6th Sustainability and Resilience of Coastal Management (SRCM 2025)

BIO Web of Conferences Volume 216 (2026)

Surabaya, Indonesia and Online
27 November 2025

ISBN: 979-8-3313-3263-1

Printed from e-media with permission by:

Curran Associates, Inc.
57 Morehouse Lane
Red Hook, NY 12571

proceedings.com

Some format issues inherent in the e-media version may also appear in this print version.

For additional information, please contact EDP Sciences – Web of Conferences
at the address below.

EDP Sciences – Web of Conferences
17, Avenue du Hoggar
Parc d'Activité de Courtabœuf
BP 112
F-91944 Les Ulis Cedex A
France

Phone: +33 (0) 1 69 18 75 75
Fax: +33 (0) 1 69 28 84 91

contact-edps@webofconferences.org

Additional copies of this publication are available from:

Curran Associates, Inc.
57 Morehouse Lane
Red Hook, NY 12571 USA
Phone: 845-758-0400
Fax: 845-758-2633
Email: curran@proceedings.com
Web: www.proceedings.com

TABLE OF CONTENTS

GEO-MARINE AND MAPPING APPLICATION FOR COASTAL AREAS

INTEGRATED PLANNING OF COASTAL AREAS

ENVIRONMENTAL AND HAZARD MITIGATION

CLIMATE CHANGE ADAPTATION AND MITIGATION

SUSTAINABLE CONSTRUCTION

ENVIRONMENTAL MONITORING AND SUSTAINABILITY

MARINE AND COASTAL AREA POLLUTION FROM WASTEWATER, NUTRIENTS, AND MARINE LITTER

SRCM 2025

Joint International Conference on Sustainable Coastal Management

The 6th Sustainability and Resilience of Coastal Management

Nopember 27th , 2025

HYBRID CONFERENCE

KEYNOTE SPEAKERS

Prof. Dr. Ir. Suhono Harso Supangkat, M.Eng.
Institut Teknologi Bandung

Prof. Dr. Lin-Chi Wang
National Kaohsiung University of Science and Technology

Prof. Nicholas Hutchins
The University of Melbourne

Prof. Dr. -Ing. Pierre Grussenmeyer
INSA Strasbourg (France)

Hepi Hapsari Handayani, S.T., M.Sc., Ph.D.
Institut Teknologi Sepuluh Nopember

CALL FOR PAPER

SRCM 2025's Sub Themes

- Integrated planning of coastal area;
- Technology and management related to marine and fisheries resources;
- Geo-marine and mapping application for coastal area;
- Marine and coastal area pollution from wastewater, nutrient, and marine litter;
- Artificial Intelligence (AI) and Internet of Things (IoT) for Climate and Disaster Resilience;
- Conservation of coastal area;
- Climate change adaptation and mitigation;
- Environmental and hazard mitigation;
- Community resilience and empowerment;
- Green and clean technology for coastal area;
- Sustainable construction;
- Disaster risk management;
- Environmental monitoring and sustainability;
- Creative Media for Disaster Mitigation;

SCHEDULE TIMELINE

TIMELINE	Abstract Submission	Acceptance Notification	Payment Deadline	Full Paper Submission	Oral Presentation (MAIN EVENT)	Full Revision Submission (Reviewed by Scientific Committee)	Conference Indexed by Scopus
Batch 1	September 30th, 2025	October 1st, 2025	October 8th, 2025	November 17th, 2025	November 27th, 2025	December 12th, 2025	
Batch 2	October 10th, 2025	October 14th, 2025	October 28th, 2025				
EXTEND	October 31st , 2025	November 1st , 2025	November 4th, 2025				

CONFERENCE & PUBLICATION FEE

PAYMENT INFORMATION

BNI

Name : DRPM - 25-SM-0005 SRCM
VA Number : 8257711030950056
Institut Teknologi Sepuluh Nopember (ITS)

	Presenting Participant	Non-Presenting Participant
Non Student	Local: IDR 2.500.000 International: USD 153	Local: IDR 500.000 International: USD 35
Student	Local: IDR 2.000.000 International: USD 123	Local: IDR 300.000 International: USD 20

FURTHER INFORMATION

Contact Person:
MKPI ITS +62 851-1781-2525

Email:
srcm.its2025@gmail.com

REGISTER HERE!

Presenting Participant
its.id/m/RegistSRCM2025

Non-Presenting Participant
its.id/m/RegistSRCM2025_NP

Statement of Peer review

In submitting conference proceedings to BIO Web of Conferences, I certify to the Publisher that I adhere to the **Policy on Publishing Integrity** of the journal in order to safeguard good scientific practice in publishing.

1. All articles have been subjected to peer review administered by the proceedings editors.
2. Reviews have been conducted by expert referees, who have been requested to provide unbiased and constructive comments aimed, whenever possible, at improving the work.
3. Proceedings editors have taken all reasonable steps to ensure the quality of the materials they publish and their decision to accept or reject a paper for publication has been based only on the merits of the work and the relevance to the journal.

Title, date and place of the conference

The 6th Sustainability and Resilience of Coastal Management 2025

Proceedings editor(s):

Hepi Hapsari Handayani, S.T., M.Sc., Ph.D. - Scientific Chair

Date and editor's signature

January 22, 2026

Hepi Hapsari Handayani, S.T., M.Sc., Ph.D.
Scientific Chair

BathySAR-Net: An Artificial Neural Network-Based Model For Predicting Intertidal Zone Depth Using Sentinel-1 Synthetic Aperture Radar (SAR) Imagery

Belvana Eka Putri[1], *Muhammad* Aldila Syariz[1,*], *Lino* Garda Denaro[2], and *Manh* Van Nguyen[3]

[1]Geomatics Engineering Department, Sepuluh Nopember Institute of Technology, Surabaya, Indonesia,
[2]National Research and Innovation Agency, Bandung, Indonesia
[3]Faculty of Geography, Hanoi National University of Education, Hanoi, Vietnam

Abstract. Coastal depth monitoring plays a crucial role in navigation and environmental management but is often constrained by the high operational costs and accessibility issues of conventional surveys. To address these challenges, this study implements "BathySAR-Net," an Artificial Neural Network (ANN) model designed to predict sea depth using Sentinel-1 Synthetic Aperture Radar (SAR) imagery combined with BATNAS bathymetric data. The model utilizes a Multilayer Perceptron (MLP) architecture to capture the complex, non-linear relationships between radar backscatter coefficients and water depth. Evaluation using robust K-Fold Cross Validation yielded a Root Mean Square Error (RMSE) of 8.7060 meters and a coefficient of determination (R^2) of 0.0838. While the model demonstrated improved stability in deeper zones (10–15 meters), predictive performance in shallow intertidal areas (<5 meters) remained limited due to significant radar signal noise and data scarcity. These findings suggest that while SAR-based ANN offers a promising, accessible alternative for bathymetry, further integration of hydrodynamic variables is essential to resolve nearshore dynamics effectively.

1. Introduction

Use Coastal zones, serving as a critical transition between terrestrial and marine environments, harbor exceptionally rich ecosystems [1]. These include natural formations such as coral reefs, mangrove forests, estuaries, and deltas, as well as artificial ecosystems like aquaculture ponds, many of which are under threat. However, these natural resources are under intense pressure from anthropogenic activities, such as overfishing driven by market demand, which often involves destructive practices [2]. Concurrently, the impacts of global climate change are a major contributing factor to abrasion and erosion, which significantly alter bathymetric contours and threaten the stability of coastal aquatic ecosystems [3].

* Corresponding author: aldilasyariz@its.ac.id

Monitoring bathymetric changes is therefore a crucial step for understanding coastal environmental dynamics [4]. Historically, these measurements have been conducted using two primary approaches: direct methods, such as diver surveys and lead-lines, and indirect methods that rely on advanced acoustic instruments like Single Beam Echosounder (SBES) and Multi Beam Echosounder (MBES) [5]. While capable of yielding high-accuracy data, the field application of these methods is largely inefficient. They are time-consuming, expensive, highly dependent on favorable weather conditions, and require extensive expert teams [6]. These constraints are exacerbated in shallow intertidal zones, where the accessibility of survey vessels is severely limited by their draft, precluding safe and optimal operation.

To overcome these limitations, remote sensing technology offers an alternative that enables large-scale bathymetric mapping without direct physical contact with the target [7]. Within the field of remote sensing, satellite imagery from active sensors, such as Synthetic Aperture Radar (SAR), demonstrates superior potential over passive sensors, which are dependent on solar illumination. The fundamental advantage of SAR lies in its ability to emit its own microwave pulses and record the backscatter, enabling effective operation regardless of cloud cover, fog, or darkness [8]. Consequently, data from satellites such as Sentinel-1 SAR are ideal for monitoring highly dynamic and constantly changing intertidal zones.

Although SAR technology consistently provides surface data, interpreting the radar backscatter signal to determine water depth is not a straightforward task. The relationship between radar signal characteristics and water depth is highly complex and non-linear, influenced by multiple factors including sea surface roughness and seabed material composition [9]. Addressing this challenge requires an analytical method capable of identifying hidden patterns within the data. This study proposes the use of an Artificial Neural Network (ANN), a machine learning method inspired by the information processing capabilities of the human brain [10]. By training on a given dataset, an ANN model can learn to approximate these non-linear relationships autonomously [11]. Therefore, the objective of this research is to develop and evaluate the performance of an ANN model for predicting intertidal zone depth along the coast of Bawean Island, Gresik, using Sentinel-1 SAR data. The model's accuracy will be quantitatively assessed using the Root Mean Square Error (RMSE) and the Coefficient of Determination (R^2)

2. Materials and Methods

2.1 Study Area & Datasets

This research was conducted on Bawean Island, which is administratively part of Gresik Regency, East Java, and located in the Java Sea. The specific study area is defined by the coordinates 5°49'4.8" to 5°50'27.6" South latitude and 112°44'2.4" to 112°46'1.2" East longitude.

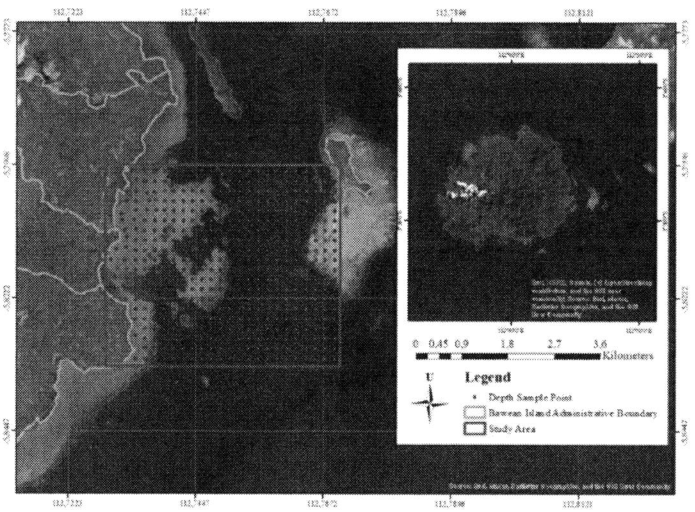

Fig. 1. Map of the study area

The southern coastal region of Bawean Island encompasses diverse habitats, including sandy, vegetated, and rocky shores, supported by critical ecosystems such as mangrove forests, seagrass beds, and coral reefs. However, these ecosystems face significant pressure from anthropogenic activities, including illegal logging and destructive fishing practices, leading to reduced coverage of coral reefs, mangroves, and seagrass. Oceanographically, the area is influenced by ocean waves with an average height of 1.41 meters and exhibits a diurnal tidal pattern. While the coastal abrasion rate is relatively low, at approximately 0.5 meters per year, the impacts of sea-level rise—such as coastal flooding (locally known as *rob*) and saltwater intrusion—are already affecting the local community [12].

The dataset consists of Sentinel-1A SAR imagery acquired on 29 May 2021 at 10:50 (UTC+7) in Interferometric Wide Swath mode with VV and VH polarizations, ascending orbit. The imagery was obtained from the Copernicus Open Access Hub. Bathymetric data were derived from the National Bathymetric Chart (BATNAS) produced by the Geospatial Information Agency (BIG) through the integration of gravity anomaly inversion and sounding data from multiple institutions. The dataset covers 110–115° E and 10–5° S at a spatial resolution of 6 arc-seconds (±180 m) referenced to the Mean Sea Level (MSL) datum, and was accessed via Ina-Geoportal.

2.2 Methodology

The research workflow commences with the acquisition of Sentinel-1 Ground Range Detected (GRD) SAR imagery and the National Bathymetry (BATNAS) dataset. This is followed by the extraction of depth soundings from the BATNAS data and a comprehensive pre-processing pipeline for the SAR imagery. The pre-processing stage yields calibrated backscatter values ($\beta°$, $\gamma°$, $\sigma°$) for both VV and VH polarizations. These two datasets are then merged, normalized, and labelled. To ensure a robust model evaluation, a 10-Fold Cross-Validation scheme is implemented. The prepared data is used to train and validate the proposed BathySAR-Net (Artificial Neural Network) architecture. Finally, the trained model is employed for depth prediction across the study area, with the final output presented as a bathymetric map.

2.2.1 Data Pre-Processing

The primary dataset utilized in this study is the Sentinel-1 Level-1 Ground Range Detected (GRD) Synthetic Aperture Radar (SAR) imagery. This dataset was selected due to its high spatial resolution and its ability to acquire data independently of weather conditions and daylight availability. The preprocessing stage is essential to enhance the quality of the SAR imagery and prepare it for further analysis. The preprocessing workflow begins with subsetting, which limits the analysis to the specific study area. This is followed by the Apply Orbit File process, which corrects the satellite's positional accuracy by incorporating updated orbital information from the metadata. Next, Thermal Noise Removal is conducted to eliminate thermal noise artifacts that can interfere with the backscatter signal, thereby improving the reliability of the image. The Border Noise Removal step follows, which removes noise along the image edges and filters out invalid pixels. Subsequently, Radiometric Calibration is applied to convert Digital Number (DN) values into calibrated backscatter coefficients, expressed as sigma naught (σ^0), in decibel (dB) units. This step ensures that the backscatter values are physically meaningful and comparable across different scenes. Speckle Filtering is then performed to reduce speckle noise, a common granular noise in SAR data, thereby enhancing image clarity. Following this, Range Doppler Terrain Correction is implemented to rectify geometric distortions caused by topographic variation and sensor geometry. The output of this preprocessing stage includes sigma naught values and incidence angle data, which are essential for subsequent feature extraction. The final preprocessing step is the conversion to decibels (dB), which transforms the calibrated backscatter values into logarithmic scale to facilitate interpretation and analysis [13]. After completing the SAR preprocessing workflow, the next phase involves extracting depth information from the BATNAS bathymetric dataset within the study area. This process yields 429 spatially distributed depth data points, ranging from 0 to 32 meters. Following the integration of SAR-derived features and depth labels from BATNAS, the dataset undergoes a data preparation stage prior to model training. This includes data normalization using the Min-Max Scaling method, which serves to standardize the range of all input features. This normalization ensures that variables with larger value ranges do not disproportionately influence the learning process [14]. Min-Max Scaling transforms each feature value into a new scale—typically between 0 and 1—according to the following formula:

$$x' = \frac{x - \min}{\max - \min} \quad (1)$$

where x' is the normalized value, x is the original data value, and *min* and *max* represent the original minimum and maximum values, respectively. Once all feature values have been normalized, the prepared dataset is validated using a 10-Fold Cross Validation scheme. In this method, the dataset is divided into ten equally sized subsets or "folds." The training and validation process is performed iteratively ten times, where in each iteration, one fold is used as the validation set while the remaining nine folds are used for training. This technique ensures that each data point is used exactly once for validation and nine times for training.

As a result, this approach provides a more robust and unbiased estimate of model performance by minimizing the variability caused by any single train-test split.

2.2.2 BathySAR-Net Network Structure

In this study, the proposed Artificial Neural Network (ANN) model is referred to as BathySAR-Net. This model is developed using a Multilayer Perceptron (MLP) architecture implemented with the Keras Sequential API, specifically designed to learn the non-linear relationship between backscatter coefficients derived from Synthetic Aperture Radar (SAR) data and corresponding water depth values. The selection of input features is grounded in the hydrodynamic modulation theory of SAR bathymetry. Since microwave radar cannot penetrate the water column, depth estimation relies on the indirect imaging mechanism where underwater topography modulates the surface current velocity. This interaction alters the sea surface roughness, which in turn modifies the radar backscatter intensity through Bragg resonance [15]. To capture these nuances, the model incorporates Sigma Nought ($\sigma°$) as the primary indicator of surface roughness, alongside Beta Nought ($\beta°$) and Gamma Nought ($\gamma°$) to account for radar reflectivity and geometric incidence angle corrections. Both VV and VH polarizations are utilized to distinguish between surface scattering (VV-dominant) and volume scattering or noise (VH-dominant). Therefore, the model input consists of six discrete features: $\beta°$ VH, $\beta°$ VV, $\gamma°$ VH, $\gamma°$ VV, $\sigma°$ VH, and $\sigma°$ VV.

The network architecture is configured with a single hidden layer containing 64 neurons, followed by an output layer consisting of a single neuron responsible for predicting the water depth value directly. The model was trained using the Adam optimizer with a learning rate of 0.001, minimizing the Mean Squared Error (MSE) loss function. Training was conducted for a maximum of 300 epochs with a batch size of 64. To prevent overfitting and enhance generalization, three regularization techniques were applied. First, the ReLU activation function was used in the hidden layer, followed by a Dropout layer with a rate of 0.3, meaning that 30% of neurons were randomly deactivated at each training step. Second, L2 regularization with a lambda coefficient of 0.01 was implemented to control model complexity. Third, an Early Stopping mechanism with a patience value of 20 was employed to automatically halt training if no improvement was observed in validation performance for 20 consecutive epochs. The final performance of the model was evaluated using the Root Mean Squared Error (RMSE) metric to assess the accuracy of depth predictions.

2.2.3 Performance Evaluation of BathySAR-Net

To evaluate the predictive performance of the proposed ANN model, this study employs two primary statistical metrics: Root Mean Squared Error (RMSE) and the Coefficient of Determination (R^2). These metrics are selected for their ability to provide a comprehensive assessment of model accuracy and explanatory power. RMSE is particularly useful because it quantifies the average magnitude of prediction error in the same units as the target variable (i.e., depth in meters), making the results intuitive and easy to interpret. A lower RMSE value indicates better predictive accuracy.In contrast, R^2 complements the evaluation by indicating the proportion of variance in the observed depth values that is explained by the model. An R^2 value closer to 1 signifies a strong correlation between the predicted and actual values, suggesting a well-fitted model. The RMSE is calculated using the following formula:

$$RMSE = \sqrt{\frac{1}{n}\sum_{i=1}^{n}(yi - \hat{yi})^2} \quad (2)$$

where n is the total number of data points, y_i is the actual depth value, and $\hat{y_i}$ is the depth value predicted by the model. The Coefficient of Determination (R^2) is calculated by comparing the sum of squared residuals to the total sum of squares, using the following formula:

$$R^2 = \frac{SSR}{SST} = \frac{\sum_{i=1}^{n}(\hat{yi} - \bar{y})^2}{\sum_{i=1}^{n}(yi - \bar{y})^2} \quad (3)$$

In this formula, $\hat{y_i}$ is the predicted value, y_i is the actual value, and $\bar{y}$ is the mean of all actual depth values.

3. Results

To evaluate the performance of the ANN model in estimating sea depth, this research conducted testing using the 10-Fold Cross-Validation method on the BathySAR-Net model as an initial case study. The results from each fold were recorded to observe the variability of the model's performance against the data division. The evaluation metrics used are Root Mean Squared Error (RMSE) and R-squared (R^2). RMSE provides information regarding the average magnitude of prediction error in meters, which in this research represents the deviation of the predicted depth from the actual value. Meanwhile R^2 indicates the proportion of variance in the depth data that can be explained by the model, with a higher value indicating a better fit of the model to the observation data. The following is a statistical visualization of the depth distribution for each fold:

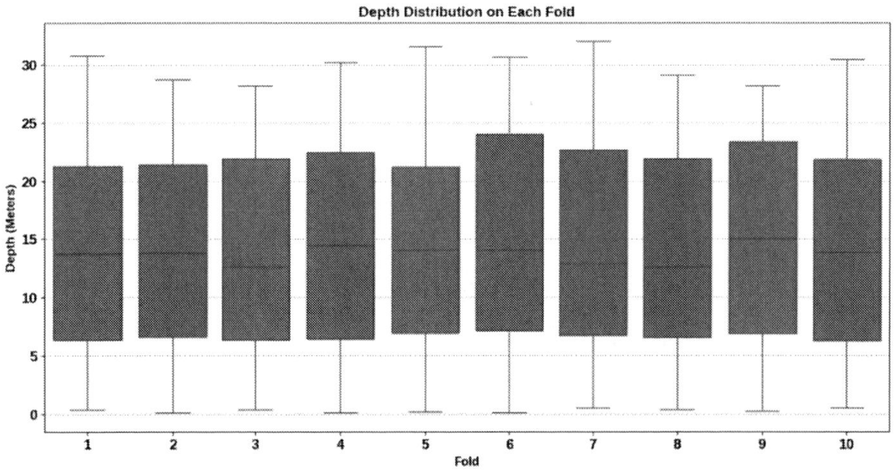

Fig. 2. Depth Distribution on Each Fold

Statistical analysis of depth distribution in each fold indicates that the data partitioning in the 10-fold cross-validation was conducted in a representative and balanced manner. The mean depth for each fold ranged from 14.019 meters to 14.779 meters, with a maximum difference of only 0.760 meters, reflecting equality in central tendency. The median depth was also consistent, ranging from 12.619 meters to 15.016 meters, indicating stability in the central value. The standard deviation, as an indicator of data dispersion, exhibited a uniform pattern across all folds, ranging from 8.800 meters to 9.244 meters with a difference of only 0.444 meters. Depth ranges were similarly consistent, with minimum values approaching 0 meters in all folds and maximum values varying from 28.149 meters to 32.009 meters. Although Fold 3 and Fold 8 displayed slightly lower medians (around 12.6 meters) and Fold 9 showed a more concentrated data distribution with the highest median, lowest standard deviation, and lowest maximum value, these variations are minor and statistically insignificant. Overall, this uniform distribution provides a fair and robust foundation for evaluating the ANN model's performance. This consistency ensures that evaluation results can be interpreted objectively, as they are influenced purely by the model architecture's performance rather than imbalances in data distribution.

To assess the performance of the BathySAR-Net model for identifying sea depth, this research employed a Multilayer Perceptron (MLP) architecture configured with a single hidden layer containing 64 neurons. This architecture was designed to efficiently capture the non-linear relationship between the six input features— $\beta°$ VV+VH, $\gamma°$ VV+VH, and $\sigma°$VV+VH —and the single output value representing depth. The training process for the model employed an identical set of optimized hyperparameters, including activation function, optimizer, and learning rate, to ensure robust and stable learning. The final performance of the BathySAR-Net model was quantitatively assessed using the average RMSE and R^2 values obtained from the 10-Fold Cross-Validation, as presented in Table 1.

Table 1. Visualization of ANN Model Architectures

Model	RMSE (m)	R^2 (m)
BathySAR-Net (64 Neurons)	8.7060	0.0838

Based on the 10-Fold Cross-Validation, the BathySAR-Net model demonstrated a Root Mean Squared Error (RMSE) of 8.7060 meters and a coefficient of determination (R^2) of 0.0838. The RMSE value indicates that, on average, the model's depth estimation deviates by approximately 8.7 meters from the actual depth value. Furthermore, the low R^2 value suggests that the model only explains a small proportion of the total variance in the observed depth data. This outcome, despite the consistent training process, indicates that the model's performance is primarily constrained by the limited information content in the input features (Sentinel-1 backscatter values). This suggests that the relationship between the radar backscatter and depth, as captured by this particular set of features, does not substantially dominate the variance in the depth data. BathySAR-Net Model was therefore employed in the subsequent stage to generate bathymetric predictions for the study area, as illustrated in Figure 3.

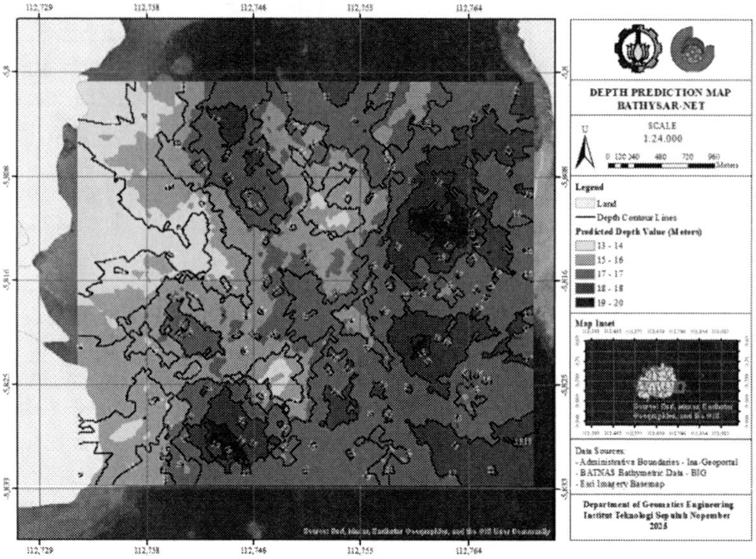

Fig. 3. Predicted Bathymetry Map using the BathySAR-Net Model

The predicted bathymetry map reveals a diverse depth distribution, ranging from shallow waters of approximately 13 meters, represented in light blue, to depths exceeding 19 meters, shown in dark blue. Spatially, deeper areas are consistently located in the southeastern part of the study area, extending toward the northeast, while shallower waters dominate the western and northern regions adjacent to Bawean Island's coastline. From a seafloor morphology perspective, the tightly spaced contours in the southeast indicate steep slopes, likely representing submarine escarpments where depth changes occur rapidly. In contrast, the wider contour spacing in the west and north suggests a gently sloping and relatively flat seabed. The presence of dark blue to purple areas in the southeast further suggests possible geomorphological features such as localized depressions or small troughs. These patterns appear consistent rather than random, indicating the potential presence of significant underwater structures that could be subjects of further investigation.

However, as this map is the output of an ANN-based model derived from radar imagery, it is important to note that the spatial patterns strongly depend on the quality and spatial resolution of the input data. In areas with noise or unrepresentative SAR texture—caused, for instance, by high waves or exceptionally calm sea surfaces—the model may produce less accurate predictions. Therefore, field verification remains essential to confirm the identified geomorphological features.

4. Discussion

This section presents an in-depth analysis of predicted depth variations compared with actual depths across the ranges of 0–5 meters, 0–10 meters, 0–15 meters, and 0–20 meters, as discussed in Section 4.1, followed by the application of the model to a different water area in Section 4.2.

4.1 Comparative Analysis of Predicted and Actual Depths Across Depth Ranges

To evaluate the model's performance in predicting bathymetry across different depth ranges, visualizations were generated for 0–5 meters, 0–10 meters, 0–15 meters, and 0–20 meters. This analysis aimed to specifically assess the model's ability to represent varying depth levels, particularly in shallow waters, which are generally more sensitive to variations in radar signals. The visualizations compare actual depth data from BATNAS with the model's predictions for each range. By progressively limiting the depth intervals, this approach provides a more detailed understanding of the model's spatial accuracy and predictive value distribution, while also identifying potential biases or error patterns at specific depths. The following figures present the visualization results for each depth range, illustrating spatial comparisons between actual data and the ANN model's bathymetric predictions.

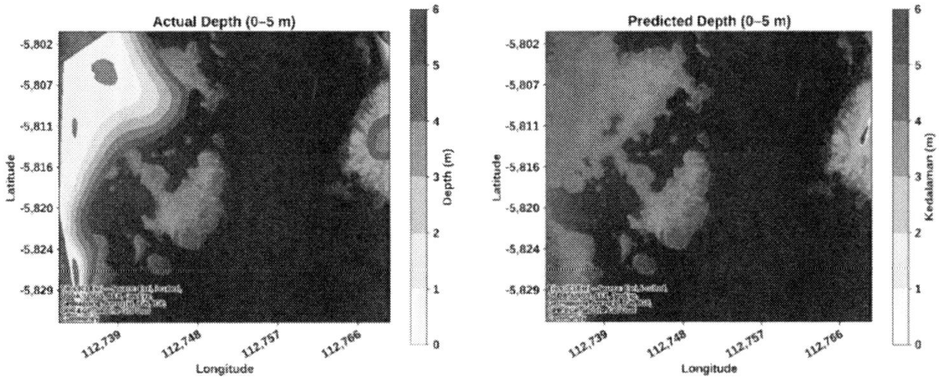

Fig. 4. Comparison of Actual and Predicted Depth Visualizations for the 0–5 Meter Range

In the first depth range (0–5 meters), the ANN model failed to produce any predictions, as indicated by the absence of depth values in the visualization. This is a critical issue, considering that this very shallow zone corresponds to the intertidal area, which is a primary focus of the study.

The results suggest that the model was unable to recognize SAR backscatter patterns representative of extremely shallow waters. This critical limitation in the 0-5 m range is primarily attributed to the physical complexity of the intertidal zone and data scarcity, rather than the ANN architecture itself. Physically, the intertidal zone is a highly dynamic environment characterized by breaking waves, extreme surface turbulence, and rapid tidal fluctuations. In these conditions, the radar backscatter is dominated by non-coherent noise and specular reflections from exposed seabeds during low tide, which mask the subtle hydrodynamic modulation required for bathymetric extraction. Furthermore, the training dataset contained a disproportionately low number of samples for the 0-2 m range, preventing the model from learning a stable representation of these chaotic signal patterns. Consequently, this study suggests that for operational intertidal mapping, the proposed SAR-based method cannot be used as a standalone solution and requires integration with auxiliary data, such as tidal gauges or optical imagery, to resolve nearshore ambiguities.

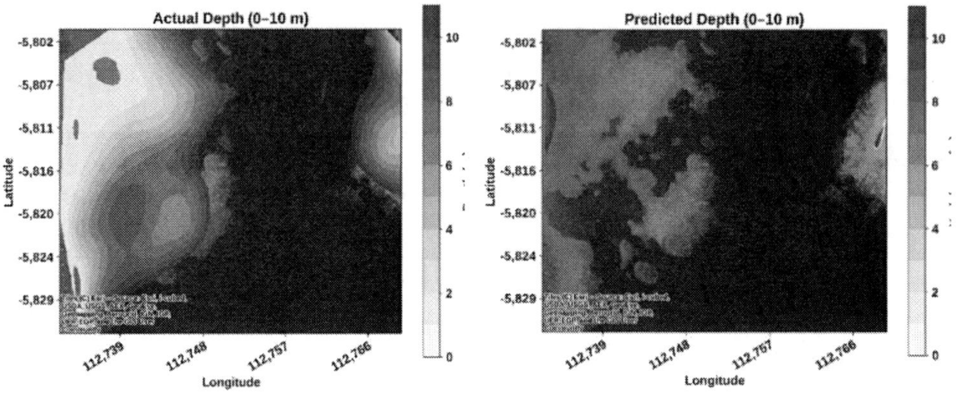

Fig. 5. Comparison of Actual and Predicted Depth Visualizations for the 0–10 Meter Range

The figure above compares actual depth maps and ANN model predictions for the 0–10 m range. Overall, the ANN model shows limitations in mapping shallow waters, especially within and around the intertidal zone. Two key observations can be made. First, in the actual depth map, the northwestern and southwestern intertidal areas display a broad, well-defined 0–5 m contour zone, while in the predicted map these areas are poorly detected, with only small portions classified within the correct range. Second, predicted depths are scattered irregularly without forming clear contour patterns, as seen in the central part of the map, where depths of 8–10 m are predicted in areas that are actually shallower or outside the 0–10 m range.

Spatial inconsistencies are also apparent where areas with uniform actual depths contain prediction anomalies or artifacts. These issues indicate the model's difficulty in capturing the spatial and textural relationships in shallow-water backscatter data. Likely causes include:

1. High environmental dynamics in the intertidal zone, such as substrate variability, tides, and shallow-wave effects influencing backscatter.
2. Limited training data for very shallow depths (0–2 m), reducing the model's ability to distinguish relevant signals.
3. The ANN's current architecture being less effective at extracting spatial features for shallow depths, performing better for deeper, more stable patterns.

Improving performance in this zone may require adding shallow-water training samples and enhancing the model architecture or integrating additional features such as texture indices or tidal data.

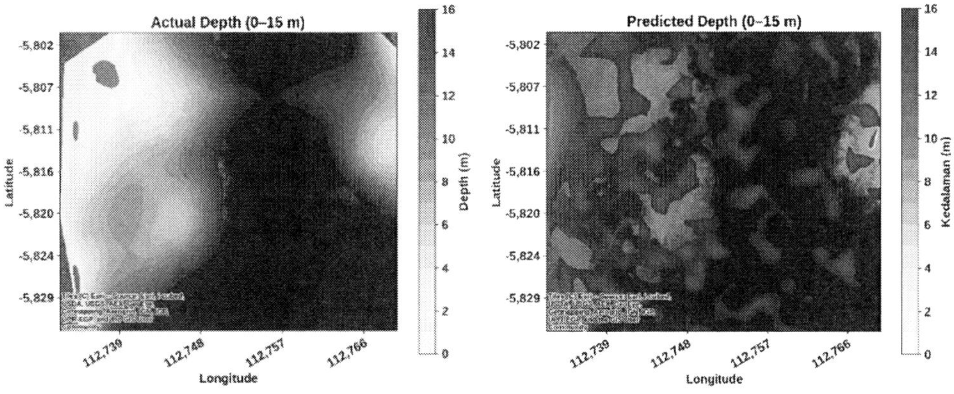

Fig. 6. Comparison of Actual and Predicted Depth Visualizations for the 0–15 Meter Range

The visualization above compares the actual depth map with the ANN model's predicted depth map for the 0–15 m range. Compared to the 0–10 m range, the ANN model demonstrates a broader spatial coverage in its predictions. However, several notable limitations remain in its performance.

A major issue lies in the misclassification of shallow-water areas. Several regions with depths of less than 5 m in the actual data are predicted to fall within the 10–16 m range, visually represented by dark blue contours near the shoreline. The distribution of these predictions also appears spatially inconsistent, with similar depth values scattered in non-adjacent locations and failing to form natural contour patterns. This indicates a tendency toward overestimation in shallow zones, likely due to the model's difficulty in distinguishing SAR backscatter signals for shallow to intermediate depths.

In the central and northeastern portions of the prediction map, areas of high predicted depth appear in locations that do not match the smooth depth gradient observed in the actual map. This suggests that the model has not fully captured the spatial transitions of depth across the study area.Potential contributing factors to these errors include:

1. An imbalance in training data representation between shallow and intermediate depths.
2. The complexity of SAR backscatter reflectance in mixed-substrate environments (sand, mud, submerged vegetation), which is challenging for conventional ANN models to differentiate.
3. The model's limited ability to incorporate spatial context between neighboring pixels, resulting in fragmented depth patterns.

While the predictions for this range show improved spatial coverage compared to the 0–10 m case, accuracy in separating shallow and intermediate depths remains an area for improvement.

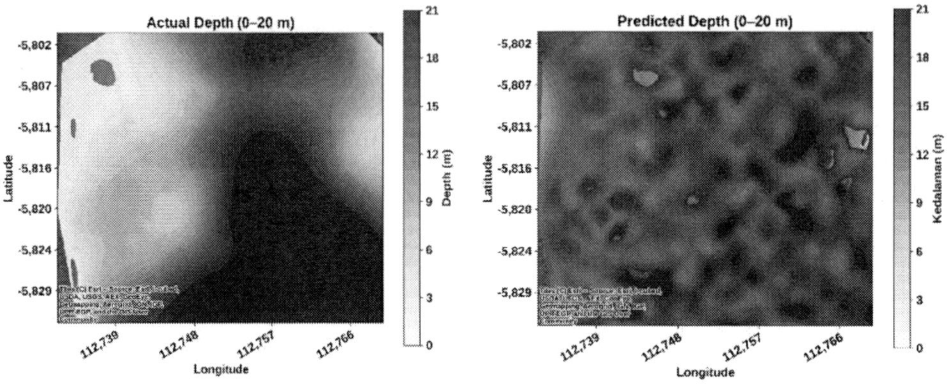

Fig. 7. Comparison of Actual and Predicted Depth Visualizations for the 0–20 Meter Range

A closer examination of the two maps reveals stark differences in how depth is distributed. The actual depth map displays smooth and orderly contour gradients, with depth increasing gradually from the shoreline toward the open sea. This pattern aligns with the typical morphology of coastal seabeds, which generally feature gentle slopes and consistent contour lines. In contrast, the predicted map exhibits irregular and fragmented depth patterns, with random spatial distributions that fail to reflect natural contour gradients. Areas of high (dark) and low (light) depth values appear scattered without a coherent spatial structure.

The mismatch becomes even more evident when examining the shallow-depth patterns in the predicted map. Numerous low-depth patches (0–6 m) appear inconsistently within regions that should be considerably deeper (10–20 m). Such patterns suggest potential local overfitting or noise during model training, where the ANN captures non-representative variations in the training data. The presence of artifact-like spots further indicates that the model struggles to capture spatially coherent depth patterns.

This spatial inconsistency points to a fundamental weakness in the model's ability to recognize the natural structure of seafloor contours. Compared to actual conditions, the ANN's predictions fail to capture continuous bathymetric structures, revealing its limited capacity to generalize in areas underrepresented in the training set. This lack of generalization poses a major challenge for applying ANN-based SAR imagery to bathymetric prediction.

Possible factors contributing to the poor predictive accuracy for this depth range include:

- The absence of explicit spatial features in the model's input (the ANN processes individual pixel values without accounting for spatial relationships between pixels).
- The increasing complexity of SAR backscatter signals at depths >10 m due to radar energy attenuation.
- Imbalanced training data distribution, particularly in the mid-depth (10–15 m) and deep (15–20 m) classes, causing prediction bias toward certain depth categories.

These findings indicate that, although the ANN model's predictions cover the full extent of the study area, the accuracy of its bathymetric representation remains low. Further steps are needed to enhance prediction accuracy, including data augmentation, outlier filtering and spatial smoothing, and integration with spatially-aware models.

Synthesizing the performance across all depth ranges, the overall low coefficient of determination ($R^2 = 0.0838$) warrants a fundamental physical explanation. Unlike optical bathymetry methods, which directly measure light attenuation through the water column and typically achieve higher correlations ($R^2 > 0.7$) in clear waters, SAR bathymetry relies on an indirect imaging mechanism. The radar signal does not penetrate the water; instead, it captures the modulation of sea surface roughness induced by the interaction between tidal currents and underwater topography. The results of this study suggest that relying solely on backscatter intensity features (σ^0) without explicit hydrodynamic input data (e.g., current velocity) limits the model's explanatory power to resolve this complex, indirect relationship. Consequently, the primary scientific contribution of this work is in delineating the performance boundaries of pure backscatter-based ANN models, highlighting that machine learning architectures alone cannot fully compensate for the lack of hydrodynamic variables in SAR bathymetry.

4.2 Application of the Model to a Different Coastal Area

As part of the extended analysis of BathySAR-Net's performance, an additional prediction test was conducted using Sentinel-1 SAR imagery from a location situated near the previous study area, but slightly shifted to the south. The objective of this test was to evaluate the model's ability to generalize across both familiar areas (previously included in training) and new, unseen regions. The dataset used for this evaluation was derived from Sentinel-1 SAR imagery of the Bawean Island region, acquired in May 2025. The imagery underwent the same pre-processing steps as the 2021 dataset used for model training, ensuring consistency in input preparation and comparability of results. The following figure presents the predicted bathymetry map generated by BathySAR-Net based on the newly acquired imagery.

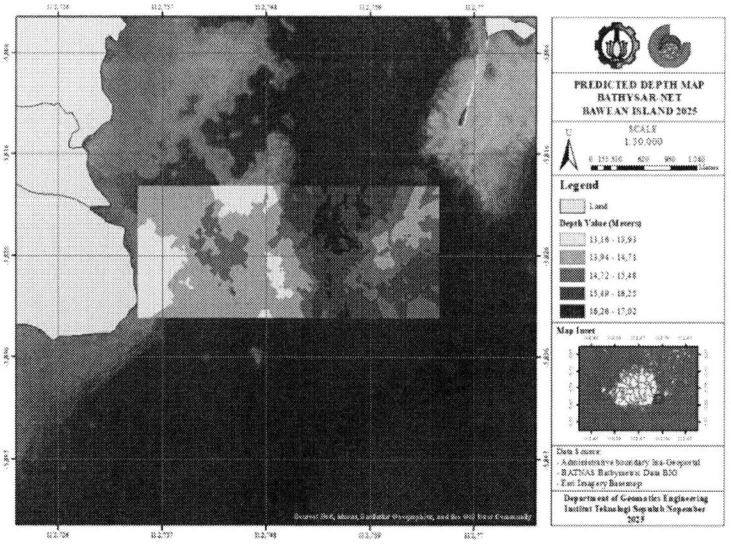

Fig. 8. Predicted Bathymetry of Bawean Island (2025)

To facilitate evaluation, the predicted bathymetric values generated by the model were grouped into five depth ranges: 0–5 m, 5–10 m, 10–15 m, 15–20 m, and >20 m. Accuracy assessment was then carried out for each range using two primary metrics: the Root Mean Square Error (RMSE), which measures the absolute prediction error, and the relative RMSE, calculated by comparing the RMSE value to the maximum depth of each range. This approach was adopted because each range has a different depth span, making the interpretation of error values more proportional. In other words, the same absolute error may have a different impact depending on the maximum depth of the range. The model's performance evaluation results for each depth range are presented in Table 2.

Table 2. RMSE Values Across Depth Ranges for the BathySAR-Net Model (2025)

Depth Range	RMSE (m)	Relative RMSE (%)	Number of Data Points
0 – 5	11,652	389,66	9
5 – 10	7,249	102,90	41
10 – 15	2,830	22,67	14
15 – 20	4,025	22,65	20
> 20	11,154	42,83	93

The highest RMSE value was observed in the 0–5 m depth range, both in absolute terms (11.652 m) and relative terms (389.66%). This high percentage error occurs because the maximum depth in this range is only about 5 m, meaning that even a small deviation of a few meters in prediction can result in a disproportionately large relative error. This substantial inaccuracy is likely due to the complex characteristics of shallow waters, such as tidal influence, wave disturbances, turbidity changes, and specular reflections from the water surface, which introduce significant noise into the SAR backscatter signal.

Model performance improves slightly in the 5–10 m range; however, the relative RMSE remains high at 102.9%. This indicates that the intertidal to shallow subtidal zones remain a significant challenge for SAR-based bathymetry modeling, particularly because radar imagery has limited penetration through the water column and is more sensitive to surface conditions.

In contrast, model performance improves considerably in the 10–15 m and 15–20 m depth ranges, with relative RMSE values around 22%. This suggests that the model begins to effectively learn and identify patterns linking SAR backscatter values to water depth in these mid-depth zones. At these depths, SAR imagery tends to exhibit more stable backscatter values and is less affected by surface dynamics, allowing the model to generalize more effectively.

In the >20 m depth range, the absolute RMSE increases again to 11.154 m, yet the relative RMSE remains at a moderate 42.83% due to the larger scale of depth values in this range. In other words, while the absolute error is substantial, it is proportionally more acceptable for deep-water estimation. Moreover, this range contains the largest number of evaluation points (93), making the performance assessment more representative.

Spatially, the predicted results are visualized as a depth distribution map classified into five ranges. Light blue represents shallow areas, while dark blue denotes deeper waters. The visual pattern shows that the eastern and southeastern parts of Bawean Island are dominated by greater depths, which is consistent with the seafloor topography indicated in previous bathymetric maps.

Although the final visualization does not yet exhibit smooth and detailed depth contours comparable to interpolated conventional bathymetric surveys, the model successfully produces a logical depth distribution pattern and spatial depth gradient. This is evident from the consistent transition of colors from shallow to deep zones, albeit still in coarse segmented blocks rather than continuous contours. Such visual coarseness may be attributed to:

- Limited observation points in certain ranges, resulting in suboptimal spatial interpolation.
- High local variability not fully captured by the model due to SAR's spatial resolution limitations or inherent data noise.

While limitations persist, particularly in shallow-water zones, these results demonstrate that BathySAR-Net has promising potential as an alternative approach for bathymetric mapping using SAR imagery. The model's ability to predict logical depth patterns in mid- to deep-water zones suggests it could be operationally viable, especially in areas difficult to access with conventional bathymetric surveys. Future model development could focus on expanding the quantity and diversity of training data, integrating tidal corrections, implementing spatial models such as CNNs, and conducting temporal and cross-location validation to ensure stronger generalization.

Overall, BathySAR-Net's predictions for 2025 Sentinel-1 SAR imagery of Bawean Island indicate a reasonably good generalization capability in mid- and deep-water depths, although performance in shallow zones still requires improvement. The spatial visualizations suggest that the model is beginning to capture realistic depth patterns, albeit with coarse contours. With further development and improved data quality, this approach could serve as an efficient alternative for radar-based bathymetric mapping.

5. Conclusion

Based on the findings and discussions presented in the previous chapters, it can be concluded that the BathySAR-Net model, which utilizes a Multilayer Perceptron (MLP) with a single hidden layer and 64 neurons, still falls short of practical applicability, particularly in shallow waters or intertidal zones. Despite undergoing systematic training and evaluation, the model achieved an average Root Mean Squared Error (RMSE) of 8.7060 meters with a coefficient of determination (R^2) of only 0.0838, indicating a very weak correlation between predicted and actual depths. Performance was notably uneven across depth ranges, with the model performing best at depths greater than 20 meters (RMSE: 4.061 meters) but failing to accurately predict depths below approximately 11 meters. While performance improved slightly when tested across a broader depth range with more data—where RMSE increased to 11.729 meters—the inability to capture the characteristics of intertidal zones remained a major limitation. This result suggests that for SAR backscatter data, which typically exhibits weak signals and high noise, the chosen MLP architecture provides the most effective balance, as overly complex models are more prone to overfitting by learning noise rather than meaningful patterns. This performance gap is primarily attributed to the physical limitations of the SAR backscatter signal itself, which has an inherently weak and non-linear relationship with water depth. To address this limitation and enhance predictive performance in future developments, it is recommended to integrate external environmental data, such as wind speed and ocean currents, into the modeling process. Since SAR backscatter is fundamentally modulated by sea surface roughness, incorporating these hydrodynamic variables could significantly strengthen the model's ability to distinguish between depth-induced patterns and surface noise caused by environmental dynamics.

Acknowledgment
The authors would like to express their sincere gratitude to the European Space Agency (ESA) for providing access to the Sentinel-1 imagery, and to the Geospatial Information Agency of Indonesia (Badan Informasi Geospasial, BIG) for supplying the tidal observation data in the vicinity of Bawean Island. These datasets were essential for the implementation and validation of this study.

Conflict of Interest
The authors declare no conflict of interest.

References

[1] S. A. Mukhtar, G. Handoyo, and H. Setiyono, "Studi Batimetri Dan Pasang Surut Untuk Rekomendasi Alur Pelayaran Di Dermaga Pelabuhan Marunda," *Indones. J. Oceanogr.*, vol. 6, no. 4, pp. 375–382, 2024, doi: 10.14710/ijoce.v6i4.16541.

[2] M. Postacchini and A. Romano, "Dynamics of the Coastal Zone," *J. Mar. Sci. Eng.*, vol. 7, no. 12, pp. 7–9, 2019, doi: 10.3390/jmse7120451.

[3] Z. A. Khamis, R. Kalliola, and N. Käyhkö, "Geographical characterization of the Zanzibar coastal zone and its management perspectives," *Ocean Coast. Manag.*, vol. 149, pp. 116–134, 2017, doi: 10.1016/j.ocecoaman.2017.10.003.

[4] R. Bishop-Taylor, S. Sagar, L. Lymburner, and R. J. Beaman, "Between the tides: Modelling the elevation of Australia's exposed intertidal zone at continental scale," *Estuar. Coast. Shelf Sci.*, vol. 223, no. March, pp. 115–128, 2019, doi: 10.1016/j.ecss.2019.03.006.

[5] Osayi Philip Igbinenikaro, Oladipo Olugbenga Adekoya, and Emmanuel Augustine Etukudoh, "Review of Modern Bathymetric Survey Techniques and Their Impact on Offshore Energy Development," *Eng. Sci. Technol. J.*, vol. 5, no. 4, pp. 1281–1302, 2024, doi: 10.51594/estj.v5i4.1018.

[6] P. Westfeld, H. G. Maas, K. Richter, and R. Weiß, "Analysis and correction of ocean wave pattern induced systematic coordinate errors in airborne LiDAR bathymetry," *ISPRS J. Photogramm. Remote Sens.*, vol. 128, pp. 314–325, 2017, doi: 10.1016/j.isprsjprs.2017.04.008.

[7] M. Ellis, R. Formanek, and N. Townsend, "Satellite Derived Bathymetry AusSeabed Community Guidelines: Satellite Derived Bathymetry," p. 13, 2022.

[8] H. Cruz, M. Véstias, J. Monteiro, H. Neto, and R. P. Duarte, "A Review of Synthetic-Aperture Radar Image Formation Algorithms and Implementations: A Computational Perspective," *Remote Sens.*, vol. 14, no. 5, pp. 1–30, 2022, doi: 10.3390/rs14051258.

[9] R. M. Asiyabi, A. Ghorbanian, S. N. Tameh, M. Amani, S. Jin, and A. Mohammadzadeh, "Synthetic Aperture Radar (SAR) for Ocean: A Review," *IEEE J. Sel. Top. Appl. Earth Obs. Remote Sens.*, vol. 16, pp. 9106–9138, 2023, doi: 10.1109/JSTARS.2023.3310363.

[10] T. Sagawa, Y. Yamashita, T. Okumura, and T. Yamanokuchi, "Shallow Water Bathymetry Derived by Machine Learning and Multitemporal Satellite Images," *Int. Geosci. Remote Sens. Symp.*, pp. 8222–8225, 2019, doi: 10.1109/IGARSS.2019.8899043.

[11] D. Ushizima, *Artificial neural networks*. 2023. doi: 10.1201/9781003359593-7.

[12] Z. Hidayah, A. Romadhon, and Y. Witjarnoko, "Penilaian Kerentanan Wilayah Pesisir Selatan Pulau Bawean terhadap Kenaikan Muka Air Laut Vulnerability Analysis of Sea Level Rise In The Southern Coast of Bawean Island," vol. 20, no. 2, pp. 87–94, 2018, doi: 10.22146/jfs.36109.

[13] F. Filipponi, "Sentinel-1 GRD Preprocessing Workflow," p. 11, 2019, doi: 10.3390/ecrs-3-06201.

[14] Y.-S. Kim, M. K. Kim, N. Fu, J. Liu, J. Wang, and J. Srebric, "Investigating the Impact of Data Normalization Methods on Predicting Electricity Consumption in a Building Using different Artificial Neural Network Models.," *Sustain. Cities Soc.*, vol. 118, no. June 2024, p. 105570, 2024, doi: 10.1016/j.scs.2024.105570.

[15] H. Liu, H. Zeng, and H. Huang, "Bragg resonant re fl ection of surface waves from deep water to shallow water by a fi nite array of trapezoidal bars," *Appl. Ocean Res.*, vol. 94, no. October 2019, p. 101976, 2020, doi: 10.1016/j.apor.2019.101976.

The influence of bidirectional tidal flow on scour mechanisms and morphological changes around the Suramadu bridge piers

Zidny Omnu Nurilhaq[1], *Candida Aulia De Silva* Nusantara[1*], *Khomsin*[1], *Danar Guruh* Pratomo[1], and *Irena Hana* Hariyanto[1]

[1]Department of Geomatics Engineering, Faculty of Civil Planning and Geo Engineering, Institut Teknologi Sepuluh Nopember, Sukolilo, Surabaya 60111, Indonesia

*Corresponding author: candida.nusantara@its.ac.id

Abstract. The Suramadu Bridge is a vital infrastructure in the Madura Strait, facilitating regional economic connectivity. However, complex oceanographic characteristics—notably bidirectional currents and seasonal variations—pose significant scouring risks to bridge piers, potentially threatening structural integrity. This study models current patterns and sediment transport to analyze scouring dynamics around the piers. Using hydrodynamics model, simulations were conducted for 2024, incorporating bathymetry, tides, wind, and sediment data. Model validation against tidal data showed high accuracy (MAE: 0.1473 m; RMSE: 0.1771 m). Results reveal a tidal-driven west–east bidirectional flow with stable average velocities (0.21–0.25 m/s) across monsoons. Interaction between currents and piers causes upstream flow deceleration and lateral acceleration, triggering deposition and erosion, respectively. Piers P46, P47, and P49 experienced the most significant scouring, with maximum depths reaching -0.250 m, while P37, P41, and P56 remained stable. Erosion trends were spatially and temporally consistent across both west and east monsoons. These findings establish critical monitoring priorities to ensure the bridge's structural sustainability in alignment with SDG Goal 9.

Keywords: current patterns, hydrodynamic modelling, scouring, sediment transport, sustainable infrastructure

1. Introduction

The Surabaya–Madura Bridge, widely known as the Suramadu Bridge, is the longest bridge in Indonesia with a total length of 5,438 meters. Its construction began under Presidential Decree No. 55 of 1990, later revised by Presidential Decree No. 79 of 2003, and was officially opened on 10 June 2009 [1]. The bridge connects Surabaya on Java Island with Bangkalan on Madura Island, crossing the Madura Strait. As the only direct land connection between Java and Madura, it plays a crucial role in enhancing regional connectivity and supporting socio-economic development. Since becoming operational, the bridge has significantly improved mobility, facilitated the movement of goods and people, and contributed to the economic growth of the four regencies in Madura [2].

However, the location of the bridge within Madura Strait introduces considerable challenges. The strait exhibits complex oceanographic characteristics, including a bidirectional current system driven by tidal forcing and seasonal monsoonal patterns. During the west monsoon, the current flows from the Java Sea toward the Madura Strait, while during the east monsoon it reverses toward the Bali Sea [3]. Recorded current velocities range from 0.03 to 0.41 m/s [4], and when combined with varying sediment types such as sandy clay and silt, they contribute to potential seabed alterations around the bridge structure [5]. One of the most critical risks associated with these conditions is local scour, a localized erosion process around bridge foundations driven by flow-induced sediment entrainment. Its mechanisms often involve three-dimensional flow structures and horseshoe vortices forming at the base of piers [6]. If not properly mitigated, scouring can compromise foundation stability and lead to structural failure.

Several studies have attempted to model scouring phenomena at the Suramadu Bridge, including applications of HEC-18 and Hanco methods [7]. However, previous research relied on single-direction flow assumptions and lacked comprehensive seasonal integration [8]. Addressing these limitations, this study models current dynamics and sediment transport around the bridge piers by incorporating bidirectional tidal flow and seasonal variability through a year-long numerical hydrodynamic simulation. This approach provides a more complete understanding of scour behavior and supports resilient and sustainable infrastructure management strategies aligned with Sustainable Development Goal (SDG) 9.

2. Research method

2.1 Study area

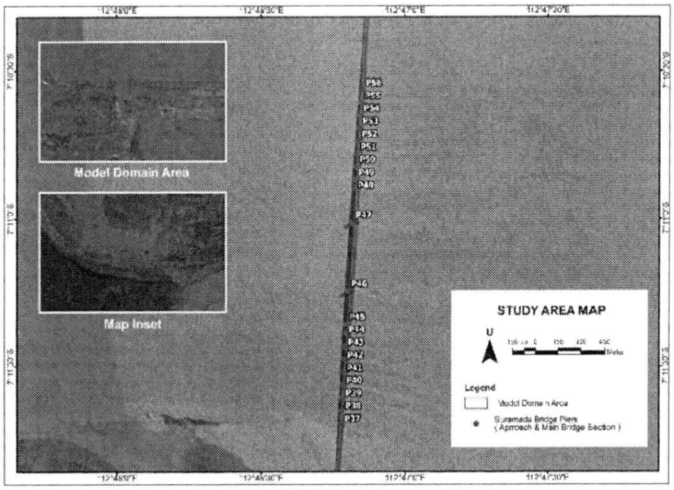

Fig. 1. Study location map in the Madura Strait waters.

The study area is located within the waters of the Madura Strait surrounding the Suramadu Bridge. The modeling domain encompasses the administrative marine boundaries of Surabaya City, Bangkalan Regency, and Gresik Regency, covering an area of approximately 1,514.63 km², as illustrated in Figure 1. The geographical extent of the study area ranges from 6°52'4.45" S to 7°20'40.36" S and from 112°36'10.40" E to 112°52'4.45" E. The focus of the scour analysis is highlighted by red markers indicating the coordinates of the Suramadu

Bridge piers, consisting of both the approach bridge and main bridge sections, with detailed information provided in Table 1.

Table 1. Coordinates of the Suramadu Bridge Piers (Approach and Main Bridge Sections)

Piers Number	Section	Latitude	Longitude
P37	*Approach Bridge*	-7°11'41.55"	112°46'46.60"
P38	*Approach Bridge*	-7°11'38.92"	112°46'46.71"
P39	*Approach Bridge*	-7°11'36.32"	112°46'46.83"
P40	*Approach Bridge*	-7°11'33.71"	112°46'46.95"
P41	*Approach Bridge*	-7°11'31.12"	112°46'47.06"
P42	*Approach Bridge*	-7°11'28.49"	112°46'47.18"
P43	*Approach Bridge*	-7°11'25.92"	112°46'47.32"
P44	*Approach Bridge*	-7°11'23.29"	112°46'47.46"
P45	*Approach Bridge*	-7°11'21.44"	112°46'47.57"
P46	*Main Bridge*	-7°11'14.72"	112°46'47.98"
P47	*Main Bridge*	-7°11'0.77"	112°46'48.91"
P48	*Approach Bridge*	-7°10'53.99"	112°46'49.38"
P49	*Approach Bridge*	-7°10'52.13"	112°46'49.51"
P50	*Approach Bridge*	-7°10'49.51"	112°46'49.72"
P51	*Approach Bridge*	-7°10'46.92"	112°46'49.92"
P52	*Approach Bridge*	-7°10'44.36"	112°46'50.15"
P53	*Approach Bridge*	-7°10'41.69"	112°46'50.38"
P54	*Approach Bridge*	-7°10'39.12"	112°46'50.60"
P55	*Approach Bridge*	-7°10'36.54"	112°46'50.82"
P56	*Approach Bridge*	-7°10'33.93"	112°46'51.05"

2.2 Data and equipment

This study utilized various instruments and software to support primary data acquisition, data processing, and hydrodynamic modelling. The equipment consisted of a Single Beam Echosounder (SBES) for obtaining bathymetric data around the Suramadu Bridge piers and a grab sampler for collecting sediment samples at six locations.

The datasets used in this research comprised model domain construction data, model input data, and model validation data. The model domain was developed using shoreline administrative data for East Java Province from the Geospatial Information Agency (BIG) and the pier dimension data obtained from CCCC Highway & Bridge Consultants Co., Ltd. [9]. Bathymetry data were sourced from direct field measurements carried out from 1–7 July 2024, global bathymetry from BATNAS, and digitized Indonesian nautical charts. The model input included tidal forcing for 2024, global wind data for 2024 derived from ERA5 reanalysis [10], and river discharge data from the Kalimas River obtained from previous studies [11]. Sediment characteristics were obtained from laboratory analyses of samples collected from 1–7 July 2024, while wave parameters were extracted from ERA5 [10] for sediment transport modelling. Model validation was conducted using tidal observations from the BIG tide gauge station SRBY for the year 2024.

2.3 Research workflow

The study began with the identification of key issues related to bridge pier scouring, followed by an extensive literature review on hydrodynamic modeling approaches and the relevant oceanographic parameters within the study area. Subsequently, both primary and secondary datasets were collected and processed to support the development of the hydrodynamic and sediment transport models. These steps include domain construction, numerical model setup, calibration and validation, and analysis of current patterns, sediment transport, and scour formation.

2.3.1 Domain construction

The land boundary was defined by following the coastline derived from the Rupa Bumi Indonesia (RBI) shoreline dataset for East Java Province, covering the coastal areas of Surabaya City, Gresik Regency, and Bangkalan Regency. The open boundaries were assigned along the northern edge of the domain, with a length of approximately 23 km, and along the eastern edge, extending about 10 km. An additional boundary was applied at the Kalimas River, with the width set to 100 m to represent the river mouth.

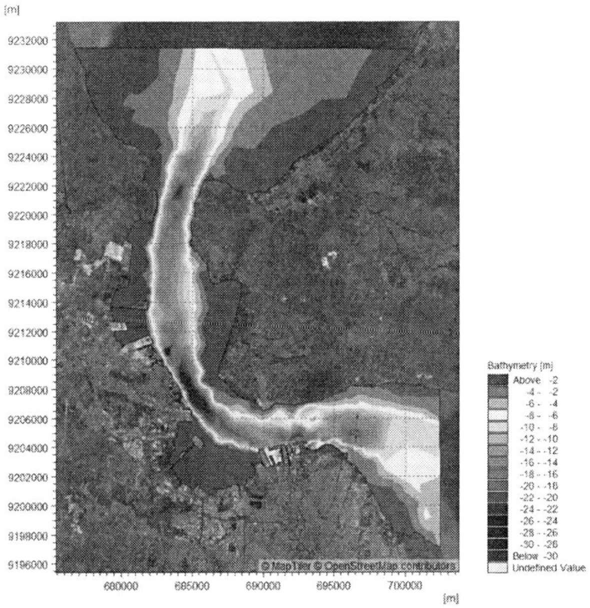

Fig. 2. Hydrodynamic model domain.

In constructing the model domain, an unstructured triangular grid (flexible mesh) was applied, combined with multiple bathymetric datasets. High-resolution depth data were obtained from direct field measurements using a Single Beam Echosounder (SBES) around the Suramadu Bridge piers to enhance detail within the focus area. These data were integrated with digitized Indonesian nautical charts and the National Bathymetry (BATNAS) dataset, which provided wider-area depth coverage referenced to Mean Sea Level (MSL). The resulting interpolated bathymetry, illustrated in **Fig. 2**, shows significant depth variability in the Madura Strait, ranging from shallow nearshore waters at 0 m to deeper sections along the main navigation channel reaching 30.445 m. This spatial depth distribution plays an essential role in influencing current patterns, sediment transport dynamics, and scouring processes around the bridge piers.

2.3.2 Numerical model setup

The hydrodynamic and sediment transport simulations were conducted using hydrodynamic model. Wind and wave input parameters were extracted using Ocean Data View prior to being incorporated into the model. The extracted wind and wave datasets, along with tidal forcing, river discharge, and sediment grain size (D50), were then used as boundary and initial conditions for the hydrodynamic and sediment transport simulations. The model was run for a full one-year period, from January 2024 to December 2024. The temporal parameters used in the simulation are summarized in **Table 2**.

Table 2. Simulation time parameters

Parameter	Value
Simulation period	373 days
Simulation soft start	7 days
Start time	24 December 2023
Stop time	31 December 2024
Time step interval	1 second

The simulation was carried out over a total duration of 373 days, starting on 24 December 2023 and ending on 31 December 2024. A soft start period of 7 days was applied at the beginning of the simulation to gradually stabilize the model and prevent numerical shocks, ensuring a smooth transition toward realistic hydrodynamic conditions. This soft start phase is crucial for minimizing initial transients and improving the accuracy of the subsequent results. A time step interval of 1 second was selected due to the use of an unstructured grid, in which element sizes vary considerably across the model domain. A larger time step could violate numerical stability constraints and lead to unrealistic outputs or model instability. The chosen 1-second interval is sufficiently small to maintain the Courant number below critical thresholds while remaining computationally efficient for a year-long simulation.

2.3.3 Calibration and validation

Before analyzing the model results, a validation procedure was conducted to assess how well the model represents real-world conditions. This validation compared the simulated tidal water levels with observational tidal data recorded by the Geospatial Information Agency (BIG) at the Surabaya tide gauge station. Statistical performance was evaluated using the Mean Absolute Error (MAE) and Root Mean Square Error (RMSE) metrics. The formulations for MAE and RMSE are presented in **equations (1)** and **(2)** [12].

$$MAE = \frac{1}{N}\sum_{i=1}^{N} |OBS_i - SIM_i| \tag{1}$$

$$RMSE = \sqrt{\frac{1}{N}\sum_{i=1}^{N}(OBS_i - SIM_i)^2} \tag{2}$$

where N represents the number of data points (observations), OBS_i denotes the actual observational value, and SIM_i denotes the model-predicted value. The error categories for MAE and RMSE are presented in **Table 3.**

Table 3. Interpretation of MAE and RMSE value [12]

MAE and RMSE value	Error level
0 – 0.299	Small
0.30 – 0.599	Medium
0.60 – 0.899	Large
>0.90	Very large

2.3.4 Current pattern and sediment transport analysis

The current and sediment transport parameters serve as the basis for evaluating the scouring mechanism around the bridge piers. Current patterns were analyzed to identify the dominant hydrodynamic forces responsible for mobilizing seabed sediments and their interaction with the pier structures. Meanwhile, sediment transport patterns were examined to determine the spatial distribution of sediment movement and its response to the prevailing flow conditions. The analysis of current and sediment transport patterns was conducted for both the west and east monsoon seasons under spring and neap tide conditions, along with an assessment of current distribution and its interaction with the bridge piers.

2.3.5 Scouring analysis

The scouring analysis was carried out by examining the interaction between current patterns and sediment transport around the bridge piers, which leads to the development of local scour at the Suramadu Bridge. The scouring assessment involved comparing the cumulative erosion values observed during the west and east monsoon seasons around each pier. This analysis was further extended by calculating the annual scour depth and scour rate using a linear regression approach, as formulated in **equation (3)** [13].

$$\hat{y} = a + bx \tag{3}$$

where $\hat{y}$ represents the dependent variable, x denotes the independent variable, a is the intercept, and b is the gradient that describes the change in $\hat{y}$ per unit change in x.

3. Result and discussion

3.1 Model validation

Validation is a crucial stage in the numerical modeling of currents and sediment transport. This step plays an essential role in assessing the model's ability to accurately represent the actual physical conditions occurring in the field. The validation process is performed by comparing the model simulation results with observational data, allowing the degree of agreement between the model output and real-world measurements to be evaluated. Through this process, the reliability and accuracy of the model can be assessed, while also identifying potential discrepancies that may arise. The parameter used for validation in this study is tidal water level, comparing the model-generated tide data with observations recorded by the Geospatial Information Agency (BIG) at the Surabaya tide gauge station. The comparison between simulated and observed tidal elevations is presented in **Fig. 3** and **Fig. 4**.

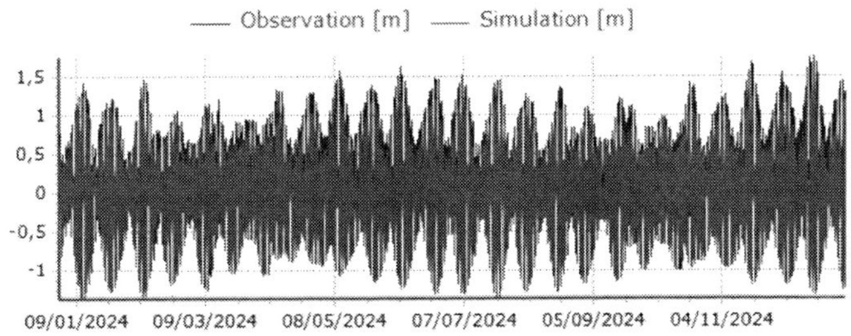

Fig. 3. Comparison of tidal parameters between the model results and BIG observational data.

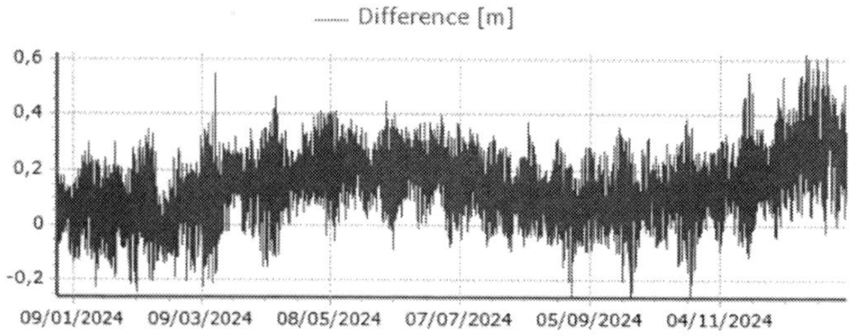

Fig. 4. Differences in tidal elevations between the model results and BIG observational data.

Fig. 3 and **4** sequentially present the tidal elevation plots and the corresponding differences between the model simulation and the BIG observational data from January to December 2024. The largest discrepancy in surface elevation was recorded at 0.62 m on 13 December 2024 at 15:00 UTC. The validation results using the Mean Absolute Error (MAE) and Root Mean Square Error (RMSE) methods are summarized in **Table 4**.

Table 4. Interpretation of MAE and RMSE value

Method	Value
Mean Absolute Error	0.1473 m
Root Mean Square Error	0.1771 m

The statistical evaluation shows that the differences between the two-validation metrics are not significant, indicating that the model errors are relatively uniform and exhibit good predictive stability. Based on the validation results and the classification criteria in **Table 3**, the MAE and RMSE values obtained for each season fall within the *small error* category (< 0.299). Therefore, the model can be considered valid, and the simulation results are suitable for further analysis.

3.2 Current patterns result from the model

The analysis of the hydrodynamic simulation results was carried out under spring tide and neap tide conditions for both the west and east monsoon seasons, with particular focus on the flow characteristics around the Suramadu Bridge piers. This approach was used to identify the maximum current velocities generated under each extreme tidal condition. The determination of spring and neap tide periods for each season was based on the lunar phase and the modeled tidal elevation data. The simulated tidal curves, along with the identification of spring and neap tide conditions, are presented in **Fig. 5.**

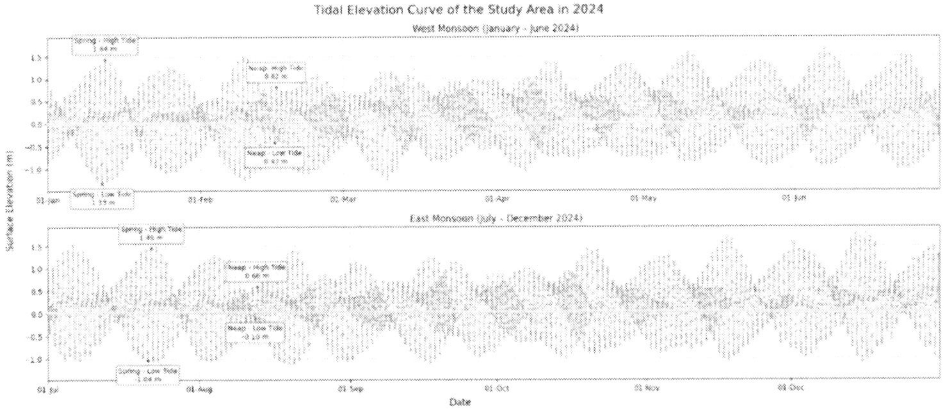

Fig. 5. Tidal elevation graph of the modeled study area in 2024.

Fig. 5 presents the tidal elevation curve for the study area throughout 2024. The blue line illustrates the variability of sea surface elevation, while the yellow line represents the Mean Sea Level (MSL) used as the reference datum. Red markers indicate spring tide conditions, and green markers denote neap tide conditions. During the west monsoon, the spring tide occurred on 12 January 2024, characterized by a minimum water level of −1.33 m at 00:00 UTC and a maximum of 1.44 m at 16:00 UTC on the same day. The neap tide was identified on 16 February 2024, with a minimum elevation of −0.43 m at 01:00 UTC and a maximum of 0.82 m at 08:00 UTC. In the east monsoon, the spring tide occurred on 21 July 2024, marked by a minimum elevation of −1.04 m at 11:00 UTC and a maximum of 1.46 m at 04:00 UTC on 22 July 2024. The neap tide for this season was observed on 12 August 2024, with a minimum level of −0.10 m at 11:00 UTC and a maximum of 0.60 m at 21:00 UTC. These extreme tidal conditions were used as references in determining the simulation scenarios for the hydrodynamic model. By incorporating spring and neap tide conditions from both monsoon seasons, the analysis of current patterns captures both the maximum and minimum energy states associated with tidal forcing, enabling a more comprehensive hydrodynamic assessment.

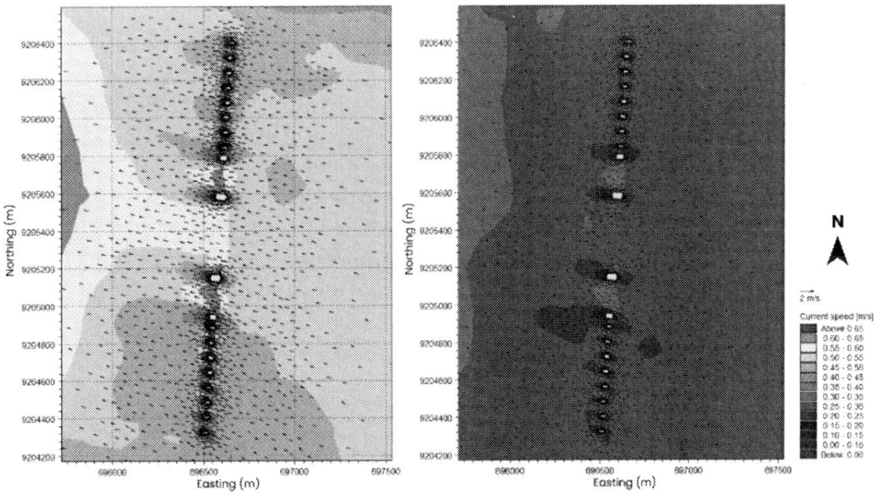

(a) (b)

Fig. 6. Current pattern in west monsoon (a) spring tide (b) neap tide.

During the west monsoon, **Fig. 6(a)** shows the distribution of current velocity under extreme spring tide conditions on 12 January 2024 at 05:00 UTC. The currents are predominantly represented by green shades, indicating velocities ranging from 0.40 to 0.55 m/s. The maximum velocities, ranging from 0.60 to 0.65 m/**s**, are shown in orange and are widely distributed, particularly along the northern and southern sides of the bridge pier structure. In contrast, under the extreme neap tide condition illustrated in **Fig. 6(b)** on 16 February 2024 at 08:00 UTC, the current velocities are generally weaker, dominated by blue shades that represent speeds below 0.30 m/s. The maximum velocities during this neap tide condition fall within the range of 0.30 to 0.35 m/s.

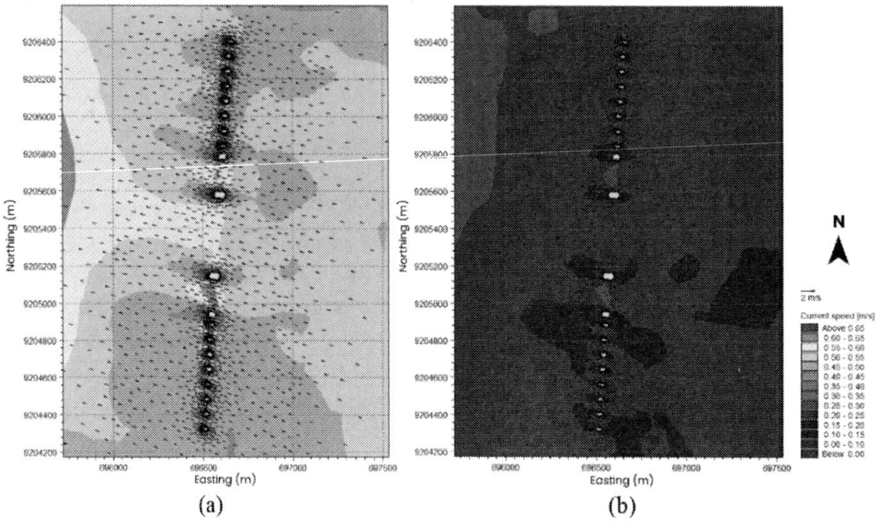

Fig. 7. Current pattern in east monsoon (a) spring tide (b) neap tide.

Meanwhile, during the east monsoon, the distribution of current velocity under extreme spring tide conditions is shown in **Fig. 7(a)** for 21 July 2024 at 16:00 UTC. The current pattern exhibits characteristics similar to those in the west monsoon, with green shades dominating the flow field, indicating velocities between 0.40 and 0.55 m/s. The maximum velocities, ranging from 0.60 to 0.65 m/**s**, appear along the northern and southern sides of the bridge pier structure. Under the extreme neap tide condition, illustrated in **Fig. 7(b)** on 12 August 2024 at 21:00 UTC, the current velocities are noticeably weaker, dominated by values below 0.25 m/s, with maximum velocities ranging from 0.25 to 0.30 m/s. The bidirectional flow pattern observed in both tidal phases reflects the strong influence of tidal forcing in driving water mass movement in the study area. During flood tide, the generated energy drives the flow westward, while during ebb tide, the flow reverses eastward.

Fig. 8 illustrates the interaction between the current patterns and the P46 bridge pier structure during both monsoon seasons. As described previously, the flow in the study area exhibits a bidirectional pattern, alternating between eastward and westward directions under the dominance of tidal forcing. When the incoming flow encounters the bridge pier, the structure acts as a physical barrier, forcing the water to decelerate. This results in a pronounced reduction in current velocity on the upstream side of the pier both west and east depending on the flow direction creating a stagnation point, highlighted by the red-circled region.

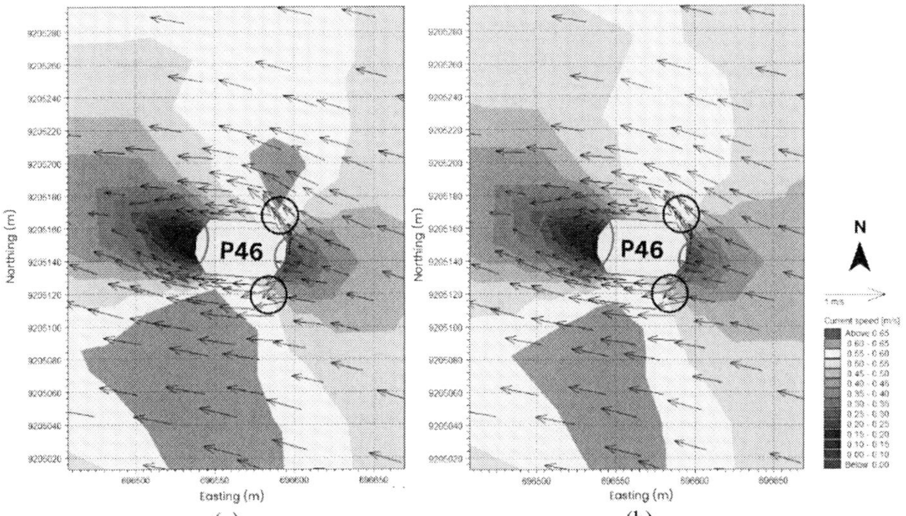

(a) (b)

Fig. 8. Interaction between current patterns and bridge piers (a) west monsoon (b) east monsoon.

The presence of the pier also obstructs the natural flow path and constricts the available flow area, causing the water mass to accelerate along the elongated sides of the pier. This flow constriction generates significant velocity amplification along the northern and southern sides, marked by the black-circled region. Once the flow passes the pier, the obstruction is removed and the current gradually returns toward its ambient state; this recovery is evident downstream on the western side of the pier, where the flow begins to spread and velocity decreases toward background levels.

For both monsoon seasons, the reduction in velocity at the stagnation zone ranges from 0.00 to 0.20 m/**s**, represented by blue to purple shades. Meanwhile, the lateral acceleration around the northern and southern sides reaches 0.50 to 0.65 m/**s**. Although no major differences are observed between the two seasons in terms of the flow structure interaction mechanism, the west monsoon exhibits a wider distribution of high-velocity zones, particularly north of pier P46.

3.3 Sediment transport result from the model

Sediment transport modeling represents a subsequent stage of the hydrodynamic simulation and aims to understand the dynamics of sediment movement within both the water column and the seabed. The transport process is influenced by interactions among several physical parameters, including current velocity, wave conditions, and sediment characteristics. In this study, sediment transport modeling was conducted to evaluate the potential for both sedimentation and scouring around the bridge piers. The model integrates hydrodynamic outputs with wave conditions and sediment properties as the primary inputs for simulating sediment transport. The sediment characteristics in the vicinity of the Suramadu Bridge piers were classified based on laboratory analyses of sediment samples, as presented in **Table 5**.

Table 5. Laboratory results of sediment sample analysis

Sample Code	Gravel (%)	Sand (%)	Silt + Clay (%)
35 T	0.00	21.49	78.51
29 T	2.55	37.48	59.97
17 T	1.90	94.57	3.53
35 B	1.75	12.22	86.03
29 B	0.00	25.93	74.07
21 B	5.26	88.67	6.07

Based on **Table 5**, the sediment composition is dominated by two major fractions: sand and fine particles (silt + clay). The gravel content is relatively low in all samples, with the highest value recorded at only 5.26% (sample 21 B). Samples such as 17 T (94.57% sand) and 21 B (88.67% sand) are predominantly composed of sand. In contrast, samples 35 T (78.51% silt + clay), 35 B (86.03% silt + clay), 29 B (74.07% silt + clay), and 29 T (59.97% silt + clay) exhibit a dominance of fine sediments.

By combining the sediment fraction characteristics with the seasonal current patterns, the resulting sediment movement around the Suramadu Bridge piers during the west and east monsoons is illustrated in **Fig. 9(a)** and **Fig. 9(b)**.

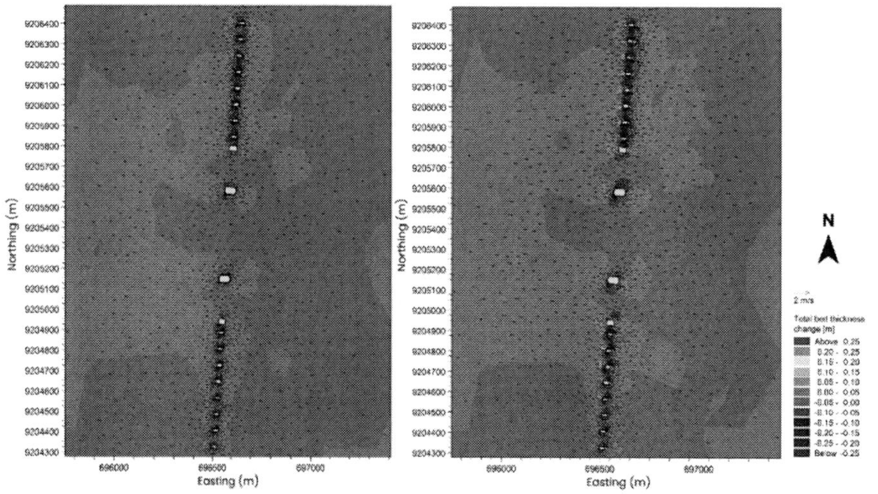

(a) (b)

Fig. 9. Sediment transport pattern (a) west monsoon (b) east monsoon.

In both monsoon seasons, the current patterns identified in earlier discussions show that the flow is predominantly driven by tidal forcing, resulting in a bidirectional east–west movement. The interaction between this oscillating flow and the seabed sediment consisting of both cohesive and non-cohesive materials drives the sediment transport processes in the study area. These processes involve the movement of material as bed load and suspended load, producing spatially variable zones of sedimentation and erosion around the Suramadu Bridge piers.

As shown in **Fig. 9(a)**, the cumulative sedimentation erosion pattern during the west monsoon indicates that much of the area around the bridge piers is dominated by erosion, primarily within the range of 0.0 to −0.05 m, while sedimentation mainly occurs within the range of 0.0 to 0.05 m. A similar pattern is observed in the east monsoon, depicted in **Fig. 9(b)**, where erosion remains dominant within the 0.0 to −0.05 m range and sedimentation remain within the 0.0 to 0.05 m range. Although both seasons exhibit generally similar magnitudes of erosion and sedimentation, differences arise near the pier structures, which will be discussed in the subsequent section. Nonetheless, closer inspection of **Fig. 9(a)** and **Fig. 9(b)** shows that sedimentation becomes more widespread and evenly distributed during the east monsoon compared to the west monsoon, while erosion becomes deeper and expands further around the pier structures. These patterns indicate that sedimentation and erosion processes are continuous between the two monsoon seasons, supported by the similar flow directions and hydrodynamic conditions prevalent throughout the year.

3.4 Bridge pier scour analysis

The reduction in current velocity within the stagnation zone results in a significant decrease in kinetic energy, which in turn lowers the flow's ability to keep sediment particles in suspension or transport them further. In this zone, the current velocity falls below the critical threshold required for sediment entrainment. Consequently, sediment particles lose their carrying force, begin to settle from the water column, and accumulate on the seabed, leading to sedimentation. Conversely, in the lateral zones, the flow is forced to constrict around the pier, resulting in accelerated velocities that increase the kinetic energy and bed shear stress exerted on the seabed. When the applied shear stress exceeds the critical shear stress of the sediment, the bed material becomes mobilized and is carried away by the flow, resulting in erosion.

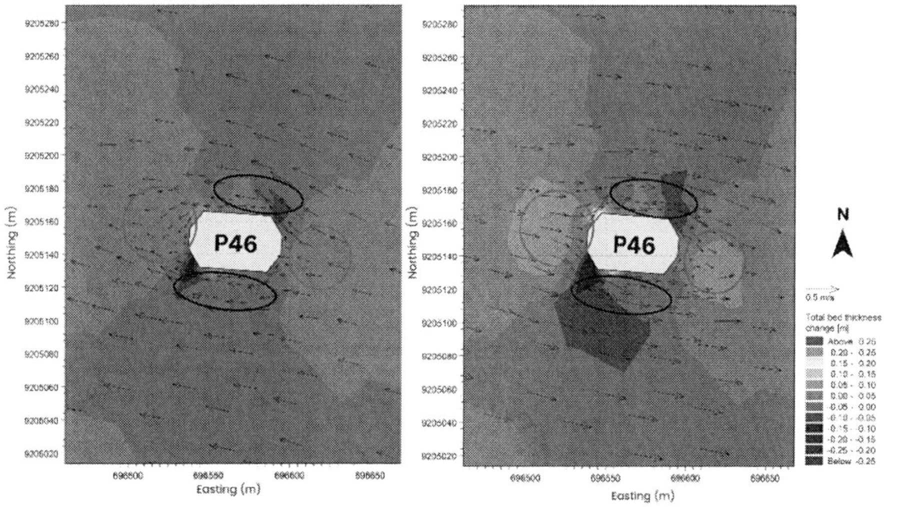

(a) (b)

Fig. 10. Sedimentation and erosion patterns around the bridge pier structure (a) west monsoon (b) east monsoon.

During the west monsoon, as shown in **Fig. 10(a)**, erosion occurs along the lateral sides of the pier, reaching values between −0.05 and −0.15 m, while sedimentation on the upstream and downstream sides ranges from 0.0 to 0.05 m. In the east monsoon, the processes of erosion and sedimentation continue, as illustrated in **Fig. 10(b).** Areas experiencing sedimentation gradually expand over time, and the same applies to zones affected by erosion. This can be observed from the widening extent of both sedimentation and erosion throughout the east monsoon. Sedimentation values around the pier reach 0.05 to 0.10 m, while erosion depths increase to −0.20 to −0.25 m. These conditions indicate that the relatively consistent seasonal flow patterns during both the west and east monsoons produce stable sediment transport behaviour across the study area and around the pier structure. The recurring seasonal flow directions and magnitudes create persistent zones of potential erosion and deposition. Consequently, despite seasonal variations, the similar hydrodynamic forcing in both monsoons promotes the continuous development of sedimentation and erosion in the same regions over time. This explains why the seabed around the pier exhibits cumulative elevation changes either deposition or scouring throughout the east monsoon.

To identify the pier that experienced the greatest scour depth during the simulation period, the changes in seabed elevation at all bridge piers were compared. The results are presented through the plot in **Fig. 11** and in **Table 6**, which show the maximum scour values at each pier based on elevation changes from the beginning to the end of the 2024 simulation period.

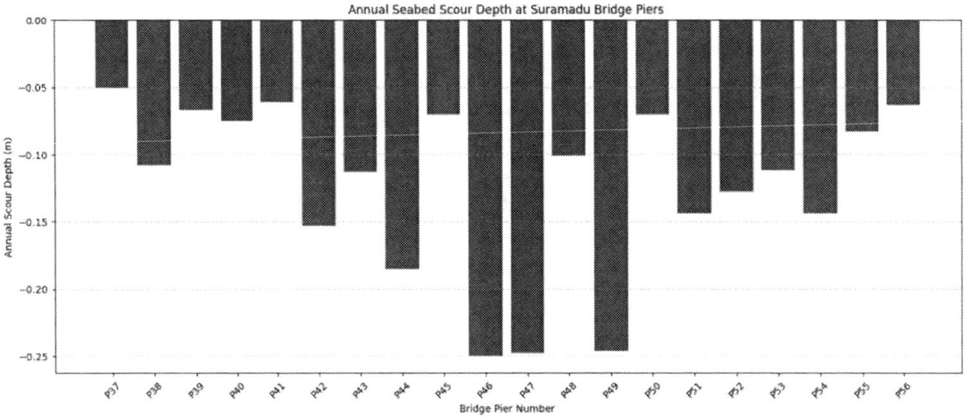

Fig. 11. Comparison of scouring depths at each Suramadu Bridge pier.

Table 6. Comparison of scouring depths at the Suramadu Bridge piers

Piers Number	Scour Depth (m)	Piers Number	Scour Depth (m)
P37	-0.050	P47	-0.248
P38	-0.108	P48	-0.101
P39	-0.067	P49	-0.246
P40	-0.075	P50	-0.070
P41	-0.061	P51	-0.144
P42	-0.153	P52	-0.128
P43	-0.113	P53	-0.112
P44	-0.185	P54	-0.144
P45	-0.070	P55	-0.083
P46	-0.250	P56	-0.063

Based on the table above, pier P46 exhibits the greatest scour depth over the one-year simulation period, with a maximum value of −0.250 m, followed by piers P47 and P49, which recorded scour depths of −0.248 m and −0.246 m, respectively. These piers fall into the category of high-scour structures when compared to the other bridge piers. Pier P37 shows the smallest scour depth at −0.05 m, while other piers with relatively low scour include P41, P56, and P39, each with scour depths less than −0.07 m. As illustrated in **Fig. 11**, the piers that experienced the deepest scour are located near the central section of the bridge, where the water depth is relatively greater. In contrast, the piers with minimal scour are situated in shallower areas. Although not entirely uniform, the results indicate a general tendency:

shallower pier locations exhibit lower scour potential, whereas deeper pier locations are more susceptible to higher scour intensity.

After identifying the piers with the greatest maximum scour depth, the next step is to evaluate the scour rate, which represents the temporal rate of bed degradation around the bridge piers. The scour rate is computed from the slope of the linear regression line fitted to the seabed elevation data for each pier. A more negative regression slope indicates a higher rate of erosion. The regression slopes, and thus the scour rates for each pier, are summarized in **Table 7**.

Table 7. Comparison of scouring rates at the Suramadu Bridge piers

Piers Number	Linear Regression Equation	Regression Slope (Scour Rate)
P37	$y = -0.000138x + 0.002460$	−0.000138
P38	$y = -0.000344x + 0.015517$	−0.000344
P39	$y = -0.000182x + 0.002703$	−0.000182
P40	$y = -0.000214x + 0.006885$	−0.000214
P41	$y = -0.000163x + 0.002825$	−0.000163
P42	$y = -0.000404x - 0.006058$	−0.000404
P43	$y = -0.000298x + 0.001019$	−0.000298
P44	$y = -0.000494x + 0.002646$	−0.000494
P45	$y = -0.000187x + 0.002129$	−0.000187
P46	$y = -0.000641x - 0.023369$	−0.000641
P47	$y = -0.000630x - 0.016320$	−0.000630
P48	$y = -0.000273x - 0.002264$	−0.000273
P49	$y = -0.000626x - 0.032602$	−0.000626
P50	$y = -0.000184x + 0.002551$	−0.000184
P51	$y = -0.000386x + 0.010842$	−0.000386
P52	$y = -0.000339x - 0.003369$	−0.000339
P53	$y = -0.000297x + 0.005101$	−0.000297
P54	$y = -0.000435x + 0.008826$	−0.000435
P55	$y = -0.000217x - 0.001259$	−0.000217
P56	$y = -0.000164x - 0.002892$	−0.000164

Based on the table above, the scour rate for each pier is determined from the slope of its linear regression equation. Pier P46 exhibits the highest scour rate at −0.000641 m/day, followed by piers P47 and P49, with scour rates of −0.000630 m/day and −0.000626 m/day, respectively. These piers fall into the category of high-scour structures when compared to the others. Pier P37 has the lowest scour rate, at −0.000138 m/day, followed by P41 and P56, with scour rates of −0.000163 m/day and −0.000164 m/day, respectively. Taken together both the maximum scour depth and scour rate evaluation piers P46, P47, and P49 represent the structures most significantly affected by scour during the one-year simulation period. The high scour rates at these piers indicate hydrodynamic conditions that promote continuous bed degradation around the pier foundations. Conversely, piers P37, P41, and P56 exhibit relatively stable conditions, characterized by minimal scour depth and low scour rates. Furthermore, the analysis shows that scour locations identified during the west monsoon generally persist into the east monsoon, with consistent erosion patterns observed throughout the one-year simulation. This evaluation serves as an important basis for prioritizing monitoring efforts to ensure the structural integrity of the Suramadu Bridge against long-term scour-related risks.

4. Conclusion

Based on the numerical hydrodynamic modeling conducted in this study, a comprehensive understanding of the current patterns and scour dynamics around the Suramadu Bridge piers has been obtained by incorporating seasonal variability. The flow regime in the study area is predominantly driven by tidal forcing, generating a bidirectional oscillating current with a dominant west–east orientation. Although peak velocities differ between spring and neap tide conditions, the seasonal average current velocities remain relatively stable. Average inflow velocities reached 0.21 m/s on the eastern side of the pier and 0.25 m/s on the western side. The interaction between the incoming flow and the bridge piers produces upstream deceleration and lateral acceleration due to flow constriction. After passing the pier, the flow gradually recovers toward ambient conditions. Velocity reductions of 0.00–0.20 m/s occur at the stagnation zones, while lateral accelerations reach 0.50–0.65 m/s. These hydrodynamic features directly influence the sediment transport processes, generating sedimentation in low-velocity zones and erosion in regions where shear stress exceeds the critical threshold. Sedimentation values range from 0.05 to 0.10 m, while erosion depths reach −0.20 m to −0.25 m.

The one-year simulation further indicates that the most significant scour occurs at piers P46, P47, and P49, with maximum erosion depths of −0.250 m, −0.248 m, and −0.246 m, respectively, accompanied by high scour rates ranging from −0.000641 to −0.000626 m/day. In contrast, piers P37, P41, and P56 exhibit relatively stable conditions, with minimum erosion depths between −0.050 and −0.063 m and low scour rates of −0.000138 to −0.000164 m/day. These results suggest that piers located in shallower waters tend to experience lower scour intensity, whereas deeper-water piers are more susceptible to stronger bed degradation. The scour process is found to be progressive and continuous, as evidenced by consistent erosion patterns across both monsoon seasons. For example, pier P46 experienced −0.075 m of scour during the west monsoon and −0.175 m during the east monsoon, resulting in a cumulative erosion depth of −0.250 m by the end of the simulation. The persistence of negative scour rates throughout the year indicates that the seasonal flow regime plays a major role in maintaining stable and progressive scour zones around the pier foundations over time.

Acknowledgement

The authors gratefully acknowledge financial support from the Institut Teknologi Sepuluh Nopember for this work, under project scheme of the Publication Writing and IPR Incentive Program (PPHKI) 2025. The authors further acknowledge PT. Hidronav Teknikatama, PT. Geosolution Pratama Nusantara and the Hydro Fieldcamp and Hydrography 2024 team for their assistance in the data collection process for this study.

References

[1] Presiden Republik Indonesia, Keputusan Presiden (Keppres) Nomor 55 Tahun 1990 tentang Pembangunan Jembatan Surabaya–Madura, Pemerintah Pusat (1990). https://peraturan.bpk.go.id/Details/63993/keppres-no-55-tahun-1990

[2] M. Effendi, Dampak Pembangunan Jembatan Suramadu Terhadap Perekonomian Pulau Madura (Studi Kasus Kabupaten Bangkalan), Universitas Diponegoro (2013). http://eprints.undip.ac.id/42107/

[3] I.A. Dumatubun, W.S. Pranowo, A. Sartimbul, J. Setiyadi, S.H. Sari, F.O. Setyawan, Karakteristik arus permukaan laut pada selatan Madura. J. Chart Datum **10**, 1 (2024)

[4] A. Siswanto, Sebaran Total Suspended Solid (TSS) pada profil vertikal di perairan Selat Madura Kabupaten Bangkalan. J. Kelautan **8**, 26–32 (2015).

[5] Chomaedhi, Pelaksanaan pondasi bored pile 240 cm Main Bridge – Jembatan Suramadu. Seminar Nasional Aplikasi Teknologi Prasarana Perkotaan, Surabaya (2008).

[6] A. Dictanata, L. Lutjito, Pengaruh penempatan tirai satu baris pada pilar jembatan terhadap kedalaman gerusan. J. Tek. Sipil Arsitektur **12**, 2 (2016). https://doi.org/10.21831/inersia.v12i2.12587

[7] H. Armono, A. Budipriyanto, Study on local scouring at Suramadu Bridge piers for structural integrity monitoring. Proc. 7th Int. Conf. Asian and Pacific Coasts (APAC), Bali, 107–112 (2013).

[8] F.H. Putra, D.G. Pratomo, Analisis arus dan transpor sedimen menggunakan pemodelan hidrodinamika 3 dimensi (Studi kasus: Teluk Ambon, Kota Ambon, Maluku). J. Tek. ITS **8**, 2 (2019).

[9] Consortium of Chinese Contractors, Detailed Design of Main Bridge for Indonesian Suramadu Bridge Project, (2005).

[10] H. Hersbach, B. Bell, P. Berrisford, G. Biavati, A. Horányi, J.M. Sabater, J. Nicolas, ERA5 hourly data on single levels from 1940 to present. C3S Climate Data Store (2023). https://doi.org/10.24381/cds.adbb2d47

[11] U. Lasminto, Studi potensi tampungan air sebagai sumber air baku Kota Surabaya. Simposium I Jaringan Perguruan Tinggi untuk Pembangunan Infrastruktur Indonesia (2016), No. 5 (2017). https://doi.org/10.12962/J23546026.Y2017I5.3113

[12] M.K. Khotimah, Validasi Tinggi Gelombang Signifikan Model Gelombang Windwave-5 Dengan Menggunakan Hasil Pengamatan Satelit Altimetri Multimisi, Universitas Indonesia (2022).

[13] N. Roustaei, Application and interpretation of linear regression. Med. Hypothesis Discov. Innov. Ophthalmol. **13**, 3, 151–159 (2024). https://doi.org/10.51329/mehdiophthal1506

A Comparison of Window Radius In Convolutional Neural Network for Estimating Chlorophyll-A in Water Bodies: Case In Laguna Lake, Philippines

Audina Maghfiroh[1*], *Januar* Kailani Suaeb[2]

[1]Department of Geomatics Engineering, Institut Teknologi Sepuluh Nopember, Surabaya, Indonesia

[2]Department of Informatics, Institut Teknologi Sepuluh Nopember, Surabaya, Indonesia

Abstract. Estimating chlorophyll-a in tropical inland waters is difficult because of complex optical conditions, limited field data, and frequent failures in atmospheric correction (over 70%). Traditional algorithms (R^2 below 0.45) do not perform well in Case-2 waters such as Laguna Lake in the Philippines. In this study, we introduce a two-stage transfer learning approach using 3D Convolutional Neural Networks (3D-CNNs). We use simulated pre-training with 126,000 samples, optimize spatial context with patch sizes from 5×5 to 11×11, and apply geometric augmentation to increase the dataset size by six times. Our process includes quality filtering with six Water Quality and Science Flags, per-band Z-score normalization, and stratified sampling to evaluate Sentinel-3 OLCI 16-band images at 300 m resolution. The best results came from the 9×9 patch model, which reached $R^2 = 0.5315$, RMSE = 0.6870, and MAE = 0.3221 log[10 μg/L] on 21,135 test samples. This improved baseline performance by 17.9% and outperformed traditional methods by 18 to 28%. Transfer learning was key, giving a 40% R^2 increase over direct training, and the two-stage method (simulated pre-training, head adaptation, full fine-tuning) led to further improvements. These findings show that deep learning with transfer learning and spatial context optimization (using a 9×9 patch) can greatly improve chlorophyll-a estimation in complex tropical lakes.

1 Introduction

Inland waters provide drinking water, fisheries, flood control, and cultural services for billions of people, making water quality a crucial element of global water security and ecosystem health. Deterioration of rivers, lakes, and reservoirs is now recognized as one of the greatest threats to human water security and aquatic biodiversity, caused by increased nutrient inputs, urbanization, and climate change [1]. These issues are clearly reflected in the United Nations Sustainable Development Goals, particularly SDG 6 (Clean Water and Sanitation), which aims to provide safe and affordable water, improve water quality, and protect and restore freshwater ecosystems. However, the SDG 6 Synthesis Report highlights that many countries,

*Corresponding author: 6016231019@student.its.ac.id

particularly in the Global South, still lack the monitoring capacity to track the status and trends of inland water quality, necessitating scalable, observation-based monitoring approaches.

Optically, Laguna Lake falls into the Case-2/inland water category, dominated by phytoplankton, suspended sediment, and colored dissolved organic matter (CDOM), making the chlorophyll-a signal difficult to distinguish from other constituents [4]. Research in productive turbid waters has shown that variability in bio-optical parameters such as phytoplankton absorption coefficients, particle backscatter, and CDOM concentration, and changes in phytoplankton community composition can lead to significant uncertainties in chlorophyll-a estimates, especially when using empirical or semi-analytical algorithms that assume uniformity [5]. The shallow, polymictic nature of Laguna Lake further complicates the situation because varying bottom reflectance and shallow water depths can significantly influence the sensor-detected radiance, allowing satellite observations to capture not only the water column but also the bottom substrate [6]. Even with the advent of new generations of high-resolution multispectral sensors, recent studies have emphasized that developing truly robust and transferable chlorophyll-a algorithms across complex optical lakes remains a significant challenge [7].

On the atmospheric side, Laguna Lake's tropical monsoon climate and proximity to the urban-industrial areas surrounding Metro Manila result in high cloud cover, aerosol loading, and bright land adjacency effects, which significantly degrade the quality of surface reflectance products. Studies have shown that atmospheric corrections in coastal and inland waters often fail or yield negative, noisy reflectance values, particularly in turbid and bright waters, when using standard, non-adapted approaches [8]. Furthermore, cloud cover, sun glint, and aerosol model uncertainties significantly reduce the number of usable images for chlorophyll-a monitoring, resulting in a much lower effective observation frequency compared to nominal satellite revisits [9]. At the same time, many algorithms and evaluations of chlorophyll-a products in inland waters are reportedly supported by only hundreds of in situ measurements that are unevenly distributed in space and time, making them susceptible to location bias and specific hydrological conditions [10]. Sampling strategies that focus on easily accessible points tend to result in station clusters near coastlines and water transport routes, even though the spatial-temporal variability of phytoplankton production in shallow lakes can be so high that many trophic events and conditions are not captured [9]. As a result, the distribution of training and test data for model development is often skewed towards meso-eutrophic conditions, while oligotrophic and other extreme classes are underrepresented, which can affect the accuracy of metrics and the generalizability of chlorophyll-a algorithms in complex tropical waters such as Laguna Lake.

To address the limitations of empirical algorithms, semi-analytical approaches and neural network-based models derived from radiative transfer simulations have been developed, including the MERIS Case-2 Water Algorithm and its successor, C2RCC, for Sentinel-2 and Sentinel [11]. These processors integrate atmospheric correction and neural network-based optical inversion trained on large-scale synthetic datasets, making them theoretically better able to handle complex optical variations. Technical documents and independent evaluations generally report the performance of C2RCC and similar algorithms with R^2 values of 0.4–0.5 for chlorophyll-a in coastal and inland waters, and often exhibit significant bias or error under highly turbid or shallow conditions [12]. Therefore, these generic semi-analytical/neural-network algorithms provide a strong starting point for comparison but do not fully address the issues of accuracy and robustness in highly variable tropical waters.

Further developments have involved the use of machine learning methods such as random forests, support vector machines (SVMs), and ensemble learning to study the nonlinear relationship between multispectral reflectance and chlorophyll-a using data-driven approaches.

Several studies in lakes across North America, Europe, and China have shown that random forests, gradient boosting, or multi-sensor combinations can outperform band-ratio or semi-analytical algorithms, with R^2 values typically ranging from 0.5–0.6 [13]. However, this high performance is often achieved in temperate lake systems with relatively abundant in-situ and imagery data, and still presents limitations in terms of transferability when applied across lakes or across regions in very different optical environments.

With advances in computer vision, deep learning methods are now being applied to estimate chlorophyll-a from Sentinel-2 and Landsat-8 imagery, primarily through the use of convolutional neural networks (CNNs) capable of simultaneously extracting spatial and spectral patterns. Recent research has shown that CNNs and deep neural network architectures can achieve or slightly exceed the performance of classical machine learning methods, with R^2 values remaining in the range of 0.5–0.6 for a variety of large lakes and reservoirs in temperate regions [14].

A convolutional neural network-based model, termed [15], has demonstrated efficacy in the end-to-end processing of Sentinel 3 satellite imagery for the purpose of monitoring chlorophyll-a levels in a lake located in the Philippines. WaterNet functions as an end-to-end model that seamlessly integrates feature extraction, band expansion, and chlorophyll-a estimation within a neural network framework. The model employs a 3D convolutional kernel, thereby fully leveraging both spectral and spatial information present in the imagery. Consequently, WaterNet is capable of managing artificial objects in the water, such as aquaculture cages, and mitigating the impact of satellite instrument errors. Additionally, data augmentation techniques are employed, and data rebalancing is proposed during data preprocessing to enhance the labeled in situ data by accounting for the variability and balance of chlorophyll-a concentration data for model training. Despite its capabilities, WaterNet still presents opportunities for optimization, as its accuracy remains suboptimal. By adjusting the window radius, the model can extract data with a broader and more detailed perspective than currently achieved by WaterNet.

This study makes several novel contributions that distinguish it from previous research. First, this study is among the first applications of a 3D-CNN architecture to Sentinel-3 OLCI data for estimating chlorophyll a in a tropical lake with complex optics, in which the spatial and spectral dimensions are integrated and learned to capture chlorophyll patterns at multiple spatial scales. Unlike most previous deep learning studies that focused on Sentinel-2 or Landsat-8 imagery in temperate lakes, this study targets tropical environments characterized by high turbidity, shallowness, and extreme spatial-temporal variability, thus broadening the geographic and optical scope of deep learning applications for water quality. Second, this study systematically optimizes the spatial context by comparing four patch sizes (5×5, 7×7, 9×9, 11×11), thus not only reporting the best model performance, but also providing a conceptual understanding of the most relevant context scales for a large shallow lake like Laguna Lake.

2 Methodology

2.1 Study area

Laguna Lake (Laguna de Bay), as shown in **Fig. 1,** is the largest lake in the Philippines, covering approximately 900 km^2. It is located southeast of Metro Manila and surrounded by dense urban, agricultural, and industrial areas. The lake is shallow, with an average depth of approximately 2.8 meters, and has complex hydrodynamics due to its horseshoe shape, numerous bays, and large river estuaries. This makes it highly susceptible to high turbidity, sediment resuspension, and nutrient input from a catchment area inhabited by approximately 12 million people. The combination of shallowness, sediment input, and anthropogenic pressures makes Laguna Lake highly sensitive to changes in water quality and difficult to monitor using remote sensing, owing to the strong contributions of suspended sediment and bottom substrate to the reflectance signal.

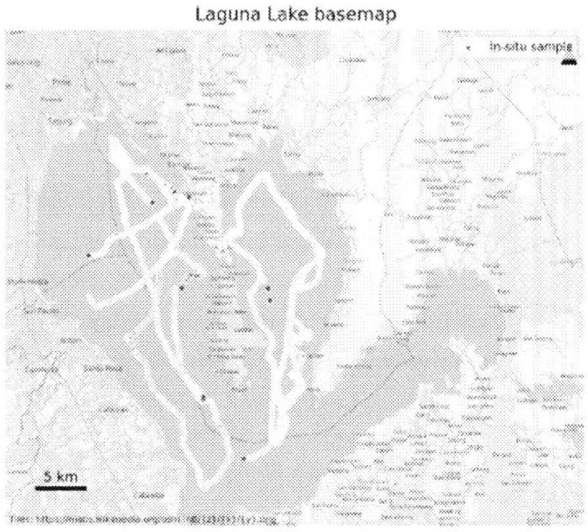

Fig. 1. Laguna lake area and map of Insitu collection area

Water quality in Laguna Lake generally ranges from mesotrophic to eutrophic, with chlorophyll-a levels reported at around 0.5–150 μg/L depending on season, location, and hydrological conditions. Increased phosphorus and nitrogen loads originate from point sources, such as industry and wastewater treatment plants, and nonpoint sources, including runoff from agriculture, settlements, and livestock. This is exacerbated by intensive fish farming activities in floating net cages. In some areas near river mouths and industrial areas, high concentrations of suspended sediment and nutrients cause the water to become turbid with a greenish-brown color, indicating strong spatio-temporal variability. This makes Laguna Lake a representative case study for testing satellite-based chlorophyll-a models in tropical lakes stressed by human activities.

Water quality monitoring in Laguna Lake has traditionally faced challenges such as access, infrastructure, security, and the cost of in-situ surveys, making field measurements infrequent, patchy, and poorly able to capture rapid changes in chlorophyll-a that can occur over days to weeks. In this regard, Sentinel-3 OLCI, with its 300 m spatial resolution and effective weekly revisit frequency, provides near-uniform spatial coverage of the lake surface and a much higher observation frequency at no additional data acquisition cost, thanks to the European Union's open data policy. Utilizing historical archives dating back to 2016 and combining relatively cloud-free satellite imagery from 2018–2020 with six in-situ campaigns from 2018–2019, this study designed a study period that represents seasonal variations and trophic conditions ranging from oligotrophic to eutrophic, while reflecting the practical constraints of monitoring in tropical lakes and the need for a realistic and replicable modeling framework for water resource management agencies.

2.2 Datasets

This study uses satellite data from the Ocean and Land Color Instrument (OLCI) aboard the Sentinel-3A and Sentinel-3 B satellites. OLCI produces color imagery of the ocean and land with a spatial resolution of 300 m (Full Resolution product) and 21 spectral channels spanning approximately 400–1020 nm. In this study, 16 main channels relevant for water quality applications are used, especially the blue, green, red, and red-edge channels that are sensitive to changes in chlorophyll-a, turbidity, and colored dissolved substances (CDOM). The product used is Level-2 Water Full Resolution (L2 WFR), processed through the EUMETSAT marine processing chain, providing reflectance and water-quality parameters ready for analysis after additional quality filtering. The spatial subset was defined by restricting the study area to the Laguna Lake boundary polygon obtained and cleaned from OpenStreetMap data, retaining only pixels within the lake. From each Sentinel-3 scene, only about 27.5% of pixels passed the quality check due to stringent filtering for clouds, land masses, sensor saturation, and atmospheric correction failure, which are common characteristics of inland water remote sensing in the tropics. Nevertheless, the number of valid pixels per scene (approximately 3,000 lake pixels) was sufficient for spatial patch extraction and model training, particularly when combined with multiple images from the 2018–2020 study period.

In-situ chlorophyll-a data collection was conducted through six field survey campaigns from November 2018 to March 2019, with each campaign covering approximately 35–50 stations and a total of 240 measurements successfully matched with Sentinel-3 imagery. Surface water samples were taken from a depth of 0–0.5 m and analyzed using a fluorometer instrument (such as Turner Designs) to obtain laboratory-calibrated chlorophyll-a concentrations. Measurement points were distributed across various morphological and anthropogenic stress zones in Laguna Lake, including areas near river mouths, fish-farming regions, and relatively cleaner open waters. To maintain consistency with satellite data, each in situ measurement was matched to the nearest Sentinel-3 pixel using a time-matching criterion of ±3 hours relative to the overpass time and a maximum spatial distance of 1 pixel (±300 m). Extremely deviant data were removed by applying a $\pm 3\sigma$ outlier criterion to the log-chlorophyll-a distribution, so that only physically plausible observations were retained. After this matching and cleaning process, a dataset representative of the seasonal and spatial variation of chlorophyll-a in Laguna Lake was obtained, which was then used as the basis for training, validating, and testing the 3D-CNN model. A summary of the dataset used in this study is provided in **Table 3.1**.

Table 2.1 Dataset description summary

Item	Description
Study area	Laguna Lake, Philippines
Period	November – March 2019
Satellite	Sentinel-3A and Sentinel-3B
Sensor	Ocean and Land Color Instrument (OLCI)
Spatial resolution	300 m
Temporal resolution	Daily
Field campaigns	6 campaigns
Field measurements	240 measurements
Satellite-field matches	52.5
Pixel QC pass rate	~27.5%
Avg valid pixels	~3100 pixels

2.3 Preprocessing

The data preprocessing phase aims to convert the raw Level-2 Sentinel-3 OLCI product into a clean, consistent, patch-based dataset ready for training a 3D CNN. The initial step is spatial subtraction, where each OLCI scene is cropped using Laguna Lake boundary polygons constructed from OpenStreetMap data and cleaned of small island artifacts and digitization errors, like Fig. 2. A point-in-polygon process is applied to each pixel to ensure that only pixels that are actually above the lake surface are retained, thus reducing the number of irrelevant pixels and lowering the computational burden compared to processing the entire satellite scene.

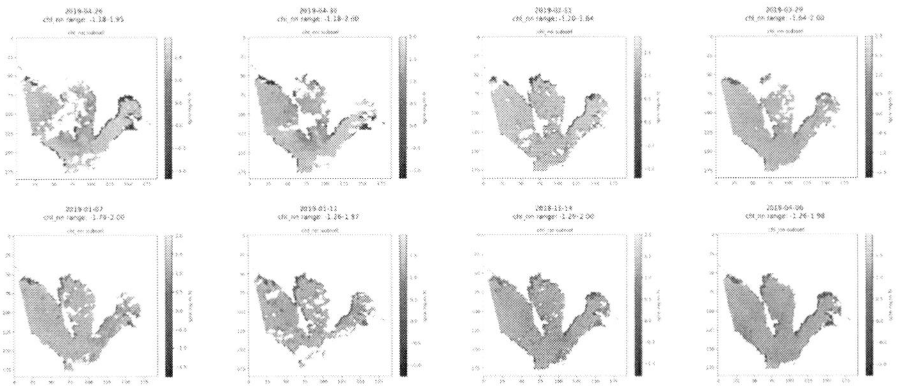

Fig. 2. Laguna Lake image data after preprocessing

The next step is quality filtering using a bitmask applied to the Water Quality Status Flag (WQSF) in the OLCI product. Several key bits considered problematic, such as LAND (land contamination), CLOUD (cloud), INVALID (invalid data), SATURATED (saturated sensor), AC_FAIL (atmospheric correction failure), and RWNEG_O (negative water irradiance), were combined into a single "bad" mask, and only pixels without these bits enabled were considered valid. This stringent filtering eliminated approximately 72.5% of the lake pixels but was crucial for maintaining the integrity of the spectral signal for modeling. From these valid pixels, spatial patches of 5×5, 7×7, 9×9, and 11×11 pixels were extracted, centered on the in situ measurement location or simulation sample point, provided that all pixels within the patch met the quality criterion of no cloud or land contamination within the spatial window.

After patch extraction, all spectral channels were normalized using a per-channel Z-score. In this process, each reflectance value was transformed to $z = (x - \mu) / \sigma$, using the mean (μ) and standard deviation (σ) calculated only from the training data. The same statistics were then consistently applied to the validation and test data to prevent information leakage. This normalization is essential for equalizing the scales across channels and stabilizing neural network training. Furthermore, data augmentation was performed geometrically through rotations (90°, 180°, 270°) and horizontal and vertical flips, which remain physically valid for isotropic water bodies. This combination of preprocessing steps resulted in a larger, statistically balanced set of patches free from major quality artifacts, thus improving the ability of the 3D-CNN model to learn the spectral-spatial patterns of chlorophyll-a in Laguna Lake.

2.4 Model Architecture

This study uses a three-dimensional convolutional neural network (3D-CNN) as its primary model, designed to simultaneously extract spectral and spatial information from Sentinel-3 OLCI image patches. Each input sample is represented as a four-dimensional tensor with an array (batch, spectral channels = 16, height, width), where the height and width dimensions are determined by the patch size (5x5, 7x7, 9x9, or 11x11 pixels). The model's core architecture is optimized for 9x9 patches. It comprises three consecutive convolutional blocks that extract features from low to high levels, followed by a global pooling layer, an intermediate fully connected layer with dropout, and a single output neuron to predict chlorophyll-a concentration. This design was chosen because it balances model complexity with the risk of overfitting and is flexible enough to handle variations in patch size by adjusting the input spatial dimensions without changing the number of feature channels or the layer structure.

In practice, all patch configurations (5x5, 7x7, 9x9, and 11x11) use a 3D-CNN architecture identical to that of the 9x9 patch model, in terms of layer order, number of convolutional channels, and the regression head structure at the end of the network. The primary difference between the configurations lies solely in the size of the spatial field of view extracted from OLCI imagery, allowing the models to learn the same patterns across different spatial contexts (local to mesoscale). This approach ensures that performance comparisons across patch sizes reflect the effects of spatial context alone, without being influenced by differences in model capacity or depth. Therefore, when results show that 9x9 patches provide the best compromise between accuracy and complexity, the conclusion can be attributed more confidently to the optimal choice of spatial context rather than to variations in neural network design.

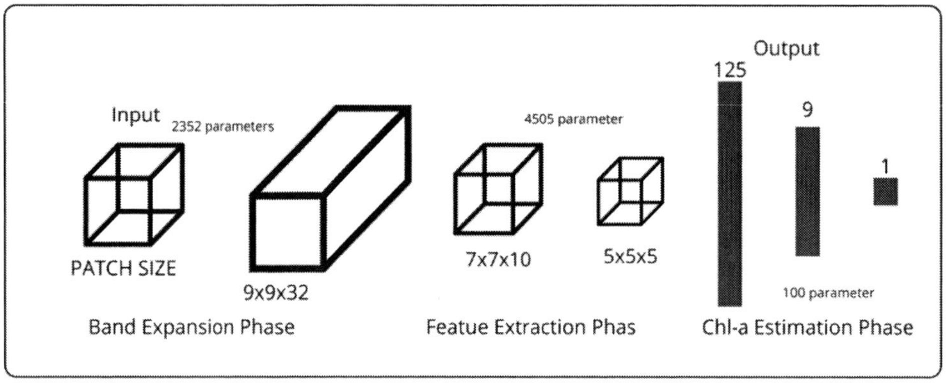

Fig. 3. Model Architecture

Technically, the first convolutional block converts the 16 input spectral channels into 32 features, followed by ReLU activation and batch normalization to stabilize the activation distribution during training. The second and third blocks sequentially increase the number of feature channels to 64 and 128, respectively, enabling the network to capture increasingly complex spectral-spatial combinations, including smooth gradients and heterogeneous patterns of chlorophyll-a in Laguna Lake. After the feature extraction stage, an adaptive average pooling layer reduces the spatial dimensionality to 1×1, generating a global feature vector that is then passed to a fully connected layer with an intermediate number of neurons (e.g., 128 to 64), followed by dropout to reduce overfitting. A final layer of single neurons without nonlinear (linear) activation is used to generate continuous predictions of chlorophyll-a concentration. This consistent architectural scheme across patch sizes simplifies the implementation of the three-stage transfer learning framework and facilitates model replication at other sensors or locations with minimal adjustments.

2.5 Training Strategy

The training strategy implemented in this study was designed to optimize the use of limited simulated and in situ datasets through a three-stage transfer-learning framework. In Stage 1 (simulated pre-training), the 3D-CNN model was trained with approximately 126,000 simulated samples (after augmentation) generated from bio-optical modeling with a wide range of chlorophyll-a, turbidity, CDOM, and water depth. All network parameters were made trainable using the Mean Squared Error (MSE) loss function, the Adam optimizer, an initial learning rate of 0.001, and approximately 50 epochs until convergence on the simulated data. This stage aimed to build a robust, generalizable spectral-spatial feature extractor, such that the network's initial weights already contained a realistic representation of chlorophyll-a patterns before being exposed to more limited field data.

In Phase 2 (in-situ head adaptation), the model previously trained on simulated data is fine-tuned on the Laguna Lake domain using an in-situ dataset augmented to approximately 174,000 patches. In this phase, the convolutional backbone is locked, leaving only the final regression head layer trainable. This is intended to fine-tune the mapping from features to observed chlorophyll-a values without "forgetting" the general features learned in Phase 1. Training lasts for approximately 30 epochs with a higher learning rate (0.002) to accelerate adaptation, while still using the MSE loss function and the Adam optimizer. This two-stage approach helps mitigate the risk of overfitting on the relatively small in-situ dataset while addressing the domain differences between the simulated data and the actual optical conditions in Laguna Lake.

Phase 3, known as full fine-tuning, is the phase in which all network layers, including the backbone and head, are retrained using the same in-situ dataset. At this stage, the learning rate is reduced substantially (e.g., to 0.0001) to enable finer adjustments without causing gradient instability. Training lasts for approximately 50 epochs, with early stopping based on the R^2 metric applied to the validation set and a patience limit of approximately 15 epochs to prevent overfitting. Overall, the data are split into 70% for training, 15% for validation, and 15% for testing, stratified by chlorophyll-a class, and weighted sampling is used to address the imbalance in the rare oligotrophic class. The entire training process is performed using the PyTorch framework on an NVIDIA GPU with 16 GB of memory, requiring approximately 20–30 hours per model to complete all three training stages.

2.6 Evaluation

The performance of the 3D-CNN model in predicting chlorophyll-a concentration was evaluated using a combination of quantitative metrics and visual diagnostic analysis. This was done to ensure that the model not only achieved good numerical accuracy but also exhibited stable and physically plausible behavior. The four main regression metrics used were the coefficient of determination (R^2), Root Mean Square Error (RMSE), Mean Absolute Error (MAE), and Mean Absolute Percentage Error (MAPE). The R^2 value is used to measure how much of the observed chlorophyll-a variation can be explained by the model, while RMSE and MAE provide an indication of the magnitude of the average error in log-chlorophyll-a units relative to the target. MAPE was used cautiously to assess relative error, noting that its interpretation becomes less reliable at very small chlorophyll-a values. The combination of these four metrics allows for a balanced assessment of absolute and relative accuracy, both at low and high concentration ranges.

3 Result

3.1 Model Performance Comparison

This subsection compares the performance of 3D-CNN models across four spatial patch sizes (5×5, 7×7, 9×9, and 11×11) using key regression metrics on the test set. Overall, all patch configurations demonstrated competitive performance, like in **Table 3.1,** in predicting chlorophyll-a in Laguna Lake, with R^2 values ranging from 0.50 to 0.53 and relatively similar RMSE and MAE values among the models. However, the 9×9 patch stood out as the configuration with the best balance between accuracy and model complexity, producing the highest R^2 value (≈0.5315) and the lowest RMSE among all models. The 11×11 patch, despite having the largest number of parameters, yielded the smallest MAE, indicating an advantage in reducing mean error, but did not yield a statistically significant improvement over the 9×9 patch. Meanwhile, the 5x5 patch performed only slightly worse than the 9x9 patch despite being the lightest architecture, whereas the 7x7 patch underperformed the other three configurations, suggesting a mismatch between model capacity and available spatial context.

Table 3. 1 Result comparison each Patch Size

Patch Size	RMSE	R^2	MAE	MAPPE
5x5	0.5168	0.6948	0.3502	104.6%
7x7	0.5025	0.6992	0.3428	96.6%
9x9	0.5315	0.6870	0.3221	132%
11x11	0.5191	0.6889	0.3139	125.7%

Statistical analysis using ANOVA and post hoc tests, such as Tukey's HSD, revealed that patch size significantly affected model performance ($p < 0.001$). The 9x9 patch proved significantly superior to the 5x5 and 7x7 patches in several key metrics. However, the difference between the 9x9 and 11x11 patches was not statistically significant, suggesting that the 9x9 patch is more advantageous with respect to computational efficiency and model size. These results suggest that adding spatial context around the central pixel is essential to some extent. Still, beyond the optimal range, the additional benefits become limited and may even lead to overfitting or the introduction of irrelevant spatial noise. These findings underpin the selection of the 9x9 patch as the primary configuration for further analysis in this study.

3.2 Transfer Learning Effectiveness

The effectiveness of the three-stage transfer learning framework was assessed by comparing the performance of several alternative training strategies, including direct training using only in-situ data without pre-training, simulation-only pre-training, a two-stage combination (simulation + head adaptation), and the complete three-stage framework (simulation + head adaptation + full fine-tuning). The results showed that models trained without pre-training tended to overfit, with lower validation and test R^2s. In contrast, models relying solely on simulation pre-training struggled to adapt to differences in optical characteristics between simulated data and real-world conditions in Laguna Lake. In contrast, the two-stage (Stages 1 and 2) and especially the full three-stage (Stages 1,2, and 3) strategies yielded significant improvements in R^2 on both the validation and test sets, with the complete framework achieving the highest R^2 and the most stable training convergence.

Quantitatively, the three-stage transfer learning approach improved R^2 by approximately 40% compared to the scenario without pre-training, while reducing the need for extensive in-situ data. A pre-training stage on a simulated dataset is essential for establishing standard spectral-spatial features. In contrast, a head adaptation stage allows for rapid calibration to the local chlorophyll-a distribution and optical characteristics of Laguna Lake without disrupting the learned features. A complete fine-tuning stage then refines all layers to adapt to the spectral nuances and spatial patterns specific to this tropical lake. The gradual improvement pattern from Stage 1 to Stage 3 demonstrates that each stage makes a clear contribution to performance gains, and that a combination of all three is necessary to achieve optimal results in complex, data-poor aquatic environments.

3.3 Data Augmentation Impact

The effect of data augmentation on model performance was evaluated by comparing two training configurations: one without augmentation (using approximately 29,000 original patches) and one with geometric augmentation, which expanded the training set to approximately 174,000 patches. The results showed that augmentation increased R^2 on the test set and consistently reduced MAE, while also reducing the performance gap between the training, validation, and test sets. The configuration without augmentation tended to exhibit symptoms of overfitting, in which the model learns overly specific patterns from a limited sample and fails to generalize well to new samples. In contrast, augmentation provided additional spatial variation that remained physically valid (because the water body was assumed to be isotropic), allowing the network to learn a representation of the chlorophyll-a pattern that was more robust to changes in patch position and orientation.

3.4 Comparison with the Literature

To evaluate the relative performance of the proposed approach compared with existing methods, the 9x9 patch 3D-CNN model was compared with various algorithms reported in the literature, including traditional band-ratio-based algorithms, semi-analytical algorithms such as C2RCC, and machine learning models such as Random Forest, as shown in **Table 3.2**. The comparison results show that the model in this study is among the top performers, with R^2 and MAE values comparable to or better than those of many previous studies, particularly given that Laguna Lake is a tropical body of water with complex optics and relatively limited in situ data. Compared with simple band ratio and C2RCC, the R^2 improvement is substantial (18–40%), whereas compared with Random Forest and WaterNet models, the performance improvement is still observed, albeit in a more moderate range.

Table 3.2 Comparison with previous methods

Category	Methods	R^2
Semi-analytical	C2RCC	0.4 – 0.5
Classical Machine Learning	Random Forest, Gradient Boosting, multi-sensor ML	0.5 – 0.6
Deep Learning	Ours	0.68

This comparison highlights two important points. First, the integration of spatial context through 3D-CNN and optimal patch selection (9x9) provides a clear advantage over approaches that rely solely on single-pixel information or simple spectral combinations. Second, the three-stage transfer learning framework and data augmentation developed in this study achieve competitive performance relative to studies of temperate lakes, which typically have more extensive field data. Thus, the results obtained not only demonstrate the scientific feasibility of this approach but also position it as a promising candidate for adoption or further development in the context of inland water quality monitoring in tropical regions facing similar data limitations and optical challenges.

4 Discussion

This chapter explores in depth why the 9×9 patch is the best configuration for estimating chlorophyll-a in Laguna Lake, and how a transfer learning framework and data augmentation improve its performance. Spatially, the 2.7 km window covered by the 9×9 patch proved to be an effective "sweet spot" for capturing chlorophyll autocorrelation patterns at mesoscopic scales (1–3 km), providing a broad enough context to distinguish water quality gradients without introducing excessive noise from unrepresentative pixels. In terms of model capacity, the 3D-CNN architecture with approximately 119,000 parameters on this patch offers a good balance between complexity and the risk of overfitting: richer in representation than the 5×5 and 7×7 configurations, which tend to underfit, yet still more lightweight and stable than the 11×11 configuration, which potentially captures local, non-generalized patterns. Computationally, the model size of approximately 4.7 MB and the fast inference time make the 9×9 configuration realistic for operational deployment and integration into routine monitoring systems.

The model's success was further enhanced by a three-stage transfer-learning strategy that addressed the scarcity of in situ data and the varying optical characteristics of tropical waters. Pre-training on a simulated dataset enabled the network to learn general and robust chlorophyll-a spectral-spatial patterns, significantly reducing the need for field samples. A head-adaptation stage (Stage 2) then focused on learning to adapt to the real-world data distributions of Laguna Lake without destroying the established underlying features, while a full fine-tuning stage (Stage 3) refined the entire network to adapt to complex field conditions. The combination of these three stages resulted in approximately a 40% performance improvement over direct training without pre-training, while also suppressing overfitting. Statistically validated geometric augmentation amplified this effect by increasing the dataset size sixfold without altering the signal's physical properties, thereby providing the spatial diversity necessary for the model to learn more general patterns that are insensitive to image orientation.

However, the discussion highlights several challenges and operational implications of this approach and opens the door to generalization to other tropical lakes. The high pixel-rejection rate due to atmospheric correction failures (approximately 72.5%) creates a trade-off between spatial coverage and data quality; therefore, a multi-date composite approach is recommended to ensure reliable weekly or biweekly maps. Decreased performance in the very low and very high chlorophyll ranges suggests the need for more targeted sampling campaigns to enrich samples at the tail of the distribution, while issues with temporal and spatial generalization indicate the need for future cross-temporal and multi-lake testing. From an operational and regional perspective, the combination of 3D-CNN, three-stage transfer learning, and geometric augmentation provides a methodological framework that can be transferred to other tropical lakes at relatively low additional cost, provided a small amount of in-situ data is available for adaptation, thus supporting broader and more sustainable water quality monitoring in developing countries.

5 Conclusion

This research successfully developed a robust deep learning-based chlorophyll-a estimation system for Laguna Lake, addressing key challenges in water quality monitoring in complex and data-poor tropical inland waters. Through a three-stage transfer learning framework involving pre-training on simulated data, adaptation to in-situ data, and full fine-tuning, a 3D-CNN model with 9x9 patches achieved an R^2 of 0.5315 on test data, an improvement of approximately 17.8% over standard algorithms and comparable performance to state-of-the-art methods in temperate regions. Spatial context optimization showed that 9x9 patches (2.7 km in size) provided the best compromise between accuracy and computational efficiency, making the approximately low memory size model operationally ready for deployment on typical computing infrastructure used by water resource management agencies. Scientifically, this study provides several important contributions, including the first demonstration of the application of 3D-CNN to Sentinel-3 OLCI data for tropical inland waters, empirical evidence that spatial context optimization through patch size comparison can generate clear design recommendations, and proof that a three-stage transfer learning framework can improve performance by up to 40% compared to direct training without pre-training.

References

1. C. J. Vörösmarty, P. B. McIntyre, and M. O. Gessner, "Global threats to human water security and river biodiversity," *Nature*, vol. **467**, pp. 555–561, 2010.
2. C. B. Mouw *et al.*, "Aquatic color radiometry remote sensing of coastal and inland waters: challenges and recommendations for future satellite missions," *Remote Sens Environ*, vol. **160**, pp. 15–30, 2015.
3. G. Dall'Olmo and A. A. Gitelson, "Effect of bio-optical parameter variability on the remote estimation of chlorophyll-a concentration in turbid productive waters: experimental results," *Appl Opt*, vol. **50**, no. 15, pp. 2111–2125, 2011.
4. T. Kutser, "The possibility of using remote sensing to monitor shallow lakes," *Hydrobiologia*, vol. **698**, pp. 25–42, 2012.
5. K. Shi, Y. Zhang, G. Zhu, B. Qin, D. Pan, and C. Huang, "Challenges and perspectives in remote sensing of inland water quality in the era of high-resolution multispectral satellites," *Remote Sens Environ*, vol. **280**, p. 113182, 2022.
6. Q. Vanhellemont, "Adaptation of the dark spectrum fitting atmospheric correction for aquatic applications of the Landsat and Sentinel-2 archives," *Remote Sens Environ*, vol. **225**, pp. 175–192, 2019.
7. N. Pahlevan, S. Sarkar, B. A. Franz, S. V Balasubramanian, and J. He, "Satellite-based remote sensing of water quality in lakes and reservoirs: Challenges and opportunities," *Water (Basel)*, vol. **9**, no. 7, pp. 1–24, 2017.
8. D. Odermatt, A. Gitelson, V. E. Brando, and M. Schaepman, "Review of constituent retrieval in optically deep and complex waters from satellite imagery," *Remote Sens Environ*, vol. **118**, pp. 116–126, 2012.
9. K. Toming, T. Kutser, L. Tuvikene, M. Viik, T. Tamm, and T. Nõges, "Spatio-temporal variability of phytoplankton primary production in a large shallow lake," *Hydrobiologia*, vol. **780**, pp. 185–199, 2016.
10. R. Doerffer and H. Schiller, "The MERIS Case 2 water algorithm," *Int J Remote Sens*, vol. **28**, no. 3–4, pp. 517–535, 2007.
11. EUMETSAT, "OLCI Algorithm Theoretical Basis Document (OLCI ATBD) – Neural network atmospheric correction and in-water processing (C2RCC)," EUMETSAT, Darmstadt, Germany, 2017.
12. G. Tilstone, "Performance of ocean colour chlorophyll-a algorithms for the coastal and shelf seas of the North East Atlantic," *Remote Sens Environ*, vol. **189**, pp. 1–19, 2017.
13. J. Chen, Y. Li, and Y. Zhang, "Estimation of chlorophyll-a concentration in inland waters using Sentinel-2 MSI imagery and ensemble learning," *Remote Sens (Basel)*, vol. **12**, no. 3, p. 441, 2020.
14. Y. Zhang, K. Shi, and B. Qin, "Deep learning-based retrieval of chlorophyll-a concentration in large shallow lakes from Sentinel-2 MSI," *Science of the Total Environment*, vol. **776**, p. 145928, 2021.
15. M. Syariz, C.-H. Lin, M. Nguyen, L. Jaelani, and A. Blanco, "WaterNet: A Convolutional Neural Network for Chlorophyll-a Concentration Retrieval," *Remote Sens (Basel)*, vol. **12**, p. 1966, Jun. 2020, doi: 10.3390/rs12121966.

Initial Assessment of Hydrocarbon Potential in the Alor Basin Using Integrated Hydrographic Data

Irena Hana Hariyanto* and *Alvina* Viasty

Department of Geomatics Engineering, Institut Teknologi Sepuluh Nopember, Surabaya, 60111, Indonesia

Abstract. Identification of hydrocarbons is a preliminary step in oil and gas exploration. By using hydro-acoustic data, this study aims to assess the hydrocarbon potential of the Alor Basin. Several data were acquired during the Jala Citra III Expedition in the Flores Sea in 2023, including Multibeam Echosounder and Marine Magnetometer. Bathymetric processing indicates that the Alor Basin reaches a maximum depth of 4,777.68 meters. The seabed is predominantly composed of clay-rich sediments, inferred to result from the deposition of fine-grained marine biogenic material or volcanic ash. This sediment analysis indicates depositional environments that may influence hydrocarbon generation and accumulation processes. Marine magnetometer data revealed residual magnetic anomalies ranging from −1,031.022 nT to 361.214 nT, with negative anomalies corresponding to intense tectonic activity. This interpretation is supported by fault structures analysis along the seismic profiles from the Marine Geological Research and Development Centre (P3GL), which may act as structural traps or migration pathways for hydrocarbons. Integrated bathymetric, magnetic, and seismic interpretations explain the basin's geodynamic evolution and its correlation to hydrocarbon system formation. The fault patterns, seabed morphology, and magnetic anomalies all indicate active tectonic processes, suggesting that the Alor Basin holds significant potential for hydrocarbon accumulation and future exploration.

1 INTRODUCTION

Ocean basins are crucial for storing natural resources, including oil and gas. Currently, there are 128 oil and gas basins in Indonesia, and approximately 53% of these remain unexplored [1]. These basins typically form in areas of interplate subduction that result in volcanic activity. This creates back-arc basins and fore-arc basins along volcanic belts. Back-arc basins are underwater basins formed behind island arcs and can produce large quantities of crude oil decades later [2].

Research in Eastern Indonesian waters has revealed significant geological structures, revealing the existence of seven basins in the Flores Sea and its surroundings. One of the identified basins is the Alor Basin in the eastern Flores Sea. The Alor Basin is located in the back-arc region of the Lesser Sunda Islands (LSI) and contains submarine volcanoes. This leads to the formation of back-arc thrust faults, which at certain points can lead to plate collisions, leading to deformation and the formation of strong thrust faults. This situation is closely related to sedimentation, which can store hydrocarbons used in oil and gas exploration [3].

Petroleum, or hydrocarbons, is a general term for organic compounds found in crude oil and natural gas [4]. Hydrocarbons in the ocean are formed from the accumulation of organic matter, such as plankton and marine plants, trapped by seafloor sediments over millions of years. In geophysics, these traps are referred to as reservoir rocks. Hydrocarbon identification is generally performed through seismic analysis to determine the geological structure and reservoir potential before drilling [5]. This study used a combination of hydrographic survey data and seismic sections to identify the presence of hydrocarbons in the Alor Basin.

This study utilizes various hydrographic instruments for further processing and analysis of the potential presence of hydrocarbons in the Alor Basin. Sounding data from a multibeam echosounder can show water depth and backscatter data can determine the characteristics of sediment in the first layer or seabed [6]. Magnetometer data is used to identify magnetic anomalies in the study area [7]. Areas suspected of containing hydrocarbons will have different magnetic

* Corresponding author: irena@its.ac.id

values from their surroundings. Hydrographic surveys are conducted as an innovation and support in the hydrocarbon identification process. This is in line with Sustainable Development Goals point nine regarding industry, innovation, and infrastructure.

2 METHOD

2.1 Research Location

This research location is in the eastern Flores Sea, specifically in the Alor Basin. Figure 1 shows the eastern Flores Sea, Indonesia. The area is located between 7°01'51" to 8°23'56" South and 122°54'35" to 124°40'55" East. This water is classified as tropical water and is famous for its rich marine natural resources.

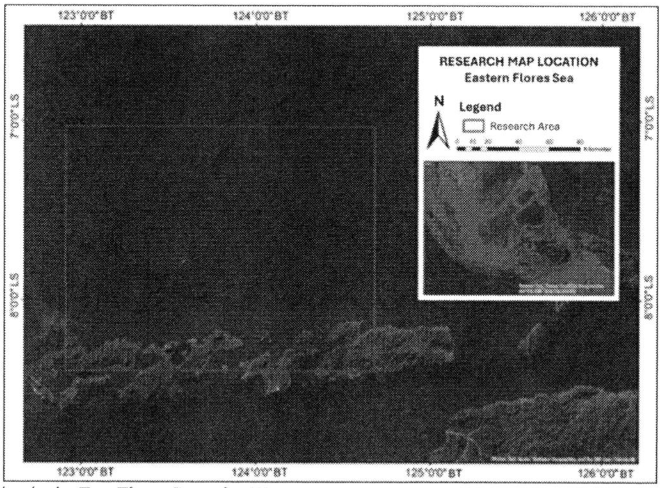

Fig. 1. Research Location in the East Flores Sea, Alor Waters

2.2 Data and Equipment

The data used in this study comes from the Jala Citra III Expedition, acquired from March to May 2023. The data was installed on the KRI SPICA 934, owned by the Indonesian Navy's Center for Hydro-Oceanography (Pushidrosal), which was used during the expedition. The details of the data utilized in this research are described in the table below.

Table 1. Setting Word's margins.

No	Data	Information
1	Raw Bathymetry and Backscatter Strength	Kongsberg EM 302 Multibeam Echosounder
2	Tides	TPXO 8.0 model
3	Sound Velocity	Sound Velocity Profiler (SVP)
4	Magnetic Value	G-882 Cesium Marine Magnetometer

In addition, this study also used declination, inclination, and IGRF correction data from the National Geophysical Data Center (NGDC) website of the National Oceanic and Atmospheric Administration (NOAA), accessed through: https://www.ngdc.noaa.gov/geomag/calculators/magcalc.shtml?useFullSite=true

To support the research, all data were processed using several software programs that are compatible with spatial data processing. These include Qimera, a bathymetry data processing software; Fledermaus FMGT, a backscatter data processing software; Oasis Montaj, a magnetic anomaly data processing software; and other general spatial data software.

2.3 Data Processing

The data processing phase for generating an analysis of the potential presence of hydrocarbons in the study area consists of several steps. Preparation for multibeam echosounder data processing includes patch test data calibration, raw data file preparation, tidal correction data, and sound velocity correction data. Patch test calibration using Qimera software is performed to correct the instrument's sensors in multibeam echosounder data processing. Patch test calibration includes roll, pitch, and yaw (heading). Raw multibeam echosounder data (*.all) is corrected using tidal data (*.tid) and sound velocity data (*.asvp). The corrected data is cleaned of noise and spikes. Data cleaning is performed using Qimera software and produces output in the form of depth data (*.asc) and other formats (*.gsf) used in backscatter processing.

The depth data (*.asc) is then layout-processed using spatial data processing software to produce a bathymetric map. Data in *.gsf format is used in Fledermaus FMGT software to perform mosaic and then perform ARA analysis, which will help in classifying the types of seabed sediment based on the characteristics of the acoustic signal reflection that occurs [8]. Magnetometer data processing is carried out with daily data correction (diurnal) and IGRF correction.

Daily data correction (diurnal) is carried out to reduce the impact of differences in acquisition time and solar radiation [9].

$$\Delta H = H_{total} \pm \Delta h_{\text{daily}} \quad (1)$$

information:

H_{total} : Total daily correction value
Δh_{daily} : Difference in daily values

IGRF correction is carried out to reduce the influence of the Earth's main magnetic field on the acquisition data [9].

$$\Delta H = H_{total} \pm \Delta h_{\text{daily}} \pm H0 \quad (2)$$

information:

H_{total} : Total daily correction value
Δh_{daily} : Difference in daily values
$H0$: International Geomagnetic Reference Field (IGRF)

After that, the Gridding process is carried out to obtain an overview of the total magnetic field anomaly value. Reduction to the equator is carried out on the residual anomaly data to form a magnetic anomaly pattern into a monopole, thus simplifying the analysis process [10]. The separation of regional and residual magnetic field anomalies from the total magnetic field is carried out using a Butterworth filter. The separation produces regional anomaly data and residual anomaly data.

3 RESULT AND DISCUSSION

3.1 Depth Determination

The bathymetric data comes from the 2023 Jala Citra III Expedition. The area surveyed in May 2023 included 90 lanes within the Alor Basin. Data processing was performed using Qimera software. The process began with sensor corrections, including patch tests and data corrections, utilizing sound velocity and tidal data. This processing used a resolution of 0.000898 ° × 0.000898 ° due to the large processing area. Afterward, the data were cleaned to remove noise and spikes to produce a corrected bathymetric map.

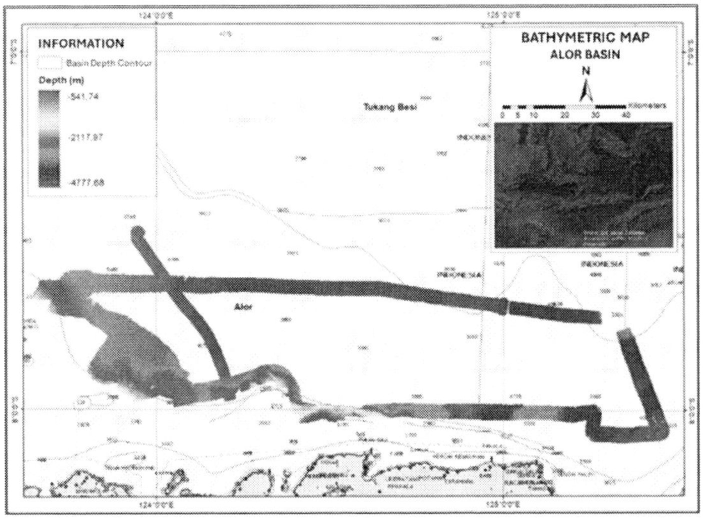

Fig. 2. Bathymetric Map of the Research Location

The processing yields varying depths and marine morphology visible from various angles. Overall, the data indicate that the waters in the Alor Basin are deep waters, namely bathyal and abyssal zones [11]. The average depth value of the measurement area is 3643 m, with the maximum depth value in the southeast showing a depth of up to 4777.68 m. On the east side, there is a steep slope due to the presence of seamounts outside the basin area, namely Baruna Komba and Abang Komba, and in the southern part of the basin is closer to the Flores Islands. Meanwhile, the central area to the eastern side shows a gentler depth but with deeper values.

3.2 Acoustic Backscatter Value

Backscatter data processing was performed using the corrected data. The data was processed using FMGT software. Processing began with configuration settings and the creation of mosaics containing intensity values. These intensity values will be used in Angular Response Analysis (ARA) to classify sediment types in the Alor Basin.

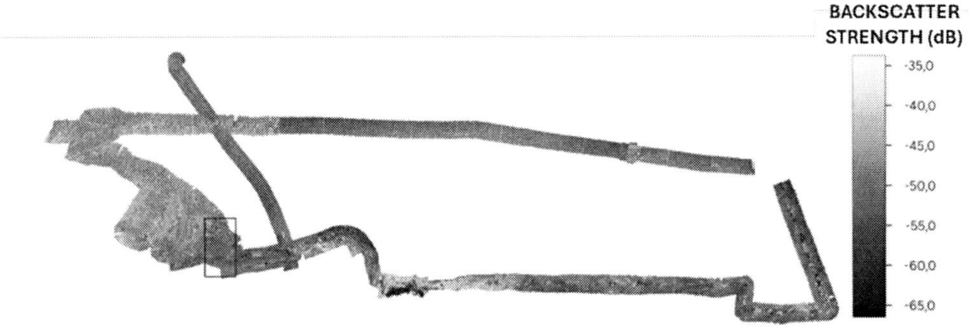

Fig. 3. Backscatter Mosaic Building Results

The mosaic creation results are shown in Figure 3. The resulting mosaics have a range of values indicating variations in backscatter intensity across the area. The study area has an intensity between -0.65 and -0.35 dB. Backscatter mosaics with high energy are indicated by light colors, while mosaics with dark colors indicate low energy.

ARA processing indicates that almost the entire study area has similar sediment types. The visualization in Figure 4 shows that clay sediment types dominate the study area. The Angular Response Curve was used to assess the level of uniformity of seabed material on the surface and identify the suitability of the mosaic data between the right and left sides of the nadir. The blue curve in Figure 6 shows a symmetrical trend in average reflectance with a peak around the nadir angle (90°) and decreases towards smaller angles. The red and green curves represent actual data from the left and right sides of the nadir. Fluctuations in both curves reflect variations in seabed surface reflectance classified as clay with patterns that experience slight deviations due to surface irregularities or local conditions. This is consistent with the Seabed Surface Sediment Map published by the Center for Marine Geology Research and Development (2010), which reveals mud sediment types, primarily silt and clay containing 20% sand, including blue mud, green mud and clay, black mud and clay, coral mud, volcanic glass, volcanic mud, gray clay, and sandy mud [12].

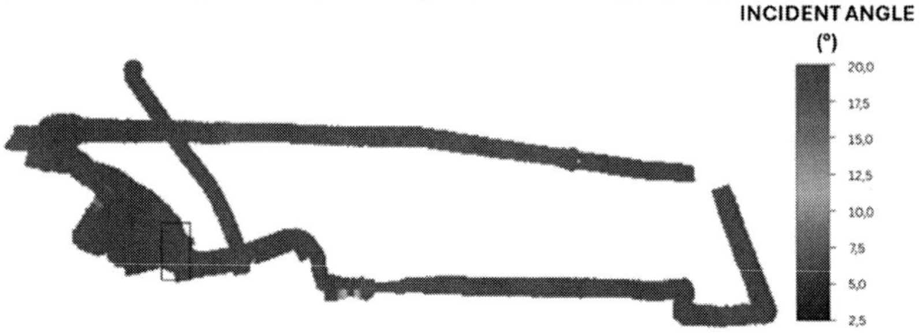

Fig. 4. The Result of *Angular Response Analysis*

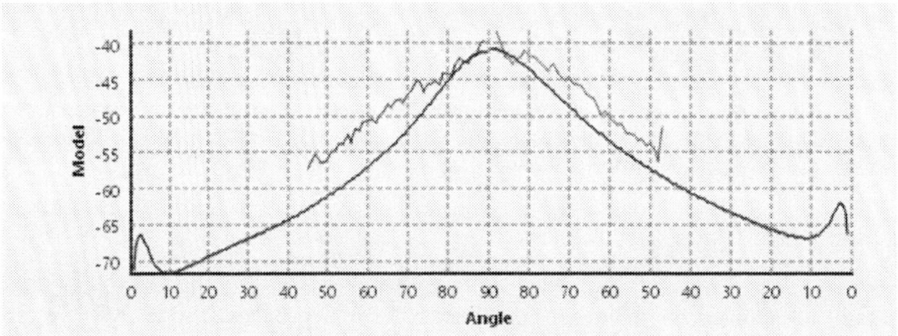

Fig. 5. *Angular Response Curve*

A variety of geological information can be obtained from sediment analysis, reflecting the geological processes and dynamics of a region. The Alor Basin is part of a deep marine basin characterized by low energy and slow sedimentation rates, which strongly support the deposition of fine particles such as clay. Clay can accumulate with other deep-sea materials such as radiolarians and foraminifera. Clay is thought to originate from volcanic ash transformed into clay minerals such as montmorillonite and illite, as well as from the weathering of rocks on land that are then carried to the sea by rivers or ocean currents. Furthermore, hydrothermal activity on the seafloor can also produce certain types of clay, such as smectite.

3.3 Magnetic Anomaly

Magnetic data was processed using the Oasis Montaj software. These magnetic field variations are caused by the contrast or distribution of magnetic properties between rock types within the Earth's crust. This results in a non-uniform Earth's magnetic field, known as a magnetic anomaly. The total magnetic field intensity map was obtained after performing diurnal correction and IGRF correction. Figure 6(a) displays the total magnetic field intensity results. This map still has a dipole shape, requiring reduction toward the equator to make it a monopole for easier interpretation. Figure 6(b) shows the total magnetic field intensity results after being reduced toward the equator and converted to a monopole. The resulting total magnetic anomaly values range from -2307.669 nT to 197.810 nT. However, this value is a combination of regional and residual magnetic values.

Separating regional and residual anomalies requires depth information from the measured data. Therefore, depth estimation was performed using the radially averaged power spectrum method. The regional zone has a depth of almost 4000 m, the residual zone is at a depth of around 1500 m, while white noise appears at a depth of around 1000 m. This corresponds to the seabed depth values obtained in the area. The regional zone is characterized by a sharply decreasing graph, while the residual zone shows a more gradual graph. The gradient shift indicating the boundary between the zones occurs on the x-axis at a value of 0.45 km^{-1} and 1.12 km^{-1}.

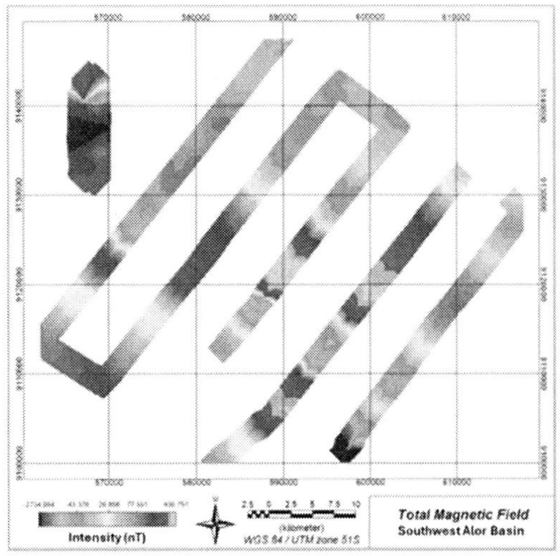

(a)

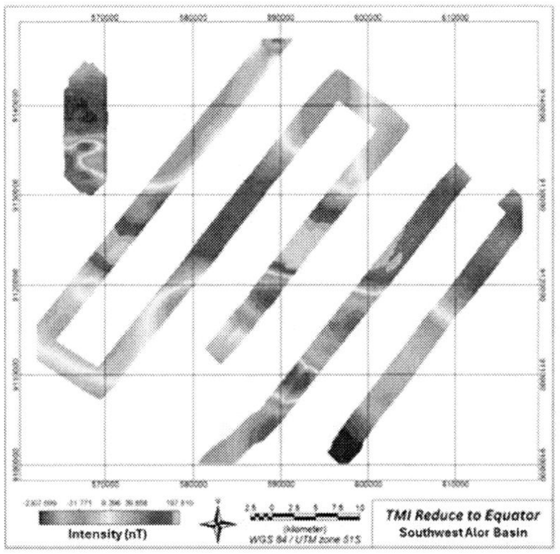

(b)

Fig. 6. *Total Magnetic Intensity* and Equatorial Reduced TMI

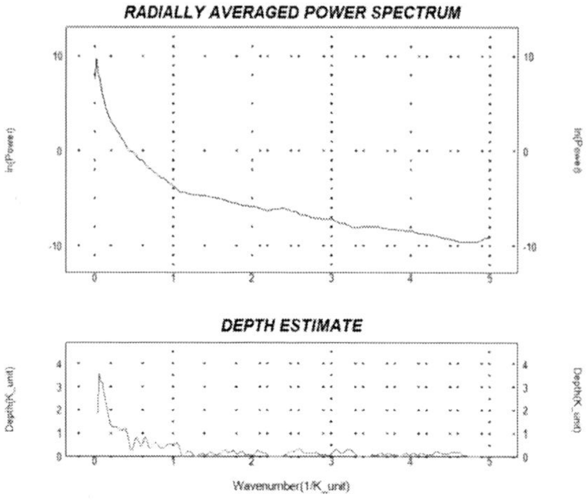

Fig. 7. The Graph of Radially Averaged Spectrum and Depth Estimate

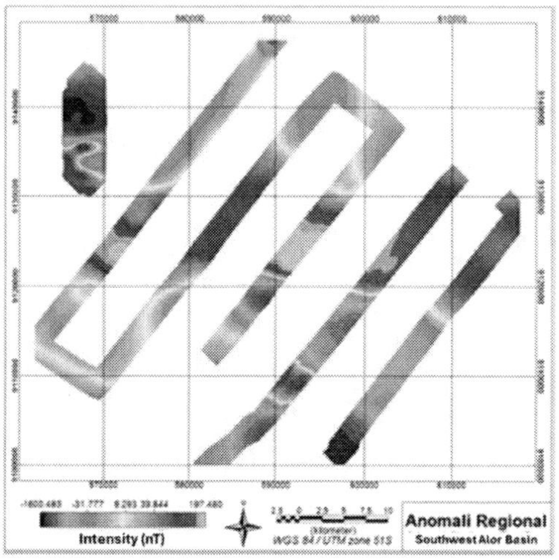

Fig. 8. Regional Anomaly Separation Results

Anomaly separation was performed using a Butterworth filter of degree 8. The choice of central wavelength depended on the resulting graph. For regional anomalies, a central wavelength of 1540 was used, and for residual anomalies, a central wavelength of 2860 was used. Figure 8 visually demonstrates the differences between regional and residual anomalies. Regional anomalies represent regional subsurface conditions, covering a large area and at significant depths, resulting in homogenous rock characteristics with minimal striking geological structural contrasts. In contrast to regional anomalies, residual anomalies provide geological information at shallower depths. This presents heterogeneity in the rock composition and the possibility of a more diverse range of geological structures.

The similar colors in Figures 8 and 9 for regional and residual anomalies do not necessarily reflect the same magnetic field values. The magnetic field values for regional anomalies range from -1600.483 nT to 197.480 nT. Meanwhile, the magnetic field values for residual anomalies range from -1031.022 nT to 361.214 nT. Both are nearly identical. Residuals have a smaller negative range and a larger positive range than regional anomalies. A magnetic field value that tends to be negative indicates the presence of lithology with a low level of magnetization. Another possibility is that the area was previously composed of highly magnetized rocks, but underwent hydrothermal alteration or certain chemical reactions that caused demagnetization in the rocks. Conversely, a positive range of magnetic field values reflects the presence of ferromagnetic or highly magnetized materials and the possibility of an increase in magnetization in the host rock (country rock). Anomaly boundaries have been identified in the residuals, with positive anomalies indicating the presence of mining or mineral deposits, while negative anomalies indicate the presence of petroleum deposits [13].

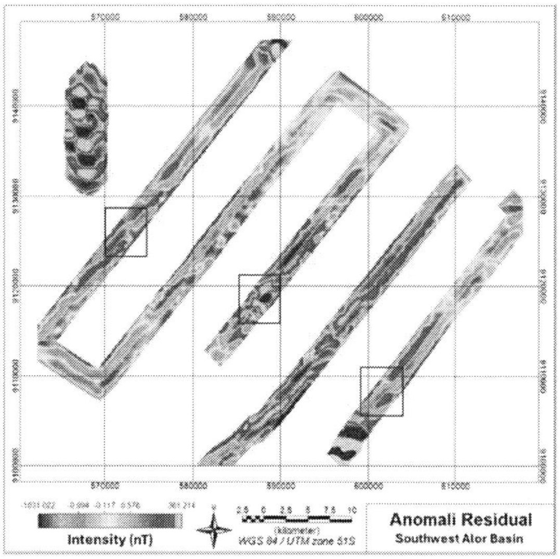

Fig. 9. Residual Anomaly Separation Results

Striking differences in values occur in several areas marked with color. Figure 9 shows significant magnetic activity in the three areas shown. The differences in magnetic values found indicate tectonic activity. This is in line with the discovery of faults in previous research [14]. The faults were identified as NW-SE oriented, which is in line with the formation of the seamounts in the area, namely Baruna Komba and Abang Komba.

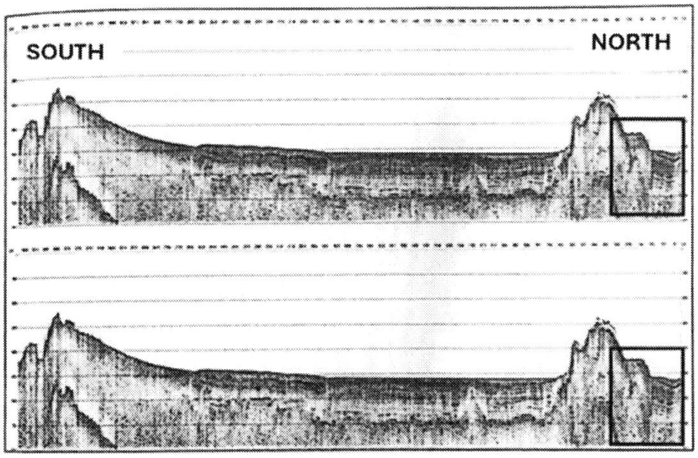

Fig. 10. Seismic Cross-Section Results of Track 6 Post-Stack Migration Results (Top) and Pre-Stack Migration Results (Bottom) [15]

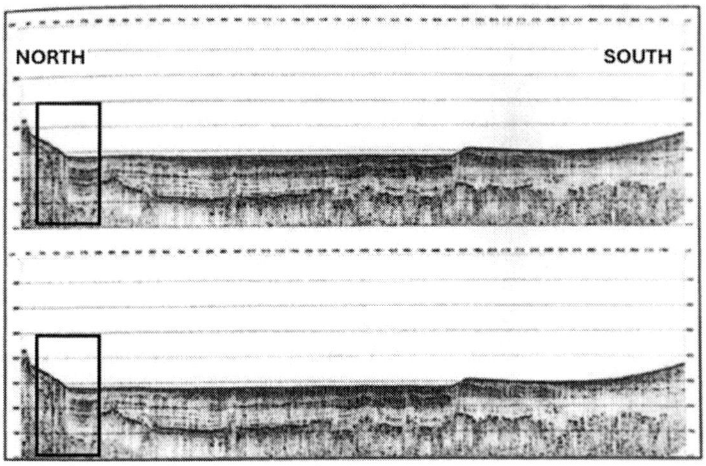

Fig. 11. Seismic Cross-Section Results of Track 8 Post-Stack Migration Results (Top) and Pre-Stack Migration Results (Bottom) [15]

3.4 Seismic Analysis

The seismic data used were processed by the Center for Marine Geology Research and Development (P3GL) in 2014 [15]. The data were collected in Alor waters at LP 2308 and LP 2309, which are located in the same location as this research area. Tracks 6 and 8 in the P3GL report are similar to the magnetometer data collection area. The seismic data oriented south-north and north-south, respectively, revealing rock layers. One of these layers exhibits a fault, supporting the analysis of the magnetic anomaly data. Figure 11 shows a steep morphology in the north and irregularities and shifts in the reflector layer on the right side of the cross-section, indicating the presence of a fault structure. This fault is characterized by a discontinuous layer and indicates vertical displacement. Meanwhile, Figure 12 shows a relatively flat seabed surface, but on the north side, there is a shift in the reflector layer, also indicating the presence of a fault. The presence of this fault can function as a migration route or even a structural trap for hydrocarbons. The relatively flat and consistent reflective layers in the center of the section likely represent reservoir rock layers, while the shift in the fault could create a closed zone that allows hydrocarbons to accumulate. Thus, the fault structure observed in this seismic section not only demonstrates the dynamics of the underwater geology but also serves as an important indicator in assessing the potential for hydrocarbon deposits in the area.

4 CONCLUSION

The Alor Basin has an identified depth of up to 4,777.68 m, with the topmost sedimentary layer being clay. The Alor Basin is part of a deep marine basin characterized by low energy and slow sedimentation rates, thus favoring the deposition of fine particles such as clay. Clay can accumulate with other deep-sea materials such as radiolarians and foraminifera. The presence of clay in the Alor Basin has the potential to support petroleum systems. Clay can act as a source rock if it contains organic matter or as a cap rock due to its impermeable nature. The differences in magnetic anomaly intensity recorded in the area are quite significant due to indications of tectonic activity. This is supported by research by P3GL in 2014 using seismic sections. The presence of these faults can function as migration pathways or even structural traps for hydrocarbons. Faults can act as migration pathways for hydrocarbons from source rock to reservoir rock. Furthermore, faults can also create complex geological conditions that support hydrocarbon accumulation, such as fault traps or fault-cut anticlines.

The results of this study still require further research. This study only indicates that magnetic anomalies exist beneath the seabed of the Alor Basin due to tectonic activity. Integration of other geophysical data, such as regional geology and the structure of the rocks forming it, as well as surface and subsurface geological analysis, is needed to confirm the existence of structures that could potentially act as hydrocarbon traps.

ACKNOWLEDGEMENT

The author would like to thank all parties who assisted and supported the Geomarine Laboratory lecturers who assisted in conducting this research. He also thanks the Indonesian Navy's Hydrological Research Center (Pushidros TNI AL) and the Indonesian National Marine Research and Development Agency (BBSPGL) for providing limited data to assist in this research. The authors also gratefully acknowledge financial support from the Institut Teknologi Sepuluh Nopember for this work, under the project scheme of the Publication Writing and IPR Incentive Program (PPHKI) 2025.

REFERENCE

[1] A. Cahyono, "Masifkan Eksplorasi, 68 Cekungan Migas Simpan Potensi Besar," Kementerian Energi dan Sumber Daya Mineral. Diakses: 5 Januari 2025. [Daring]. Tersedia pada: https://www.esdm.go.id/id/media-center/arsip-berita/masifkan-eksplorasi-68-cekungan-migas-simpan-potensi-besar
[2] G. A. S. Nayoan, "Offshore Hydrocarbon Potential of Indonesia," Energy, vol. 11, no. 6, hlm. 1225–1246, 1981, doi: 10.1016/0360-5442(81)90034-7.
[3] R. W. V Bemmelen, The Geology of Indonesia: General Geology of Indonesia and Adjacent Archipelagoes, vol. IA. The Hague: Government Printing Office, 1949.
[4] P. G. McLeroy dan J. P. Riva, "Origin of Hydrocarbons," Britanica. Diakses: 5 Januari 2025. [Daring]. Tersedia pada: https://www.britannica.com/science/petroleum/Origin-of-hydrocarbons
[5] Wahyudi, "Aplikasi Mikroseismik untuk Memindai dan Mengidentifikasi Keberadaan Hidrokarbon (Microseismic Aplication for Direct Hydrocarbon Detection and Identification)," Berkala MIPA, vol. 18, no. 2, hlm. 114–123, 2008, Diakses: 5 Januari 2025. [Daring]. Tersedia pada: https://jurnal.ugm.ac.id/bimipa/article/view/33495
[6] K. Akbar, D. G. Pratomo, dan Khomsin, "Analisis Nilai Hambur Balik Sedimen Permukaan Dasar Perairan Menggunakan Data Multibeama Echosounder EM302," Jurnal Teknik ITS, vol. 6, no. 2, hlm. 2337–3520, 2017.
[7] Moh. Ryan, J. R. Husain, dan H. Bakri, "Studi Anomali Magnetik Total untuk Pencarian Daerah Prospek Hidrokarbon Daerah Pulau Buru Provinsi Maluku," Journal Geomine, hlm. 17–21, 2015, doi: https://dx.doi.org/10.33536/jg.v1i1.5.
[8] R. C. Hasan, D. Lerodiaconou, L. Laurenson, dan A. Schimel, "Integrating Multibeam Backscatter Angular Response, Mosaic and Bathymetry Data for Benthic Habitat Mapping," PLoS One, vol. 9, no. 5, Mei 2014, doi: 10.1371/journal.pone.0097339.
[9] Ab. T. Insani, Aplikasi Side Scan Sonar dan Magnetometer untuk Pemetaan Sebaran Anomali Magnetik Dasar Laut. Malang: Institut Teknologi Nasional Malang, 2015.
[10]A. H. Pinandita dan W. Sutresno, "Karakteristik Anomali Magnetik dari Metode Reduksi ke Kutub dan Reduksi ke Ekuator pada Struktur Sesar," Globe: Publikasi Ilmu Teknik, Teknologi Kebumian, Ilmu Perkapalan, vol. 2, no. 3, hlm. 250–257, 2024, doi: 10.61132/globe.v2i3.515.
[11]S. Puryono, S. Anggoro, Suryanti, dan I. S. Anwar, Pengelolaan Pesisir dan laut Berbasis Ekosistem. Badan Penerbit Universitas Diponegoro Semarang, 2019.
[12]A. H. Satyana, "Geologi Kelautan Indonesia," 2022.
[13]Diva Prameswari, Analisis Pemodelan Inversi Data Geomagnet untuk Pemetaan Bawwah Permukaan Cekungan Bone Sulawesi. 2024.
[14]L. Sarmili dan R. Arief Troa, "Keberadaan Sesar dan Hubungannya dengan Pembentukan Gunung Bawah Laut di Busur Belakang Perairan Komba, Nusa Tenggara," Jurnal Geologi Kelautan, vol. 12, no. 1, 2014.
[15]Subarsyah, Laporan Pengolahan Data Seismik 2D Multichannel Perairan Alor LP 2308 dan 2309. 2014.

High-Resolution Coastal Topographic Model Using UAV Based Multi Georeferencing Method for Enhancing Sustainable and Resilient Coastal Zone Management

Mochamad Zidan Alfirmansyah[1], *Achmad* Fahriza[1], *Candida* Aulia De Silva Nusantara[1,*], *Putra* Maulida[1], and *Khomsin*[1]

[1]Department of Geomatic Engineering, Faculty of Civil Engineering, Planning and Geosciences, Institute Technology Sepuluh Nopember, Surabaya, Indonesia

*E-mail: candida.nusantara@its.ac.id

Abstract. Coastal zones are dynamic and vulnerable environments that require high-resolution topographic data to support sustainable and resilient management. This study aims to develop a high-resolution coastal topographic model using UAV photogrammetry with a multi-georeferencing approach, comparing Ground Control Points (GCP), Post-Processed Kinematic (PPK), and a combined GCP–PPK method. The research was conducted in the Kenjeran coastal area near the Suramadu Bridge, Surabaya, Indonesia using UAV survey. UAV surveys were performed to generate orthomosaics and digital terrain models (DTMs). Each georeferencing method was applied independently to evaluate its influence on spatial accuracy. Accuracy assessment used twelve Independent Check Points (ICPs), with horizontal and vertical accuracy evaluated through Circular Error (CE) and Linear Error (LE). The results indicate that the GCP-based model achieved the highest accuracy, with LE and CE values of 0.149 m and 0.264 m, respectively, outperforming both the PPK and combined methods. While PPK offers greater operational efficiency by reducing field control requirements, GCP-based georeferencing remains the most reliable for centimeter-level coastal mapping. The combined method provides a balanced trade-off between accuracy and efficiency. This study supports SDGs 13 (Climate Action) and 14 (Life Below Water) by enhancing climate-resilient coastal management.

1 Background

Coastal zones, encompassing the dynamic interface between terrestrial and marine ecosystems, accommodate approximately 40% of the global population within 100 km from shorelines and contribute significantly to economic prosperity through diverse ecosystem services valued at USD 28 trillion annually [1]. However, these critical areas face unprecedented challenges from climate-induced sea-level rise (SLR) projected at 0.43-0.84 m by 2100 under moderate emission scenarios, coupled with accelerating coastal erosion rates reaching 0.5-2.0 m/year in vulnerable regions [2]. The compounding effects of anthropogenic pressures and natural hazards have resulted in substantial degradation of coastal ecosystems worldwide, necessitating urgent implementation of data-driven coastal zone management strategies [3].

Traditional coastal topographic mapping methodologies, including terrestrial surveys and satellite-based remote sensing, demonstrate significant limitations in capturing fine-scale geomorphological features essential for comprehensive coastal vulnerability assessments [4]. Satellite imagery, despite broad spatial coverage, typically provides horizontal resolutions of 0.5-10 m and vertical accuracies ranging from 2-15 m, which are insufficient for detailed coastal erosion monitoring and hazard mapping in dynamic shoreline environments [5]. Conversely, terrestrial surveying techniques, while achieving centimeter-level accuracy, require extensive field time and substantial financial investment, limiting their applicability for large-scale or repeated coastal monitoring programs [6].

The emergence of Unmanned Aerial Vehicle (UAV) photogrammetry has revolutionized coastal mapping capabilities by enabling high-resolution topographic data acquisition with sub-decimeter horizontal accuracy and vertical precision at significantly reduced operational costs compared to conventional methods [4,6]. This technological advancement has positioned UAV-based Structure-from-Motion (SfM) photogrammetry as a leading tool for detailed coastal morphodynamic analysis, erosion assessment, and hazard mapping.

The accuracy of UAV-derived topographic products is fundamentally governed by georeferencing quality, where Ground Control Point (GCP) configuration, density, and measurement precision play a critical role in determining final model reliability [7]. Previous studies indicate that reliance on single georeferencing strategies either GCP-based or Real-Time Kinematic (RTK) positioning can result in systematic horizontal and vertical errors exceeding acceptable thresholds in complex coastal terrains characterized by vegetation cover and heterogeneous topography [5]. Recent advances suggest that integrated multi-georeferencing approaches, combining differential GNSS measurements, RTK-UAV positioning, and photogrammetric bundle adjustment, can substantially improve positional accuracy and model robustness [7].

In archipelagic nations with extensive coastlines, the demand for accurate coastal topographic data is further intensified by high population exposure, economic dependency on coastal resources, and escalating erosion risks [8]. National coastal management frameworks mandate the availability of high-accuracy topographic information to support detailed spatial planning, infrastructure protection, and ecosystem conservation [9]. Moreover, alignment with the Sustainable Development Goals particularly SDG 13 (Climate Action) and SDG 14 (Life Below Water) requires scientifically robust baseline datasets to inform evidence-based adaptation and mitigation strategies [10].

Despite rapid advancements in UAV platforms and photogrammetric processing algorithms, critical knowledge gaps persist regarding the optimal integration of multi-georeferencing techniques that balance accuracy, operational efficiency, and cost-effectiveness for large-scale coastal mapping applications. The absence of standardized validation frameworks tailored to heterogeneous coastal environments constrains the broader operational adoption of UAV-based geospatial technologies. Addressing these gaps is essential for enhancing the reliability of high-resolution coastal topographic models and strengthening sustainable coastal zone management under increasing climate and anthropogenic pressures.

2 Method

This research was conducted along the coast of Kenjeran Beach, Sukolilo District, Surabaya. The research area included residential areas, beaches, vacant land, vegetation, and cultural heritage sites (Kedung Cowek Fort) with an area of 43.4 hectares, as shown in Figure 1. The research area has a generally flat topography, typical of the northern coast of Java, with an elevation ranging from 0 to 5 meters above mean sea level (MSL).

Figure 1. Study Area Marked with Yellow Border

Data acquisition was performed using a UAV DJI Mavic 3 Enterprise quadcopter model ware equipped with DRTK to produce high-resolution aerial photographs. The sensor installed on the UAV was a non-metric camera sensor with a resolution of 20 MP RGB. More specifications of the DJI Mavic 3 Enterprise UAV are presented in Table 1.

Table 1. UAV Specification

Spesification	Notes
Type UAV	Quad-copter
Camera	Wide Camera
Resolution	20 MP
Width Sensor	4/3 CMOS
Focal Length	24mm
Geolocation	GNSS DRTK Module

Source : [11]

The research stages carried out included preparation, data acquisition, processing, map accuracy testing, and analysis, which were conducted systematically as shown in Figure 2.

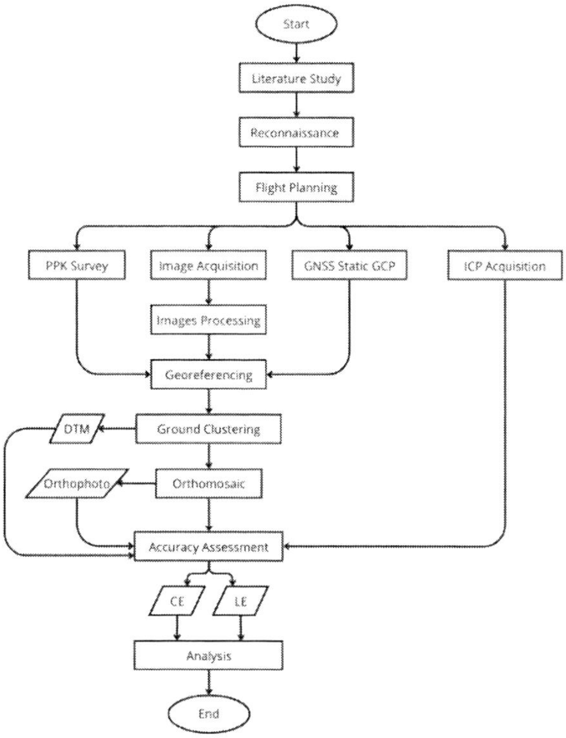

Figure 2. Research Flow Chart

During the preparation stage, aerial photo survey planning was carried out, including reconnaissance and acquisition route planning. The reconnaissance process was carried out before starting field work using the desk study method on Google Earth. This stage was to ensure that the study area is safe from tall objects such as towers, buildings, SUTT, and others. In addition, wind patterns and speeds in the study area are analyzed to ensure that wind speeds are safe for UAV flights below 12 m/s.

The acquisition route planning is using DJI GO software on the UAV remote control. The side-lap, front-lap, and flight altitude parameters were set at 80%, 70%, and 80 meters above ground level, respectively, to produce a GSD value of $\leq$ 5 cm in accordance with the requirements of PERKA BIG Number 1 of 2020 for a map scale of 1:2,500 [12]. The acquisition process was carried out in 3 flights as shown in Figure 3. with a total flight time of 31 minutes 10 seconds, resulting in 1,235 photos. Each area has 5% overlap to combine each acquisition data mission. The acquisition was carried out from 7 to 10 a.m. in clear weather conditions and wind speeds below 10 m/s.

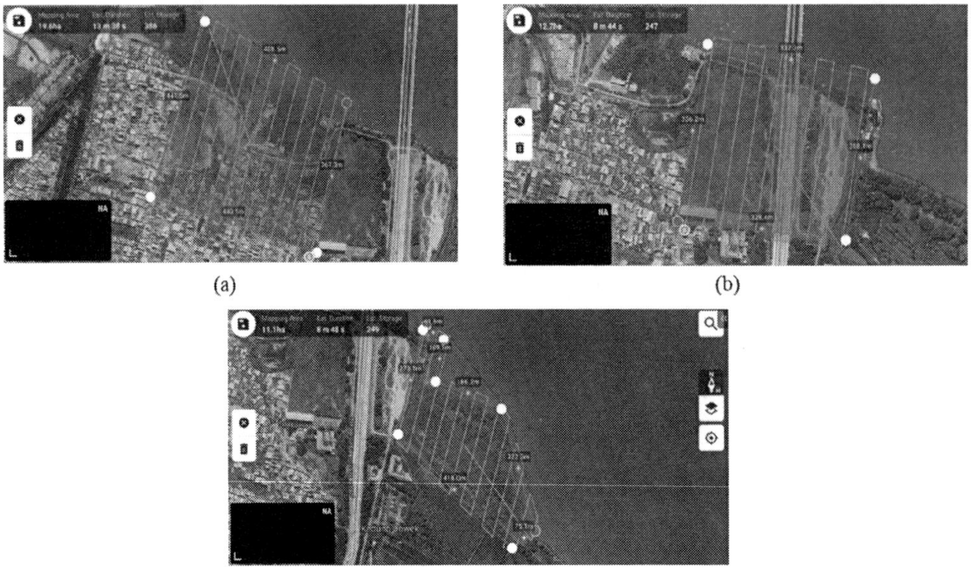

(a) (b)

(c)

Figure 3. Flight Plan (a) First Flight, (b) Second Flight, and (c) Third Flight

In this study, the georeferencing process used three methods, namely orthorectification using Ground Control Points (GCP), Post Processing Kinematic (PPK), and a combination of both. These methods were used to produce the most accurate orthophoto. A total of 8 GCPs were used, 8 GCPs for GCP method and 4 GCP for combined method, were distributed throughout the study area, as shown in Figure 2. The GCPs were marked with 1x1 meter pre-marks with black and white patterns. Each GCP was measured using the GNSS static method for 6 hours, which was processed using baseline adjustment against the 4 nearest CORS stations.

After all data was collected, aerial photos were processed using the Structure of Motion (SfM) method in Agisoft Metashape software. The processing was carried out in stages: photo alignment, georeferencing, point cloud extraction, DEM, and orthophoto. The DEM data obtained from the processing was then used for ground classification to obtain topography through the process of separating ground objects from other objects such as vegetation and building. To determine the accuracy of the result, an accuracy test was conducted using 12 Independent Check Points (ICP) spread evenly throughout the study area, as shown in Figure 3. The accuracy value of the result was calculated based on the CE90 and LE90 values in accordance with PERKA BIG Number 6 of 2018 for a map scale of 1:2,500 [13].

$$RMSEr = \sqrt{\frac{\sum_{i=1}^{n}(\Delta X_i^2 + \Delta Y_i^2)}{n}} \text{ and } RMSEz = \sqrt{\frac{\sum_{i=1}^{n}(\Delta Z_i^2)}{n}} \quad (1)$$

$$CE90 = 1,5175 \, x \, RMSEr \quad (2)$$

$$LE90 = 1,6499 \, x \, RMSEz \quad (3)$$

Map accuracy is determined based on the accuracy class provisions for 1:2,500 scale base maps presented in Table 2.

Table 2. Class Accuracy Threshold

First Class		Second Class		Third Class	
CE90 (in meters)	**LE90 (in meters)**	**CE90 (in meters)**	**LE90 (in meters)**	**CE90 (in meters)**	**LE90 (in meters)**
0,5	0,5	0,75	0,75	1,25	1,25

Map accuracy is qualified if the CE90 and LE90 values are below the specified thresholds [13]

3 Result

The results of processing GNSS observation data for GCP points obtained coordinates referenced to WGS 1984 and MSL Datum from tidal observations. The results are presented in Table 3. The GCP point coordinates were then used to georeference aerial photographs in the next stage.

Table 3. GCP Coordinates

GCP Name	X (m)	Y (m)	Z MSL (m)	Z Elliposoid (m)
G01	695826.128	9202890.998	1.007	31.769
G06	695934.062	9202655.503	1.406	32.168
G39	696103.464	9202789.646	1.861	32.623
G16	696412.064	9202639.908	1.981	32.743
G23	696765.280	9202361.855	1.415	32.177
G29	696718.440	9202556.797	1.161	31.923

Table 4. presents the coordinates of the ICP points as test points for aerial photo accuracy

Table 4. ICP Coordinates

ICP Name	X (m)	Y (m)	Z MSL (m)	Z Elliposoid (m)
ICP01	696180.4915	9202558.51	1.050	31.792
ICP02	696376.7276	9202803.32	1.545	32.287
ICP03	696404.0039	9202547.88	1.943	32.685
ICP04	695990.0575	9202811.088	1.513	32.255
ICP05	696429.98	9202802.14	1.246	31.988
ICP06	696552.17	9202780.04	2.566	33.308
ICP07	696000.9351	9202822.412	3.149	33.891
ICP08	696056.0731	9202609.992	2.941	33.683
ICP09	695847.4541	9202960.414	2.419	33.161
ICP10	696627.491	9202605.121	1.984	32.726
ICP11	696560.9821	9202594.012	1.148	31.890
ICP12	696559.5813	9202618.894	1.498	32.240

The processing results from three methods: GCP, PPK, and a combination of both (GCP+PP) were tested for accuracy using the CE and LE equations in accordance with PERKA BIG Number 6 of 2018, as presented in Table 5 until
Tabel 7.

Table 5. Accuracy GCP Method

ICP Name	ΔX (m)	ΔY (m)	ΔZ (m)	$\Delta X^2 + \Delta Y^2$	ΔZ^2
ICP01	-0.142	-0.150	0.089	0.043	0.008
ICP02	0.042	0.010	-0.194	0.002	0.038
ICP03	-0.064	0.000	0.002	0.004	0.000
ICP04	0.002	0.022	0.102	0.000	0.010
ICP05	0.000	-0.001	-0.022	0.000	0.000
ICP06	-0.028	-0.112	0.357	0.013	0.128
ICP07	0.051	-0.024	0.182	0.003	0.033
ICP08	0.016	-0.077	-0.160	0.006	0.026
ICP09	0.098	0.030	0.162	0.010	0.026
ICP10	0.064	0.051	-0.143	0.007	0.020
ICP11	-0.140	0.045	0.143	0.022	0.020
ICP12	0.012	0.098	-0.167	0.010	0.028
		RMSE (m)		**0.098**	**0.160**
		CE90 (m)		**0.149**	
		LE90 (m)		**0.264**	

Table 6. Accuracy PPK Method

ICP Name	ΔX (m)	ΔY (m)	ΔZ (m)	$\Delta X^2 + \Delta Y^2$	ΔZ^2
ICP01	-0.726	-0.593	0.369	0.878	0.136
ICP02	0.131	0.454	-0.302	0.224	0.091
ICP03	0.253	-0.365	-0.202	0.197	0.041
ICP04	0.230	-0.036	-0.095	0.054	0.009
ICP05	-0.909	0.174	0.245	0.857	0.060
ICP06	0.172	-0.112	0.357	0.042	0.128
ICP07	-0.194	-0.190	0.282	0.074	0.080
ICP08	0.058	-0.154	-0.330	0.027	0.109
ICP09	0.059	-0.265	-0.473	0.074	0.224
ICP10	-0.250	0.298	-0.368	0.151	0.136
ICP11	-0.201	0.147	0.305	0.062	0.093
ICP12	0.178	-0.535	-0.179	0.318	0.032
		RMSE (m)		**0.497**	**0.335**
		CE90 (m)		**0.753**	
		LE90 (m)		**0.553**	

Table 7. Accuracy Combine Method

ICP Name	ΔX (m)	ΔY (m)	ΔZ (m)	$\Delta X^2 + \Delta Y^2$	ΔZ^2
ICP01	-0.164	-0.174	0.170	0.057	0.029
ICP02	0.074	0.093	-0.212	0.014	0.045
ICP03	-0.169	0.075	0.053	0.034	0.003
ICP04	0.048	0.151	0.188	0.025	0.035
ICP05	0.111	0.2	0.020	0.052	0.000
ICP06	-0.116	-0.125	0.532	0.029	0.183
ICP07	0.094	-0.087	0.296	0.016	0.088
ICP08	0.241	-0.197	-0.166	0.097	0.028
ICP09	0.147	0.059	0.269	0.025	0.072
ICP10	0.197	0.086	-0.143	0.046	0.020
ICP11	-0.161	0.124	0.243	0.041	0.059
ICP12	0.156	0.159	-0.276	0.050	0.076
		RMSE (m)		**0.202**	**0.239**
		CE90 (m)		**0.306**	
		LE90 (m)		**0.395**	

The GCP method obtained the highest accuracy values with a CE value is 0.149 m and 0.246 for LE value, then combined method with a CE value of 0.305 m and an LE value of 0.246 m, and the lowest accuracy is PPK method with a CE value of 0,753 m and an LE of 0,553 m.

Accuracy levels refer to geometric errors represented by RMSE values, which indicate the error values at each test point. The comparison and spatial distribution of errors represented by RMSE values are shown in Figure 4 and Figure 5. Figure 4. presents the distribution pattern of georeferencing errors calculated based on Independent Check Point (ICP) test points in the study area. Spatial visualization of RMSE is used to identify variations in horizontal-vertical positional accuracy and assess the consistency of geometric transformation results across the entire study area. Through this approach, areas with high errors are shown in red and areas with low errors are shown in green, allowing for analysis of their relationship with the distribution and density of control points.

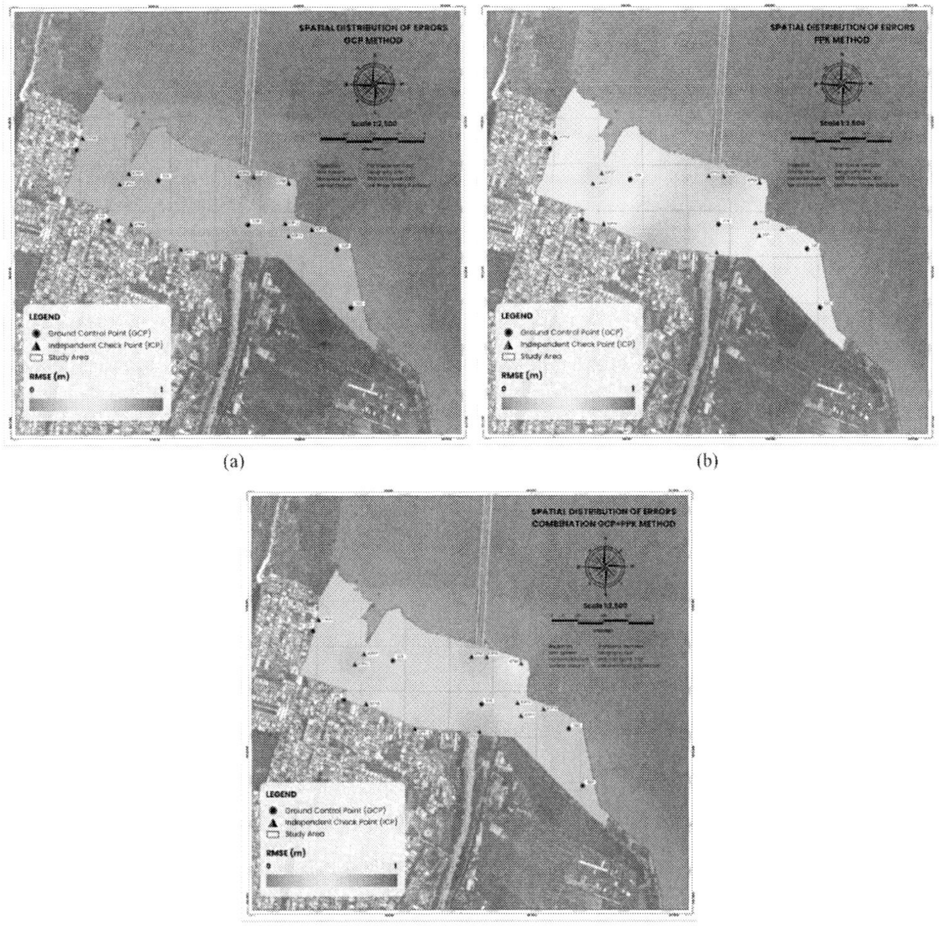

(a) (b)

(c)

Figure 4. Distribution of Errors (a) GCP Method, (b) PPK Method, and (c) combined method

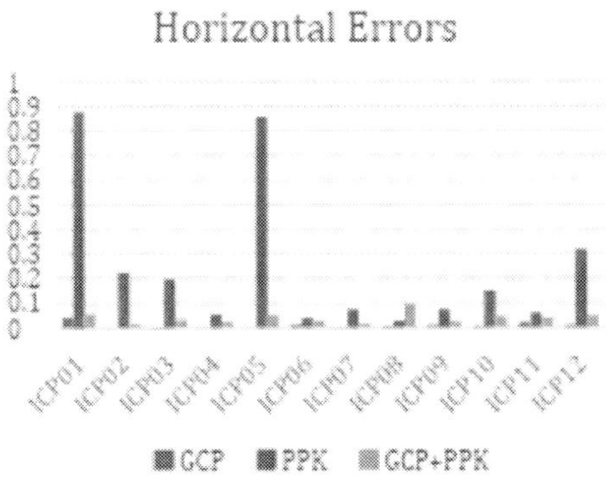

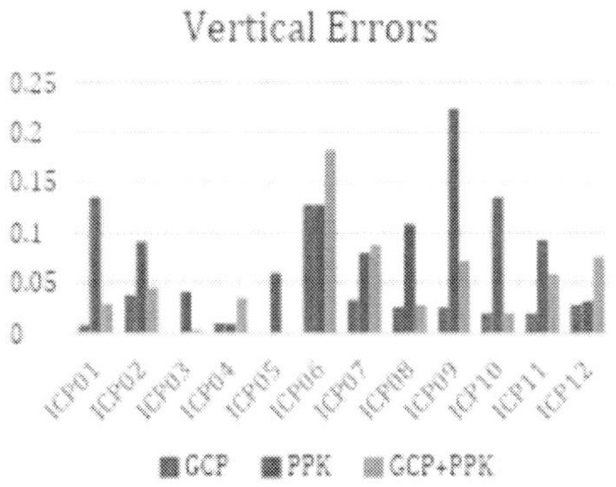

Figure 5. Comparation of ICP Errors

Based on Figures 4 and 5, the distribution of errors both horizontally and vertically shows differences in geometric accuracy between georeferencing methods. The GCP method produces the smallest and most stable errors at each test point both horizontally and vertically, as indicated by the dominance of green colors. An increase in error values occurs in ICP06 with an increase of up to 3.5 times greater than the average vertical error. This condition is caused by location of ICP06, which is far from the GCP points used. This can also be seen in the combination method, which has fewer GCPs than the GCP method, resulting in a relatively higher error value than GCP but remains stable, as indicated by the greenish-yellow color spread throughout the study area except at ICP06. The increase in error values in the combination method was due to the use of fewer GCPs to improve data acquisition and processing efficiency, although it still met the class 1 standard in accordance with PERKA BIG Number 6 of 2018. Meanwhile, the PPK method produced a more random error distribution, as indicated by the fluctuations in the graph and color patterns on the map. Errors in the PPK method are more accumulated in the central area, which is far from the DRTK base station. The accuracy at each test point is influenced by image stitching errors in coastal areas, base-rover distance, and signal quality during the acquisition process.

The results of processing using the GCP method consist of orthophoto maps presented in Figure 6 and DTM maps presented in Figure 7.

Figure 6. Digital Terrain Model

Figure 7. Orthophoto

4 Discussion

Grounded in the results yielded by this research, processing using the GCP-based georeferencing method has the best accuracy compared to the PPK method and the combination of GCP and PPK. This superior accuracy is also reflected in the spatial distribution of errors, where both horizontal and vertical errors values show a homogeneous and stable pattern across the study area. Spatial error maps indicate that most areas are dominated by low error values, while localized increases occur at ICP locations situated farther from the surrounding GCP network. This highlights the strong influence of GCP spatial configuration in constraining the bundle adjustment process and minimizing geometric distortion throughout the model.

The accuracy values are shown by the CE90 and LE90 values, which are 0.149 and 0.264, respectively. These values meet the accuracy threshold for class 1 maps at a scale of 1:2,500, as specified in PERKA BIG No. 6 of 2018. The high accuracy of the GCP method is influenced by the even distribution of GCPs and the measurement of GCPs using the static GNSS method for six hours, which provides stable and reliable reference coordinates.

Instead, the PPK method showed an increase of error with a CE90 value of 0.753 and a LE90 value of 0.553. The increase of error in the PPK method results from the absence of GCPs to constrain the SfM adjustment, height estimation limitations of onboard GNSS sensors, and residual multipath effects and systematic biases that are not fully corrected during post-processing. These factors contribute to less stable spatial accuracy and a more heterogeneous error distribution. Although the PPK method provides the lowest accuracy among the tested approaches, it offers superior flexibility and faster data acquisition, making it suitable for rapid surveys in areas with limited accessibility.

The combination of both methods, PPK+GCP, provides improved accuracy compared to the PPK-only method. The accuracy values obtained by the PPK+GCP method are CE90 of 0.306 and LE90 of 0.395, which satisfy the class 1 accuracy requirements for 1:2,500 scale maps. However, these values remain lower than those obtained using the full GCP-based approach due to the reduced number of GCPs intentionally employed to improve operational efficiency. A smaller GCP set weakens the geometric constraints during bundle adjustment, particularly in areas located farther from control points, resulting in slightly higher residual errors. This pattern aligns with the results reported by Zhang et al. (2019), who demonstrated that while PPK-based UAV photogrammetry combined with a limited number of GCPs that can significantly improve an accuracy and reduce vertical bias, the highest positional accuracy is consistently achieved when a dense and well-distributed GCP network is used [14]. Thus, the PPK+GCP approach represents a trade-off between accuracy and efficiency rather than a replacement for full GCP deployment.

Overall, this study demonstrates that three georeferencing strategies GCP, PPK, and PPK combined with GCP can be effectively applied to generate high-resolution coastal topographic models. Each method offers different advantages depending on accuracy requirements, field conditions, and resource availability. While GCP-based georeferencing provides the highest accuracy, PPK offers rapid data acquisition with reduced field effort, and the combined approach serves as a balanced alternative that reduces GCP requirements while still meeting national accuracy standards. The availability of these methodological options is expected to support the production of reliable coastal topographic data for climate-resilient coastal management and marine ecosystem protection. Furthermore, accurate coastal mapping contributes directly to the achievement of the Sustainable Development Goals (SDGs), particularly SDG 13 (Climate Action) and SDG 14 (Life Below Water), by enabling data-driven mitigation, adaptation, and sustainable marine spatial planning.

5 Conclusion

This study aims to build a high-resolution coastal topographic model using an Unmanned Aerial Vehicle (UAV) with a multi-georeferencing approach that combines GCP Georeferencing, PPK, and both Ground Control Points (GCPs) and Post-Processed Kinematic (PPK) methods. The accuracy of the three processing results was tested using 12 ICP points spread across the study area. The accuracy values were calculated based on the provisions stipulated in PERKA BIG Number 6 of 2018, including CE90 and LE90 values. The accuracy values were used to determine the accuracy class of the map at a scale of 1:2,500.

The accuracy test results for the three georeferencing methods used showed that the GCP method had the highest accuracy value with a CE90 value of 0.149 and an LE90 value of 0.264, which is classified as accuracy class 1 for a 1:2,500 scale map. In contrast, the PPK method had the lowest accuracy value with a CE90 value of 0.753 and an LE90 value of 0.553, which is classified as accuracy class 2 for a 1:2,500 scale map. Meanwhile, the combination of both methods (GCP+PPK) produced moderate results with a CE90 value of 0.306 and an LE90 value of 0.395, as well as better acquisition process efficiency than the GCP method.

The three methods applied in this study were able to produce orthophoto and DEM results with accuracy that needed to be adjusted to the requirements. The acquisition and processing strategies used will affect the results obtained, especially in coastal areas.

Acknowledgements

This study is supported by Department of Geomatics Engineering ITS, PT. Geosolution Pratama Nusantara, TSS Ocean, PT. Hidronav Tehnikatama, Print n Print, Be Collection, Larita, and Cheers.

Conflict of Interest

The authors declare no conflicts of interest

References

1. B. Neumann, A.T. Vafeidis, J. Zimmermann, R.J. Nicholls, Future coastal population growth and exposure to sea-level rise and coastal flooding. *PLoS One 10*, e0118571 (2015). https://doi.org/10.1371/journal.pone.0118571

2. M.I. Vousdoukas, R. Ranasinghe, L. Mentaschi, T.A. Plomaritis, P. Athanasiou, A. Luijendijk, L. Feyen, Sandy coastlines under threat of erosion. *Nat. Clim. Chang.* 10, 260–263 (2020). https://doi.org/10.1038/s41558-020-0697-0
3. C.M. Duarte, I.J. Losada, I.E. Hendriks, I. Mazarrasa, N. Marbà, The role of coastal plant communities for climate change mitigation and adaptation. *Nat. Clim. Chang.* 3, 961–968 (2018). https://doi.org/10.1038/nclimate1970
4. J.A. Gonçalves, R. Henriques, UAV photogrammetry for topographic monitoring of coastal areas. *ISPRS J. Photogramm. Remote Sens.* 104, 101–111 (2015). https://doi.org/10.1016/j.isprsjprs.2015.02.009
5. Y. Taddia, C. Corbau, E. Zambello, A. Pellegrinelli, UAVs for Structure-From-Motion coastal monitoring. *Sensors* 19, 1717 (2019). https://doi.org/10.3390/s19071717
6. I.L. Turner, M.D. Harley, C.D. Drummond, UAVs for coastal surveying. *Coast. Eng.* 114, 19–24 (2016). https://doi.org/10.1016/j.coastaleng.2016.03.011
7. T. Zimmerman, K. Jansen, J. Miller, Analysis of UAS flight altitude and ground control point parameters on DEM accuracy along a complex, developed coastline. *Remote Sens.* 12, 2305 (2020). https://doi.org/10.3390/rs12142305
8. Ministry of Marine Affairs and Fisheries, *Marine and fisheries in figures 2020* (Center for Data, Statistics and Information, Jakarta, 2020).
9. Government of Indonesia, Law No. 27 of 2007 on the management of coastal areas and small islands (State Gazette of the Republic of Indonesia, 2007).
10. United Nations, Transforming our world: The 2030 agenda for sustainable development, UN General Assembly Resolution A/RES/70/1 (2015).
11. DJI, Mavic 3 Enterprise Series - Specifications, https://enterprise.dji.com/mavic-3-enterprise/specs (2024)
12. Geospatial Information Agency of Indonesia, Head of Geospatial Information Agency Regulation Number 1 of 2020 concerning Technical Guidelines for Geospatial Information Presentation at Map Scale 1:2,500 (Geospatial Information Agency, Jakarta, 2020)
13. Geospatial Information Agency of Indonesia, Head of Geospatial Information Agency Regulation Number 6 of 2018 concerning Control Point and Geospatial Data Acquisition at Map Scale 1:2,500 (Geospatial Information Agency, Jakarta, 2018)
14. H. Zhang, E. Aldana-Jague, F. Clapuyt, F. Wilken, V. Vanacker, K. Van Oost, Evaluating the potential of post-processing kinematic (PPK) georeferencing for UAV-based structure-from-motion (SfM) photogrammetry and surface change detection. *Earth Surf. Dyn.* 7, 807–827 (2019). https://doi.org/10.5194/esurf-7-807-2019

Innovations in Precast Concrete Technology for Giant Sea Wall Construction in Indonesia: A Literature Review

Candra Irawan[1*], *Made* Yujiro Semita[1]

[1]Department of Civil Engineering, Institut Teknologi Sepuluh Nopember, Surabaya, Indonesia

Abstract. Coastal erosion is a major challenge for Indonesia, where extensive shorelines are exposed to high-energy waves and anthropogenic pressures. This study introduces Hexaloc, a novel concrete armor unit with six interlocking arms designed to enhance stability and ease of installation in breakwater structures. The research focuses on the innovation of concrete molds and production systems, enabling scalable fabrication of Hexaloc units with consistent quality. Laboratory-scale trials were conducted to validate compressive strength, density, and durability using Ultrasonic Pulse Velocity (UPV), hammer test, and compressive strength testing. Results confirm that Hexaloc units achieve the targeted strength of 30 MPa and demonstrate effective interlocking performance under simulated wave conditions. The prototype development reached Technology Readiness Level (TRL) 5, with potential for industrial-scale application and community-based coastal protection programs. This innovation contributes to sustainable coastal infrastructure by offering a practical, durable, and environmentally adaptive solution for shoreline defense in tropical regions.

1 Introduction

Indonesia's northern Java coastline is increasingly exposed to compound hazards, including recurrent tidal flooding (rob), accelerated land subsidence, and the projected impacts of global sea-level rise. These phenomena threaten densely populated urban areas such as Jakarta, Semarang, and Surabaya, where critical infrastructure and economic activities are concentrated. In response, the Government of Indonesia has proposed the construction of a Giant Sea Wall (GSW), envisioned as a multi-hundred-kilometer coastal defence system extending from Banten to East Java (shown in Fig. 1). While the project promises large-scale protection, it also raises concerns regarding feasibility, cost, ecological impacts, and social acceptance among coastal communities

(a) (b) (c)

Fig. 1. Jakarta's GSW Construction using: (a) Corrugated Concrete Sheet Piles, (b) Pile Construction Combined with Armour Units, (c) Cylindrical Piles [1]

International precedents demonstrate both the potential and limitations of massive seawall projects. The Afsluitdijk in the Netherlands, completed in 1932, remains an iconic example of coastal engineering that has safeguarded the

* Corresponding author: chandra@ce.its.ac.id

nation for nearly a century [2]. Japan's post-2011 tsunami barriers, extending over 395 km, illustrate the integration of structural defences with ecological measures such as large-scale coastal afforestation [3]. Similarly, innovations in Hong Kong highlight the role of precast concrete technologies in accelerating construction and improving durability [4]. These cases provide valuable lessons for Indonesia in balancing engineering performance with environmental and social sustainability.

Precast concrete systems offer distinct advantages for GSW construction, including factory-controlled quality, rapid installation, and scalability for mass production [5]. However, their application in marine environments requires careful attention to durability against chloride penetration [6] and sulfate attack [7], hydraulic performance under extreme wave loading [8], and long-term ecological compatibility [9]. Recent advances in reinforced precast armour units, eco-engineered block geometries, and protective coatings suggest promising pathways to address these challenges [10].

This paper consolidates lessons learned from Indonesian and international seawall projects, identifies material and structural challenges specific to precast applications in Indonesia, and proposes a precast-centered construction concept for the GSW. The study emphasizes strategies that combine engineering robustness with ecological integration, aiming to deliver a resilient and sustainable coastal defence system aligned with the United Nations Sustainable Development Goals (SDGs).

2 Lessons learned and structural models

2.1 Indonesian experience and observed issues

Recent coastal reinforcement projects in Indonesia, particularly in Jakarta, have revealed recurring challenges that complicate large-scale implementation. The most pressing issues include the exceptionally high construction costs, significant social impacts on fishing communities, and ongoing debates regarding the balance between engineering effectiveness and ecological or social trade-offs. These concerns underscore the complexity of designing coastal defences that are not only technically robust but also socially equitable and environmentally sustainable. The discourse surrounding these projects reflects the need for integrated approaches that consider hydrodynamic performance, long-term durability, and community acceptance as inseparable dimensions of success.

To address these challenges, several structural concepts have been tested in local contexts. Corrugated Concrete Sheet Piles (CCSP) have been employed for vertical barriers, pile-supported systems combined with armour units have been used to dissipate wave energy, and large-diameter cylindrical piles have provided foundation stability in soft marine soils. Each of these models demonstrates technical feasibility and offers valuable insights into construction practices under Indonesian coastal conditions. However, they also highlight persistent stakeholder concerns, particularly regarding extended construction schedules and the scalability of solutions for national-level projects such as the Giant Sea Wall. These lessons emphasize the importance of adopting innovative precast concrete technologies that can accelerate construction, improve durability, and reduce socio-environmental conflicts in future coastal defence initiatives.

2.2 International precedents

Historical examples such as the Dutch Afsluitdijk and Japan's extensive tsunami barriers show long-term protection benefits but also underline the need for integrated ecosystem measures. Precast armour-unit systems (e.g., Accropode™, Core-loc®, Xbloc) and applications in South Africa, Hong Kong (GJPRCP no-dredge techniques), and Taiwan illustrate diverse technical solutions adapted to site-specific hydraulic loads and environmental constraints.

Historical precedents such as the Dutch Afsluitdijk and Japan's extensive tsunami barriers demonstrate the long-term protective benefits of large-scale coastal infrastructure. The Afsluitdijk, completed in 1932, has safeguarded the Netherlands from the North Sea for over 90 years, while also serving as a multifunctional structure for transport and recreation [2]. Similarly, Japan's tsunami barriers, constructed after repeated catastrophic events, highlight the importance of robust engineering in reducing disaster risks [3]. However, these examples also underline the necessity of integrating ecosystem-based measures, as rigid structures alone can disrupt sediment transport, fisheries, and coastal biodiversity. Modern coastal protection strategies increasingly emphasize hybrid approaches that combine hard infrastructure with ecological restoration to achieve resilience and sustainability.

In terms of technical innovation, precast armour-unit systems such as Accropode™, Core-loc®, and Xbloc have become widely adopted in breakwater and seawall construction. These units are designed to dissipate wave energy efficiently while reducing material use compared to traditional rubble mound structures. The Xbloc system, developed in the Netherlands in 2001, has been extensively tested and applied worldwide, offering resilience against waves up to twice the design height [11]. Such precast solutions provide modularity, rapid installation, and durability in harsh marine environments, making them attractive for large-scale projects like Indonesia's proposed Giant Sea Wall. Their adaptability to different hydraulic conditions ensures that engineers can tailor designs to local site-specific challenges, balancing structural stability with environmental considerations.

Applications across diverse regions further illustrate the versatility of precast systems. In Hong Kong, the GJPRCP no-dredge technique has enabled construction without extensive seabed disturbance, reducing ecological impacts [4]. These international experiences highlight that while the engineering principles are universal, successful implementation requires sensitivity to local hydraulic loads, sediment dynamics, and ecological constraints. These methods are shown on Fig. 2. For Indonesia, learning from these cases can inform a more integrated approach—combining advanced precast technologies with protective coatings, high-performance concrete mixes, and ecological design features to ensure both durability and environmental compatibility.

Fig. 2. (a) Afsluitdijk in the Netherlands [2] (b) Japan's Tsunami Barriers [3] (c) GJPRCP Seawall in Hong Kong [4]

2.3 Structural typology for precast seawalls

Vertical and curved solid walls constructed from precast panels represent one of the most straightforward solutions for coastal defense, offering high resistance against direct wave impact and ease of modular installation [2]. In addition, perforated or cellular precast units are widely applied to dissipate wave energy, reducing reflection and turbulence while enhancing structural stability [3]. These systems are particularly effective in environments with high wave energy, as they combine hydraulic efficiency with reduced material requirements. Pile-supported precast caps also provide flexibility in areas with soft seabed conditions, enabling load transfer to deeper strata while minimizing settlement issues [4].

Discrete armour-unit layers, such as Accropode™, Core-loc®, and Xbloc, are another practical option, designed to interlock and resist displacement under severe hydraulic loads [11]. The choice among these alternatives depends strongly on local wave climate, seabed geotechnical properties, and operational constraints such as dredging allowances and construction access. International applications demonstrate how site-specific conditions dictate the optimal combination of precast technologies. For Indonesia's Giant Sea Wall initiative, integrating these lessons ensures that structural resilience is balanced with constructability and environmental compatibility.

3 Material and durability considerations for marine precast concrete

3.1 Key degradation mechanisms

In coastal environments, concrete structures are subjected to multiple aggressive deterioration processes that significantly reduce their service life if not properly mitigated. Chloride ingress and sulfate attack are among the most critical chemical mechanisms, as they penetrate the concrete matrix and initiate reinforcement corrosion, ultimately compromising structural integrity. Quantitative thresholds such as chloride concentration at the reinforcement level exceeding 0.4% by weight of cement or sulfate concentrations above 1500 mg/L in seawater are commonly used indicators of durability risk [6]. Mechanical degradation also plays a major role: abrasion and attrition caused by suspended sediments and repeated wave impacts can erode armour units and reduce hydraulic stability [14], [21]. Research suggests that abrasion rates of 0.1–0.5 mm/year under high sediment loads and wave pressures exceeding 50 kN/m² can significantly shorten service life [8]. Together, these factors highlight the necessity of adopting marine-specific design standards, including low-permeability concrete mixes (<1000 Coulombs), corrosion-resistant reinforcement, and protective coatings, to ensure that precast seawall elements achieve the resilience required for long-term coastal defence [10]

3.2 Effects of marine constituents and curing practices

The use of seawater as mixing water or sea sand as fine aggregate in concrete, when applied without mitigation strategies, has been shown to reduce long-term compressive strength and significantly increase the risk of chloride-induced reinforcement corrosion [6]. Elevated chloride concentrations penetrate the concrete matrix, accelerating steel deterioration and compromising structural integrity in marine environments [7]. In addition to chemical attack, curing regimes and initial exposure conditions exert a critical influence on carbonation depth and overall durability. Studies have demonstrated that improper curing, particularly with seawater, can intensify carbonation and weaken the protective capacity of the concrete cover, thereby shortening service life . These findings underscore the necessity of adopting marine-grade design standards, including low-permeability mixes, supplementary cementitious materials, and controlled curing practices, to ensure that precast seawall elements maintain resilience under aggressive coastal conditions [8].

Comparative research highlights that curing under moist air conditions has been shown to result in higher carbonation rates than water curing, with carbonation depths nearly double after one year of exposure. The use of supplementary cementitious materials (SCMs) such as fly ash and silica fume, combined with extended curing, has been reported to lower permeability and delay the onset of reinforcement corrosion [6]. These findings underscore the necessity of adopting marine-grade design standards, including low-permeability mixes (<1000 Coulombs), SCM incorporation, and controlled curing practices, to ensure that precast seawall elements maintain resilience under aggressive coastal conditions [9].

3.3 Recommended material strategies

To enhance the durability of precast concrete elements in aggressive marine environments, several material and structural strategies must be adopted. Cement selection plays a critical role, with sulfate-resistant cement or Portland-slag cement (PSC) recommended to mitigate chemical attack from chlorides and sulfates [7]. Supplementary cementitious materials such as silica fume and controlled proportions of fly ash are effective in reducing permeability and improving long-term strength [5]. Partial replacement options using locally sourced materials, for example clamshell powder, may also be explored to lower embodied carbon while maintaining performance, provided that rigorous testing validates their suitability. Protective systems which include durable surface coatings, corrosion inhibitors, and densifiers are essential to slow the ingress of aggressive ions and extend service life [10]. Finally, reinforcement strategies should be carefully considered: armour units and slender precast elements benefit from reinforcement or fibre enhancement to limit catastrophic breakage, while unreinforced designs require thorough verification to ensure hydraulic stability and structural integrity [8]. Collectively, these measures provide a comprehensive framework for producing precast seawall components that balance durability, sustainability, and resilience under coastal loading conditions.

Findings from reviewed studies indicate that PSC and blended cements incorporating slag or fly ash consistently reduce chloride penetration depth by 25–40% compared to ordinary Portland cement [7]. Silica fume additions of 8–10% have been shown to lower permeability by up to 50% and increase compressive strength retention under marine exposure [5]. Experimental trials with clamshell powder as partial cement replacement demonstrated comparable strength development to fly ash mixes, while reducing embodied carbon by approximately 15% [12]. Protective coatings and corrosion inhibitors further extend service life, with field studies reporting a 20–30 year delay in reinforcement corrosion initiation when such systems are applied [10].

Material replacement strategies provide both durability and sustainability benefits when carefully validated. The combined use of PSC, silica fume, and fly ash offers proven reductions in permeability and improved resistance to chloride ingress, while innovative local materials such as clamshell powder present promising pathways for lowering carbon footprints without compromising performance. Protective systems and reinforcement enhancements complement these measures, ensuring that precast seawall components achieve resilience under severe coastal loading. Future research should focus on long-term field validation of alternative binders and replacement materials, quantifying durability gains and lifecycle carbon reductions to establish robust design standards for marine infrastructure.

3.4 Quality control and production

Precast concrete production offers significant advantages for coastal defence applications by enabling high-quality casting under controlled factory conditions. Unlike in-situ construction, precast systems allow precise regulation of curing regimes, ensuring optimal hydration and strength development [5]. Factory environments also facilitate the application of protective systems such as surface coatings, corrosion inhibitors, and densifiers, which enhance resistance to chloride ingress and abrasion [10]. These controlled processes reduce variability across units, improve reproducibility, and allow performance certification prior to deployment, thereby ensuring that each element meets marine-specific durability standards [8]. As a result, precast production not only accelerates construction schedules but also provides a reliable pathway to achieving long-term resilience and sustainability in large-scale seawall projects [11].

4 Proposed precast-based Giant Sea Wall concept and implementation strategy

4.1 System concept overview

A modular precast seawall system offers a versatile and constructible solution for large-scale coastal defence, combining structural robustness with ecological sensitivity. The foundation is established through a pile grid composed of large-diameter precast piles or driven precast piles, providing stability in soft marine soils and ensuring long-term load-bearing capacity. Above this, precast capping beams and vertical panels can be assembled to form a continuous barrier where uninterrupted protection is required, while reinforced precast armour units or interlocking concrete elements are deployed along seaward slopes to dissipate wave energy and reduce hydraulic stresses. To further enhance ecological outcomes, perforated or cellular elements [13] may be incorporated in selected stretches, reducing wave reflection and simultaneously promoting habitat formation for marine organisms as shown in Fig. 3. This modular approach not only accelerates construction through standardized components but also enables performance certification, adaptability to site-specific conditions, and integration of ecosystem-based measures, positioning it as a sustainable pathway for Indonesia's future coastal defence initiatives.

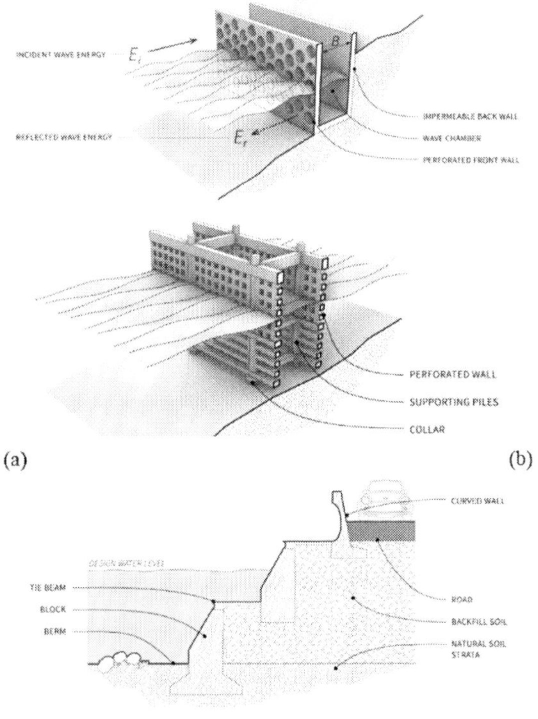

(c)

Fig. 3. Illustrations of: (a) Perforated Vertical Breakwater, (b) Vertical Cellular Walls, (c) Thin-Walled Hollow Seawall [13]

4.2 Design principles

The successful implementation of modular precast seawalls requires a holistic design framework that balances robustness, durability, constructability, and environmental integration. Robustness must be achieved by designing for local extreme wave and current loads with redundancy to ensure resilience under rare but severe events. Durability is enhanced by targeting characteristic concrete strengths above 35 MPa [6], employing low-permeability mixes, and applying protective coatings in factory settings to resist chloride ingress and abrasion [10]. Constructability is optimized through extensive off-site production and the use of barge or heavy-lift installation methods, thereby shortening on-site works and reducing disruption; where seabed disturbance is constrained, no-dredge or low-dredge techniques such as grouted jetted precast piling (GJPRCP) should be adopted [4]. Finally, environmental integration is essential, with hybrid nature-engineering measures which include mangrove planting, vegetated berms, and intertidal habitats embedded in cavities that can be introduced to improve biodiversity and foster social acceptance [9]. Collectively, these strategies provide a pathway for seawall systems that are technically resilient, sustainable, and socially inclusive.

4.3 Materials and element specifications (practical guidance)

Precast element design for coastal defence requires careful material and structural specification to ensure long-term resilience under aggressive marine exposure. Portland-slag cement (PSC) or sulfate-resistant cement combined with silica fume is recommended to mitigate chemical attack, with target 28-day cube strengths in the range of 35 – 40 MPa depending on element function [6]. Reinforcement strategies should prioritize corrosion-resistant options, including epoxy-coated or stainless steel bars in critical zones, with internal cathodic protection considered for key elements subject to severe chloride ingress [10]. As presented in Armour units, particularly those with slender geometries, benefit from reinforced or fibre-reinforced designs to limit catastrophic breakage, while interlocking geometries are preferred to minimize displacement and hydraulic instability [8]. Factory-applied offshore-grade coatings, supplemented by sacrificial layers in high-wear zones, provide an additional barrier against abrasion and chloride penetration, extending service life and reducing maintenance requirements [5]. Collectively, these measures establish a robust framework for precast seawall elements that balance durability, constructability, and environmental performance.

(a
)

(b
)

Fig. 4. Illustrations of Armour Units: (a) Damaged Cube (Left) and Dolos (Right), (b) Accropode (Left), Core-loc (Middle), and Xbloc (Right) [11]

4.4 Construction sequencing and logistics

Phased implementation is essential for the successful deployment of modular precast seawalls, beginning with pilot stretches that validate structural, hydraulic, and social performance under local conditions before scaling to full alignment. Locating factory production facilities close to harbour areas enables efficient logistics, with roll-on/roll-off systems and heavy-lift barges facilitating the transport of precast elements and minimizing on-site construction time [4]. Equally important is the integration of community engagement plans, particularly for fishing communities whose livelihoods are directly affected by coastal infrastructure. Corporate social responsibility (CSR) programs, inclusive access designs, and participatory planning frameworks can foster trust, improve social acceptance, and align engineering interventions with local socio-economic priorities [14]. This combined approach ensures that seawall projects achieve technical resilience while maintaining legitimacy and sustainability in the eyes of affected stakeholders.

5 Discussion and Conclusions

5.1 Benefits and trade-offs

Precast systems provide notable advantages for coastal defence infrastructure, particularly in terms of accelerated construction schedules, improved quality control, and potential long-term cost savings through reduced maintenance requirements [5]. Factory-controlled casting and curing enable reproducibility and performance certification prior to deployment, thereby enhancing durability and reliability compared to conventional in-situ methods [10]. However, these benefits are balanced by trade-offs: initial capital investment for precast facilities and the scaling of supply chains represent significant challenges, especially in regions with limited industrial capacity. In addition, social impacts which include restricted fishing access, altered coastal aesthetics, and community concerns regarding livelihood disruption require proactive mitigation through inclusive design features and structured stakeholder engagement programs [14]. Addressing these trade-offs through integrated planning ensures that precast seawall systems achieve both technical resilience and social legitimacy.

5.2 Sustainability and SDG alignment

The proposed modular precast seawall approach is designed to align with the United Nations Sustainable Development Goals (SDGs), ensuring that coastal defence strategies contribute to broader global sustainability objectives. By protecting vulnerable coastal communities from flooding and erosion, the system directly supports SDG 11 on sustainable cities and communities. Through the adoption of resilient infrastructure solutions—including precast elements with enhanced durability, factory-controlled quality assurance, and redundancy against extreme hydraulic loads—the approach advances SDG 9 on industry, innovation, and infrastructure. Furthermore, the integration of hybrid nature-engineering features (illustrated on) such as mangrove planting, vegetated berms, and perforated habitat modules preserves marine ecosystem services and biodiversity, thereby contributing to SDG 14 on life below water. Collectively, these measures demonstrate how engineering innovation can be harmonized with social and ecological priorities to deliver coastal infrastructure that is both technically robust and globally sustainable [15].

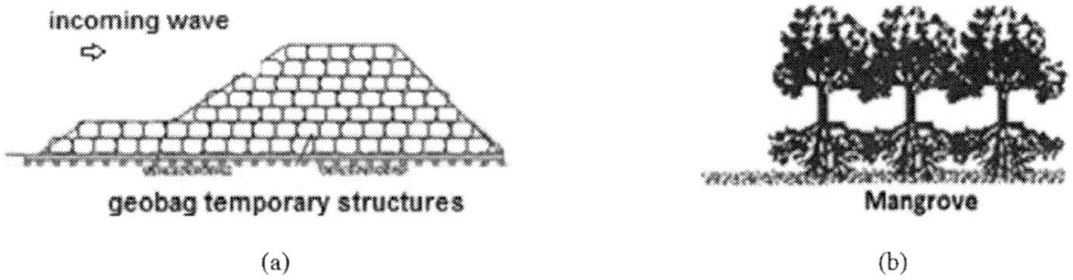

(a) (b)

Fig. 5. Coastal Protection Structures using (a) Temporary Geobags and (b) Mangrove Plants [15]

5.3 Limitations and research needs

Site-specific geotechnical and hydrodynamic investigations are indispensable for the design of modular precast seawalls, as they provide the foundation for load assessments, pile installation feasibility, and hydraulic stability under local extreme events. Long-term performance data are required for locally adapted precast concrete mixes, including innovative partial replacements such as clamshell powder, to validate strength, permeability, and durability under Indonesian marine exposure. Full-scale trials of reinforced armour units in regional seas are equally critical to assess interlocking behavior, breakage resistance, and hydraulic performance under site-specific wave climates [8]. Beyond engineering validation, socioeconomic studies must accompany technical plans to evaluate the displacement of fishing activities, ensure equitable compensation, and foster community acceptance through participatory frameworks [14]. Integrating these technical and social dimensions ensures that seawall projects achieve resilience, sustainability, and legitimacy in the Indonesian context.

5.4 Conclusions

Precast concrete technologies have the potential to serve as the backbone of Indonesia's Giant Sea Wall initiative, offering a constructible, durable, and more sustainable alternative to conventional in-situ methods. By leveraging factory-controlled production and modular installation, precast systems can accelerate construction schedules while ensuring consistent quality across large-scale coastal defence structures. Their adaptability to marine-specific standards makes them particularly suited to withstand the harsh conditions of Indonesia's coastal environment.

The success of such an approach depends on several technical imperatives. Durable, low-permeability concrete mixes are essential to resist chloride and sulfate attack, while reinforced or fibre-enhanced armour units can provide greater resilience against hydraulic loads and mechanical abrasion. Protective factory-applied coatings further extend service life by limiting chemical ingress, and construction methods that minimize seabed disturbance reduce ecological impacts during installation. Together, these strategies position precast concrete as a viable pathway to deliver a Giant Sea Wall that balances engineering robustness, environmental sustainability, and long-term socio-economic benefits for coastal communities.

5.5 Recommendations

To ensure the successful implementation of the Giant Sea Wall, it is recommended to initiate pilot projects spanning 1–3 km that employ the proposed precast system, thereby validating both design parameters and logistical feasibility under real site conditions. Parallel to this, the development of local precast supply chains with strict quality control and marine-grade concrete standards will be essential to guarantee durability and scalability. Engineering design should be complemented by ecosystem-based measures such as mangrove planting and the incorporation of habitat cavities to enhance ecological resilience. Early engagement with affected coastal communities is critical, including the provision of compensatory programs and access arrangements to secure social acceptance. Finally, targeted research on the use of local materials, long-term durability performance, and comprehensive lifecycle environmental impact assessments will provide the evidence base needed to refine strategies and ensure that the Giant Sea Wall achieves both technical robustness and sustainable outcomes.

References

[1] CNN Indonesia, "FOTO: Wajah Pembangunan Tanggul Pengaman Pesisir Jakarta," CNN Indonesia, 9 Sep. 2025. [Online]. Available: https://www.cnnindonesia.com/ekonomi/20250909063344-94-1271458/foto-wajah-pembangunan-tanggul-pengaman-pesisir-jakarta

[2] Ministerie van Infrastructuur en Waterstaat, "Project Afsluitdijk," Rijkswaterstaat. [Online]. Available: https://www.rijkswaterstaat.nl/water/projectenoverzicht/afsluitdijk

[3] Geoengineer.org, "Balancing Coastal Risk: Japan's 395–430 km Tsunami Seawall with Nature-Based Measures and Localised Opposition," Geoengineer.org, 6 Jul. 2025. [Online]. Available: https://www.geoengineer.org/news/balancing-coastal-risk-japans-395430-km-tsunami-seawall-and-nature-based-measures-and-localised-opposition

[4] Q. Z. Q. Yue, "Grouted jetted precast reinforced concrete piling technology for no-dredge seawall and land reclamation in Hong Kong," in HKIE Geotechnical Division 36th Annual Seminar Proceedings, 2016, pp. 59–71. (Conference paper — PDF available).
[5] S. H. Kosmatka and M. L. Wilson, Design and Control of Concrete Mixtures, 15th ed., Portland Cement Association, Skokie, IL, 2011.
[6] W. Zhuo, Q. Yan, Z. Yang, S. Lin, K. Lin, and F. He, "Chloride penetration in coastal concrete structures: field investigation and model development," Advances in Materials Science and Engineering, Article ID 4537283, 2019. doi:10.1155/2019/4537283.
[7] C. Zhang, J. Li, M. Yu, Y. Lu, and S. Liu, "Mechanism and performance control methods of sulfate attack on concrete: a review," Materials 17, 4836 (2024). doi:10.3390/ma17194836.
[8] G. Scaravaglione, J.-P. Latham, J. Xiang, A. Francone, and G. R. Tomasicchio, "Historical overview of the structural integrity of concrete armour units," Coastal and Offshore Science and Engineering 1, pp. 68–98 (2022). doi:10.53256/COSE_220105.
[9] M. A. Reidenbach et al., "Performance evaluation of natural and nature-based features for coastal protection and co-benefits," Annual Review of Marine Science 18, 8.1–8.29 (2025). doi:10.1146/annurev-marine-040423-023251.
[10] R. E. Melchers, "Recent developments of reinforcement corrosion in concrete structures in the ocean," Corrosion Engineering, Science and Technology: The International Journal of Corrosion Processes and Corrosion Control, (2025). doi:10.1177/1478422X251358291.
[11] M. Muttray and B. Reedijk, Design of Concrete Armour Layers, Delta Marine Consultants, Gouda, Netherlands, Oct. 2008. (Technical/report).
[12] A. L. da Silva, E. R. Kohlman Rabbani, and M. Shakouri, "Seashell powder as a sustainable alternative in cement-based materials: a systematic literature review," Sustainability 17, 592 (2025). doi:10.3390/su17020592.
[13] B.-L. Dang, H. Nguyen-Ngoc, T. D. Hoang, H. Nguyen-Xuan, and M. Abdel Wahab, "Numerical investigation of novel prefabricated hollow concrete blocks for stepped-type seawall structures," Engineering Structures 198, 109558 (2019). doi:10.1016/j.engstruct.2019.109558.
[14] M. Esteban et al., "Adaptation to sea level rise: learning from present examples of land subsidence," Ocean & Coastal Management 189, 104852 (2020). doi:10.1016/j.ocecoaman.2019.104852.
[15] N. Yuanita, A. Kurniawan, H. Setiawan, F. Hasan, and M. Khasanah, "Physical model of natural coastal protection system: wave transmission over mangrove seedling trees," Journal of Coastal Research, Special Issue 91, pp. 176–180 (2019). doi:10.2112/SI91-036.1.

Spatial Analysis of Forest Fire Hotspots in Kalimantan Using Inhomogeneous Neyman-Scott Cluster Area

Berlian A. Kindhi[1*], *Rifqi* A. Mahmada[2], *Muhammad* F. Zaqi[2], *Dimas* A. Saputra[2], and *Narsischa* S. Ningrum[2]

[1]Department of Electrical Automation Engineering, Institut Teknologi Sepuluh Nopember, Surabaya, Indonesia
[2]Department of Statistics, Institut Teknologi Sepuluh Nopember, Surabaya, Indonesia

Abstract. Forest and land fires in Kalimantan pose a serious environmental challenge with significant impacts on ecosystems, public health, and the achievement of the Sustainable Development Goals (SDGs). In recent years, fire intensity has increased due to the combined effects of human activities, peatland characteristics, and extreme climate conditions associated with El Niño. Spatial analysis shows that fire hotspots are unevenly distributed, with a strong concentration in peatland areas. This study applies the Inhomogeneous Neyman–Scott Cluster Process (NSCP) to model the spatial clustering of hotspots under heterogeneous conditions by incorporating covariate information. The objective is to identify hotspot distribution patterns in Kalimantan during 2022–2024 and to generate fire risk maps to support targeted mitigation and environmental management policies. Model comparison using the Akaike Information Criterion (AIC) indicates that the Thomas Cluster Process with distance to BPBD facilities as a covariate provides the best fit, with the lowest AIC value (-456,944.7). The results reveal a negative relationship between distance to BPBD facilities and hotspot intensity, indicating that fire occurrences decrease with increasing distance. These findings confirm that hotspot distribution in Kalimantan is not random and highlight the importance of spatially informed fire management strategies.

1 Introduction

Forest and land fires are a serious environmental problem in Indonesia, particularly in Kalimantan, which is known as one of the world's tropical forest centres. This issue is closely related to the Sustainable Development Goals (SDGs), particularly SDG 13 on climate action and SDG 3 on good health and well-being. In recent years, forest fires in Kalimantan have become increasingly widespread. As reported in [1], during the period from January to April 2024, East Kalimantan was the most severely affected region, with more than 18.000 hectares of land burned, the largest amount compared to other provinces in Indonesia.

The main factors causing forest fires include peatland conversion, land clearing practices involving burning, and extreme climatic conditions such as El Niño, which prolongs the dry season [2]. Spatio-temporal analysis using hotspot data shows that dry peatlands are highly flammable and accelerate the spread of fire, thereby increasing the risk of fires in surrounding areas. As a result, forest fires have widespread impacts, ranging from ecosystem damage and loss of biodiversity to threats to public health due to haze pollution.

In addition to these widespread impacts, forest fires in Kalimantan tend to be concentrated in areas with high peat density. Several previous studies have examined forest fire patterns using various approaches. For example, satellite image-based analysis and hotspot data have been used to map fire-prone areas and identify environmental factors that influence the spread of fire [2, 3]. Other studies have conducted spatio-temporal analyses to understand the dynamics of fire events over time and the interactions between hotspots in different locations [4]. These results show that hotspots tend to form spatial clusters, where fires in one location can increase the risk of fires in surrounding areas.

*Corresponding author: berlian@its.ac.id

The pattern of fires forming clusters is relevant for analysis through a point process framework such as the Cox process, in which large fires can increase the likelihood of subsequent fires in the surrounding area due to the spread of heat or embers. Therefore, the Neyman-Scott Cluster Process (NSCP) is used, as this model is able to capture spatial intensity variations and provide a more accurate representation of fire patterns. Study [4] shows that NSCP is effective in mapping the frequency, intensity, and level of damage caused by fires in Kalimantan's peatlands, thus providing a basis for the development of more targeted forest fire risk maps. Previous studies have demonstrated the relevance of this approach. Study [5] developed a non-homogeneous Cox process to model the distribution of large earthquakes in Sulawesi and Maluku using geological variables. In the context of forest fires, existing studies are still limited. Study [6] examined fire hotspot patterns in Kalimantan using descriptive analysis based on SNPP-VIIRS satellite imagery. However, this study did not use a point process approach or produce forest fire risk maps. Furthermore, study [7] conducted a combined spatio-temporal analysis of hotspots, burned area, and carbon emissions with drought indicators; however, this study emphasised the relationship between climate and forest fires rather than point process-based spatial intensity modelling.

2 Literature Review

2.1 Fire Forest

In Indonesia, forest fires are a common environmental calamity, especially in Kalimantan, which is home to extensive tropical forests and peatland ecosystems. Study [8] states that peatland fires are more severe because they spread underground and are challenging to put out, while these fires harm vegetation and biodiversity in addition to degrading the environment through increased carbon emissions, soil damage, and health issues from smoke. Haze lowers air quality, causes respiratory illnesses, interferes with community activities, and even has cross-border effects, according to socio economic perspectives. Climate variability, particularly the El Niño phenomenon that lowers water levels and prolongs dryness, exacerbates human activities like land clearance for plantations, peatland drainage, and land burning [3, 4]. Thus, forest fires in Kalimantan are regarded as both ecological and social calamities, necessitating community empowerment and sustainable land management in addition to technological suppression. As of October 2023, the affected area in Indonesia was 994.313 hectares, with Kalimantan being the most affected, according to KLHK data. This led to a number of prevention and monitoring initiatives, including the SiPongi+ portal and policies under Regulation No. 32 of 2016 on Forest and Land Fire Control, which place an emphasis on law enforcement, integrated patrols, and monitoring.

2.2 Spatial Point Process

Spatial Point Process (SPP) is a stochastic framework for representing the positions of events as a collection of points in R^2. This analysis aims to describe the intensity (first order), the pattern of dependence between points (second order), and the relationship between point patterns and spatial variables. Where the realization of this process is represented as a set of points $\{xi\}$, so this model is suitable for various ecological and environmental applications because it intuitively distinguishes between regular patterns (inhibition), clustering, and random patterns (Poisson) [9]. First-order statistics focus on intensity (x), which is defined such that for a small area B around location x.

$$\mathbb{E}[N(B)] = \lambda(x).\ |B| \tag{1}$$

In a homogeneous process, $\lambda(x)$ is constant with respect to (λ), while the intensity of an inhomogeneous process varies depending on location and is often modeled using regression against spatial covariates, $\lambda(x) = \exp(\beta^T z(x))$. Prior to second-order analysis, it is generally necessary to estimate (x) (e.g., smoothing kernel or GLM/GAM model) to prevent misinterpretation of clustering patterns that are caused by intensity inhomogeneity [10].

Second-order statistics focus on interactions between points. The probability of finding points in a limited area around x and y at the same time is estimated by $\int x \int x\rho^2 (x, y)\, dx\, dy$, which is the definition of the second order product density $\rho^2 (x, y)$. The pair correlation function (PCF) is obtained from ρ^2 as below.

$$g\,(x, y) = \frac{\rho^2(x,y)}{(\lambda x)\cdot\lambda(y)} \tag{2}$$

This PCF becomes a function of distance $r = \|x - y\|$ for isotropic/stationary processes denoted by (r). If $(r) > 1$, there are more pairs of points at distance r (clustering), whereas is $(r) < 1$, there are fewer pairs (inhibtion).

The Ripley's K-Function is a popular second-order summary besides PCF. For stationary processes, this function is defined as follow.

$$K(r) = \frac{1}{\lambda}\, \mathbb{E}[N(B)] \tag{3}$$

Where the expected complete spatial randomness CSR is $(r) = \pi r2$. The relationship between (r) and (r) in two dimensions is expressed below.

$$g\,\frac{(r) = K\prime(r)}{2\pi r} \tag{4}$$

To prevent the actual clustering signal from being obscured by uneven intensity patterns, it is important to use the appropriate intensity-reweighted version of the K-Function and an estimate of $g(r)$ that takes into account the variation $\lambda(x)$ for inhomogeneous data [11].

The inhomogeneous Kfunction was developed to address the limitation of Ripley's K-function, which assumes that point intensity is constant across space. In real-world cases like forest fire hotspots, this assumption is unrealistic since fire risks vary significantly due to factors such as peatland distribution, precipitation, and human activity. Introduced by study, the inhomogeneous K function adjusts for this non-stationarity by weighting point pairs with an estimated intensity, $\hat{\lambda}(x)$. This correction makes it possible to separate first-order effects (environmental influences on fire density) from second-order effects (spatial clustering that reflects fire spread or local interactions). Inhomogeneous K-function $K_{\text{inhom}}(r)$ for the Neyman–Scott family in the case where the parent (cluster centers) process is homogeneous (constant intensity κ_0), then the simplification that occurs when parents are homogeneous, and give the Thomas (Gaussian) closed form as an example plus practical note. For a point process if $N \sim Poiss(\mu)$ then $v = \mu 2$ and the prefactor simplifies to $\frac{1}{k_0}$. So with intensity function (x) we can expresses.

$$K_{inhom}(r) = \pi r\,2 + \frac{1}{k_0} \int_{\|h\| \le r} (f * f)(h)\, dh \tag{4}$$

Where $(f * f)(h)$ denotes the convolution of the offspring density with itself, cap turing the contribution of point pairs that originate from the same cluster. The first term $\pi r\,2$ represents the theoretical K-function of a completely spatially random (CSR) process, while the second term introduces an additional component due to clustering. In empirical applications, (x) and its estimate $\hat{\lambda}\,(x)$ represent the spatial intensity function and its nonparametric estimate, respectively, and W denotes the observation window within which the process is observed. Estimating $K_{inhom}(r)$ thus involves correct ing for spatial variation in point intensity by weighting each observed pair inversely by the product $\hat{\lambda}(x)\hat{\lambda}(y)$, ensuring that the result reflects second-order clustering rather than large-scale intensity inhomogeneity.

2.3 Inhomogeneuous Neyman-Scott

An extensive overview of the theory of point processes and Neyman–Scott processes is presented, among others, by study [11]. In essence, a Neyman–Scott process X can be described as a union of clusters $\cup_{c\epsilon}$, where C is a Poisson process of centers of clusters (parent process). Given C, the clusters X_c, $c \in C$ are independent Poisson processes, with an intensity function $ak(u - c, \omega)$, where $k(u,\omega)$ is a probability density function parameterized by ω, which determines the spread of daughter points around its mother and a is the expected number of daughters per cluster. If $(u.)$ is the density of a symmetric normal distribution, then the process is called a modified Thomas Process [12].

An inhomogeneous Neyman–Scott process is often formulated by thinning the daughter points, producing what is known as a second order intensity reweighted stationary model. Our formulation differs by introducing inhomogeneity at the cluster center level C, treating C as an inhomogeneous Poisson process governed by an inten sity function $fb\ (u) : \kappa \exp(\beta^T z(u))$. Where $z = (z_1, \ldots, z_k)$ is a vector of covariates and $\beta = (\beta_1, \ldots, \beta_k)$ is a regression parameter.

The intensity of the Neyman-Scott point process with inhomogeneous cluster centers is then.

$$(u) = \int_{\mathbb{R}2} (c)\mathbb{E}[Nc\,]f(u - c)dc$$

$$= \mu \int (c)(x - c)dc,\ u \in \mathbb{R}^2$$

$$= (\kappa \times f)(x) \tag{5}$$

where $(\kappa \times f)(u)$ denotes the spatial convolution of the parent intensity function kappa (κ) with the offspring kernel f. This expression shows that the intensity of the observed process is a smoothed version of the underlying parent intensity, with the degree of smoothing determined by the spread of the offspring kernel f. Consequently, spatial variation in (c) is transferred to the offspring process through convolution, producing inhomogeneous clustering that reflects both the spatial heterogeneity of parent centers and the dispersal properties of the offspring distribution.

Thomas Cluster Process, Fire spread patterns are distributed around the ignition sources following a bivariate normal with variance parameter σ^2 (isotropic) [13]. The density function is given by

$$f(u) = \frac{1}{2\pi\sigma\ 2} \exp\ (-\ \|u\|\ 2\ 2\sigma\ 2\ ) \tag{6}$$

Where σ^2 denotes the variance of the Gaussian distribution, and $\|u\|$ is the Euclidean distance between an offspring and its parent. This kernel ensures that offspring are symmetrically distributed around the parent center with expected radial dispersion proportional to σ. Smaller values of σ indicate that offspring points are more tightly concentrated near their parent, forming denser clusters, whereas larger σ values correspond to broader, more diffuse clustering patterns. Since (u) integrates to one over the plane, it serves as a valid spatial probability density governing the offspring displacement mechanism in the homogeneous parent Neyman–Scott process. For the Thomas cluster process, the pair correlation function is expressed as

$$g(r, \sigma) = 1 + \frac{1}{4\pi\kappa\sigma\ 2} \exp\ (\frac{r2}{4\sigma 2})\ ,\ r = \|u - v\| \tag{7}$$

while the K-function is given by

$$K(r) = \pi r^2 + \frac{1}{\kappa}\{1 - exp\ (-\frac{r2}{4\sigma 2})\} \tag{8}$$

To estimate the estimation of inhomogeneous clustered processes is the first-order composite log-likelihood (treating observed points as an inhomogeneous Poisson pro cess with intensity $(x;,)$ and ignoring interactions). When parametrize $(c; \beta)$ (parent inhomogeneity parameters β) and cluster parameters θ appear in $(u; \theta)$ and $\mu(\theta)$, the firstorder composite log-likelihood is.

$$\ell_{c1}(\beta) = {}_{\Sigma xi\epsilon X\cap W}\, lo(xi\,;\beta) - \textstyle\int_W \lambda(u;\beta)du \tag{9}$$

To maximize (7) we can use Hessian and the entries (8).

$$\partial\theta^2{}_{j\theta k}\,\ell_{c1} = \Sigma i{=}1\; n\; \partial\theta j\, \frac{\partial\theta j\; \lambda(xi\,)\partial\theta k\; \lambda(xi)}{\lambda\,(Xi)2} - \Sigma i{=}1\; n\, \frac{\partial\theta 2\; j\theta k}{\lambda(xi\,)}$$

$$\textstyle\int_W \partial^2\, \theta_j\theta_k\,(u)du \tag{10}$$

captures both the local curvature at observed data points and the compensating contribution over the observation window W.

Matérn Cluster Process (MCP) is a special case of the Ney man–Scott family of cluster processes, where offspring points are uniformly distributed within a fixed radius around their parent points [14-15]. In this model, parent points are generated from a homogeneous Poisson process with intensity κ, and each parent independently produces a Poisson-distributed number of offspring with mean μ. The offspring points are then placed uniformly within a circular region of radius R centered at their parent location. This uniform displacement mechanism implies that all locations within the cluster radius are equally likely to contain offspring, resulting in clusters with homogeneous intensity within their boundaries. The probability density function governing the offspring distribution is expressed as below.

$$f(u) = \{ \frac{1}{\pi R2\,,} if\; \|u\| \leq R,\; 0, \tag{11}$$

$$oth\,erwise$$

Here, R defines the spatial extent of the cluster, and $\|u\|$ represents the Euclidean dis tance between an offspring and its parent. The uniformity of (u) ensures that the off spring are evenly scattered within the disk, creating clusters with sharp boundaries compared to the smoother Gaussian spread in the Thomas process [19].

The pair correlation function of the Matérn cluster process declines from a finite value at short distances to one as r increases, reflecting spatial aggregation limited to the neighborhood defined by R. The corresponding K-function takes the form (9).

$$K(r) = \pi r^2 + \frac{1}{\kappa} k(r, R) \tag{12}$$

Where $(r,)$ represents the clustering contribution within radius R. Model estimation for inhomogeneous Matérn cluster processes typically relies on the first-order composite likelihood, treating observed points as arising from an inhomogeneous Pois son process with intensity $\lambda(x; \beta,\theta)$ while ignoring higher-order interactions. Parameter estimation is then achieved by maximizing this likelihood with respect to both the parent process parameters β and cluster parameters θ. The Matérn process is particularly suitable for modeling spatial phenomena where clusters exhibit uniform density within a finite spatial extent, such as vegetation patches, animal nests, or fire hotspots confined within localized regions.

2.4 Model Evaluation

Model selection plays a crucial role in spatial analysis, especially in point process modelling, to ensure that the selected model can accurately represent the spatial distribution of points. The Akaike Information Criterion (AIC) is commonly applied to achieve a balance between model fit and model complexity to minimize the risk of overfitting [19,20].

The AIC is defined by (10). $AIC = -2L\max + 2p$ (13) Where p is the number of estimated parameters and L_{max} denotes the maximum value of the likelihood function for the fitted model. In the case of a Poisson process model, Lmax is obtained from the first-order composite likelihood, while for a Cox process model, it is derived from the second-order composite likelihood. The model with the lowest AIC value is considered the most suitable, as it indicates minimal information loss and provides a more accurate and efficient spatial representation.

3 Methodology

3.1 Study Area and Data Descriptions

This study focuses on the island of Kalimantan, Indonesia, which consists of five provinces: West Kalimantan, Central Kalimantan, South Kalimantan, East Kalimantan, and North Kalimantan. Kalimantan is selected due to its extensive peatland coverage and its recurrent forest and land fire events, particularly during prolonged dry seasons influenced by El Niño phenomena. The primary dataset consists of forest fire hotspot locations detected during the dry season of 2024. Only hotspots with high confidence levels were retained to reduce false detections and improve spatial reliability. Each hotspot is represented as a spatial point in a two-dimensional geographical coordinate system (longitude, latitude). In addition to the point pattern data, spatial covariates were incorporated to model intensity inhomogeneity. These covariates include:

1. Distance to the nearest **Regional Disaster Management Agency (BPBD)** office,
2. Distance to the nearest **water source (river)**.

Distances were computed using Euclidean distance metrics based on spatial layers obtained from Google Maps. The response variable is the spatial location of fire hotspots, while covariates are treated as continuous spatial predictors.

Table 1. Research Variables

Variable	Descripton	Label/Scale	Source of Data	Year
X	Forest fire locations (hotspots)	Degree cooordinate (latitude, longitude)	Ministry of Forestry	2024
Covariates	Distance to water sources	Ratio	Google Maps	2024
	Dinstance to BPBD	Ratio	Google Maps	2024

3.2 Data Structure

In our purpose method, X represents forest fire point data, which is a coordinate point consisting of latitude and longitude. Then, z is the covariate, namely the distance from the point to the nearest river and the distance from the point to the nearest BPBD. Let

$$X = \{x_i = (x_{i1}, x_{i2}) : i = 1, \dots, n\} \qquad (13)$$

denote the observed spatial point pattern of forest fire hotspots within an observation window $W \subset R^2$, where x_{i1} and x_{i2} represent longitude and latitude, respectively. Each point x_i is associated with a vector of spatial covariates:

$$z(x_i) = z_1(x_i), z_2(x_i)). \tag{14}$$

where:

1. $z_1(x_i)$ denotes the distance to the nearest river,
2. $z_2(x_i)$ denotes the distance to the nearest BPBD office.

The following is the data structure used in this study.

Table 2. Data Structure

i	***Longtitude*** **($X_{long,i}$)**	***Latitude*** **($X_{lat,i}$)**	**Z_1 (x_i)**	**Z_2 (x_i)**
1	$X_{long,1}$	$_{lat,1}$	$Z_{1,1}$	$Z_{2,1}$
2	$X_{long,2}$	$_{lat,2}$	$Z_{1,2}$	$Z_{2,2}$
..	....	....	...	....
n	$X_{long,n}$	$_{lat,n}$	$Z_{1,n}$	$Z_{2,n}$

3.3 . Exploratory Spatial Data Analysis

Prior to model fitting, exploratory spatial data analysis (ESDA) was conducted to assess the global structure and potential spatial dependence of hotspot locations. Quadrat counting analysis was employed to test the hypothesis of Complete Spatial Randomness (CSR). The observation window was partitioned into several subregions, and a chi-square test was applied to compare observed and expected counts under a homogeneous Poisson process.

To further examine spatial dependence, Ripley's K-function and the inhomogeneous K-function were computed. The homogeneous K-function was used as an initial diagnostic to identify clustering tendencies, while the inhomogeneous K-function was employed to account for spatial intensity variation induced by covariates. This step is crucial to distinguish true second-order interaction (clustering) from first-order intensity inhomogeneity. The estimated intensity function λ was evaluated on a regular spatial grid covering the study area to produce a continuous hotspot risk map. Regions with higher estimated intensity indicate areas with greater susceptibility to forest and land fires, providing spatially explicit information for mitigation planning and early-warning strategies.

3.4 Analysis Steps

The analysis begins with the acquisition and preprocessing of forest fire hotspot data and spatial covariates. Only hotspots with high confidence levels were retained to ensure the reliability of detected fire events. Spatial covariates, including the distance to the nearest BPBD office and distance to water sources, were computed using geographic information system (GIS) tools based on Euclidean distance. All spatial layers were projected into a common coordinate reference system, and the study region was defined as a bounded observation window $W \subset R^2$. Data cleaning procedures were applied to remove duplicated points and ensure spatial consistency.

Exploratory spatial analysis was conducted to assess the global structure and heterogeneity of the hotspot distribution. Initial visualization through spatial point maps revealed substantial variation in point density across Kalimantan. To formally evaluate the assumption of Complete Spatial Randomness (CSR), a quadrat counting test was applied by partitioning the study area into several subregions and comparing observed hotspot counts with those expected under a homogeneous Poisson process. The rejection of CSR indicated strong spatial inhomogeneity, suggesting that fire occurrence is influenced by underlying spatial factors.

To investigate spatial dependence beyond first-order effects, second-order statistics were employed. Ripley's K-function was first used to detect clustering patterns in the hotspot distribution. Because the intensity of fire occurrences varies spatially, an inhomogeneous K-function was subsequently applied to account for non-stationary intensity. The persistence of clustering after intensity correction indicated the presence of genuine spatial interaction among hotspot events, thereby justifying the use of cluster process models.

Based on the exploratory findings, the hotspot pattern was modeled using an Inhomogeneous Neyman–Scott Cluster Process (NSCP), where cluster centers follow an inhomogeneous Poisson process with intensity expressed as a log-linear function of spatial covariates. Two alternative offspring dispersion mechanisms were considered: the Thomas cluster process, which assumes Gaussian dispersion of events around parent locations, and the Matérn cluster process, which assumes uniform dispersion within a fixed radius. These two models represent different spatial fire spread behaviors and clustering scales.

Model parameters were estimated using the first-order composite likelihood approach, treating the observed process as an inhomogeneous Poisson process while incorporating clustering effects through spatial convolution. The likelihood integral over the observation window was approximated using the Berman–Turner device to ensure computational efficiency. Backward variable elimination was applied to identify the most parsimonious covariate structure for each cluster model.

Model selection was performed using the Akaike Information Criterion (AIC) to compare competing specifications. The model with the lowest AIC value was selected as the optimal representation of the spatial distribution of forest fire hotspots. The estimated regression coefficients and cluster parameters were then interpreted to assess the influence of spatial covariates on hotspot intensity and to characterize the strength and spatial extent of clustering.

Finally, the selected NSCP model was used to generate spatial predictions by evaluating the estimated intensity function over a regular grid covering Kalimantan. The resulting intensity surface was visualized as a forest fire risk map, where higher intensity values indicate areas with greater susceptibility to fire occurrence. This risk map provides spatially explicit information to support targeted fire prevention, monitoring, and mitigation strategies.

4 Results and Discussion

4.1 Exploratory Spatial Pattern of Forest Fire Hotspot

The initial analysis focuses on the spatial distribution of forest fire hotspots across Kalimantan in 2024. Visual inspection of the hotspot map reveals a highly uneven spatial pattern, with clear concentrations in specific regions, particularly in the western and central parts of Kalimantan. These areas exhibit dense aggregations of hotspot points, while other regions show sparse or scattered occurrences. Such strong spatial heterogeneity indicates that forest fire events are unlikely to follow a random spatial process.

The quadrat counting analysis further supports this observation. The extremely large chi-square statistic and the corresponding p-value far below conventional significance levels strongly reject the hypothesis of Complete Spatial Randomness (CSR). This result confirms that hotspot occurrences are influenced by spatially varying factors, such as environmental conditions and human activities, rather than arising from a homogeneous Poisson process. Consequently, modelling approaches assuming spatial homogeneity would be inappropriate and potentially misleading for this dataset.

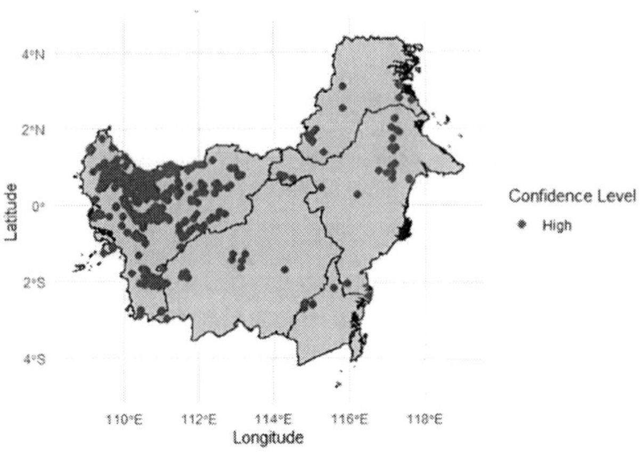

Fig. 1. Distribution of forest fire hotspots with confidence in Kalimantan in 2024.

The distribution of hotspots on the map (Fig. 1) is generally somewhat irregular or uneven across the island. This is indicated by extreme variations in point density. The western part of Kalimantan is the area with the most striking and massively concentrated hotspots density, with a main focus in the 110° E until 112° E longitude and 0° until 2° S latitude zones. This area is the most significant hotspots zone, as indicated by its high concentration. In contrast, other regions on the island of Kalimantan show much lower hotspot densities or are scattered sporadically. This distribution pattern clearly shows a tendency toward clustering, indicating that the formation of hotspots is not random but is influenced by certain local factors. Therefore, this visual evidence strongly indicates that the hypothesis of Complete Spatial Randomness (CSR) should most likely be rejected for this hotspot data. However, for further verification, a quadrat counting analysis was conducted to observe spatial trends, and then a K-function plot was used to detect interactions between points. To perform quadrat testing under the null hypothesis of Complete Spatial Randomness (CSR), the observation window was divided into nine irregular quadrats to assess whether the number of hotspot points within each sub-region differs significantly from what would be expected under a homogeneous Poisson process. The test result yielded a chi-square statistic of 1174,7 with 8 degrees of freedom and a p value $< 2,2 \times 10{-}16$, indicating a strong rejection of the CSR hypothesis. This suggests that the hotspot pattern is not randomly distributed but exhibits clear spatial variation across the study area. Such a pattern implies the presence of spatial inhomogeneity, possibly due to geological and spatial factors.

Then, we will use the K-function to see the spatial correlation. Fig. 2 presents the result of the K-function analysis for hotspot locations. The empirical K-function lies above the theoretical Poisson reference line, indicating a clear clustering pattern among the hotspots. The sudden drop in the empirical K-function near the right end of is characteristic of edge effects or a lack of data coverage at those large distances, r. This suggests that the hotspots are not randomly distributed but tend to occur in groups, possibly influenced by environmental or geological factors. Next, we consider the inhomogeneous K-function (Fig. 3) to assess whether the clustering persists after accounting for spatial variation in intensity. The dashed blue line represents the K-function of an inhomogeneous Poisson process. In general, the edge corrected inhomogeneous K-function still indicates the presence of clustering, although the degree of clustering is weaker than in the homogeneous case. This implies that some clustering can be explained by spatial inhomogeneity, but a significant level of spatial dependence remains. Hence, assuming a stationary process may overlook important spatial trends affecting the hotspot distribution.

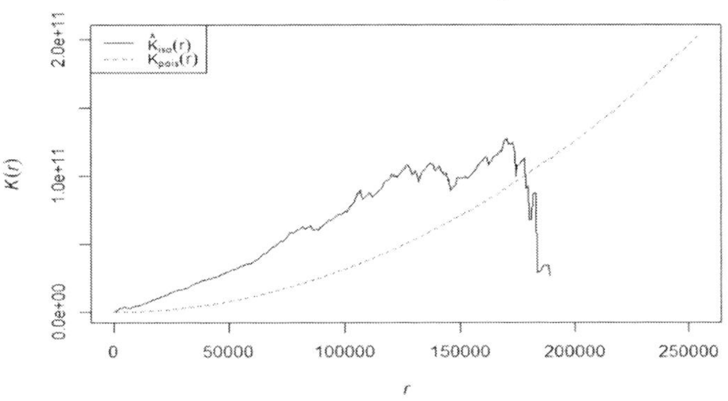

Fig. 2. The K-function plot for distribution of forest fire hotspots in Kalimantan in 2024.

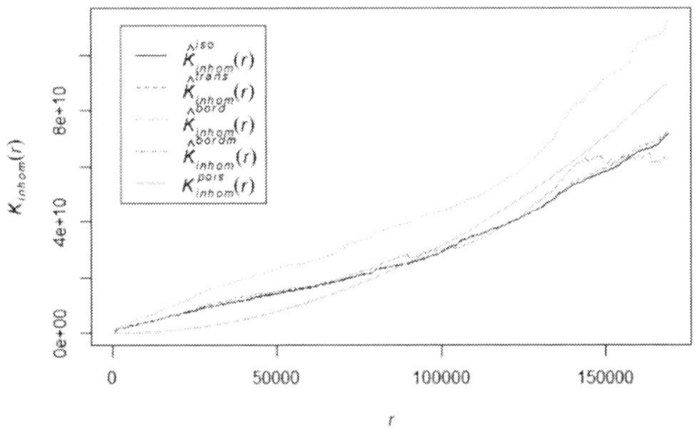

Fig. 3. Inhomogeneous Kfunction plot for distribution of forest fire hotspots in Kalimantan in 2024.

4.2 Model Selection

The AIC is a model selection metric used to compare competing statistical models by balancing goodness of fit and model complexity. Lower AIC values indicate a model that provides a better trade-off between fit and parsimony. In the context of spatial point process modeling, AIC helps identify which cluster process (Thomas or Matérn) and which spatial covariate best describe the observed pattern of events. Table 3 presents AIC values and spatial parameters for the Thomas and Matérn cluster process models, each fitted with a different spatial covariate below.

Table 3. Thomas and Matérn cluster process models parameter architecture

	Thomas			Matérn		
	Distance between BPBD and river	**Distance to BPBD**	**Distance to river**	**Distance between BPBD and river**	**Distance to BPBD**	**Distance to river**
K	$2,766.10^{-10}$	$2,628 \times 10^{-10}$	$2,652 \times 10^{-10}$	$2,652 \times 10^{-10}$	$2,523 \times 10^{-10}$	$2,604 \times 10^{-10}$
ω	$9,621 \times 10^{3}$	9,88x103	1,993x104	2,041x104	2,041x104	2,011x104
AIC	-451714,3	-456944,7	-452044,5	-451665,9	-456901,1	-451997,9

Based on AIC comparison, the Thomas process incorporating the Distance to BPBD covariate yields the most parsimonious fit to the observed spatial pattern (AIC = −456944,7). This indicates that the spatial distribution of events is best explained by a Gaussian-type cluster process in which intensity declines with increasing distance from BPBD sites. The Thomas model's consistently have lower AIC values relative to the Matérn process further suggest that event clustering follows a smoother, more continuous decay pattern rather than a sharply bounded cluster structure.

In the Thomas model, the estimated κ parameter ranges from $2{,}6\times10^{-10}$ to 2,8 × 10−10, while the scale (weight) parameter varies between 9,6 × 103 and 9,9 ×103. The relatively small κ value indicates a high degree of spatial clustering, suggesting that the parent process operates with low intensity. Meanwhile, the smaller scale value implies that the aftershock points tend to be concentrated in closer closer to the main event centers. In comparison, the Matérn model yields slightly smaller κ estimates (approximately 2,5 × 10−10 to 2,7×10−10) but considerably larger scale values (around 1,99 ×104 to 2,04 × 104). This suggests that the Matérn process represents a more dispersed cluster ing structure, where events are distributed over a wider spatial range surrounding the mainshock centers.

Overall, the Thomas model exhibits a stronger clustering pattern with a lower parent process intensity compared to the Matérn model. Based on the spatial parameter estimates and the corresponding AIC values, it can be concluded that the Thomas model incorporating the Distance to BPBD covariate provides the most appropriate representation of the spatial structure observed in the data

4.3 Model Interpretation

An in-depth analysis of the calculated parameters was conducted based on the model selection results, which showed that the Thomas Cluster Process Model with the BPBD Distance variable provided the most economical fit to the spatial pattern of hotspots. The Thomas model was able to capture the clustering effect and the intensity trend of the cluster center (λp) based on spatial covariates. The parameter that controls this in tensity is β, which is modeled in the loglinear function $\ln(\lambda p(\boldsymbol{u})) = \beta 0 + \beta 1\ (u)$, where $\beta 1$ is for BPBD Distance.

Table 4. Estimated Parameters

	β	**exp(β)**
Intercept	-20,572	$1{,}16 \times 10^{-9}$
Z_1	$-2{,}199 \times 10^{-6}$	0,999998

The spatial model estimation results show that the estimated value of the intercept parameter ($\beta 0$) is −20,572 and $\beta 1$is −2,119 × 10−6. These estimates were obtained using the maximum composite likelihood approach, which aims to describe the effect of distance on the spatial distribution of fire hotspot occurrences in the study area.

The negative value of $\beta 1$ indicates that the closer a point is to the BPBD office, the higher the tendency for a hotspot to occur. Conversely, as the distance from the BPBD increases, the likelihood of hotspot occurrence decreases. In other words, hotspots are more likely to be found in areas relatively close to BPBD locations. Based on the $\beta 1$ value, it can be calculated that for every one kilometer decrease in distance toward the BPBD, the intensity of hotspot occurrence increases by approximately exp ($\beta 1$) ≈ 0,999998, meaning that the change in risk is relatively small per unit of distance. However, this effect becomes more evident over larger distances. For example, in an area located 100 km closer to the BPBD, the probability of hotspot occurrence increases by approximately 1,00021 times compared to more distant regions. This finding suggests that BPBD offices, which are typically located in areas with high human activity and accessibility, may be associated with a greater frequency of hotspots due to anthropogenic factors. The proximity to human settlements and infrastructure likely contributes to a higher fire risk or hotspot concentration.

Meanwhile, the intercept ($\beta 0$) represents the baseline log intensity under the condition where the distance to the BPBD is zero (i.e., exactly at the BPBD location). Based on this value, the baseline intensity of hotspots (λ) can be calculated as.

$$\lambda = e(\beta_0) \times \kappa = (1,16 \times 10^{-19}) \times (2.628 \times 10^{-10} = 3,06 \times 10^{-19} \quad (14)$$

The value $\lambda = 3,06 \times 10^{-19}$ indicates that when the distance to the BPBD is zero, the intensity of hotspot occurrence is extremely low. This implies that even though the location is very close to BPBD, the probability of hotspot occurrence remains minimal. This condition may be attributed to stronger monitoring and fire-control measures implemented around BPBD offices. These results confirm that the distance to BPBD has a negative influence on the spatial intensity of hotspots, but the effect is relatively mild and tends to become more apparent over larger spatial scales. Overall, these findings support the idea that the distribution of hotspots in Kalimantan is determined by a clustering process that is strongly influenced by the relative distance to BPBD facilities.

4.4 Model Prediction

The modeling results using the Non-Stationary Cluster Process (NSCP) with the Thomas cluster model approach and covariates of river proximity and BPBD location produced a predicted map of hotspot intensity distribution across Kalimantan, as shown in Fig. 4. This map presents an estimation of hotspot density (intensity) based on the developed spatial model, where lighter colors indicate areas with higher hotspot occurrence intensity.

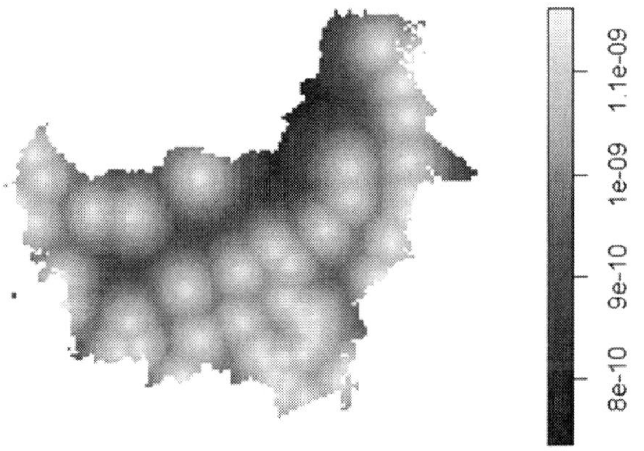

Fig. 4. Predicted hotspot intensity using the Thomas NSCP model in Kalimantan.

The regions with the highest intensity are identified in the central and eastern parts of Kalimantan, along with several smaller concentrations in the southwestern and southern areas. This distribution pattern indicates that forest and land fire activities tend to cluster in specific areas, most likely influenced by proximity to triggering variables such as river distance and BPBD location. The predicted intensity map remains indicative and potentially biased. This condition is attributed to the limited number and uneven distribution of hotspot data used, as well as possible spatial imbalances in the covariate variables. The interpretation of this map should therefore be conducted with caution, and further validation is recommended using additional observational data or alternative modeling approaches.

5 Conclusion and Future Works

Based on the model comparison using Akaike Information Criterion (AIC), the Thomas Cluster Process model incorporating the "Distance to BPBD" covariate provides the best and most parsimonious fit to the observed spatial pattern of hotspot occurrences. With the lowest AIC value (AIC = −456944,7), this model effectively captures the spatial clustering behavior of events, indicating that hotspot intensity decreases smoothly with increasing distance from BPBD facilities. The Thomas process demonstrates stronger spatial clustering ($\kappa = 2,6\times10^{-10}$) and smaller cluster spread ($\omega = 9,9\times10^{3}$) compared to the Matérn process, which exhibits a more dispersed clustering structure with larger scale values ($\approx 2.0\times10^{4}$). This implies that hotspot events tend to concentrate more tightly around central points under the Thomas model, while the Matérn process represents a wider, less cohesive distribution.

From the log-linear intensity model $\ln(\lambda\square(u)) = \beta_0 + \beta_1 (u)$, the intercept ($\beta_0 = -20,572$) and slope ($\beta_1 = -2,119 \times 10^{-6}$) reveal that hotspot intensity is inversely related to the distance from BPBD locations. In other words, areas closer to BPBD facilities are more likely to become cluster centers for fire events, while regions farther away show a significant reduction in clustering intensity.

In summary, the analysis concludes that the Thomas Cluster Process with the BPBD Distance covariate is the most suitable model to describe the spatial structure of hotspot occurrences in Kalimantan. The model reflects a Gaussian type clustering mechanism influenced strongly by proximity to BPBD facilities, suggesting that the spatial distribution of fire hotspots is not random but rather governed by localized clustering dynamics and accessibility to disaster management infrastructure.

Some suggestions for further research are to add several spatial covariates directly related to the location of the fire, such as land cover and soil type, or add some temporal covariates such as rainfall and wind direction throughout the year. In addition, because the data used is very large, analysis methods and tools that are robust and tested on high-volume data can be used.

References

[1] Madani Berkelanjutan, "40 Ribu Hektare Karhutla Diindikasi Terjadi di Indonesia Selama Januari-April 2024, Terbesar di Kalimantan Timur," madani berkelanjutan.id. Available:
https://madaniberkelanjutan.id/40-ribu-hektare-karhutla-diindikasiterjadi-di-indonesia-selama-januari-april-2024-terbesar-dikalimantan-timur/. [Accessed: Feb. 10, 2025].

[2] L. Syaufina and I. S. Sitanggang, "Peatland fire detection using spatio-temporal data mining analysis in Kalimantan, Indonesia," J.
Trop. For. Sci., vol. 30, no. 2, pp. 154–162, 2018. doi:
10.26525/jtfs2018.30.2.154162.

[3] A. J. Horton et al., "Identifying key drivers of peatland fires across Kalimantan's Ex-Mega Rice Project using machine learning," Earth and Space Sci., vol. 8, no. 9, Art. no. e2021EA001873, 2021. doi: 10.1029/2021EA001873.

[4] A. Schmidt et al., "Fire frequency, intensity, and burn severity in Kalimantan's threatened peatland areas over two decades," Front. Forests Global Change, vol. 7, Art. no. 1221797, 2024. doi:
10.3389/ffgc.2024.1221797.

[5] A. Choiruddin, Aisah, F. Trisnisa, and N. Iriawan, "Quantifying the effect of geological factors on distribution of earthquake occurrences by inhomogeneous Cox processes," Pure Appl. Geophys., vol. 178, pp.
1579–1592, 2021.

[6] R. Kumalawati et al., "Hotspot spatial patterns using SNNP-VIIRS for fire potential monitoring," Int. J. Forestry Res., vol. 2023, Art. no. 3121862, 2023.

[7] S. Nurdiati, A. Sopaheluwakan, P. Septiawan, and M. R. Ardhana, "Joint spatio-temporal analysis of various wildfire and drought indicators in Indonesia," Atmosphere, vol. 13, no. 10, Art. no. 1591, 2022.

[8] F. Bioresita, D. E. Pongdatu, N. Hayati, U. W. Deviantari, and C. B. Pribadi, "Estimation of forest fire areas in Palangka Raya, Central Kalimantan, Indonesia using NBR2 and its impact on environment," Indonesian J. Forestry Res., vol. 12, no. 1, pp. 1–12, 2025. doi: 10.59465/ijfr.2025.12.1.1-12.

[9] Y. Wang, A. Degleris, A. Williams, and S. W. Linderman, "Spatiotemporal clustering with Neyman-Scott processes via connections to Bayesian nonparametric mixture models," J. Amer. Statist. Assoc., vol. 119, no. 547, pp. 2382–2395, 2024. doi: 10.1080/01621459.2023.2257896.

[10] HA Sidharta, B Al Kindhi, E Mulyanto, MH Purnomo ,"Semantic Segmentation of Pedestrian Groups Based on Directional-oriented Density Features with Shallow U-net Architecture", International Journal of Intelligent Engineering & Systems, 18, 1, 2025

[11] A. J. Baddeley, J. Møller, and R. Waagepetersen, "Non- and semiparametric estimation of interaction in inhomogeneous point patterns," Stat. Neerl., vol. 54, no. 3, pp. 329–350, 2000.

[12] J. Illian, A. Penttinen, H. Stoyan, and D. Stoyan, Statistical Analysis and Modelling of Spatial Point Patterns. Chichester, U.K.: Wiley, 2008.

[13] A. Baddeley, E. Rubak, and R. Turner, Spatial Point Patterns: Methodology and Applications with R. New York, NY, USA: Chapman and Hall/CRC, 2015.

[14] S. Portet, "A primer on model selection using the Akaike Information Criterion," Infect. Disease Modelling, vol. 5, pp. 111–128, 2020. doi: 10.1016/j.idm.2019.12.010.

[15] J. Zhang, Y. Yang, and J. Ding, "Information criteria for model selection," WIREs Comput. Stat., vol. 15, no. 5, Art. no. e1607, 2023. doi: 10.1002/wics.1607.

Analysis of the Effects of Mesoscale Convective System on the Enhancement of Wind and Significant Wave Height in Indonesian Maritime Areas

Fajar Setiawan[1,2*], and *Gede* Gangga Wisnawa[1,2]

[1]Department of Geomatics Engineering, Sepuluh Nopember Institute of Technology, Surabaya, Indonesia

[2]Marine Meteorological Station of Tanjung Perak Surabaya, Surabaya, Indonesia

Abstract. Extreme weather characterized by strong winds and high waves poses serious risks to maritime safety in the Indonesian Maritime Continent. This study comprehensively quantifies the impact of Mesoscale Convective Systems (MCS) on Significant Wave Height and Wind Speed enhancements using Himawari-8/9 imagery and multi-mission satellite altimetry. By utilizing a robust 300-km spatial filter to isolate standalone offshore events, the research employs quantile-based and multi-classification approaches to evaluate spatial and temporal anomalies. The results indicate that ocean-atmosphere coupling is highly non-linear; multilinear regression models consistently failed to predict environmental anomalies (R^2<0.03). However, quantile analysis revealed critical threshold behaviors. Spatially, intense systems consistently drove higher positive anomalies. Significantly, a multi-classification interaction analysis demonstrated that the synergistic combination of the "strongest" MCS parameters (largest area, coldest temperatures, and highest top cloud) generated wave impacts eight times greater than the "weakest" systems. Conversely, temporal responses relative to a 24-hour baseline were complex and non-monotonic, suggesting the dominance of lag effects over instantaneous intensity. These findings provide quantitative evidence that MCS modulates air-sea interactions through complex non-linear mechanisms, demonstrating that operational maritime forecasting must adopt advanced probabilistic modeling rather than relying on simple linear parameterizations to capture these hazardous events.

1 Introduction

Indonesian Maritime Continent (IMC) represents one of the most convectively active regions on Earth, acting as a primary heat engine for the global atmospheric circulation [1]. Within this complex archipelago, Mesoscale Convective Systems (MCSs) which is known as organized clusters of thunderstorms with the scales extending hundreds of kilometers are one of the dominant weather phenomena, accounting for a significant proportion of total

* Corresponding author: fajar.setiawan@bmkg.go.id

precipitation and extreme weather events over tropical regions [2]. In certain condition, MCS could produce a wide range damaging and widespread windstorms with the mean speed up to 23.6 ms^{-1} (45.9 knots), and responsible for the hazardous weather including tornadoes and flash floods which is called as Mesoscale Convective Complexes [3]. While the atmospheric characteristics of MCSs, such as their diurnal cycle, propagation, and rainfall contribution, have been extensively studied, most existing literature in the IMC has focused on precipitation dynamics, where MCSs are reported to contribute up to 45-70% of total rainfall in certain regions [4]. However, their coupling with the ocean surface dynamics remains a critical yet under-explored frontier. To date, quantitative assessments regarding the impact of MCSs on the sea surface state in Indonesian waters, particularly concerning the localized enhancement of wave height are virtually non-existent in the published literature. Specifically, the quantitative hydrodynamic response of the sea surface manifested as abrupt changes in Significant Wave Height (ΔSWH) and surface Wind Speed (ΔWS) to these intense convective forcing events is not fully quantified.

The interaction between MCSs and the ocean surface is governed by complex physical mechanisms. As deep convection develops, strong downdrafts transport high-momentum air from the mid-troposphere to the surface, creating divergent outflows and gust fronts [2]. These transient wind fields can generate steep and high-frequency wind waves that differ significantly from the background swell. However, standard meteorological and oceanographic models often struggle to capture the localized and transient nature of these interactions. Previous studies have predominantly relied on linear assumptions to scale storm intensity with environmental impacts. Yet, the hydrodynamic response to convection is likely highly non-linear [5], dependent on a synergistic combination of storm size, cloud-top height, and thermal intensity rather than any single parameter alone.

A major challenge in quantifying this impact lies in the observational gap. Unfortunately, In-situ buoy networks in the IMC are sparse due to very expensive funding and vandalism constraints, making it difficult to capture the spatial heterogeneity of MCS impacts over open waters. Furthermore, coastal effects often contaminate the signal of pure ocean-atmosphere coupling. This study focuses strictly on the offshore environment, following the findings of land-sea breeze circulation effects can extend up to 30 km from the coastline, this research isolates the analysis to open waters beyond this buffer zone to ensure that the observed anomalies are driven by convective systems rather than local land-sea breeze or thermal regimes [6].

A synergistic multi-mission satellite approach is used to detect MCS and its effects on sea. A high-resolution infrared imagery from Himawari-8/9 which is utilized with Advanced Himawari Imager (AHI) has capabilities to precisely resolve cloud parameters at a 2 km resolution to detect and characterize MCS properties which are MCS area (area), Cloud Top Temperature (CTT), and the estimation of Cloud Top Height (CTH) based on multichannel infrared difference [7]. A combination with sea state measurements from a constellation of altimeter satellites is used to gain the physical sea surface data (SWH and WS) with high accuracy and low bias in the open ocean [8].

The primary objective of this research is to comprehensively quantify the impact of MCSs on the enhancements of significant wave height (ΔSWH) and wind speed (ΔWS) in Indonesia offshore region. This study represents a significant advancement (state-of-the-art) in Indonesian maritime meteorology, as it is the first to explicitly decouple convective-driven wave anomalies from background conditions using a multi-mission satellite approach. We employ several methodologies that includes a spatial sensitivity analysis to empirically determine the effective radius of MCS influence, a dual-perspective analysis comparing spatial anomalies (inside vs. outside MCS) and temporal anomalies

(MCS vs. non-MCS baseline), and also a quantile-based multi-classification approach to decouple the effects of MCS size, intensity, and vertical extent. By moving beyond linear approximations, it aims to provide new insights into the of ocean-atmosphere interactions in the Maritime Continent.

2 Study Area and Data

2.1 Study Area

The geographical scope of this research encompasses the offshore waters of the Indonesian Maritime Continent, defined in Figure 1 within the coordinates of 93°E to 147°E longitude and 17°S to 17°N latitude. To ensure the analysis focuses strictly on open-ocean convective interactions and is unaffected by local diurnal wind regimes, a spatial filter was applied to mask coastal areas. Specifically, observations located within 30 km of the coastline which indicates that the influence of local land–sea breeze circulation were excluded from the dataset. By removing this coastal band, the study isolates the ocean-atmosphere coupling mechanisms driven by Mesoscale Convective Systems (MCSs) from near-shore local effects.

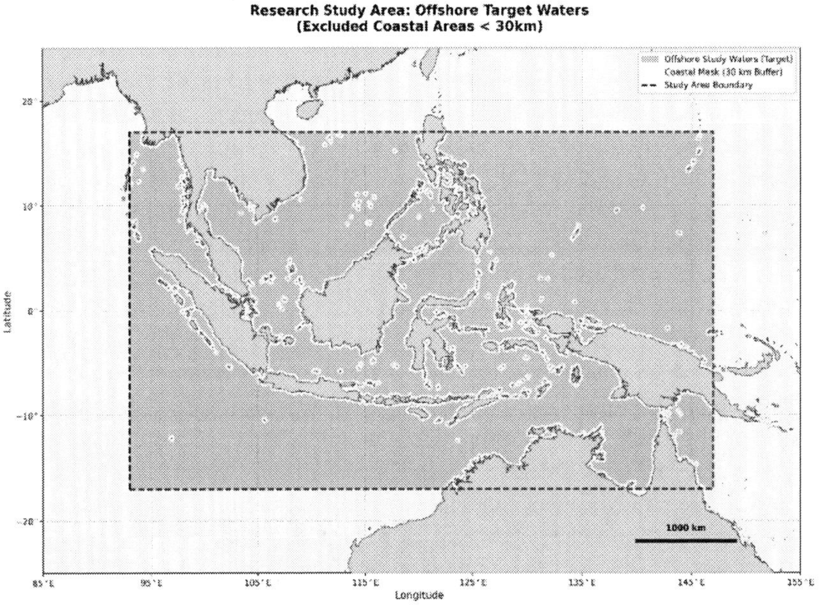

Fig. 1. Offshore study area for Indonesia Maritime Continent

2.2 Data Acquisition

A dataset that combines high-spatiotemporal-resolution geostationary satellite imagery with multi-mission satellite altimetry is used to gain the air-sea interaction effects. The observation period spans from January 1, 2021 to August 31, 2025, capturing a wide range of convective activity over the Indonesian Maritime Continent. By merging these complementary datasets, we aim to simultaneously resolve the atmospheric evolution of

Mesoscale Convective Systems (MCSs) and the corresponding variations in ΔSWH and ΔWS.

2.2.1 Himawari-8/9 Satellite Imagery

MCS detection and characterization were performed using data from the Advanced Himawari Imager (AHI) onboard the Himawari-8/9 satellites. it was identified using infrared Channel 13 (10.4μm) to delineate convective cloud clusters which was defined based on standard thresholds with a cloud-top brightness temperature (CTT) less than 235 K and a contiguous area of ≥ 10,000 km^2 [7, 10]. To determine the MCS's Cloud Top Height (CTH), the Split-Window technique was employed using the brightness temperature difference (BTD) between the water vapor absorption band (Channel 9, 6.9μm) and the clean infrared window band (Channel 14, 11.2μm) [9].

2.2.2 Multi-Mission Satellite Altimetry

Significant Wave Height (SWH) and surface Wind Speed (WS) data are served as the dependent variables for this analysis. These were derived from a comprehensive constellation of 10 satellite altimetry missions to ensure higher spatiotemporal coverage. The constellation includes: CryoSat-2, SARAL/AltiKa, Jason-3, Sentinel-3A, Sentinel-3B, CFOSAT, HY-2B, HY-2C, Sentinel-6 Michael Freilich, and SWOT. The data retrieval was automated using the Marine Copernicus Python module, ensuring access to standardized and quality-controlled oceanographic measurements.

3 Methodology

To evaluate the coupling between atmospheric convection and ocean surface dynamics, a multi-step processing workflow was developed from selecting potential MCS events which were identified and rigorously filtered to retain the only isolated systems occurring over the open sea (unaffected by sea-land breeze). These verified events were matched spatiotemporally with altimetry tracks to calculate the deviation of wind and wave parameters relative to background conditions. Finally, the relationship between MCS physical properties and the observed oceanic response was modeled using a combination of multiclassification techniques and multilinear regression.

3.1 Detection of Standalone Offshore MCS

Detection of Mesoscale Convective Systems (MCSs) was conducted using Cloud Top Temperature (CTT) data from the Himawari-8/9 Advanced Himawari Imager (AHI) using infrared bands where it has a 2 kilometers spatial resolution [7]. The MCS identification followed the objective threshold-based approach widely used in tropical convection studies, in which contiguous cold cloud-top areas with CTT values below 235 K (-38°C) were classified as convective clusters with a minimum area exceeding 10,000 km^2 to ensure that the detected features represent truly MCSs [10]. This classification will create a polygon of MCS with an area more than 10,000 km^2.

Spatial independence was enforced to prevent the interaction signals from neighboring storms. The required separation distance was determined by analyzing the spatial decay of the wave impact relative to the MCS's size and shape. An iterative analysis was performed by creating multiple buffer zones around the MCS polygons, expanding from the cloud shield edge in steps of 50 km. For each buffer interval, the wave height enhancement (anomaly) was evaluated. The effective zone of influence was defined as the distance from the MCS edge where the wave anomaly ceases to be positive. The threshold identified through this analysis was subsequently used to filter the dataset, ensuring that only single (standalone) systems were located within each other's sphere of influence.

A land–sea mask derived from the Indonesian coastal polygon database was also applied to isolate the offshore MCSs. Only MCS entirely located over oceanic surfaces and showing no intersection with 30 kilometers from land were retained. This step ensures that the analyzed convective systems developed over the open sea and were unaffected by local wind regimes, such as land–sea breeze circulations. The resulting dataset represents standalone offshore MCS events, which provide more ideal basis for analyzing pure ocean–atmosphere coupling mechanisms.

3.2 Computation of Spatio-Temporal Differences of Altimetry Data

To evaluate the ocean surface response to MCS activity, a spatial and temporal (Spatio-Temporal) filtering procedure was implemented to extract relevant observations of significant wave height (SWH) and surface wind speed (WS) from multi-mission satellite altimetry data. Each MCS detection timestamp (T_{MCS}) was used as a temporal reference to define a ±30-minute observation window centered around the MCS occurrence time. Spatially, all altimetry measurements located within the geographical extent of the MCS polygon were selected to represent the "MCS area". 150 km buffer zone surrounding the MCS polygon was defined to represent the background (non-convective) oceanic condition in the near-same time. This filtering method allows for the separation of altimetry-derived parameters between regions directly influenced by convective activity and adjacent unaffected ocean areas to capture the immediate mesoscale environment, often referred to as the 'near-environment', while avoiding contamination from far-field and macroscale synoptic features [11].

Following the spatio-temporal filtering process, two categories of differences were computed to quantify the oceanic response associated with MCS events which are spatial differences and temporal differences. The spatial difference was calculated as the deviation between the mean altimetry measurements inside the MCS polygon and those within the surrounding buffer zone. This parameter represents the instantaneous effect of convective forcing on the ocean surface relative to nearby unaffected regions.

The temporal difference was derived by comparing altimetry observations within the MCS polygon during its active phase (T_{MCS} ± 30 minutes) against measurements acquired over the same region during non-convective periods within ±24 hours of the event. This temporal comparison captures the time-evolving impact of MCS passage on wave and wind conditions. Both difference parameters were calculated for SWH and WS, resulting in a comprehensive dataset that describes the spatial and temporal variability of ocean surface responses to mesoscale convective activity.

3.3 Statistical, Quantile Based Multiclassification, and Multilinear Analysis

To assess the quantitative influence of MCS characteristics on ocean surface dynamics, a combination of descriptive statistics, quantile-based multiclassification analysis, and multilinear regression was applied.

3.3.1 Quantile-Based Classification

Prior to regression analysis, three fundamental MCS properties derived from satellite data namely MCS area (area), mean cloud-top temperature (CTT), and mean cloud-top height (CTH) were categorized based on its quartile distribution. The quartile method divides the dataset into four equal intervals (Q1–Q4), representing the increasing magnitudes of each parameter to generate meaningful clusters [12]. For example, Q1 represents the lowest 25% of observations (smaller, warmer, and lower MCSs), whereas Q4 represents the top 25% (largest, coldest, and highest systems). This classification allows the investigation of systematic trends in the enhancement of significant wave height (SWH) and surface wind speed (WS) according to MCS intensity levels. The mean and standard deviation of ΔSWH and ΔWS were computed for each quartile class to assess how stronger convective systems produce greater oceanic responses. The quantile-based analysis thus provides a discrete, physically interpretable framework prior to continuous regression modeling.

3.3.2 Multiclassification Analysis

In addition to individual quartile evaluations, a multiclassification approach was introduced to capture the combined influence of multiple MCS attributes. Each MCS event was assigned a categorical label based on the joint quartile class of all three parameters (area, CTT, CTH). For instance, a system classified as (Q4–Q4–Q4) represents a large, cold, and high-topped MCS, while (Q1–Q1–Q1) denotes a weakest convective system.

By grouping events according to these combined categories, the mean ΔSWH and ΔWS values were evaluated across all multiclass combinations. This enables identification of the most influential combination of MCS properties responsible for the greatest wave and wind enhancement. The results from this multiclassification were later compared to the multilinear model outcomes to validate and interpret the statistical dependencies identified by regression.

3.3.3 Multilinear Regression Analysis

To assess the quantitative influence of MCS characteristics on ocean surface dynamics, a multilinear regression analysis was applied by using MCS parameters derived from satellite data (area, CTT and CTH). These parameters respectively represent the horizontal extent, convective intensity, and vertical development of each system, while dependent variables were defined as the spatial and temporal differences of SWH and WS obtained from the altimetry filtering results. The relationship between these variables was expressed as:

$$\Delta Y = \beta_0 + \beta_1 X_1 + \beta_2 X_2 + \beta_3 X_3 + \epsilon \qquad (1)$$

where ΔY corresponds to either ΔSWH or ΔWS. X_1, X_2, X_3 as independent variables (area, CTT, CTH). β_0 is the baseline value of ΔY. $\beta_1, \beta_2, \beta_3$ are coefficients of regression, representing the mean change in Y for a one-unit change in the corresponding X variable, while holding all other predictors constant. ϵ as the model error term (residual).

The model was applied to the four distinct datasets (spatial wave, temporal wave, spatial wind, temporal wind), while coefficient of determination (R^2) and Root Mean Square Error (RMSE) were used to assess model performance. R^2 was selected to quantify the proportion of variance in the oceanic response that can be explained by the linear combination of MCS parameters. It indicates the strength of the relationship and is calculated as:

$$R^2 = 1 - \frac{\sum_{i=1}^{n}\left(y_i - \hat{y}_i\right)^2}{\sum_{i=1}^{n}\left(y_i - \bar{y}\right)^2} \quad (2)$$

where y_i is the observed value, $\hat{y}_i$ is the predicted value from the regression model, and $\bar{y}$ is the mean of the observed data. A higher R^2 indicates a model that better fits the observed variability. Complementarily, the RMSE was used to measure the average magnitude of the model's prediction errors. Unlike R^2, RMSE provides an error metric in the same physical units as the dependent variable (meters for waves, m/s for wind), making it intuitive for interpreting the accuracy of the model estimates. Furthermore, RMSE penalizes larger errors more heavily, which is critical for assessing performance in extreme weather scenarios. It is defined as:

$$RMSE = \sqrt{\frac{1}{n}\sum_{i=1}^{n}\left(y_i - \hat{y}_i\right)^2} \quad (3)$$

This multilinear framework provides a statistical means to interpret how variations in convective system structure and intensity modulate the ocean surface, particularly in terms of wave height amplification and wind speed enhancement beneath MCS area.

3.4 Workflow Summary

The complete workflow is summarized in Figure 2. The methodology begins with the identification of MCS events using Himawari-8 AHI imagery, followed by spatial filtering to retain the only standalone offshore systems. These events are then matched with multi-mission altimetry data through a spatio-temporal filtering process to isolate measurements within and around MCS regions. Subsequent calculations of spatial and temporal differences enable the quantification of MCS-induced changes in sea surface conditions. Finally, the statistical and multilinear analyses integrate all derived parameters to evaluate the relationship between MCS properties and their corresponding impacts on wave and wind enhancement.

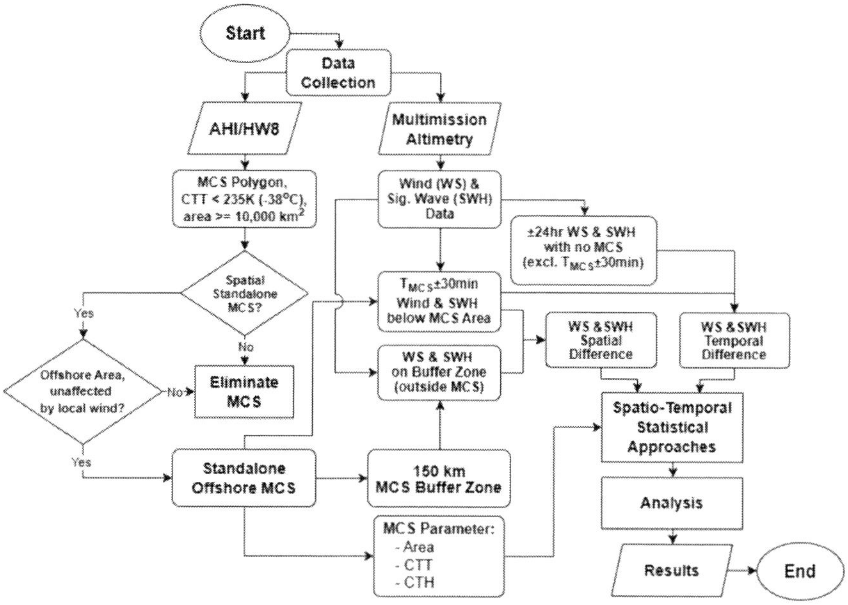

Fig. 2. Flow chart showing the process to analyze the effects MCS to wind and waves enhancements

This systematic approach allows for a comprehensive understanding of air-sea interactions driven by mesoscale convective systems, particularly in offshore environments that are minimally affected by local terrestrial dynamics.

4 Results and Discussion

Based on the sensitivity analysis described in methodology, the empirical assessment as shown in Table 1 revealed that the MCS-induced SWH amplification is confined within the average of 200 km buffer zone, after which the wave height difference (ΔSWH) values transition to negative. This suggests that the constructive influence of the MCS ceases beyond this threshold. To provide a robust safety margin and guarantee that the analyzed systems were fully isolated from neighboring convective influence, the exclusion radius was extended to 300 km.

Table 1. Significant wave height difference based on distance from MCS

Buffer Zone (km)	n-data	Mean ΔSWH	Std. Dev.
0-50	146	0.0865	0.2859
50-100	117	0.0621	0.2646
100-150	151	0.0611	0.2931
150-200	195	0.0120	0.2544
200-250	200	-0.0025	0.2339
250-300	86	-0.0227	0.2576

The interpretations of statistical analysis of MCS impacts on ΔSWH and ΔWS is structured into two primary methodologies which are a spatial analysis comparing conditions inside the MCS to a simultaneous external buffer zone and a temporal analysis comparing conditions during an MCS event to a baseline non-MCS period.

4.1 General Statistical Overview and Variability

A comparative descriptive analysis of all four study datasets presented in Table 2 gives a critical initial finding that MCS events are associated with an increase in both wave height and wind speed on average. Mean values in both spatial and temporal for ΔSWH (0.0369 m and 0.0351 m, respectively) and ΔWS (0.6434 m/s and 0.6013 m/s, respectively) are positive. However, the most prominent characteristic of the data is the extreme variability. Standard deviation for all four datasets is exceptionally large, often 4 to 5 times greater than its mean (e.g., ΔSWH mean 0.0369 vs. ΔSWH standard deviation 0.1666). This high variance indicates that while the average impact is positive, individual MCS events can produce a wide spectrum of effects, from strong localized suppression to significant amplification.

Table 2. Descriptive Statistics of Wave and Wind Difference under the existence of MCS

Analysis	Variable	n-data	Mean	Std. Dev.	Median (50%)	Min	Max
Spatial	ΔSWH	1951	0.0369	0.1666	0.0297	−1.4424	1.7158
Temporal	ΔSWH	1512	0.0351	0.2937	0.0225	−1.1523	1.8202
Spatial	ΔWS	1734	0.6434	1.4456	0.5028	−5.1916	11.7817
Temporal	ΔWS	883	0.6013	2.1988	0.3586	−6.9534	9.8016

A linear regression analysis performed using MCS parameter (area, CTT, CTH) to predict the magnitude of spatial and temporal anomalies for both SWH and WS gave the results that this regression analysis has the coefficient of determinations which are universally negligible, ranging from 0.0011 to 0.0204. This indicates that less than 3% of the variability in wind and wave anomalies can be explained by a simple linear combination of MCS size, temperature, and height. This lack of linear correlation is visually confirmed in Figure 3 (Comprehensive Scatterplot Matrix). The regression lines for all parameters are nearly horizontal, and the data points exhibit a high degree of dispersion (heteroscedasticity) rather than clustering around a trend line.

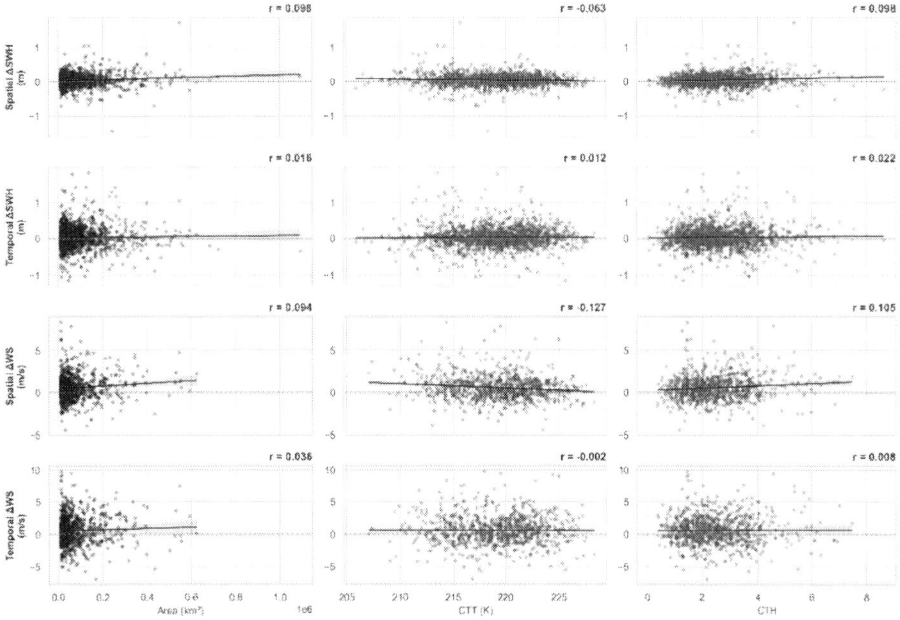

Fig. 3. Distributions of MCS parameter correlation to winds and waves enhancement

4.2 Justification for Quantile Based Analysis

A preliminary step using linear correlation analysis performed between three MCS parameters (area, CTT, CTT) and the dependent variables (ΔSWH and ΔWS) has revealed universally weak correlation coefficients for all parameters in both spatial and temporal contexts. The strongest Pearson correlation was merely 0.098 (area). This finding demonstrates that a simple multilinear regression model is fundamentally very weak and unsuited for capturing the complex and non-linear dynamics of MCS impacts to environmental changes. This weakness necessitates a more robust, non-parametric approach, justifying the use of the following quantile-based classification and interaction analysis to uncover the underlying trends.

4.2.1 Single-Parameter Quantile-Based Classification

To investigate non-linear trends, the mean impact of each MCS parameter was analyzed after stratification into quartiles (Q1: weakest/smallest, Q4: strongest/largest). The results for single-parameter quantile-based classification visualized in Figure 4 show the trends for every MCS Parameter in general. Spatial impact analysis between SWH and WS over MCS area vs. Buffer Zone has revealed several distinct trends where ΔSWH results a clear and positive monotonic relationship in both area and CTH which progressively increased from the smallest area (Q1: 0.013 m) to the largest (Q4: 0.061 m). Similarly, the impact from the lowest-top MCSs (Q1: 0.021 m) was three times weaker than that from the highest-top MCSs (Q4: 0.065 m). Meanwhile, the strongest trend of ΔWS had an inverse relationship with CTT. The coldest MCS produced the largest ΔWS (0.915 m/s), an effect that systematically weakened as cloud-top temperatures increased (Q4: 0.477). A positive (though non-monotonic) trend was also seen with CTH, where the highest-top storms (Q4: 0.864) had a significantly stronger impact than the Q2 (0.469). Spatially, the hypothesis that "stronger storms produce stronger impacts" holds true. Larger, higher, and colder (more intense) MCSs generate a significantly stronger wind and wave field within the storm's footprint compared to the immediate surrounding environment. This suggests that convective intensity and storm size are primary drivers of the spatial differential.

In sharp contrast, the temporal analysis (MCS vs. Non-MCS Baseline) revealed far more complex and often non-monotonic relationships. ΔSWH results the trends for area and CTT were non-linear. The impact peaked at Q3 (0.056) for area parameter and decreased slightly for Q4 (0.055). Meanwhile, U-shaped pattern was observed in CTT where the strongest impacts found at the extremes: Q1 (0.042) and Q4 (0.051). More chaotic trend is shown in ΔWS, no parameter produced a simple monotonic relationship. For instance, MCSs area impact was the weakest for Q2 (0.428), while CTT showed a no clear trend. The complexity of the temporal results suggests that the cumulative change from a baseline condition is not governed by peak intensity alone. Non-linear trends imply that different physical mechanisms might be at play during ±24 comparation.

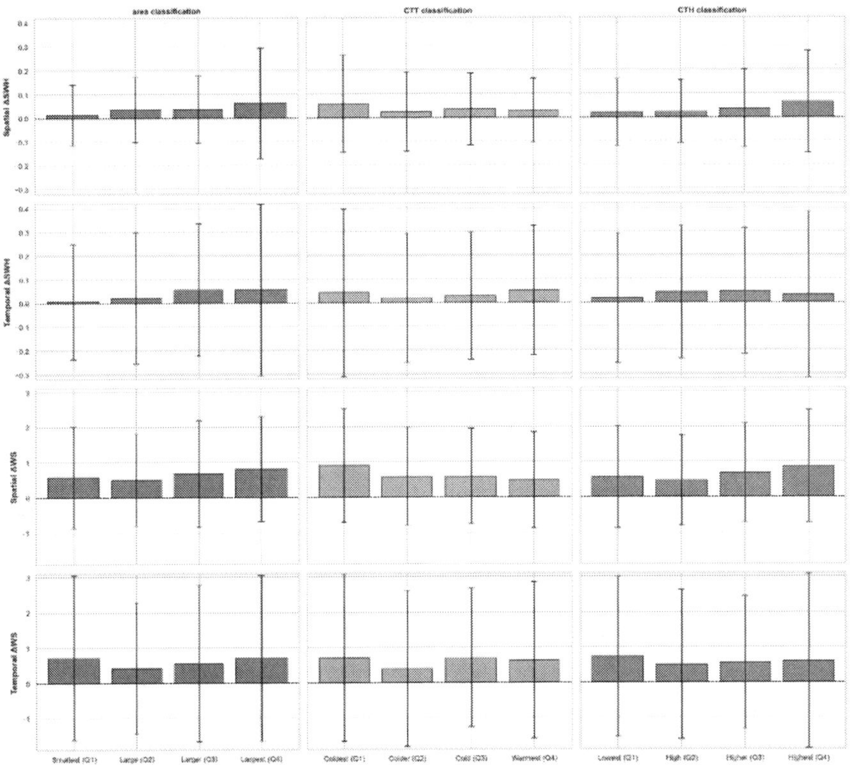

Fig. 4. Spatial and temporal Quantile based distributions of MCS parameter

4.2.2 Quantile-Based Multi-Classification Interaction Analysis

One of the core finding of this study emerged from the multi-classification analysis which evaluated the combined interaction of all three parameters by comparing the "Strongest" MCS combination (Q4 area, Q1 CTT, Q4 CTH) against the "Weakest" configuration (Q1 area, Q4 CTT, Q1 CTH). Table 3 shows the results where interaction effect of Spatial ΔSWH is dominant. A "Strongest" combination increases mean wave impact that is 8.0 times greater than the "Weakest" combination. This demonstrates that the synergistic combination of large area, high intensity (cold tops), and high altitude is the primary driver of extreme spatial wave generation. This "stronger is stronger" logic holds for Temporal ΔSWH (2.1x) and Spatial ΔWS (1.7x), confirming that the "Strongest" configuration is a robust generator of environmental impact.

A major anomaly was discovered for Temporal ΔWS. In this context, the "Strongest" MCS combination was less impactful than the "Weakest" combination, resulting in an inverse (0.8x) ratio. This is a profound finding, directly contradicting the logic of the other three analyses which strongly suggests that small, warm-topped, and low-altitude systems may induce a larger cumulative change in the wind field over time compared to non-MCS baseline. This could be due to longer durations, different propagation characteristics, or a dominance of large-scale stratiform wind fields rather than localized convective downdrafts.

Table 3. Multi-classification comparation of "Strongest" MCS against "Weakest" MCS

Variable	"Strongest"		"Weakest"		Impact Ratio
	n-samples	Mean	n-samples	Mean	
Spatial ΔSWH	165	0.0869 m	96	0.0108 m	8.0x
Temporal ΔSWH	126	0.0534 m	84	0.026 m	2.1x
Spatial ΔWS	148	0.9724 m/s	90	0.5636 m/s	1.7x
Temporal ΔWS	73	0.9899 m/s	51	1.2123 m/s	0.8x

4.3 Multilinear Regression Analysis

To investigate the potential for a direct linear prediction of environmental anomalies based on MCS parameters, a multiple linear regression (MLR) model was applied. The model utilized MCS parameter (area, CTT, CTH) as independent predictors for the spatial and temporal changes in SWH and WS. The statistical summary of the regression models is presented in Table 4.

Table 4. Multilinear regression equations analysis for Spatial and Temporal difference

Analysis Type	R^2	RMSE	Regression Equation
$\Delta SWH_{Spatial}$	0.0126	0.1743	$y=0.221+(1.07\times10^{-7})area-(0.0010)ctt+(0.0079)cth$
$\Delta SWH_{Temporal}$	0.0011	0.2938	$y=-0.481+(1.92\times10^{-8})area+(0.0023)ctt+(0.0078)cth$
$\Delta SWH_{Spatial}$	0.0204	1.4416	$y=8.799+(8.04\times10^{-7})area-(0.0383)ctt+(0.0418)cth$
$\Delta SWH_{Temporal}$	0.0015	2.1989	$y=0.041+(1.12\times10^{-6})area+(0.0025)ctt-(0.0274)cth$

The primary finding from this analysis is the negligible predictive power of the linear models across all scenarios. The Coefficient of Determination (R^2) values are consistently below 0.03, indicating that less than 3% of the variance in ΔSWH and ΔWS can be explained by a linear combination of MCS physical parameters. This lack of linear correlation is consistent with previous studies suggesting that convective-scale interactions are inherently non-linear and governed by stochastic processes rather than simple scaling laws [5]. The spatial models performed marginally better than the temporal ones, though still poorly. Model achieved the highest R^2 (0.0204) for ΔWS. The signs of the coefficients in the spatial equations align with physical expectations where negative coefficient for CTT and positive coefficient for area and CTH suggest that colder, wider and higher convective clouds tend to drive stronger positive anomalies. However, the high RMSE (1.44 m/s for wind) relative to typical anomaly values confirms that this linear trend is overwhelmed by other variability. The temporal models showed essentially no linear correlation with R^2 values approaching zero (ΔSWH: 0.0011, ΔWS: 0.0015). The regression coefficients for CTT even flipped to positive values in temporal models, contradicting the expected thermodynamic relationship where colder storms are typically more intense. This near-zero correlation in temporal models likely reflects the mismatch between instantaneous satellite snapshots and the delayed response of the ocean surface, a phenomenon known as the 'fetch-limited' or 'time-lagged' growth of waves under transient storm forcing [13].

The failure of the MLR analysis to yield significant correlations provides a crucial insight into the ocean-atmosphere coupling mechanism during MCS events. It demonstrates that the impact of an MCS on the sea surface does not follow a simple scaling law. Literature on tropical convection suggests that environmental responses often exhibit "threshold behaviors" where significant impacts only materialize once a storm reaches a critical level of organization or intensity [14]. A storm that is twice as large or twice as cold does not linearly produce twice the wave height or wind speed anomaly. Instead, the interaction is likely characterized by threshold behaviors and complex non-linear synergies between storm size, intensity, and the background environmental state. As noted by [15], the momentum transfer from convective downdrafts to the sea surface is highly dependent on the stability of the boundary layer, which cannot be captured by cloud-top parameters alone. Consequently, relying solely on linear parameterizations to predict MCS impacts in marine forecasting would be insufficient.

5 Concluding Remarks

This study provides a comprehensive quantification of the ocean surface response to Mesoscale Convective Systems (MCSs) in the Indonesian Maritime Continent. By integrating geostationary imagery with multi-mission satellite altimetry, we establish that the hydrodynamic impact of MCSs on Significant Wave Height (ΔSWH) and Wind Speed (ΔWS) is inherently non-linear and governed by complex interaction effects. Our findings demonstrate that simple linear models are insufficient for predicting convective-driven sea state anomalies, as extreme responses depend on a synergistic combination of storm size, intensity, and vertical development rather than single parameters alone.

Furthermore, the study highlights a significant divergence between instantaneous spatial anomalies and time-lagged temporal evolutions. These insights underscore the necessity for operational forecasting systems to shift from linear parameterizations toward non-linear, probabilistic, or machine-learning-based frameworks to accurately mitigate maritime hazards associated with deep convection in tropical offshore environments.

The Acknowledgements: The authors would like to express their sincere appreciation to Indonesian Meteorology, Climatology, and Geophysics Agency (BMKG) for providingmdata and necessary facilities that supported this research.
Fundings: This research was financially supported by the Indonesia Endowment Fund for Education (LPDP).
Data availability: This study has been conducted using E.U. Copernicus Marine Service Information; 10.48670/moi-00176; 10.48670/moi-00183

References

1. Yang, S., Zhang, T., Li, Z. et al. Climate Variability over the Maritime Continent and Its Role in Global Climate Variation: A Review. J Meteorol Res 33, 993–1015 (2019). https://doi.org/10.1007/s13351-019-9025-x
2. Houze, R. A. (2004). Mesoscale convective systems. In Reviews of Geophysics (Vol. 42, Issue 4, pp. 1–43). https://doi.org/10.1029/2004RG000150
3. Maddox, R. A. (1983). Large-Scale Meteorological Conditions Associated with Midlatitude, Mesoscale Convective Complexes. *Monthly Weather Review*, *111*(7), 1475-1493. https://doi.org/10.1175/1520-0493(1983)111<1475:LSMCAW>2.0.CO;2
4. Crook, J., Morris, F., Fitzpatrick, R. G. J., Peatman, S. C., Schwendike, J., Stein, T. H., Birch, C. E., Hardy, S., & Yang, G. Y. (2024). Impact of the Madden–Julian oscillation and equatorial waves on tracked mesoscale convective systems over southeast Asia. Quarterly Journal of the Royal Meteorological Society, 150(760), 1724–1751. https://doi.org/10.1002/qj.4667
5. Drignei, D., Forest, C. E., & Nychka, D. (2008). Parameter estimation for computationally intensive nonlinear regression with an application to climate modeling. Annals of Applied Statistics, 2(4), 1217–1230. https://doi.org/10.1214/08-AOAS210
6. Pokhrel, R., & Lee, H. (2011). Estimation of the effective zone of sea/land breeze in a coastal area. Atmospheric Pollution Research, 2(1), 106–115. https://doi.org/10.5094/APR.2011.013
7. Chen, D., Guo, J., Yao, D., Lin, Y., Zhao, C., Min, M., et al. (2019). Mesoscale convective systems in the Asian monsoon region from Advanced Himawari Imager: Algorithms and preliminary results. Journal of Geophysical Research: Atmospheres, 124, 2210–2234. https://doi.org/10.1029/2018JD029707
8. Mitsopoulos, P., & Peña, M. (2023). Characterizing Coastal Wind Speed and Significant Wave Height Using Satellite Altimetry and Buoy Data. Remote Sensing, 15(4). https://doi.org/10.3390/rs15040987
9. Silalahi, F. S. A., Farda, N. M., & Nurjani, E. (2024). Uji Perbandingan Metode Estimasi Curah Hujan Menggunakan Satelit Himawari: Metode Konvensional dan Machine Learning. Jurnal Geosains Dan Remote Sensing, 5(2), 131-141. https://doi.org/10.23960/jgrs.ft.unila.333
10. Rafati, S., Karimi, M. Assessment of mesoscale convective systems using IR brightness temperature in the southwest of Iran. Theoretical and Applied Climatology 129, 539–549 (2017). https://doi.org/10.1007/s00704-016-1797-7
11. Maddox, R. A. (1980). An Objective Technique for Separating Macroscale and Mesoscale Features in Meteorological Data. *Monthly Weather Review*, *108*(8), 1108-1121. https://doi.org/10.1175/1520-0493(1980)108<1108:AOTFSM>2.0.CO;2
12. Goswami, S., & Chakrabarti, A. (2012). Quartile Clustering: A quartile based technique for Generating Meaningful Clusters. *ArXiv, abs/1203.4157*
13. Hwang, P. A. (2016). Fetch- and Duration-Limited Nature of Surface Wave Growth inside Tropical Cyclones: With Applications to Air–Sea Exchange and Remote Sensing. Journal of Physical Oceanography, 46(1), 41-56. https://doi.org/10.1175/JPO-D-15-0173.1
14. Meerveld, H.J., & McDonnell, J.J. (2006). Threshold relations in subsurface stormflow: 1. A 147-storm analysis of the Panola hillslope. Water Resources Research, 42.

15. Powell, M., Vickery, P. & Reinhold, T. Reduced drag coefficient for high wind speeds in tropical cyclones. Nature 422, 279–283 (2003). https://doi.org/10.1038/nature01481

Early Afterslip Deformation Captured by Sub-Daily Kinematic GPS following the 2010 Mw 7.8 Mentawai Earthquake

Anandiko Savieros Sujanto[1], *Putra Maulida*[1*], *Mokhamad Nurcahyadi*[1], *Ira Mutiara Anjasmara*[1], *Agnis Triahadini*[2]

[1]Department of Geomatics Engineering, Institut Teknologi Sepuluh Nopember, Surabaya, Indonesia
[2]Department of Geophysics Engineering, Institut Teknologi Sepuluh Nopember, Surabaya, Indonesia

*Corresponding author: putra@its.ac.id

Abstract. On 25 October 2010, an Mw 7.8 earthquake occurred west of the Mentawai Islands, releasing energy from a shallow offshore fault and generating complex ground deformation patterns. Although surface deformation has been monitored for over a decade, the precise geometry of the slipping zone remains unresolved. This study investigates the event using sub-daily kinematic GPS observations from the Sumatra GPS Array (SuGAr) to analyze coseismic and early postseismic deformation following the rupture. The resulting displacement field reveals limited horizontal motion of approximately 20–25 cm and vertical subsidence of about 4–6 cm near the islands, suggesting that most slip was concentrated offshore toward the trench. In the subsequent days, the GPS time series displays a gradual decay in motion, which can be modeled with a logarithmic function and characteristic timescales of roughly five days. This transient response is interpreted as an early afterslip evolving downdip of the main rupture area. These findings underscore the heterogeneous nature of the plate interface, where seismic and aseismic processes occur concurrently over short timescales. Future work will estimate early afterslip using a time-dependent inversion model to further characterize the rupturing zone and clarify the transition between seismic and aseismic deformation.

1 Introduction

1.1 Background

The subduction zone where the Indo-Australian plate dives beneath the Sunda block is recognized globally as a high-risk tectonic boundary, historically producing massive megathrust earthquakes [1]. The Mentawai segment is particularly notable within this system for its potential to generate "tsunami earthquakes" rare events that cause tsunamis far larger than their seismic magnitude would suggest [2]. This threat became reality in 2010 when a shallow section of the megathrust ruptured west of the Pagai Islands (Mw 7.8) [3]. Unlike typical powerful earthquakes, this event was marked by slow rupture propagation and weak high-frequency radiation, yet it unleashed a destructive tsunami with run-ups reaching 16 meters [3]. Such ruptures within the shallow accretionary prism challenge the prevailing assumption that the updip end of the seismogenic zone is predominantly aseismic or resistant to slip, underscoring the need for improved understanding of the frictional properties of the shallow plate boundary [4]. Following the mainshock's rapid release of accumulated strain, the resulting stress change typically induces significant aseismic movement, known as afterslip, which can persist for months or years [5]. This postseismic period is critical for elucidating fault zone properties, as it often accounts for a substantial portion of total displacement and can increase stress on adjacent fault segments [6]. Previous geodetic studies of the Sumatra-Andaman system have demonstrated that afterslip is a principal postseismic process. For instance, afterslip following the 2007 magnitude 8.4 Bengkulu earthquake increased stress on the adjacent Mentawai segment, potentially accelerating the 2010 event [7]. While the long-term decay of afterslip has been extensively studied using daily GPS data, the evolution of this process during the initial minutes to hours after rupture the early postseismic phase remains poorly understood [8].

The main obstacle to characterizing this early phase is the limited temporal resolution of standard geodetic techniques. Conventional processing typically utilizes daily static GPS solutions, which average coordinates over 24-

hour windows to maximize precision [9]. Although this approach is effective for tracking long-term tectonic rates, the temporal averaging acts as a low-pass filter, effectively masking the rapid transient motions that occur right after an earthquake [10]. Recent studies using high-rate, sub-daily kinematic GPS have shown that substantial displacement occurs within this initial window. In standard daily processing, this early motion is often incorrectly lumped into the coseismic estimate [8]. For example, analyses of the 2010 Maule and 2015 Illapel earthquakes revealed that ignoring this early phase can lead to an underestimation of the postseismic slip budget and a misunderstanding of the fault's frictional heterogeneity [8]. Accurately resolving these displacements is crucial for assessing the full energy budget of large earthquakes and characterizing the transition from seismic to aseismic slip [8].

This observational gap is particularly relevant for the 2010 Mentawai event, which ruptured a region previously considered to be weakly coupled [11]. To overcome these resolution limitations, this study employs sub-daily kinematic GPS observations from the Sumatra GPS Array (SuGAr) to reconstruct high-frequency displacement waveforms associated with the 2010 earthquake. By utilizing a kinematic processing approach instead of daily solutions, the analysis aims to pinpoint the precise onset of afterslip and distinguish it from the true coseismic offset. Furthermore, the Principal Component Analysis-based Inversion Method (PCAIM) is applied to model the spatiotemporal evolution of slip [12]. This methodology facilitates the detection of rapid logarithmic decay signals that are often aliased in conventional processing, thereby providing new constraints on the frictional properties of the shallow megathrust and enhancing the understanding of the slip budget associated with tsunami earthquakes [13].

2 Materials and methods

2.1 Geodetic data

Continuous GPS observations were obtained from the Sumatra GPS Array (SuGAr), a network jointly operated by the Earth Observatory of Singapore (EOS) and BRIN [3], [7]. This array offers ideal near-field coverage of the Mentawai segment. We focused our analysis on 16 stations situated on Sipora, North Pagai, and South Pagai Islands (including BSAT, PRKB, SLBU), as these locations lie directly above the rupture's downdip limit. By utilizing high-rate sampling (15 seconds), we were able to capture sub-daily crustal motions that are typically smoothed out in standard 24-hour geodetic solutions [8].

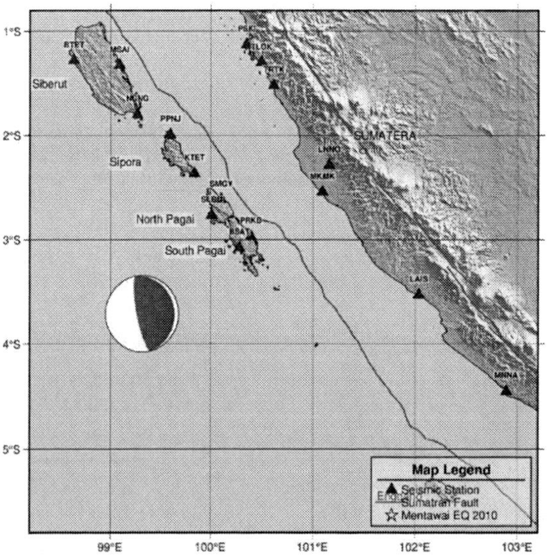

Fig. 1 Study Area

2.2 Kinematic PPP processing

To retrieve instantaneous station positions, we processed the raw GPS observables using the GipsyX software package from NASA's Jet Propulsion Laboratory (JPL) [14]. Instead of using network-based double-difference methods, we applied the Precise Point Positioning (PPP) technique [15], which enables the autonomous and precise determination of coordinates for individual stations.

Data were processed using the GipsyX software [14] employing the Precise Point Positioning (PPP) strategy [15]. We utilized JPL's final orbits and clock products, applied single-receiver ambiguity resolution, and aligned coordinates to ITRF2014 with VMF1 tropospheric corrections.

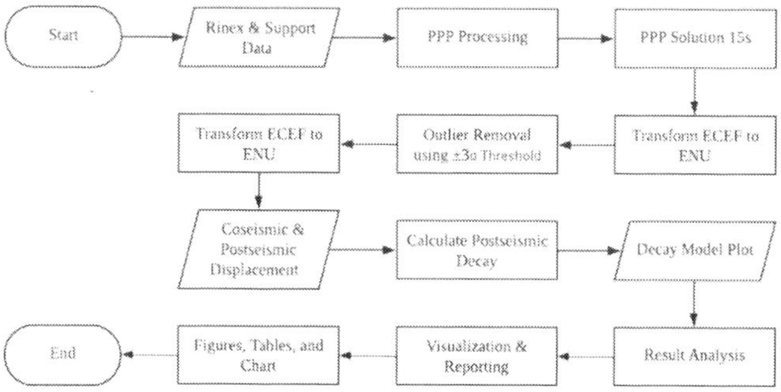

***Fig.** 2 Research Workflow*

2.3 Filtering and signal extraction

Kinematic time series inherently contain high-frequency noise and outliers caused by multipath effects and receiver instability. To extract the tectonic signal, we implemented a post-processing filtering strategy. A sidereal filter was applied to remove repeating multipath patterns based on the satellite constellation's repeat cycle. Additionally, spatial filtering was used to mitigate common-mode errors (CME) that affect the regional network. To ensure data integrity, outliers exceeding three times the standard deviation (3σ) were iteratively removed. This cleaning process effectively eliminates non-geophysical artifacts, yielding a clean displacement waveform suitable for inversion and ensuring that the modeled decay represents true ground deformation rather than observational noise [8].

2.4 Displacement and Decay Visualization

The cleaned kinematic time series reveals three distinct components of deformation: long-term secular tectonic motion, an abrupt coseismic jump, and transient postseismic relaxation. To separate these signals, we first calculated the displacement from the preseismic baseline to the coseismic step, and then tracked the position change from the coseismic moment through the minutes, hours, and days of early postseismic deformation. The temporal evolution of this postseismic motion was modeled using the following equations:

$$u(t) = c + a \ln\left(1 + \frac{t}{\log \tau}\right) \quad (1)$$

$$u(t) = c + a\left(1 - e^{-\frac{t}{\tau exp}}\right) \quad (2)$$

Here, u(t)denotes the position at time t (meters), c is the coseismic offset (m), a is the amplitude of the transient decay, t is the time elapsed since the mainshock (days), log τ represents the characteristic timescale for logarithmic decay, and τ_{exp} is the exponential decay constant. The logarithmic term describes the early afterslip mechanism, defined by its decay amplitude and characteristic time τ [5]. In this analysis, we explicitly differentiate rapid near-field afterslip from long-term viscoelastic relaxation, as previous studies have demonstrated that logarithmic decay offers a better fit for early postseismic deformation in the days immediately following an earthquake [13].

3 Result and Discussion

3.1 Kinematic Time Series and Decay Analysis

To characterize the deformation associated with the 2010 Mentawai earthquake, we first analyzed the 15-second kinematic position estimates derived from the SuGAr network. Unlike standard daily static solutions, which provide only a single coordinate averaged over a 24-hour period, our sub-daily kinematic time series reveal the continuous evolution of crustal motion from the precise moment of rupture (t=0) through the early postseismic phase. The displacement time series for the North and East components at important near-field stations, including SLBU, BSAT, KTET, and LNNG, display a clear tectonic signal, as seen in Fig. 3. Despite the high-frequency scatter inherent in sub-daily solutions attributable to unmodeled short-term tropospheric variability and multipath effects the tectonic trend remains unmistakable and clearly distinguishable from observational noise.

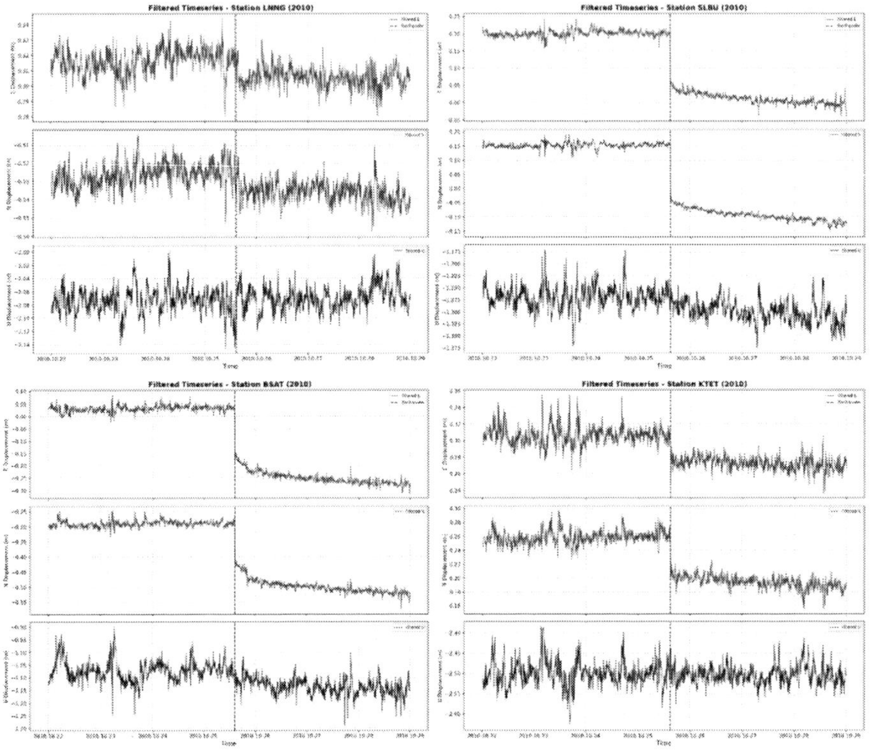

Fig. 3 *Time series of kinematic displacement for the vertical, east, and north components for particular near-field locations (e.g., SLBU, BSAT, KTET, LNNG). The red,green,blue scatter points represent 15-second kinematic east, north, uplift position estimates. capturing the instantaneous coseismic offset at t=0 and the subsequent rapid postseismic decay. Note the distinct 4–6 cm subsidence (downward shift) in the vertical component, confirming the offshore nature of the rupture.*

The coseismic rupture is indicated by a dramatic, immediate offset that occurs precisely at the moment of the earthquake, as shown by the horizontal components of the time series. This is immediately followed by a rapid, non-linear transient motion that continues in the same direction but decays in velocity over time. For example, the East component at station SLBU shows a sudden jump followed by a smooth logarithmic curve that persists for days. This ability to track the post-rupture motion second-by-second allows us to separate the true coseismic step from the subsequent rapid afterslip, a distinction that is often blurred in conventional daily processing where the first 24 hours of motion are averaged into a single data point.

Crucially, the vertical (normal) time series at the Pagai Island stations provide a definitive geometric constraint on the rupture source. As illustrated in the vertical component plots in Fig. 3, stations SLBU and BSAT reveal an instantaneous and stable subsidence, or downward displacement, of approximately 4 to 6 cm. This vertical signal is stable throughout the postseismic phase and is a critical piece of evidence regarding the earthquake's mechanism. The location of the station in relation to the slip patch determines whether the ground moves upward or downward in the context of subduction zone seismicity.

Surface subsidence at the location of the Pagai Islands confirms that the primary slip patch was located offshore, well to the southwest of the islands. If the rupture had extended downdip beneath the islands, geodetic theory predicts that we would have observed surface uplift. Therefore, the widespread subsidence recorded in the kinematic time series provides strong, independent evidence that the 2010 event was a "tsunami earthquake," characterized by a rupture confined to the shallow, compliant sedimentary wedge near the trench, rather than a deep megathrust event.

3.2 Coseismic Displacement Field and Rupture Characteristics

The coseismic offsets extracted from the kinematic time series exhibit a coherent spatial pattern consistent with a shallow subduction rupture. The stations on the North and South Pagai Islands, which are situated precisely above the downdip border of the rupture patch, recorded the greatest displacements, as shown in Fig. 4. The vector map indicates that the main motion occurred in the direction of the trench, southwest. Specifically, stations SLBU and BSAT experienced horizontal trench-ward motions ranging from 20 to 25 cm. These values are substantial but relatively moderate for a magnitude 7.8 event, further suggesting that the main energy release was distant from the islands.

The quantitative values for these offsets are detailed in Table 3. For instance, station BSAT recorded a coseismic eastward displacement of -25.43 cm and a northward displacement of -18.31 cm. Similarly, in the East and North components, station PRKB recorded -13.87 cm and -9.59 cm, respectively. These precise static offsets serve as the input for any finite fault inversion and define the starting point for the postseismic relaxation analysis. By using kinematic estimation, we ensure that these values reflect the true elastic rebound at the moment of the quake, uncontaminated by the rapid afterslip that occurred in the subsequent hours.

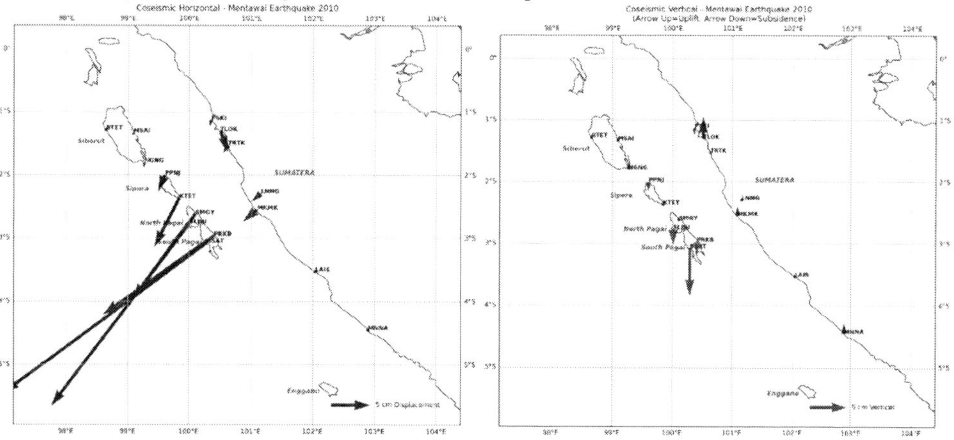

***Fig. 4** Horizontal (left) and vertical (right) coseismic displacement fields of the 2010 Mw 7.8 Mentawai earthquake derived from sub-daily kinematic PPP solutions. The vectors indicate moderate trench-ward motion and subsidence at the Pagai islands, consistent with a shallow offshore rupture.*

The vertical displacement field, also shown in Fig. 4 (right Fig.), reinforces the interpretation of a shallow source. The widespread subsidence of 4 to 6 cm across the Pagai Islands is consistent across multiple stations, ruling out local site effects. This pattern of moderate horizontal motion coupled with consistent subsidence is the geodetic fingerprint of a shallow rupture that did not propagate significantly downdip. It suggests that the slip was focused in the megathrust's updip section, where the contact usually displays frictional behavior that weakens with velocity.

Furthermore, the deformation field displays a rapid spatial decay in magnitude with distance from the trench. Stations located farther inland on the island of Sipora (e.g., KTET) and on the Sumatran mainland register progressively smaller displacements. For example, station KTET recorded only about 3 cm of horizontal motion, and mainland stations showed negligible offsets. Unlike conventional megathrust ruptures that cause extensive deformation along the whole forearc and mainland, this strong localization of deformation in the outer forearc wedge demonstrates that the 2010 Mentawai earthquake was a highly localized shallow event.

3.3 Early Postseismic Deformation and Decay Analysis

Following the instantaneous coseismic step, the kinematic solutions reveal that the crust did not immediately reach equilibrium but instead underwent a significant and rapid transient deformation phase. Fig. 5 (top Fig.) visualizes the cumulative postseismic displacement vectors over increasing time intervals, from the first 30 minutes up to 3 days. Spatially, the postseismic motion generally aligns with the azimuth of the coseismic slip but demonstrates a distinct spatial extension. The vectors at the Pagai Islands point trench-ward, but their significant magnitude suggests that the deformation zone migrated downdip from the coseismic rupture patch into the deeper plate interface beneath the islands.

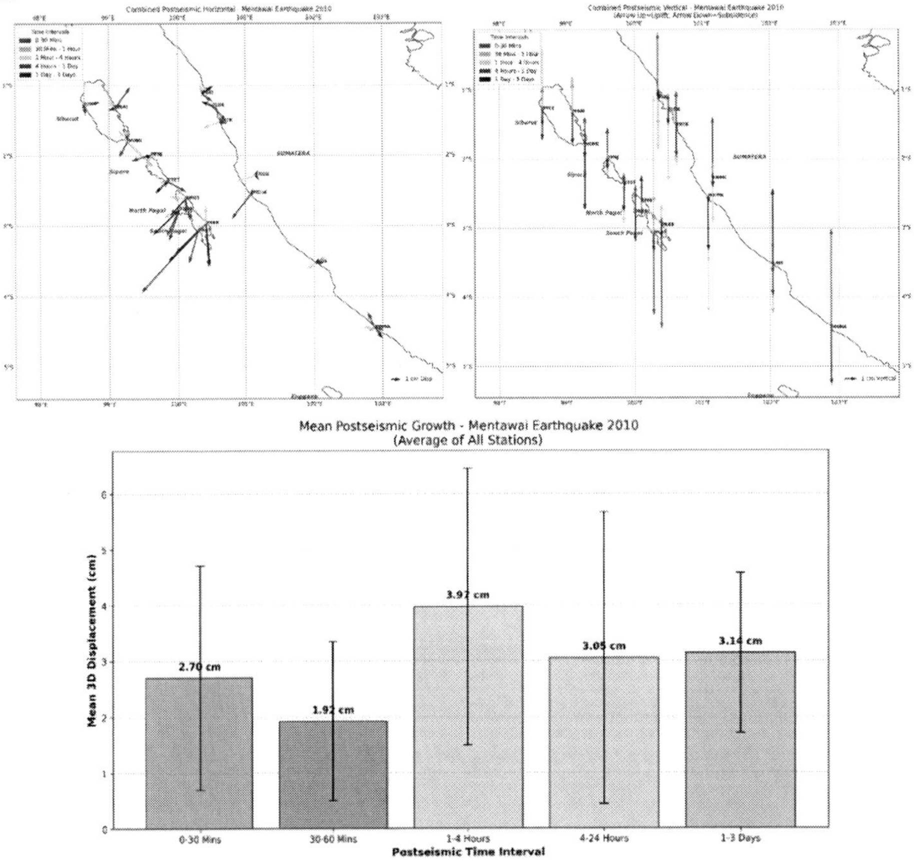

***Fig. 5** Spatiotemporal evolution of early postseismic deformation. (Top) Cumulative postseismic displacement vectors over varying time intervals. (Bottom) Mean 3D cumulative displacement growth across all stations, highlighting the rapid accumulation within the first 4 hours.*

 https://doi.org/10.1051/bioconf/202621603003

The rapidity of this early postseismic phase is quantitatively analyzed in the mean 3D displacement growth chart shown in Fig. 5 (bottom Fig.). This analysis reveals that a substantial portion of the total afterslip occurred almost immediately. On average, a displacement of 2.70 cm was accumulated within the first 30 minutes post-rupture. This was followed by an additional 1.92 cm in the subsequent 30-minute interval and a massive 3.97 cm over the next 1–4 hours. This means that the fault continued to slide rapidly even after the seismic waves had subsided, releasing a significant amount of stored strain energy in the form of aseismic creep.

Comparing these intervals reveals the critical importance of sub-daily observation. The displacement accumulated in the first 4 hours alone (approximately 8.6 cm cumulative) is significantly larger than the total displacement observed over the entire subsequent 1–3 day period, which amounted to only 3.14 cm. This finding indicates that the decay of afterslip velocity is extremely steep; the fault moves fastest immediately after the break. If one were to rely solely on daily GPS solutions, this entire 4-hour period of rapid motion would be averaged out, leading to a severe underestimation of the initial postseismic velocity.

This spatiotemporal evolution suggests a mechanical continuity between the seismic rupture and the aseismic afterslip. The stress released by the shallow rupture was transferred to the adjacent downdip segment of the fault. Because this deeper segment has different frictional properties specifically, it is velocity-strengthening it did not rupture seismically but instead responded by sliding stably at a high rate. The border between the locked zone that broke and the creeping zone that slid silently is successfully mapped by the kinematic data.

3.4 Decay Mechanism and Physical Interpretation

To understand the physical process driving this rapid transient motion, we modeled the displacement time series using two competing decay functions: logarithmic (Eq. 1) and exponential (Eq. 2). Determining which mathematical model best fits the data is essential because different models correspond to different rheological mechanisms. Logarithmic decay is typically associated with frictional afterslip on a fault interface, whereas exponential decay is indicative of linear viscoelastic relaxation of the mantle. We fitted both models to the 15-second kinematic position estimates for all near-field stations.

***Table 1** Estimated parameters and RMS misfits for the Logarithmic decay model.*

Site	Logarithmic				RMS
	a_{East} (cm)	a_{North} (cm)	C_{east} (cm)	C_{north} (cm)	
BSAT	-2.995	-2.683	-17.134	-42.782	0.841
BTET	-0.065	-0.023	-3.203	-66.888	0.518
KTET	-0.267	-0.631	27.987	21.125	0.728
LAIS	-0.203	-0.056	17.8	-19.301	0.631
LNNG	-0.18	-0.107	30.974	-53.291	0.529
MKMK	-0.102	-0.034	18.9	-51.398	0.552
MNNA	-0.101	-0.168	12.407	4.478	0.742
MSAI	-0.049	-0.124	30.241	6.335	0.682
NGNG	0.031	-0.312	42.22	5.756	1.112
PPNJ	-0.07	-0.124	29.418	-60.657	0.59
PRKB	-1.488	-1.075	23.935	-42.469	1.292
PSKI	-0.065	0.018	8.198	-33.397	0.607
SLBU	-1.711	-2.288	5.688	-3.181	0.622
SMGY	-0.709	-0.957	32.508	16.302	0.753
TLOK	-0.863	0.301	134.662	-196.4	2.177
TRTK	-0.064	-0.153	5.337	-72.901	0.452

The statistical results, summarized in Table 1 and Table 2, unequivocally favor the logarithmic model. We used the Root Mean Square (RMS) misfit to evaluate the goodness-of-fit. For example, at station BSAT, the exponential model yields an RMS misfit of 0.926 cm, but the logarithmic model achieves an RMS misfit of 0.841 cm. A similar pattern is observed at station PRKB, where the logarithmic fit (RMS 1.292 cm) consistently outperforms the exponential fit (RMS 1.300 cm). Across the entire network, the logarithmic function provides a closer approximation to the observed data, minimizing the residuals between the modeled curve and the actual GPS observations.

Table 2 *Estimated parameters and RMS misfits for the Exponential decay model.*

Site	Exponential				RMS
	a_{East} (cm)	a_{North} (cm)	C_{east} (cm)	C_{north} (cm)	
BSAT	-8.774	-7.808	-18.805	-44.313	0.926
BTET	-0.208	-0.047	-3.227	-66.914	0.518
KTET	-0.81	-1.872	27.858	20.788	0.728
LAIS	-0.654	-0.141	17.729	-19.349	0.631
LNNG	-0.534	-0.281	30.877	-53.373	0.53
MKMK	-0.342	-0.122	18.873	-51.401	0.551
MNNA	-0.314	-0.499	12.362	4.388	0.742
MSAI	-0.168	-0.358	30.23	6.263	0.683
NGNG	0.095	-0.979	42.235	5.625	1.109
PPNJ	-0.23	-0.364	29.395	-60.726	0.59
PRKB	-4.312	-3.136	23.072	-43.079	1.3
PSKI	-0.211	0.054	8.177	-33.387	0.607
SLBU	-5.088	-6.776	4.787	-4.408	0.632
SMGY	-2.09	-2.833	32.121	15.789	0.755
TLOK	-2.956	1.054	134.46	-196.344	2.173
TRTK	-0.191	-0.426	5.305	-73	0.453

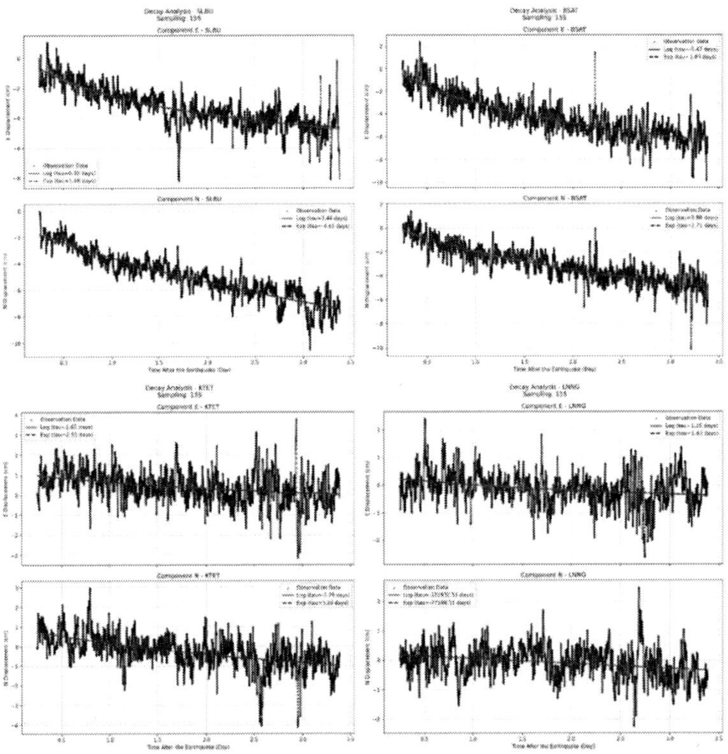

Fig. 6 *Kinematic displacement time series (black/grey dots) for North and East components. The best-fit logarithmic model is shown by red line, and the exponential model is shown by blue line. Note how the red line better fits the rapid initial motion.*

Visually, this preference is evident in Fig. 6, where the best-fit curves are overlaid on the kinematic time series. The logarithmic curve (indicated by the solid red line) tracks the dense cloud of data points with high fidelity, particularly during the critical first 24 hours where the curvature of the displacement signal is sharpest. In contrast, the exponential curve (indicated by the blue dashed line) often fails to reproduce this rapid onset, tending to underestimate the initial velocity and overestimate the later deformation. The visual mismatch of the exponential model confirms that viscoelastic processes are too slow to explain the immediate post-earthquake deformation.

The parameters derived from the logarithmic modeling further support the afterslip hypothesis. As detailed in Table 1, the estimated characteristic decay times (τ) are extremely short, ranging from approximately 0.05 to 0.3 days. For example, the East component at station SLBU exhibits a decay time of roughly 0.30 days. This extremely short timescale is a hallmark of rate-and-state friction mechanics, where the fault response to a stress step is immediate and decays inversely with time. Therefore, we conclude that frictional afterslip on the megathrust interface specifically in the velocity-strengthening zone downdip of the rupture was the dominant driver of deformation in the days following the 2010 Mentawai earthquake.

3.5 Comparison of Coseismic and Postseismic Budgets

A direct comparison of the coseismic offsets and the cumulative early postseismic displacements highlights the significant contribution of afterslip to the total deformation budget. By integrating the postseismic motion over a 3-day interval, we can assess how much additional slip occurred relative to the mainshock. The comparative values for all stations are summarized in Table 3. While the coseismic slip dominates the overall magnitude of displacement, the early afterslip contributes a remarkably large fraction, ranging from 10% to 40% of the total surface displacement at near-field stations.

The most significant postseismic contributions are observed at stations located closest to the downdip edge of the rupture. For example, Table 3 shows that station BSAT recorded a massive postseismic displacement of -11.77 cm in the East component and -11.34 cm in the North component. When compared to its coseismic offset of -25.43 cm (East) and -18.31 cm (North), the afterslip represents a substantial continuation of the fault motion. Similarly, station SLBU recorded a postseismic northward motion of -9.41 cm, which corresponds to approximately 42% of its coseismic jump (-22.24 cm). These high ratios indicate that the earthquake was not a singular event but rather the beginning of a prolonged slip episode.

***Table 3** Comparison of Coseismic offsets versus cumulative Early Postseismic displacements.*

Site	Coseismic		Postseismic	
	E (cm)	N (cm)	E (cm)	N (cm)
BSAT	-25.43	-18.31	-11.77	-11.34
BTET	0.1	0.07	-0.58	0.76
KTET	-3.14	-5.98	-0.64	-3.51
LAIS	-0.02	-0.2	-0.57	-0.52
LNNG	-1.18	-1.01	-1.15	-0.76
MKMK	-1.74	-1.47	-1.07	-1.81
MNNA	0.07	0.23	0.43	0.25
MSAI	-0.19	-0.42	-1.03	1.05
NGNG	-0.12	-0.71	0.02	-0.63
PPNJ	-0.86	-2.07	0.03	-0.66
PRKB	-13.87	-9.59	-4.94	-7.36
PSKI	-0.17	-0.8	-0.03	0.24
SLBU	-17.28	-22.24	-6.49	-9.41
SMGY	-7.93	-10.43	-2.39	-4.48
TLOK	0.96	-2.6	-0.59	-0.76
TRTK	-0.6	-0.99	0.24	-1.73

The spatial distribution of these high-amplitude postseismic signals is visualized in Fig. 7. The comparison map shows that while the black coseismic vectors are large, the red postseismic vectors are far from negligible, particularly at the Pagai Islands. This implies that creeping patches downdip of the main slip zone were instantly activated by the stress changes brought on by the shallow seismic rupture. This downdip region's high afterslip magnitude suggests that a sizable amount of the plate convergence budget is released aseismically.

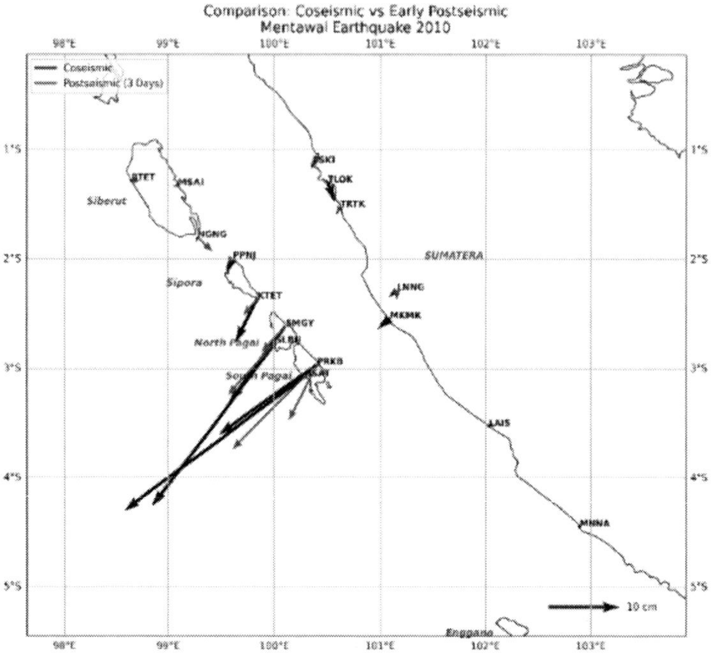

Fig. 7 Comparison of coseismic (black vectors) and early postseismic (red vectors) displacement fields.

Conversely, far-field stations such as LAIS and MKMK exhibit minimal postseismic influence, with displacements generally less than 1 cm. This sharp contrast confirms that the rapid afterslip observed in this study is a near-field phenomenon, confined to the segment of the megathrust immediately adjacent to the rupture. These results paint a picture of a heterogeneous frictional landscape along the Mentawai segment, characterized by a sharp transition from a shallow velocity-weakening zone that ruptures in tsunamigenic earthquakes to a deeper velocity-strengthening zone that creeps silently in their aftermath.

4 Conclusion

This study utilized high-rate kinematic GPS observations to characterize the deformation field of the 2010 Mw 7.8 Mentawai tsunami earthquake, effectively resolving both the instantaneous rupture and the rapid afterslip phase. The primary results are:

- The coseismic displacement field shows up to 20–25 cm of horizontal motion and 4–6 cm of vertical subsidence, confirming a very shallow rupture concentrated near the trench.
- Early postseismic deformation, typically obscured in daily GPS solutions, was clearly detected and spans the first minutes to days after the earthquake. These signals reach several centimeters in amplitude and evolve smoothly.
- Logarithmic decay models fit the postseismic trajectories better than exponential models, indicating that frictional afterslip dominates the early deformation phase. Characteristic decay times are on the order of a few days.
- The spatial pattern suggests that afterslip occurred downdip of the main rupture, highlighting a strongly heterogeneous fault interface where seismic rupture in the shallow prism transitions rapidly to aseismic slip at depth.

These insights significantly advance our knowledge regarding the mechanics of tsunami earthquakes and clarify the complex interaction between seismic rupture and aseismic creep along the Sunda subduction zone.

Acknowledgement

We extend our sincere gratitude to Institut Teknologi Sepuluh Nopember (ITS) for funding this work under the research grant titled "Spatiotemporal Variations of Coseismic and Early Afterslip Deformation of >M7 Earthquakes Along the Sumatra-Java Subduction Zone Using Kinematic GNSS." This financial assistance was pivotal in ensuring the successful execution and publication of this research.

References

1. K. Sieh, "The Sunda megathrust - Past, present and future," Journal of Earthquake and Tsunami, vol. 1, no. 01, pp. 1–19, 2007.
2. A. V. Newman, G. Hayes, Y. Wei, and J. Convers, "The 25 October 2010 Mentawai tsunami earthquake, from real-time discriminants, finite-fault rupture, and tsunami excitation," Geophysical Research Letters, vol. 38, no. 5, 2011.
3. E. M. Hill et al., "The 2010 Mw 7.8 Mentawai earthquake: Very shallow source of a rare tsunami earthquake determined from tsunami field survey and near-field GPS data," Journal of Geophysical Research: Solid Earth, vol. 117, no. B6, 2012.
4. R. Cattin et al., "Stress change and effective friction coefficient along the Sumatra-Andaman-Sagaing fault system after the 26 December 2004 (Mw = 9.2) and the 28 March 2005 (Mw = 8.7) earthquakes," Geochemistry, Geophysics, Geosystems, vol. 10, no. 3, 2009.
5. C. J. Marone, C. H. Scholz, and R. Bilham, "On the mechanics of earthquake afterslip," Journal of Geophysical Research: Solid Earth, vol. 96, no. B5, pp. 8441–8452, 1991.
6. H. Perfettini et al., "Seismic and aseismic slip on the central Peru megathrust," Nature, vol. 465, no. 7294, pp. 78–81, 2010.
7. L. L. H. Tsang et al., "Afterslip following the 2007 Mw 8.4 Bengkulu earthquake in Sumatra loaded the 2010 Mw 7.8 Mentawai tsunami earthquake rupture zone," Journal of Geophysical Research: Solid Earth, vol. 121, pp. 9034–9049, 2016.
8. K. Liu et al., "Rapid early afterslip characteristics of the 2010 moment magnitude (Mw) 8.8 Maule earthquake determined with sub-daily GPS solutions," Satellite Navigation, vol. 5, no. 23, 2024.
9. C. Twardzik, M. Vergnolle, A. Sladen, and A. Avallone, "Unravelling the contribution of early postseismic deformation using sub-daily GNSS positioning," Scientific Reports, vol. 9, p. 1775, 2019.
10. Y. Bock and D. Melgar, "Physical applications of GPS geodesy: A review," Reports on Progress in Physics, vol. 79, no. 10, 2016.
11. M. Chlieh et al., "Heterogeneous coupling of the Sumatran megathrust constrained by geodetic and paleogeodetic measurements," Journal of Geophysical Research: Solid Earth, vol. 113, no. B5, 2008.
12. A. P. Kositsky and J.-P. Avouac, "Inverting geodetic time series with a principal component analysis-based inversion method," Journal of Geophysical Research: Solid Earth, vol. 115, no. B3, 2010.
13. R. Raharja et al., "Long aseismic slip duration of the 2006 Java tsunami earthquake based on GPS data," Earthquake Science, vol. 29, no. 5, pp. 291–298, 2016.
14. W. Bertiger et al., "GipsyX/RTGx: A new tool for geodetic analysis," GPS Solutions, vol. 24, no. 4, 2020.
15. J. F. Zumberge, M. B. Heflin, D. C. Jefferson, M. M. Watkins, and F. H. Webb, "Precise point positioning for the efficient and robust analysis of GPS data from permanent stations," Journal of Geophysical Research: Solid Earth, vol. 102, no. B3, pp. 5005–5017, 1997.

Public participation and willingness to participate in marine debris management in Batam City: A preliminary assessment using an open-ended contingent valuation approach

I Wayan Koko Suryawan[1,2,3,4], *Nova* Ulhasanah[1,2,3], *Mega Mutiara* Sari[1,2,3], Evi Siti Sofiyah[1,2], and *Ari* Rahman[1,2*],

[1] Department of Environmental Engineering, Faculty of Infrastructure Planning, Universitas Pertamina, Jalan Sinabung II, Terusan Simprug, Jakarta, 12220, Indonesia
[2] Center for Environmental Solution (CVISION), Universitas Pertamina, Jalan Sinabung II, Terusan Simprug, Jakarta, 12220, Indonesia
[3] Graduate Program of Science in Sustainability, Faculty of Infrastructure Planning, Universitas Pertamina, Jalan Sinabung II, Terusan Simprug, Jakarta, 12220, Indonesia
[4] Center for Interdisciplinary Research on Ecology and Sustainability, College of Environmental Studies and Oceanography, National Dong Hwa University, Hualien 97401, Taiwan, ROC

*E-mail: ari.rahman@universitaspertamina.ac.id

Abstract. Marine debris has become a growing environmental challenge in Batam City, driven by rapid coastal development, increasing tourism, and cross-border waste transport within the Singapore Strait. Understanding community engagement is critical for designing effective waste reduction strategies, yet empirical evidence on local participation remains limited. This preliminary study assesses public willingness to participate (WtP) in marine debris management using an open-ended contingent valuation method (CVM). A total of 70 respondents from five coastal zones Nongsa, Marina, Tanjung Pinggir, Barelang, and Batu Ampar were surveyed in September 2025 through face-to-face interviews. WtP was measured in terms of hours per month that individuals were willing to contribute to beach clean-ups, environmental monitoring, educational campaigns, and related activities. The results show an average willingness to participate of 5.68 hours/month/person, with a median and mode of 2 hours, indicating a highly skewed distribution driven by a small group of highly motivated individuals. Chi-Square analysis reveals that age is the only demographic factor significantly associated with participation levels ($p = 0.008$), while gender, occupation, and income show no significant relationship. The participation curve follows an exponential decline, and the marginal willingness to participate evaluated at the mean is approximately 0.80 persons per additional hour, suggesting that higher time commitments sharply reduce overall community engagement.

1 Introduction

Marine debris has emerged as an increasingly pressing environmental issue in numerous coastal regions globally [1], with Batam City exemplifying this challenge. Situated within the heavily trafficked Singapore Strait, Batam, as a rapidly developing island, experiences significant maritime traffic, industrial activities, tourism, and inter-island mobility. These factors render Batam particularly susceptible to the leakage of land-based waste, littering from port areas, tourism hotspots, and transboundary marine debris from adjacent coastal jurisdictions. The accumulation of debris jeopardizes coastal ecosystems, diminishes tourism potential, disrupts fisheries, escalates local government expenditures on clean-up operations, and adversely affects the health and well-being of coastal residents [2]. Addressing these multifaceted issues necessitates not only the implementation of appropriate infrastructure and regulatory frameworks but also a comprehensive understanding of community engagement, as public participation is often the most effective mechanism for promoting long-term coastal stewardship. Despite Batam's strategic significance, empirical research concerning marine debris management in the region remains limited, particularly with respect to its social dimensions. Existing studies on coastal waste issues in Indonesia predominantly focus on biophysical assessments, the sources and composition of plastic waste, ecological impacts, and technical waste management solutions [3]. While these studies underscore the scale of the problem, they frequently overlook community perceptions of marine debris challenges and the willingness of residents to engage in mitigation efforts. This deficit in social and behavioral insights creates a gap in the formulation of participatory coastal governance strategies. Absent an understanding of community members' readiness to invest time, effort, and labor, government-led initiatives may lack the sustained impact necessary for effective change.

Another notable gap in literature pertains to valuation approaches. Most valuation studies in environmental management primarily utilize monetary willingness to pay (WTP) as the chief indicator of community support [4–6]. However, monetary contributions do not consistently reflect the actual level of environmental stewardship in communities where time availability, social norms, and a culture of volunteerism significantly influence environmental behavior more than financial resources [7]. In numerous coastal communities, particularly those characterized by diverse socio-economic backgrounds, non-monetary contributions [8,9], including volunteer hours, may offer a more accurate representation of local commitment to environmental initiatives. Limited research within Indonesia has employed time-based willingness to participate (WtP) metrics, and an even smaller number of studies have utilized WtP as the principal valuation instrument for marine debris management. This underscores a methodological deficiency that the current study aims to address.

The originality of this research lies in its implementation of an open-ended contingent valuation method (CVM) to evaluate willingness to participate in terms of hours volunteered per month, rather than focusing on monetary contributions. By prioritizing time, the study captures an alternative dimension of community value that is more closely aligned with the labor-intensive nature of marine debris mitigation activities, which include beach clean-ups, environmental monitoring, and community education. The open-ended format enables respondents to express the amount of time they are willing to contribute freely, thereby capturing nuanced variations in participation that are often obscured in closed-response formats. At the same time, this approach serves as an exploratory valuation tool, acknowledging that open-ended CVM responses may exhibit considerable variance and skewness. In this context, integrating CVM with a time-based WtP framework provides a coherent methodological justification, as it enables the study to directly connect individual participation behavior with broader dimensions of adaptive capacity that shape community preparedness and responses to environmental challenges.

A further innovative element is the incorporation of adaptive capacity indicators into the survey's design and interpretation. Rather than treating participation as an isolated behavioral variable, the study contextualizes it within Batam's larger governance framework, encompassing the availability of coastal waste bins, green port initiatives, risk communication, educational outreach, enforcement mechanisms, and community-based programs. These indicators elucidate how community willingness to participate interacts with governance performance, infrastructure deficiencies, and institutional readiness. Although this preliminary study does not conduct a comprehensive adaptive capacity index or structural modeling, it establishes a conceptual foundation for a more extensive valuation framework in subsequent research phases.

2 Method

This study adopts a preliminary quantitative approach to examine public participation and willingness to participate (WtP) in marine debris management in Batam City. Willingness to participate is defined as the number of hours per month that community members are willing to allocate to activities such as beach clean-ups, environmental monitoring, educational campaigns, or community-based waste initiatives. This measure of non-monetary contribution provides an early understanding of community engagement potential, which is often a stronger predictor of long-term environmental stewardship than financial willingness alone. The analytical foundation of this study is rooted in the adaptive capacity framework, in which public participation is influenced by key dimensions such as assets, flexibility, learning, organization, and agency. The status quo and recommended improvements for each of these dimensions were evaluated using a set of indicators previously compiled through field observations and expert judgment, as shown in Table 1.

Table 1. Adaptive capacity dimensions, status quo conditions, and proposed improvement scenarios for marine debris management in Batam City.

Attribute	Attribute Description (Adaptive Capacity Dimension)	Status Quo (Current Condition in Batam)	Alternative
Asset Availability	Asset availability reflects the community's access to physical resources needed for marine debris management. These include waste disposal facilities, cleaning tools, and operational infrastructure such as coastal waste stations and marine waste collection systems. Higher asset availability enhances a community's ability to take action and reduces constraints that prevent participation.	Coastal areas such as Nongsa, Barelang, and Batu Ampar lack designated coastal waste bins and disposal points, resulting in scattered trash along the shoreline. Residents and volunteers do not receive waste bags, gloves, or cleaning tools, and there is no regular system of marine waste collection using boats to retrieve debris from nearby islands.	The improved scenario provides organized coastal waste disposal infrastructure, including segregated waste bins and accessible 3R coastal stations. Residents and volunteers are supplied with suitable cleaning tools such as reusable bags, gloves, and protective equipment distributed through CSR programs. A scheduled marine waste-collection boat service operates routinely to remove debris from surrounding islets before it accumulates or disperses into the open sea.

Flexibility (Time Availability)	Flexibility reflects the community's ability to adjust schedules and allocate time for environmental activities despite work and household commitments. Strong flexibility enables residents to participate because activities align with their available time and are supported by social and workplace structures.	Most residents work full-time and lack the free time necessary to join environmental programs. Clean-up events are irregular and not aligned with times when residents are available, and employers generally do not grant time allowances or recognize environmental volunteerism.	The improved scenario introduces regular beach-cleaning activities held during weekends and national holidays when residents are more available. Community environmental activities become incorporated into neighborhood agendas, and employers support environmental participation by granting volunteer leave or adjusting work schedules to accommodate involvement in marine debris management.
Learning (Knowledge & Awareness)	Learning reflects the community's access to environmental knowledge, training, and awareness about marine debris. Strong learning capacity enables residents to understand environmental risks, adopt proper waste practices, and engage in informed participation.	Residents have limited awareness of the sources, impacts, and solutions related to marine debris. There is minimal dissemination of educational content through digital media, and no regular workshops are conducted by universities, NGOs, or government agencies. As a result, knowledge of environmentally responsible behavior remains low.	The improved scenario provides widespread digital learning materials such as videos, infographics, online modules, and interactive content explaining marine debris issues in simple and accessible formats. Universities, NGOs, and government institutions conduct workshops, school programs, and community outreach activities. Residents also indicate the number of hours they are willing to devote each month to marine debris management, reflecting increased environmental awareness and willingness to learn.
Organization (Collective Action & Governance)	Organization reflects the presence of structured community groups, leadership systems, and collaborative networks that allow coordinated, sustained environmental action. Strong organizational capacity helps communities maintain long-term programs, mobilize volunteers, and build partnerships.	There is no formal community structure for coastal environmental management. Activities remain dependent on government-led initiatives, and there are no dedicated committees, planned schedules, or collaborative partnerships with industrial areas, tourism operators, or NGOs. As a result, participation is inconsistent and short-lived.	The improved scenario includes the formation of community environmental committees responsible for organizing clean-up programs and monitoring coastal conditions. These groups maintain structured schedules, hold regular meetings, and coordinate volunteers. Collaboration is established with industrial zones, port authorities, tourism operators, and NGOs, resulting in long-term

			support, funding, and shared responsibilities.
Agency (Responsibility & Initiative)	Agency represents the community's sense of responsibility, motivation, and willingness to take independent action. Strong agency reflects empowerment, initiative, and the belief that individual and collective actions can improve environmental outcomes.	Many residents believe marine debris management is solely the responsibility of the government, resulting in low personal motivation and limited community-driven initiatives. Youth engagement and local leadership are also minimal.	In the improved scenario, residents show a greater sense of responsibility and actively participate in clean-up activities and environmental programs. Youth and community leaders initiate and organize events, share educational content, and mobilize participation through social media and local networks. Residents voluntarily monitor coastal conditions, report illegal dumping, and contribute to ongoing environmental efforts without relying solely on government directives.

Data collection was conducted in September 2025 across five major coastal zones in Batam City, namely Nongsa, Marina, Tanjung Pinggir, Barelang, and Batu Ampar. These areas were selected due to their high exposure to marine debris accumulation and their representation of diverse socio-economic and functional coastal settings, including tourism corridors, fishing settlements, port and industrial zones, and mixed residential areas. Previous studies have documented significant coastal degradation and waste accumulation in several parts of Batam, particularly in Tanjung Uma, where a once-flourishing marine ecosystem has deteriorated due to persistent pollution and unmanaged waste. Similar challenges have been reported in densely populated small islands such as Pulau Lengkang in Belakang Padang, where urban waste pressure has intensified coastal environmental degradation [10]. Additional studies from Penawar Rindu and surrounding small islands in Batam further highlight weaknesses in solid waste management systems and their implications for marine environments [11].

A total of 70 respondents were interviewed using accidental sampling. This technique was appropriate for a preliminary study whose primary aim was to capture initial perceptions from coastal users who experience marine debris firsthand in their daily routines. The respondents represented a broad cross-section of the coastal population, including fishermen, traders, housewives, students, manual laborers, and office workers. The survey instrument consisted of three components. The first section gathered demographic information, including gender, age, occupation, and monthly income. The second section introduced respondents to the current challenges of marine debris in Batam, referring to the previously defined governance indicators coastal waste bin availability, green port initiatives, environmental education programs, cross-border data sharing, community clean-up activities, incentive mechanisms, risk communication, and law enforcement. These indicators reflect the operationalization of adaptive capacity in the marine debris context and were presented with short descriptions, photographs, and simple explanations to ensure respondents clearly understood the local environmental problems.

The third section focused on measuring willingness to participate. Respondents were asked an open-ended question in which they stated the number of hours per month they were willing to dedicate to marine debris management activities. The open-ended approach, reflected in the format illustrated in Figure 1, allowed for unrestricted responses and captured subtle variations that fixed-category questions might overlook. This format is particularly valuable for preliminary studies because it reveals the full distribution of participation levels including very low and very high values providing a detailed picture of community commitment.

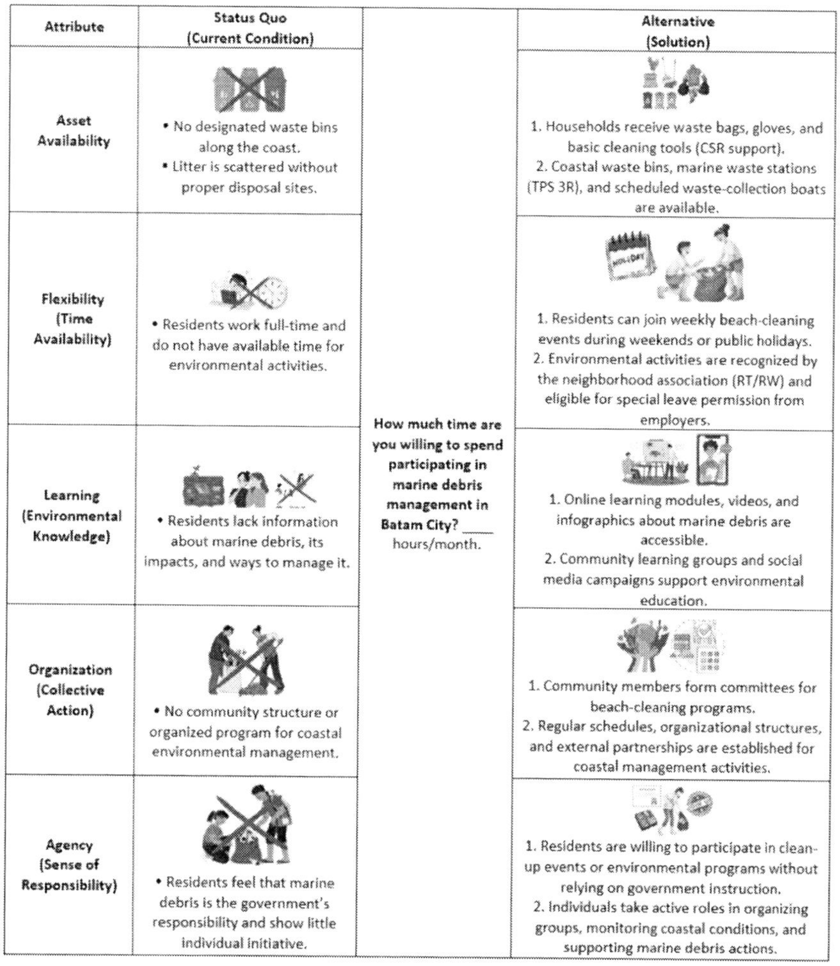

Fig. 1. Conceptual illustration of adaptive capacity dimensions and WtP scenarios for marine debris management in Batam City.

Data collection was carried out through face-to-face interviews in public coastal areas, fishing docks, small business areas, and residential shoreline zones. Enumerators followed a standardized script to avoid interviewer bias and ensure consistent explanation of "participation hours," particularly regarding the types of activities included. All respondents were informed that the study was exploratory, participation was voluntary, and no personal identifying information would be recorded. Ethical protocols were followed throughout the process. The dataset was analyzed using descriptive statistics and simple regression. Descriptive measures such as mean, median, mode, standard deviation, minimum and maximum values, and quartiles were used to summarize the distribution of participation hours. This analysis provided an overview of general trends in community participation, revealing whether willingness levels clustered around low, moderate, or high contributions and identifying outliers that may indicate strong leadership potential or exceptional willingness.

To explore the socio-economic determinants of participation, a simple regression model was applied, using willingness to participate (hours per month) as the dependent variable. The independent variables included age, income, occupation type, and gender. The regression was intentionally kept basic to match the exploratory nature of the study, aiming to detect early associations rather than build a complex predictive model. These initial patterns can help researchers refine hypotheses and inform the design of more advanced statistical models in future studies. The methodological design emphasizes clarity, practicality, and suitability for a preliminary assessment. Unlike valuation studies requiring economic bids or complex utility modeling, willingness to participate measures community commitment directly and can be easily understood by respondents. The insights generated from this preliminary research will support the development of more comprehensive frameworks for community-based marine debris management in Batam, including the integration of behavioral drivers, motivation factors, and long-term monitoring in subsequent phases of the research.

3 Results and discussion

Table 2 summarizes the descriptive statistics for the community's willingness to participate (WtP) in marine debris management activities in Batam City, expressed in hours per month. The results show that, on average, respondents are willing to contribute 5.68 hours per month, although participation levels vary widely across the sample. The median and mode values are both 2 hours, indicating that most respondents tend to offer only a few hours each month, while a smaller number of highly motivated individuals contribute substantially more time. This variation is reflected in the relatively large standard deviation of 9.6 hours, with participation ranging from 0 hours at the minimum to 60 hours at the maximum. The percentile distribution further clarifies the spread of responses: 25% of respondents are willing to participate for 1 hour, 50% for 2 hours, and 75% for 6 hours per month. These descriptive patterns provide an important baseline for understanding community engagement potential and the diversity of participation levels among coastal residents in Batam.

Table 2. Descriptive statistics of community WtP in marine debris management activities

Statistic	Value
Mean	5.68 hours/month
Standard Error of Mean	1.16
Median	2 hours
Mode	2 hours
Standard Deviation	9.6
Minimum	0 hours
Maximum	60 hours
25th Percentile (P25)	1 hour
50th Percentile (P50)	2 hours
75th Percentile (P75)	6 hours

Figure 2 illustrates the distribution of respondents' WtP in marine debris management activities, expressed in hours per month. The histogram shows that participation levels are heavily concentrated at the lower end of the scale, with most respondents contributing only a small number of hours. The highest frequencies occur around 1–2 hours per month, consistent with the median and mode values reported earlier (both 2 hours). This confirms that while many residents are willing to participate, their available time for environmental activities is limited. The distribution is positively skewed, with a long tail extending toward higher participation levels. A small number of respondents reported substantially higher contributions, ranging from 12 hours up to 60 hours per month, indicating the presence of highly motivated individuals or active community leaders who consistently engage in coastal clean-up or environmental activities. An exponential trendline was fitted to the data, represented by the function $y = 12.306e^{-0.186}x$, which closely follows the decreasing pattern of the histogram. The coefficient of determination ($R^2 = 0.4608$) indicates a moderate level of explanatory fit, suggesting that an exponential decay pattern reasonably captures the overall decline in participation frequency as hours increase. This pattern is common in early-stage participation studies, where the majority of the population contributes minimally, and only a small proportion offers substantial time commitments.

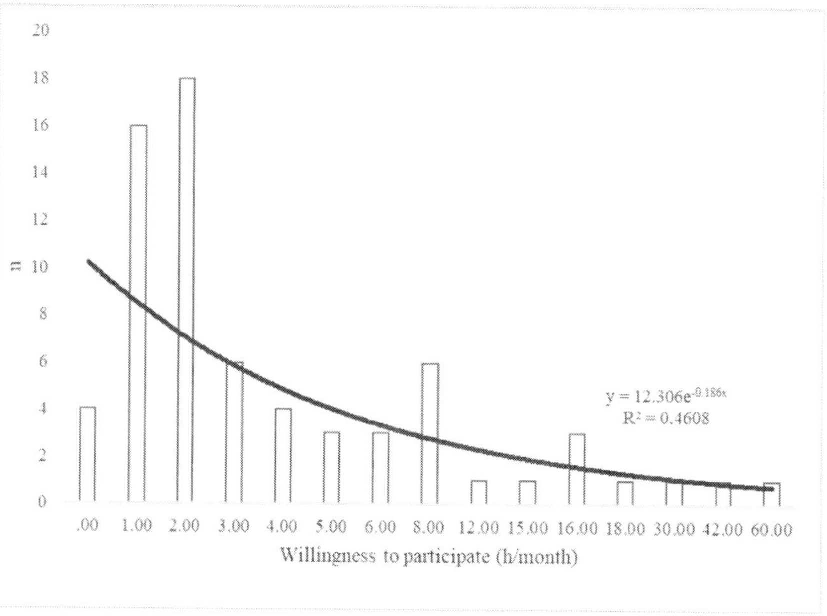

Fig. 2. Distribution of community WtP in marine debris management activities with fitted exponential decay curve.

To further interpret the distribution of participation hours, the marginal willingness to participate (MWtP) was derived from the exponential decay function fitted to the histogram in Figure 2. The relationship between participation frequency and participation hours was expressed as (Equation 1):

$$y = 12.306\, e^{-0.186x} \quad (1)$$

where y represents the number of respondents and x denotes willingness to participate in hours per month. The marginal rate of change was obtained by differentiating the function with respect to x (Equation 2):

$$\frac{dy}{dx} = -0.186 \times 12.306\, e^{-0.186x}, \quad (2)$$

which simplifies to (Equation 3):

$$\frac{dy}{dx} = -2.290\, e^{-0.186x}. (3)$$

Because the derivative is always negative, the number of people willing to participate decreases as the required level of participation (hours per month) increases. This reflects the expected decline in engagement when time commitments become more demanding. To estimate the marginal willingness to participate at the average participation level, the derivative was evaluated at the mean of 5.68 hours/month/person. Substituting this value yields (Equation 4):

$$\frac{dy}{dx}\Big|_{x=5.68} = -0.796 \ \ (4)$$

This value indicates that for each additional hour of expected participation at the mean, the number of individuals willing to contribute decreases by approximately 0.8 hour/month/person. In other words, requiring one more hour of participation reduces the pool of willing participants by nearly one person. This highlights the strong sensitivity of community engagement to time availability and suggests that marine debris programs should prioritize low-time-burden activities to sustain broad participation.

Table 3 presents the results of the Chi-square tests examining whether demographic characteristics are associated with differences in community willingness to participate in marine debris management activities in Batam City. The variables analyzed include gender, age group, occupation, and income, providing a preliminary assessment of the socio-economic factors that may influence participation behavior. The results indicate that most demographic variables do not exhibit statistically significant associations with participation levels. Gender ($\chi^2 = 13.773$, $p = 0.467$), occupation ($\chi^2 = 32.495$, $p = 0.995$), and income ($\chi^2 = 51.371$, $p = 0.650$) all yield p-values greater than 0.05, suggesting no meaningful relationship with willingness to participate. In contrast, age group shows a statistically significant association with participation behavior ($\chi^2 = 67.277$, $p = 0.008$), indicating that willingness to participate varies across age categories. This finding suggests that age is the only demographic factor that meaningfully influences participation in this preliminary study. Understanding age-related differences is important for designing targeted community engagement strategies, as different age groups may exhibit varying capacities, motivations, and time availability for involvement in marine debris reduction initiatives.

Table 3. Results of Chi-square tests examining associations between demographic characteristics and willingness to participate in marine debris management.

Variable	Pearson Chi-Square (χ^2)	df	p-value (Asymptotic Sig.)	Interpretation
Gender	13.773	14	0.467	Not significant
Age Group	67.277	42	0.008	Significant association
Occupation	32.495	56	0.995	Not significant
Income	51.371	56	0.65	Not significant

The findings of this preliminary study provide early insights into how communities in Batam City perceive their role in marine debris management and how demographic characteristics, participation patterns, and governance gaps shape their adaptive capacity. Although the study is exploratory, several implications emerge for strengthening policy design and developing a more comprehensive adaptive capacity valuation model tailored to coastal urban settings.

The WtP data indicate that community engagement is present but highly uneven. With an average participation of 5.68 hours/month/person and a median of 2 hours, the distribution suggests that a core group of highly committed individuals is compensating for a broader population that contributes minimally. This skewed distribution is common in early-stage environmental stewardship studies, where participation is often concentrated among those with stronger awareness or proximity to environmental risks [12]. The declining marginal willingness to participate further confirms that additional time requirements sharply reduce overall engagement, indicating that future programs should prioritize low-burden, flexible participation options.

In terms of socio-demographic influences, the Chi-Square analysis reveals that age is the only significant predictor of willingness to participate. Younger groups (18–29 years) show higher levels of engagement, likely due to increased environmental awareness, digital exposure, and participation in school or university activities. In contrast, gender, income, and occupation did not significantly influence participation levels. This finding suggests that marine debris management in Batam can be designed as an inclusive program across social groups provided that activities remain accessible, time-efficient, and supported by clear benefits to community well-being. From an adaptive capacity perspective, the results align with the notion that capacity emerges from the interaction of assets, learning, organizational structures, flexibility, and agency [13–15].

Policy implications from this study center on strengthening governance arrangements that both enhance institutional performance and leverage community willingness. First, because younger respondents demonstrate the highest levels of engagement, outreach and participation programs should be anchored in schools, universities, and youth communities, employing gamified and technology-based engagement strategies to sustain interest and long-term involvement. Second, given the high perceived importance but low performance of several governance indicators, municipal agencies and port authorities should prioritize investments in coastal waste management infrastructure, particularly in zones with high exposure to marine debris such as Nongsa and Barelang. Third, the non-significant influence of income suggests that participation is not primarily constrained by financial capacity, reinforcing the relevance of non-monetary engagement pathways, including time contributions, voluntary monitoring, and citizen science initiatives. Finally, this preliminary study establishes a foundation for developing an adaptive capacity valuation model that integrates time-based WtP, demographic characteristics, and governance gaps. Future research can build on this foundation by expanding sample size, incorporating behavioral predictors, evaluating program effectiveness, and examining longitudinal changes over time. Such an approach can ultimately support policymakers in designing participatory, equitable, and resilient marine debris management systems in Batam.

4 Conclusion

This preliminary study provides an initial assessment of public participation and time-based willingness to participate (WtP) in marine debris management in Batam City using an open-ended contingent valuation approach. The findings indicate that the community demonstrates a foundational level of engagement, with an average willingness to contribute 5.68 hours/month/person. However, the highly skewed distribution reflected in a median mode of only 2 hours suggests that meaningful participation is concentrated among a relatively small group of motivated individuals. From a policy perspective, this pattern implies that marine debris programs in Batam should be designed to minimize individual time burdens while maximizing collective impact. Practical programmatic responses include organizing short, regular clean-up activities, integrating participation into weekend or community events, and embedding marine debris actions into existing neighborhood or school-based programs to sustain broad engagement without requiring excessive time commitments. The significant influence of age on participation further highlights the need for age-responsive engagement strategies. In practice, this finding supports prioritizing youth-centered programs through schools, universities, and digital platforms, where environmental awareness and time flexibility tend to be higher. At the same time, the absence of significant associations for gender, occupation, and income indicates that participation is not structurally limited by socio-economic status. This reinforces the relevance of non-monetary policy instruments, such as volunteer-based monitoring, citizen science initiatives, and community reporting mechanisms, as inclusive tools for marine debris management across diverse population groups.

The results point to the early-stage nature of Batam's adaptive capacity. While community members recognize the urgency of marine debris problems, their willingness to contribute time remains constrained by uneven access to knowledge, limited coastal waste infrastructure, and weak organizational and institutional support. Translating WtP into effective action therefore requires governance interventions that reduce participation costs and enhance enabling conditions, including improved coastal waste facilities, clearer coordination between municipal agencies and port authorities, and sustained environmental education efforts. In this regard, the study provides a valuable empirical baseline for designing structured participation programs that align community time availability with institutional capacity.

Several limitations must be acknowledged. First, the relatively small sample size of 70 respondents, while appropriate for exploratory analysis, limits the generalizability of the findings. Second, the use of accidental sampling constrains representativeness, as respondents were selected based on availability rather than probabilistic methods. Third, although the open-ended CVM format effectively captures the full range of participation hours, it is sensitive to extreme values and may generate skewed distributions, as observed in this study. Finally, the analysis relies on descriptive statistics and simple regression and does not account for contextual dynamics such as seasonal variation, tourism intensity, or cross-border waste flows, despite Batam's exposure to transboundary marine debris in the Singapore Strait. Building on this initial assessment, future research should employ larger and more diverse samples to improve statistical power and representativeness and extend data collection across seasons and additional coastal zones. The next phase of this research will apply a choice experiment (CE) approach to estimate preference structures and trade-offs among marine debris management attributes, including waste infrastructure, environmental education, green port initiatives, community incentives, and enforcement mechanisms. Integrating CE with time-based WtP will allow for more robust policy simulations and the estimation of marginal values for both monetary and non-monetary contributions. Future studies should also incorporate psychological and behavioral factors, such as environmental attitudes, risk perception, and social norms, which may influence participation more strongly than demographic characteristics alone.

Funding

This research was funded by the Directorate General of Higher Education, Research, and Technology (Direktorat Jenderal Pendidikan Tinggi, Riset, dan Teknologi – DIKTI), Ministry of Education, Culture, Research, and Technology of the Republic of Indonesia, through the 2025 DIKTI Research Grant. The study was conducted under the Rector's Decree of Universitas Pertamina (No. 241/UPER-R/SK/HK.01/VII/2025).

Author Contributions

I. W. K. Suryawan contributed to the conceptualization of the study, research design, methodological development, data interpretation, and manuscript drafting and revision. N. Ulhasanah contributed to the development of the conceptual framework, data validation, and interpretation of results. M. M. Sari contributed to data collection, preliminary analysis, and manuscript preparation. E. S. Sofiyah contributed to survey design, field coordination, and data verification. A. Rahman contributed to methodology design, statistical analysis, supervision of data processing, and critical review and editing of the manuscript. All authors have read and approved the final version of the manuscript.

Data Availability Statement

The data supporting the findings of this study are available from the corresponding author upon reasonable request. The data are not publicly available due to ethical considerations related to respondent confidentiality.

Acknowledgments

The authors extend their sincere appreciation to the Directorate General of Higher Education, Research, and Technology (Direktorat Jenderal Pendidikan Tinggi, Riset, dan Teknologi – DIKTI), Ministry of Education, Culture, Research, and Technology of the Republic of Indonesia, for providing financial support through the 2025 DIKTI Research Grant. This study was carried out under the Rector's Decree of Universitas Pertamina (No. 241/UPER-R/SK/HK.01/VII/2025). The authors also thank all respondents and local stakeholders in Batam City for their participation and support during the data collection process.

Declaration of Competing Interest

The authors declare that they have no known competing financial interests or personal relationships that could have appeared to influence the work reported in this paper.

References

1. F. Ayala, K. Castillo-Morales, and S. Cárdenas-Alayza, Mar. Pollut. Bull. **174**, 113281 (2022)
2. G. G. N. Thushari and J. D. M. Senevirathna, Heliyon **6**, e04709 (2020)
3. M. R. Cordova, T. Purbonegoro, R. Puspitasari, R. Subandi, M. T. Kaisupy, S. P. A. Wibowo, Nurjamin, Suparmo, and S. Sapulete, J. Ecol. Eng. **22**, 131 (2021)
4. Y. Kuang and B. Lin, Sustain. Cities Soc. **67**, 102741 (2021)
5. T. R. Walker, E. McGuinty, S. Charlebois, and J. Music, Humanit. Soc. Sci. Commun. **8**, 80 (2021)
6. B. Maskey and M. Singh, Environments **4**, (2017)
7. R. L. Ryan, R. Kaplan, and R. E. Grese, J. Environ. Plan. Manag. **44**, 629 (2001)
8. R. K. Turner, J. Mar. Syst. **25**, 447 (2000)
9. J. D. Lau, C. C. Hicks, G. G. Gurney, and J. E. Cinner, Ecosyst. Serv. **35**, 219 (2019)
10. G. Prajati and S. Widiantoro, in *2019 2nd Int. Conf. Appl. Eng.* (IEEE, 2019), pp. 1–6
11. D. P. Safitri, Y. Adicita, N. Ulfah, and A. S. Purba, in *Int. Conf. Soc. Marit. Bord. Area (SHIMBA 2023)* (Atlantis Press, 2023), pp. 21–27
12. W. N. Adger, in *Clim. Chang. Adapt. Capacit. Dev.* (PUBLISHED BY IMPERIAL COLLEGE PRESS AND DISTRIBUTED BY WORLD SCIENTIFIC PUBLISHING CO., 2003), pp. 29–49
13. N. L. Engle, Glob. Environ. Chang. **21**, 647 (2011)
14. P. J. Cohen, S. Lawless, M. Dyer, M. Morgan, E. Saeni, H. Teioli, and P. Kantor, Ambio **45**, 309 (2016)
15. L. W. Robinson and F. Berkes, Glob. Environ. Chang. **21**, 1185 (2011)

Evaluating community priorities for marine debris reduction in Batam City using importance–performance analysis: A preliminary study

Ari Rahman[1,2], *Nova* Ulhasanah[1,2,3], *Mega Mutiara* Sari[1,2,3], *Evi Siti* Sofiyah[1,2], and *I Wayan Koko* Suryawan[1,2,3,4*]

[1] Department of Environmental Engineering, Faculty of Infrastructure Planning, Universitas Pertamina, Jalan Sinabung II, Terusan Simprug, Jakarta, 12220, Indonesia

[2] Center for Environmental Solution (CVISION), Universitas Pertamina, Jalan Sinabung II, Terusan Simprug, Jakarta, 12220, Indonesia

[3] Graduate Program of Science in Sustainability, Faculty of Infrastructure Planning, Universitas Pertamina, Jalan Sinabung II, Terusan Simprug, Jakarta, 12220, Indonesia

[4] Center for Interdisciplinary Research on Ecology and Sustainability, College of Environmental Studies and Oceanography, National Dong Hwa University, Hualien 97401, Taiwan, ROC

*Corresponding author: i.suryawan@universitaspertamina.ac.id

Abstract. Marine debris has become an escalating environmental challenge in Batam City, driven by rapid coastal development, port activities, tourism pressure, and waste leakage from surrounding settlements. As part of an early scoping effort, this preliminary study examines community priorities for improving marine debris management using the Importance–Performance Analysis (IPA) framework. Eight key indicators of coastal waste governance: coastal waste bin provision, green port and marine waste transport, marine debris education, cross-border data sharing, community beach clean-ups, participation incentive schemes, marine debris risk information, and enforcement of anti-littering were evaluated by 70 respondents across major coastal areas of Batam in September 2025. Descriptive statistics were used to summarize respondent characteristics, while IPA mapped perceived importance against perceived performance for each indicator. The results reveal several priority gaps, with some indicators rated as highly important yet insufficiently implemented, highlighting areas that require immediate policy attention. Infrastructure availability, environmental education, and regulatory enforcement emerged as relative strengths, whereas risk communication, institutional collaboration, and community engagement showed notable performance challenges. The findings provide an initial evidence base to guide more targeted interventions and support the development of adaptive, community-centered marine debris reduction strategies in Batam City.

1 Introduction

Marine debris has emerged as one of the most persistent and complex environmental challenges confronting coastal cities across Southeast Asia [1]. Batam City, located in the Riau Islands and situated at a critical maritime intersection between Indonesia and Singapore, faces increasing pressures stemming from population growth, tourism expansion, intensive maritime traffic, industrial development, and rapid coastal urbanization. These dynamics contribute to significant volumes of unmanaged waste entering the marine environment, adversely affecting not only ecological systems but also public health, economic activities, and coastal livelihoods [2]. Prior research in Indonesia has indicated that inadequate waste infrastructure, limited public awareness, and weak institutional coordination are core factors driving the accumulation of marine debris in many coastal regions [3]. As an emerging coastal hub, Batam reflects these national challenges while also possessing unique socio-ecological conditions that necessitate localized assessment. The urgency of addressing marine debris in Batam is heightened by its geographical position within an international maritime corridor. Floating waste in Batam's waters frequently originates from both domestic sources and transboundary flows from surrounding regions. The presence of shipyards, ports, fishing zones, and tourism sites means that various stakeholders engage daily with the coastal environment, each contributing to or being affected by marine debris in distinct manners. This complexity underscores the necessity for governance approaches that are not only technically sound but also socially inclusive and responsive to community perceptions.

Despite the growing concern regarding marine debris, empirical studies examining community-level perspectives on marine debris governance in Batam remain scarce. Much of the existing literature focuses on ecological impacts, waste composition, marine spatial planning, or large-scale waste management strategies. However, few studies investigate how local residents perceive key governance indicators such as infrastructure availability, green port initiatives, environmental education, institutional cooperation, or enforcement mechanisms. Understanding these perceptions is critical, as community support and involvement can significantly influence the efficacy of coastal waste management interventions [4]. Without the incorporation of local views, programs may fail to align with public expectations or may not achieve long-term sustainability [5]. A notable gap in the current body of research is the absence of structured tools to identify which aspects of marine debris management are deemed most important by residents and how well these aspects are currently being implemented. Prior research has not systematically mapped the relationship between perceived importance and perceived performance of marine debris governance indicators in Batam. Consequently, policymakers lack a practical evidence base to prioritize interventions and allocate resources more effectively. This gap indicates the necessity for diagnostic tools that can swiftly elucidate areas of strategic weakness, community dissatisfaction, or governance mismatch.

The novelty of the present study resides in its application of the Importance–Performance Analysis (IPA) framework, which has been extensively utilized in tourism, public services, and environmental management, yet has rarely been employed to assess marine debris governance in Indonesian coastal cities. By applying IPA, this study introduces a structured and visually intuitive method for identifying priority indicators that require immediate improvement, as well as indicators that should be maintained, minimized, or redesigned [6,7]. Furthermore, unlike many ecological assessments, this study prioritizes community perceptions, acknowledging that successful coastal management depends not solely on infrastructure and policy but also on local engagement and social acceptance. Another innovative aspect is the inclusion of eight governance indicators specifically tailored to the Batam context: coastal waste bin provision, green port and marine waste transport, marine debris education, cross-border data sharing, community beach clean-ups, participation incentive schemes, marine debris risk information, and enforcement of anti-littering regulations. These indicators reflect a comprehensive socio-institutional approach that transcends traditional "clean-up only" perspectives. They encapsulate the multi-dimensional nature of marine debris governance, encompassing infrastructure, education, incentives, regulation, and international cooperation.

2 Method

This study employed a preliminary quantitative survey design to understand community perceptions of marine debris management in Batam City, Indonesia. The overall purpose of the study was to identify how residents evaluate a set of governance-related indicators that are essential for improving coastal cleanliness, strengthening community-based environmental management, and supporting adaptive capacity in coastal areas. Marine debris has become a persistent challenge in Batam due to high maritime traffic, tourism activity, inter-island mobility, and waste leakage from residential and industrial zones. To capture early insights into community responses, the research adopted the Importance–Performance Analysis (IPA) approach, which enables the identification of priority indicators that require immediate intervention [8-12].

The context for this study includes eight indicators selected through literature review, field observation, and local policy relevance (Table 1). These indicators were coastal waste bin provision, green port and marine waste transport, marine debris education, cross-border data sharing, community beach clean-ups, participation incentive schemes, marine debris risk information, and enforcement of anti-littering regulations. The selection of these indicators reflects multi-dimensional considerations of marine debris governance, including infrastructure readiness, institutional initiatives, community engagement, regional cooperation, communication strategies, and regulatory enforcement. The descriptions and operational definitions of all indicators used in this study were adapted directly from the indicator table provided in Table 1. Because this was a preliminary assessment, the study aimed to generate baseline understanding rather than test complex theoretical models.

Data collection took place in September 2025, covering several coastal zones in Batam, namely Nongsa, Marina, Tanjung Pinggir, Barelang, and Batu Ampar. These sites were selected because they represent diverse coastal typologies touristic beaches, local residential shorelines, fishing areas, and economic zones connected to port activities. Each site faces different forms of marine debris pressures, making them suitable for capturing a wide range of perceptions. The field team used accidental sampling, approaching individuals present in coastal public spaces, fishing docks, small business strips, and local community clusters. Although non-probability sampling, accidental sampling was considered appropriate for an exploratory study focusing on rapid data acquisition and early-stage indicator validation.

A total of 70 respondents participated in the survey. Respondents included community residents, fishermen, informal sector workers, formal employees, traders, students, and housewives. The diversity of participants ensured that the data reflected the perspectives of individuals with different levels of coastal interaction and environmental exposure. Prior to data collection, the questionnaire was reviewed by subject-matter experts to ensure clarity, relevance, and comprehension, especially for respondents with varying educational backgrounds.

Table 1. Description of marine debris management indicators used in the study

No.	Indicators	Description
1	Coastal waste bin provision	Measures the availability and adequacy of coastal waste bins, garbage sacks, and collection points installed across Batam's major beaches (Nongsa,

		Barelang, Batu Ampar). This indicator captures infrastructure readiness by the Sanitation Department to reduce shoreline littering and improve on-site waste handling.
2	Green port & marine waste transport	Refers to the establishment of Batu Ampar Port as Indonesia's first green port and the provision of dedicated marine waste-collection vessels serving surrounding small islands. This indicator reflects institutional investment in pollution control technologies and port-based waste retrieval operations.
3	Marine debris education	Assesses educational initiatives delivered by environmental educators, universities, local NGOs, and coastal community groups through social media campaigns, school programs, community workshops, and outreach activities to increase public understanding of marine debris management.
4	Cross-border data sharing	Measures the use of international marine debris databases (e.g., UNEP, IOC) and collaborative research between Indonesia and Singapore for data exchange, monitoring, hotspot identification, and regional knowledge dissemination regarding transboundary marine litter flows.
5	Community beach clean-up	Represents routine weekend-based coastal clean-up programs organized in areas such as Nongsa, Marina, and Tanjung Pinggir, where local communities, youth groups, and volunteers engage in shoreline waste removal linked with eco-tourism and environmental awareness events.
6	Participation incentive scheme	Evaluates CSR-supported incentive schemes (rewards, vouchers, cashback) provided by industrial zones (e.g., Batamindo) and port operators to encourage residents and fishermen to participate in beach clean-ups and marine litter reduction programs.
7	Marine debris risk information	Measures the extent of information dissemination concerning the health risks, tourism losses, and cross-border diplomatic tensions caused by marine debris. This indicator focuses on public communication materials, campaigns, and hazard warnings.
8	Enforcement of anti-littering	Captures the consistency and strictness of law enforcement toward illegal dumping at sea by communities, fishermen, coastal industries, and tourism operators. Includes monitoring, penalties, and coastal patrol activities.

The questionnaire was structured into three sections. The first section covered demographic characteristics such as gender, age group, occupation, and income. This information was essential for determining whether demographic factors influenced perception patterns within the community. The second section contained importance ratings for each indicator. Respondents were asked to evaluate how important each marine debris governance indicator was for improving coastal environmental conditions. The third section asked respondents to assess the current performance of the same indicators in Batam. Both importance and performance were measured using a five-point Likert scale ranging from 1 (very low) to 5 (very high). This symmetry of measurement allowed for direct comparison between perceived importance and perceived implementation. Face-to-face data collection ensured that respondents fully understood each item being evaluated. This was especially important because several indicators, such as cross-border data sharing or green port initiatives, may be unfamiliar to some community members. Enumerators provided brief explanations when necessary but avoided influencing responses. Participation was voluntary, and respondents were informed that their answers would remain anonymous. No personal identifying information was collected.

After completing data collection, responses were coded and entered into a statistical software package for analysis. Descriptive statistics, including means, standard errors, and standard deviations, were computed to summarize respondent evaluations and demographic distributions. IPA was then applied by plotting each indicator onto a two-dimensional grid. Indicators with high importance and low performance were considered priority areas requiring immediate attention, while those with high importance and high performance were regarded as strengths that should be maintained [8]. Indicators falling into low-importance quadrants were interpreted accordingly but with caution due to the exploratory nature of the study.

Chi-square tests were conducted to examine whether demographic variables were associated with significant differences in importance and performance ratings. This analysis provided insight into the potential variation in perceptions among different groups, such as younger versus older respondents or fishermen compared with office workers. These statistical procedures ensured that the findings were based on systematic analysis rather than anecdotal impressions. Throughout the study, ethical considerations were strictly followed. All respondents participated willingly after receiving a clear explanation of the study's purpose. No financial incentives were provided, and confidentiality was fully guaranteed. The study was conducted as an initial step to support more comprehensive research on adaptive marine debris governance in Batam and to provide evidence for designing future community-based interventions.

3 Results and discussion

The demographic composition of the respondents reflects a diverse segment of Batam's coastal community, with characteristics that help contextualize the perception patterns captured in this study (Table 2). A total of 70 individuals participated in the survey, consisting of 65.7% male (n = 46) and 34.3% female (n = 24), indicating that men were more frequently encountered in accessible coastal and public activity areas during the September 2025 data collection period. In terms of age, the majority of respondents were relatively young, with 55.7% falling within the 18–29 year range (n = 39), followed by 17.1% aged 40–49 years (n = 12), 14.3% aged 50 years or older (n = 10), and 12.9% aged 30–39 years (n = 9). The occupational structure further illustrates the heterogeneity of coastal users in Batam. Formal-sector employees constituted the largest group at 24.3% (n = 17), while other categories such as housewives (12.9%), students (11.4%), manual laborers (8.6%), traders (5.7%), fishermen (2.9%), and farmers/laborers (2.9%) were also represented. A substantial portion, 37% (n = 26), categorized themselves under "others," indicating a mix of informal workers, small business owners, and residents with varied economic engagements. This demographic distribution provides important contextual grounding for interpreting perceptions of marine debris initiatives across age, gender, and occupational groups.

Table 2. Demographic Characteristics of Respondents (n = 70)

Category	Sub-category	n	%
Gender	Male	46	65.7
	Female	24	34.3
Age Group	18–29 years	39	55.7
	30–39 years	9	12.9
	40–49 years	12	17.1
	≥ 50 years	10	14.3
Occupation	Others	26	37
	Employee/Office worker	17	24.3
	Housewife	9	12.9
	Student	8	11.4
	Manual labour (Buruh)	6	8.6
	Trader	4	5.7
	Fisherman	2	2.9
	Farmer/Labourer	2	2.9

The assessment of marine debris management indicators elucidates distinct priorities as perceived by the surveyed community (Table 3). Respondents ascribed the highest importance score to green port and marine waste transport, with a mean rating of 4.500 (SE = 0.099; SD = 0.830), underscoring the public's strong recognition of the critical role that port operations and waste-collection vessels play in mitigating marine debris. This was closely followed by coastal waste bin provision, which attained an important score of 4.457 (SE = 0.113; SD = 0.943), and marine debris education, yielding an importance mean of 4.429. These metrics suggest that accessible infrastructure and public awareness are regarded as fundamental components of coastal waste governance.

However, performance ratings indicate significant gaps in implementation. Marine debris risk information was perceived as the best-performing indicator, with a mean score of 3.229 (SE = 0.157; SD = 1.310), implying that communication efforts regarding environmental risks are relatively prominent within the community. Conversely, indicators such as green port and marine waste transport, as well as cross-border data sharing, received considerably lower performance scores of 2.857, reflecting community concerns that these institutional or cross-jurisdictional initiatives remain limited or inadequately recognized at the local level. Coastal waste bin provision exhibited moderate performance at 3.214, indicating partial implementation while still leaving substantial room for enhancement. Other indicators, including community beach clean-ups (3.071) and enforcement of anti-littering regulations (3.057), were assessed as fair but fell below optimal standards. Collectively, the disparity between high importance ratings and lower performance values highlights a pressing need for more coordinated, resource-supported, and community-integrated initiatives to enhance marine debris management in Batam.

Table 3. Importance–performance scores for marine debris management indicators in Batam City

Indicator	Importance (Mean; SE; SD)	Imp. Rank	Performance (Mean; SE; SD)	Perf. Rank
Coastal waste bin provision	4.457 (0.113; 0.943)	2	3.214 (0.169; 1.413)	2
Green port & marine waste transport	4.500 (0.099; 0.830)	1	2.857 (0.162; 1.354)	7
Marine debris education	4.429 (0.107; 0.894)	3	3.100 (0.167; 1.395)	3
Cross-border data sharing	4.300 (0.121; 1.012)	7	2.857 (0.168; 1.407)	7
Community beach clean-up	4.357 (0.112; 0.933)	5	3.071 (0.161; 1.344)	4
Participation incentive scheme	4.214 (0.139; 1.166)	8	2.900 (0.163; 1.364)	6
Marine debris risk information	4.314 (0.125; 1.043)	6	3.229 (0.157; 1.310)	1
Enforcement of anti-littering	4.414 (0.114; 0.955)	4	3.057 (0.153; 1.284)	5

Figure 1 illustrates the results of the IPA for eight governance indicators pertinent to marine debris management in Batam City, derived from standardized (z-score) values of importance and performance. By centering both axes at the grand means, the IPA framework facilitates a relative comparison of community expectations against perceived implementation outcomes [9,10]. This methodology emphasizes structural imbalances in governance performance rather than assessing absolute effectiveness, rendering it particularly adept at identifying priority intervention areas in preliminary assessments. The IPA results demonstrate a distinct differentiation between indicators that yield visible, community-oriented outcomes and those that are institutionally focused yet less perceptible at the local level. Indicators situated in the *Keep Up the Good Work* quadrant specifically, coastal waste bin provision, marine debris education, and enforcement of anti-littering regulations represent governance actions that residents highly value and perceive as functioning effectively. These indicators share a crucial characteristic: they produce tangible and immediately observable benefits, including cleaner beaches, clearer behavioral guidance, and a visible regulatory presence. Their positioning suggests that investments in physical infrastructure, awareness-raising, and enforcement have achieved a level of maturity sufficient to meet baseline community expectations. Nonetheless, the maintenance of these gains necessitates sustained institutional commitment; any decline in service quality or enforcement visibility could rapidly undermine public confidence and participation.

Conversely, the *Concentrate Here* quadrant underscores the most significant governance gaps, encompassing green port initiatives and marine waste transport systems. These indicators are regarded as highly important due to Batam's strategic role as a major port city and its vulnerability to transboundary marine debris flows in the Singapore Strait. The low performance scores associated with these indicators imply that, although institutional programs may exist, their outcomes are not sufficiently visible or effective from the community's perspective. This discrepancy suggests a disconnection between upstream, port-based interventions and the downstream coastal conditions experienced by residents. Analytically, this points to a coordination challenge rather than a deficiency in policy intent. Enhancing inter-agency collaboration, improving operational efficiency, and increasing transparency regarding port-related waste management outcomes could facilitate the translation of institutional investments into tangible coastal improvements and restore public confidence in these initiatives.

The *Low Priority* quadrant includes community beach clean-ups, participation incentive schemes, and cross-border data sharing. Despite the frequent emphasis on these measures in policy discourse, their placement indicates that the public currently perceives them as peripheral to effective marine debris management. While community clean-ups hold symbolic significance, they may be perceived as reactive and inadequate in addressing systemic waste inflows. Participation incentive schemes appear to exert limited motivational influence, suggesting that extrinsic rewards are less effective than intrinsic motivations such as environmental concern and civic responsibility. Cross-border data sharing, although pertinent to transboundary pollution, is likely regarded as remote and technical, operating beyond the immediate awareness of local communities. This reflects a visibility and communication gap rather than an absence of importance at the institutional level, indicating that these initiatives may require reframing to underscore their relevance to local environmental outcomes. The *Possible Overkill* quadrant encompasses marine debris risk information, characterized by relatively strong performance yet comparatively lower perceived importance. This suggests that risk communication efforts, such as informational signage, warnings, or public campaigns, are present and functioning, but may not fully align with community priorities. From an analytical perspective, this does not imply that risk information is unnecessary; rather, it indicates diminishing returns from standalone communication strategies. Enhancing the effectiveness of risk information may necessitate closer integration with actionable measures, such as linking health risks to specific waste sources or coupling informational campaigns with infrastructure improvements and enforcement actions.

The IPA results reveal a governance structure in which community-facing interventions are generally more successful than institutionally complex or transboundary measures. The analysis emphasizes the necessity of rebalancing governance efforts by fortifying high-importance but underperforming domains, particularly in relation to port-related waste management and marine transport systems, while sustaining effective practices in infrastructure provision, education, and enforcement. By elucidating both performance gaps and perceptual mismatches, Figure 1 provides a coherent analytical foundation for prioritizing policy interventions and enhancing Batam's adaptive capacity in managing marine debris.

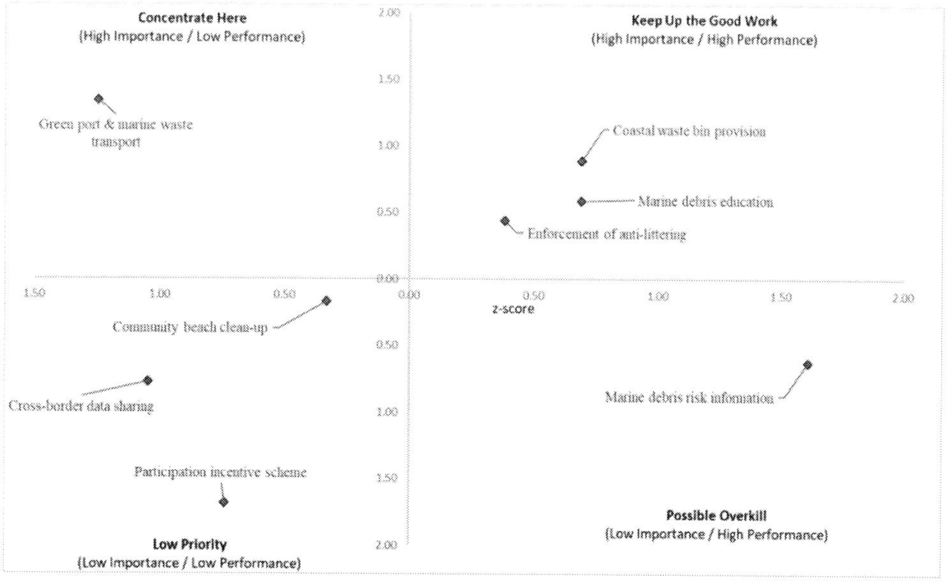

Fig. 1. IPA quadrant grid based on standardized (Z-Score) importance and performance values

The "Low Priority" quadrant, characterized by low importance and low performance, includes community beach clean-ups, participation incentive schemes, and cross-border data sharing. Community beach clean-up programs, with an important score of 4.357 but performance score of 3.071, lies close to the quadrant boundary, suggesting that while respondents recognize their value, these activities are not viewed as urgent compared to infrastructure- or enforcement-based approaches. Participation incentive schemes (importance 4.214, performance 2.900) appear further from the center, suggesting that the public does not prioritize financial or reward-based motivation as strongly as other strategies. This aligns with findings from coastal community studies showing that intrinsic motivations such as environmental concern, stewardship, and community identity often outweigh extrinsic incentives [13,14]. The placement of cross-border data sharing (importance 4.300, performance 2.857) indicates moderate recognition of its relevance; however, its technical nature may make it less visible to the general public. Since transboundary waste dynamics between Indonesia and Singapore are complex and often managed at institutional levels, respondents may not fully understand or directly perceive these processes. The "Possible Overkill" quadrant reflects low importance but high performance. Here, green port and marine waste transport, with an important score of 4.500 but a performance score of 2.857, appears unexpectedly in this region due to the relative placement of the grand means. Although respondents considered this indicator to be important, its performance is perceived lower relative to others, causing its position to appear within or near the quadrant boundary. This placement suggests a mismatch: institutional initiatives such as port greening and marine waste-collection boats are visible but may not directly translate into improvements that residents experience on the coastline. It is possible that the community perceives port-based waste management as an institutional matter that does not immediately address household-level or beach-level waste concerns. This type of mismatch has been observed in other coastal governance studies where institutional investments fail to generate direct public satisfaction [15].

The IPA results highlight the need for targeted improvement, particularly in areas that the community deems important but insufficiently implemented. Strengthening risk communication, expanding infrastructure provisioning, enhancing educational programs, and ensuring consistent enforcement represent high-priority strategies. Meanwhile, less urgent indicators such as incentive schemes and technical cross-border cooperation may be improved gradually or supported indirectly through increased public awareness. The preliminary findings underscore the necessity of multi-level collaboration between government, port authorities, NGOs, and communities to develop an adaptive, inclusive, and sustainable marine debris management system for Batam.

4 Conclusion, limitation, and future research

This preliminary study provides early insights into community perceptions of marine debris management in Batam City by evaluating eight key governance indicators through Importance–Performance Analysis (IPA). The results highlight that residents generally view infrastructure, education, and enforcement as highly important components of effective coastal waste governance. While several indicators such as coastal waste bins, anti-littering enforcement, and public education are perceived to perform reasonably well, others show clear gaps between importance and implementation. Risk communication, community programs, and several institutional initiatives require greater attention to meet community expectations. Overall, the findings emphasize the need for stronger coordination between government agencies, port authorities, industries, NGOs, and local communities to develop a more adaptive and community-centered marine debris management system in Batam.

This study has several limitations that should be acknowledged. First, the sample size was relatively small (n = 70) and collected using accidental sampling, which limits the generalizability of the results beyond the surveyed communities. Second, the indicators used in the IPA were adapted for preliminary assessment and may not fully capture the complexity of Batam's marine debris governance landscape. Third, responses relied on self-reported perceptions, which may be influenced by respondent awareness, personal experiences, or seasonal variations in marine debris conditions. Lastly, the study did not include direct environmental measurements or observational data, which could provide stronger triangulation with community perceptions.

Future studies should expand the sample size and adopt probability-based sampling methods to improve representativeness across Batam's coastal districts. Additional indicators may be incorporated to reflect broader governance dimensions such as funding mechanisms, stakeholder coordination, and behavior change interventions. Mixed-method approaches, including in-depth interviews, focus groups, and field observations could offer deeper insight into the drivers of community perceptions and institutional performance. Longitudinal research would also be valuable for monitoring changes over time, particularly as Batam implements new policies related to port environmental management and transboundary waste issues. Finally, comparative studies involving nearby islands or cross-border cooperation with Singapore could further enrich understanding of regional marine debris dynamics.

Author Contributions

A.R. contributed to the study conceptualization, methodological design, statistical analysis, and interpretation of results, and provided critical revisions to the manuscript. N.U. contributed to the development of the conceptual framework, data validation, and interpretation of findings. M.M.S. contributed to data collection, preliminary data processing, and manuscript preparation. E.S.S. contributed to survey instrument development, field coordination, and data verification. I.W.K.S. contributed to the overall study design, supervised data collection and analysis, led manuscript drafting and revision, and served as the corresponding author. All authors have read and approved the final version of the manuscript.

Funding

This research was funded by the Directorate General of Higher Education, Research, and Technology (Direktorat Jenderal Pendidikan Tinggi, Riset, dan Teknologi – DIKTI), Ministry of Education, Culture, Research, and Technology of the Republic of Indonesia, through the 2025 DIKTI Research Grant. The study was conducted under the Rector's Decree of Universitas Pertamina (No. 241/UPER-R/SK/HK.01/VII/2025).

Data Availability Statement

The data supporting the findings of this study are available from the corresponding author upon reasonable request. The data are not publicly available due to ethical considerations and the need to protect respondent confidentiality.

Ethics Approval and Consent to Participate

This study involved human participants and was conducted in accordance with ethical research standards. Participation was voluntary, informed consent was obtained from all respondents prior to data collection, and no personally identifiable information was recorded.

Declaration of Competing Interest

The authors declare that they have no known competing financial interests or personal relationships that could have appeared to influence the work reported in this paper.

Acknowledgments

The authors gratefully acknowledge the Directorate General of Higher Education, Research, and Technology (Direktorat Jenderal Pendidikan Tinggi, Riset, dan Teknologi – DIKTI), Ministry of Education, Culture, Research, and Technology of the Republic of Indonesia, for financial support through the 2025 DIKTI Research Grant. This study was carried out under the Rector's Decree of Universitas Pertamina (No. 241/UPER-R/SK/HK.01/VII/2025). The authors also thank all respondents and local stakeholders in Batam City for their valuable participation and cooperation during the field survey.

References

1. L. C. M. Omeyer, E. M. Duncan, K. Aiemsomboon, N. Beaumont, S. Bureekul, B. Cao, L. R. Carrasco, S. Chavanich, J. R. Clark, M. R. Cordova, F. Couceiro, S. M. Cragg, N. Dickson, P. Failler, G. Ferraro, S. Fletcher, J. Fong, A. T. Ford, T. Gutierrez, F. Shahul Hamid, J. G. Hiddink, P. T. Hoa, S. I. Holland, L. Jones, N. H. Jones, H. Koldewey, F. M. Lauro, C. Lee, M. Lewis, D. Marks, S. Matallana-Surget, C. G. Mayorga-Adame, J. McGeehan, L. F. Messer, L. Michie, M. A. Miller, Z. F. Mohamad, N. H. M. Nor, M. Müller, S. P. Neill, S. E. Nelms, D. F. L. Onda, J. J. L. Ong, A. Pariatamby, S. C. Phang, R. Quilliam, P. E. Robins, M. Salta, A. Sartimbul, S. Shakuto, M. W. Skov, E. B. Taboada, P. A. Todd, T. C. Toh, S. Valiyaveettil, V. Viyakarn, P. Wonnapinij, L. E. Wood, C. L. X. Yong, and B. J. Godley, Sci. Total Environ. **841**, 156704 (2022)
2. F.-C. Mihai, S. Gündoğdu, L. A. Markley, A. Olivelli, F. R. Khan, C. Gwinnett, J. Gutberlet, N. Reyna-Bensusan, P. Llanquileo-Melgarejo, C. Meidiana, S. Elagroudy, V. Ishchenko, S. Penney, Z. Lenkiewicz, and M. Molinos-Senante, Sustainability **14**, (2022)
3. S. D. A. Smith and E. Bernal, Ocean Coast. Manag. **202**, 105433 (2021)
4. J. Jambeck, R. Geyer, C. Wilcox, T. R. Siegler, M. Perryman, A. Andrady, R. Narayan, and K. L. Law, Mar. Pollut. **347**, 768 (2015)
5. Y. Rydin and M. Pennington, Local Environ. **5**, 153 (2000)
6. L. Barth, L. Schweiger, R. Benedech, and M. Ehrat, Decis. Anal. J. **9**, 100347 (2023)
7. A. Taghipour, A. Padash, V. Etemadi, M. Khazaei, and S. Ebrahimi, Sustainability **16**, (2024)
8. J. A. Martilla and J. C. James, J. Mark. **41**, 77 (1977)
9. J. A. Durlak and E. P. DuPre, Am. J. Community Psychol. **41**, 327 (2008)
10. E. Frauman and S. Banks, Tour. Manag. **32**, 128 (2011)
11. M. L. Tseng, Environ. Monit. Assess. **172**, 171 (2011)
12. A. K. Das, J. Mukherjee, and U. Chatterjee, Innov. Infrastruct. Solut. **7**, 187 (2022)
13. H. Li, H. Liu, and W.-Y. Chang, Environ. Manage. **75**, 1520 (2025)
14. A. J. Dean, K. Uebel, T. Schultz, K. S. Fielding, E. Saeck, H. Ross, and V. Martin, People Nat. **6**, 1452 (2024)
15. O. Aburto-Oropeza, V. Platzgummer, E. M. Ferrer, C. López-Sagástegui, R. de G. A. Mirabent, A. Ávalos Galindo, F. Favoretto, A. Giron-Nava, I. Mendoza Camacho, and C. Núñez Sañudo, Front. Mar. Sci. **11**, 1491483 (2025)

Comparative analysis of shoreline dynamics in Surabaya and Sidoarjo under climate variability

Eko Hadi Santoso[1,2*] *and Lalu Muhamad* Jaelani[1]
[1]*Department of Geomatics Engineering, FTSPK, ITS, Surabaya, East Java, Indonesia*
[2]*Maritim Meteorological Station of Tanjung Perak, BMKG, Surabaya, East Java, Indonesia*
**Corresponding author: eko.santoso@bmkg.go.id*

Abstract. This study investigates the influence of hydroclimatic variability on shoreline dynamics along two contrasting coastal systems in East Java, Indonesia, Surabaya's engineered urban coast and Sidoarjo's natural tidal-flat coast, during 2015–2025. Shorelines were extracted from multi-temporal Landsat imagery using NDWI, shoreline change rates were calculated with DSAS, and rainfall data from BMKG were correlated with mean End Point Rate (EPR) to assess hydroclimatic impacts on coastal morphodynamics. Results reveal contrasting shoreline trajectories driven by coastal typology. Surabaya exhibits semi-cyclic shoreline adjustments controlled by reclamation barriers and restricted sediment pathways, resulting in localized accretion despite declining rainfall. In contrast, Sidoarjo shows rainfall-responsive erosion–accretion cycles, where open tidal channels enable rapid sediment redistribution during high-rainfall periods. Rainfall–EPR correlations are strong at the local scale ($r = 0.76$ in Surabaya; $r = 0.80$ in Sidoarjo) but weak when aggregated ($r \approx 0.10$). This indicates that integrating hydroclimatic indicators with geomorphic context does not increase overall correlation strength, but improves interpretation of typology-specific shoreline responses obscured in aggregated analyses. The NDWI–DSAS–rainfall workflow provides a scalable framework for monitoring shoreline dynamics in tropical monsoonal environments.
Keywords: shoreline dynamics; rainfall variability; NDWI; EPR; Landsat imagery; climate variability, coastal management

1. Introduction

Coastal zones represent dynamic interfaces where natural processes and human interventions continuously reshape geomorphic configurations [1]. Among the various indicators of coastal change, shoreline dynamics are particularly significant, reflecting the cumulative effects of hydrodynamic, climatic, and anthropogenic factors over time [2]. Globally, nearly one-quarter of sandy shorelines are retreating due to sea-level rise, storms, and land-use modification [3]). However, in tropical monsoonal regions such as Indonesia, rainfall variability exerts an additional and often dominant control on sediment transport and coastal morphology [4].

Rainfall influences shoreline evolution primarily through its control on sediment delivery via river discharge and surface runoff, which together determine the magnitude of sediment available for coastal deposition or erosion [5]. Intense precipitation enhances sediment supply to estuarine and coastal zones, promoting accretion, while prolonged dry conditions reduce sediment input and increase susceptibility to tidal erosion. However, this relationship has rarely been quantified explicitly, particularly across contrasting coastal typologies.

In Indonesia, Surabaya and Sidoarjo exemplify two distinct coastal systems shaped by differing geomorphic and anthropogenic conditions. Surabaya's urban coastline is characterized by extensive port development, reclamation, and engineered drainage networks that constrain sediment transport [6]. Conversely, Sidoarjo's tidal-flat coast remains more natural, dominated by sediment-laden inflows from the Porong River and tidal mixing within aquaculture zones. These contrasting conditions provide an ideal natural laboratory for examining how climatic variability interacts with coastal morphology under different levels of human modification.

Remote sensing has transformed shoreline monitoring by providing synoptic, repeatable, and cost-effective observations across time [7]. Landsat imagery, with its consistent temporal archive since the 1980s, offers a valuable resource for detecting decadal-scale shoreline trends [8]. This study utilizes Landsat 8 Operational Land Imager (OLI) and Landsat 9 OLI-2 surface reflectance products, selected for their radiometric consistency, moderate spatial resolution (30 m), and suitability for long-term shoreline analysis at regional scales. The Normalized Difference Water Index (NDWI), proposed by McFeeters (1996), enhances spectral contrast between land and water, enabling robust shoreline delineation under variable environmental conditions. When integrated with the Digital Shoreline Analysis System (DSAS) developed by the U.S. Geological Survey [9], quantitative metrics such as the End Point Rate (EPR) can be derived to assess erosion–accretion dynamics over time.

Despite technological advances, tropical coastal studies often focus on sea-level rise or anthropogenic stressors, with limited integration of hydroclimatic parameters like rainfall [10]. As a result, the interaction between rainfall-driven hydrological inputs and coastal morphodynamics in monsoonal environments remains poorly understood, especially where river discharge variability and tidal mixing strongly influence sediment pathways. This gap limits predictive modeling capacity and the formulation of climate-informed management strategies.

This study addresses that gap through an integrative, data-driven framework that combines NDWI-based shoreline extraction, DSAS EPR computation, and rainfall correlation analysis to quantify rainfall-induced shoreline dynamics in East Java. Specifically, this study aims to assess decadal shoreline change rates (2015–2025) in Surabaya and Sidoarjo using multi-temporal Landsat data, evaluate the statistical relationship between rainfall variability and shoreline movement, and derive typology-specific insights to support climate-resilient coastal management.

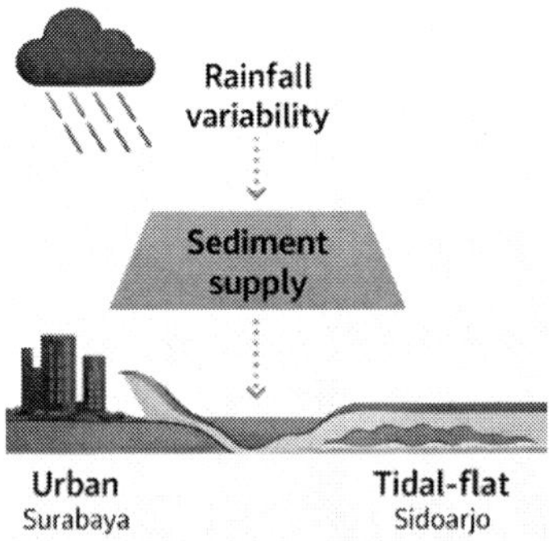

Fig. 1. Conceptual framework linking rainfall variability, sediment supply, and shoreline response in urban (Surabaya) and tidal-flat (Sidoarjo) coastal systems.

The research contributes conceptually and methodologically by embedding rainfall—a climatic proxy—within the geospatial analysis of shoreline change. By contrasting urban and tidal-flat coasts, the study enhances understanding of how hydroclimatic and anthropogenic forces jointly influence tropical coastal evolution. Figure 1 presents the conceptual framework linking rainfall variability, inferred sediment redistribution processes, and shoreline response. Within this framework, rainfall variability is represented by BMKG rainfall records, shoreline response is quantified using NDWI-derived shorelines and End Point Rate (EPR) metrics from DSAS, while sediment-related processes are interpreted conceptually based on spatial erosion–accretion patterns and coastal typology differences.

Through this framework, the paper bridges geomorphological theory, remote sensing analytics, and applied climate adaptation, offering actionable insights for sustainable coastal governance in data-limited tropical regions.

2. Materials and Methods

2.1 Study Area

The study focuses on two contrasting coastal systems located along the northern coast of East Java, Indonesia: Surabaya (urban coast) and Sidoarjo (tidal-flat coast). Both are situated between 7°19′–7°58′S and 112°45′–112°88′E (Figure 2). Surabaya's shoreline has been heavily modified by reclamation, harbor dredging, and industrial expansion, resulting in fragmented sediment pathways. In contrast, Sidoarjo's coast comprises tidal flats and aquaculture ponds that remain geomorphically active, shaped by sediment input from the Porong River and monsoonal tides. The region experiences a tropical monsoon climate with a wide range of annual rainfall (up to ~2,500 mm/year), as reported by BMKG (2024) for Java. BMKG further notes that high precipitation in East Java is linked with large-scale climate drivers, including the Asian–Australian monsoon and ENSO variability [11-12].

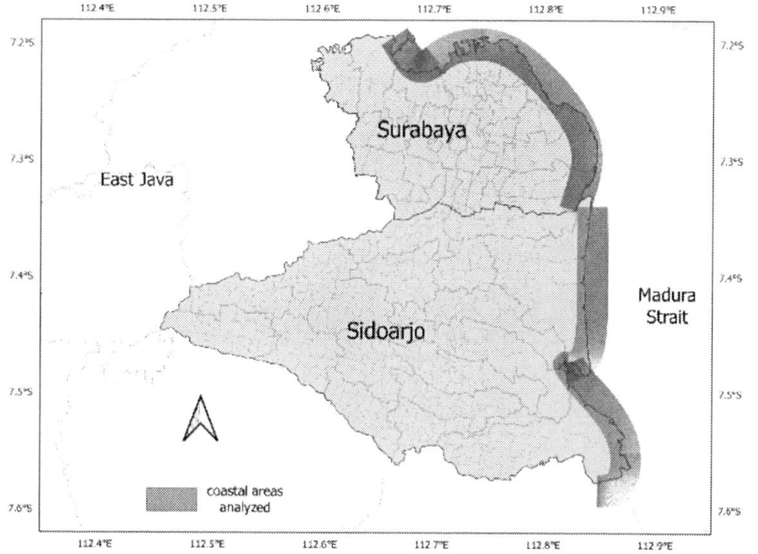

Fig. 2. Study area of Surabaya (urban coast) and Sidoarjo (tidal-flat coast) along the northern coast of East Java, Indonesia.

2.2 Data Acquisition

Table 1. Summary of datasets and their applications.

Data Type	Source	Resolution	Period	Application
Landsat 8 OLI	USGS EarthExplorer	30 m	2015-2021	Shoreline extract
Landsat 9 OLI/TIRS	USGS EarthExplorer	30 m	2022-2025	Shoreline extract
Rainfall data	BMKG (Tanjung Perak, Juanda	Monthly	2015-2025	Hydroclimatic correlation

Three main datasets were used (Table 1):

a. Landsat 8 OLI (2015, 2020) and Landsat 9 OLI/TIRS (2025) imagery were obtained from USGS EarthExplorer at 30 m spatial resolution. The analysis utilized visible and near-infrared spectral bands, specifically Band 2 (Blue), Band 3 (Green), Band 4 (Red), and Band 5 (Near-Infrared), which are commonly employed for shoreline extraction and land–water discrimination. Images were selected under cloud cover less than 10%, based on USGS metadata, and atmospherically corrected prior to analysis.
b. Rainfall data from BMKG's Tanjung Perak (Surabaya) and Juanda (Sidoarjo) stations, representing urban and tidal-flat rainfall regimes from 2015–2025. Monthly rainfall was aggregated into annual totals.
c. Supporting datasets, including shoreline baselines and administrative boundary maps, were obtained from open-access national geospatial sources, specifically the Geospatial Information Agency of Indonesia (Badan Informasi Geospasial, BIG). These datasets were used for georeferencing, spatial alignment, and interpretation of shoreline changes.

2.3 Shoreline Extraction using NDWI

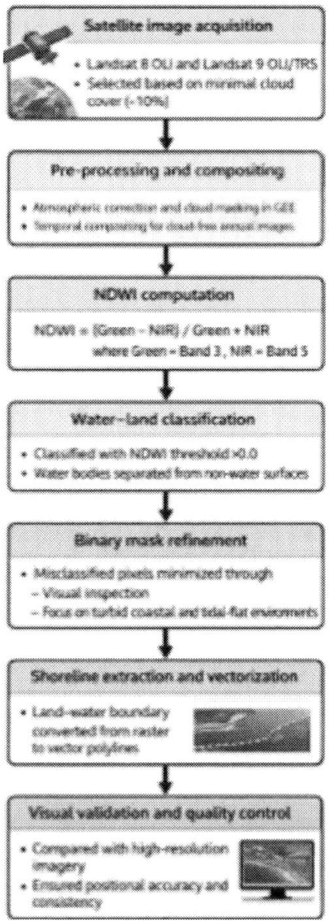

Fig. 3. Detailed workflow of NDWI-based shoreline extraction and validation using Google Earth Engine, including image pre-processing, NDWI thresholding, shoreline vectorization, and visual quality control.

Shoreline positions were delineated using the Normalized Difference Water Index (NDWI) [13]:

$$NDWI = \frac{Green - NIR}{Green + NIR} \quad (1)$$

where Green and NIR represent surface reflectance values from Landsat Band 3 and Band 5. Processing was implemented on Google Earth Engine (GEE), ensuring radiometric consistency and cloud-free composites. Pixels were classified using an NDWI threshold of > 0.0, a commonly adopted criterion for distinguishing water from non-water surfaces in Landsat-based coastal studies, as positive NDWI values indicate the dominance of water reflectance characteristics. The threshold was empirically verified through visual calibration against high-resolution reference imagery to ensure robust land–water separation under varying tidal and atmospheric conditions. The extracted shorelines were visually validated against high-resolution imagery to ensure positional reliability.

2.4 Shoreline Change Rate Analysis

The Digital Shoreline Analysis System (DSAS) v6.0 was used to compute shoreline change rates via the End Point Rate (EPR) method [9]:

$$EPR = \frac{D_2 - D_1}{T} \quad (2)$$

where D_1 and D_2 are shoreline positions (m) at times t_1 and t_2, and T is the elapsed time (years). A baseline was generated parallel to the coast, and transects were spaced every 30 m — totaling 1,015 in Surabaya and 984 in Sidoarjo. EPR values (m/year) were categorized into erosion (negative), accretion (positive), or stable (±0.5 m/year).

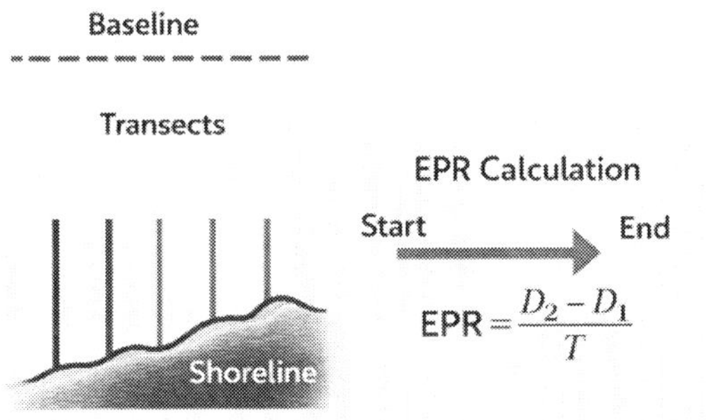

Fig. 4. DSAS analytical framework showing baseline, transects, and EPR calculation workflow.

2.5 Rainfall–Shoreline Correlation Analysis

To quantify the climatic influence, Pearson's correlation coefficient (r) was applied to assess the relationship between annual rainfall totals and mean EPR values (2015–2025):

$$r = \frac{\sum(x_i - \bar{x})(y_i - \bar{y})}{\sqrt{\sum(x_i - \bar{x})^2 \sum(y_i - \bar{y})^2}} \quad (3)$$

where x represents annual rainfall and y represents EPR. Statistical analysis was performed using Excel.

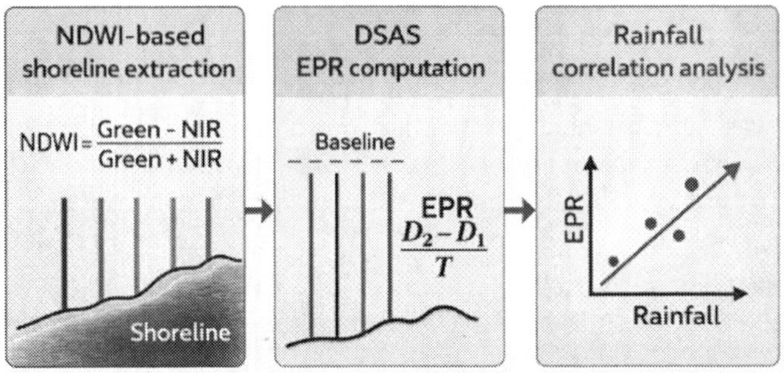

Fig. 5. Research workflow integrating NDWI-based shoreline extraction, DSAS EPR computation, and rainfall correlation analysis.

3. Results and Discussion

3.1 Shoreline Dynamics Overview (2015–2025)

Multi-temporal analysis revealed distinct morphodynamic responses between the two coasts.

- Surabaya: EPR values ranged from −67.5 to +120 m/year (mean = +4.5 m/year). Approximately 28% of transects indicated erosion, 41% accretion, and 31% stability. Erosional hotspots were concentrated near Benowo (Tambak Oso Wilangun), Krembangan (Perak Barat), and Pabean Cantian (Perak Utara), as highlighted in Figure 6, while localized accretion prevailed along Bulak and Sukolilo Baru. The spatial distribution of these erosion–accretion patterns reflects the combined influence of coastal infrastructure development, modified drainage networks, and sediment trapping by nearshore structures.

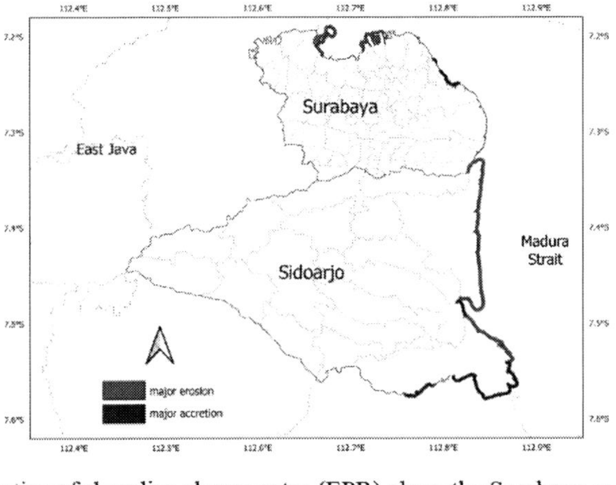

Fig. 6. Spatial distribution of shoreline change rates (EPR) along the Surabaya and Sidoarjo coast from 2015 to 2025.

- Sidoarjo: EPR values ranged from −179 to +189 m/year (mean = −1.6 m/year). Nearly 47% of transects experienced erosion, especially in Sedati (Kalanganyar and Segorotambak) and Jabon (southern Kupang), while 32% accretion occurred in northern Kupang and Kedungpandan, reflecting sediment accumulation within tidal channels but insufficient to offset net coastal retreat.

Surabaya's engineered shoreline promotes localized progradation due to sediment trapping behind coastal structures. In contrast, Sidoarjo's open tidal plain remains highly susceptible to erosion because hydrodynamic energy and reduced sediment supply act more freely across the system. These contrasting morphologies reflect how human modification in Surabaya constrains sediment mobility, while tidal processes in Sidoarjo amplify natural morphodynamic variability [2], [5].

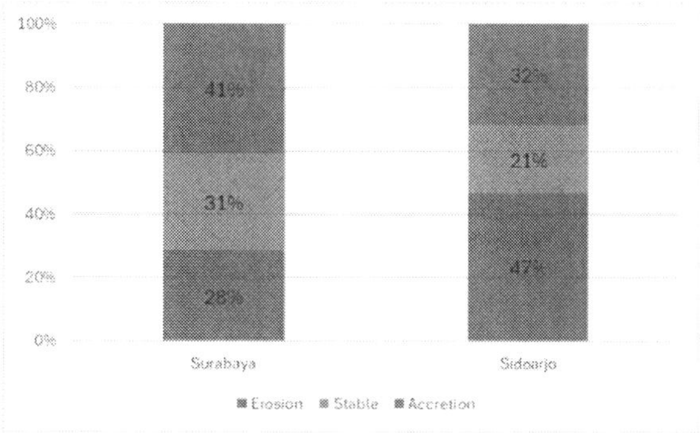

Fig. 7. Comparative distribution of erosion, accretion, and stable shoreline segments in Surabaya and Sidoarjo (2015–2025).

3.2. Rainfall Variability and Shoreline Response

Annual rainfall ranges from 1,371 mm/year (2023–2025) in Surabaya to 2,370 mm/year (2015–2017 and 2019–2020) in Sidoarjo. Correlation analysis revealed coast-specific relationships between rainfall variability and shoreline change, as illustrated in Figure 8. For Surabaya, the scatter distribution shows a clear upward trend, where higher annual rainfall generally corresponds to positive EPR values, indicating net accretion ($r = 0.76$). Similarly, Sidoarjo exhibits a stronger positive association ($r = 0.80$), characterized by a steeper trend and wider dispersion of EPR values under high-rainfall conditions, reflecting the sensitivity of tidal-flat environments to rainfall-driven sediment redistribution.

In contrast, when data from both regions are combined, the scatter points in Figure 8 become widely dispersed around a near-horizontal regression line ($r = 0.10$), indicating the absence of a consistent rainfall–shoreline response across the study area. This divergence highlights the role of local coastal controls, such as urbanized shore protection in Surabaya and tidal-channel dynamics in Sidoarjo, which modulate how rainfall signals are translated into shoreline change.

Periods of high rainfall corresponded with net accretion (2015–2017, 2019–2021), while dry phases (2017–2019, 2023–2025) coincided with erosion. This finding confirms that rainfall variability acts as a hydroclimatic driver modulating sediment redistribution, though its effect is filtered by coastal typology [4].

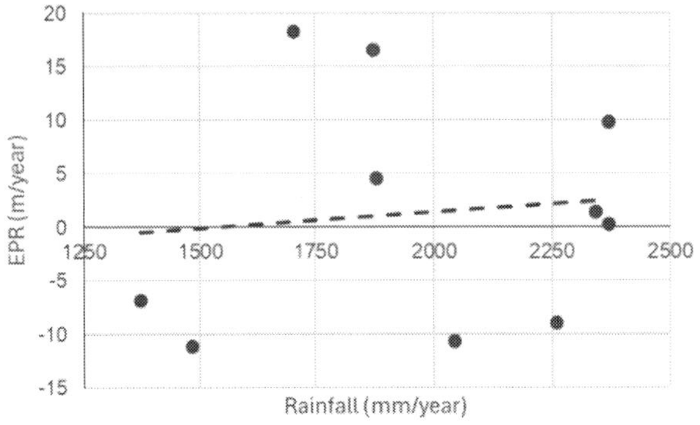

Fig. 8. Scatter plot illustrating the relationship between annual rainfall and shoreline change rates (EPR) for the Surabaya and Sidoarjo coasts during 2015–2025.

In Surabaya, rainfall-driven sediment flux was weakened by engineered drainage and reclamation barriers, leading to localized accretion near confined basins but resulting in overall morphological rigidity. In Sidoarjo, the open tidal system responded more directly—high rainfall enhanced fluvial sediment input, promoting widespread accretion, whereas low rainfall reduced sediment supply, increasing tidal erosion [14]. This spatially divergent behavior explains why rainfall–EPR correlations remain strong locally but lose coherence when both coasts are analyzed together.

3.3. Spatiotemporal Morphodynamic Patterns

Temporal segmentation (2015–2017, 2017–2019, 2019–2021, 2021–2023, 2023–2025) revealed alternating erosion–accretion phases (Figure 9).

- Surabaya exhibited a semi-cyclic behavior, defined here as a recurring but non-periodic alternation between accretion and erosion phases over successive time segments rather than a regular oscillatory cycle. Accretion dominated during 2015–2017 and 2019–2023, whereas erosion prevailed in 2017–2019 and 2023–2025. As shown in Figure 9, these phase transitions coincide with shifts in rainfall regimes and are further modulated by major port reclamation activities along the Surabaya waterfront between 2016 and 2021, which altered sediment pathways and promoted localized sediment trapping behind coastal structures. Consequently, net accretion dominates despite persistent erosion in western sectors.
- Sidoarjo demonstrated a more synchronous response, meaning that shoreline changes occurred consistently and coherently with rainfall variability across the analyzed periods. As illustrated in Figure 9, strong accretion systematically corresponds to high-rainfall phases (2015–2017 and 2019–2023), while pronounced erosion aligns with drier intervals. This close temporal alignment reflects the dominance of tidal-channel–controlled sediment transport and the limited buffering capacity of engineered coastal structures in the Sidoarjo tidal-flat environment.

These results underscore that Surabaya's morphodynamic pattern is primarily engineering-driven, while Sidoarjo's evolution remains rainfall-driven, producing opposite shoreline tendencies—net accretion versus net erosion. These cycles demonstrate that rainfall–sediment coupling is more effective in natural tidal systems than in engineered coasts [5-6].

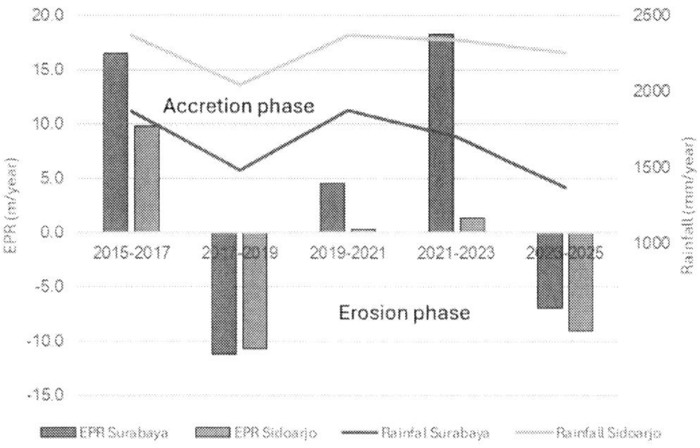

Fig. 9. Temporal evolution of mean EPR and rainfall anomalies showing alternating erosion–accretion phases.

As illustrated in Figure 9, the temporal evolution of mean EPR exhibits a clear alternation between accretion and erosion phases that broadly co-varies with rainfall variability. During 2015–2017, both Surabaya and Sidoarjo experienced positive mean EPR values, coinciding with elevated rainfall, indicating a net accretion phase. This pattern shifted in 2017–2019, when reduced rainfall corresponds to negative mean EPR values at both sites, marking a transition to erosion-dominated conditions.

The accretion phase re-emerged during 2019–2021 and intensified in 2021–2023, particularly in Surabaya, where peak positive EPR values align with above-average rainfall. In contrast, the subsequent period (2023–2025) shows a decline in rainfall accompanied by negative mean EPR values, signaling a renewed erosion phase. The consistent temporal alignment between rainfall peaks and positive EPR, as well as rainfall minima and negative EPR, supports the interpretation of alternating erosion–accretion cycles driven by rainfall-modulated sediment redistribution, with local coastal settings influencing the magnitude of the response.

3.4. Comparative Morphodynamic Interpretation

Table 2. Comparative morphodynamic indicators of Surabaya and Sidoarjo (2015–2025).

Parameter	Surabaya (Urban)	Sidoarjo (Tidal-flat)
Dominant driver	Anthropogenic (reclamation, drainage control)	Hydroclimatic (rainfall, tide, etc)
Mean EPR (m/year)	+4.5	-1.6
Rainfall-EPR Correlation (r)	0.76	0.80
Sediment connectivity	Low (confined by infrastructure)	High (open tidal channels)
Morphodynamic behavior	Localized, delayed response	Immediate, widespread response

The comparative assessment emphasizes that geomorphic openness and human modification jointly determine how rainfall signals manifest in shoreline morphology. As summarized in Table 2, Surabaya shows a net accretion trend, whereas Sidoarjo exhibits net erosion. In the urbanized coast of Surabaya, structural confinement reduces sediment mobility, creating localized accretional zones and overall morphological persistence. In contrast, Sidoarjo's open tidal system permits more immediate responses to hydroclimatic forcing, translating rainfall decline into accelerated erosion.

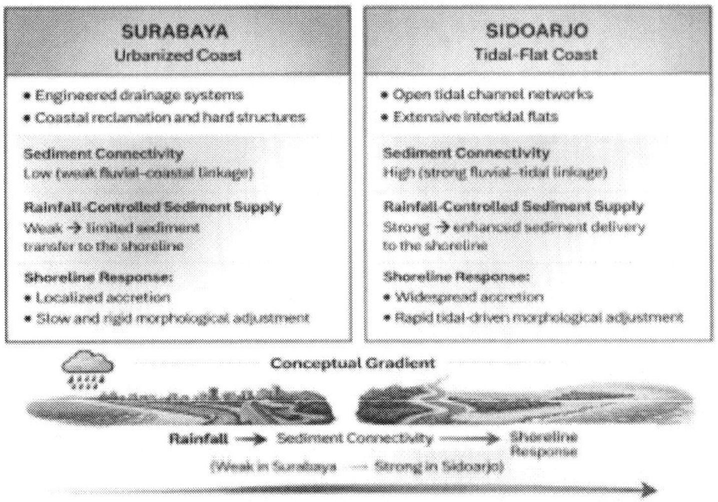

Fig. 10. Conceptual comparison of sediment pathways and rainfall influence in Surabaya vs. Sidoarjo.

While the urban coast gains structural stability at the expense of ecological flexibility, the tidal-flat coast retains dynamic resilience but faces persistent spatial volatility. These opposing behaviors confirm that anthropogenic confinement promotes accretion but suppresses adaptive sediment redistribution, whereas geomorphic openness amplifies erosional risk under reduced rainfall conditions. These contrasts are further clarified through a simplified conceptual comparison illustrating how sediment pathways and rainfall sensitivity differ between the two systems (Figure 10), reinforcing the need for differentiated coastal management strategies.

Specifically, structurally reinforced and maintenance-oriented interventions—such as breakwater optimization, controlled reclamation, and periodic sediment management—are more appropriate for Surabaya, where shoreline behavior is dominated by anthropogenic confinement. In contrast, nature-based and adaptive approaches, including mangrove rehabilitation, tidal-flat conservation, and the preservation of sediment exchange pathways, are better suited for Sidoarjo, where shoreline dynamics respond more directly to rainfall-driven sediment fluctuations. This distinction consistently appears across all temporal segments and underpins the contrasting shoreline behaviors observed in both regions [1], [3].

3.5. Integrated Conceptual Model

Figure 11 synthesizes the observed relationships among rainfall variability, sediment flux, and shoreline response. In Surabaya, the influence of rainfall is strongly mediated by engineered infrastructure and reduced drainage connectivity, producing compartmentalized sediment dynamics and spatially confined accretion. In Sidoarjo, rainfall directly governs sediment transport through tidal channels, intensifying erosion when precipitation declines. This model supports the broader climate–morphodynamic framework, highlighting that hydroclimatic variability is a major but spatially filtered determinant of coastal evolution in tropical regions. Accordingly, the degree of human alteration dictates whether rainfall acts as a stabilizing (urban) or destabilizing (tidal-flat) agent within the shoreline system [4]. However, this apparent stabilization along Surabaya's urban coast may introduce long-term vulnerabilities, as rigid coastal structures can limit landward sediment accommodation and increase the risk of coastal squeeze under future sea-level rise, potentially reducing the system's adaptive capacity.

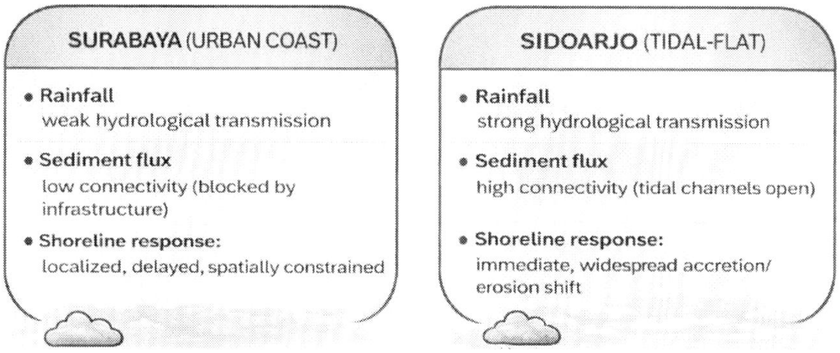

Fig. 11. Integrated conceptual model of rainfall–sediment–shoreline interactions in urban and tidal-flat coastal systems.

4. Conclusions

This study examined shoreline dynamics in two contrasting coastal systems—Surabaya's urban coastline and Sidoarjo's tidal-flat coastline—over the period 2015–2025 using multi-temporal Landsat imagery, NDWI-based shoreline extraction, DSAS-derived change rates, and rainfall correlation analysis. The results reveal fundamentally different morphodynamic behaviors governed by geomorphic setting and the degree of human modification.

Surabaya exhibits a predominantly accretional shoreline response (mean EPR = +4.5 m/year), reflecting the influence of reclamation structures, engineered drainage networks, and coastal protection measures. These interventions produce what is referred to here as semi-stable geomorphic conditions, characterized by persistent net accretion combined with reduced short-term shoreline mobility and a dampened response to rainfall variability across successive temporal segments. In contrast, Sidoarjo displays a predominantly erosional regime (mean EPR = –1.6 m/year), where open tidal-flat morphology and active channel networks maintain strong sediment connectivity, allowing rainfall variability to translate directly into alternating erosion–accretion phases.

Correlation analysis further confirms rainfall as a significant climatic driver at the individual-coast scale, with strong positive relationships observed in Surabaya ($r = 0.76$) and Sidoarjo ($r = 0.80$). However, when both coastlines are analyzed together, the relationship weakens substantially ($r \approx 0.10$), demonstrating that aggregation across contrasting geomorphic typologies can obscure hydroclimatic signals. This finding highlights the importance of spatial context when interpreting climate–shoreline interactions in heterogeneous coastal systems.

Overall, the integrated NDWI–DSAS–rainfall framework proved effective for decadal-scale shoreline assessment, enabling robust detection of morphodynamic cycles and sediment–rainfall interactions under data-limited conditions. The study underscores that hydroclimatic variability alone cannot fully explain shoreline evolution in tropical coastal environments; instead, shoreline responses emerge from the combined influence of rainfall forcing, geomorphic openness, sediment connectivity, and human alteration. These insights reinforce the need for coastal management strategies that are explicitly tailored to local morphodynamic settings rather than uniformly applied across regions.

The findings highlight three key points:

- Coastal typology governs sensitivity to rainfall: engineered coasts exhibit morphological stability, whereas tidal-flat coasts demonstrate amplified responses to rainfall fluctuations.
- Rainfall–shoreline coupling is scale-dependent: correlations are strong at local scales but diminish when contrasting coasts are aggregated.
- Sediment connectivity mediates climate forcing: structural confinement reduces system responsiveness, while geomorphic openness increases susceptibility to erosion during low-rainfall years.

5. Implications for Sustainable Coastal Management

The implications of this study span policy, management, and methodological dimensions, reflecting the distinct morphodynamic behavior of Surabaya and Sidoarjo. In highly urbanized coastal settings such as Surabaya, sediment management actions shown in Figure 12 do not imply large-scale natural sediment bypassing. Instead, feasibility is limited to localized and managed sediment redistribution, implemented through periodic dredging at river mouths and port areas followed by controlled placement of sediments to downdrift coastal segments. This approach integrates with existing port maintenance activities and avoids major land-use conflicts while partially restoring sediment continuity.

a. Policy Implications

Urban coasts such as Surabaya require integrated planning that aligns coastal development with sediment system functioning. Reclamation zoning, adaptive drainage design, and sediment bypassing mechanisms are essential to prevent sediment imbalance caused by structural confinement. In highly dense urban settings, sediment bypassing is feasible primarily as a localized and managed intervention, implemented through periodic dredging at river mouths, port entrances, or engineered drainage outlets, followed by controlled sediment placement to downdrift coastal segments. Given that urban sediment connectivity is weak and rainfall–shoreline coupling is limited, policies should prioritize maintaining residual sediment pathways rather than relying on hydroclimatic inputs alone. Incorporating rainfall variability into coastal risk assessments enhances long-term planning under climate uncertainty [6].

b. Ecosystem-Based Management

In natural tidal systems like Sidoarjo, ecosystem-based adaptation (EbA) and nature-based solutions (NbS) can strengthen resilience to rainfall-driven erosion. Strategies may include mangrove rehabilitation, tidal channel conservation, and controlled sediment redistribution. Because sediment connectivity is strong and rainfall–EPR correlation is high, management must preserve hydrological openness to allow sediment transport to continue buffering erosional phases during low-rainfall years. This approach supports dynamic stability without restricting natural morphodynamic processes [14].

c. Methodological Advancement

The integration of NDWI-derived shorelines, DSAS rate metrics, and rainfall analysis demonstrates a scalable and cost-efficient monitoring framework for tropical coasts. The weakened combined correlation ($r \approx 0.10$) indicates that regional shoreline forecasting should avoid aggregating geomorphically different coastlines into unified models; instead, typology-specific modelling improves interpretability and predictive performance.Future work should include wave climate, tidal energy, land subsidence, and anthropogenic sediment interventions, with machine learning approaches (e.g., Random Forest, hybrid neural models) to enhance predictive skill [10].

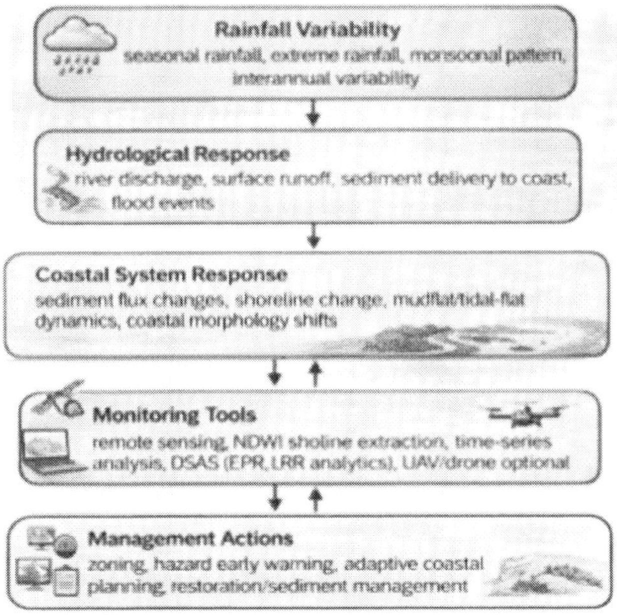

Fig. 12. Framework for integrating rainfall variability into coastal management and monitoring.

This framework highlights that rainfall-driven shoreline responses must be interpreted according to coastal typology, as urban engineered coasts and natural tidal flats react differently to the same hydroclimatic forcing. By linking rainfall anomalies, sediment connectivity, and morphodynamic behavior, it directs managers toward typology-specific adaptation pathways.

The weak combined correlation ($r \approx 0.10$) further indicates that geomorphically distinct coasts should not be merged in regional forecasting; rainfall must be incorporated separately for each coastal type to ensure diagnostic reliability. This integrative approach supports evidence-based decision-making consistent with the Sendai Framework and SDG 13, strengthening climate-resilient coastal governance for Indonesia's northern shoreline.

6. Limitations and Future Research Directions

The NDWI–DSAS–rainfall framework provides a robust basis for comparing shoreline dynamics across contrasting coastal typologies; however, several methodological and interpretative limitations should be acknowledged. The 30 m spatial resolution of Landsat imagery constrains the detection of fine-scale shoreline changes, particularly within wide intertidal zones. Positional uncertainty associated with shoreline extraction may range from approximately 10 to 30 m, depending on tidal stage, shoreline slope, and image acquisition conditions. This uncertainty directly affects the calculated End Point Rate (EPR), potentially dampening short-term erosion–accretion signals while remaining less influential on decadal-scale trends. Consequently, while absolute EPR magnitudes should be interpreted cautiously, the dominant accretional and erosional tendencies identified for Surabaya and Sidoarjo, respectively, are considered robust at the temporal scale examined.

Residual tidal-phase inconsistencies among multi-temporal images may further influence shoreline positioning, with a greater impact in Sidoarjo's gently sloping tidal flats where small vertical tidal differences translate into larger horizontal shoreline shifts. In contrast, Surabaya's engineered coastline exhibits sharper geomorphic boundaries, making shoreline position more sensitive to spatial resolution than to tidal variability.

Although rainfall was treated as the primary hydroclimatic driver, the weak correlation observed when both coastlines were aggregated ($r \approx 0.10$) reflects not merely a limitation of the correlation approach, but a physical dominance of contrasting coastal controls. Aggregation masks rainfall signals because shoreline behavior is filtered by fundamentally different geomorphic and anthropogenic settings: engineered confinement and restricted sediment connectivity in Surabaya versus open tidal channels and rainfall-sensitive sediment fluxes in Sidoarjo. This finding underscores that correlation-based analyses are highly context-dependent and must be interpreted within typology-specific frameworks rather than regional aggregation alone.

Future research should therefore prioritize: (i) higher-resolution datasets (e.g., Sentinel-2, PlanetScope, UAV imagery) to reduce positional uncertainty and improve EPR reliability; (ii) coupled hydrodynamic–morphodynamic modeling to explicitly represent wave climate, tidal prisms, and river discharge, particularly for open tidal-flat systems; and (iii) typology-specific machine learning models that integrate multiple physical drivers while avoiding distortions introduced by pooled datasets. Strengthening these approaches will improve understanding of how hydroclimatic variability interacts with geomorphology and human modification, supporting more reliable shoreline prediction and adaptive coastal management in tropical regions.

Acknowledgments

The author gratefully acknowledges the BMKG for providing financial and institutional support that made this study possible. This work was developed as part of the ongoing research collaboration between BMKG and ITS, focusing on the integration of geospatial technologies, coastal monitoring, and climate–ocean interaction studies to enhance coastal resilience and adaptive management.

References

[1] R. J. Nicholls and A. Cazenave, "Sea-level rise and its impact on coastal zones," *Science*, vol. 328, no. 5985, pp. 1517–1520, 2010. https://doi.org/10.1126/science.1185782

[2] A. Luijendijk, G. Hagenaars, R. Ranasinghe, F. Baart, G. Donchyts, and S. Aarninkhof, "The state of the world's beaches," *Scientific Reports*, vol. 8, no. 1, Art. no. 6641, 2018. https://doi.org/10.1038/s41598-018-24630-6

[3] M. I. Vousdoukas *et al.*, "Sandy coastlines under threat of erosion," *Nature Climate Change*, vol. 10, no. 3, pp. 260–263, 2020. https://doi.org/10.1038/s41558-020-0697-0

[4] R. Almar *et al.*, "Influence of El Niño on the variability of global shoreline change," *Nature Communications*, vol. 14, Art. no. 38742, 2023. https://doi.org/10.1038/s41467-023-38742-9

[5] T. Sultana, M. T. Islam, M. S. Rahman, A. B. Siddique, and A. N. M. S. Huda, "Evaluating the long-term geomorphic process in relation to hydrodynamics in the central coastal zone of Bangladesh," *Heliyon*, vol. 9, no. 6, Art. no. e17368, 2023. https://doi.org/10.1016/j.heliyon.2023.e17368

[6] G. Le Cozannet *et al.*, "Sea level change and coastal climate services: The way forward," *Journal of Marine Science and Engineering*, vol. 5, no. 4, Art. no. 49, 2017. https://doi.org/10.3390/jmse5040049

[7] E. H. Boak and I. L. Turner, "Shoreline definition and detection: A review," *Journal of Coastal Research*, vol. 21, no. 4, pp. 688–703, 2005. https://doi.org/10.2112/03-0071.1

[8] J. E. Pardo-Pascual, J. Almonacid-Caballer, L. A. Ruiz, and J. Palomar-Vázquez, "Automatic extraction of shorelines from Landsat TM and ETM+ multi-temporal

images with subpixel precision," *Remote Sensing of Environment*, vol. 123, pp. 1–11, 2012. https://doi.org/10.1016/j.rse.2012.02.024

[9] E. R. Thieler, E. A. Himmelstoss, J. L. Zichichi, and A. Ergul, *Digital Shoreline Analysis System (DSAS) Version 4.0—An ArcGIS Extension for Calculating Shoreline Change*, U.S. Geological Survey, Open-File Report 2008–1278, 2009. https://doi.org/10.3133/ofr20081278

[10] M. Sengupta, M. R. Ford, P. S. Kench, and G. L. W. Perry, "Exploring the drivers of reef island shoreline change using machine learning models," *Scientific Reports*, vol. 15, Art. no. 16735, 2025. https://doi.org/10.1038/s41598-025-00136-w

[11] BMKG, *Catatan Iklim dan Kualitas Udara Indonesia 2024*, Badan Meteorologi, Klimatologi, dan Geofisika, 2024. [Online]. Available: https://iklim.bmkg.go.id/bmkgadmin/storage/buletin/Catatan%20Iklim%20dan%20Kualitas%20Udara%202024%20BMKG.pdf

[12] A. Bershadskii, "Hamiltonian distributed chaos in the Asian–Australian monsoons and in the ENSO," *arXiv preprint* arXiv:1805.00397, 2018. [Online]. Available: https://arxiv.org/abs/1805.00397

[13] S. K. McFeeters, "The use of the Normalized Difference Water Index (NDWI) in the delineation of open water features," *International Journal of Remote Sensing*, vol. 17, no. 7, pp. 1425–1432, 1996. https://doi.org/10.1080/01431169608948714

[14] F. Xing, Y. P. Wang, and J. Jia, "Hydrodynamics and sediment transport patterns on intertidal flats along middle Jiangsu coast," *Anthropocene Coasts*, vol. 5, Art. no. 12, 2022. https://doi.org/10.1007/s44218-022-00012-4

Sustainable Boatbuilding: Evaluating the Feasibility of Recycled HDPE as a Structural Material

Adristi Nisazarifa[1,*], *Desrilia* Nursyifaulkhair[1], *Amanda* Rosalina[2], *I Putu Arta* Wibawa[1], *Muhammad Rizki Andika* Saputra[1]

[1]Shipbuilding Engineering Department, Shipbuilding Institute of Polytechnic Surabaya, Surabaya, Indonesia
[2]Marine Engineering Department, Shipbuilding Institute of Polytechnic Surabaya, Surabaya, Indonesia
* Corresponding author: adristinisazarifa@ppns.ac.id

Abstract. The boatbuilding industry is increasingly shifting toward environmentally sustainable materials. High-Density Polyethylene (HDPE) has become a highly viable alternative material because of its outstanding chemical and impact resistance, as well as its low-carbon manufacturing process. This makes it a practical substitute for traditional materials such as wood and fiberglass-reinforced plastic for constructing small vessels. However, the substantial generation of HDPE waste poses a significant challenge in maritime manufacturing. In response, this study presents a solution for the plastic waste produced during boatbuilding by recycling HDPE sourced from the Non-Metal Workshop at the Shipbuilding Institute of Polytechnic Surabaya (PPNS) and evaluating its suitability for use as a manhole cover. The compliance of recycled HDPE (r-HDPE) as a structural material was assessed based on the BKI (Biro Klasifikasi Indonesia) Standards for Thermoplastic Vessels, Volume 2, 2023 Edition. The mechanical testing of r-HDPE showed a tensile strength of 24.50 MPa and a flexural strength of 41.93 MPa. Finite Element Analysis using ANSYS showed that increasing the cover thickness from 6.5 mm to 12.5 mm significantly reduced deformation and stress, resulting in a safety factor of 8.29 under a maximum load of 1470 N. Therefore, 12.5 mm thick r-HDPE is structurally viable and supports sustainable shipbuilding by reducing waste.

Keywords: Ansys, Bending Test, HDPE, Manhole, Tensile Test.

1 Introduction

Small vessels measuring up to 30 meters in length are still widely used in Indonesian waters for many purposes, including fishing vessels, cargo and passenger vessels for inter-island crossings, water ambulances, patrol boats, and others. However, the availability of materials for boatbuilding poses a significant challenge to the sustainability of these small vessels in Indonesia. While wood remains the primary material, sourcing high-quality wood that meets the necessary dimensions for construction is becoming increasingly difficult [1]. Additionally, deforestation due to wood exploitation is widely recognized as a contributor to global warming[2]. Conversely, Fiber-Reinforced Polymer (FRP), which has been introduced more than a decade ago as an alternative material, presents its own environmental challenges [1]. FRP presents considerable recycling difficulties, and its degradation results in microplastics that are harmful to aquatic ecosystems. The production processes for FRP boats also involve health risks, as toxic substances are used during manufacturing. In Indonesia, the majority of FRP boat production employs the hand lay-up method in open-air workshops, where operators are often provided with inadequate personal protective equipment. Furthermore, waste from FRP boatyards is poorly managed, despite its classification as hazardous waste. Currently, there is no effective solution for handling FRP waste in shipyards. It is usually sent to licensed B3 (hazardous and toxic) waste handlers, yet in many facilities throughout Indonesia, this waste is left to accumulate or is burned openly.

Nowadays, High-Density Polyethylene (HDPE) is increasingly employed as an alternative in small vessel construction, influenced by environmental concerns associated with the use of wood and FRP [3]. Patrol vessels within the Indonesian National Armed Forces (TNI) have started utilizing HDPE, and fishing vessels in various regions are also beginning to adopt HDPE material. This shift is due to HDPE's outstanding properties, which include remarkable impact strength, resistance to chemicals and corrosion, and a low-carbon emission production process [4]. Furthermore, HDPE provides an environmentally friendly material because its thermoplastic properties allow repeated melting and remolding, thereby facilitating recycling several times [5]. Plastic recycling also plays a crucial role in environmental preservation, primarily by reducing the consumption of fossil fuels required to make virgin plastic materials [6]. Moreover, recycling HDPE waste significantly reduces the accumulation of non-biodegradable plastics and subsequently offers a practical approach to addressing environmental challenges.

A significant challenge is that recycled HDPE (r-HDPE) often exhibits reduced mechanical properties and lower chemical stability than its virgin state (v-HDPE). This degradation is attributed to the service life, natural aging, and thermo-mechanical stresses during recycling [7]. A previous study showed that the mechanical performance of r-HDPE decline after several cycles of mechanical recycling. [8]. Despite the limitation, r-HDPE remains a promising alternative for small vessel applications, especially when its mechanical properties approximate those of virgin HDPE [9]. One potential application of r-HDPE in small vessels is for manufacturing manhole covers. These covers serve as access points for fuel and ballast tanks, as well as various ship compartments. Additionally, they enable crew members to perform maintenance, inspections, and repairs.

This research aims to assess the feasibility of recycling HDPE waste into manhole covers for a small boat. Mechanical testing of r-HDPE, including tensile and flexural testing, was conducted and compared with the BKI (Biro Klasifikasi Indonesia) standard (2023). Finite element analysis was also conducted out to assess the distribution of stress and deformation of the r-HDPE manhole cover under applied loading conditions.

2. Materials and Methods

2.1 Manufacturing of HDPE sheets

The HDPE waste used in this research was sourced from the waste produced during ship fabrication at the Non-Metal Workshop of the Shipbuilding Institute of Polytechnic Surabaya. Virgin HDPE sheets (v-HDPE) with a thickness of 6 mm originated from Agru. The recycled HDPE (r-HDPE) was first crushed into smaller pieces to facilitate the melting and compression process, then washed to eliminate any remaining contaminants, and subsequently dried overnight. A total of 250 g of waste HDPE flakes was placed into an aluminum mold to produce r-HDPE sheets with dimensions of 200 × 200 × 30 mm. The fabrication process for the r-HDPE sheet is illustrated in Figure 1. The r-HDPE sheet was fabricated using a manual hot compression molding process under a load of 88.2 N at a temperature of 200 °C for 15 minutes, utilizing a Memmert Laboratory Oven located in the Metallurgy Laboratory of the Department Mechanical Engineering at Sepuluh Nopember Institute of Technology. Following the melting process, the pressure was continuously maintained during cooling to ensure solidification and to prevent distortion, resulting in a solid structure of r-HDPE.

Fig 1. Fabrication process of the r-HDPE sheet

2.2 Mechanical Testing

The r-HDPE and v-HDPE sheets were respectively cut and prepared for mechanical properties testing in accordance with the ASTM D638 standard. The specimen dimensions were classified as Type I for the r-HDPE material and Type III for the v-HDPE material. The tensile test was conducted to measure the yield strength (MPa), elongation (%), and modulus of elasticity (MPa) for both r-HDPE and v-HDPE materials. The specimens were evaluated using a Universal Testing Machine in the Mechanical Engineering Department Laboratory at Sepuluh Nopember Institute of Technology, Surabaya. Additionally, flexural (bending) testing was performed in accordance with ISO 178, using the three-point bending method to measure flexural strength (MPa) for both r-HDPE and v-HDPE.

2.3 Manhole Structural Analysis

The structural simulation was conducted using ANSYS software on a manhole model measuring 600 mm × 400 mm. Figure 2 shows the geometry of the manhole cover, with all dimensions specified in millimeters. The thickness of the manhole cover was varied to 6.5 mm, 9.5 mm, and 12.5 mm, in accordance with the recommended range specified in ASTM

F-1144. The 3D model was created as a solid body, with the base of the manhole cover representing the section attached to the ship's deck and defined as a fixed support.

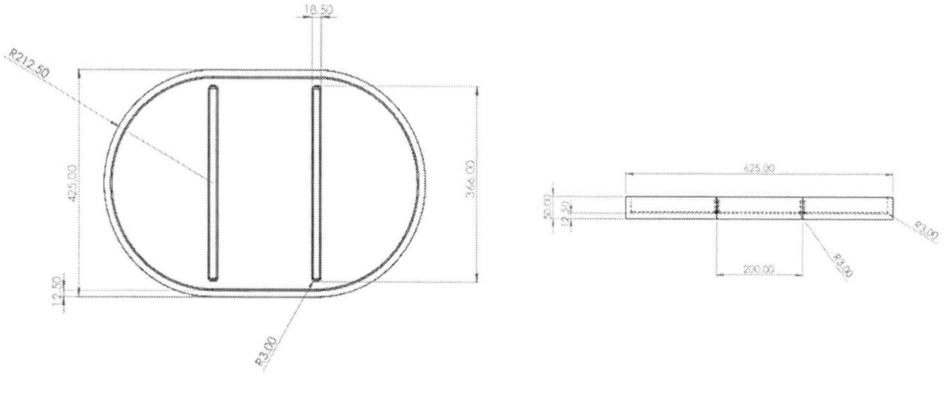

a. Top view b. Side View

Fig 2. Geometry of the manhole cover

The properties of r-HDPE were determined through experimental mechanical tests and were modeled as homogeneous, isotropic, and linear elastic materials. Key parameters included an elastic modulus of 347 MPa, a yield strength of 24.502 MPa, a Poisson's ratio of 0.35, and a density of 0.96 g/cm³, all of which were incorporated into the ANSYS material library.

Boundary conditions were applied to simulate the attachment of the cover to the deck by defining the base region of the manhole cover as a fixed support. A uniformly distributed load was applied to the top surface of the cover at two levels, 980 N and 1470 N, corresponding to the weight of one to two individuals.

A grid independence test (GIT) was performed to ensure that the finite element results are independent of the selected mesh configuration. The mesh was systematically refined by decreasing the overall element size from 10 mm to 1 mm. As a result, the 2 mm mesh with 515.704 elements, as presented in Table 2, was selected as the optimal choice for all subsequent analyses. This selection ensured computational efficiency while maintaining result accuracy within the convergence criterion of a change of less than 0.5%.

Table 1. Grid Independence Test

No.	Mesh size (mm)	Elements	Max deformation (mm)	Error (%)
1	10	16,161	15.338	—
2	9	18,829	15.414	0.49
3	8	24,064	15.606	1.23
4	7	31,556	15.881	1.73
5	6	43,867	15.926	0.28
6	5	70,290	16.022	0.60
7	4	99,962	16.095	0.45
8	3	201,689	16.173	0.48
9	**2**	**515,704**	**16.225**	**0.32**
10	1	2,268,841	16.261	0.22

The primary simulation outputs were equivalent (Von Mises) stress and total deformation, which evaluate stress concentration zones and maximum deflection under human loading. These parameters provide a direct assessment of structural integrity by comparing peak stresses with material limits and deflections against acceptable serviceability thresholds. The safety factor (SF) is the key metric for determining structural adequacy, as shown in Equation 1.

$$SF = \frac{\sigma_y}{\sigma_{max}} \quad (1)$$

3. Result and Discussion

3.1 Mechanical Properties Evaluation

Tensile and bending tests were used to assess the mechanical properties of r-HDPE and v-HDPE samples. The tensile tests provided values for tensile strength, elongation, and elastic modulus, which are detailed in Table 2. Meanwhile, the bending tests offered data on flexural strength, reported in units of MPa or N/mm^2.

Table 2. Tensile and Bending Test Results

Materials	Yield Strength (MPa)	Modulus of Elasticity (MPa)	Flexural Strength (MPa)	Elongation (%)
r-HDPE	24.502	347	41.926	10.027
v-HDPE	24.484	246.301	40.372	198.770

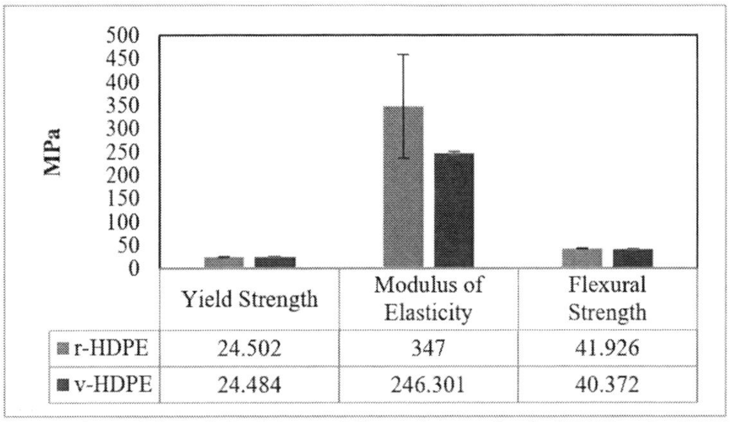

Fig 3. The yield strength, modulus of elasticity, and flexural strength of r-HDPE and v-HDPE

According to Fig 3, the yield strength of r-HDPE (24.502 MPa) and v-HDPE (24.484 MPa) is nearly identical. This similarity can be attributed to an effective recycling process, which allows recycled HDPE to maintain its mechanical properties. The recycling process can alter the crystallinity of materials, occasionally leading to a slight increase in their tensile strength, but might also lead to brittleness as the recycling process continues [10] This behavior aligns with findings from research conducted by Chrysafi et al. [11], in which the r-HDPE exhibits higher tensile strength than v-HDPE. Additionally, they also observed that r-HDPE tends to become more brittle compared to v-HDPE following the recycling process.

The most notable difference was observed in elongation at break, as illustrated in Fig 4. The v-HDPE demonstrated a very high elongation of 198.770%, indicating its ductility. In contrast, the r-HDPE exhibited significantly lower elongation (10.027%), indicating greater brittleness. This reduction in ductility properties was likely due to the recycling effect, where the heating process can break polymer chains and alter the crystallinity, thereby affecting brittleness [10].

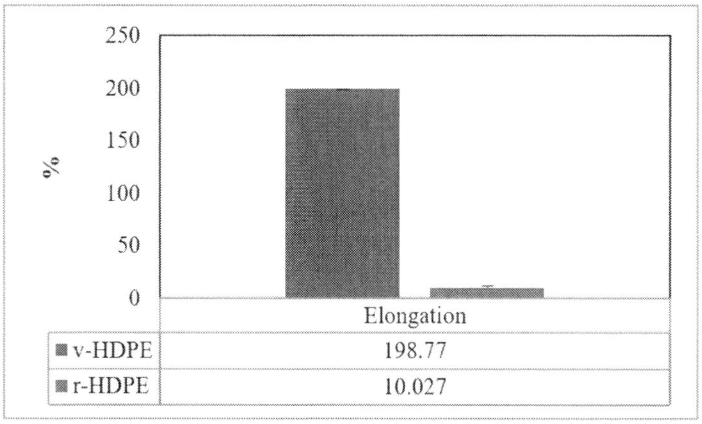

Fig 4. Elongation at break

The r-HDPE demonstrates a higher elastic modulus and flexural strength, with the values of 347 MPa and 41.926 MPa, respectively, in comparison to v-HDPE, which has values of 246.301 MPa and 40.372 MPa. The increased modulus of elasticity of r-HDPE indicates that the material is stiffer and more rigid. The increased rigidity may also result from the recycling process, which could alter the crystalline structure.

According to the BKI standard [12], "Guidelines for Thermoplastic Vessels, Volume 2, 2023 Edition," the minimum tensile yield strength required for HDPE to qualify as a structural material for the hull is 17 N/mm² or MPa. The r-HDPE achieved a tensile yield strength of 24.502 MPa, as seen in Table 3, surpassing the minimum standard by 44.13%. It indicates that the r-HDPE possesses adequate strength for these applications. For tensile elongation at break, the BKI standard allows a wide range from 10% to 1500%. The v-HDPE exhibits a higher elongation at 198.770%, demonstrating excellent ductility. In contrast, the r-HDPE exhibits an elongation of 10.027%, meeting the minimum required level standard. Additionally, the flexural strength also shows a similar trend, with a value of 41.926 MPa for r-HDPE, slightly above the minimum requirement of 40 MPa. This implies that r-HDPE is more brittle and has a lower capacity to stretch before fracturing. It aligns with its lower elongation, indicating a less flexible material structure.

Table 3. Mechanical Properties of r-HDPE and HDPE (BKI)

No	Properties	Unit	r-HDPE	HDPE (BKI)
1.	Tensile Yield Strength	MPa	24.502	17
2.	Tensile elongation at break	%	10.027	10 to 1500%
3.	Flexural Strength	MPa	41.926	40

3.2 Manhole Structural Analysis

The finite element method (FEM) simulation was performed using ANSYS software. The analysis was performed on the manhole design shown in Figure 2 by applying load variations of 980 N and 1470 N to different manhole thicknesses of 6.5 mm, 9.5 mm, and 12.5 mm. A FEM simulation within ANSYS Workbench was performed to assess the mechanical behavior of a manhole composed of r-HDPE under tensile loading. The main goals of this simulation were to assess the distribution of Von Mises stress, total deformation, stress concentration areas, and the material's safe working limit prior to failure. The material properties of r-HDPE obtained through tensile testing were used as simulation parameters. These included an elastic modulus of 347 MPa, a yield strength of 24.502 MPa, a Poisson's ratio of 0.35, and a density of 0.96 g/cm³. The results of the ANSYS simulation are presented in Table 4 and Figure 3. The finite element analysis (FEA) provides a structural representation of the r-HDPE manhole under variations of thickness and applied loading. The primary analysis parameters include total deformation contours and equivalent Von Mises stress, which serve as the basis for evaluating deformation characteristics and the resulting stress distribution from human-induced loading.

Table 4. ANSYS Simulation Results on r-HDPE Manhole Cover

Thickness (mm)	Load (N)	Maximum Total Deformation (mm)	Minimum Equivalent Stress (Von Mises) (MPa)	Maximum Equivalent Stress (Von Mises) (MPa)
6.5	980	10.841	5.254×10^{-4}	5.195
9.5		4.691	3.820×10^{-4}	2.869
12.5		2.467	3.334×10^{-4}	1.970
6.5	1470	16.261	7.880×10^{-4}	7.793
9.5		7.037	5.071×10^{-4}	4.277
12.5		3.700	5.001×10^{-4}	2.955

The simulation results indicate that increasing the recycled HDPE plate thickness from 6.5 mm to 9.5 mm and 12.5 mm results in a significant reduction in the equivalent (Von Mises) stress under all loading conditions tested. This phenomenon suggests that increasing plate thickness improves stress distribution and reduces stress concentration in critical areas. Previous studies have reported that increasing the cross-sectional dimensions, particularly thickness, can enhance the mechanical stability of recycled plastic panels and reduce stress responses to external loads [13]. Furthermore, other research has highlighted that thicker r-HDPE sections provide better structural performance [14]. Based on the simulation results, the 12.5 mm thick r-HDPE plate configuration is considered the most effective and suitable option for use as a manhole cover plate within the loading range of 980 N to 1470 N.

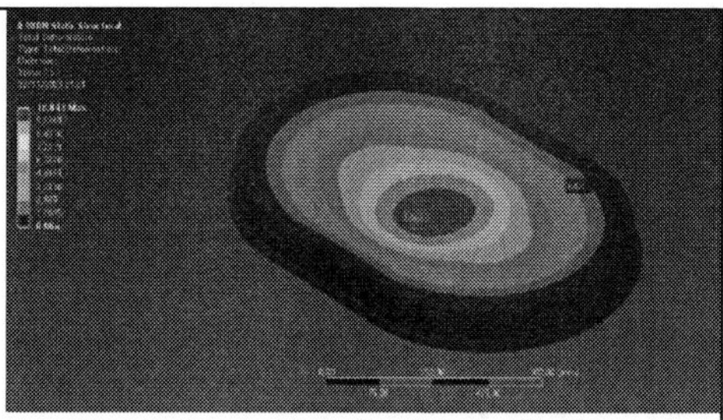

a. Total deformation for 6.5 mm in thickness and a load of 980 N

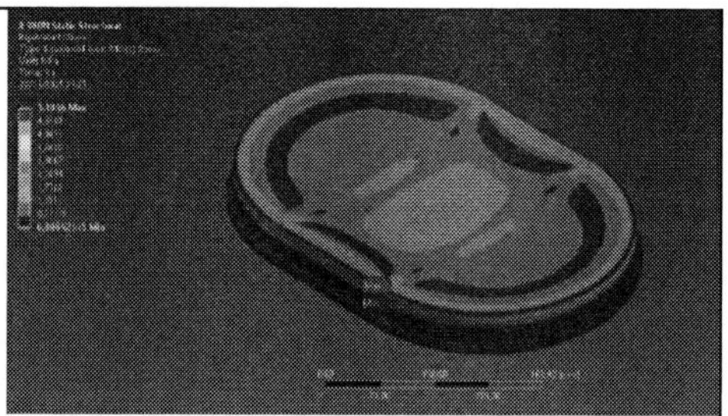

b. Von Mises stress distribution for 6.5 mm in thickness and a load of 980 N

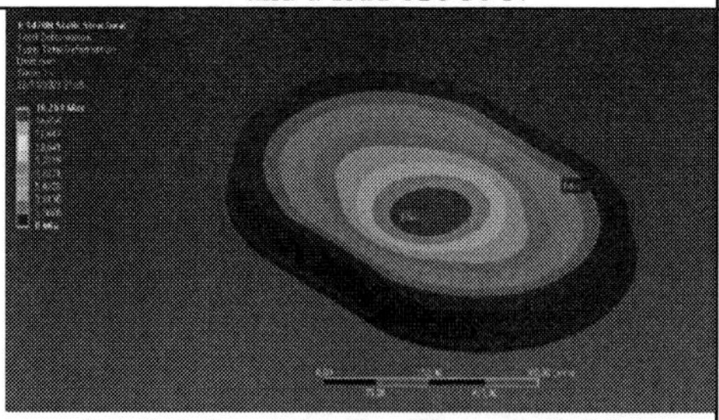

c. Total deformation for 6.5 mm in thickness and a load of 1470 N

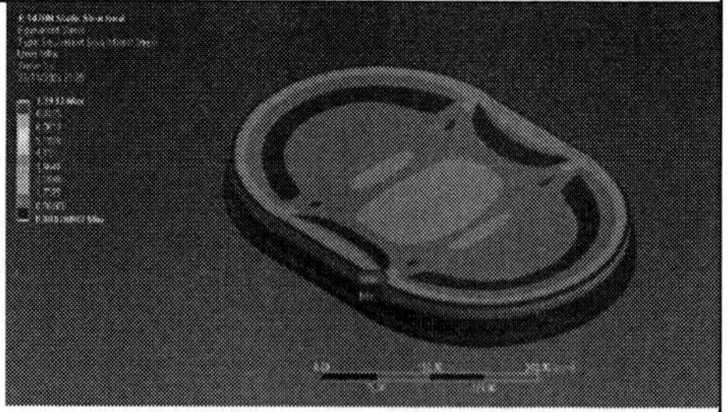

d. Von Mises stress distribution for 6.5 mm in thickness and a load of 1470 N

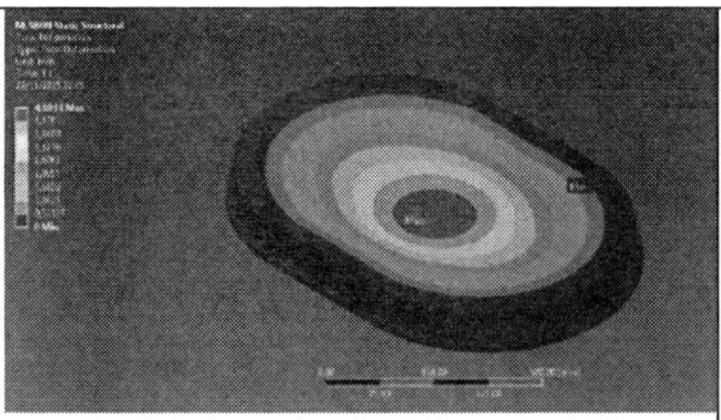

e. Total deformation for 9.5 mm in thickness and a load of 980 N

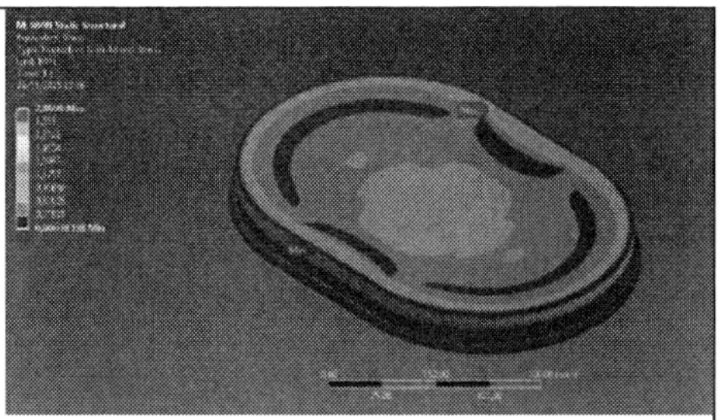

f. Von Mises stress distribution for 9.5 mm in thickness and a load of 980 N

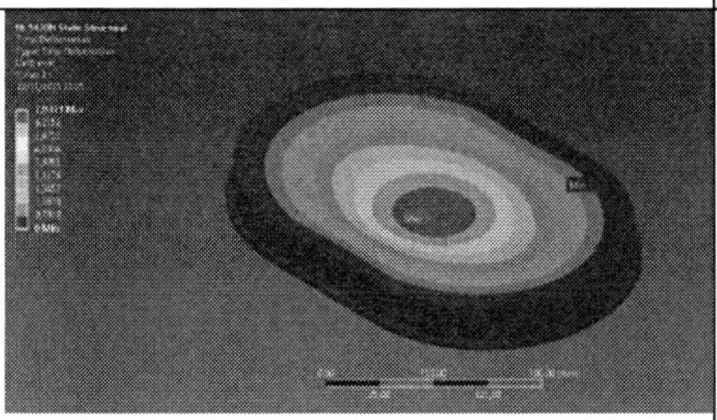

g. Total deformation for 9.5 mm in thickness and a load of 1470 N

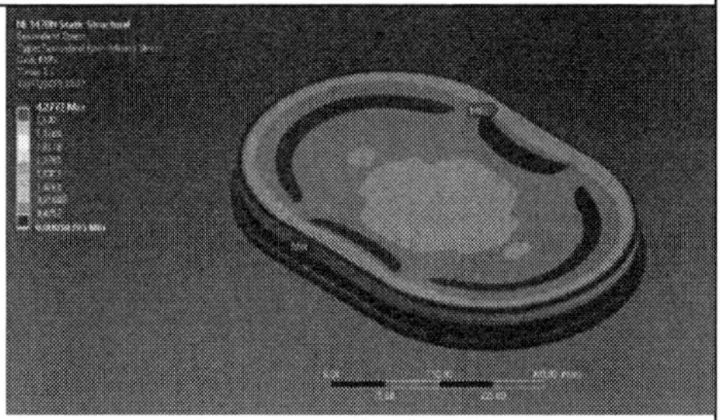

h. Von Mises stress distribution for 9.5 mm in thickness and a load of 1470 N

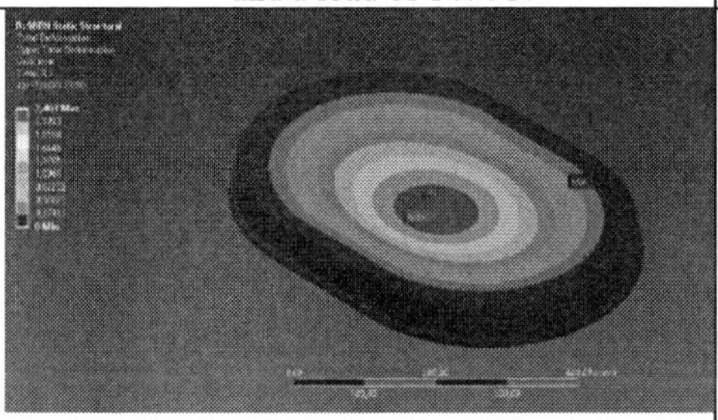

i. Total deformation for 12.5 mm in thickness and a load of 980 N

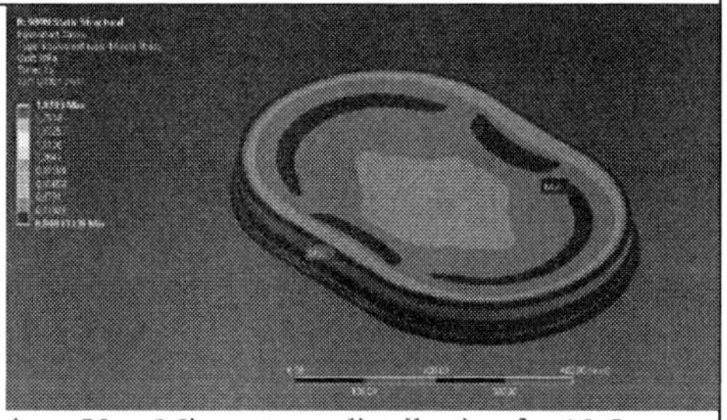

j. Von Mises stress distribution for 12.5 mm in thickness and a load of 980 N

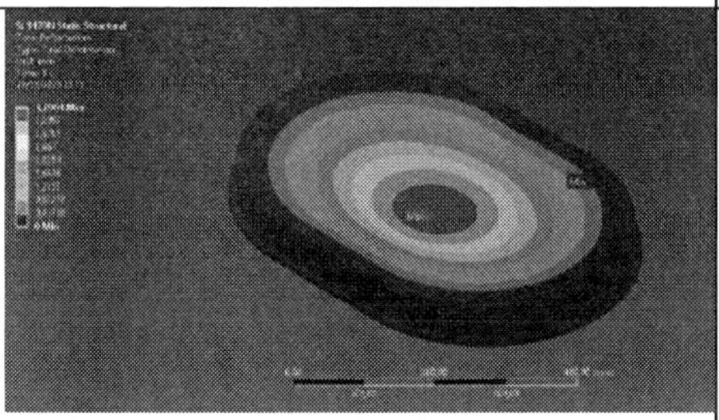

k. Total deformation for 12.5 mm in thickness and a load of 1470 N

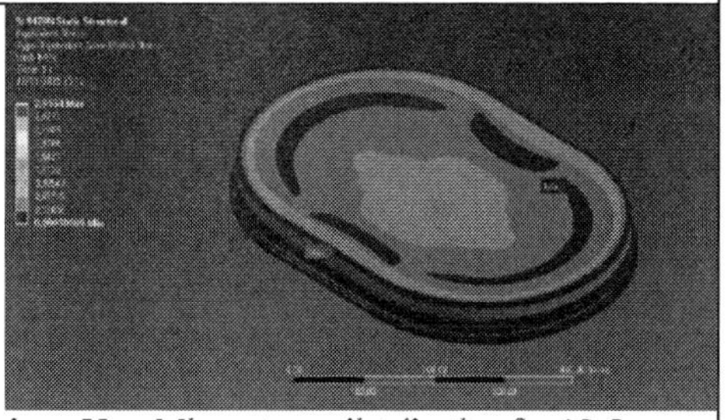

l. Von Mises stress distribution for 12.5 mm in thickness and a load of 1470 N

Fig 5. Total Deformation and Von Mises Stress Distribution of r-HDPE Manhole Cover

The analysis of deformation and stress distribution demonstrates a significant influence of plate thickness on structural performance. In the case of a 6.5 mm thick plate, maximum deformation occurs at the load application area, with contour colors transitioning from dark blue (demonstrating minimum deformation) to red (demonstrating maximum deformation), as illustrated in Figures 5a and 5c. This indicates considerable deformation values, particularly under a 1470 N load. The contour pattern displays a color gradient from dark blue in the support area, representing the minimum deformation, progressing to yellow and red around the loading point, where the maximum deformation reaches 16.261 mm. This distribution pattern is consistent with thin-plate theory, which suggests that deflection is most significant at the center of the span and approaches zero at the supports. Additionally, the Von Mises stress is concentrated around the support region and the geometric transition zones, as depicted in Fig 5b and 5d. Notably, the maximum stress is greater under the 1470 N load compared to the 980 N load, measuring at 7.793 MPa. Red-to-orange stress contours represent these critical areas, while the remaining regions fall within the blue-to-green range. These simulation results are consistent with the theory of geometric discontinuities and plate behavior under bending conditions, which states that maximum stresses typically occur in the regions of direct contact .

As the thickness increased to 9.5 mm, deformation decreased substantially, as evidenced by predominantly blue-green contour colors (Fig 5e). This indicates lower deflection and a more uniform distribution of deformation, with the maximum deformation measuring 7.097 mm (1470 N). The Von Mises stress also decreases, and the stress contour pattern is predominantly represented by blue-to-green colors (Fig 5d and 5f), indicating a more uniform stress distribution.

For a thickness of 12.5 mm, deformation is minimized, with contour colors dominated by blue-green (Fig 5i and 5k). This configuration demonstrates significantly improved load distribution and a notable increase in plate stiffness. Fig 5k shows a significant reduction in deformation to 3.700 mm, with contours predominantly colored blue to green, while yellow appears only locally around the loading point. This condition aligns with principles of structural mechanics, where increasing the plate thickness results in a cubic enhancement in flexural stiffness [15]. Increasing the thickness of panels made from recycled plastic effectively reduces deflection under loading [13]. Additionally, the maximum Von Mises stress diminishes in comparison to other configurations, reaching a lower maximum value of 2.955 MPa. This indicates a safe and stable state under operational loads. This stress reduction suggests that increasing the thickness enhances the load distribution area and reduces local stress concentration. Panels made from recycled plastic with greater thickness demonstrate improved structural performance in terms of stress distribution [13].

Table 5. Safety Factor of r-HDPE Manhole Covers

Thickness (mm)	Load (N)	Maximum Stress (MPa)	Safety Factor	Remarks
6.5	980	5.1956	4.716	Structurally safe in terms of stress
9.5		2.8699	8.538	Safe, suitable for light loading conditions
12.5		**1.9703**	**12.436**	**Highly safe, recommended configuration**
6.5	1470	7.7933	3.144	Safe, but at the lower limit of acceptability, less optimal
9.5		4.2772	5.729	Safe under moderate loading
12.5		**2.9554**	**8.291**	**Safest configuration, recommended design**

The safety factor calculation results, presented in Table 5, indicate that all plate thickness variations made from r-HDPE exhibit safety factor values greater than 3. Under the highest loading condition of 1470 N, the lowest SF value is observed at a thickness of 6.5 mm, with a value of 3.14. An SF value of ≥ 2 is considered to meet the structural safety requirements for both non-structural and semi-structural components under static loading conditions [15]. From a material strength perspective, all tested plates are considered safe against failure due to stress. It has been demonstrated that increasing the plate thickness enhances the safety factor. For example, under a load of 980 N, the safety factor increases from 4.72 to 12.44 as the thickness rises from 6.5 mm to 12.5 mm. A similar tendency is observed under the 1470 N load, where the SF increases from 3.14 to 8.29 with the same thickness variation. This observation aligns with the theory of mechanics of materials, which states that increasing the cross-sectional dimensions enhances stiffness and reduces local stress distribution [13]. The mechanical performance of r-HDPE improves when applied in configurations with greater cross-sectional thickness [14]. Recycled plastic panels can perform optimally when the panel thickness is sufficiently designed to ensure stress stability under light to moderate loading conditions [13]. Consequently, although the 6.5 mm thickness theoretically meets the safety factor requirement (SF > 2), the 12.5 mm thickness is optimal because it yields the highest SF, providing enhanced safety and improved design reliability to accommodate variations in operational loads and potential long-term material degradation.

4. Conclusion

The results of this study suggest that recycled HDPE (r-HDPE) has strong potential as a suitable material for manhole covers in small vessels. Mechanical testing revealed that r-HDPE exhibits superior tensile strength, elastic modulus, and flexural strength compared to virgin HDPE, highlighting the increased rigidity resulting from the recycling process. Although a significant reduction in elongation was observed, indicating greater brittleness, r-HDPE demonstrated a yield strength of 24.502 MPa, exceeding the minimum requirement specified by BKI (17 MPa), thus confirming its appropriateness as a lightweight structural material. Additionally, the flexural strength of virgin HDPE was 40.372 MPa, whereas that of r-HDPE was 41.926 MPa, slightly above the minimum threshold of 40 MPa. This indicates that r-HDPE is more brittle and has a lower capacity for stretching before fracture, as evidenced by its lower elongation of 10.027%, suggesting a less flexible material structure.

The FEM simulation performed with ANSYS showed that increasing the plate thickness from 6.5 mm to 12.5 mm significantly decreases both deformation and Von Mises stress, while simultaneously increasing the safety factor from 3.14 to 8.29 under a maximum load of 1470 N. Although all thickness variations were considered safe (SF > 2), the 12.5 mm configuration was identified as the most optimal design. The use of r-HDPE contributes to sustainable shipbuilding practices by reducing non-biodegradable plastic waste generated during construction and encouraging the use of environmentally friendly materials within the maritime sector. For future studies, it is advisable to evaluate the long-term performance of r-HDPE in marine environments, analyze its behavior under cyclic dynamic loads, and explore potential enhancements to its structural properties through the addition of fillers or hybrid composite configurations.

References

[1] I. P. A. Wibawa and R. W. Birmingham, "Wood vs FRP, Sustainable Material for Indonesian Fishing Vessels Based on Fishers' Perspectives," in *Proceeding the 9th International Conference on Marine Technology*, 2014.

[2] S. S. Kumar, D. Das P. H., P. N. Jha, B. M.V., L. Edwin, and M. Lekshmi N., "Assessment of Life Cycle and Environmental Impact Hotspots Associated with the Construction and Disposal of Fiberglass Reinforced Plastic Fishing (FRP) Boats in the Small-Scale Fishing Sector of India," *Fishery Technology*, vol. 61, no. 4, Oct. 2024, doi: 10.56093/ft.v61i4.153466.

[3] T. Firmanda, S. Anggara, M. R. Hariadi, and A. R. Prakasa, "HDPE as a New Alternative Material for Small Vessel Boat Strength," in *Proceedings of the 5th International Conference on Applied Science and Technology on Engineering Science*, 2022, pp. 473–477.

[4] J.-M. Lee, "Feasibility study of shell element-based elastic FE approach for welding-induced thermal distortion prediction in HDPE welded structures," *International Journal of Naval*

Architecture and Ocean Engineering, vol. 15, p. 100559, 2023, doi: 10.1016/j.ijnaoe.2023.100559.

[5] A. Petrovič *et al.*, "Comparative Analysis of Virgin and Recycled Thermoplastic Polymer based on Thermochemical Characteristics," *CHEMICAL ENGINEERING TRANSACTIONS*, vol. 94, pp. 1321–1326, 2022.

[6] K. Ragaert, L. Delva, and K. Van Geem, "Mechanical and chemical recycling of solid plastic waste," *Waste Management*, vol. 69, pp. 24–58, Nov. 2017, doi: 10.1016/j.wasman.2017.07.044.

[7] L. Dong, C. Dong, W. Zhi, P. Zhang, W. Li, and B. Gu, "A comprehensive approach to physical recycling of hazardous waste HDPE industrial drums: Overcoming challenges in safety and foam interference," *Chemical Engineering Journal*, vol. 497, p. 154502, Oct. 2024, doi: 10.1016/j.cej.2024.154502.

[8] P. Oblak, J. Gonzalez-Gutierrez, B. Zupančič, A. Aulova, and I. Emri, "Processability and mechanical properties of extensively recycled high density polyethylene," *Polymer Degradation and Stability*, vol. 114, pp. 133–145, Apr. 2015, doi: 10.1016/j.polymdegradstab.2015.01.012.

[9] D. A. I. Malyuta, K. L. Matteson, M. P. Berry, D. Bajwa, and C. Ryan, "Experimental statistical modeling of tensile properties and flexural stiffness of recycled high-density polyethylene (rHDPE) thermoplastic using response surface methodology (RSM)," *Results in Materials*, vol. 20, p. 100472, Dec. 2023, doi: 10.1016/j.rinma.2023.100472.

[10] D. A. I. Malyuta, K. L. Matteson, C. Ryan, M. P. Berry, and D. Bajwa, "An investigation into the tensile properties of recycled high-density polyethylene (rHDPE) blended with talc filler," *Results in Materials*, vol. 17, p. 100377, Mar. 2023, doi: 10.1016/j.rinma.2023.100377.

[11] I. Chrysafi *et al.*, "Characterization of the Thermal, Structural, and Mechanical Properties of Recycled HDPE," *Macromolecular Symposia*, vol. 405, no. 1, Oct. 2022, doi: 10.1002/masy.202100224.

[12] Biro Klasifikasi Indonesia, *Guidelines for Thermoplastic Vessels, Volume 2, 2023 Edition*. 2023.

[13] A. J. Das and M. Ali, "Multiscale Evaluation of Recycled Plastic Corrugated Panels for Sustainable Construction," *Buildings*, vol. 15, no. 14, p. 2423, Jul. 2025, doi: 10.3390/buildings15142423.

[14] S. Ahmed and M. Ali, "Tensile and Flexural Properties of Recycled HDPE for Application in Building Products," in *MATEC Web of Conferences*, A. T. Jafry, Ed., Jun. 2024, p. 01038. doi: 10.1051/matecconf/202439801038.

[15] J. M. Gere and S. P. Timoshenko, *Mechanics of Materials*. Boston, MA: Springer US, 1997. doi: 10.1007/978-1-4899-3124-5.

Integration of photogrammetry data and 3D modelling GIS for analyzing spatial utilization intensity based on Jambi city government regulations in sustainable spatial planning (Case study: Pasar Jambi Sub-district)

Dira Muvianti Warman[1*], *Hepi Hapsari* Handayani[2], and *Cahyono* Susetyo[3]

[1] Department of Geomatics Engineering, Faculty Of Civil, Planning, and Geo-Engineering, Institut Teknologi Sepuluh Nopember, Surabaya, Indonesia

[2] Department Urban and Regional Planning, Faculty Of Civil, Planning, and Geo-Engineering, Institut Teknologi Sepuluh Nopember, Surabaya, Indonesia

**Corresponding author: diramuvianti9@gmail.com*

Abstract. Increasing urbanization in metropolitan areas has led to the emergence of many densely populated areas and illegal buildings [1]. To control this growth, local governments have made various efforts to control the intensity of development or land use. This study evaluates spatial utilization intensity in Pasar Jambi District to address the challenges of rapid urbanization and the proliferation of non-compliant structures. By integrating 3D photogrammetric modeling and Geographic Information Systems (GIS), the research assesses the compliance of Building Coverage Ratio (BCR), Floor Area Ratio (FAR), Green Coverage Ratio (GCR), and Building Setback Lines with Jambi City Regulation No. 3/2015 and Draft Regulation No. 5/2024. The findings indicate that spatial utilization in the study area remains suboptimal, characterized by excessive BCR, deficient GCR, and frequent setback violations, which effectively "lock" land resources and hinder functional adaptation. Conversely, FAR values generally remain below the maximum allowable thresholds, suggesting significant potential for controlled vertical intensification. Shifting urban growth from a horizontal to a vertical trajectory is recommended to alleviate pressure on land availability and ensure the implementation of sustainable spatial planning.

1. Introduction

Under Law No. 26 of 2007, district and city governments have the authority to implement spatial planning, which includes regional spatial planning, spatial utilization, and control of spatial utilization in districts/cities [2]. Based on these regulations, the Jambi City government issued a regional regulation governing spatial planning, as stipulated in Jambi City Regional Government Regulation No. 3 of 2015 concerning harmonizing development, and the draft Jambi City Regional Regulation No. 5 of 2024 on the spatial plan for the Jambi City area for 2024-2044 [3][4]. Therefore, the government needs to implement a more scientific and rational spatial layout and optimize land allocation in order to improve the utilization of existing land resources [5]. The purpose of this study is to analyze and provide input to the government on how regulations align with practices in the field. Based on this, an Intensity of Space Utilization (ISU) analysis was conducted. ISU provides an overview of how space is used and utilized to maintain urban balance, as well as indicating whether the land is being managed optimally in terms of infrastructure load, environmental quality, and urban comfort [6].

The operational variable of spatial utilization intensity refers to spatial planning guidelines in Indonesia. In accordance with the appendix to Minister of Agrarian Affairs and Spatial Planning Regulation No. 16 of 2018, spatial utilization intensity includes Building Coverage Ratio (BCR), Floor Area Ratio (FAR), Green Coverage Ratio (GCR), Basement footprint, Building Setback Line, and Building Density. In this study, the operational variables used to indicate the intensity of space utilization include Building Coverage Ratio (BCR), Floor Area Ratio (FAR), Green Coverage Ratio (GCR), Building Setback Line, and Building DensityIn implementing the Detailed Spatial Plan (RDTR) to understand the provisions for spatial utilization, a technique was used to combine 2D and 3D spatial patterns [8]. Where 3D mapping requires other supporting basic data in the form of elevation models, building footprint polygons that can be extracted from orthophoto data as basic data for the formation of Level of Detail (LoD) 2.1 in 3D mapping, as well as polygons of land use zones. The creation of this 3D modelling was carried out using aerial photographs processed using photogrammetry principles, which were then converted into point clouds and subsequently into DTM and DSM data.

The combination of these two methods is expected to meet the need for building information, both vertical and horizontal, generated from the modelling that has been carried out. The resulting 3D model can assist in solving problems related to calculating space utilization intensity variables. BCR, FAR, GCR, Building Setback Line, and Building Density affect space utilization intensity in Pasar Jambi District and have a significant impact on space utilization. The BCR values in the trade and service areas and residential areas indicate that most blocks are in the dense to very dense category, so that the building footprint dominates the land surface and leaves limited open space. Therefore, this research is very important to evaluate the effectiveness of the existing spatial utilization intensity policy in the existing conditions in the Pasar Jambi District with applicable standards, namely based on Jambi City Government Regulation No. 3 of 2015 concerning harmonizing development, and the draft Jambi City Regional Regulation No. 5 of 2024 concerning the 2024-2044 spatial plan for the city of Jambi.

2. Materials and Methods

2.1 Study Area

This research was conducted in Pasar Jambi Subdistrict, one of the subdistricts in Jambi City. This subdistrict covers an area of approximately 168.8 hectares and is one of the urban areas that plays a strategic role in economic, trade, and service activities in the city of Jambi. The characteristics of the area, which is dominated by urban functions, make the Pasar Jambi Subdistrict a relevant location for studying the dynamics of spatial utilization, building density, and urban development patterns.

Geographically, Pasar Jambi Subdistrict is bordered by Palayangan Subdistrict to the north, Jelutung Subdistrict and Telanaipura Subdistrict to the west, Jelutung Subdistrict to the south, and Jambi Timur Subdistrict to the east. These boundaries indicate that Pasar Jambi District is located in the core area of the city, surrounded by other districts with relatively similar urban characteristics, resulting in fairly intensive interaction between regions. This condition has implications for high pressure on land use and the need for integrated spatial control.

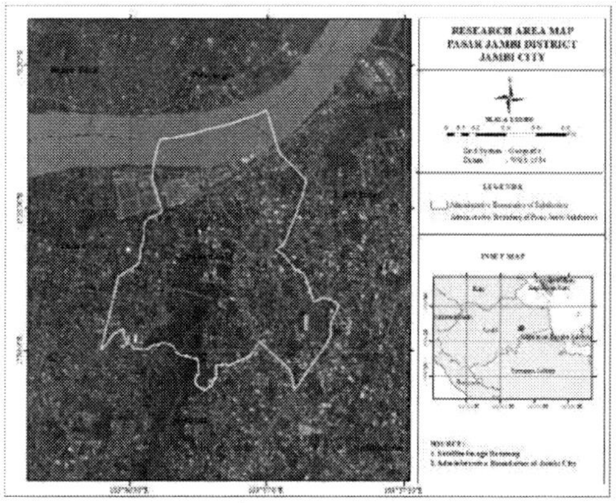

Fig 1. Research area Map

2.2 Data Sources

This study applies a geospatial approach to evaluate the intensity of existing land use in Pasar Jambi District. The primary data used comes from aerial photo mapping acquired in October 2024, resulting in orthophoto data with a resolution of 3.3 cm and a horizontal accuracy of 32.5 cm. Digital Terrain Model (DTM) and Digital Surface Model (DSM) data with a resolution of 12 cm and an accuracy of 49.5 cm. This data was further processed to produce a 3D City Model (Level of Detail/LOD-2) where the resulting model has more detailed geometry, including roof representation. The model was created based on the digitization of building roofs and Normalized Digital Surface Model (NDSM) calculations, resulting in a LoD 2 building model. The results of the spatial utilization intensity analysis were then identified and evaluated for compliance based on the spatial planning regulations in force in the City of Jambi (Local Regulation No. 3 of 2015 and Draft Local Regulation No. 5 of 2024).

2.3 Methods

2.3.1 Research Framework

This study was conducted to analyze the intensity of space utilization by developing a 3-dimensional city model using DTM, DSM, and Orthophoto data generated from photogrammetric processing. The parameters used to determine the intensity of spatial utilization include BCR, FAR, GCR, Building Setback Line, and Building Density. The 3D city model was built at Level of Detail/LOD-2 by processing DTM and DSM data to produce a Normalized Digital Surface Model (NDSM), then manually segmenting building roofs based on orthophoto data. In addition, land use grouping was also carried out in existing conditions, resulting in maps of land use, green open spaces (RTH), roads, and water bodies. The creation of building footprints, roof segmentation, and land use maps was carried out through visual classification using on-screen digitization techniques. To ensure data accuracy, field validation was conducted using sampling methods to test the conformity between digitized results and actual conditions in the field.

The nDSm value obtained is used to determine the height of the building, then the building footprint data is added to produce a LOD-2 3D city model. The accuracy of the resulting 3D model is tested to determine the level of precision of the model. Building sampling tests are conducted with reference to the American Society for Photogrammetry and Remote Sensing Edition 2, Version 2 ASPRS Positional Accuracy Standards for Digital Geospatial Data [10]. Where the total number of samples for Horizontal Accuracy and Non-Vegetated Vertical Accuracy is 30 samples for the area, 1000 KM^2. Where the accuracy value of the building based on CityGML uses the Root Mean Square Error (RMSE) value as a reference. The RMSE value produced in the LOD 2 building model is in accordance with the CityGML standard, which is < 2 meters [11].

The results of the 3D building formation that has met the standards are then analyzed using existing land use area data, and an ISU analysis is carried out in accordance with RDTR regulations. This analysis is related to BCR, FAR, GCR, BSL, and Building Density. The results of this analysis will then be compared with the provisions of the 2024-2044 RDTR for the City of Jambi. This evaluation of compliance will reveal which areas violate or do not violate IPR regulations based on the RDTR regulations.

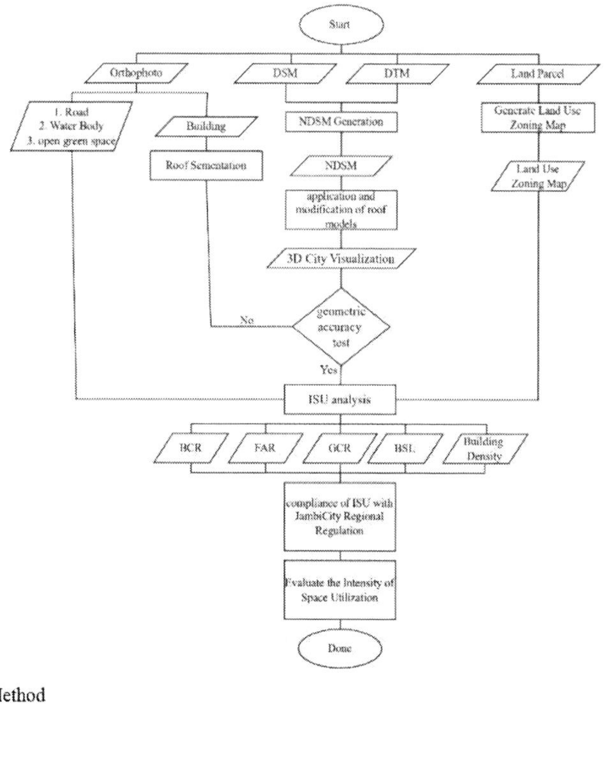

Fig. 2. Research Flow Method

2.3.2 Intensity of Space Utilization (ISU)

The intensity of space utilization is analyzed to determine how often the space is used by humans, so that the direction of development and land use can be regulated [6]. According to Edward Ng (2010), provisions related to Building Floor Area Ratio (BFAR) and Green Floor Area Ratio (GFAR) in an urban context tend to be mechanical in nature, but need to be understood contextually within the overall physical spatial structure. This includes aspects of the urban system and the external environment. The physical form resulting from the control of BCR, GCR, and FAR remains within the framework of the unity of the urban spatial structure to ensure spatial integration [12]. o calculate these variables, you can use the following equation:

$$BCR = \frac{Building\ Footprint\ Area}{Total\ land\ Area} \times 100\% \quad (1)$$

$$FAR = \frac{Total\ Floor\ Area\ Building}{Total\ land\ Area} \quad (2)$$

$$GCR = \frac{Green\ Open\ Space\ Area}{Total\ land\ Area} \times 100\% \quad (3)$$

In calculating the BCR value, a comparison is made between the built-up area and the total land area. The FAR is calculated based on the ratio of the total floor area of the building to the total land area, and the GCR is calculated by comparing the open space area to the total land area. Meanwhile, the BSL determines the minimum distance between a building and the boundary of the land or public facilities such as roads, riverbanks, railroad tracks, and high-voltage power lines. The BSL provisions are made with consideration of safety and health aspects to ensure that buildings are at a safe distance from these elements to avoid risks that could endanger residents and the surrounding environment. The calculation of building density in Dangnga's dissertation (2002) explains that the identification of settlement density is done by calculating the total area of the building (building coverage) with the area of the settlement block [13].

2.3.3 Legal Framework for Regional Spatial Utilization in Jambi City

Regulations on spatial planning in Jambi City can focus on two main legal instruments: Jambi City Regional Regulation No. 3 of 2015 concerning Buildings and Draft Jambi City Regional Regulation No. 5 of 2024 concerning the 2024–2044 Spatial Plan for Jambi City. These two regulations form the legal basis for analyzing and evaluating the suitability of land use in the Jambi City area. The 2015 Perda regulates technical and administrative provisions related to buildings, including parameters for land use intensity such as Building Coverage Ratio (BCR), Floor Area Ratio (FAR), and Green Coverage Ratio (GCR). Meanwhile, the 2024 Draft Regional Regulation provides new guidelines on regional spatial structure and patterns, zoning, and building density standards that will serve as a reference for sustainable development in the future. Based on this, the maximum limits for BCR, FAR, and GCR values based on land use zones can be seen in Table 1.

Table 1. Maximum limits of the BCR, FAR,and GCR values.

No	Land Use Zone	BCR	FAR	GCR
1	Commercial and Services Zone	50%	10.0	10%
2	Residential Zone	70%	2.0	5%
3	Public Facilities Zone	60%	4.2	10%
4	Park and Urban Green Space Zone	15%	0.2	85%
5	Tourism Zone	50%	1.0	20%
6	Industrial Zone	50%	1.0	30%
7	Cemetery Zone	5%	0.1	5%
8	Defense and Security Zone	50%	1.5	10%

Based on PUPR Regulation No. 28/PRT/M/2015 concerning the Determination of River Boundaries and Lake Boundaries [14] and PUPR Regulation No. 06 of 2021 concerning the classification of minimum building limits for public facilities as can be seen in Table 2.

Table 2. Building Setback Line Classification

Variable	Classification	Criteria
Building Setback Line (BSL)	Building distance from the road	Arterial roads: 5 – 7 meters
		Collector roads: 3 – 5 meters
		Local roads: 1.5 – 3 meters
	Building distance from the river	Minor (tributary) river: 10 meters
		Medium river: 15 – 20 meters
		Large river: > 25 meters

The calculation of building density in Dangnga's dissertation (2002) explains that the identification of settlement density is carried out by calculating the total area of buildings (building coverage) with the area of the settlement block [13]. Where the Director General of Human Settlements in 1979 classified building density levels, as can be seen in Table 3.

Table 3. Building Density Variable

	Classification	Criteria
Building Density Variable	BCR <40%	Low
	BCR 40% - 60%	Moderate
	BCR 60% - 80%	Dense
	BCR > 80%	Very Dense

2.3.2 3D City Model

The creation of this 3D city model was carried out to digitally represent the earth's surface and surrounding objects, such as the ground, vegetation, buildings, and so on. The 3D modeling was created based on Level of Detail (LoD) 0 to LoD3, but this study used LoD-2 modeling, where the model has more detailed building geometry than LoD-1, but already has a roof shape. 3-dimensional geospatial data defines locations in 3-dimensional space determined by the X, Y, and Z axes. Spatial objects in the real world are regular and irregular in shape. In GIS, these spatial objects are represented in the form of points, lines, and surfaces. There are two categories of spatial object representation, namely surface-based and volume-based. Surface-based object representation is the representation of objects represented by the shape of the primitive object surface. One of the simplest methods of forming 3D objects is by extrusion. Extrusion is the stretching of a 2D flat plane into a 3D object. The footprint or flat plane as the base (initial 2D data) on each side will be pushed upwards according to the specified height to create a polyhedron [12].

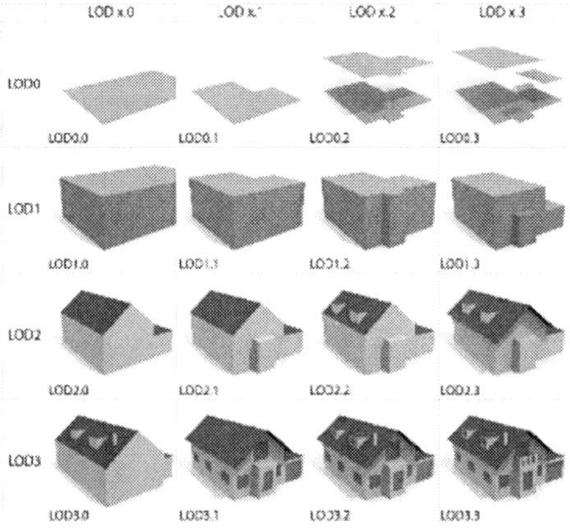

Fig. 3. grouping of levels of detail

3. Result

3.1. Land Use Zone

To regulate the intensity of land use, zoning regulations are implemented based on specific land use zones. These zoning regulations are designed to ensure that each zone is used in accordance with its designated function as stipulated in the RDTR regulations. Land use zones are delineated based on existing land use conditions in the field. Each land use zone in Pasar Jambi Subdistrict can be seen in Figure 2. These land use zoning regulations are listed in the Detailed Spatial Plan document. Zoning regulations in Jambi City are regulated in Jambi City Regulation Number 5 of 2024 concerning the 2024-2044 Jambi City Spatial Plan.

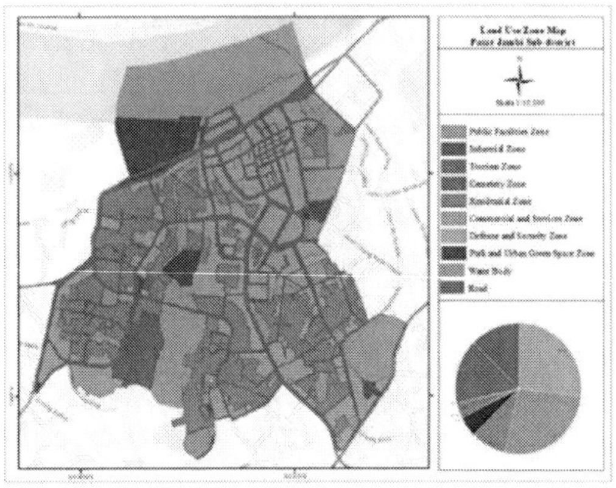

Fig. 4. Landuse Zona Map

Generally, the Pasar Jambi District is dominated by residential areas (26.78%) and trade and service areas (26.83%). The high concentration of trade and service functions is in line with the strategic location of the Pasar Jambi subdistrict, which is often the center of commercial activity. Meanwhile, the smallest land use zone is the tourism zone, at 0.25%. In addition, the land use zones in the Pasar Jambi subdistrict are designated as public facility zones, city park zones, industrial zones, cemetery zones, and defense and security zones. Based on Figure 3, the Pasar Jambi subdistrict is an area with fairly high economic activity, where residential areas, trade, and services dominate half of the total area. This spatial pattern is in accordance with the Detailed Spatial Plan (RDTR) which designates this area for residential and commercial use. This conformity is reinforced by empirical indicators in the form of high levels of economic activity. Based on the types of activities that frequently occur in this area due to the presence of traditional markets, terminals, shopping centers/malls, hotels, schools, places of worship, and government offices.

3.2. 3D Model Visualization

This 3D building model is a representation of existing buildings on the earth's surface in the research area. This model was obtained based on the overlay of DTM, DSM, and building footprint data, as well as the extraction of building roofs, resulting in a LOD-2.1 3D city model. This LOD-2 model presents buildings in the form of simplified outer shells. The 3D modeling of buildings in the Jambi Market area is dominated by residential, commercial, and service areas, where the majority of tall buildings are shops, malls, and hotels. The modeling results can be seen in Figure 5.

https://doi.org/10.1051/bioconf/202621605002

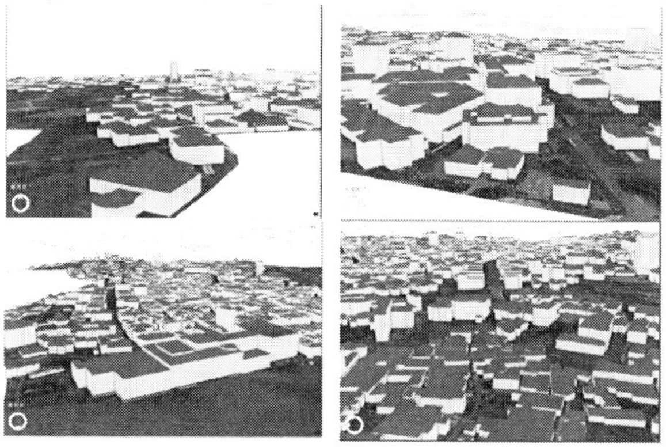

Fig. 5. 3D City Visualization

Based on land use zones, this area is dominated by commercial and service areas, resulting in many shop buildings, most of which are three stories high. In addition, high-rise buildings are dominated by hotels and malls. The roofs of the buildings formed from this modelling are dominated by flat roofs, while there are also buildings with dome, hip, gambrel, spherical, and vaulted roofs. Residential buildings usually have hip and gambrel roofs.The existing modelling was used to test geometric accuracy by reviewing the length and height segments of the model. In this case, the building length and height segment data will be tested and then the percentage of error and Root Mean Square Error (RMSE) will be calculated.

Table 4. geometric accuracy test

Number of Validation Samples	50
Number of Roof Segments	3579
Highest Error Difference	3,13354
Lowest Error Difference	-1,45200
Average Error	0,1951265227
Model Error Percentage	1,16240%
RMSE	0,4417312789

From the model that was formed, 50 samples of building length and height were used, where the distribution of these building samples was evenly spread across various building types, including tall, medium, and low buildings. Based on this data, an RMSe value of 0.441 was obtained, which does not exceed 2 meters. This indicates that the resulting model meets the standards and has good spatial quality and is suitable for use in further mapping and planning analyses, where in this study, the height of the model greatly affects the FAR value that will be obtained, and the length of the model will affect the existing BCR value. Therefore, the model obtained must be of good quality so that it can be used for further ISU analysis.

3.3. Intensity of Space Use Analysis (ISU)

The results of 3D visualization of the city and land use zone maps can be used to calculate the intensity of space utilization in existing conditions, which can then be compared with applicable standards, namely Local Regulation No. 3 of 2015 and Draft Local Regulation No. 5 of 2024 of the City of Jambi. Spatial Utilization Intensity includes Building Coverage Ratio (BCR), Floor Area Ratio (FAR), Green Coverage Ratio (GCR), Building Setback Line, and Building Density.

3.3.1 Building Coverage Ratio (BCR)
Building Coverage Ratio (BCR) is a useful tool for controlling land coverage by buildings. This is done to maintain a balance between the parts of the land covered by buildings and the parts that are still open, as well as to absorb water. The BCR value percentage is obtained from the ratio of building area to land area in the land use zone. The resulting BCR percentage values vary considerably, ranging from 0% to 100%, with 45% of all blocks falling within the 70% to 100% range. Based on the BCR values obtained, a comparison was then made between the existing BCR percentage values and the Jambi city RDTR design, as shown in Figure 6.

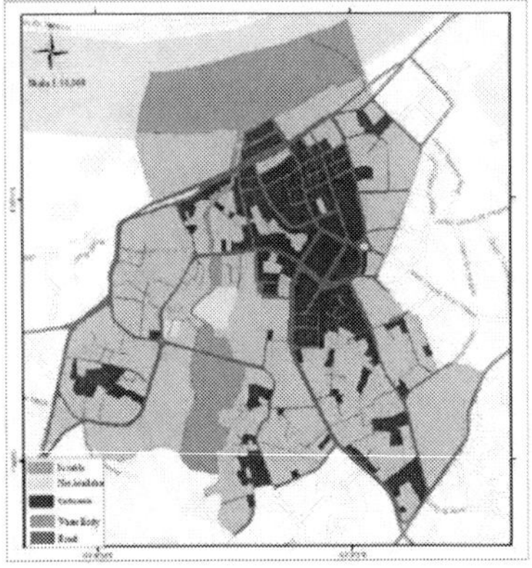

Fig. 6. BCR Compliance Level

Although dominated by compliance, a significant percentage of zones were categorized as Non-Compliant with KDB, namely 42.73%. These non-compliant areas are densely populated, dominated by commercial and service zones. The high level of non-compliance in this zone is due to the presence of retail buildings (shopping centers) where almost all of the land is used for buildings. These results indicate the need for a review and stricter enforcement of regulations in trade and service zones to ensure a balance between economic needs and environmental functions in accordance with the Regional Spatial Plan (RTRW).

3.3.2 Floor Area Ratio (FAR)
Floor Area Ratio (FAR) is the ratio of the total floor area of a building to the land area. The results of the 3D city model can be used to obtain the height of each building, which can then be used to calculate the FAR value. The FAR value in this area ranges from 0 to 5.4. The FAR values in this area are predominantly in the range of 0.6 to 1.0. The obtained FAR values are then compared with the existing conditions and applicable regulations, as shown in Figure 7.

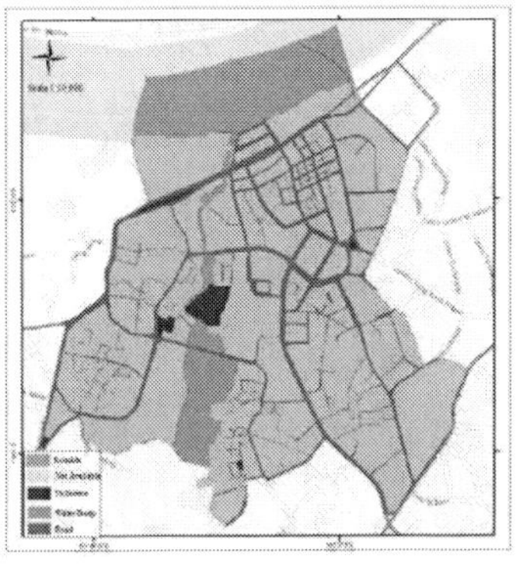

Fig. 7. FAR Compliance Level

Based on the FAR compliance level, the Pasar Jambi subdistrict has a compliance level of 95.91%, where the existing FAR value does not exceed the applicable provisions. And 4.09% of the zone does not have an FAR value. So, looking at the FAR value in the existing conditions, the FAR in the Pasar Jambi area is still far from the maximum FAR value, so this area can still be developed vertically. In general, there are no issues related to building height in the Pasar Jambi District, so vertical development is still possible in this area, as the FAR value obtained is still far from the maximum height threshold and the FAR value in this area. However, in several hotel buildings, there is a row of buildings with contrasting heights compared to other buildings in the vicinity. Therefore, there is a need for building layout guidelines that regulate building height at the micro level.

3.3.3 Green Coverage Ratio (GCR)

Green Coverage Ratio (GCR) is a percentage figure comparing the total area of open space outside buildings to the total area of land parcels. The green base coefficient is inversely proportional to the building base coefficient. The GRC percentage in this area ranges from 0% to 96.52%, with a GCR value of 0% or none dominating 41.82% of all existing zone blocks. A GCR value of 0% indicates that there is no green open space in the block. This is due to significant development pressure and limited green open space. The existing GCR values are compared with the applicable regulations, as shown in Figure 8.

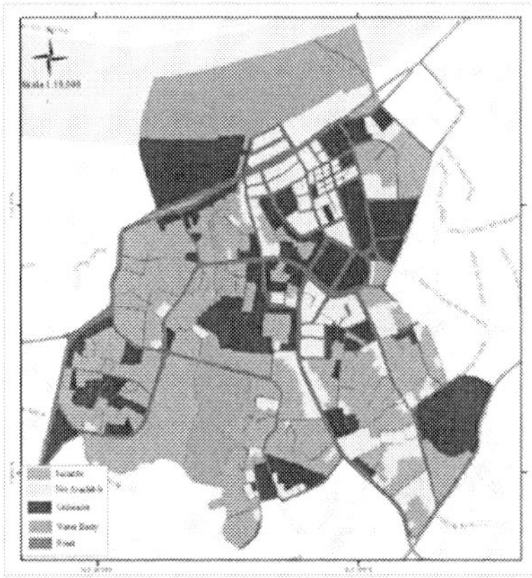

Fig. 8. GCR Compliance Level

Based on the results of comparing GCR compliance levels, 37.27% of zones meet GCR values, while 20.45% of zones do not meet GCR values, and 42.27% of zones do not have GCR values. It can be seen that in the trade and services sector, there are still many zones that do not have GCR values. The GCR aspect requires special attention. In the trade and service zones, there are still many blocks that do not have GCR or do not meet the applicable GCR standards, so that the proportion of green open space in these intensive activity areas is very limited. The rows of shop houses or shops show a tendency towards minimal greening in this zone. This condition has the potential to reduce environmental quality, reduce water absorption, and reduce thermal and visual comfort for area users. Therefore, in the future, the local government can develop green open areas, especially in trade and service areas.

3.3.4 Building Setback Line (BSL)

GSB determines the minimum distance between buildings and land boundaries or public facilities such as roads, riverbanks, railroad tracks, and high-voltage power lines. These GSB provisions are made with safety and health aspects in mind, to ensure that buildings are at a safe distance from these elements in order to avoid risks that could endanger occupants and the surrounding environment. Based on PUPR Regulation No. 28/PRT/M/2015 concerning the Determination of River and Lake Boundary Lines[14] and PUPR Regulation No. 06 of 2021 concerning the classification of minimum building distances from public facilities [15]. The level of compliance of building distances from public facilities can be seen in Figure 8.

Where the level of compliance of BSL that meets the minimum distance requirements for buildings to roads and rivers is 34%, while the distance of buildings that do not meet the requirements for the distance of buildings to roads and rivers is 61.8%.

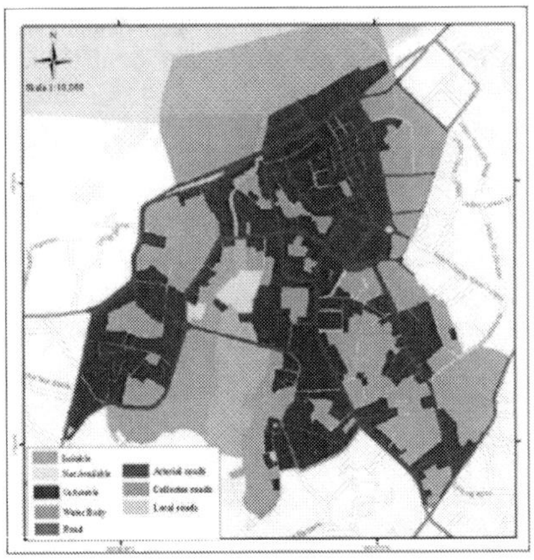

Fig. 9. BSL compatibility level

2.4 Build Density

The calculation of building density in Dangnga's dissertation (2002) explains that the identification of settlement density is carried out by calculating the ratio between the total area of buildings (building coverage) and the area of the settlement block, where, based on the 1979 Director General of Human Settlements, building density is grouped into four categories, namely low density, medium density, high density, and very high density. This can be seen in Figure 10.

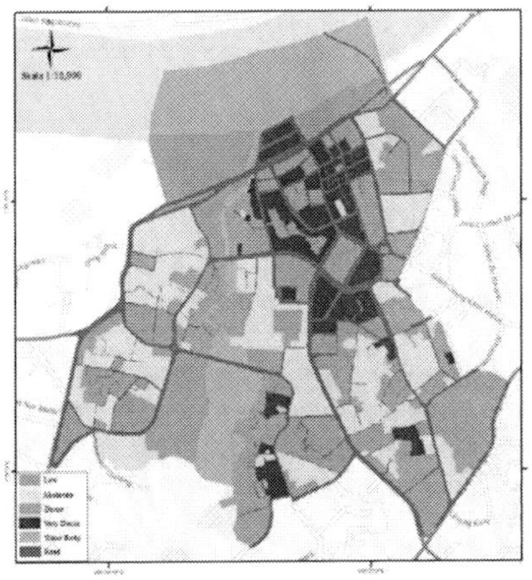

Fig. 10. Building Density Level

Based on the results of building density classification, it can be seen that several buildings located in residential zones and commercial and service zones have relatively high density levels. The high density of these buildings is closely related to the intensity of land use, where almost all available land has been used for development. This phenomenon is mainly caused by the proliferation of shop buildings that dominate the area, leaving very limited open space.

Table 5. Building Density Classify

Criteria	Classification	Number of blocks	Persentase
<40%	Low	51	23.18%
40% - 60%	Moderate	49	22.27%
60% - 80%	Dense	71	32.17%
> 80%	Very Dense	49	22.17%

The data in Table 5 Building Density Classification shows that Pasar Jambi Subdistrict has a very high building density. This can be seen in Table 5 and Figure 9. Overall, Pasar Jambi Subdistrict has a very high building density. This can be seen from the dominance of the dense and very dense categories. When combined, these two categories cover 120 blocks, which are dominated in the central to northern to southern parts. This shows that the majority of the area in this subdistrict has developed with high building intensity. This indicates the need for spatial control and optimization of the green base coefficient to balance building density and environmental sustainability.

4. Conclusions

Based on the processing and analysis of the results, it was found that 3D modeling of buildings with LoD-2.1 used aerial photo and orthophoto data with a semi-automatic method. From the model formed based on 50 measured length and height samples, the resulting RMSe value was 0.441, which did not exceed 2 meters. This result shows that the model obtained has a good spatial pattern. Thus, the model can be used for further analysis related to mapping and planning for ISU analysis. ISU analysis in Pasar Jambi Subdistrict, based on the influence of Building Coverage Ratio (BCR), Floor Area Ratio (FAR), Green Coverage Ratio (GCR), Building Setback Line, and Building Density on ISU in Pasar Jambi Subdistrict, shows a significant influence on space utilization. The BCR values in the trade and service areas and residential areas indicate that most blocks are in the dense to very dense category, so that the building footprint dominates the land surface and leaves limited open space. This condition is further reinforced by the high building density, which indicates that most of the land has been "locked" into buildings, leaving little flexible space for rearrangement, functional adaptation, or the addition of green open space.

On the other hand, FAR, which is generally still far below the maximum provisions, shows that vertical intensity can actually be increased in a controlled manner, so there is an opportunity to shift some of the growth from horizontal to vertical in order to reduce pressure on the land base. The level of compliance with spatial utilization in Pasar Jambi Subdistrict is still not optimal, as indicated by the high building density, the incompatibility of existing BSLs, and the lack of green open spaces in Pasar Jambi Subdistrict. Overall, the combination of high BCR, potential for increased FAR, low GCR, numerous BSL violations, and building density that has exceeded half of the block area illustrates high but unbalanced spatial utilization intensity. The high building density identified requires a special planning strategy, so that as a policy consideration, ISU management techniques such as the use of land use zones can be applied. This mechanism can be used as an instrument to balance building density with environmental sustainability, thereby providing additional incentives for building intensity for developers who provide green open spaces or public facilities.

References

[1] E. Rustiadi, A. E. Pravitasari, Y. Setiawan, S. P. Mulya, D. O. Pribadi, and N. Tsutsumida, "Impact of continuous Jakarta megacity urban expansion on the formation of the Jakarta-Bandung conurbation over the rice farm regions," *Cities*, vol. 111, no. October, p. 103000, 2021.

[2] Undang-Undang Republik Indonesia, "Undang-Undang Republik Indonesia No.26 Tahun 2007 Tentang Penataan Ruang," 2007.

[3] Perda Kota Jambi, "Peraturan Daerah Kota Jambi No.3 Tahun 2015 Tentang Bangunan," 2015.

[4] Perda Kota Jambi, "Peraturan Daerah Kota Jambi Nomor 5 Tahun 2024 tentang Rencana Tata Ruang Wilayah Kota Jambi Tahun 2024-2044," 2024.

[5] W. Yuewu, G. Yinxi, W. Faliang, C. Changyu, and Y. Yanqin, "Method on the Evaluation of Urban Territorial Space Intensive Utilization," vol. XLVIII, no. April, pp. 6–11, 2025.

[6] J. Gabriel, J. I. Kindangen, and R. Prijadi, "Kajian Intensitas Pemanfaatan Ruang di Kecamatan Wenang Study of Space Utilization in Wenang Sub-District," *J. Sabua*, vol. 13, no. 2, 2024.

[7] E. N. Wijayanti and H. Sutanta, "Visualisasi 3D Rencana Detail Tata Ruang Kota Yogyakarta Dengan Cesium," vol. 3, no. 2, pp. 125–134, 2020.

[8] ASPRS, "The ASPRS Positional Accuracy Standards , Edition 2 : The Geospatial Mapping Industry Guide to Best Practices," vol. 89, no. 10, pp. 581–588, 2023.

[9] G. Gröger, T. H. Kolbe, C. Nagel, and K.-H. Häfele, "Open Geospatial Consortium OGC City Geography Markup Language (CityGML) En- coding Standard," 2012.

[10] Edward Ng, *Designing High-Density Cities for Social and Environmental Sustainability.* Earthscan, 2010.

[11] M. S. Dangnga, "Pengaruh Interaksi Antara Pertumbuhan Penduduk. Perumahan. dan Kualitas Lingkungan Terhadap Sarana dan Prasarana Perumahan dan Faktor Faktor Kualitas Lingkungan.," Pascasarjana. Institut Pertanian Bogor., 2002.

[12] Permen PUPR, "Permen PUPR No 28/PRT/M/2015 Tentang Penetapan Garis Sepadan Sungai dan Garis Sepadan Danau," no. 2, 2015.

[13] F. Biljecki, J. Stoter, H. Ledoux, S. Zlatanova, and A. Çöltekin, "Geo-Information Applications of 3D City Models : State of the Art Review," pp. 2842–2889, 2015.

[14] GDI-DE, "Modeling Guide for 3D Objects Part 2 : Modeling of Buildings (LoD1 , LoD2 and LoD3)," no. November, pp. 1–51, 2014.

[15] A. B. Purnomo, "Koefisien Dasar Bangunan (KDB) dan Koefisien Lantai Bangunan (KLB) Sebagai Dasar Gubahan Massa," pp. 111–121, 2021.

Evaluation of Infrastructure Condition in the Transmigration Area to Support Regional Development (Case Study: Labangka Transmigration Area)

Berlian A. Kindhi[1*], *Nazilla* F. Novianti[2], *Daniyal* W. Athaurakhman[3], *Taufanni* Y. Fadlilah[1], and *Fatih* K. Ramadhan[1]

[1]Department of Automation Electrical Engineering, Institut Teknologi Sepuluh Nopember, Surabaya, Indonesia
[2]Department of Urban and Regional Planning, Institut Teknologi Sepuluh Nopember, Surabaya, Indonesia
[3]Department of Statistics, Institut Teknologi Sepuluh Nopember, Surabaya, Indonesia

Abstract. This study evaluates the condition of public facilities and infrastructure in the Labangka Transmigration Area as a basis for strengthening area development grounded in local potential. The evaluation was conducted through descriptive analysis, spatial analysis, content analysis, and SWOT and PESTEL approaches to identify existing conditions, service accessibility, and strategic intervention needs. The findings indicate that educational and health infrastructure in the area remains insufficient, characterized by limited facilities at the secondary education level and the absence of referral-level health services. In addition, most Posyandu facilities and primary education infrastructure are in substandard to unfit condition. The road network also exhibits substantial structural gaps, with a predominance of severely damaged and unpaved segments, particularly in Labangka Village, Suka Mulya Village, and UPT Prode SP III, thereby hindering accessibility and reducing logistical efficiency. The accessibility analysis reveals pronounced spatial disparities in education and health services, which may further exacerbate social inequality. These findings underscore the need for place-based prioritization of basic infrastructure development to foster a more livable, productive, and sustainable region.

1 Introduction

The development of transmigration areas in Indonesia is currently directed not only as a population relocation program but also as a strategic effort to establish new economic growth centers based on local resources. The transmigration program was first regulated under Law No. 15 of 1997 on Transmigration, positioning transmigration as part of national development aimed at supporting equitable development and population redistribution. Historically, from the post-independence period through the New Order era, this policy facilitated the movement of people from densely populated regions such as Java and Bali to less populated areas including Sumatra, Kalimantan, Sulawesi, Maluku, and Papua [1].

The transmigration program has long contributed to national development, particularly through population redistribution and the establishment of new settlement areas. Transmigration zones are intended not only as residential areas but also as emerging economic centers grounded in local potential (Ministry of Village, Development of Disadvantaged Regions, and Transmigration, 2020). In practice, however, the development of transmigration areas continues to face persistent challenges, especially inadequate infrastructure, which remains a major barrier to local economic activities. Priyadi's study highlights that infrastructure significantly influences the development of transmigration areas [2].

Facilities and infrastructure constitute key determinants of economic growth, particularly in newly developed regions such as transmigration areas. Road networks, bridges, clean water systems, electricity, markets, and

*Corresponding author: berlian@its.ac.id

irrigation infrastructure have direct impacts on production, distribution, and marketing of local commodities. Adequate infrastructure reduces logistical costs, enhances accessibility, and expands economic participation opportunities [3].

Infrastructure plays a central role in opening economic pathways for transmigrant communities [4]. Road and bridge networks, for example, not only facilitate the distribution of key commodities to markets but also significantly reduce logistical expenses [5]. Access to electricity and irrigation similarly forms a fundamental foundation for supporting production and distribution activities. Du and Jiao (2023) demonstrate that rural infrastructure development improves environmental governance, subsequently stimulating local economic growth [6]. Furthermore, improvements in irrigation systems have been shown to produce measurable impacts on agricultural productivity; strategies that enhance total factor productivity in rice through irrigation rehabilitation and road improvement significantly increase rice output.

Strengthening basic infrastructure such as clean water systems, electricity, and rural bridges generates multiplier effects that promote broader local economic development. A study in Banjarnegara Regency found that water and electricity infrastructure significantly contributes to local Gross Regional Domestic Product (GRDP) [7]. With adequate basic facilities, transmigrant communities gain improved access to markets and production inputs while optimizing the potential of local resources [8]. Consequently, investments in infrastructure within transmigration areas should be viewed as essential drivers of regional transformation toward sustainable and equitable economic growth.

2 Methodology

2.1 Data Analysis Methods

Various analytical methods were employed in accordance with the expected outputs of the TEP program. For the output describing the General Overview of the Area, Descriptive Analysis [9], Spatial Analysis (for geographical and administrative conditions) [10], and Content Analysis (for the historical development of the area) [11] were utilized. For the Social and Institutional Conditions output, the applied methods included Stakeholder Mapping (Salience/Power–Interest Matrix) [12], Descriptive Analysis, and Content Analysis. The Economic Conditions of the Area were examined through Descriptive Analysis, Content Analysis, and SWOT Analysis [13]. For Ecological Conditions and Basic Services, Carrying Capacity Analysis [14], Land Suitability Analysis, Spatial Analysis, Zoning and Land Designation Analysis, and Infrastructure Needs Assessment were conducted. Field findings and their synthesis were analyzed using Descriptive Analysis, SWOT Analysis for transmigration areas, and sector-based PESTEL Analysis. Finally, Strategic Recommendations were formulated using Benchmarking Analysis (policies, best practices, and community aspirations) and Descriptive Analysis [15].

2.2 Cross-Sector Coordination and Stakeholder Engagement

To ensure the effectiveness of the TEP program, cross-sector coordination and stakeholder engagement were essential components. The team was required to establish systematic collaboration with various strategic actors at the national, regional, and local levels. This collaborative mechanism involved cooperation with Local Government (through potential mapping, development program integration, and utilization of regional data), partnership with Higher Education Institutions (through placement of students/alumni and involvement of faculty members as research supervisors), synergy with Patriot Scholarship Recipients (through the role of the Expedition Team as field mentors and co-development of projects), and engagement with Local Communities (through participatory research training and regular forums for sharing research findings).

3 Result and Discussion

3.1 Existing Condition of Infrastructure in the Study Area

This subsection presents an in-depth evaluation of the existing public facilities and infrastructure (Sarpras) in the Labangka Transmigration Area. The assessment focuses on three critical pillars of infrastructure that determine the social and economic sustainability of the region: educational facilities, health facilities, and road network conditions.

Educational and health facilities constitute essential social infrastructure for improving Human Resources (HR) quality and the Human Development Index (HDI) in Labangka, which in turn supports long-term economic productivity. Meanwhile, the road network functions as the economic backbone of the area, playing a vital role in connecting agricultural production centers to markets, influencing logistical costs, and determining overall accessibility.

Table 1. Number of Educational Facilities in the Study Area
Source: Kecamatan dalam Angka 2025

Village	PAUD	TK	SD/MI	SMP/ MTs	SMA/ SMK	PT
Jaya Makmur	2	0	1	0	0	0
Suka Damai	4	1	2	3	0	0
Suka Mulya	3	0	2	1	0	0
Sekokat	2	0	1	0	0	0
Labangka	2	0	2	1	1	0
UPT Prode SP I	2	0	1	1	0	0
UPT Prode SP II	1	0	1	0	0	0
UPT Prode SP III	1	1	1	1	0	0

The evaluation indicates that the availability of educational facilities in the Labangka Transmigration Area is dominated by early childhood institutions, with a total of 17 PAUD units distributed evenly across the region, whereas only 2 kindergarten units are available, creating an access gap at the formal preschool level. At the primary level, the availability of SD/MI is relatively adequate, with 11 units present across all villages/UPTs; however, accessibility declines sharply at the lower secondary level, with only 6 SMP/MTs units operating in four out of eight villages. The most significant disparity appears at the upper secondary and tertiary levels, where only one SMA/SMK unit exists—concentrated in Labangka Village—and no higher education facilities are present throughout the entire area. This pattern indicates that the higher the education level, the more limited the accessibility for the residents of Labangka.

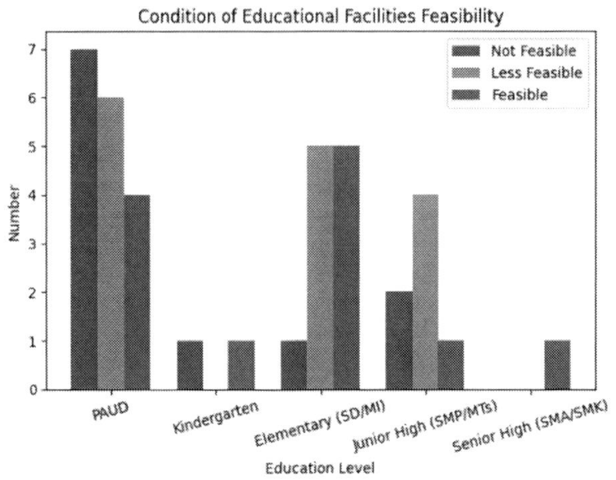

Fig. 1. Diagram of Educational Facility Conditions

The overall condition of educational facilities in the Labangka Transmigration Area is dominated by units requiring substantial rehabilitation. Early childhood education facilities (PAUD) exhibit the most critical condition, with 13 units categorized as Unfit (7 units) and Poor (6 units). At the lower secondary level, the majority of SMP/MTs facilities also fall into the Poor category (4 units), significantly outnumbering those deemed Adequate (1 unit). Although SD/MI facilities show a more balanced distribution between Poor and Adequate conditions (5 units each), one unit remains categorized as Unfit. This pattern indicates that most educational infrastructure—except for the single Adequate SMA/SMK unit and the relatively balanced condition of kindergarten facilities—requires immediate rehabilitation to ensure functional performance in supporting human capital development within the area.

Table 2. Number of Health Facilities in the Study Area
Source: Kecamatan dalam Angka 2025

Village	Hospital	Health Center	Posyandu	Pharmacy	Village Health Post	Auxiliary Health Center
Jaya Makmur	0	0	5	0	1	1
Suka Damai	0	1	6	0	1	0
Suka Mulya	0	0	3	0	1	1
Sekokat	0	0	4	0	1	1
Labangka	0	0	6	1	1	1
UPT Prode SP I	0	0	1	0	1	1
UPT Prode SP II	0	0	1	0	1	1
UPT Prode SP III	0	0	1	0	1	1

The evaluation reveals that the Labangka Transmigration Area does not possess a hospital or pharmacy in most villages, except for Labangka Village which has one pharmacy, while a single health center (Puskesmas) is available only in Suka Damai. Primary healthcare facilities are more widely distributed, with one Village Health Post (Poskesdes/Polindes) in every village/UPT (8 units total) and six Auxiliary Health Centers (Pustu). This distribution reflects a system emphasizing community-level and primary care services. Posyandu constitutes the most abundant health facility (27 units), with the highest concentrations in Suka Damai and Labangka (6 units each), highlighting substantial efforts toward maternal and child health services. However, the absence of hospitals and limited availability of health centers indicate restricted access to referral care and pharmaceutical services.

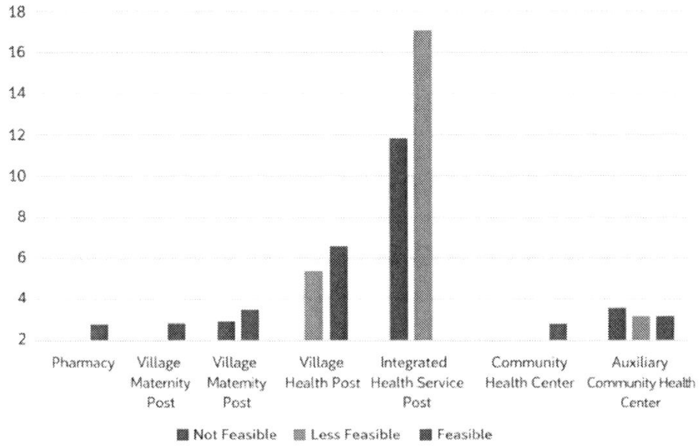

Fig. 2. Diagram of Health Facility Conditions

The condition of healthcare facilities in the Labangka Transmigration Area demonstrates severe adequacy issues, particularly among community-level facilities. Most of the 27 Posyandu units fall into the Poor (17 units) and Unfit (10 units) categories, suggesting that nearly all maternal and child health posts require comprehensive rehabilitation. While the condition of Auxiliary Health Centers shows a more balanced distribution, Village Health Posts exhibit relatively better adequacy, with a majority classified as Fit. Overall, the primary healthcare infrastructure, especially Posyandu, operates at low functional capacity, posing significant barriers to providing optimal primary healthcare services.

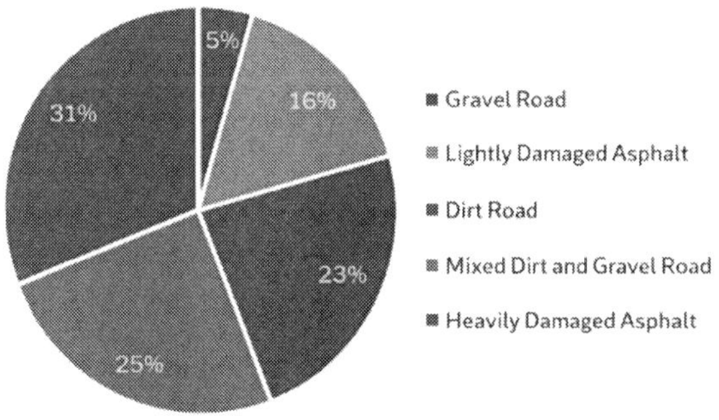

Fig. 3. Diagram of Pavement Conditions in the Study Area

Analysis of road accessibility gaps indicates that the quality of basic transportation services in Labangka is in critical condition, with 79% of problematic road segments characterized by severe structural damage and inadequate pavement (dirt or gravel roads). These conditions substantially hinder mobility and increase logistical costs for

https://doi.org/10.1051/bioconf/202621605003

transmigrant households. In several settlement units, asphalt roads have deteriorated to the point that full reconstruction is urgently required to restore the area's livability and economic function.

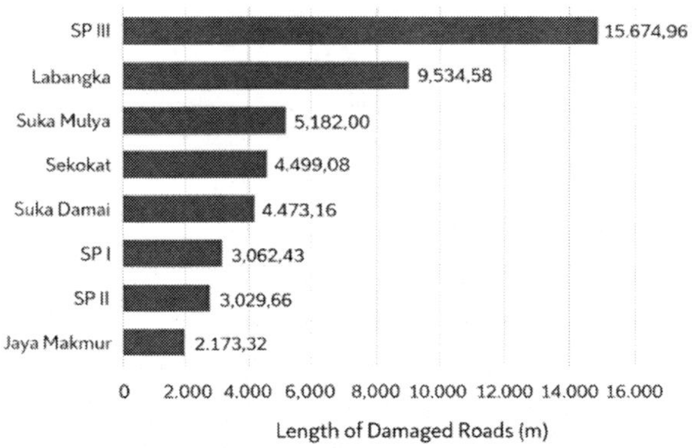

Fig. 4. Diagram of Road Damage Length in the Study Area

Further analysis of road damage highlights stark disparities in infrastructure quality, with the worst structural failures and unpaved segments concentrated in Labangka Village, which recording the highest number of severely damaged roads, followed by Suka Mulya and Sekokat. This uneven distribution underscores the need for priority allocation of road rehabilitation resources to these areas, which serve as key economic corridors within the region.

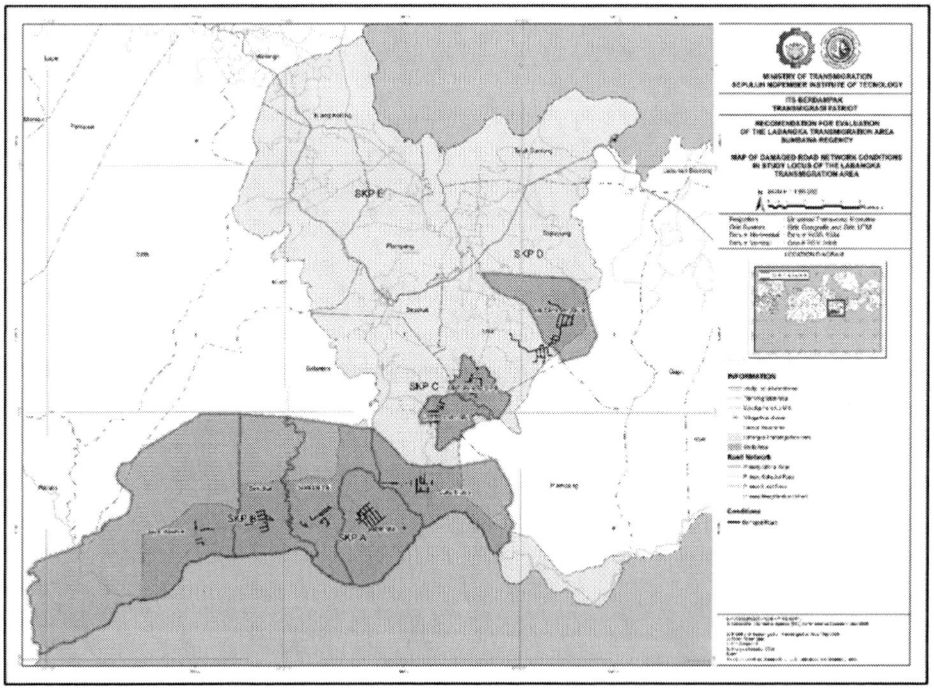

Fig. 5. Map of Damaged Road Network in the Study Area

3.2 Accessibility of Public Facilities

This subsection presents a detailed assessment of service accessibility for existing public facilities in the Labangka Transmigration Area. The analysis focuses on spatial disparities and quality conditions across three critical infrastructure pillars: Educational Facilities, Health Facilities, and Road Improvement Requirements.

Assessing the accessibility of educational and healthcare facilities is essential, as both serve as fundamental drivers of human capital development and improvements in the Human Development Index (HDI). Spatial inequity in access to these services directly contributes to social disparities and reduced economic productivity. Meanwhile, the assessment of road improvement needs is crucial because road networks serve as the economic backbone of the area, connecting production centers with markets and shaping logistical efficiency and daily mobility.

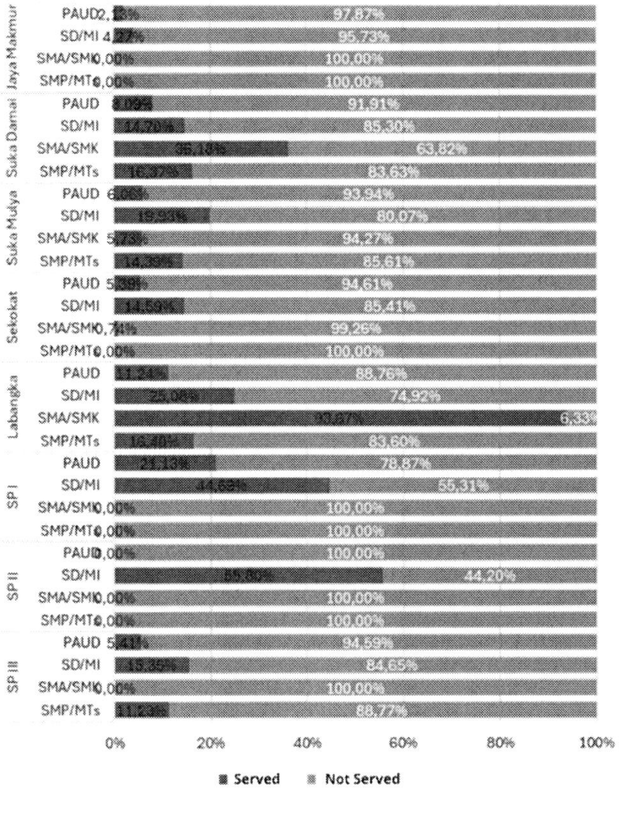

Fig. 6. Diagram of Educational Facility Accessibility

The visualization illustrates significant spatial disparities in access to educational facilities across the area. While certain villages—such as Sekokat and Suka Mulya—demonstrate high levels of service coverage (exceeding 90% of the area served), critical gaps persist in major locations such as Jaya Makmur (approximately 70% unserved) and Labangka (approximately 60% unserved). The predominance of unserved areas reflects suboptimal facility placement and underscores the need for place-based construction of new schools to ensure equitable access to basic education. A map illustrating this accessibility is provided in the following figure.

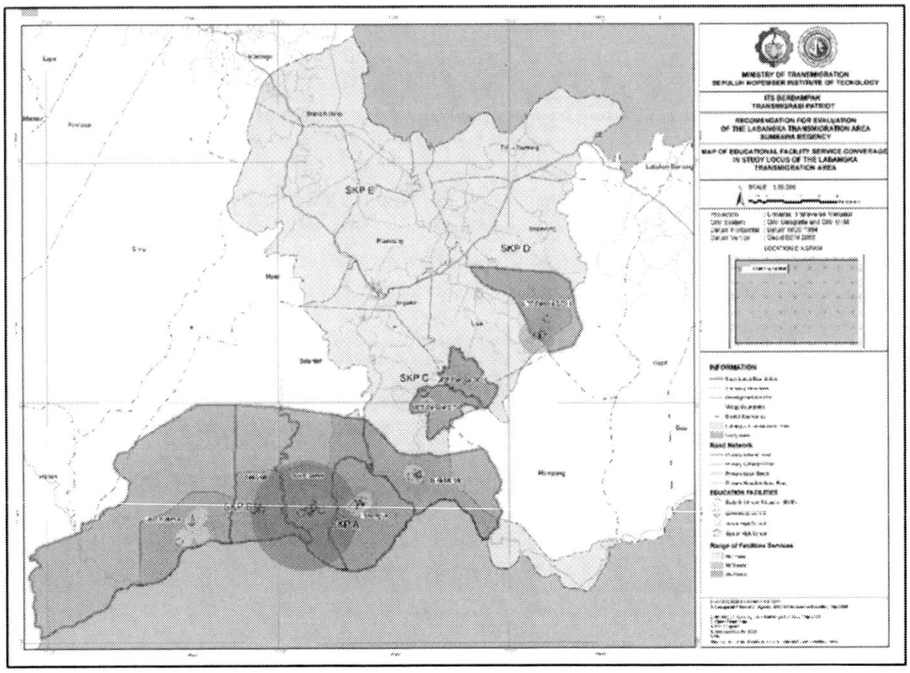

Fig. 7. Map of Educational Facility Accessibility in the Study Area

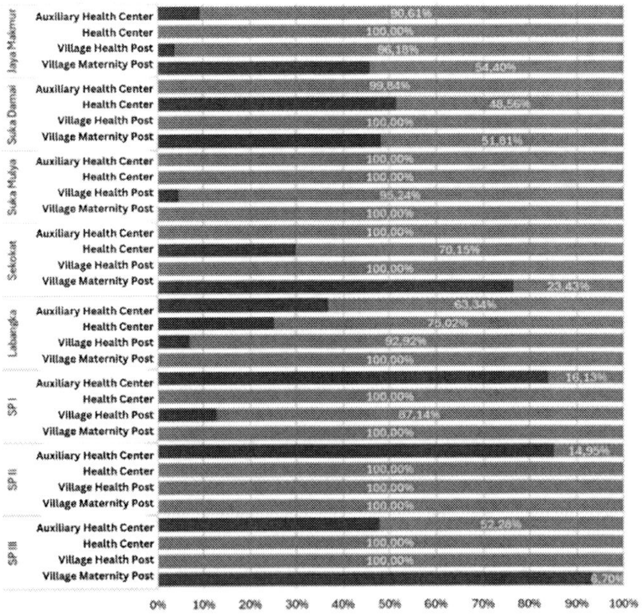

Fig. 8. Diagram of Health Facility Accessibility

The visualization reveals alarming spatial inequities in healthcare service coverage, with substantial areas across the region remaining unserved. Although Sekokat, Suka Mulya, and SP1 exhibit relatively good coverage (over 80% served), critical deficits appear in Jaya Makmur (85–90% unserved), Plampang (around 80% unserved), and Labangka (approximately 75% unserved). These substantial service gaps highlight major risks of social inequity and low facility utilization, thus necessitating strengthened primary healthcare networks through the construction of additional Auxiliary Health Centers or the deployment of mobile clinics.

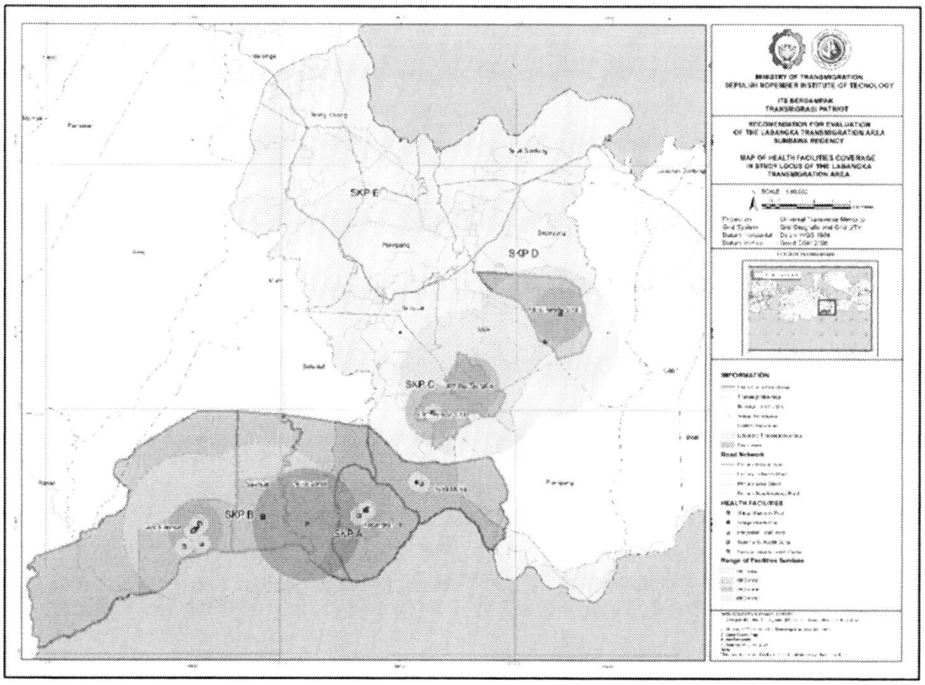

Fig. 9. Map of Health Facility Accessibility in the Study Area

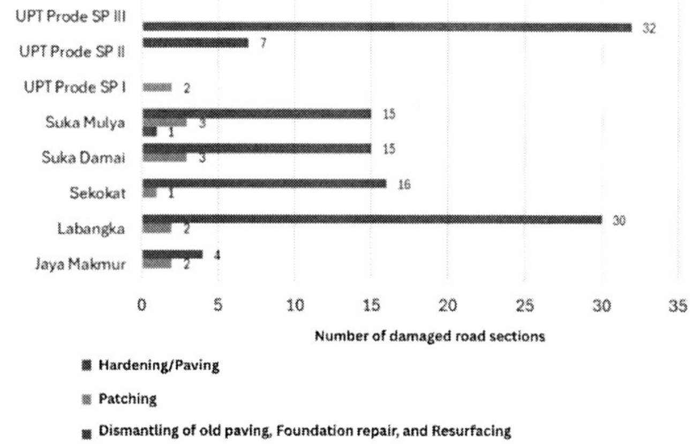

Fig. 10. Diagram of Road Damage Length in the Study Area

The visualization categorizes repair needs across damaged road segments, clearly indicating that the primary issue in the road network is not routine maintenance but fundamental structural deficiency and the absence of basic pavement. Data show that UPT Prode SP III contains the highest number of severely damaged segments requiring full reconstruction. This settlement unit represents a critical hotspot within the region, where comprehensive road rehabilitation is an urgent prerequisite for restoring mobility, reducing logistical burdens, and improving the overall viability of the transmigration area.

4 Conclusion

This study demonstrates that the condition of infrastructure in the Labangka Transmigration Area remains far from adequate to support the acceleration of socio-economic development. The availability of educational and healthcare facilities reveals imbalances across levels and locations, while the quality of these facilities is often categorized as Poor or Unfit. The severely damaged and underdeveloped road network represents the most significant barrier to population mobility and the distribution of local commodities. Furthermore, major spatial disparities in accessibility to essential services highlight structural inequities that hinder human capital development and local productivity.

These findings underscore the importance of targeted infrastructure interventions, particularly in rehabilitating educational and healthcare facilities, improving road quality, and redesigning the spatial allocation of public services based on actual accessibility needs. Implementing these recommendations is expected to strengthen the region's function as an emerging growth center and enhance the long-term welfare of transmigrant communities.

References

[1] M. N. S. NEGARA and R. INDONESIA, "UNDANG-UNDANG REPUBLIK INDONESIA NOMOR 15 TAHUN 1997 TENTANG KETRANSMIGRASIAN," 1997.

[2] I. Priyadi, "Optimalisasi Peran Daerah Tujuan Guna Mendukung Peningkatan Ekonomi Dan Status Perkembangan Kawasan Transmigrasi," vol. 6, pp. 79–85, 2024.

[3] U. N. ESCAP, "ASIA AND THE PACIFIC SDG PROGRESS REPORT," Bangkok, 2020.

[4] A. K. M. A. Ullah, S. Naushin, and S. A. Muse, "Transmigration village development : the state and community organizations in rural Indonesia," no. October, pp. 1–11, 2024, doi: 10.3389/fpos.2024.1441393.

[5] K. A. Saragih, R. Maharddhika, and M. I. Shaleh, "The Effect of Rural Bridge Infrastructure on Agricultural Economic Growth in Indonesia," vol. 1, no. 1, pp. 1–11, 2024.

[6] X. Du and F. Jiao, "How the rural infrastructure construction drives rural economic development through rural living environment governance — case study of 285 cities in China," no. September, pp. 1–18, 2023, doi: 10.3389/fenvs.2023.1280744.

[7] G. Rediansyah, R. Khoirudin, and D. Yuniarti, "Pengaruh Infrastruktur , Air dan Listrik Terhadap Perekonomian Daerah di Kabupaten Banjarnegara," pp. 1–10, 2023.

[8] T. Wahyuni, "Pengembangan Infrastruktur dalam Usaha Meningkatkan Perekonomian Masyarakat Transmigran Desa Parit Keladi 1 , Pontianak : Studi Fenomenologi," vol. 7, no. 2, pp. 163–173, 2024.

[9] D. M. Mertens, *Research and Evaluation in Education and Psychology: Integrating Diversity with Quantitative, Qualitative, and Mixed Methods Approaches*, 5th ed. California: SAGE, 2020.

[10] J. Malczewski and C. Rinner, *Multicriteria Decision Analysis in Geographic Information Science*, 1st ed. Heidelberg: Springer Berlin Heidelberg, 2021. doi: https://doi.org/10.1007/978-3-540-74757-4.

[11] S. E. Shannon, "Three Approaches to Qualitative Content Analysis," vol. 15, no. 9, pp. 1277–1288, 2005, doi: 10.1177/1049732305276687.

[12] S. Olander, *External Stakeholder Analysis in Construction Project Management*, no. September. 2006.

[13] B. S. Utomo and J. Timur, "Analysis of the Quality of Economic Growth in Districts / Cities in East Java," vol. 3, no. 3, pp. 699–712, 2024.

[14] Y. Wei, C. Huang, P. T. I. Lam, Y. Sha, and Y. Feng, "Using Urban-Carrying Capacity as a Benchmark for

Sustainable Urban Development: An Empirical Study of Beijing," pp. 3244–3268, 2015, doi: 10.3390/su7033244.
[15] D. R. Andjarweni, R. Musyafa, F. Silalahi, and F. Maharani, "Menggali Potensi Ekonomi Desa Lubuk Sao , Nagari Dalko , Kabupaten Agam : Analisis PESTEL & SWOT untuk Pengembangan Berkelanjutan," 2025.

Evaluation of the Floater N219 Structure with CEVL Material in Response to Random Wave Excitation

Windha Umi Alifia[1],*, Aries Sulisetyono[1], Bimo Hadi Putro[1]

[1] Department of Naval Architecture, Faculty of Marine Technology, Institut Teknologi Sepuluh Nopember, Surabaya, Indonesia
* Corresponding author: windha.alifia@its.ac.id

Abstract. Amphibious aircraft are seaplanes fitted with dual floats attached to the fuselage, allowing for landing and take-offs on aquatic surfaces. This research presents a method for evaluating the structural integrity of amphibious aircraft subjected to stochastic wave load stimulation through a probabilistic framework. The wave loads on the aircraft are assessed using the panel approach in a time domain simulation with ANSYS AQWA. Aircraft operations are simulated under three wave height: 0.5 m, 1.0 m, and 1.5 m, with three variations in relative wave direction: 90°, 180°, and 0°, within a wave frequency range of up to 2 rad/s. The simulation of the floater model attempts to predict the vertical bending moment experienced by the structure; this value is subsequently utilized as input in static load modelling through the finite element method to determine the maximum stress value. A probabilistic approach was employed to account for the stochastic characteristics of wave loads, with all potential loads represented as a probability density function (PDF). Moreover, the structural reliability evaluation, which ascertains the likelihood of structural failure, was estimated by combining the load PDF with the strength PDF, derived from CEVL material testing. The evaluation results indicate that the probability of structural failure is -0.34, -0.23 and -0.029 for wave heights of 0.5 m, 1.0 m, and 1,5 m, respectively. The reliability of the floater structure might be enhanced by diminishing the stress induced by wave loads by reinforcement of floater's longitudinal structure and/or the fortification of the CEVL material.

1. Introduction

N219 known as Indonesia aircraft as utility aircraft. This aircraft has 2 twin-engines with 17 until 19-seater, this transport aircraft that designed for multi-purpose missions in remote areas in Indonesia, which is as archipelago country. The N219 can carry loads up to 7,030 kg at take-off and 6,940 kg on landing operation. The operating empty weight of the aircraft (without payload) is 4,039 kg. Type Certificate for N219 aircraft is obtained after flying for 340 hours. The N219 aircraft can fly with a maximum speed of 210 knots, and a lowest speed of up to 59 knots (PTDI, 2022). The urgencies of conversion the aircraft to amphibious as seaplane or floatplane configuration are connecting each island and flexibility of her to landing and take-off not only on land but also in waterfront areas. The conversion will replace wheel on two side below the fuselage with two demi hull floaters. With these reasons, is essential to design and analyze of reliability structure floater itself, especially vertical bending moment that occur when operational of floatplane.

In the context of amphibious conversion, the floater functions as a critical structural component that directly interacts with stochastic wave environments during take-off, landing, and floating operations. Random wave excitation generates vertical bending moments along the longitudinal axis of the floater, which may govern the structural safety of the floatplane. Previous investigations on the N219 floatplane floater have shown that variations in wave height and wave heading significantly influence the magnitude of bending moments and the resulting stress distribution within the structure, with head-sea conditions producing the most critical response. Consequently, a purely deterministic strength assessment is insufficient to represent the inherent uncertainties associated with wave loading and material strength. A structural reliability-based approach, incorporating random wave loading analysis, finite element stress evaluation, and probabilistic assessment using Monte Carlo simulation, provides a more representative framework for evaluating the structural performance and failure probability of the N219 floater under realistic operational conditions (Putro et al., 2025).

In previous studies, the structural reliability of the N219 floatplane floater has been investigated to account for the uncertainties associated with wave-induced loads and material strength. (Meiliana *et al.*, 2025) demonstrated that vertical bending moments generated by random wave excitation significantly affect the reliability index of the floater structure, particularly under adverse sea conditions. Their findings indicate that deterministic stress-based evaluation may overestimate structural safety, as reliability levels can decrease to critical values even when stresses remain below allowable limits. These results highlight the necessity of incorporating probabilistic reliability analysis in the structural assessment of N219 floatplane floaters to ensure adequate safety margins under realistic operational wave environments.

2. Methodology

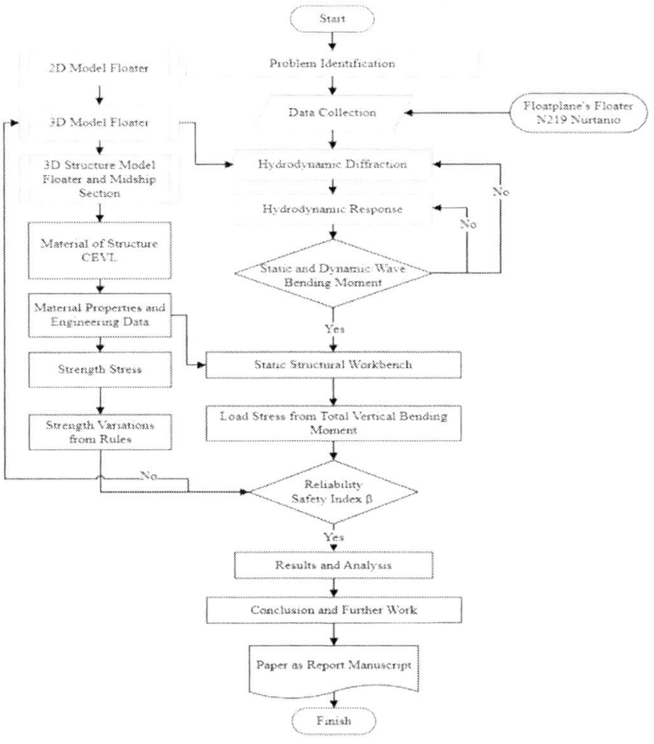

Diagram 1. Flow-Chart of Methodology Study

The research begins with problem identification, focusing on the interaction between ocean waves and the aircraft's floater during waterborne operations. Particular attention is given to the generation of wave-induced loads and their influence on vertical bending moments acting on the floater structure. Following this, a comprehensive data collection stage is conducted. This includes geometric data of the floater, material properties of the structural components, and relevant engineering parameters. A two-dimensional (2D) geometric model is first developed to support preliminary hydrodynamic characterization, followed by a detailed three-dimensional (3D) floater model to accurately represent the actual geometry. In addition, a 3D structural model of the floater and midship section is constructed to enable structural stress analysis and load transfer evaluation.

Material properties of the structure are defined based on certified engineering data and reference standards. These properties include elastic modulus, Poisson's ratio, density, and allowable stress limits, which are essential inputs for the subsequent structural strength assessment. The hydrodynamic analysis is performed through diffraction analysis, which evaluates the interaction between incident waves and the floater body. This analysis yields wave-induced pressures and forces acting on the floater surface. The results are then used in the hydrodynamic response analysis to determine the floater's response characteristics in the frequency domain. These responses form the basis for calculating both static and dynamic wave-induced bending moments.

When the calculated bending moments are found to be significant, the analysis proceeds to the static structural assessment using a static structural workbench. In this stage, the total vertical bending moment, derived from hydrodynamic loading, is applied to the structural model. The resulting stress distribution within the floater structure is evaluated to identify critical regions subjected to maximum stress levels. The computed stresses are then compared against strength criteria derived from applicable design rules and standards. If the stress levels do not satisfy the prescribed limits, structural modifications or reassessments are required.

Conversely, if the criteria are fulfilled, the study advances to the reliability evaluation phase. Structural reliability is assessed through the calculation of the safety index (β), which quantifies the margin between structural capacity and applied loads. This index serves as an indicator of the probability of failure and the overall safety level of the floater structure under wave-induced loading conditions.

Finally, the results are compiled and discussed in the results and analysis section, emphasizing the relationship between hydrodynamic response, structural stress behavior, and safety performance. The study concludes with a summary of key findings and recommendations for future work, and the complete analysis is documented in the form of a scientific report manuscript.

2.1 N219 Floatplane and Floater Lines Plan Configuration

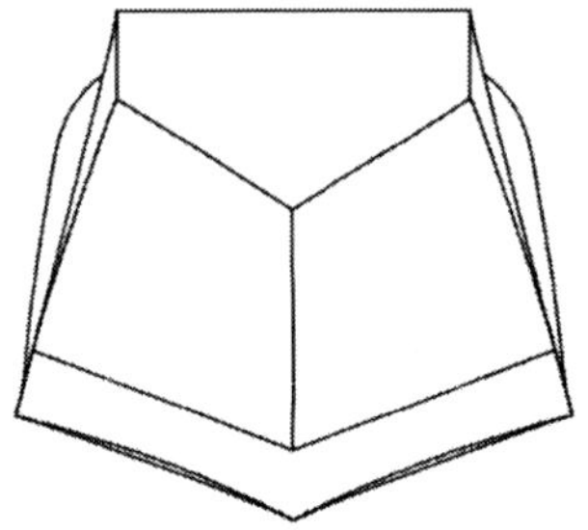

Fig. 1. After body of Floater

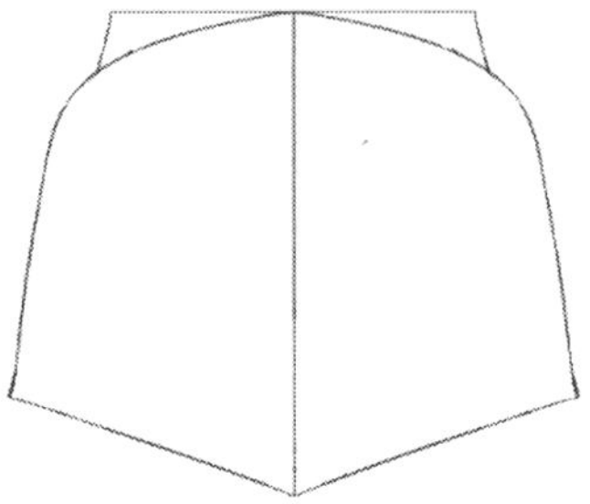

Fig. 2. Fore body of Floater

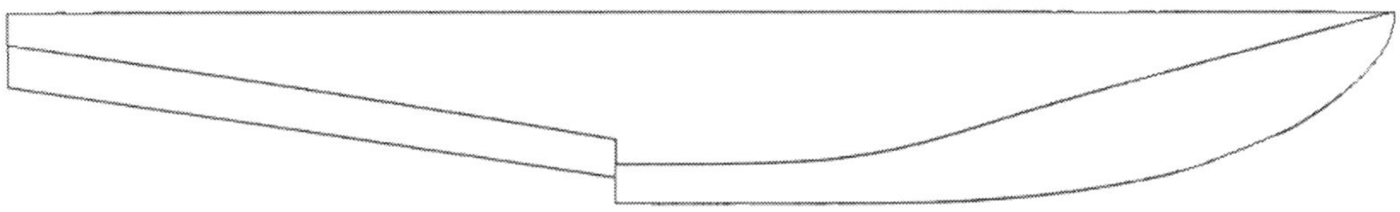

Fig. 3. Sheer Plan of Floater

The design of floater according to (Aliffrananda, 2021) and (Washoya, 2019) used for this study to analyzed the reliability and to design the structural that suitable and reliable during bending moment that applied in operational itself.

Table 1. Main Dimension of Floater

Principal Dimension	Magnitude
Length Overall (LOA)	9.60 m
Length of Waterline (LWL)	9.448 m
Breadth of Demi hull (B)	1.25 m
Height (H)	1.10 m
Draught (T)	0.77 m
Space between demi hull (S)	3.95 m
Forebody length (L_f)	5.4 m
Afterbody length (L_a)	4.2 m

Deadrise angle	20°
Displacement (Δ)	7.330 ton

2.2 Floater Structure and Material

Material used for this study is called Carbon Vinyl Ester Laminate. Known as suitable in application aircraft, coatings, pipes, tanks, windmill blades, carbon vinyl ester laminate (CEVL) with outstanding properties related to chemical and corrosion resistance (Wypych, 2016). Constraints of structural design in floater are weight and strength. The weight of total demi-hull should not exceed the requirement allowed, which is 2.99 ton.

Table 2. Material Properties of CEVL

Material Property	Magnitude	Unit
Density (ρ)	1600	kg/m^3
Young Modulus (E)	55	GPa
Tensile Strength (X_T)	890	MPa
G_{12}	4.5	GPa

Source: (Gargano et al., 2019)

From Table 2, X_T represents the maximum tensile stress that the material can withstand along the primary fibre direction before failure, as for G_{12} represents the material's resistance to shear deformation within the laminate plane, specifically in the plane defined by the fibre direction (1-direction) and the transverse direction (2-direction). Material Properties and engineering data input in static structural workbench for material Carbon-Vinyl Ester Laminate can be seen in Table 3.

Table 3. Material Properties CEVL for Input Calculations

Material Properties CEVL			
Property		**unit**	
Density	ρ	1600	kg/m^3
Young's Modulus	E	55	GPa
Tensile Strength	X_T	890	MPa
Poisson Ratio	υ	0.25	

Table 4. Weight Floater Calculation

Scantling Structure with Carbon Vinyl Ester Laminate CEVL			
Part	**area (m^2) or volume (m^3)**	**Thickness (m)**	**Weight (ton)**
Shell	29.15	0.014	1.30
Bulkhead	2.25	0.012	0.08
Longitudinal	0.011		0.036
	Demi hull	∑weight (ton)	1.42

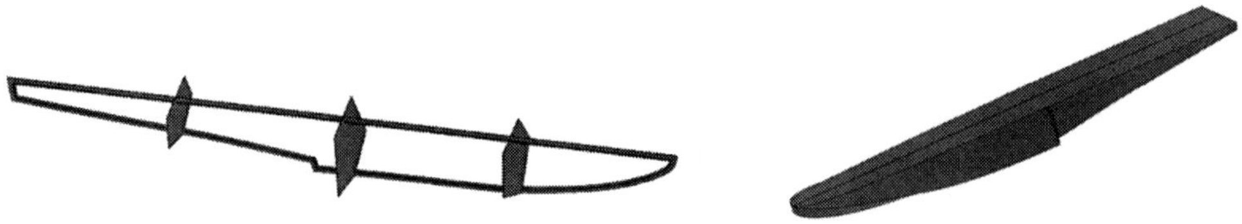

Fig. 4. Bulkhead Configuration Structure and 3D Model Floater

The minimal requirement of watertight compartment that must be in each hull of floater is 4 (four) units. This means each floater should have 3 watertight bulkheads. The calculation of weight estimation of hull structure above with tabulation manually measures volume of scantling times with density of CEVL. With material define as CEVL, Table 5 below detailed scantling that used for structure of floater.

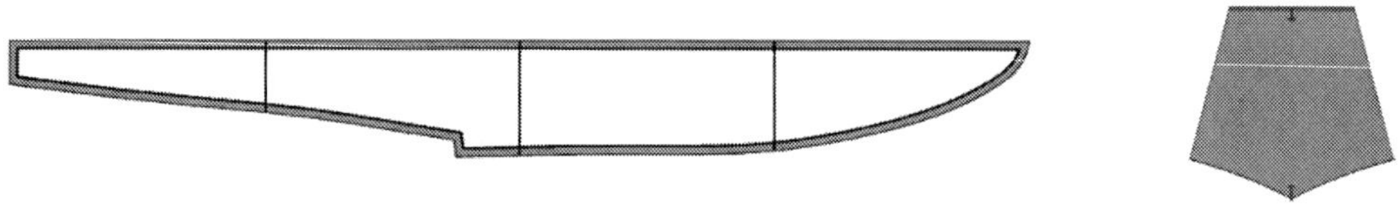

Fig. 5. Bulkhead Arrangement and Midship Section Structure

Table 5. Scantling Floater

Part		Thickness (m)
Shell	Deck	0.014
	Side	0.014
	Bottom	0.014
	Step	0.014
	Transom	0.014
Bulkhead	BHD 1	0.012
	BHD 2	0.012
	BHD 3	0.012
Longitudinal	Web	0.06
	Face	0.04

2.3 Ship Motions

Grid independence conducted to find optimum configuration. With several types of grids or meshing, found optimum configuration as shown in Figure 6. Comparison value of element or mesh size with minimum response in Mesh element size used 0.075 m for Hydrodynamic Diffraction analysis with amount number of element and nodes are 60,959 and 153,725, respectively.

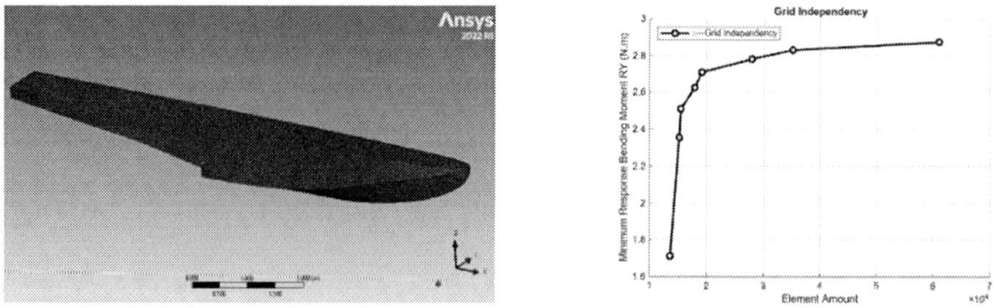

Fig. 6. Mesh model floater Hydrodynamic Diffraction and Grid Independency Study

https://doi.org/10.1051/bioconf/202621605004

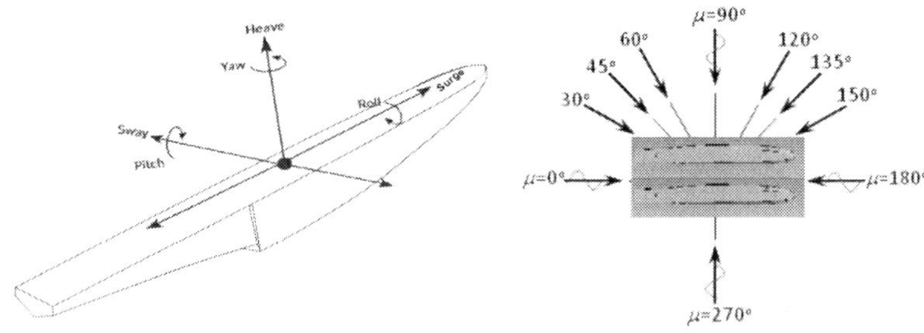

Fig.7. Degrees of Freedom and Wave Direction to Body

Based on Figure 7 shown that degrees of freedom for floating structure and wave direction along of demi hull of floatplane's floater. Ship motion formulation as general defines with formulation of $(M + A(x))\ddot{x} + B(x, \dot{x})\dot{x} + C(x)x = F(x, \dot{x}, t)$. This study analyzed vertical bending moment in floater with 3 wave directions head seas, beam seas and following seas.

Fig. 8. Wave Direction to Floater

The definition by (Bretschneider, 1959) at about wave spectrum is represents the wave energy (or wave amplitude) distribution of individual wave frequencies in a stationary sea state. Pierson-Moskowitz (PM) known as a type of wave spectra that optimum with 2 parameters.

$$S(\omega) = \frac{5}{16} H_s^2 \frac{\omega_p^4}{\omega^5} \exp\left[-\frac{5}{4} \left(\frac{\omega_p}{\omega} \right)^4 \right]$$

T_p with ITTC definition as the wave or the period of another response like vertical bending moment, in s second, with most energy in the wave or response spectrum, i.e. the most probable maximum wave or response in a short-term sea state. The wave frequency for ω with unit rad/s and ω_p define as spectral peak frequency in rad/s. With this parameter, this research used the formulation of Pierson-Moskovitz's formula as below:

$$S(\omega) = \frac{1}{2\pi} \frac{H_s^2}{4} \left(\frac{2\pi}{T_z} \right)^4 \omega^{-5} \exp\left[-\left(\frac{2\pi}{T_z} \right)^4 \omega^{-4} \right]$$

Where S (ω) is Pierson-Moskovitz spectrum (m2/(rad/s)), H_s known as significant wave height, and T_Z as average zero up-crossing wave period (s), also ω wave frequency (rad/s). From plot of graphics results shown in Figure 9 which range of wave frequency (rad/s) that implied significant value for spectral energy. There were 4 range of frequency, 0.1 to 1.0, 1.0 to 2.0, 2.0 to 3.0 to 4.0. Frequency of wave 0.1 until 1.0 observed give significant value for Hs 0.5 m, with range 0.4 1.2 for Hs 1.0 m. This study analyzed and found from 0.1 until 2.0 rad/s for significant range period in hydrodynamic response analysis.

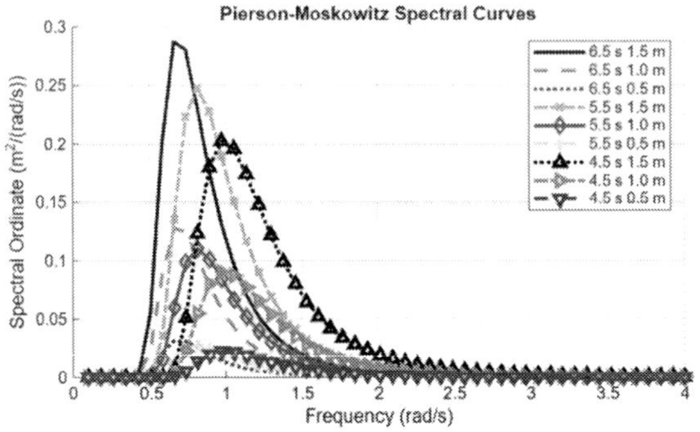

Fig. 9. Spectrum Wave for Significant Wave (Hs) variations

This study conducted for area of java seas. The java seas have surface area of the sea is 320,000 square kilometers and the average depth is 46 m. The maximum length and maximum width of the sea are 1600 km and 380 km respectively. The average depth of the sea is just 46 meters because it is a shallow sea as shown in Figure 10 (The Editors of Encyclopedia Britannica, 2019).

The analysis attended to several values of H_s and peak period T_z that have value more than 500 for the probabilities. This wave scatter diagram as seen in Table 6, consists of several cells. Each cell of wave scatter diagram which are short-term descriptions of the sea in terms of joint probability of occurrence of a significant wave height Hs and a characteristic period. These cells contain 3 minimum data items below:

a. The significant wave height Hs (usually in meter),
b. The characteristic wave period (in seconds),
c. The number of occurrences for the sea state.

Table 6. Wave scatter Diagram of Java Sea

Hs (m)	Wave Periods (s)													Sum
	3.5	4.5	5.5	6.5	7.5	8.5	9.5	10.5	11.5	12.5	13.5	14.5	15.5	
0.5	94	321	161	11	19	14	7	5	4	2	1			639
1	227	1197	1388	712	316	285	166	89	57	27	8	3	1	4476
1.5		185	680	408	527	213	162	120	67	31	11	5	2	2411
2			71	229	404	323	66	37	34	19	10	9	2	1204
2.5				33	185	290	86	20	12	7	2	2	1	638
3				1	33	218	97	10	4	2			1	366
3.5					1	60	81	8	1	1				152
4						7	45	10	1					63
4.5							20	16	1					37
5							2	7	1					10
5.5								3	1					4

Sum	321	1703	2300	1394	1485	1410	732	325	183	89	32	19	7	10000

Source: (Wibowo, 2012; Kadhafi, 2016)

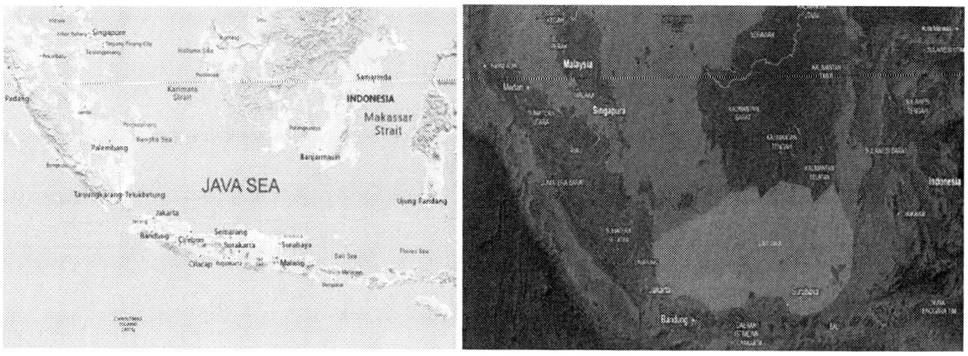

Fig. 10. Java Seas in Map
Source: (The Editors of Encyclopedia Britannica, 2019)

The peak period should be defined for calculated in Hydrodynamic response. The range of significant frequency starts from 0.1 to finish frequency 2.0 rad/s with T_z are 4.5 s, 5.5 s and 6.5 s. From Table 7, the value of periods that are used for zero up-crossing period in wave The calculation with wave direction 0°, 90°, and 180° and forward speed of floater used 9.625 m/s each Tz and wave directions, this study calculated for bending moment static and dynamic has been done with several results. As explained for static vertical bending moment, conducted calculations for RY components in ANSYS AQWA TM. After Statical Bending Moment found, Dynamic Bending Moment used the workbench Hydrodynamic Diffractions, which got several values for each variation shown in Figure 12, Figure 13 and Figure 14.

Table 7. Wave Periods and its Frequencies from Java Seas Wave Scatter

Hs(m)	Wave Periods (s)			Sum
	4.5	**5.5**	**6.5**	
0.5	321	161	11	493
1.0	1197	1388	712	3297
1.5	185	680	408	1273
Sum	**1703**	**2229**	**1131**	**5063**
Hs (m)	**Frequencies (rad/s)**			**Sum**
	1.39	**1.14**	**0.96**	
0.5	321	161	11	493
1.0	1197	1388	712	3297
1.5	185	680	408	1273
Sum	**1703**	**2229**	**1131**	**5063**

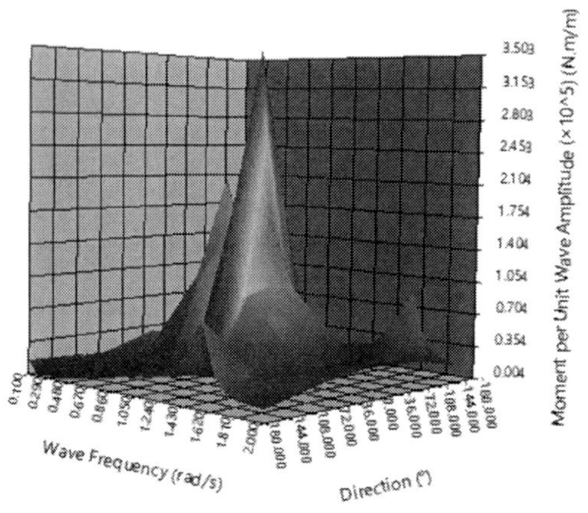

Fig. 11. Typical of Bending Moment (RAO) for RY Vertical

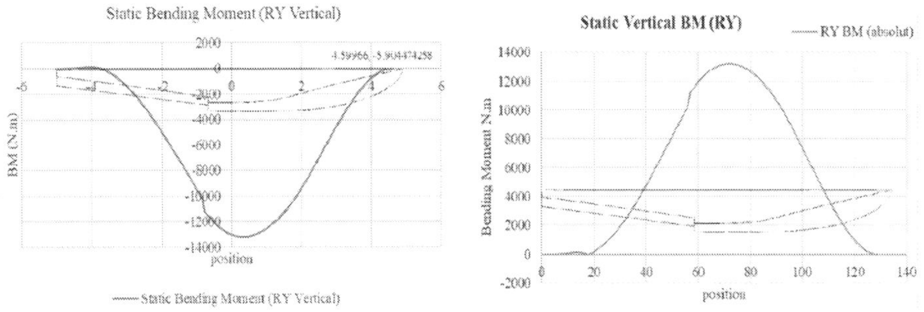

Fig. 12. Static Bending Moment Floater for Vertical Bending Moment (RY)

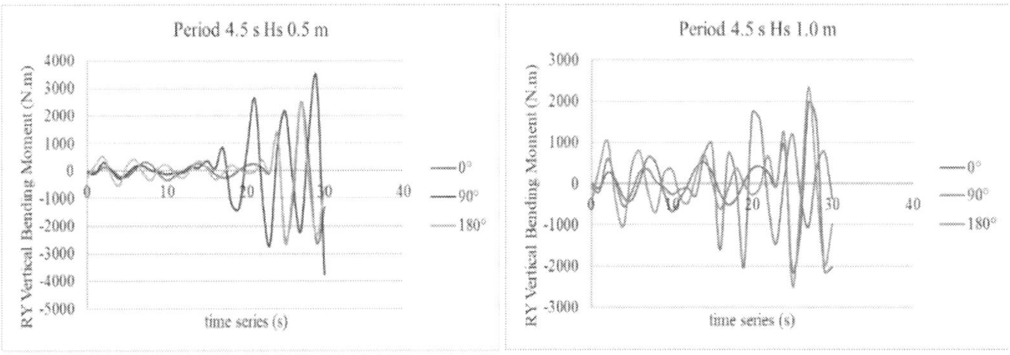

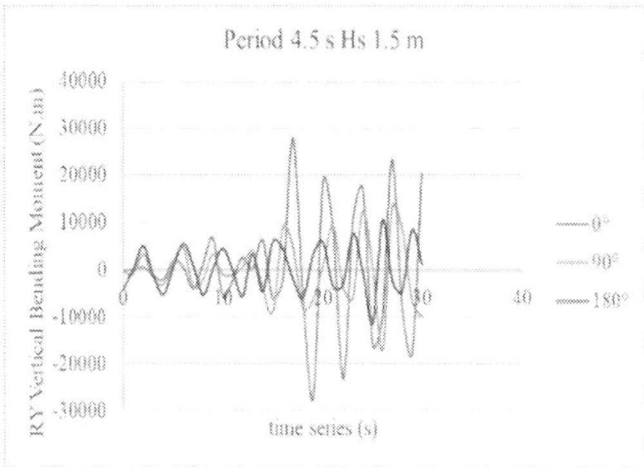

Fig. 13. Dynamic Vertical Bending Moment Hs 0.5 m

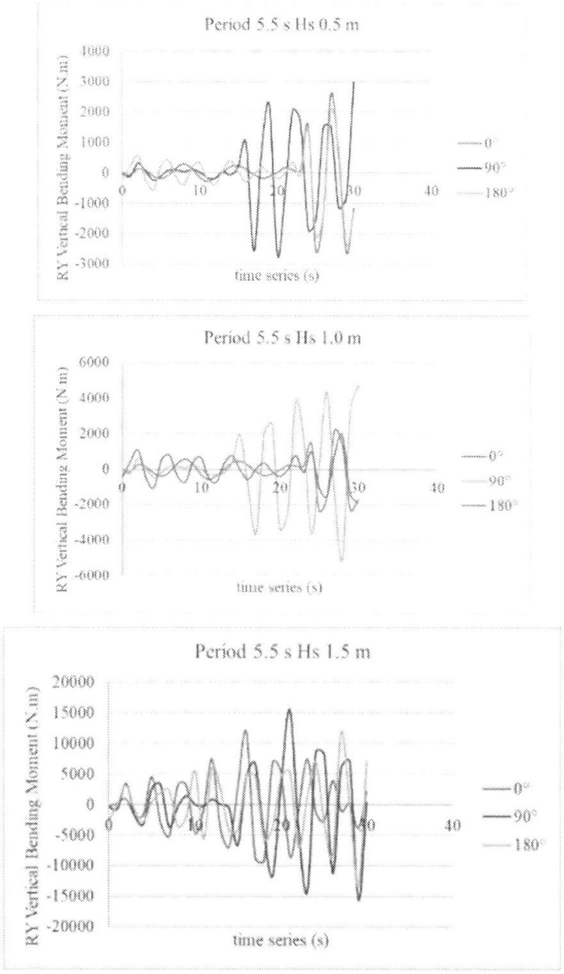

Fig. 14. Dynamic Vertical Bending Moment Hs 1.0 m

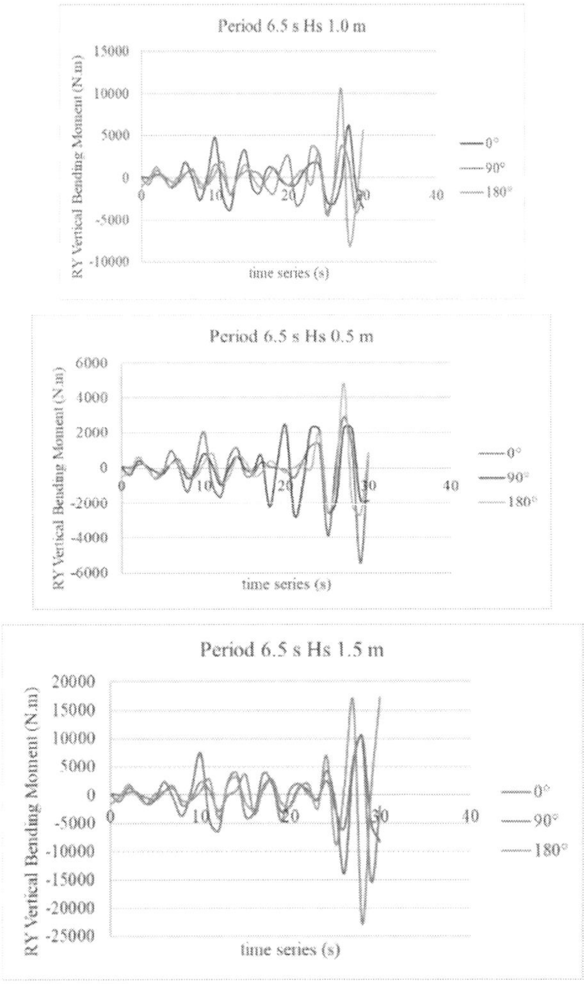

Fig. 15. Dynamic Vertical Bending Moment Hs 1.5 m

2.4 Structural Stress Analysis

Structural stress analysis is performed to quantify the stress response of the amphibious aircraft floater under wave-induced loading conditions. The analysis is conducted using the Finite Element Method (FEM), which is widely applied in marine and offshore structural engineering for evaluating the response of floating structures subjected to hydrodynamic loads. The adopted methodology follows established practices recommended by the Ship Structure Committee (SSC) for global structural strength assessment under wave excitation.

Wave-induced loads acting on the floater are obtained from time-domain hydrodynamic simulations using ANSYS AQWA. The hydrodynamic response is expressed in terms of vertical bending moment, which represents the dominant global load component affecting the longitudinal strength of floating structures. According to SSC guidelines, global wave loads may be transferred to the structural model in the form of equivalent bending moments when assessing overall structural response.

The structural analysis is performed using a quasi-static FEM approach, assuming that the wave encounter frequencies are significantly lower than the natural frequencies of the floater structure. Under this assumption, dynamic amplification effects are considered negligible, and the structural response is governed primarily by static equilibrium. This approach is commonly applied in wave-induced structural analyses of ships and floating bodies and provides an efficient and sufficiently accurate representation of global stress behavior.

The floater structure is discretized into finite elements connected at nodal points, forming a numerical model capable of capturing global deformation and stress distribution. The governing equilibrium equation of the FEM formulation is expressed as:

$$Ku = F$$

where K denotes the global stiffness matrix, u is the nodal displacement vector, and F represents the external load vector derived from the applied vertical bending moment. Material properties and geometric characteristics of the floater are incorporated into the stiffness formulation.

The displacement field obtained from the FEM solution is used to compute strain and stress distributions within the structure based on linear elastic constitutive relations. The structural stress response is evaluated using the von Mises equivalent stress, which is adopted as the representative stress parameter for assessing structural performance. Stress evaluation is focused on longitudinal structural components, as these members primarily resist global bending loads induced by wave action.

Considering the stochastic nature of wave excitation, the resulting structural stress is treated as a random variable. The probabilistic distribution of stress is obtained by propagating the wave load probability density function through the FEM-based structural response model. This procedure ensures that the variability of the wave environment is reflected in the structural stress response.

Structural reliability is assessed by combining the stress probability density function with the material strength probability density function obtained from CEVL material testing. Failure is defined using a limit state function:

$$g(R,S) = R - S$$

where R represents material resistance and S denotes applied structural stress. Structural failure is assumed to occur when $g(R,S) < 0$. The probability of failure is determined by evaluating the overlap between the stress and strength distributions for each wave condition.

The adopted structural stress analysis framework provides a consistent basis for evaluating the reliability of amphibious aircraft floaters operating in stochastic wave environments. The results indicate that structural reliability may be improved through reduction of wave-induced stress, particularly by reinforcing longitudinal structural members and enhancing material strength, in accordance with SSC-based structural design recommendations.

2.5 Load Stress Calculation

The analysis of stress that floating body experiences could be several types combinations of load, but in this study only analyzed about vertical bending moment. By definition of about bending moment define as R or rotation component in AQWA™ (Manual Aqwa 14.5, n.d.). This study conducted only vertical bending moments from coupled motions heave and pitch. From data static and dynamic bending moment, applied in static structural workbench to found and calculated for bending moment total (MT).

Table 8. Vertical Bending Moment

Variation	Wave Period (s)	Wave Height (m)	Wave Directions (°)	Ms Static (N.m)	Mw Wave or Dynamic (N.m)	MT Total (N.m)
V1	4.5	0.5	Following Seas 0°	13201.10	2585.67	15786.78
V2			Beam Seas 90°		3755.42	16956.53
V3			Head Seas 180°		2645.05	15846.14
V4		1.0	Following Seas 0 °		2110.30	15311.40
V5			Beam Seas 90°		2017.14	15218.25

Variation	Wave Period (s)	Wave Height (m)	Wave Directions (°)	Ms Static (N.m)	Mw Wave or Dynamic (N.m)	MT Total (N.m)
V6			Head Seas 180°		2474.99	15676.09
V7			Following Seas 0 °		28091.93	41293.04
V8		1.5	Beam Seas 90°		17160.94	30362.05
V9			Head Seas 180°		11589.88	24790.99
V10			Following Seas 0 °		2637.71	15838.81
V11		0.5	Beam Seas 90°		2987.92	16189.03
V12			Head Seas 180°		2320.70	15521.80
V13			Following Seas 0 °		2335.98	15537.08
V14	5.5	1.0	Beam Seas 90°		5024.60	18225.71
V15			Head Seas 180°		1906.92	15108.03
V16			Following Seas 0 °		11986.41	25187.52
V17		1.5	Beam Seas 90°		15495.11	28696.22
V18			Head Seas 180°		13185.63	26386.74
V19			Following Seas 0 °		5442.77	18643.87
V20		0.5	Beam Seas 90°		2681.49	15882.60
V21			Head Seas 180°		4825.26	18026.37
V22			Following Seas 0 °		6203.50	19404.61
V23	6.5	1.0	Beam Seas 90°		4424.65	17625.76
V24			Head Seas 180°		10543.95	23745.06
V25			Following Seas 0 °		13853.17	27054.28
V26		1.5	Beam Seas 90°		15199.5	28400.65
V27			Head Seas 180°		22654.03	35855.14

Remoted displacement at 3 points used for calculated and analyzed in workbench as boundary condition. With several load and moment with remoted displacement applied as boundary conditions, stress value found for every variation of case. From Table 8 above applied to static structural as moment at 2 position AP and FP from bending moment total each variation with several value static or constant:

a. Hydrostatic Load with value generated by workbench,
b. Load Body with value as weight load from mass displacement of floater type found 35953.65 N with gravitational acceleration 9.81 m/s^2,
c. Remoted Displacement for boundary conditions.

From static structural workbench analysis, found that distribution of load stress accumulation centralized in area transitional along from forebody to after body. Another area is simply evaluated as evenly distributed from stress distributions. From vertical bending moments and other loads that applied in through floater structure, shown in Figure 18 that deformation and stress highly occurred in area of stepped plate with bottom stringer and bottom plate near it.

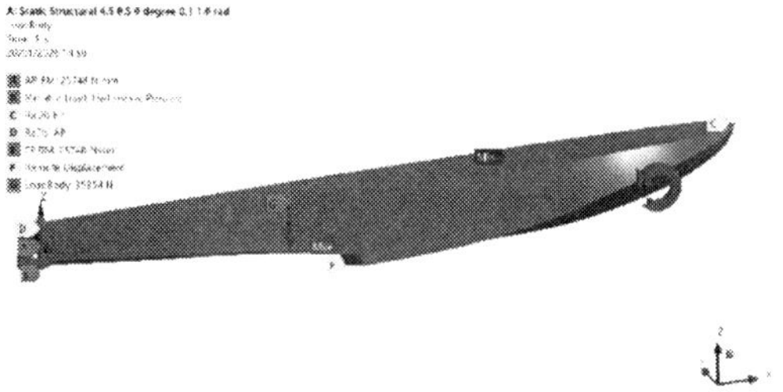

Fig. 16. AP and FP Bending Moment Applied with Hydrostatic Pressure Load and Load Body Applied in Static Structural as Boundary Condition

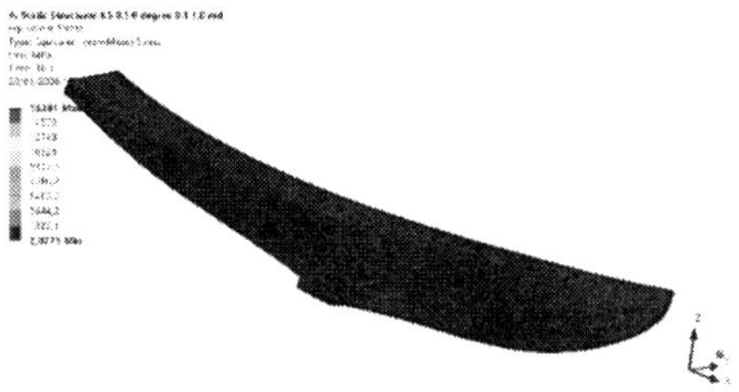

Stress in period 4.5 s and Hs 0.5 m of head seas

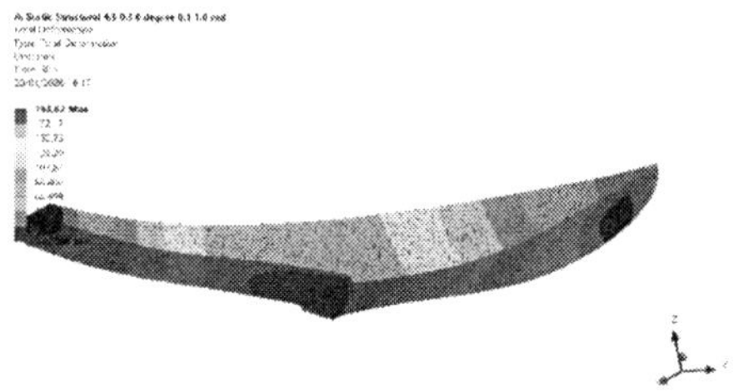

Deformation in period 4.5 s and Hs 0.5 m of head seas

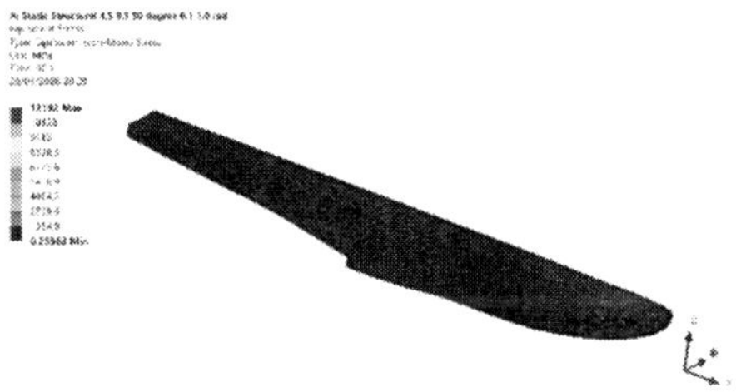

Stress in period 4.5 s and Hs 1.0 m of beam seas

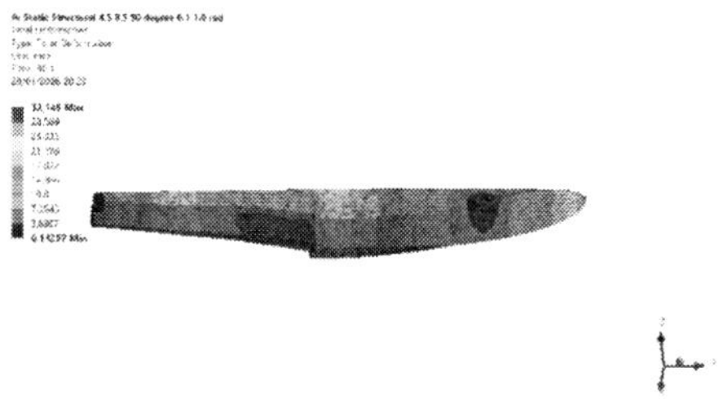

Deformation in period 4.5 s and Hs 1.0 m of beam seas

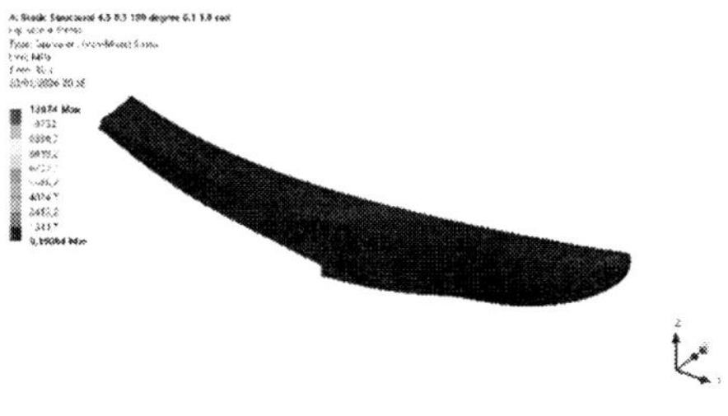

Stress in period 4.5 s and Hs 1.5 m of following seas

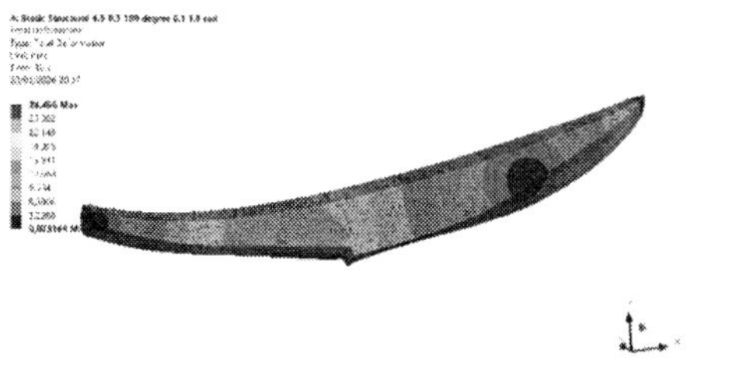

Deformation in period 4.5 s and Hs 1.5 m following seas

Fig. 17. Load Stress and Deformation Results Calculation in Static Structural

2.6 Assessment Safety Index (β) for Material Structure

For the strength of stress structure has define with reductions factor by (Biro Kasifikasi Indonesia BKI, 2021) as shown in Table 9. With the regulation and recommendation or guidelines from (Biro Klasifikasi Indonesia BKI, 2021), this study calculate and stated for variation reduction in tensile strength of structure with CEVL material shown in Table 10. Probability density function from value of variations CEVL used for strength analysis shown in Figure19. Based on the BKI *Rules For Classification and Construction*, the allowable variations for fibre-reinforced composite materials can be summarized as follows:

$$\sigma_{\text{perm}} = \frac{R_m \cdot R_1 \cdot R_2 \cdot R_3 \cdots R_8}{S} \, [\text{N/mm}^2]$$

Table 9. Reduction Factor for Laminates

Reduction Factor Description	Symbol	Value / Limitation
Statistical long-term loading (sea loads)	(R_1)	(< 0.75)
Static long-term loading (tank walls)	(R_2)	(< 0.50)
Raised temperature	(R_3)	(< 0.90)
Ageing	(R_4)	(< 0.80)
Manufacturing (hand lay-up)	(R_5)	(= 0.90)
Manufacturing (prepreg)	(R_5)	(= 1.00)
Fatigue, (N < 10^2)	(R_6)	(= 1.00)
Fatigue, (N < 10^3)	(R_6)	(= 0.80)
Fatigue, (N < 10^6)	(R_6)	(= 0.40)
Different material properties	(R_7)	(< 0.90)
Moisture	(R_8)	(< 0.80)

Table 10. Variations Reduction for Material CEVL

NO	Reduction for Laminated	Symbol	σ_{perm} A	σ_{perm} B	σ_{perm} C	σ_{perm} D	σ_{perm} E	σ_{perm} F	σ_{perm} G	σ_{perm} H	σ_{perm} I	σ_{perm} J	σ_{perm} K	σ_{perm} L
1	Statical long-term loading (sea loads)	R_1< 0.75	0.7	0.70	0.70	0.70	0.70	0.70	0.70	0.70	0.70	0.70	0.70	0.70
2	Static-long term loading (wall tanks)	R_2 < 0.5	0.45	0.45	0.45	0.45	0.45	0.45	0.45	0.45	0.45	0.45	0.45	0.45
3	Raised temperature	R_3 < 0.9	0.85	0.85	0.85	0.85	0.85	0.85	0.85	0.85	0.85	0.85	0.85	0.85
4	Ageing	R_4 < 0.8	0.75	0.75	0.75	0.75	0.75	0.75	0.75	0.75	0.75	0.75	0.75	0.75
5	Manufacturing **i.e.:**													
	- Hand Lay-up	R_5 = 0.9	1.00	1.00	1.00	0.90	0.90	0.90	0.85	0.85	0.85	0.77	0.77	0.77
	- Prepeg (human error)	R_5 = 1.0												
6	Fatigue N load:													
	N < 10^2	R_6 = 1.0	1.00	0.80	0.40	1.00	0.80	0.40	1.00	0.80	0.40	1.00	0.80	0.40
	N < 10^3	R_6 = 0.8												
	N < 10^6	R_6 = 0.4												
7	Different material properties	R_7 < 0.9	0.85	0.85	0.85	0.85	0.85	0.85	0.85	0.85	0.85	0.85	0.85	0.85
8	Moisture	R_8 < 0.8	0.75	0.75	0.75	0.75	0.75	0.75	0.75	0.75	0.75	0.75	0.75	0.75
	Strength		**103.578**	**82.863**	**41.431**	**93.220**	**74.576**	**37.288**	**88.041**	**70.433**	**35.217**	**79.237**	**63.390**	**31.695**

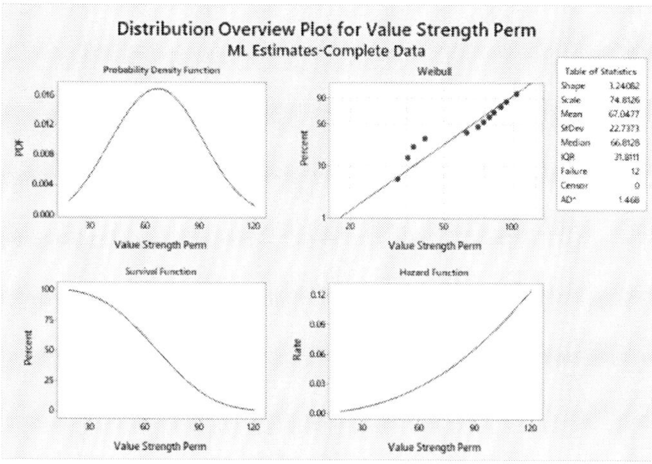

Fig. 18. Distribution of Strength of Stress Material CEVL for Structure Floater

3. Result and Analysis

With several steps of analyses in this study, to found strength stress and load stress data from design, Hydrodynamic Diffraction, Hydrodynamic Response until Static Structural as workbench analyses in ANSYS™, in this stage conducted analysis for reliability. Reliability itself define as is its ability to fulfil its design purpose for some specified reference period (Thoft-Christensen & Murotsu, 1986). Most structures have a number of possible failure modes. Therefore, in determining the reliability of a structure this fact must be taken into account (Daniel Mohammad Rosyid, 2007). The first step will usually be to estimate the structural reliability with respect to each specified failure mode, i.e. the probability that the structure will not attain each specified failure mode (ultimate or serviceability) during the specified reference period.

$$\beta = \frac{\mu_S - \mu_Z}{\sqrt{\sigma_S^2 + \sigma_Z^2}}$$

$$P_f = 1 - \beta$$

Based on (Ship Structure Committee SSC, 1990) the definitions of μ_s, μ_z as mean of strength stress and load stress, respectively, and σ_s , σ_z as standard deviations of strength stress and load stress also, respectively. The reliability index (β) and probability of failure (Pf) calculated with formulation above with each results give in Table 11. From Table 11 and Table 12, could be explained for safety index from variations of Hs and wave directions. The two tables had values for μ and σ as mean value and standard deviation and calculated for safety index and probability failure. With several of withdrawn results that shown in Figure 20 and Figure 21, typically the type of the strength distribution followed by the load distribution, with formula below for gaussian S and normal distribution formula of probability density function.

$$PDF(x) = \frac{1}{\sigma\sqrt{2\pi}} e^{-(x^2/2\sigma^2)}$$

Table 11. Recapitulation of Safety Index and Probability Failure for Hs Variations

Strength [S]		Load [Z]			
			Hs 0.5 m	Hs 1.0 m	Hs 1.5 m
μS	6.67×10^7	μz	2.03×10^7	2.16×10^7	2.54×10^7
σS	2.48×10^7	σz	2.43×10^7	2.70×10^7	3.16×10^7
		β	1.34	1.23	1.03
		P(f)	-0.34	-0.23	-0.03

Table 12. Recapitulation of Safety Index and Probability Failure for Wave Direction Variations

Strength [S]		Load [Z]			
			0°	90°	180°
μS	6.67×10^7	μz	1.76×10^7	3.12×10^7	1.88×10^7
σS	2.48×10^7	σz	2.46×10^7	3.23×10^7	2.38×10^7
		β	1.41	0.87	1.39
		P(f)	-0.41	0.13	-0.39

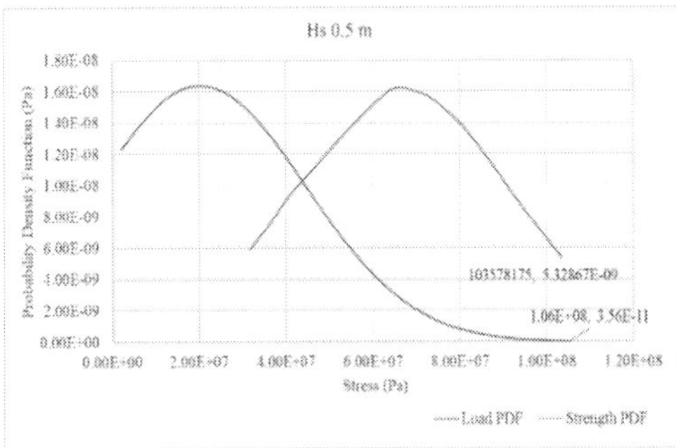

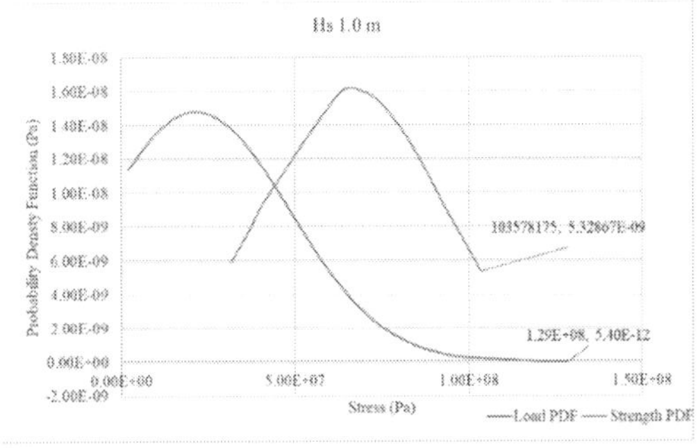

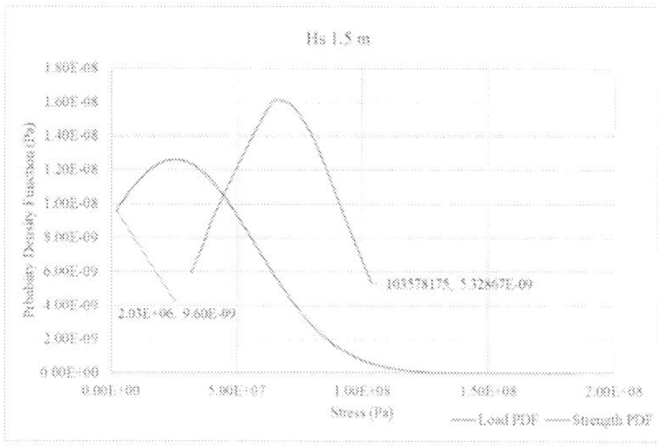

Fig. 19. Probability Density Function for each Hs variations

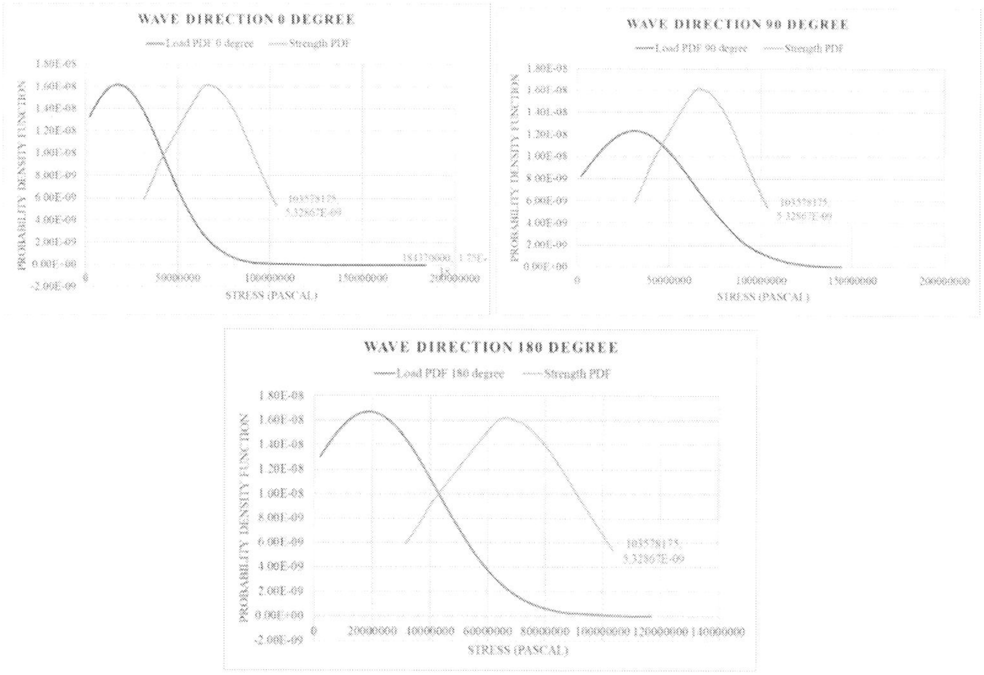

Fig. 20. Probability Density Function for each Wave Directions variations

4. Conclusion

From this study found that higher of significant wave height (Hs) give lower safety index (β), highest significant height (Hs = 1.5 m) gave lowest safety index β = 1.03. In detail wave direction gave lowest safety index is 90 degree or beam seas with values 0.87. The reliability and failure analysis indicates that the forebody–afterbody transition region requires increased material stiffness and/or larger scantling dimensions, as reflected by the decreasing reliability index (β = 1.34–1.03) and the corresponding increase in failure probability under higher wave conditions. Finally, this simply method of reliability safety index could use for analyzing reliability of body/hull of floatplane's floater N219 from vertical bending moment.

Acknowledgement

The Authors highly appreciated for Department of Naval Architecture ITS for providing research facilities supporting this research.

References

Aliffrananda, M. H. N. (2021). *Studi Ketidakstabilan Porpoising pada Floatplane dalam Operasi Lepas Landas di Perairan Tenang* [Master Thesis]. Institut Teknologi Sepuluh Nopember Surabaya.

Biro Klasifikasi Indonesia BKI. (2021). *Rules For Classification and Construction.* www.bki.co.id

Gargano, A., Das, R., & Mouritz, A. P. (2019). *Finite element modelling of the explosive blast response of carbon fibre-polymer laminates.* Composites Part B: Engineering, 177. https://doi.org/10.1016/j.compositesb.2019.107412

Muammar Kadhafi. (2016). *Analisis Umur Kelelahan (Fatigue Life) Konstruksi Kapal Perang Tipe Corvett di Perairan Indonesia* [Master Thesis]. Institut Teknologi Sepuluh Nopember ITS.

PTDI. (2022). N219 Nurtanio. PT Dirgantara Indonesia (Persero).

Ship Structure Comitte SSC. (1990). SSC-351 *An Introduction to Structural Reliability Theory.*

The Editors of Encyclopaedia Britannica. (2019, March 5). "Java Sea." Encyclopedia Britannica.Https://www.Britannica.Com/Place/Java-Sea.

Thoft-Christensen, P., & Murotsu, Y. (1986). *Application of Structural Systems Reliability Theory. In Application of Structural Systems Reliability Theory.* Springer Berlin Heidelberg. https://doi.org/10.1007/978-3-642-82764-8

Washoya, Z. A. (2019). *Analisis Beban Impact Dengan Variasi Sudut Deadrise Pada Floats Pesawat Amfibi Pada Saat Mendarat Dengan Simulasi CFD* [Undergradute Thesis]. Institut Teknologi Sepuluh Nopember Surabaya ITS.

Wypych, G. (2016). VE vinyl ester resin. Handbook of Polymers, 702–703. https://doi.org/10.1016/B978-1-895198-92-8.50214-7

Bretschneider, C. L. (1959). *Wave variability and wave spectra for wind-generated gravity waves.* U.S. Army Corps of Engineers.

ITTC – Recommended Procedures and Guidelines (Seakeeping), 2011.

Putro B. H., Rochmad A. N., Alifia W. U., Meiliana S. W. & Sulisetyono A. 2025 *Structural reliability study of the N219 floatplane's floater due to random wave loads* IOP Conference Series: Earth and Environmental Science 1473 012032 https://doi.org/10.1088/1755-1315/1473/1/012032

Meiliana S. W., Sulisetyono A. & Alifia W. U. 2025 Reliability assessment of the N219 floatplane floater under irregular wave loading Journal of Physics: Conference Series 3132 012015 https://doi.org/10.1088/1742-6596/3132/1/012015

Artificial Neural Network Backpropagation for Real-Time Coagulant Dose Prediction in Drinking Water Treatment

Hilmi Putra Pradana[1], *Berliana* Khansa Salsabila[1], *Achmad* Muzakky[1], *Mas* Agus Mardianto[1], and *Ervin* Nurhayati[1*]

[1] Department of Environmental Engineering, Faculty of Civil, Planning, and Geo Engineering, Institut Teknologi Sepuluh Nopember, Surabaya, Indonesia
* Corresponding author: ervin@its.ac.id

Abstract. A drinking water treatment plant is crucial for fulfilling the increasing water demand. As an integrated series, the coagulation unit is the most basic unit for removing particulates and reducing turbidity. However, the time gap to determine an effective coagulant dose was almost 6 h, which cannot accommodate the fluctuations in water inlet quality. A noteworthy method to reduce time is to use artificial neural network backpropagation (ANN-BP) to predict an optimum dose. The dataset consisted of five parameters (pH, temperature, conductivity, color, and turbidity) for a month of primary data sampling and historical jar test data from 2018 to 2022. The F-test result, F-value (6038,779) > F-table (2.21923), showed that one or more parameters had a statistically significant influence on the coagulant dose. Subsequently, a t-test excluded pH and temperature, with p-values lower than 0.05. Empirical models were developed through trial-and-error variations of the input layers (three and five parameters), hidden layers (2-10 nodes), and an output. The models with the lowest MSE and highest R^2 were [5-6-1] (R^2 =- 0.96051; MSE = 0.00179) and [3-4-1] (R^2 = 0.97755; MSE =0.00102). In conclusion, [3-4-1] is recommended because it has the lowest MSE and the highest R^2.

Keyword: ANN-BP, Coagulation, Drinking water, Water treatment

1. Introduction

The necessity of potable water has been rising over the years as domestic and industrial water use increases. Surabaya's water demand in 2039 was projected to be 15,528 L/s from 7,610 L/s in 2018 as the baseline value [1], while Gombi, Nigeria's water demand was forecasted to increase by 54.5% in 2031 compared to 2021 [2]. To meet the increasing demand for water quantity and the need to meet potable water quality standards, non-potable natural water can be treated using a drinking water treatment plant (DWTP), which consists of an integrated series of operations and process treatment units [3]. One of the most basic unit processes commonly and fundamentally used in DWTP is the coagulation unit paired with flocculation to remove particulates and reduce turbidity [4].

As an essential part of the treatment process, coagulation is a destabilization process of suspension or solution by adding a coagulant (a substance to give an effect of destabilization), which induces rapid turbulence to form visible floc or precipitate that will be separated by gravitational force or filtration [5]. The optimum coagulant doses are normally calculated using the jar test method, which requires a time to conduct the test of approximately half an hour to 6 hours from sampling until the outcome is obtained [6–8]. The time gap between them cannot accommodate the variety of water quality parameters that influence the coagulation-flocculation process or, in general, the water treatment process [6]. The negative effects of unfitted doses may result in increased operational costs, inefficient removal, and a risk to consumer safety [6]. Therefore, there is a need to discover a quicker technique to determine optimum coagulant doses.

A noteworthy method that has been gradually explored by many studies involves machine learning or artificial intelligence to quickly predict coagulant doses based on fluctuations in inlet water quality. Various models have been used, such as partial least squares regression (PLS), adaptive neuro-fuzzy inference system (ANFIS), multiple linear regression (MLR), and artificial neural network (ANN) [6–8]. Every study stood up by combining or comparing several methods, by using different input parameters, or various data sizes.

This study aims to determine the effectiveness of coagulant dose prediction based on a back-propagation ANN method from five regular water parameters, where the result can be obtained in near real-time from the outlet of the prior reactor or the inlet coagulation process, such as pH, temperature, conductivity, color, and turbidity. The key parameters would be extracted F-test and T-test to choose the most influential parameters. A comparison between the full set of five parameters and the statically selected parameters for building the ANN model. The performance of each model was evaluated to determine which model best represented the jar test results based on the coefficient of determination (R^2) and mean square error (MSE). Moreover, this study emphasized the use of a large historical dataset as a secondary source and primary parameter sampling to enhance the confidence and reliability of the model and minimize prediction error.

2. Methodology

2.1 Data sources

For this study, primary and secondary data were used. Primary data were acquired from real-time sampling from Karangpilang and Ngagel DWTP for a month. Samples were collected nine times a day (08.00 AM, 08.30 AM, 09.00 AM, 12.00 AM, 12.30 PM, 13.00 PM, 15.00 PM, 15.30 PM, and 16.00 PM) from the water outlet of the pre-sedimentation tank or the inlet of the coagulation unit. A 10-litre sample volume was taken for each period, then the five field parameters were measured (pH, turbidity, temperature, color, and conductivity). Subsequently, the total sampling volume was divided into 5 container sub-samples and 1 blank container for jar test analysis. Each container consisted of a 1 L sample. 1% liquid aluminium sulfate, $[Al_2(SO_4)_3 18H_2O]$, was the coagulant used in this study, as this substance is commonly utilised in Indonesia's DWTP, as well as in the location of the study object.

Jar test analysis was performed by applying 5 variations of coagulant doses separately. Mixing was arranged at 100 rpm for a minute as rapid mixing, then 40-60 rpm for 15 minutes as slow mixing phase. Finally, the settling time was set at 15 minutes. Optimum coagulant doses were chosen if the turbidity parameter was < 5 NTU; additionally, after treatment, the parameters were also tested.

Historical data, as a secondary source, were provided by the Surabaya Regional Water Company as the regulator of the primary sampling location. The data ranged from 2018 to 2022, with a total of 10282 data. Each set contained the same parameters that were obtained by primary sampling.

2.2 Determining the significant parameter affecting the dose of coagulant

Five parameters as independent variables and coagulant doses as dependent variable were statistically analyzed with F-test, while the influence of each parameter was also analyzed with pairwise t-test. Significant parameters from the individual t-test will be used as another input layer in the second ANN model, in addition to all the initial parameters in the first ANN model.

2.3 ANN model configuration

In this study, back-propagation ANN (ANN-BP) models were used. Fundamental idea of backpropagation, as primary algorithm to compute gradients efficiently in ANNs, is about updating the weights and biases of the network from the way that decreases the error or cost function by iteration process [9]. ANN-BP, as a supervised learning algorithm, consisted of three layers, namely, an input layer consisting of the values of water quality parameters, a hidden layer consisting of 2 – 10 nodes, as the exact node numbers would be determined from trial-error, and an output layer for coagulant dose values.

The ANN-BP process in this study was implemented using MATLAB. The process involved a training phase (80% of the dataset), followed by a testing phase (20% of the dataset). Each phase the input and output data would be normalized to synthesize ANN-BP model, then the results also should be denormalized. Function for process building ANN architecture in training and testing phase were 'newff' to build feed-forward backpropagation ANN, while simulation in each phase after build ANN model used 'sim' function. In the training phase, the ANN architecture parameters were 0.5 for the learning rate, 1000 for the maximum iterations (epochs), and 0,9 for the momentum parameter. Activation functions were assigned to link the layers: a sigmoid function (logsig) for the input layer and a linear function (pureline) for the output layer. The model's performance was assessed based on its accuracy in predicting the coagulant dose in comparison to the experimental jar test results from the coefficient of determination (R^2) and mean square error (MSE).

3. Result and Discussion

3.1 Significant Parameters Affected Coagulant Doses

The primary and secondary data of five parameters were statistically analysed with F-test to identify the significant effect of multiple parameters in wastewater to coagulant doses. The result shown (Table 1) that F-value (6038.779) was higher than F-table (2.21923) with significancy below α (<0.05) [10]. It can be concluded that there is one or more parameters (pH, color, temperature, conductivity, and turbidity) that have a statistically significant influence on the coagulant dose or the overall combination of all parameters affecting the coagulant dose [10].

Table 1. F-test result value

Model	df	Mean Square	F	Sig
Regression	5	1469945.479	6038.779	.000[b]
Residual	1741	243.418		
Total	1746			

After knowing the F-test result, a t-test for each parameter effect on coagulant doses was conducted. The test was applied to determine the effect of each parameter on the coagulation by comparing the change of parameter and its effect of the coagulant doses.

The statistical outcome (Table 2) revealed that the t-test significance value of temperature (0.658) and pH (0.142) parameters didn't influence coagulant doses because their significance level is higher than α (0,05). pH in this study may not influence the coagulant dose because the narrow value change on the pH range that was between 5.5 – 8.5, as the pH range is still within the optimum pH for alum [6]. This was because the type of coagulant was chosen based on the pH range of the water input of the water treatment plant. A similar effect had occurred in the temperature parameter. The slight range between 28°C - 33°C may not significantly effluence the coagulant dose in the coagulation process. Even though the change in temperature and pH can alter water viscosity and ionic charge, the increase in alum dose for optimal coagulation [11]. The variation of pH and temperature in this study was insufficient to produce a measurable difference. As a result, no statistically significant variation in coagulant doses was found.

In contrast, turbidity, color, and conductivity indicated influence of each parameter for the coagulant dose need, as those showed their significance value are lower than 0.05. The change in turbidity affects the need for coagulant dose adjustment. The higher the turbidity, the more coagulant doses are required. Increasing turbidity escalate the suspended solid that contains ionic strength with same charge, so to attract the whole suspended solid with enough opposite charge is higher comparing with water with low turbidity [12]. Interestingly, water with lower turbidity requires a higher coagulant dose ratio, while an increase in turbidity does not significantly increase the coagulant dose. It because the collision efficiency in high turbidity will increase then improving particle-aggregation caused by neutralization [13]. Color describes the presence of dissolved organic substances that are usually negatively charged, thereby increasing the consumption of coagulants through complex adsorption and bonding mechanisms. Conductivity, as an indicator of ionic strength, affects the thickness of the electrical double layer in the particles and changes the destabilization efficiency of the charge [14]. Considerable variation in these three parameters results in a significant difference in the dosage requirement of coagulants, so it is statistically significant.

Table 2. T-test result value

Model	Standardized Coefficients	t	Sig.
(constant)		-2.576	.010
Turbidity	.343	25.396	.000
Temperature	-.003	-.443	.658
pH	.008	1.467	.142
Color	.441	6.353	.000
Conductivity	.267	4.704	.000

Based on the results of this t-test, turbidity, color, and conductivity can be considered as alternative input layers in the Artificial Neural Network (ANN) model. Using only significant parameters can simplify model architecture, improve training efficiency, and reduce the risk of overfitting without sacrificing the accuracy of coagulant dose predictions.

3.2 Configurations of the ANN model

Two configuration model input layers were used as the main alternatives, as well as the neuron form hidden and output layer. The input layer variations contained five neurons as all the influent water parameters and three neurons, consisting of turbidity, color, and electric conductivity, because of the prior significant parameter test. The model parameters and properties can be seen in the Table 3.

Table 3. The model parameters and properties

Parameter	Value	Description
Input *Layer*	5 neurons	Turbidity, pH, temperature, color, electrical conductivity
	3 neurons	Turbidity, color, electrical conductivity
Hidden Layer	2-10 neurons	Trial and error
Output *Layer*	1 neuron	Optimal dose of coagulants
Learning Rate	0.5	-
Momentum	0.9	-
Epoch	1000	Minimum value
Fungsi Aktivasi	2	Logsig and purelin

The results of the model training and testing for the ANN showed variation in performance between the configuration models because of the variation in neurons in the input and hidden layers. This analysis focused on comparing the performance of all models based on two metrics, the mean squared error (MSE) and the determination coefficient (R^2), to measure the average error rate and the relationship between the input and output. The output MSE and R^2 of model configurations can be seen in Table 4 and illustrated in Figure 1a and 1b.

Table 4. The output MSE and R^2 of model configurations

Input layer	Hidden layer (node)	Output layer	Model	MSE	R^2
5 (pH, temperature, conductivity, color, turbidity)	2	1 (Coagulant dose)	[5,2,1]	0.00251	0.94422
	3		[5,3,1]	0.00820	0.80381
	4		[5,4,1]	0.00601	0.86042
	5		[5,5,1]	0.00266	0.94079
	6		[5,6,1]	0.00179	0.96051
	7		[5,7,1]	0.00870	0.79029
	8		[5,8,1]	0.00287	0.93587
	9		[5,9,1]	0.01096	0.72610
	10		[5,10,1]	0.01631	0.54459
3 (conductivity, color, turbidity)	2		[3,2,1]	0.00153	0.96631
	3		[3,3,1]	0.00152	0.96664
	4		[3,4,1]	0.00102	0.97755
	5		[3,5,1]	0.00537	0.87652
	6		[3,6,1]	0.00347	0.92201

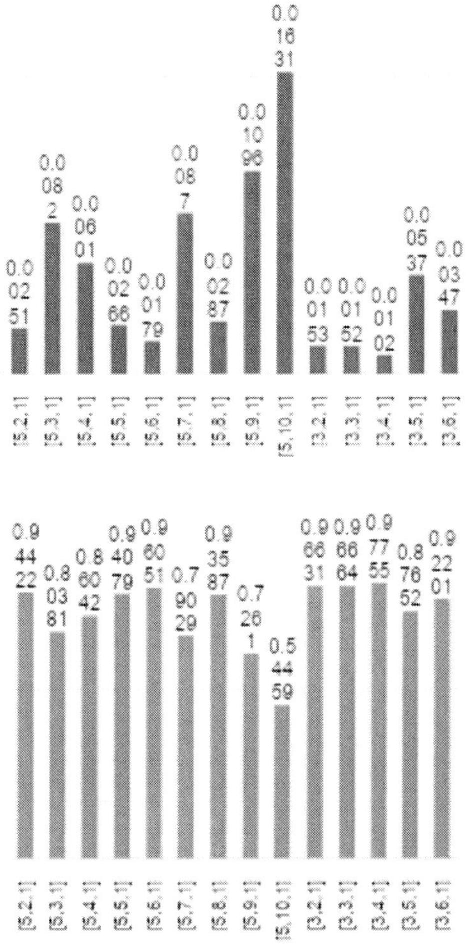

Fig. 1. The fluctuation of MSE (1a) and R^2 (1b) from the model configuration

In the configuration of the input layer with 5 neurons, the model performance shows a fluctuating pattern as the number of neurons in the hidden layer increases. The model with configuration [5,2,1] yielded an MSE of 0.00251 with an R^2 of 0.94422, indicating that the model's ability to explain data variations was quite high. However, when the number of neurons in the hidden layer was increased to 3, 4, and 5, the MSE value amplified at some point, indicating that the increase in model complexity did not necessarily result in increased accuracy and become overfitting because of increasing of hidden layer [15]. Of all the experiments with 5 input neurons, model [5,6,1] showed the best results with an MSE of 0.00179 and an R^2 of 0.96051. A low MSE value indicates a small prediction error rate, whereas a high R^2 value indicates that the model can account for approximately 96% of the variation in the output data. This means that the relationship between water quality parameters (turbidity, pH, temperature, color, and electrical conductivity) and the optimal coagulant dose can be well represented by the model. However, adding more than six neurons significantly reduced the model performance. For example, in model [5,9,1], the MSE value increases to 0.01096 and R^2 decreases to 0.72610, indicating that the model is becoming less stable and is prone to overfitting [15]

Meanwhile, in the input layer configuration with 3 neurons, the overall performance of the model showed better and stable results compared to the 5-neuron input configuration. It also shows that the use of fewer but relevant input parameters can result in more efficient models without sacrificing the prediction accuracy [15]. The best results were achieved by the model [3,4,1], with an MSE of 0.00102 and an R^2 of 0.97755. This value was the highest among all the configurations tested, indicating that the model had excellent predictive capabilities and a low error rate. An R^2 value close to 1 indicates that the model can explain almost the entire variation in the output data; in other words, the relationship between the three input parameters and the optimal coagulant dose can be very well represented by this model. In addition, these results also indicate that the addition of parameters such as pH and temperature (as in the configuration of 5 input neurons) does not provide a significant increase in accuracy and may even magnify the noise in the data and decrease the stability of the model [15].

3.3 Best ANN model and limitation model application

The Artificial Neural Network (ANN) model is used to predict the need for coagulant doses based on raw water quality parameters. Two model architectures were tested, namely the 5–6–1 model and the 3–4–1 model, which was found to be the best model configuration at the input layer of 5 neurons and 3 neurons. From these variations, both architectures demonstrated optimal performance within their respective structures, and their predictive behavior was further evaluated to determine the most reliable model. ANN-BP architectural for 5-6-1 model configuration can be seen in Fig 2.

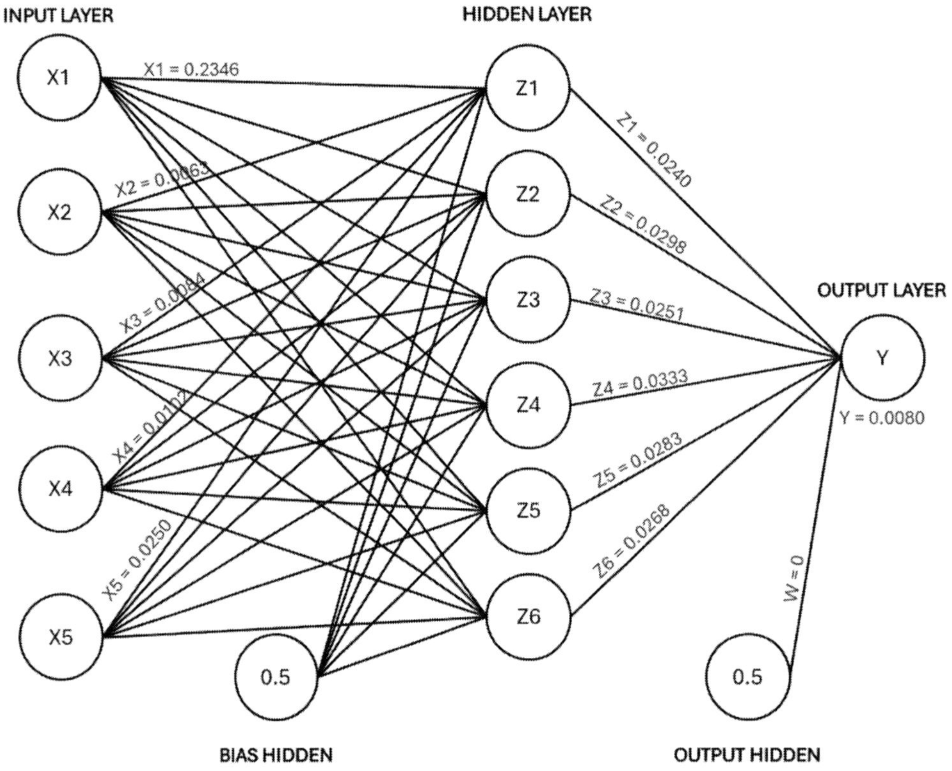

Fig. 2. ANN-BP architectural for 5-6-1 model configuration

For the 5–6–1 architecture, the model with five input neurons generates the normalized predictive equation shown in Equation (1).

$$\text{coagulant dose} = (\text{turbidity} \times 0.0268) - (\text{temperature} \times 0.0241) + (\text{pH} \times 0.0241) + (\text{color} \times 0.0242) + (\text{electrical conductivity} \times 0.0243) \quad (1)$$

After denormalization, the equation becomes (Equation 2).

coagulant dose = (turbidity x 0.298) -(temperature x 0.138) + (pH x 0.134) + (color x 0.165)

+ (electrical conductivity x 0.163) (2)

From the equation 1 and 2, turbidity is the variable with the greatest influence, indicating that increasing the turbidity of raw water will significantly increase the need for coagulants. The color and conductivity parameters have a considerable positive influence, indicating their relationship to the concentration of organic matter and the content of dissolved ions that affect the coagulation process. On the contrary, temperature exerts a negative influence, which means that rising temperatures tend to decrease the need for coagulants. This has to do with increasing the kinetic energy of particles at higher temperatures, so that the charge destabilization process can take place more effectively.

On the other hand, the result of ANN-BP architectural for 5-6-1 model configuration can be seen in Fig 3.

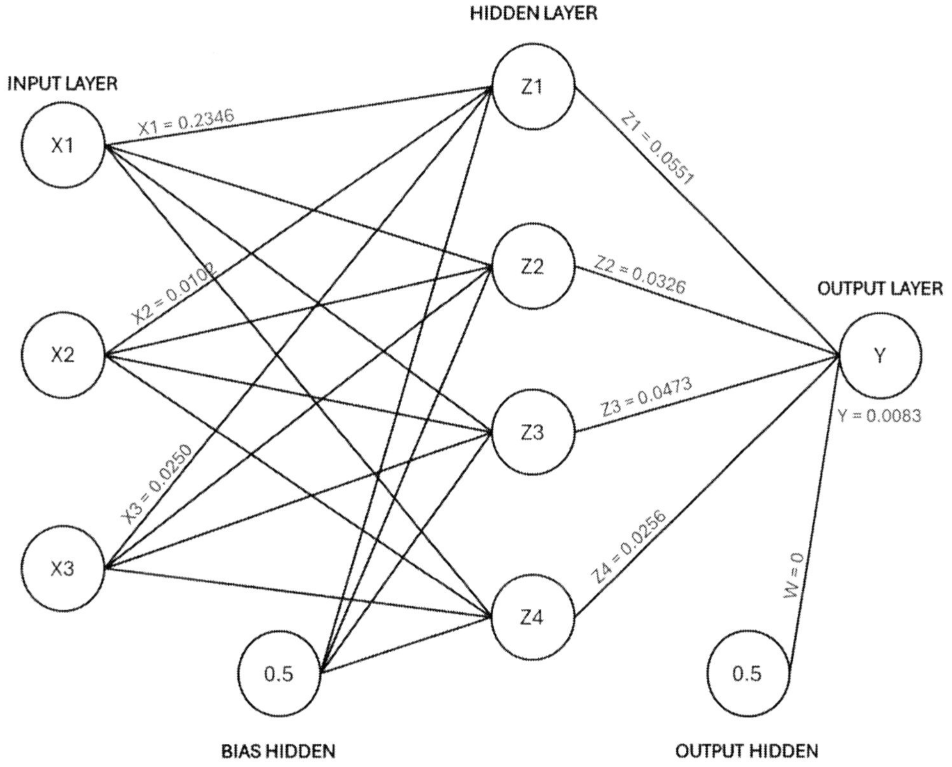

Fig. 3. ANN-BP architectural for 3-4-1 model configuration

The 3–4–1 architecture produces a simplified normalized model shown in Equation (3):

$$coagulant\ dose\ = (turbidity\ x\ 0.0414)\ +\ (color\ x\ 0.0373)\ +\ (electrical\ conductivity\ x\ 0.0376) \quad (3)$$

Following denormalization, the model becomes (Equation 4)

$$coagulant\ dose\ = (turbidity\ x\ 0.306)\ +\ (color\ x\ 0.159)\ +\ (electrical\ conductivity\ x\ 0.151) \quad (4)$$

Similar to the previous model, turbidity remains the dominant variable, with color and conductivity continuing to serve as meaningful predictors in describing organic content and total dissolved solids that influence coagulation efficiency. This simplified input structure highlights the model's ability to perform effectively even with fewer parameters, focusing on the variables most directly associated with coagulant demand.

In comparison, the [3,4,1] model has the most superior performance with the highest R^2 (0.97755) and the lowest MSE among all models, making it the most efficient and accurate model. While the model [5,6,1] still provides excellent results, it is more complex and more sensitive to data variations. This ANN model is built on data with a specific range, so its use must pay attention to limitations. This is because the model becomes accurate only when the temperature is in the range of 24°C – 33°C according to the air temperature according to BMKG (2022) and pH 5.5 – 8.5 according to the optimal pH of the use of alum-type coagulants. This limitation is important because ANN studies patterns based on available data. When the input is outside of that range, the model no longer describes the actual process conditions.

4. Conclusions

Based on the F-test result, since the p-value was 0.000 (< 0.05) and $F_{count} > F_{table}$, it can be concluded that five coagulant feed water parameters simultaneously have a significant effect on the coagulant dose. Individual t-test result presented that turbidity, color, and conductivity had significant affect to coagulant doses ($p < 0.05$), in contrast pH and temperature didn't significant influence ($p > 0.5$). Empirical models were then developed using both all five parameters and a reduced set of the three statistically significant parameters. The ANN-BP models which have lowest MSE and highest R^2 were [5-6-1] ($R^2 = 0.96051$; MSE = 0.00179) and [3-4-1] ($R^2 = 0.97755$; MSE =0.00102). In conclusion, [3-4-1] is recommended because of the lowest MSE and highest R^2.

Acknowledgement

The authors would like to acknowledge the Institut Teknologi Sepuluh Nopember for the project funding under D4-2-ITS25 - Dana Departemen/Fakultas/Unit - Batch 2

References

[1] F. F. Pradypna, B. D. Marsono, and E. S. Soedjono, A Study of Drinking Water Supply and Demand in Surabaya in the Year 2039, IOP Conf. Ser.: Earth Environ. Sci. **506**, 012028 (2020).
[2] I. D. Joshua, A. C. Salihu, A. M. Mshelia, and N. N. Ubachukwu, ANALYSIS OF COMMUNITY-BASED PATTERN OF WATER DEMAND AND SUPPLY, Jessd **6**, (2023).
[3] M. Taufik, E. Khairina, R. Hidayat, R. Kalalinggi, and M. Iqbal, Study of Government's Strategy Indonesia on Clean Water Avaiability in Indonesia, Jurnal Kesehatan Lingkungan Indonesia (JKLI) **21**, 111 (2022).
[4] N. H. Pakharuddin, M. N. Fazly, S. H. Ahmad Sukari, K. Tho, and W. F. H. Zamri, Water treatment process using conventional and advanced methods: A comparative study of Malaysia and selected countries, IOP Conf. Ser.: Earth Environ. Sci. **880**, 012017 (2021).
[5] J. Bratby, Coagulation and Flocculation in Water and Wastewater Treatment, Water Intelligence Online **15**, 9781780407500 (2016).
[6] S. Narges, A. Ghorban, K. Hassan, and K. Mohammad, Prediction of the optimal dosage of coagulants in water treatment plants through developing models based on artificial neural network fuzzy inference system (ANFIS), J Environ Health Sci Engineer **19**, 1543 (2021).
[7] Z. Shi, C. W. K. Chow, R. Fabris, J. Liu, E. Sawade, and B. Jin, Determination of coagulant dosages for process control using online UV-vis spectra of raw water, Journal of Water Process Engineering **45**, 102526 (2022).
[8] E. L. L. Tochio, B. C. Do Nascimento, and S. R. Lautenschlager, Coagulant dosage prediction in the water treatment process, Water Supply **23**, 3515 (2023).
[9] M. M. Hammad, *Artificial Neural Network and Deep Learning: Fundamentals and Theory*, arXiv:2408.16002.
[10] Robert Odek and Gordon Opuodho, F-test and P-values: A synopsis, J.M.S **13**, 59 (2023).
[11] N. Dayarathne, M. J. Angove, S. Jeong, R. Aryal, S. R. Paudel, and B. Mainali, Effect of temperature on turbidity removal by coagulation: Sludge recirculation for rapid settling, Journal of Water Process Engineering **46**, 102559 (2022).
[12] Baghvand, Optimizing Coagulation Process for Low to High Turbidity Waters Using Aluminum and Iron Salts, American Journal of Environmental Sciences **6**, 442 (2010).
[13] R. Nedjai, A. Al-Mamun, and M. Z. Alam, Effects of initial turbidity and myco-coagulant dose on the effectiveness of the coagulation process in water treatment, Appl. Chem. Eng. **7**, 1546 (2024).
[14] A. Mortadi, E. G. Chahid, A. Elmelouky, M. Chahbi, N. El Ghyati, S. Zaim, O. Cherkaoui, and R. El Moznine, Complex electrical conductivity as a new technique to monitor the

coagulation-flocculation processes in the wastewater treatment of the textile Industry, Water Resources and Industry **24**, 100130 (2020).
[15] X. Ying, An Overview of Overfitting and its Solutions, J. Phys.: Conf. Ser. **1168**, 022022 (2019).

Analysis of Urban Areas with Built-up Land Index Using ALOS PALSAR-2 Satellite Imagery (Case Study: Surabaya City)

Muhammad Nabil Hadjoe[1], *Filsa* Bioresita[1*], *Noorlaila* Hayati[1], and *Timo* Balz[2]

[1] Department Geomatics Engineering, Faculty of Civil, Planning, and Geo Engineering, Sepuluh Nopember Institute of Technology, Surabaya, Indonesia
[2] State Key Laboratory of Information Engineering in Surveying, Mapping and Remote Sensing, Wuhan University, Wuhan, China
* Corresponding author: filsa.bioresita@its.ac.id

Abstract. Surabaya, the largest metropolitan city in East Java Province, has experienced rapid urban growth driven by population increase, economic development, and migration from surrounding regions. This growth has accelerated the expansion of built-up areas, particularly in suburban zones that were previously dominated by productive land such as rice fields and forests. These land use changes present significant challenges for sustainable urban spatial management, highlighting the need for efficient and reliable monitoring techniques. This study employs Synthetic Aperture Radar (SAR) based remote sensing, which enables consistent data acquisition regardless of weather conditions or illumination. ALOS PALSAR-2 Single Look Complex (SLC) data were utilized and analyzed using the SAR-based Impervious Built-up Index (IBUIsar) method to identify and map built-up areas based on backscatter characteristics. The application of the IBUIsar method aims to produce an accurate spatial distribution map of built-up land in Surabaya. The results indicate that higher IBUIsar values are concentrated in the city center and areas with intensive development activities. This information is expected to support urban spatial planning and contribute to sustainable urban development strategies.

1 Introduction

Surabaya as the largest metropolitan city in East Java Province has experienced significant growth, becoming a center of migration for people from surrounding areas [1]. Rapid economic growth and the availability of adequate infrastructure and facilities are the main factors behind the increase in population and the expansion of built-up areas [2]. This growth has led to notable land-use changes, especially in peri-urban areas that previously consisted of agricultural fields or forested land but have now been converted into residential areas and supporting urban facilities. Rising population density continues to pressure land resources, resulting in further land-use transformation to accommodate housing and public services [3]. The expansion of urban infrastructure has also intensified land demand and accelerated changes in urban land patterns.

In the context of monitoring built-up areas, remote sensing technology offers an efficient and effective approach. Among these, Synthetic Aperture Radar (SAR) stands out for its strong capability to observe highly developed urban environments. SAR can overcome common limitations of optical imagery, such as cloud cover, atmospheric disturbances, and dependence on sunlight [4]. Its ability to operate regardless of weather conditions further enhances its utility for continuous monitoring. Moreover, SAR's geometric sensitivity allows for more accurate extraction of surface characteristics [5].

Urban area identification using SAR data can be performed through SAR-based built-up indices. One method that has shown strong potential is the Impervious Built-up Index (IBUIsar) introduced by [6]. This index relies on radar backscatter information to improve the detection of urban surfaces. The study by [6] showed that IBUIsar, which utilizes ALOS PALSAR imagery, was able to extract built-up areas in three locations with overall accuracies of 90%, 98%, and 87.6% respectively. This highlights the effectiveness of the method as a reliable approach for mapping built-up surfaces.

Detecting urban built-up areas through built-up index algorithms represents a key method in mapping urbanization patterns. Previous studies have shown that the use of SAR for detecting built-up areas still faces challenges such as speckle, backscatter variations, and limited spectral information [7], [8], [9]. Several studies have combined SAR and optical data [10], [11], [12], [13], but the results remain dependent on the quality of the optical imagery. Meanwhile, built-up index approaches such as NDBI, EBI, and UI [14], [15] are mostly effective only for optical imagery and are less optimal when applied to SAR data.

This study employs ALOS PALSAR-2 imagery using the IBUIsar built-up index, which combines dual-polarization data (HH and HV) to enhance the differentiation of urban surfaces. IBUIsar utilizes radar backscatter information to more accurately identify built-up features. This research aims to analyze urban areas using a built-up index to obtain clearer and more reliable information on urban characteristics.

2 Methods

2.1 Study Area

The research is located in Surabaya City, East Java Province, as shown in Figure 1. Geographically, The city of Surabaya borders Gresik Regency to the west and the Madura Strait to the north and east.

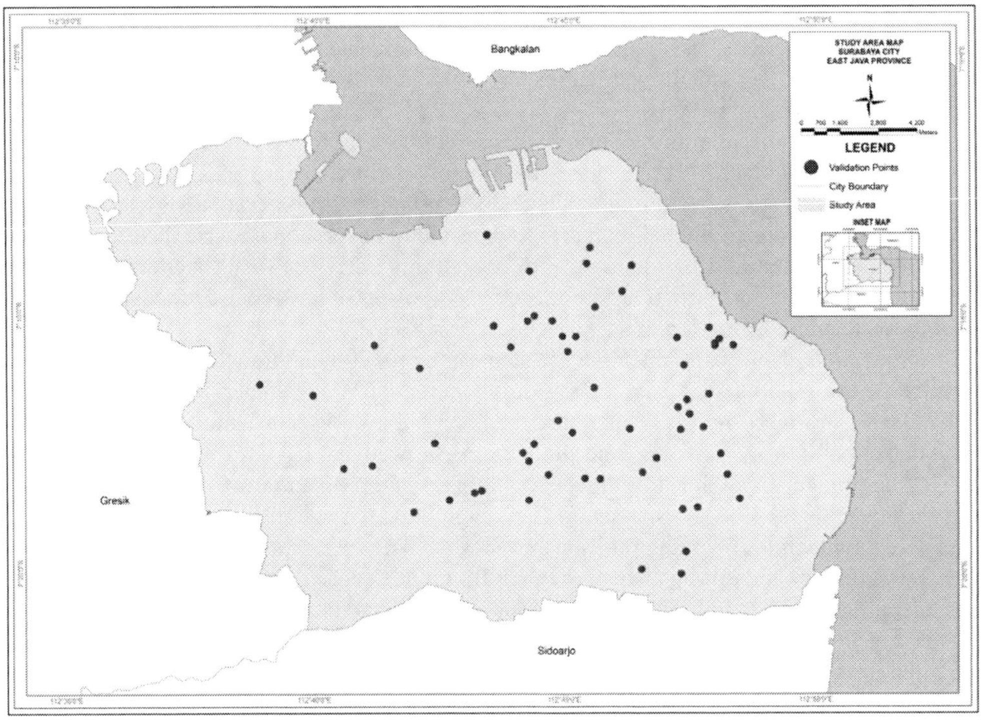

Fig. 1. Study Area

2.2 Dataset

In this study, the instruments used include satellite imagery, vector spatial data, and field observation data. The primary instrument is ALOS PALSAR-2 SLC dual-polarization (HH + HV) imagery, which provides a spatial resolution of approximately 10 m and a temporal resolution (revisit time) of about 14 days, allowing consistent observation of surface characteristics and changes over time. This data is consisting of 2 scenes acquired in 2022 and 2024. In addition to satellite data, a shapefile of the administrative boundary of Surabaya City was utilized. To validate the results of the image processing, field observations (ground truth) and Google Earth data were employed as reference sources for comparison with the built-up land index classification outcomes.

2.3 Method

The processing stage began with the use of ALOS PALSAR-2 imagery to calculate the radar-based Built-up Index (IBUIsar), which utilizes backscatter values from HH and HV polarizations to identify built-up areas. The first step involved preprocessing the ALOS PALSAR-2 data using SNAP software. The raw Single Look Complex (SLC) SAR data were converted into power or intensity data. A multi-looking operation was then applied to reduce speckle noise and enhance the signal-to-noise ratio, while slightly decreasing spatial resolution. The azimuth-to-range look ratio was set at 3:1 to optimize speckle reduction, as this ratio represents a practical compromise between reducing speckle noise and preserving spatial resolution, thereby improving radiometric stability without excessively degrading the geometric quality of urban features [14]. Radiometric calibration was subsequently performed to convert digital number values into backscatter coefficients (σ^0), ensuring that the data have consistent physical meaning across time and space. To further refine the image, a Gamma Map filter with a 3×3 moving window was applied for additional speckle suppression without compromising critical spatial details. Geocoding and terrain correction were then conducted using a Digital Elevation Model (DEM) to correct geometric distortions caused by terrain variations and to accurately project the imagery into the UTM-WGS84 coordinate system. The process employed the Range-Doppler terrain correction approach with SRTM DEM resampled to a 25 m pixel size using the nearest neighbor method. The final calibrated and corrected backscatter data were clipped to the administrative boundary of Surabaya City to prepare the dataset for subsequent classification and analysis.

The next step is to calculate the built-up land index (IBUIsar). The index value is calculated based on a mathematical combination of HH and HV backscatter values that describe the intensity of the impervious surface. The calculation involves three main components: first, parameter **a** represents the normalized difference between HH and HV backscatter (Eq. 2), emphasizing the relative contrast of surface scattering; second, parameter **b** expresses the direct ratio of HH to HV backscatter (Eq. 3), which reflects the surface roughness and orientation characteristics. These two parameters are then combined in an exponential normalization form, producing the final IBUIsar equation [14]:

$$IBUIsar = a - \left[\frac{(e^a - e^b)}{(e^a + e^b)}\right] \quad (1)$$

$$a = \left[\frac{(HH_\sigma^0 - HV_\sigma^0)}{(HH_\sigma^0 + HV_\sigma^0)}\right] \quad (2)$$

$$b = \left[\frac{(HH_{\sigma}^{0})}{(HV_{\sigma}^{0})}\right] \quad (3)$$

This formulation enhances the distinction between built-up and non built-up features. Negative or near-zero IBUIsar values typically indicate built-up areas, while positive values correspond to vegetation, bare soil, or water surfaces.

The accuracy of the built-up area classification was assessed using a confusion matrix to compare both indices against the validation data, employing 30 validation points representing built-up areas and 30 validation points representing non built-up areas. The selected points represent areas that remained unchanged in 2022 and 2024.

3 Result and Discussion

3.1 Result

The preprocessing stage of the ALOS PALSAR-2 imagery produced calibrated backscatter values (σ^0) for both HH and HV polarizations, which serve as the fundamental inputs for generating the IBUIsar index. These backscatter signals reflect the interaction between radar energy and surface features, allowing differentiation between built-up areas and other land-cover types based on their scattering characteristics.

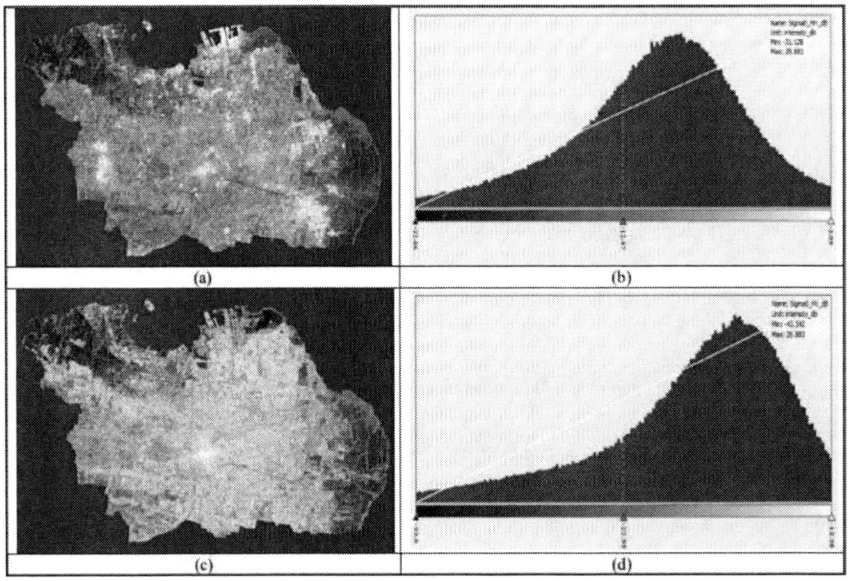

Fig. 2. ALOS PALSAR-2 backscatter in 2022, **(a)** HH polarization, **(b)** histogram of HH polarization, **(c)** HV polarization, **(d)** histogram of HV polarization

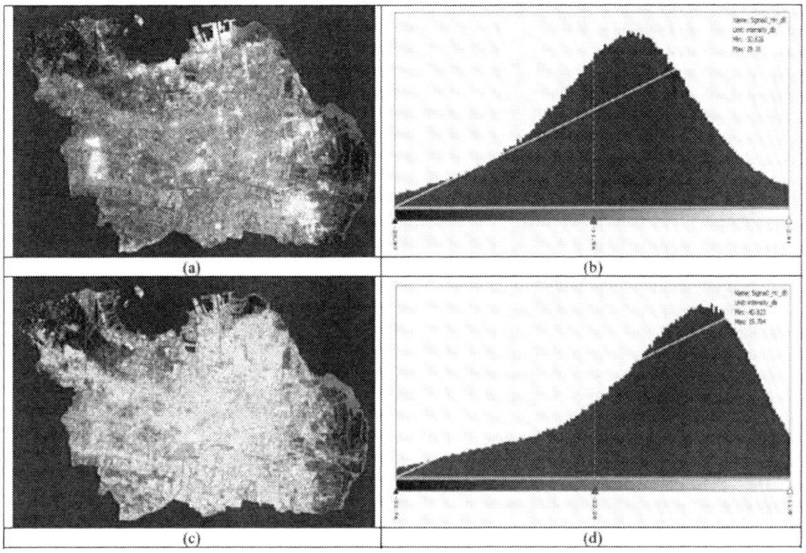

Fig. 3. ALOS PALSAR-2 backscatter in 2024, **(a)** HH polarization, **(b)** histogram of HH polarization, **(c)** HV polarization, **(d)** histogram of HV polarization

The backscatter σ^0 ALOS PALSAR-2 value range in 2022 (Figure 2) and 2024 (Figure 3) shows a clear difference in dispersion response between HH and HV polarizations, which can be directly related to surface and land-cover characteristics. In 2022, HH values ranged from –31.128 to 29.883 dB (Figure 2b), where high HH values are typically associated with strong surface and double-bounce scattering from built-up areas and other hard surfaces, while low values correspond to smooth surfaces such as water bodies or bare land. In contrast, HV values ranged from –43.342 to 26.883 dB (Figure 2d), indicating a mixture of very low returns from smooth surfaces and very high returns from hard objects or urban structures. In 2024, HH values showed a similar range, from –30.626 to 29.16 dB (Figure 3b), indicating the continued presence of stable hard surfaces and urban features. Meanwhile, HV values range decreased from –40.823 to 19.704 dB (Figure 3d), suggesting a reduced contribution of volume scattering, potentially due to changes in vegetation cover or increased dominance of surface scattering. The consistent maximum HH values across both years highlight the persistence of built-up areas, whereas the reduced HV response in 2024 indicates changes in land cover that affect vegetation-related and complex structural scattering mechanisms.

The IBUIsar algorithm is formulated using normalized combinations of HH and HV backscatter values to improve the separation between built-up and non built-up. By calculating the mean and standard deviation (SD) of the IBUIsar output across different land-cover types, the index demonstrates a strong ability to isolate built-up areas from other classes. Built-up pixels consistently show negative or near-zero IBUIsar values, while non built-up areas such as vegetation, bare soil, and water surfaces produce positive values without overlap in the SD range. This clear separation allows the classification threshold to be determined directly from the distribution of IBUIsar values. Following the approach in a previous study [14], the threshold is adopted by assigning negative IBUIsar values to built-up areas and positive values to non built-up areas. The IBUIsar algorithm result in 2022 and 2024 is shown in Figure 4.

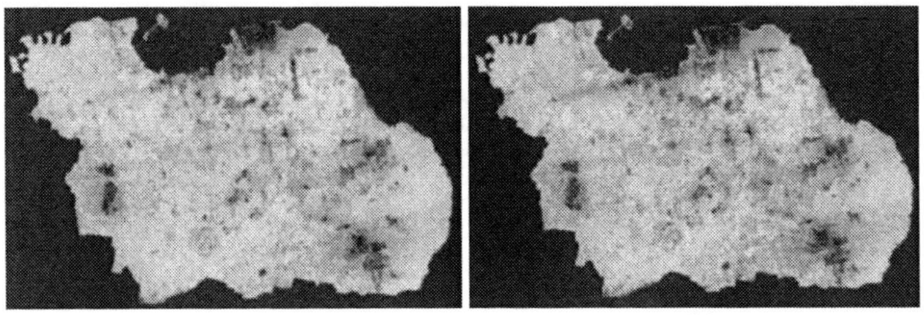

(a) (b)

Fig. 4. IBUIsar Algorithm Results in 2022 (a) and 2024 (b)

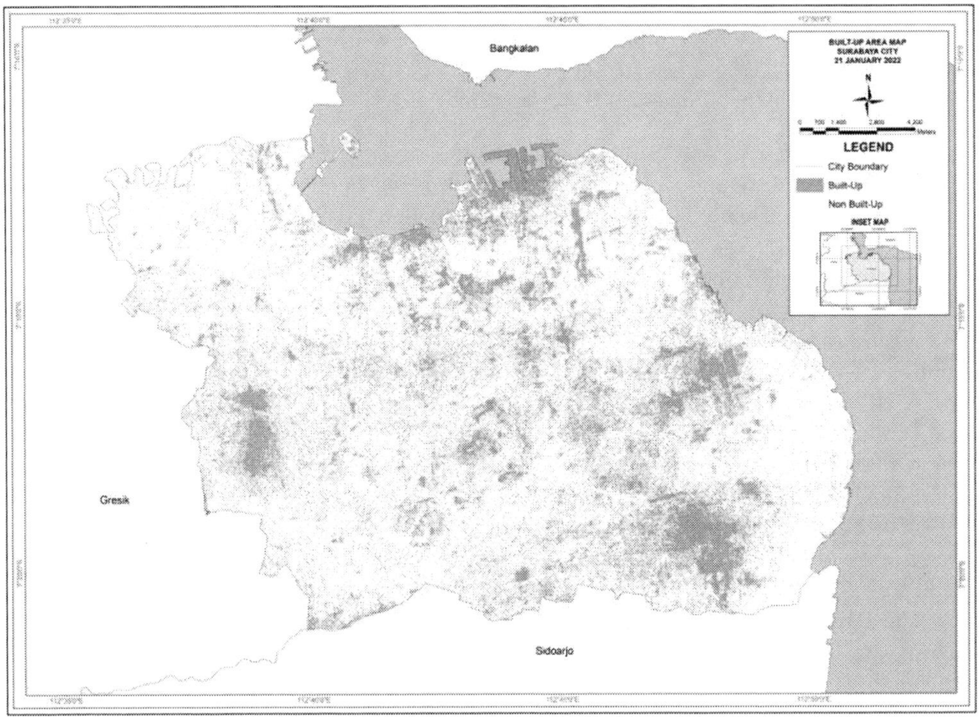

Fig. 5. Built-up Area Map of Surabaya City in 2022

Figure 5 presents the built-up area map of Surabaya City in 2022, shows two classification categories represented by distinct colours. The map is predominantly shaded in no colour, indicating that non built-up areas occupy most of the city's surface. Built-up areas, shown in orange pixel are concentrated in central, eastern, and northern parts of Surabaya, with denser patterns visible around industrial zones and the port area along the northern coastline. Smaller clusters of built-up pixels also appear in the western and southern sections, reflecting mixed land-use characteristics.

Similar to previous years, figure 6 shows a map of built-up areas in Surabaya in 2024. This map is dominated by no colour, indicating that non built-up areas still dominate most of Surabaya's land area. Built-up areas, shown in orange pixel, form dense concentrations in the center, east, and north of the city, with significant intensification around industrial areas and port complexes along the north coast. Additional clusters of built-up areas are also visible in the west and south, reflecting ongoing urban expansion and diverse land use patterns in the city's suburbs.

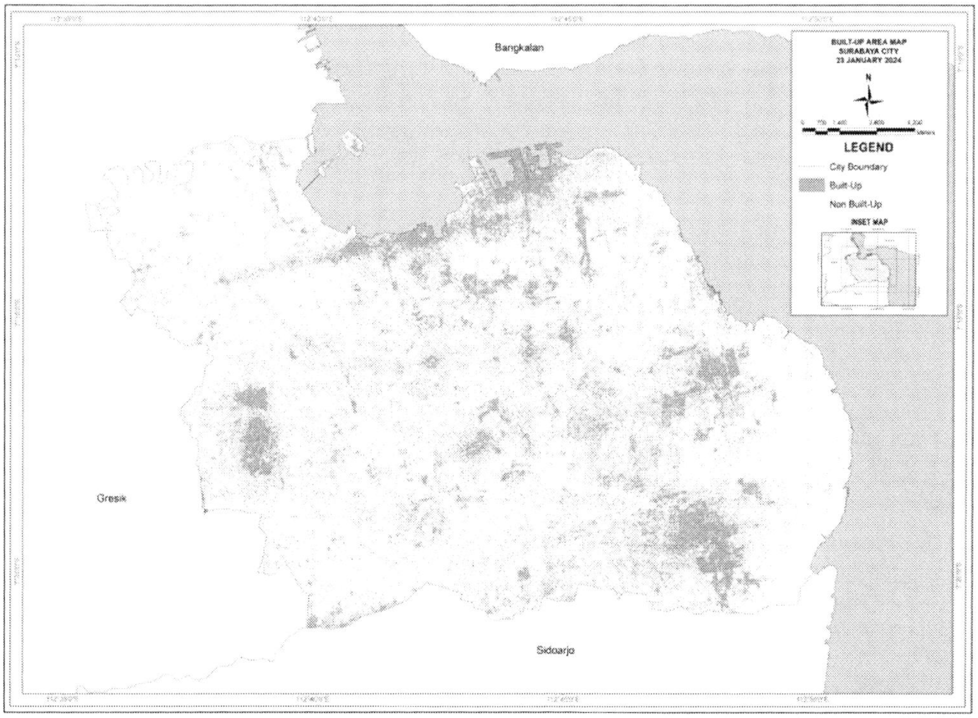

Fig. 6. Built-up Area Map of Surabaya City in 2024

A comparison of the areas in 2022 and 2024 indicates a slight change in the distribution of built-up and non-built-up areas in the city of Surabaya. In 2022, the built-up area was recorded at around 6,429.360 ha, slightly higher than in 2024, which reached 5,899.696 ha. This decrease may indicate a change in classification for several pixels that were previously detected as built-up but were later identified as non-built-up, mainly due to backscatter variations or differences in surface conditions. Meanwhile, the area of non-built-up land increased from 27,157.394 ha in 2022 to 27,684.249 ha in 2024. Overall, these changes reflect relatively stable urban surface dynamics with insignificant shifts in area over a two-year period.

The accuracy assessment was carried out using 60 validation points, consisting of 30 points representing built-up areas and 30 points representing non-built-up areas. All validation points were selected from locations that remained unchanged between 2022 and 2024 to ensure consistency in evaluating the index performance. Based on the confusion matrix shown in Table 1, the IBUIsar algorithm demonstrates a strong ability to separate built-up from non built-up areas. In 2022, the method achieved an overall accuracy of 86.67%, while in 2024 the accuracy increased to 90%, indicating improved classification reliability. The index also shows consistently better performance in detecting built-up areas, suggesting that the backscatter-based formulation effectively captures the scattering characteristics of impervious surfaces. Meanwhile, misclassifications in non-built-up areas are likely influenced by surface heterogeneity and local backscatter variations, highlighting the importance of stable validation points in ensuring reliable accuracy estimation.

Table 1. Confusion Matrix Result

Year	Predict	Actual		Total	Accuracy (%)
		Built-up	Non Built-up		
2022	Built-up	25	3	28	86.67%
	Non Built-up	5	27	32	
	Total	30	30	60	
2024	Built-up	24	0	24	90.00%
	Non Built-up	6	30	36	
	Total	30	30	60	

3.2 Discussion

The identification of built-up areas in Surabaya using ALOS PALSAR-2 imagery from 2022 and 2024 provides an overview of the city's development dynamics over two years. IBUIsar based analysis was used to distinguish between built-up and non-built-up areas, while also evaluating the consistency of the algorithm in complex urban environments. The comparison between the two years not only shows patterns of change in the city's surface, but also helps assess the extent to which IBUIsar is able to represent actual conditions in the field.

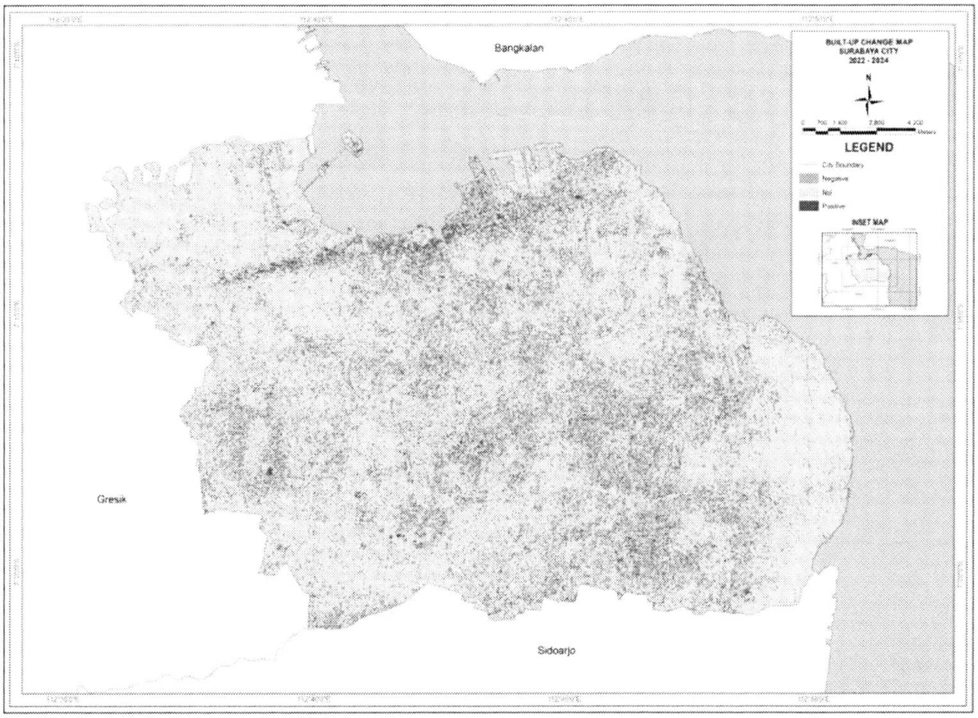

Fig. 7. Built-up Area Change Map 2022 and 2024 of Surabaya City

Figure 7 illustrates the built-up area change in Surabaya City between 2022 and 2024, derived from the IBUIsar built-up land index. The map categorizes change into three classes: positive change, negative change, and no change. Areas marked in red represent positive change, indicating locations that were classified as non built-up in 2022 but transitioned into built-up by 2024. The positive change extending across the northern part of Surabaya from east to west is likely influenced by a combination of physical surface conditions and limitations of the IBUIsar method. This area is dominated by port activities, industrial zones, and coastal infrastructure characterized by hard surfaces that generate high backscatter and are therefore prone to being detected as built-up areas. However, the clear linear pattern also suggests the influence of backscatter variations related to surface moisture, tidal conditions, or radar acquisition geometry, indicating that part of the detected positive change may reflect the sensitivity of IBUIsar to coastal and industrial environments rather than actual new urban expansion.

Areas shown in green correspond to negative change, where land previously identified as built-up in 2022 shifted to non built-up in 2024. These pixels are relatively dispersed and appear in small patches, reflecting inconsistencies likely arising from the IBUIsar classification rather than actual land-cover reduction.

Meanwhile, the yellow pixels dominate the map, representing no change between the two years. However, the overwhelming extent of this yellow class, especially in areas that are known to be highly urbanized, indicates potential misclassification in the 2024 IBUIsar output. Overall, the spatial pattern suggests that the algorithm did not fully capture the true extent of built-up development in Surabaya, as evidenced by unexpectedly large zones labeled as unchanged non built-up.

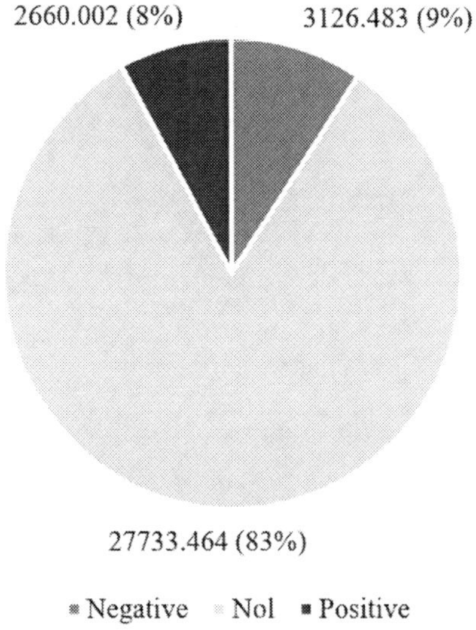

Fig. 8. Area Proportions of Built-up Change in Surabaya (2022 and 2024)

According to Figure 8, the results of the area proportion show that most of Surabaya, around 83%, did not experience any changes between 2022 and 2024 according to the IBUIsar algorithm. Areas that experienced negative changes, namely those that shifted from built-up to non built-up, covered around 9% of the total area. Meanwhile, areas with positive changes, namely changes from non built-up in 2022 to built-up in 2024, only cover about 8%. This distribution indicates that the majority of the area remains in the same condition, while the detected changes are relatively small and likely reflect the limitations of the accuracy of the IBUIsar classification in 2024.

The overall findings of this study highlight both the potential and the limitations of using the IBUIsar index derived from ALOS PALSAR-2 data for identifying built-up areas in Surabaya. While the method provides a general indication of surface conditions and captures certain patterns of urban expansion, inconsistencies particularly in the 2024 classification show that the algorithm still requires refinement to better reflect real world urban characteristics. The minimal proportion of detected changes and the unexpected dominance of non built-up classifications in heavily urbanized zones suggest that the index alone is insufficient for detailed urban monitoring. Future research would benefit from integrating additional SAR or optical datasets, adjusting thresholding methods, and incorporating more robust validation points to enhance the reliability and accuracy of built-up area detection in complex urban environments.

4 Conclusion

The IBUIsar index derived from ALOS PALSAR-2 imagery is able to provide a general overview of the distribution of built-up areas in Surabaya in 2022 and 2024, but its accuracy remains limited for more detailed urban analysis. The decrease in built-up area from approximately 6,429.360 ha in 2022 to 5,899.696 ha in 2024, the dominance of unchanged areas, and the appearance of unexpected classifications in several urban zones indicate inconsistencies in the algorithm, which are more related to weaknesses in threshold determination and the sensitivity of IBUIsar to dense urban areas rather than actual surface-cover changes. Therefore, this method requires improvements through threshold calibration, incorporation of supporting datasets, and refinement of validation samples to produce more accurate and representative built-up area mapping.

Acknowledgment

The authors would like to thank Japan Aerospace Exploration Agency (JAXA) via the 3rd Research Announcement on Earth Observation (ER3A2N511) for providing ALOS PALSAR-2 imagery data in this study. This research is funded by the Indonesian Endowment Fund for Education (LPDP) on behalf of the Indonesian Ministry of Higher Education, Science and Technology and managed under the EQUITY Program (Contract No. 4299/B3/DT.03.08/2025 & No 3029/PKS/ITS/2025.

References

[1] F. Ramdhani, "Impact of Urbanization for City Developments in Indonesia," 2013.

[2] D. Fitria Pida, K. N. Aini, and C. A. Putri, "Dampak Urbanisasi terhadap Perkembangan Kota di Indonesia: Tinjauan dari Aspek Ekonomi Pembangunan," vol. 3, no. 1, pp. 226–238, 2025, doi: 10.62383/wissen.v3i1.562.

[3] K. H. Rahmani and B. M. Sukojo, "Evaluation of the Suitability of Land Use and Building Based on BIM to Support the Spatial Planning of North Gandaria Village," vol. 19, no. 1, pp. 96–105, 2023.

[4] F. Fadlin, M. A. Thaha, F. Maricar, and M. P. Hatta, "Monitoring Perubahan Penggunaan Lahan Menggunakan Citra Satelit Sentinel 1 di DAS Wanggu Kota Kendari," *Jurnal Teknik Sumber Daya Air*, vol. 1, no. 2, pp. 77–88, Jul. 2022, doi: 10.56860/jtsda.v1i2.5.

[5] Y. Zhang, J. Zhang, H. Wu, Z. Lu, and S. Guangtong, "Monitoring of urban subsidence with SAR interferometric point target analysis: A case study in Suzhou, China," *International Journal of Applied Earth Observation and Geoinformation*, vol. 13, no. 5, pp. 812–818, 2011, doi: 10.1016/j.jag.2011.05.003.

[6] S. Sinha, A. Santra, and S. S. Mitra, "Semi-automated impervious feature extraction using built-up indices developed from space-borne optical and SAR remotely sensed sensors," *Advances in Space Research*, vol. 66, no. 6, pp. 1372–1385, Sep. 2020, doi: 10.1016/j.asr.2020.05.040.

[7] A. Maqsoom, B. Aslam, A. Yousafzai, F. Ullah, S. Ullah, and M. Imran, "Extracting built-up areas from spectro-textural information using machine learning," *Soft comput*, vol. 26, no. 16, pp. 7789–7808, Aug. 2022, doi: 10.1007/s00500-022-06794-6.

[8] T. Li *et al.*, "Built-Up Area Extraction from GF-3 SAR Data Based on a Dual-Attention Transformer Model," *Remote Sens (Basel)*, vol. 14, no. 17, Sep. 2022, doi: 10.3390/rs14174182.

[9] H. B. Makineci, "Spatio-temporal change detection of built-up areas with Sentinel-1 SAR data using random forest classification for Arnavutköy Istanbul," *Niğde Ömer Halisdemir Üniversitesi Mühendislik Bilimleri Dergis*, vol. 12, no. 2, pp. 626–636, 2023, doi: 10.28948/ngmuh.1203301.

[10] B. Shrestha, H. Stephen, and S. Ahmad, "Impervious surfaces mapping at city scale by fusion of radar and optical data through a random forest classifier," *Remote Sens (Basel)*, vol. 13, no. 15, Aug. 2021, doi: 10.3390/rs13153040.

[11] Y. Forget, M. Shimoni, M. Gilbert, and C. Linard, "Complementarity Between Sentinel-1 and Landsat 8 Imagery for Built-Up Mapping in Sub-Saharan Africa," *Polymer Preprints*, no. 2018100695, Oct. 2018, doi: 10.20944/preprints201810.0695.v1.

[12] R. Azmi, J. Chenal, H. Amar, C. S. Tekouabou Koumetio, and E. B. Diop, "A Hybrid Approach for Extracting Large-Scale and Accurate Built-Up Areas Using SAR and

Multispectral Data," *Atmosphere (Basel)*, vol. 14, no. 2, Feb. 2023, doi: 10.3390/atmos14020240.

[13] W. Wu, S. Guo, Z. Shao, and D. Li, "CroFuseNet: A Semantic Segmentation Network for Urban Impervious Surface Extraction Based on Cross Fusion of Optical and SAR Images," *IEEE J Sel Top Appl Earth Obs Remote Sens*, vol. 16, pp. 2573–2588, 2023, doi: 10.1109/JSTARS.2023.3250461.

[14] T. A. Kebede, B. T. Hailu, and K. V. Suryabhagavan, "Evaluation of spectral built-up indices for impervious surface extraction using Sentinel-2A MSI imageries: A case of Addis Ababa city, Ethiopia," *Environmental Challenges*, vol. 8, Aug. 2022, doi: 10.1016/j.envc.2022.100568.

[15] F. Samadzadegan, A. Toosi, and F. Dadrass Javan, "Automatic built-up area extraction by feature-level fusion of Luojia 1–01 nighttime light and Sentinel satellite imageries in Google Earth Engine," *Advances in Space Research*, vol. 72, no. 4, pp. 1052–1069, Aug. 2023, doi: 10.1016/j.asr.2023.05.015.

Efficiency Analysis of Coagulant and Energy Consumption of Surabaya Water Treatment Plant using Data Envelopment Analysis Method

Achmad Muzakky[1], Al Rizal Fiqri Dwi Ramdhani[1], and R. Irwan Bagyo Santoso[1], Hilmi Putra Pradana[1], Ervin Nurhayati[1]*

[1]Department of Environmental Engineering, Sepuluh Nopember Institute of Technology, 60111, Surabaya, Indonesia

*Corresponding author: ervin@its.ac.id

Abstract. The high demand for clean water in Surabaya has driven the optimisation of the water treatment process at three water treatment plants (WTPs). The primary challenge is determining the optimal dosage for coagulation and flocculation, which play a crucial role. DEA was used to assess the optimum coagulant dosage. The DEA calculates the coagulant dosage efficiency score and energy consumption in every WTP unit, with a threshold score of 1. The DEAP 2.1 software performs the assessment simulation. When the WTP score is 1, it is considered efficient in saving coagulants, and an energy A score below 1 indicates inefficiency in conserving those resources. The inputs are energy requirements and coagulant doses between 2018 and 2022. The outputs are water quality parameters such as turbidity, total dissolved solids (TDS), and colour. The highest efficiency performance was obtained by WTP 2 (46%), followed by WTP 3 (43%) and WTP 1 (29%). Overall, the recommendations for energy efficiency and coagulant dosage efficiency vary. The energy efficiency improvement recommendations are pump monitoring, maintenance of the electromotor, and supervision of the main assets. On the other hand, maintaining optimal coagulant requirements, selecting the right coagulant, and administering flocculants effectively are key.

1 Introduction

The essential resources for socioeconomic development in urban areas are water and energy. The water resources in urban environments require energy to supply, treat, and distribute. The availability of drinking water, which is affected by water scarcity, population growth, and urbanisation, will become a challenge for water utilities in the upcoming years [1].

Coagulants play an essential role in water treatment processes, particularly in the energy consumption during the coagulation process. The energy consumption efficiency in the coagulation process can be affected by several factors, such as fractal dimensions, floc growth models, floc strength, and floc settling velocity [2]. A practical method for improving coagulant dosing efficiency and energy requirements in water treatment utilities is benchmarking. Evaluating the WTP performance using various metrics and procedures has been done by several studies [1]. Life Cycle Assessment (LCA) is the most common method used to measure the energy intensity of WTP. However, this method has limitations. It focused on the energy consumption by the WTP without considering the effect of raw water quality on the quality of the product [3].

The Surabaya Municipal Water Company operates three major water treatment facilities that consume around 30–40% to the total energy utilisation in process operations. Therefore, drinking water processes contribute to more than 45 million tons of greenhouse gases annually [6]. As much as 40% of the operational costs for drinking water systems are used for energy costs [8]. This study addresses the knowledge gap of the Data Envelopment Analysis (DEA) Common Sets of Weights (CSW) approach to score coagulant dosage and energy requirements efficiency at WTPs 1, 2, and 3. Multiple inputs and outputs for each WTP unit are utilised by using the DEA method, resulting in multi-criteria analysis without requiring assumptions.

2 Methodology

This study required a field research method, collecting both primary and secondary data from WTPs 1, 2, and 3. The research framework involved several systematic stages, including problem formulation, literature review, data collection, data processing and analysis, and formulation of conclusions and recommendations. This study utilised the DEA-CSW method to overcome the limitations of previous DEA-FSW (Flexible Sets of Weights) approaches, which often resulted in multiple facilities being identified as equally efficient, thus limiting their discriminatory power [4]. The DEA-CSW approach allocates common weights to variables (inputs and outputs) for all assessed units, enabling all units to be fully discriminated and ranked in a specific sequence.

Primary data were collected through field surveys examining the condition of production pumps and equipment at all three facilities. The secondary data is using the data series between 2018 and 2022, including water quality parameters (Turbidity, Total Dissolved Solids, and Colour), coagulant dosage, and energy consumption. The input variables consisted of energy requirements and coagulant doses, whereas the output variables included turbidity, TDS, and colour measurements [5]. The efficiency maximum score (EEo) is 1. A score of 1 indicated efficient operation in saving coagulant and energy, whereas scores below 1 indicated inefficiency.

The research methodology also included calculating the optimal coagulant dosage using artificial neural network methods with a 3-4-1 network architecture and evaluating pump efficiency based on pump and electromotor specifications. Field research was conducted to identify the actual conditions of the main assets, including pumps, electromotors, and other processing equipment, at each facility. The correlation between water tariffs and pump efficiency was also analysed to understand the economic implications of the operational efficiency.

3 Result

The analysis of 1,747 data points WTP 1, 2, and 3 during 2018-2022 revealed significant differences in operational efficiency among the three facilities. Using the DEA-CSW method, WTP 2 achieved the highest average efficiency score of 46%, followed by WTP 3 at 43%, and WTP 1 at 29%. This contrasts with the DEA-FSW results, where 55 calculations achieved perfect efficiency scores of 1.0, demonstrating the limited discriminatory power of the FSW approach [7]. The DEA-CSW method successfully identified only one calculation (number 14 from WTP 2) as having the highest efficiency score of 0.994, representing the most energy-efficient coagulant dosage and energy usage among all the evaluated data.

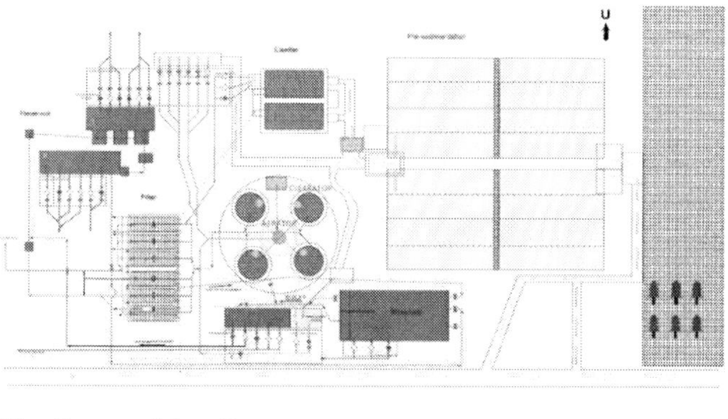

Fig. 1. Water Treatment Scheme Process

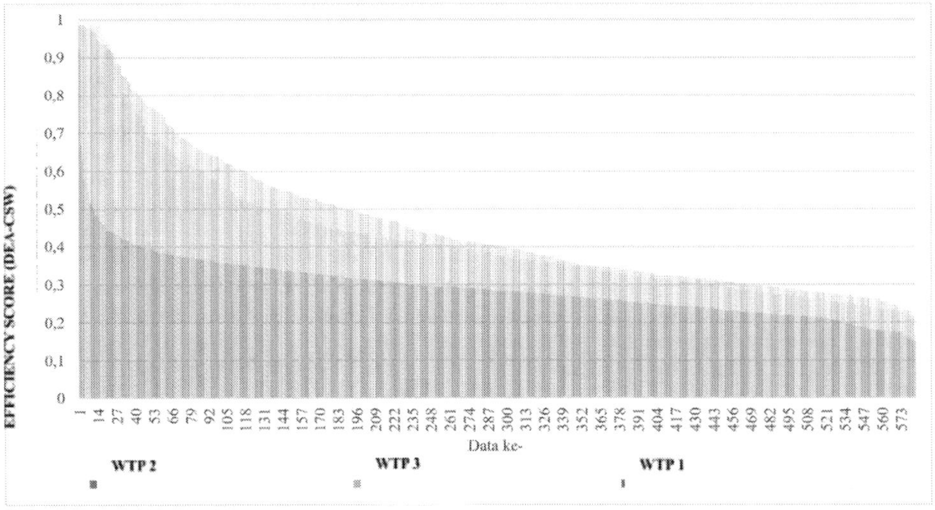

Fig. 2. DEA-CSW calculation chart based on highest to lowest scores

The superior performance of WTP 2 can be attributed to several factors identified through field studies. The facility features four pre-sedimentation basins with a gravity-based overflow settling velocity of 8.3 m/s. A primary clarifier with a 19-meter diameter for initial sedimentation by adding aluminum sulphate and an upflow accelerator clarifier also with 19-meter diameter for continued sedimentation. Additionally, WTP 2 employs 32 rapid sand filter units with dual media layers consisting of 60 cm of silica sand and 30 cm of anthracite, utilizing both air and water backwashing. Overall, WTP 2 has maintained a consistent coagulant dosage and appropriate energy.

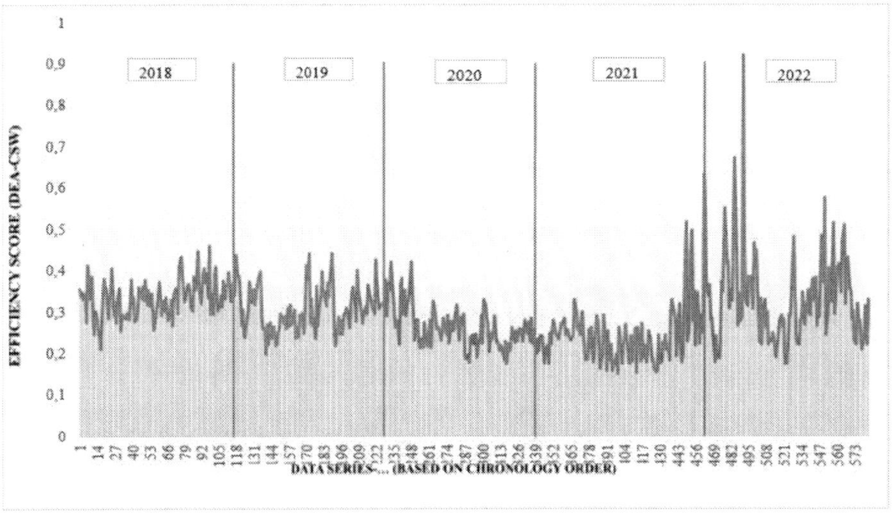

Fig. 3. DEA-CSW WTP 1 calculation chart based on chronological order

Figure 3 illustrates the DEA-CSW efficiency scores for WTP 1 from 2018 to 2022, showing moderate performance with values generally ranging between 0.25 and 0.50 during the first four years. Efficiency levels remain relatively stable but do not show significant improvement from 2018 through 2021, indicating consistent yet modest operational effectiveness. Several notable peaks emerged above 0.70 and approached 0.90, showing an impressive change in operational control, coagulant dosing optimisation, and energy management during 2022. This upward shift implies that corrective actions or process enhancements implemented at WTP 1 were successful, resulting in its strongest performance in the final year of observation.

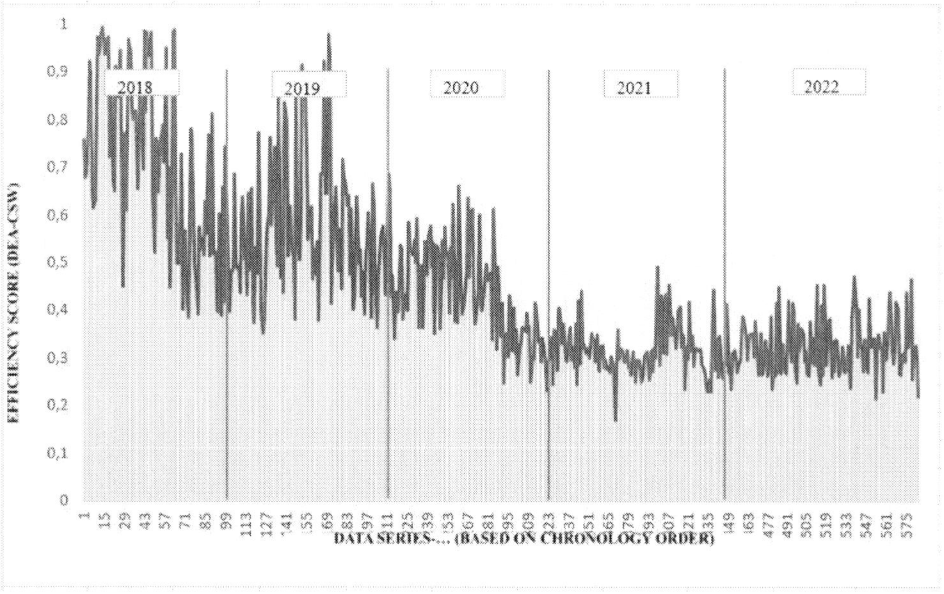

Fig. 4. DEA-CSW WTP 2 calculation chart based on chronological order

Figure 4 illustrates the efficiency trend of WTP 2. Started with the highest performance among the three plants in 2018, with many data points reaching between 0.70 and 0.95. However, a steady downward trend emerges beginning in 2019, indicating progressive declines in efficiency. This situation is related to aging infrastructure, raw water variability, or an imbalance in coagulant and energy usage. By 2020 and continuing through 2021 and 2022, the scores stabilize at significantly lower levels, generally between 0.20 and 0.40, with very few high-efficiency peaks. This sustained decline highlights deterioration in operational consistency and suggests that WTP 2 requires targeted improvements, equipment rehabilitation, and better optimisation strategies to recover its earlier high performance.

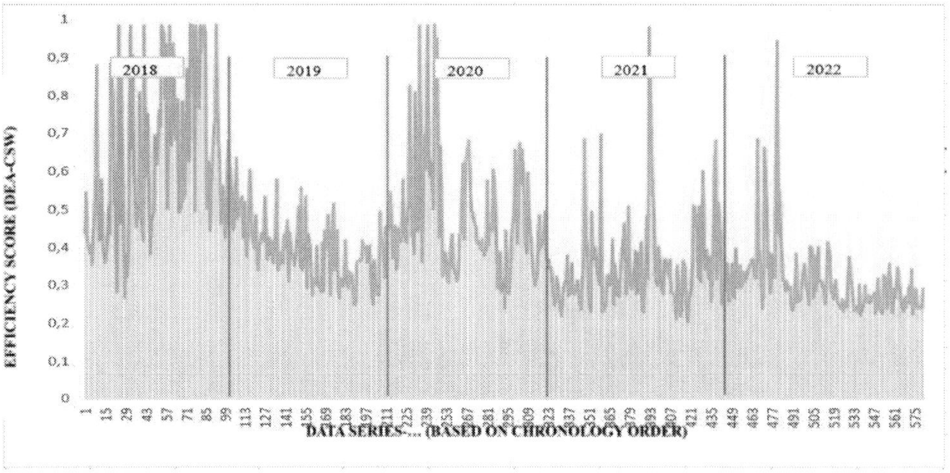

Fig. 5. DEA-CSW WTP 3 calculation chart based on chronological order

Figure 5 presents the efficiency profile of WTP 3, which exhibits extreme volatility in 2018 and 2019, with frequent spikes approaching or exceeding 0.90, indicating periods of very high performance mixed with sharp declines. This instability suggests inconsistent process conditions or fluctuating raw water quality. Beginning in 2020, the efficiency values show a pronounced downward shift, with fewer high peaks and greater clustering around lower scores. By 2021 and 2022, the efficiency stabilises in the range of 0.20 to 0.35, indicating a sustained decline in operational effectiveness. This pattern reflects deterioration in process reliability, potential equipment wear, or the need for improved coagulant dosing and energy management. Overall, WTP 3 transitions from high performance to consistently low efficiency over time.

The energy optimisation potential across all three facilities was substantial. Based on the DEA-CSW analysis, the cumulative energy that could be optimised reached 16,038,796.76 kWh/year, compared to 11,739,038.89 kWh/year using the DEA-FSW method. This energy savings potential could reduce greenhouse gas emissions by 6,704,217.05 kg CO_2 eq/year (DEA-CSW) or 4,906,918.26 kg CO_2 eq/year (DEA-FSW), equivalent to the annual emissions from 1,457 or 1,067 typical passenger vehicles, respectively. For coagulant optimisation, the cumulative potential savings reached 2,927,710 kg/year using DEA-CSW compared to 2,115,895 kg/year using DEA-FSW, representing significant operational cost reduction opportunities for WTP.

Pump efficiency analysis revealed considerable variations across facilities. Sample calculations from WTP 1 showed that southern water pumps no. 1 and 2 achieved 72.21% efficiency, while pump no. 6 achieved only 27.69% efficiency. At WTP 2, water pump no. 5 demonstrated 55.6% efficiency, whereas water pump No. 1 showed 45.5% efficiency. Primary water pumps no. 7 and 8 of WTP 3 exhibited efficiencies of 34.4% and 43.8% respectively. These findings indicate that several pumps, particularly those over 15-20 years old with efficiency scores below 50%, require replacement with high-efficiency models to optimise energy consumption.

The relationship between operational efficiency and water tariffs was analysed using data from 2016 to 2020. The average water tariff in 2020 was IDR 3,087.06 per m^3, while operational production costs showed relatively stable increases over the period, reaching IDR 842 per m^3 in 2020, representing a 3.59% increase from the previous year. The ratio of operational production costs to the average water selling price was 27.27% in 2020, up from 25.88% in 2019. This indicates that efficiency improvements in coagulant dosage, main assets, processing units, and pump operations directly impact operational production costs, ultimately affecting consumer tariffs and service quality.

4 Conclusions

This study presents a robust analysis using the DEA-CSW method to measure the energy and coagulant efficiency at three WTPs. Shortly, the potential savings in energy (16,038,796.76 kWh/year) and coagulant (2,927,710 kg/year) are significant results.

The efficiency ranking showed WTP 2 achieving the highest score of 46%, followed by WTP 3 at 43%, and WTP 1 at 29%. The superior performance of WTP 2 was attributed to its consistent operational capacity of 1,000 litres per second (lps). A comprehensive treatment process design, including four pre-sedimentation basins and dual-media rapid sand filters, followed by systematic operational practices enabling consistent coagulant dosing and appropriate energy management.

Significant optimisation potential was identified across all facilities, with a cumulative energy savings potential of 16,038,796.76 kWh/year and coagulant savings of 2,927,710 kg/year using the DEA-CSW method. These savings translate to substantial environmental benefits, equivalent to removing 1,457 passenger vehicles. Pump efficiency analysis revealed considerable variation, with some units, particularly those over 15-20 years old, achieving efficiency scores below 50%, indicating an urgent need for replacement with high-efficiency models.

Comprehensive recommendations were formulated to address both coagulant and energy efficiency through a detailed evaluation of coagulant requirements, optimal dosage determination, systematic maintenance programs, proper asset management, and strategic replacement of aging equipment. The correlation analysis between operational efficiency and water tariffs demonstrated that efficiency improvements directly impact operational production costs and consumer tariffs, emphasising the economic importance of optimisation. The implementation of these recommendations will enable the WTP facilities to achieve substantial improvements in operational efficiency, cost reduction, and environmental performance while maintaining high-quality drinking water production for the residents.

References

1. M. Rodríguez-Merchan, C. Ulloa-Tesser, C. Baeza, Y. Casas-Ledón, Evaluation of the Water-Energy nexus in the treatment of urban drinking water in Chile through exergy and environmental indicators. J. Clean. Prod. 317, 128494 (2021). 10.1016/j.jclepro.2021.128494
2. M. Molinos-Senante, C. Guzmán, Benchmarking energy efficiency in drinking water treatment plants: Quantification of potential savings. J. Clean. Prod. 176, 417-425 (2018). https://doi.org/10.1016/j.jclepro.2017.12.178
3. M. Molinos-Senante, R. Sala-Garrido, Assessment of energy efficiency and its determinants for drinking water treatment plants using a double-bootstrap approach. Energies 12(4), 759 (2019). https://doi.org/10.3390/en12040765
4. J. Wu, J. Chu, Q. Zhu, Y. Li, L. Liang, Determining common weights in Data Envelopment Analysis based on the satisfaction degree. J. Oper. Res. Soc. 67(12), 1446-1458 (2016). 10.1057/jors.2016.35
5. T. Coelli, A Guide to DEAP Version 2.1: A Data Envelopment Analysis (Computer) Program. CEPA Working Paper 96/08, Centre for Efficiency and Productivity Analysis, University of New England (2016)
6. United States Environmental Protection Agency, Energy Efficiency for Water Utilities (EPA, Washington DC, 2023)
7. A. Charnes, W.W. Cooper, E. Rhodes, Measuring the efficiency of decision making units. Eur. J. Oper. Res. 2(6), 429-444 (1978). https://doi.org/10.1016/0377-2217(78)90138-8
8. I. Contreras, A review of the literature on DEA models under common set of weights. J. Model. Manag. 15(4), 1277-1300 (2020)s. 10.1108/JM2-02-2019-0043

Feasibility Analysis of Decentralized Pressure Improvement in Peripheral Water Networks: Case Study of Zone 3, Surabaya, Indonesia

Qoyyum Imaniar[1*], *Umboro* Lasminto[1], and *Hendro Ardiansyah* Riyanto[2]

[1]Civil Engineering Department, Faculty of Civil Planning and Geo Engineering, Institut Teknologi Sepuluh Nopember, Surabaya, Indonesia.

[2]Surya Sembada Municipal Water Utility of Surabaya City, Surabaya, Indonesia.

Abstract. The study evaluates shifting Surabaya's low-pressure coastal area (Zone 3) from a centralized to a decentralized water treatment system. Current residual pressure is below 2 mH_2O, failing the 5–10 mH_2O standard, and Scenario 1 shows the existing system cannot meet 2030 demand. Scenario 2, following the utility's centralized expansion (Karang Pilang IV WTP, Putat Gede 3 Booster Pump, Mbah Ratu Reservoir), reaches 88% of customers at adequate pressure by 2030 but requires IDR 219.98 billion. A decentralized alternative with a local 500 L/s WTP and 900 m^3 reservoir dedicated to Zone 3 achieves 98% coverage at the required pressure for only IDR 100.5 billion. The study concludes that decentralized systems can more effectively and cost-efficiently boost pressure and meet growing demand where raw water sources are limited and centralized expansion is insufficient.

1 Introduction

Surabaya City is the second most populous metropolitan area in Indonesia. The city's population reached 3,018,022 in 2024, with a corresponding population density of approximately 8990 people/km² [1]. This rapid urbanization places significant strain on public infrastructure, particularly water services. Surya Sembada Municipal Water Utility of Surabaya City (PDAM Surabaya) is guided by Surabaya Regional Medium-Term Development Plan (RPJMD) 2021-2026. Which mandates achieving universal and equitable access to safe drinking water by 2030 [2]. Therefore, PDAM Surabaya plans to develop a centralized massive-scale infrastructure to improve areas that are still not covered by water, namely, zone 3 [3].

The most critical challenge is found in the northernmost coastal peripheral region, where the city's highest population density is recorded at 35,271 people/km² [1]. Furthermore, spatial planning data indicate that this coastal zone has almost no vacant land available for new construction [4], which has historically hindered the utility from establishing local treatment facilities in the area. Despite the mandate for universal access to safe drinking water by 2030, a gap remains in optimal conditions, particularly in the Zone 3 distribution network, where the low residual high water pressure and its causes are discussed in this study.

Based on the residual pressure record data from the Surabaya City PDAM, this area still has a residual pressure level of below 0.5 mH2O, namely, Simokerto sub-district and 6 sub-districts currently suffer from an average residual water pressure head below 2 mH_2O in 2025 [3], failing to meet the Ministry of Public Works' standard (No.18/RT/M/2007) of 5–10 mH_2O [5], as shown in **Fig. 1**.

* Corresponding author: qoyyumimaniah@gmail.com

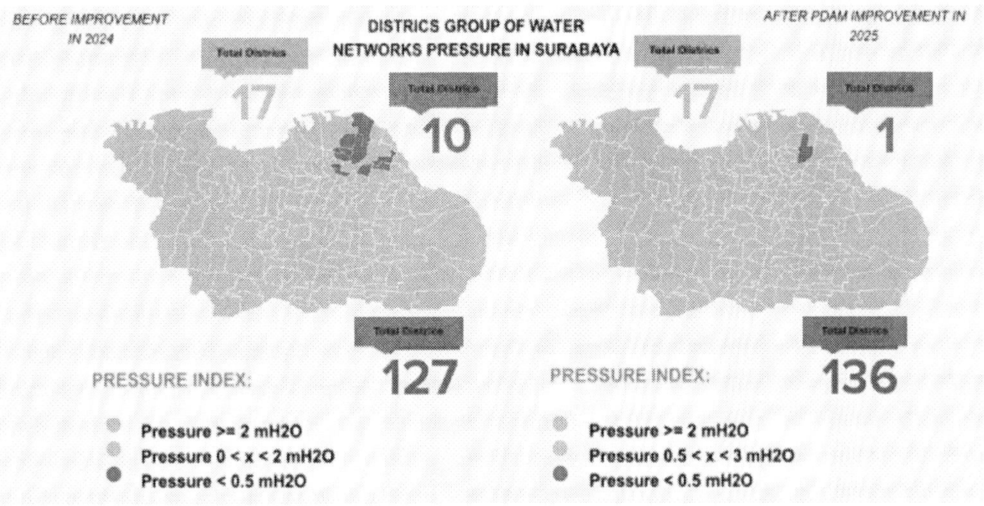

Fig. 1. Analysis of the Pressure Water Networks in Surabaya City [3].

This hydraulic failure is quantitatively validated by customer complaint data, where there is a frequent frequency of customer complaints in zone 3 per districts for the last 1 year in the form of water not flowing in Petemon, 220 complaints, Tambak Wedi, 30 complaints, Wonokusumo, as many as 28 complaints, Kapasan, 26 complaints, Kapas Krampung, 23 complaints, Bulak Banteng, 17 complaints, Bogen Baru, 1 complaint, which indicates widespread lack of water access in several coastal sub-districts.

The main cause of the low pressure in Zone 3 is its vast distance service from the main Water Treatment Plants (Karangpilang and Ngagel WTP), causing the development of centralization infrastructure to be less effective because which causes extreme frictional head loss across transmission lines [6], then This distance also increases water age within the pipes [7], leading to a decline in residual chlorine and a heightened risk of bacterial regrowth before reaching the customer [8].

The second cause, the current centralized system, is the increasing uncertainty of raw water availability. Surabaya's high-quality water supply of 4% comes from Umbulan springs and other springs in Pasuruan; however, this source is currently experiencing a significant decline in discharge quantity [3]. This reliance on a distant and diminishing source threatens the supply sustainability for peripheral zones. Moreover, local water resources in coastal areas are often non-viable due to poor surface water quality and brackish groundwater caused by extensive seawater intrusion [9].

The third cause is the high number of people in the area [1], so that the need for water is also large, and there is limited land to build clean water infrastructure in the northern coastal city of Surabaya [4] or the Zone 3 area.

While the utility's current mitigation plan focuses on centralized expansion—including the Mbah Ratu Reservoir and Putat Gede Booster with a high capital cost of IDR 219.98 billion [10]—a significant research gap remains. Comparative evidence evaluating whether this expensive centralized expansion can truly overcome the combined challenges of source uncertainty in Pasuruan and hydraulic bottlenecks in the coastal periphery is lacking. A decentralized "point-of-use" system located directly within the coastal service area could eliminate long-distance head loss and reduce dependency on diminishing external springs.

Consequently, this study aims to conduct a comparative feasibility analysis between the utility's centralized expansion plan and a proposed decentralized water treatment system. Specifically, this research aims to:

- Simulate hydraulic performance and service coverage projected for 2030.
- Evaluate technical compliance with regulatory pressure standards.
- The capital cost requirements are compared to determine the most viable and resilient solution for Surabaya's northernmost coastal peripheral water networks.

2 Materials and Methods

2.1 Study of Interest

The study was conducted in the city of Surabaya, Indonesia. More precisely, in the Zone 3 Service Unit of the Clean Water Distribution Network of Surabaya City. In the Zone 3 service unit of PDAM Surabaya City, further divided into several sub-zones, namely sub-zones 301-317, where the division of sub-zones can be seen in **Fig. 2**.

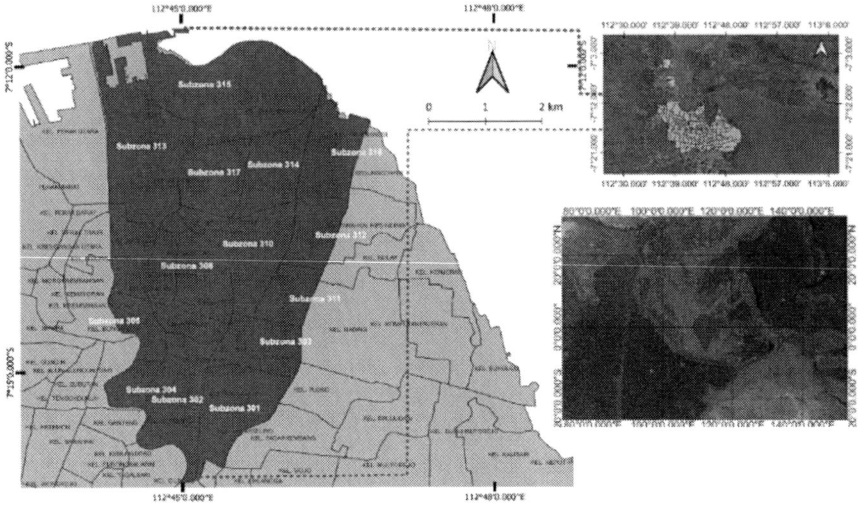

Fig. 2. Location of the Surabaya City Clean Water Distribution Network Zone 3 Services Unit.

2.2 Conceptual Study

This study was conducted by analyzing some of the problems that occur in the clean water distribution network in Surabaya, namely, being limited to peripheral areas (zone 3) that have low water head pressure only and then simulating the planned development, whether it is sufficient to meet or can be replaced by other alternatives. For more details, the conceptual diagram is found in **Fig. 3** as follows:

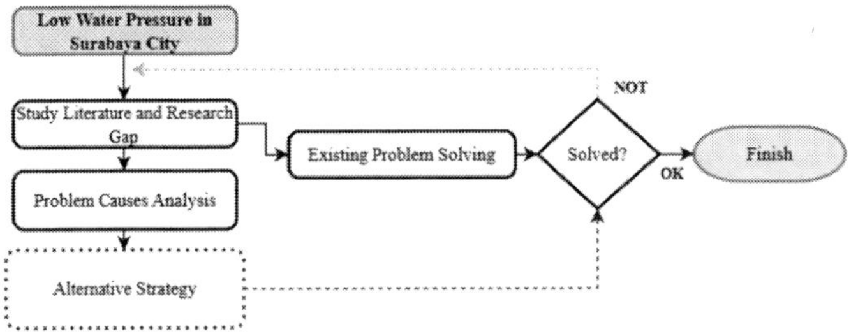

Fig. 3. Conceptual Diagram.

2.3 Methodology and Research Diagram

2.3.1 Data and analysis diagram

The data used for this study is limited to 2024, because the data from the annual book reporting has already preserved for that year, which can only be obtained at the end of 2024. The data used, sources, and usage are shown in **Table 1.**

Table 1. Data Sources and Their Uses

Data Item	Data Source	Use
City Population	Central Statistical Agency of Surabaya 2024	Calculating the amount of water per person in Surabaya, Indonesia
Total water consumption	Surabaya's Municipal Water Utility, Annual Book, 2024	Calculating the amount of water needed per person in Surabaya, divided by the total population
Water networks geospatial data	SIG Division, Surabaya Regional Water Supply	Obtaining Numerical Data on the Surabaya Pipeline
Sensor Pressure Recording	SIG Division, Surabaya Regional Water Supply	Low-pressure analysis and calibration of the numerical model
PDAM Book and E-procurement, Surabaya Infrastructure and Spatial Plan	Website/open source	To analyze the problems in the Surabaya network, the growth rate of Surabaya's population is determined, and a plan based on existing spatial conditions is chosen

For hydraulic simulation, use WaterGEMS Software recommended by Surabaya's Municipal Water Utility because it has more advanced features that make data analysis, including calibration, easier. For the research scheme and data processing to become the result of conclusions, as shown in **Fig. 4**.

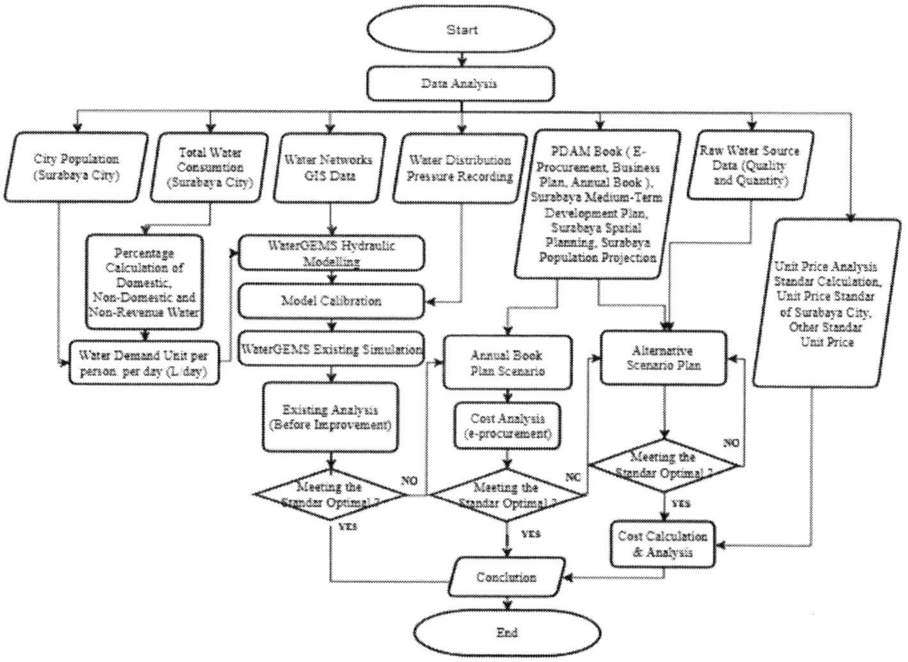

Fig. 4. Research Diagram

2.3.2 Demand Assumption

In the calculation of existing water needs using account billing data for each customer of PDAM Surabaya City in the last year, which is converted in the form of L/S, where the service coverage has reached 100%, the total water demand is the sum of domestic, non-domestic, and water loss needs, found in equation 1 by Kalensun [12] as follows:

$$Q_r = Q_d + Q_n + Q_a \tag{1}$$

Where:
Q_r = average water requirement (L/s),
Q_d = domestic water demand (L/s),
Q_n = non-domestic water needs (L/s),
Q_a = water loss (L/s).

Where the analysis of the demand approach in this planning uses the assumption that non-domestic water needs are 25% of domestic water needs according to urban clean water supply standards [11], and water loss of 31.05% in 2024 and decreases by 1% each year according to the PDAM business plan book, so that in 2030 water loss is 25.05%.

2.3.3 Hydraulic Simulation

In hydraulic analysis using WaterGEMS software, where the data needed for modeling is water demand, technical data of the clean water system of the City of Surabaya (primary and secondary pipelines, reservoirs, pumps, tanks, etc.), and field pressure data are used as calibration to adjust the simulation results to describe the actual conditions in the field. **Figure 3** shows more details of the simulation process. The hydraulic simulation itself is divided into 2 planning stages: the analysis stage of the existing year and the analysis of the projected year 2030. The demand used in the existing year is the demand from the existing billed account data. The 2030 projection uses population data multiplied by the average water usage from the existing billed account data.

2.3.4 Model calibration and validation

Data calibration aims to determine the difference between the data obtained from the simulation results and the actual data in the field. The root mean square error (RMSE) method is used to perform the calibration process [13]. The RMSE formula is as follows:

$$RMSE = \sqrt{\frac{\sum(X-Y)^2}{n}} \tag{2}$$

Where:
X = Observation value
Y = Predicted value
n = Amount of Data

2.3.5 Surabaya Regional Water Supply Utility and Service Improvement Planning

Surya Sembada Municipal Water Utility of Surabaya City is a drinking water company in Surabaya that has achieved 100% of the population of Surabaya City, with 630,491 customer connections. However, there are still areas that have small discharges with a pressure below 0.5 mH_2O, namely, in 10 villages in the north of Surabaya

City, besides a water loss of 31.06% above the regulatory standard of 25% [3], further new challenges of climate and population increase, so that the Water Utility of Surabaya City planning medium-term strategic development, and to realize the Surabaya City Regional Medium-Term Development Plan program Number 6 of 2021-2024, namely "By 2030, achieve universal and equitable access to drinking water"[2]. The rehabilitation of the pipeline network as far as 52.82 kilometers and the construction of the Karang Pilang IV Water Treatment Plant with a capacity of 1000 L/s at a cost of IDR 149,850,000,000 are among the developments planned for the medium term. Construction of the Mbah Ratu reservoir with a capacity of 6000 m^3 for IDR 36,299,552,556. Then, the Putat Gede III pump house was constructed with a capacity of 2 pumps of 375 L/s and one pump of 600 L/s, at a cost of IDR 33,834,747,495 [10].

2.3.6 Water Resources in Zone 3

Some of the raw water in Surabaya is from surface water (Surabaya River, Brantas River, and Bengawan Solo), Umbulan Springs, then used water such as Boezom Morokrembangan and Kali Mas, then groundwater but it is not recommended because the quality of Surabaya's groundwater is not good, especially in coastal areas where the city of Surabaya zone 3 has a rather brackish water condition according to groundwater salinity [10].

2.3.7 Ministry of Public Works' Water Distribution and Water Utility Services Standard

In raw water planning, there is a standard for piping distribution networks that regulates the speed of permitting and water pressure in the clean water distribution pipeline, regulated by the Ministry of Public Works' standard [5], which is found in **Table 2**. Piping distribution network standards,

Table 2. The Piping Distribution Network Standards [5].

Description	Notation	Requirements
Speed of water flow in the pipes		
a) Minimum speed	V min	0.3 - 0.6 m/s
b) Maximum speed		
PVC or ACP pipes	V max	3.0 - 4.5 m/s
Steel pipe or DCIP	Vmax	6.0 m/s
Pressure of water in the pipe		
a) Minimum pressure	h min	(0.5 – 1.0) atm at the farthest service coverage point.
b) Maximum pressure		
- PVC or ACP pipes	H Max	6 - 8 atm
- Steel pipes or DCIP	H Max	10 atm
- PE 100 pipe	H Max	12.4 MPa
- PE 80 pipe	H Max	9.0 MPa

Then, in the planning of clean water services, some standards are a reference in clean water services in urban areas regulated by the Ministry of Settlements and Regional Infrastructure [11], found in **Table 3** as follows:

Table 3. Urban Clean Water Service Standards [11].

Parameters	Metropolitan	Big Cities	Medium City	Small Town
Service Target	100%	100%	100%	80%
Water usage (l/person/day)				
Home Connection	150-210	100-150	90-100	60-100
General hydrant	30	30	30	30
Industrial Needs				
Heavy Industry (L/s)	0.5-1			
Medium Industry (L/s)	0.25-0.5			

15%-30% of the domestic needs

Parameters	Metropolitan	Big Cities	Medium City	Small Town
Light Industry (L/s)	0.15-0.25	15%-30% of the domestic needs		
Commercial				
Markets	0.1-1			
Local hotels (l/room/day)	400			
International Hotels	1000			
Social				
University (l/person/day)	22			
School	15			
Mosque (l/day)	1000-2000			
Hospital (l/room/day)	400			
Health center (l/day)	1000-2000			
Office (L/s)	0.01			
Military (l/day/ha)	10000			
Operating Hours (Hours)	24	24	24	24
Maximum water requirement	Average requirement x 1.35			
Water loss in new system	20 %			
Loss of the water legacy system	30%-40%			
Peak Hour Needs	1.65-2			

2.3.8 *Calculation of Unit Price Analysis Standard*

The BOQ calculation requires a detailed specification of the requirements per unit of work, which is derived from the engineering drawings during development analysis and planning. Material needs are considered based on the planning analysis results and adjusted to their market availability. Meanwhile, in the calculation of Capital Cost, the basis is the results of the BOQ calculation and Unit Price Analysis [14], using relevant cost data in the city of Surabaya [15] and referring to work standards.

3 Results and Discussion

3.1 Alternative scenario planning

From the problems in Zone 3, namely, low pressure and large population, the alternative planning chosen is the decentralization of production or the creation of a drinking water treatment plant in Zone 3, which is compared with the existing and planned Surabaya Municipal Water Utility. The scenarios compared are the existing conditions before improvement, the post-improvement conditions of the regional water supply utility planning/centralized scenario, and the post-Alternative planning conditions/decentralized scenario. Each scenario is analyzed with 2 stages or sub-scenarios, namely, demand in 2024 and the projected demand in 2030. The investments compared are as follows in **Table 4**:

Table 4. Scenario Planning

Scenario	Name	Sub-Scenario	Investment
1	Existing	a. Existing in 2024	- (without improvement)
		b. Existing in 2030	
2	Surabaya's Municipal Water Utility Plan	a. Surabaya's Municipal Water Utility Plan for 2024 (centralized improvement scenario)	1. Addition of a drinking water treatment plant for the Karang Pilang IV, with a capacity of 1000 L/s.

	Surabaya's Municipal Water Utility Plan	b. Surabaya's Municipal Water Utility in 2030 (centralized improvement scenario)	2. Addition of the Mbah Ratu Reservoir of 6000 m^3. 3. Addition of a Putat Gede III Boster Pump, 2 x 375 L/s + 600 L/s.
3	Alternative Plan	a. Alternative Plan for 2024 (decentralization improvement scenario)	1. Zone 3 drinking water treatment plant, Capacity 500 L/s;
		b. Alternative Plan for 2030 (decentralization improvement scenario)	2. Addition of 900 m^3 production reservoir.

3.2 Analysis of water supply and demand

In this section, the aim is to determine the water production supply and water needs compared in the existing conditions (Scenario 1) without additional supply, the conditions of adding centralized supply (Scenario 2), and adding decentralized (Scenario 3), **Table 5** shows the result of the comparison of the difference in water demand in the existing year and the projection of the three scenarios. Cumulation of water supply and demand.

Table 5. Cumulative Water Supply and Demand

Item	Explanation	Year		
		2024	**2025**	**2030**
Domestic (L/person/day)		185.76	210	210
Non-Domestic (L/person/day)	Domestic x 25 %	46.44	52.50	52.50
Normal Factor	F average = 1	232.20	262.50	262.50
Maximum day factor (MDF)	F max-day = 1.2	278.64	315.00	315.00
Peak Hour Factor	F peak = 1.8	417.96	472.50	472.50
Population (Person)		3,021,043	3,043,518	3,144,330
Non-Revenue	-1% per year	31.52%	30.52%	25.52%
Normal water demand	L/s	10,678.09	12,068.92	11,991.03
	m^3/s	10.68	12.07	11.99
Peak water demand hourly	L/s	19,220.57	21,724.06	21,583.86
	m^3/s	19.22	21.72	21.58
Production Capacity	L/s (water municipal plan)	-	1,000.00	1,000.00
	L/s (alternative plan)	-	500.00	500.00
	L/s (existing)	11,834.16	11,834.16	11,834.16
	m^3/s(existing)	11.83	11.83	11.83
	L/s (existing + water municipal plan)	11,834.16	12,834.16	12,834.16
	m^3/s (existing + water municipal plan)	11.83	12.83	12.83
	L/s (existing + alternative plan)	11,834.16	12,334.16	12,334.16
	m^3/s (existing + alternative plan)	11.83	12.33	12.33
Capacity Difference (Production and Demand)	m^3/s(existing)	1.16	-0.23	-0.16
	m^3/s (existing + water municipal plan)	1.16	0.77	0.84
	m^3/s (existing + alternative plan)	1.16	0.27	0.34

From the analysis results in **Table 5.** Water Supply and Demand Cumulation can be found that to deal with water needs in the projected year, existing conditions without additional water supply cannot face an increase in the number of population in 2025 and 2030 because water has a deficit of -0.23 m^3/s. Thus, an additional capacity of 500 L/s from Scenario 3 is sufficient and efficient to overcome water needs in 2025 and 2030 compared to an additional capacity of 1000 L/s from Scenario 2, which experiences idle capacity or excess water production and water production cost.

3.3 Hydraulic Analysis

3.3.1 Pressure calibration and validation

In the pressure calibration of the sensor records and simulations, as many as 34 points spread across the city of Surabaya were only used on primary and secondary pipes. The result of calibration pressure networks is shown in **Fig. 5**.

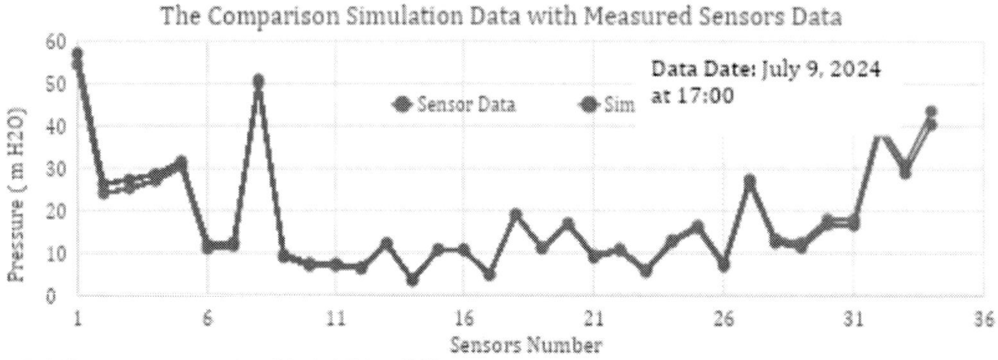

Fig. 5. Calibration pressure result on July 9, 2024, at 17:00.

Calibration is performed by adjusting the roughness value of the pipe so that the pressure can be in accordance with the field conditions. The results of the pressure adjustment indicated that the average pressure in the field data was 18.21 m H_2O, while in the simulation data, the average pressure was 18.18 m H_2O which had an average error of 0.91 m H_2O or 5% of the pressure conditions in the field with an RMSE value of 1.2 m H_2O, indicating that the error was still below 10% and valid to be used as a simulation of the actual situation.

3.3.2 Analysis of the Hydraulic Simulation Results

This hydraulics analysis is simulated at the peak hour (5:00 PM), into 3 scenarios that are Existing/without improvement in **Fig. 6**, government improvement plan/centralized scenario in **Fig. 7**, and alternative Improvement plan / decentralized scenario in **Fig. 8**, which are divided into sub-scenarios in 2024 and 2030 (Projected). Where the simulation results are explained whether the pressure is already optimal according to the regulations of the Ministry of Public Works, where the pressure is > 7 m H_2O, or not optimal, where the pressure is < 7 m H_2O [5].

1. Hydraulic analysis without improvement (Scenario 1)

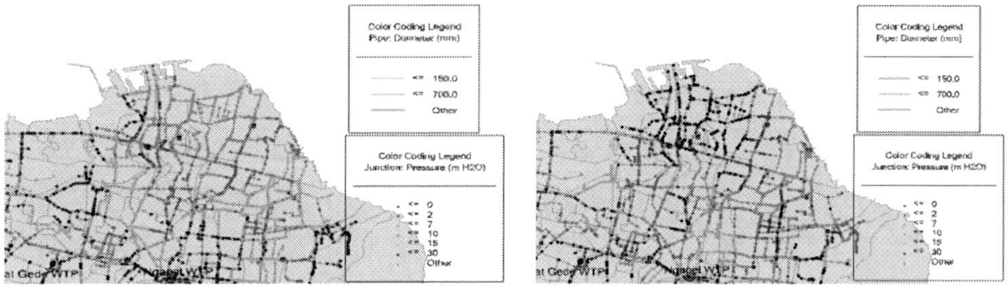

(a) (b)

Fig. 6. Hydraulic simulation of (a) 2024 year and (b) projected 2030 year without Improvement (scenario 1).

In the results of the hydraulic simulation scenario 1 or without improvement scenario in **Fig. 6**, from the analysis with a visual approach, it can be seen that in the current year 2024, the majority of nodes are orange or the pressure is below 7 mH_2O indicating that the pressure is low but not water-off. However, there are many black nodes at the time of the 2030 projection year, indicating that the water pressure is 0 mH_2O.

2. Hydraulic Analysis with Centralized Improvement (Scenario 2)

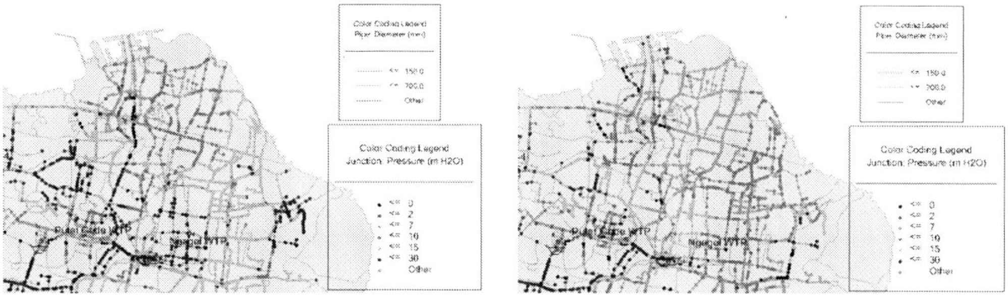

(a) (b)

Fig. 7. Hydraulic simulation of (a) 2024 year and (b) Projected 2030 year with Government Improvement Plan / Centralized Scenario

In the results of the hydraulic simulation scenario 2 / Centralized Scenario in **Fig. 7**, from the analysis with a visual approach, it can be seen that in the existing year 2024, the majority of nodes are green or the pressure is almost all optimal above 7 mH_2O but at the time of the 2030 projection year, there are many orange nodes, indicating the water pressure is below 7 mH_2O and the red nodes indicate the pressure below 2 mH_2O This indicates that scenario 2 / centralization scenario is still not optimal to increase overall pressure in 2030.

3. Hydraulic Analysis with Decentralized Improvement (Scenario 3)

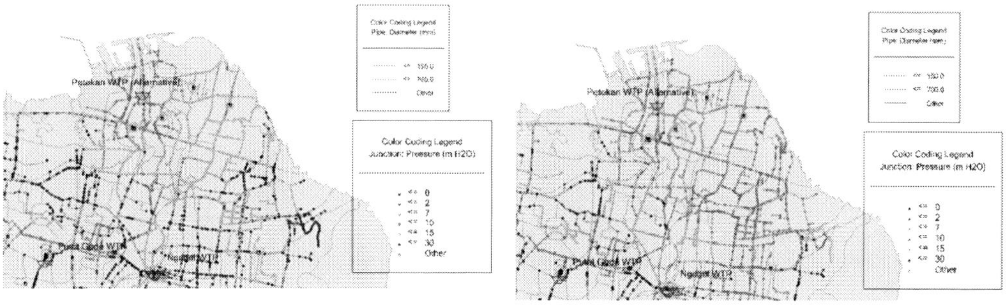

(a) (b)

Fig. 8. Hydraulic simulation of (a) 2024 year and (b) projected 2030 year with Government Improvement Plan / Decentralized Scenario

In the results of the hydraulic simulation scenario 3 / decentralized scenario in **Fig. 8**, from the analysis with a visual approach, it can be seen that in the current year 2024 the majority of nodes are light blue or the overall pressure is optimal above 15 mH_2O, but at the time of the 2030 projection year, there are many orange nodes, indicating a water pressure below 7 mH_2O and the red nodes indicate a pressure below 2 mH_2O but the pressure is not optimal compared to the previous scenario.

For more details of the output pressure per junction/node, a pressure percentage is calculated to determine the optimal amount of water pressure access, which is divided into optimal nodes with a pressure of > 7 m H_2O, non-optimal pressure < 7 m H_2O and > 2 m H_2O, and an area that cannot be watered with a low water pressure value of 2 m H_2O. The percentage of pressure calculation at the node in each scenario is shown in **Fig. 9**. Water Access Percentage per Scenario, as follows:

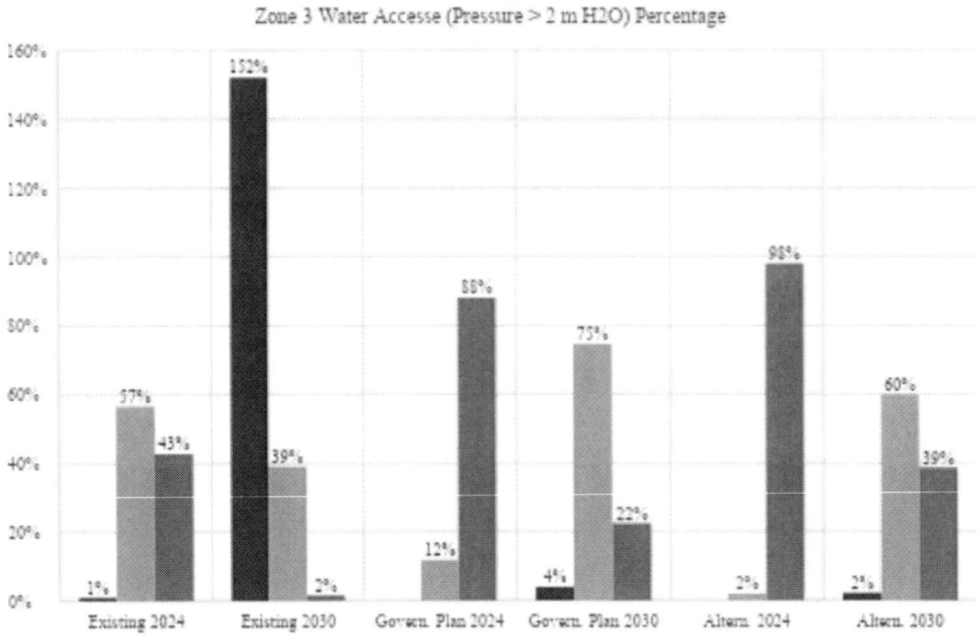

Fig. 9. Water access percentage per scenario

From the hydraulic analysis of **Figures 6, 7, and 8**, we know that the scenario without improvement cannot endure the water demand problem in 2030 at all. **Fig. 9** shows that the scenario of the government plan (centralized plan) improves to 100% water access for the existing, but 12% of the nodes still do not meet the optimal. And meet 96% accessed on the 2030 projection year. Conversely, the alternative scenario succeeds in accessing all areas for the existing year and 98% access on the 2030 projection. This indicates that in the decentralized scenario with the addition of 500 L/s supply, the optimal pressure is 12% greater than centralization with the addition of 1000 L/s supply for peripheral areas in zone 3, and the amount of water supply required in decentralization is 50% less than the centralization scenario in the case of this location.

3.4 Explanation of the Centralized Improvement Plan

Centralized planning is an alternative planned by the PDAM of Surabaya City because the main water production units, namely, Ngagel and Karang Pilang WTP, have been built, and land is still available to increase the central supply capacity. However, based on the results of the simulation using WaterGEMS in the analysis in the previous sub-chapter, complex planning and a large amount of water production are needed to meet the increased pressure in the peripheral area or Zone 3. The planning items and costs are listed in **Table 6,** with a total cost of IDR 219.98 billion.

Table 6. Cost of Centralized Improvement – PDAM's Plans

No.	Item	Total Cost (IDR)
1	Construction of the Putat Gede Pumpa House 3	33,834,747,495.00
2	Mbah Ratu Pump House and Supporting Buildings (reservoir and main distribution network)	36,299,522,566.00
3	Development of the IPAM Karang Pilang IV	149,850,000,000.00
Total Project Value		**219,984,270,061.00**

Total Project Value (Rounded)	**219,980,000,000.00**

3.5 Explanation of the alternative decentralized with WTP 500 L/s

3.5.1 Alternative Glass-Fused-to-Steel (GFS) WTP 500 L/s Preliminary Design

Based on the results of the simulation using WaterGEMS previously, it can be seen that the alternative decentralization in the form of making a water treatment plan with a capacity of 500 L/s has a significant influence on increasing pressure and saving total water compared to the centralization process by requiring a water capacity of 1000 L/s. However, in the existing conditions, there are limitations to building a decentralized WTP:

1. The quality of the existing raw water;
2. The availability of the existing land;
3. Higher cost for advanced units.

Then, the alternative is that it can save space, can treat existing river water well with low side effects, but requires slightly higher capital costs and maintenance costs than a conventional WTP structure. The Advanced Unit can be seen in **Table 7** and the layout in **Fig. 10** to determine the preliminary design of the land available.

Fig. 10. Petekan Glass-Fused-to-Steel (GFS) WTP 500 L/s (Decentralized Plan)

Table 7. Advanced WTP's Units and Dimensions

Unit	Dimension	Component
Drinking water treatment plant based on ultra-high-grade steel structure	Unit flocculation, clarifier, filtration, and reservoir	Ultra-high-grade glass-fused-to-steel
SCADA-setting system	1 set	Automatic bar screen, valve, and water level sensors
Direct Intake, Gravity Pipe	DN800, L = 15 m	Direct Intake-type intake and gravity intake
Submersible Pump	90 kW	Sewage submersible pump
Coagulation	DN500	In-Line Static Mixer

Unit	Dimension	Component
Flocculation	Dia. 8.5 m x 6 m	Vertical shaft mechanical flocculator
Clarify	Dia. 9 m x 5 m	Lamella Tube Settler
GAC Filtration	Dia. 5.5 m x 4 m	Granular Activated Carbon Filtration
Reservoir	Capacity 900 m^3	Structural concrete construction
Distribution Pump	3 x 250 L/s; 200 kW	Centrifugal Pump

Of the several units mentioned, the cost is calculated according to the standard of price calculation of the WTP capital cost per item package for 500 L/s from PDAM Surabaya. The total capital cost required for the Alternative Glass-Fused-to-Steel (GFS) WTP and the reservoir is IDR 100.5 billion. For details, see **Table 8** below:

Table 8. Capital Cost Package of the Alternative Glass-Fused-to-Steel (GFS) WTP (500 L/s)

No.	Item	Total cost (IDR)
I	Work preparation and occupational safety	900,000,000.00
II	Work of the intake unit (Vendor Estimate for 500 L/s)	5,000,000,000.00
III	Concrete foundation work for GFS WTP Structure	7,600,000,000.00
IV	Work of Process's Mechanicals	50,000,000,000.00
V	Interconnection Piping	3,000,000,000.00
VI	Electrical and Instrumentation Work	4,000,000,000.00
VII	Reservoir and Distribution Work	12,000,000,000.00
VIII	Supporting Building Work	8,000,000,000.00
Total construction cost (excluding VAT)		**90,500,000,000.00**
VAT 11%		**9,955,000,000.00**
Total Project Value		**100,455,000,000.00**
Total Project Value (Rounded)		**100,500,000,000.00**

3.5.2 *Limitations and Advantages of Decentralized WTP Design*

The limitation of decentralized WTP development is that the investment cost for a viable unit at this location is more expensive than conventional WTPs in general. Additionally, the existing water conditions are very poor, and even with advanced WTP technology to reduce the side effects of water quality, over time it can still result in high maintenance and repair costs. However, the advantage is that because it has a decentralized production system based on simulation results, it can meet lower water demands, making it more water-efficient and requiring less infrastructure, which results in a lower total capital cost for achieving higher water head pressure results compared to centralized planning.

3.6 Comparison of centralized and decentralized scenario results (offered)

Based on several analyses, both pressure results, capital costs, and the amount of water used were compared, with the comparison results are in **Table 9** and the comparison chart in **Fig. 11**.

Table 9. Capital Cost of the Alternative WTP

Scenario	Condition Networks	Result Comparisons			
		Addition of Water Production	Pressure		Capital cost (IDR)
			2024	2030	
1	Existing	-	43% Optimal	2% Optimal	-
2	Centralized	1000 L/s	88% Optimal	22% Optimal	219,980,000,000
3	Decentralized	500 L/s	98% Optimal	39% Optimal	100,500,000,000

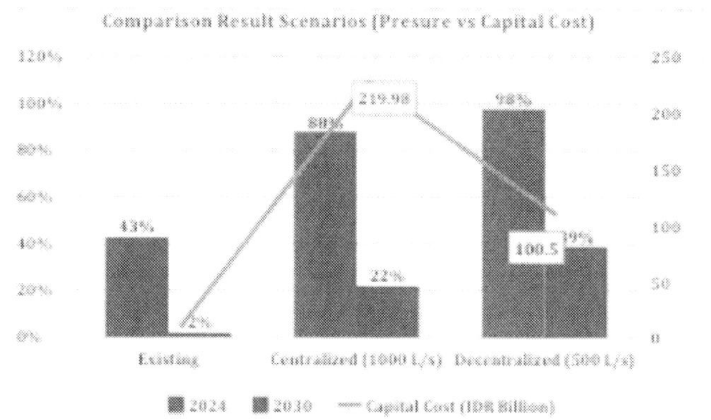

Fig. 11. Comparison of Scenario Result

From the analysis of the comparative graph of the scenario results in **Fig. 11**, it can be seen that in the existing conditions, no additional water with no cost is not able to face low pressure, both in 2024 conditions and in 2030 conditions. From the results of the analysis of the conditions of the Centralized scenario, an additional production of 1000 L/s is needed and a cost of IDR 219.98 Billion resulting in optimal pressure of 88% in 2024 and 22% in 2030, The Decentralized scenario conditions require an additional water of 500 L/s and a cost of IDR 100.5 Billion to achieve a more optimal pressure of 98% in 2024 and 39% in 2030. This indicates that the decentralized scenario has high effectiveness in increasing pressure, increasing water resistance with efficient use of the amount of water produced, and is efficient in investment costs. Therefore, increasing pressure with this decentralized concept is feasible on one hand, but on the other hand, the effect of poor raw water might still result in high maintenance and repair costs, although it does not exceed the maintenance and repair costs of a larger and more complex centralized production unit system.

4 Conclusion

From the topic of analysis of the low water pressure in Zone 3 of the clean water distribution network, it can be seen that the main cause of low water head pressure in the peripheral area in Zone 3 of the clean water distribution network of the city of Surabaya is the distance from the installation, so the alternative required is to decentralize unit water production or WTP where the water production capacity is half of the previous centralized planning by the Surabaya City PDAM or by 500 L/s, then an analysis is carried out with 3 scenarios, where it can be known that where in the analysis of scenario 1 (existing without improvement) it cannot overcome the problem of water pressure head for the projected year (2030) at all, then scenario 2 - the centralized unit production for 2024 clean water access of 100% nodes where pressure is above 2 mH_2O, but 12% of pressure head nodes is still below the optimal standard of minimum 7 mH_2O from regulations of the Ministry of Public Works, then in the 2030 projection year there will be only 88% who have access optimal drinking water or pressure above 2 mH_2O. Meanwhile, in the alternative scenario, it has managed to achieve 100% optimal water head pressure for existing and 98% optimal water head pressure for the 2030 projection year. Then, in the comparison of investment costs, improvement scenario 2 based on centralized unit production by PDAM plan of improvement data, where the investment made is the addition of the Karang Pilang IV WTP with a capacity of 1000 L/s, the addition of the Mbah Ratu Reservoir of 6000 m^3, and the addition of the Putat Gede III Boster Pump, 2 x 375 L/s + 600 L/s. Requires a total investment cost of IDR 219,984,300,051. Meanwhile, in Scenario 3, an alternative investment is made in the form of adding a Zone 3 drinking water treatment plant, a capacity of 500 L/s, and a production reservoir of 900 m^3 with a total investment of IDR 100,500,000,000, With the selection of an optimal Zone 3 installation alternative, it is hoped that, in addition, the alternative can save space, effectively treat existing river water with minimal side effects, but on the other hand, the impact of poor raw water may still result in high maintenance and repair costs, although not exceeding the maintenance and repair costs of a larger and more complex centralized production unit system.

Thus, this alternative is useful if the raw water source in the centralized unit experiences a significant decline, as it can effectively meet the increasing future water demand in an efficient and water-saving manner.

5 Future Study

There are limitations in the analysis that can be developed for further studies. First, regarding cost calculation, in addition to considering installation costs, maintenance and operational costs must also be considered. If water cannot be met in the future due to population increase or climate change, raw water can be used from other sources, namely seawater.

Acknowledgements: The author is aware of the imperfections of this writing, so criticism and suggestions can be provided via the author's email. Moreover, I thank Surya Surya Sembada Municipal Water Utility of Surabaya City for facilitating the hydraulic simulation laboratory with WaterGEMS software, guiding and providing data availability, and the Institute of Technology of Sepuluh Nopember for the implementation of this publication activity.

References

1. BPS-Statistics of Surabaya City, Kota Surabaya dalam Angka 2025 [Surabaya City in Figures 2025], (BPS-Statistics of Surabaya City, Surabaya, 2025). https://surabayakota.bps.go.id/id/publication/2025/02/28/bd1f25e59ae790cc8a7c0c07/kota-surabaya-dalam-angka-2025.html
2. Surabaya City Government, Peraturan Daerah Kota Surabaya Nomor 4 Tahun 2021 tentang Rencana Pembangunan Jangka Menengah Daerah (RPJMD) Kota Surabaya 2021-2026 [Surabaya City Regional Regulation Number 4 of 2021 concerning the Surabaya City Regional Medium-Term Development Plan 2021-2026], (Surabaya City Government, Surabaya, 2021). https://jdih.surabaya.go.id/uploads/peraturan/2021perda3578004.pdf
3. PDAM Surya Sembada Surabaya, Laporan Tahunan 2024: Air Sehat untuk Masyarakat [2024 Annual Report: Healthy Water for the Community], (PDAM Surya Sembada, Surabaya, 2025). https://www.pdam-sby.go.id/document/read/laporan-tahunan-2024
4. Surabaya City Government, Sistem Informasi Peta Peruntukan Lahan (SIPP) Kota Surabaya [Land Use Map Information System of Surabaya City], (Department of Housing, Settlement Areas, and Land (DPRKPP), Surabaya, 2025). [Online]. Available: https://www.petaperuntukan.cktr.web.id/. [Accessed: Oct. 23, 2025].
5. Ministry of Public Works and Housing, Penyelenggaraan Pengembangan Sistem Penyediaan Air Minum [Implementation of Water Supply System Development], (Ministry of Public Works and Housing, Jakarta, 2007).
6. G. Abdiganieva and U. Bakhramov, Determination of head loss in water pipes, in Proceedings of E3S Web Conf., XXVI International Scientific Conference "Construction the Formation of Living Environment" (FORM-2023).**410**,05026(2023). https://doi.org/10.1051/e3sconf/202341005026
7. R. Hasan and M. I. YUCE, Modeling of Water Age in Distribution Systems- Case Study, Int. J. Innov. Res. Eng. Manag.**11**,13–18 (2024). https://doi.org/10.55524/ijirem.2024.11.4.2
8. B. Kulmedov, E. Eroglu, and N. Kandymov, Modeling the factors affecting residual chlorine decay in the water distribution network of Abuja, Nigeria, Water Qual. Res. J.,**60**,482–497(2025). https://doi.org/10.2166/wqrj.2025.006
9. A. Nolte, S. Bender, J. Hartmann, S. Baltruschat, N. Moosdorf, and R. Reinecke, Coastal groundwater level trends reveal global susceptibility to seawater intrusion, EarthArXiv, (2024). https://doi.org/10.31223/X55457
10. PDAM Surya Sembada Surabaya, Rencana Bisnis Tahun 2023-2028 [2023-2028 Business Plan], (PDAM Surya Sembada, Surabaya, 2023).
11. Department of Settlements and Regional Infrastructure, Pedoman Petunjuk Teknis dan Manual Bagian 6. Air Minum Perkotaan [Technical Guidelines and Manual Part 6: Urban Water Supply], (Department of Settlements and Regional Infrastructure, Jakarta, 1998).

12. H. Kalensun, L. Kawet, and F. Halim, Perencanaan Sistem Jaringan Distribusi Air Bersih di Kelurahan Pangolombian Kecamatan Tomohon Selatan [Clean water distribution network system planning in Pangolombian Village, South Tomohon District], J. Sipil Statik.**4**,105-115(2016). https://ejournal.unsrat.ac.id/v2/index.php/jss/article/view/11465/11067#
13. S. Hossain, G. A. Hewa, C. W. K. Chow, and D. Cook, Modelling and Incorporating the Variable Demand Patterns to the Calibration of Water Distribution System Hydraulic Model, Water.**13**,2890(2021). https://doi.org/10.3390/w13202890
14. Directorate General of Construction Development, Lampiran IV Surat Edaran Direktur Jenderal Bina Konstruksi Nomor 30./SE/Dk/2025 tentang Tata Cara Penyusunan Perkiraan Biaya Pekerjaan Konstruksi Bidang Pekerjaan Umum dan Perumahan Rakyat: Analisis Harga Satuan Pekerjaan (AHSP) Bidang Sumber Daya Air [Appendix IV of the Circular Letter of the Director General of Construction Development Number 30./SE/Dk/2025 concerning Procedures for Preparing Construction Work Cost Estimates for Public Works and Public Housing: Work Unit Price Analysis (AHSP) for Water Resources], (Ministry of Public Works and Housing, Jakarta, 2025). https://sijkt.pu.go.id/o/SE-Dirjen-30-2025
15. Mayor of Surabaya, Mayor of Surabaya Decree Number: 100.3.3.3/11/436.1.2/2025 concerning the Fourth Amendment to the Mayor of Surabaya Decree Number 100.3.3.3/148/436.1.2/2024 concerning Regional Expenditure Unit Price Standards for Surabaya City for Fiscal Year 2025, (Surabaya City Government, Surabaya, 2025).

UTILIZATION OF LANDSAT 8 LEVEL 2 SATELLITE IMAGERY FOR IDENTIFICATION OF TEA PLANT HEALTH USING THE WEIGHTED OVERLAY METHOD (CASE STUDY: KERTAMANAH UNIT, MALABAR PLANTATION, BANDUNG)

Clara Rosanara Febryna Putri Priyanto[1], *Vivi* Cahyaningrum[1], and *Mokhamad* Nur Cahyadi[1*]

[1]Department of Geomatics Engineering, Faculty of Civil, Planning, and Earth Engineering, Institut Teknologi Sepuluh Nopember, Surabaya, Indonesia

Abstract. Monitoring tea plant health is important to maintain plantation productivity. This study aims to map the health status of tea plants in the Cinyiruan and Kertamanah divisions, Kertamanah Unit, Malabar Plantation owned by PT Perkebunan Nusantara 1 Regional 2 using Landsat 8 Level 2 satellite imagery. The parameters used were NDVI, LST, and slope from DEM. These parameters were integrated using the weighted overlay method with weights of 50% for NDVI, 30% for LST, and 20% for DEM in the Yielding Plant area. NDVI and LST processing were conducted in Google Earth Engine, while weighting and visualization were carried out in ArcMap 10.8.2. The resulting tea plant health map was classified into three classes: very healthy, quite healthy, and unhealthy. Validation using wet tea crop production data per-block for 2020–2024 produced an accuracy of 46.6%, influenced by the 30-meter image resolution and field variations such as crop age, rainfall, pests, and production records. Spatially, very healthy class dominated the central to southern areas, while unhealthy class was generally found in peripheral and steep slopes areas. These results indicate that high NDVI values, low surface temperatures, and gentle topography are closely related to better tea plant conditions.

1 Introduction

Tea has become one of the most widely known and consumed beverages in the world. This beverage generally comes from the Camellia sinensis plant, using its young leaves and buds [1]. In addition to being a beverage, tea is also used in cosmetics and medicine due to its antibacterial and antioxidant content [1]. Tea is one of the plantation commodities that has long been developed in Indonesia and plays a significant role. As one of the non-oil and gas export commodities, tea contributes to the country's foreign exchange earnings in the plantation sub-sector [2]. Furthermore, the tea plantation sector contributes to job creation, with the number of workers reaching around 41.403 people in 2018 [3].

Tea originates from subtropical regions with cool temperatures that tend to remain stable throughout the year. Tea cultivation in Indonesia needs to take environmental conditions into account, especially altitude. Altitude is one of the main ecological factors that affect microclimate conditions, such as air temperature, humidity levels, light intensity, and wind speed. As a subtropical plant, tea grows optimally at temperatures of 18–21°C with humidity above 70%, making it suitable for cultivation in mountainous areas [4]. With rainfall requirements of 1000–2500 mm per year, tea can thrive [5]. Generally, this plant is cultivated in highlands with land slopes ranging from gentle to very steep (>45%) [6].

Remote sensing using satellite or drone imagery is an efficient method for monitoring large-scale plantation conditions without having to visit the location directly. In addition, remote sensing data is also useful for analyzing temporal changes in land surface temperature, as well as identifying major atmospheric dominant gas components

* Corresponding author: cahyadi@geodesy.its.ac.id

such as (SO2) and aerosol concentration particles with Global Navigation Satellite System (GNSS) observations [7]. Positioning systems using satellite navigation have advanced significantly, with the GNSS now increasingly used to support navigation for mobile platforms [8, 9]. GNSS consists of several satellite constellations that orbit the Earth continuously. The system has advanced significantly in recent decades. Using code or pseudo-range measurements, GNSS enables users to determine latitude, longitude, and elevation with accuracy ranging from meters to millimeters [10]. Meanwhile, remote sensing technology employs sensors onboard satellites or drones to capture electromagnetic wave reflections originating from the Earth's surface. One of the most commonly used satellites is Landsat 8, which provides multispectral imagery with a ground resolution of 15–30 meters and a 16 days revisit interval [11]. These images have undergone radiometric and geometric correction, making them suitable for vegetation analysis. In tea plant health studies, one of the indices used is Normalized Difference Vegetation Index (NDVI). NDVI is calculated from the ratio between spectral reflectance in the red band and the near-infrared (NIR) band. Healthy plants tend to absorb more red light and reflect more NIR [12]. NDVI values extend from -1 to +1, where values approaching +1 represent dense and healthy vegetation [12]. NDVI is widely used as an indicator to assess vegetation health, while Land Surface Temperature (LST) reflects thermal stress on the land surface. Integrating NDVI and LST provides an opportunity to produce more comprehensive spatial mapping of areas experiencing vegetation stress [11].

One of the largest tea-producing areas is the Malabar Plantation in Bandung, West Java, which is managed by PT Perkebunan (PTPN) Nusantara 1 Regional 2. The Kertamanah unit, as part of this plantation, plays an important role in supporting national tea productivity. However, crop production levels are greatly influenced by soil quality, nutrient availability, and management. Therefore, an efficient and accurate method is needed to monitor land conditions on a regular basis to support data-driven decision making. Based on this, this study aims to utilize Landsat 8 Level 2 satellite imagery with the weighted overlay method to identify the health of tea plants in the Kertamanah Unit, Malabar Plantation, Bandung. This study aims to generate accurate and valuable information for more effective and sustainable tea plantation management.

2 Data and Methods

2.1 Locations

The study was located in Afdeling Cinyiruan and Kertamanah, Kertamanah Unit, Malabar Plantation, Bandung Regency, West Java (Fig.1). The area is dominated by sloping topography with varying degrees of inclination and a tea plantation (Camellia sinensis) managed by PT Perkebunan Nusantara I Regional 2. This analysis is limited to areas classified as Producing Plant (TM) so that the resulting analysis values can well reflect the condition of actively productive vegetation.

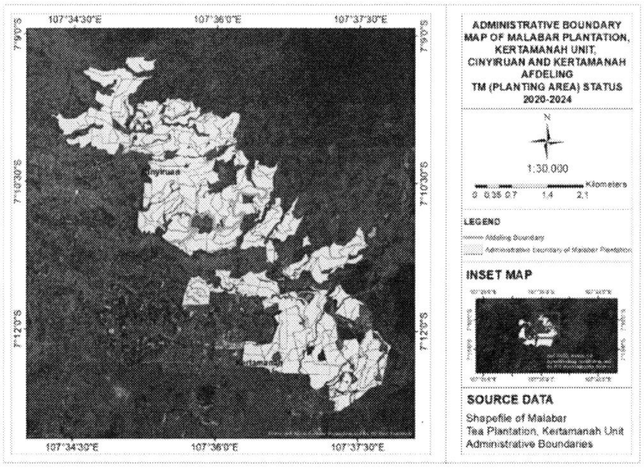

Fig.1. Kertamanah Unit Case Study Area

2.2 Data and Devices

2.2.1 Data

This study uses several types of primary data that serve as the basis for spatial analysis and validation of results. The primary data consists of Landsat 8 Level-2 images with an observation period from 2020 to 2024. These images help gather the standardized vegetation index called NDVI and the land surface temperature, or LST, which are the main factors used to evaluate how healthy the plants are. In addition, shapefile data on the boundaries of TM-status tea plantations in the Cinyiruan and Kertamanah divisions of the Kertamanah Unit, Malabar Plantation, were used to figure out the spatial boundaries of the study area. Wet production and tea plant productivity data for the 2020–2024 period obtained from PT Perkebunan Nusantara I Regional 2 were used as a reference in the validation process to compare the results of spatial analysis with actual conditions in the field.

2.2.2 Tools

Several software programs were used to support the data processing and analysis. Google Earth Engine (GEE) served as a satellite image processing platform to generate NDVI and LST parameters through a cloud-based processing system. ArcMap version 10.8.2 is used to perform advanced spatial analysis, including weighting using the weighted overlay method and compiling thematic maps that describe crop health distribution. Microsoft Excel is used to validate the analysis results by comparing crop health map values with wet production data. Furthermore, Microsoft Word is used to compile research reports in a systematic and structured manner.

2.3 Analysis Parameter Data

2.3.1 NDVI (Normalized Difference Vegetation Index)

NDVI values are calculated using the following equation.

$$NDVI = \frac{(\rho NIR - \rho Red)}{(\rho NIR + \rho Red)} \quad (1)$$

In the Landsat 8 OLI image, NIR refers to the near-infrared channel, which is Band 5, and Red refers to the red channel, which is Band 4. Rephrase The NDVI values go from -1 to +1. When the values are close to +1, it means there is a lot of vegetation. If the values are near -1, it shows places with little or no vegetation, like water or open land [13].

2.3.2 LST (Land Surface Temperature)

LST values are calculated from Brightness Temperature (BT) using the following formula:

$$LST = \frac{BT}{1+\left(\left(\frac{\lambda BT}{c2}\right)\ln ln\, e\lambda\right)} \quad (2)$$

The LST parameters show the land surface temperature measured by the TIRS thermal channel, which is Band 10 on Landsat 8 images [14].

2.3.3 Slope

Slope is determined using the Digital Elevation Model (DEM) to understand how the land's shape affects the crops' growth conditions. Steep slopes have the potential to cause erosion and reduce crop productivity [15].

2.4 Workflow

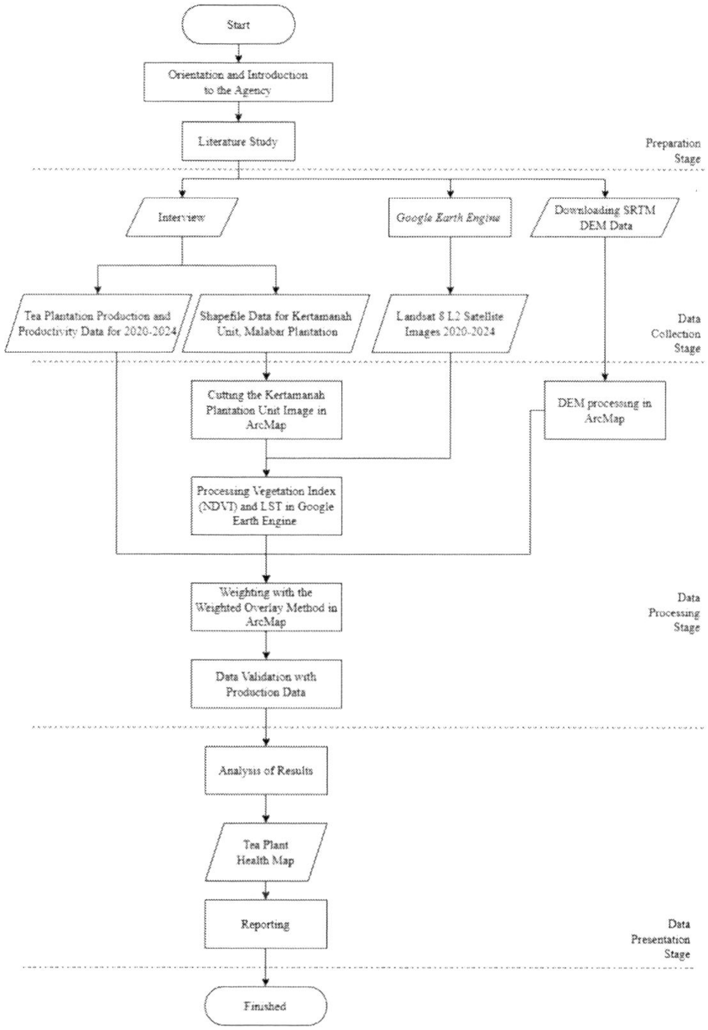

Fig.2. Work Flow Chart

The preparation stage began with an orientation at PTPN I Regional 2 Bandung to understand the plantation management system and tea production process. This activity was followed by a literature review on satellite image processing, NDVI and LST analysis, and the application of the weighted overlay method to strengthen the theoretical basis of the research.

The data collection stage included downloading Landsat 8 Level-2 images through Google Earth Engine (GEE), DEM data from online sources, and production and plantation boundary data from PTPN I Regional 2. All of this data became the main material for spatial analysis and result validation.

In the data processing stage, the images were cropped to focus on the study area using ArcMap 10.8.2. The NDVI and LST parameters were calculated through GEE, while the slope map was generated from DEM processing in ArcMap. The three parameters were combined using the Weighted Overlay method with a weight of 50% for NDVI, 30% for LST, and 20% for slope. The results were then classified into three categories: very healthy, quite healthy, and unhealthy.

The validation stage was carried out by comparing the results of the crop health map classification with wet production data per block using Microsoft Excel. Crops with high health levels generally have greater wet production than less healthy crops.

The determination of plant health categories is carried out using a statistical approach based on quartile values. Plants classified as unhealthy are determined based on wet production values below the first quartile (Q1) or the lowest 25% of data, while plants classified as healthy have production values above the second quartile (Q2), which represents the median value of the data distribution. Quartile values are calculated using equations (3) and (4).

After the class boundaries are set, validation is performed by comparing the map classification results and actual production data. The accuracy of the results is calculated using equation (5).

The final stage, which is the presentation of results, is carried out by visualizing the data in the form of a thematic map showing how healthy the tea plants are in the study area. The map and analysis results are then systematically compiled in a research report to provide a comprehensive overview of the vegetation conditions and productivity of tea plants in the study area.

3 Result

3.1 Average Normalized Difference Vegetation Index (NDVI) of Malabar Tea Plantation, Kertamanah Unit, 2020-2024

An analysis of the average NDVI values in Afdeling Cinyiruan and Kertamanah, Kertamanah Unit, Malabar Tea Plantation, Bandung, during the 2020–2024 period shows clear spatial variations in tea plant vegetation density. The analysis area focused on land with Producing Plant (TM) status to ensure the obtained NDVI values accurately represent productive vegetation conditions. TM land was chosen because it is a plant growth phase that directly contributes to tea shoot productivity and has a relatively uniform maintenance pattern and harvest cycle, thus reflecting the plant's physiological condition consistently from year to year. The average NDVI map shows three main classes that describe the level of vegetation greenness, namely: (1) low vegetation with an NDVI value of 0.190–0.609 (red), (2) moderate vegetation with NDVI values of 0.609–0.692 (yellow), and (3) high vegetation with NDVI values of 0.692–0.819 (green) (Fig.3). The spatial distribution shows that the high vegetation class dominates the central and northern areas of the plantation. These conditions indicate high canopy density and good plant physiological conditions.

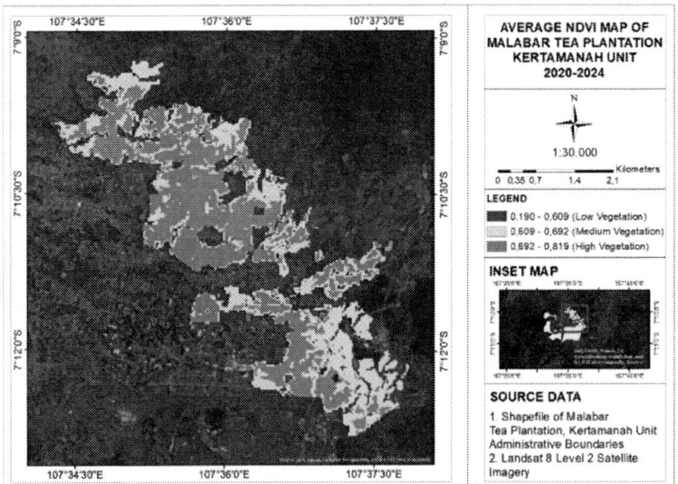

Fig. 3. Average NDVI Map of Kertamanah Unit, Malabar Tea Plantation, 2020-2024.

One representative block is TH0C26 with an NDVI value of 0.759, reflecting healthy and uniform vegetation conditions. Moderate vegetation classes are scattered around the dominant green area, acting as a transition zone between healthy and less healthy vegetation. A representative example is block TH0C38 in the southern part of the plantation with an NDVI value of 0.662. The distribution of this class indicates a phase of recovery or post-harvest vegetation growth, as well as the possibility of more intensive cultivation activities in the area. Meanwhile, low vegetation classes were identified in smaller numbers and were generally located on the northern, southern, and eastern edges of the plantation, close to residential areas and transportation routes. Block TH0C36, with an NDVI value of 0.598, is an example of an area that falls into this category. The spatial distribution of red on the map indicates potential disturbances to vegetation, whether due to human activity, land clearing, or environmental factors such as erosion and differences in land cover types.

Overall, the NDVI distribution pattern shows that vegetation in good condition is concentrated in the center of the plantation, while the edge areas show lower levels of greenness. This pattern indicates a gradient in vegetation condition influenced by topography, land management intensity, and proximity to anthropogenic activities. These findings confirm that multi-year NDVI analysis can be used as an effective indicator for monitoring the spatial and temporal variability of tea plant health across plantation areas.

3.2 Average Land Surface Temperature (LST) of Malabar Tea Plantation, Kertamanah Unit, 2020-2024

Image processing results show the average land surface temperature (LST) distribution in Afdeling Cinyiruan and Kertamanah, Kertamanah Unit, Malabar Tea Plantation, Bandung, during the period 2020–2024 (Fig.4). Spatially, the LST pattern shows clear segmentation between cold, cool, and warm zones. The LST classes are divided into three categories: cold (13.797–17.063°C, colored blue), cool (17.063–18.432°C, colored yellow), and warm (18.432–26.858°C, colored red). The cold zone (13.797–17.063°C) is widely distributed across the eastern and southern parts of the study area, with the highest concentration in areas adjacent to higher elevations. This low temperature distribution correlates with dense vegetation cover and low anthropogenic activity. The cool category (17.063–18.432°C) is indicated by yellow, which is relatively evenly distributed in the central area of the plantation. This zone acts as a transition between the cold and warm areas, often coinciding with flat land that has a mixture of moderate vegetation and active plantation activities. Meanwhile, the warm category (18.432–26.858°C) is dominated by the western and central parts of the plantation. This area is generally associated with open land, sparse vegetation, or the presence of plantation infrastructure that absorbs more heat than vegetative areas.

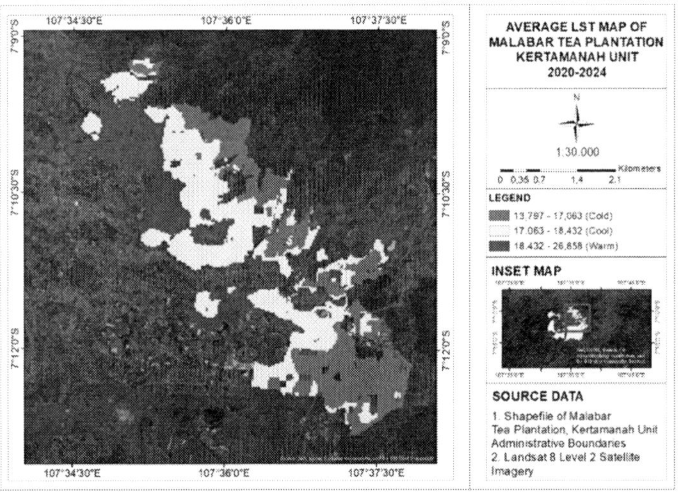

Fig. 4. Average LST Map of Kertamanah Unit, Malabar Tea Plantation, 2020-2024.

Spatial comparisons between LST and topography show a correlation between surface temperature, elevation, and slope. The results of the comparison between the LST map and the Digital Elevation Model (DEM) show that areas labeled "warm" tend to be in zones with low slope (0°–14°), which are depicted in light green on the

DEM map. This condition is common because relatively flat or gentle surfaces receive more even solar radiation and have greater heat storage capacity. Conversely, areas labeled "cool" to "cold" are mostly found on steep slopes with a gradient between 21°–58°, marked in orange to red on the DEM map. Steep slopes generally have denser vegetation, good drainage, and are located at higher elevations, resulting in relatively lower surface temperatures. In addition to the influence of slope, the aspect of the slope also contributes to temperature variation. Slopes with a west to north orientation tend to have higher temperatures in the afternoon due to greater radiation intensity, while east to south slopes are relatively cooler because they receive direct radiation in the morning for a shorter duration. Overall, the analysis results indicate a negative relationship between LST and elevation, the higher the elevation, the lower the surface temperature, and a positive correlation between LST and gentle slope, where flat or gentle areas tend to be warmer. This pattern confirms that surface temperature variations within the study area are controlled by a combination of topographic conditions and land cover characteristics, both of which play an important role in controlling surface energy balance in plantation environments.

3.3 Digital Elevation Model (DEM) of Malabar Tea Plantation, Kertamanah Unit, 2020-2024

After processing the Digital Elevation Model (DEM) data, a slope map was obtained (Fig.5). This map represents the topographic variations in the Kertamanah Unit, Malabar Plantation, particularly in the Yielding Crops (TM) area. Based on the extraction results using ArcMap software, the study area has a slope range of 0° to 58°, which is further classified into five classes, namely flat to gentle slopes (0°–8° in dark green), gentle to moderately steep slopes (8°–14° in light green), moderately steep (14°–21° in yellow), steep (21°–32° in orange), and very steep (32°–58° in red).

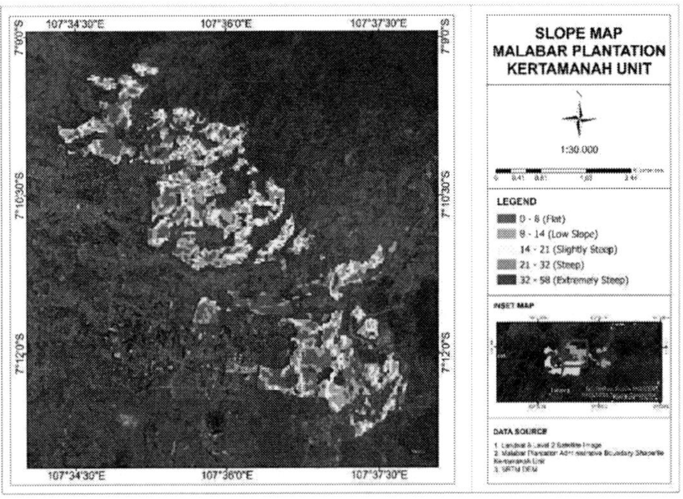

Fig. 5. Slope Map of Kertamanah Unit, Malabar Plantation.

Spatially, the distribution of slope classes shows a clear pattern. Areas with low slopes (0°–14°) are marked with dark green to light green colors and are widely spread across the central and southern parts of the study area, forming relatively flat and homogeneous topographic blocks. Medium slopes (14°–21°), colored yellow, appear as a transition zone between gentle and steep areas. Meanwhile, areas with high slopes (21°–58°) are colored orange to red and dominate the edges of the blocks, especially on the east, west, and southern ends, which are steep slopes or ridges. The presence of red areas on the southern side indicates extreme slopes that require special attention in land management.

The relationship between slope inclination and tea plant health shows that topography plays an important role in determining land suitability. Gentle to moderately steep slopes (0°–14°) are ideal conditions for tea plant growth because they have good drainage and low erosion risk. Medium slopes (14°–21°) are still suitable for tea cultivation but require soil conservation measures such as terracing. Meanwhile, areas with steep to very steep slopes

(21°–58°) have the potential for high erosion and difficulty in maintenance, which may impact plantation productivity and plant health. Thus, slope analysis based on DEM data can be used as an important parameter in evaluating land suitability and planning sustainable tea garden management.

3.4 Tea Plants Health Map at Malabar Tea Plantation, Kertamanah Unit, 2020-2024

The results of processing using the weighted overlay method produced a map showing the spatial distribution of tea plant health levels in the Planting Area (TM) in the Kertamanah Unit, Malabar Plantation (Fig.6). This analysis was conducted by considering three main parameters, namely NDVI, LST, and slope, each of which contributes to the physiological condition of plants. The parameter weights were set at 50% for NDVI, 30% for LST, and 20% for slope, with a total weight of 100%. These weights were determined based on the level of influence of each factor on plant health. NDVI plays a dominant role because it reflects the density and greenness of vegetation, which is directly correlated with photosynthetic activity. LST affects plant physiology through the regulation of surface temperature, which influences plant metabolism and stress. Meanwhile, slope plays a role in controlling water distribution, soil stability, and erosion risk, which impact plant growth.

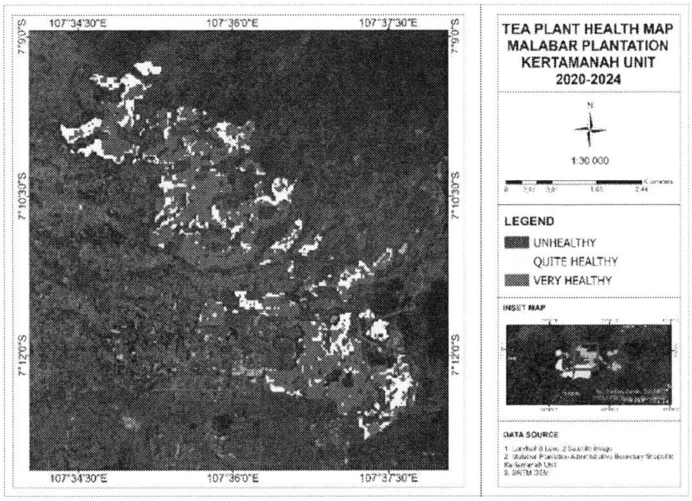

Fig. 6. Tea Plant Health Map of Kertamanah Unit, Malabar Plantation, 2020-2024

The weighted overlay map produces three main classes of tea plant health, namely very healthy (score 3, green color), quite healthy (score 2, yellow color), and unhealthy (score 1, red color). Spatially, the very healthy class dominates most of the plantation area, especially in the central to southern parts, which are characterized by high NDVI values, moderate surface temperatures, and gentle to moderate slope gradients. The quite healthy class (yellow) is scattered as a transition zone around the green area, especially across the northern, eastern, and southern parts of the plantation. For example, in the Singung plantation block with the code TH0D36, this area has moderate NDVI values, relatively high surface temperature variations, and steeper slopes compared to very healthy areas. These conditions have the potential to reduce crop productivity if not balanced with proper management. The unhealthy class (red) is distributed in limited numbers in several small spots, as seen in the Saringan block with code TH0C11. This area is generally located on steep slopes or block edges with low NDVI values, high surface temperatures, and significant erosion risk. These conditions indicate high environmental pressure and an urgent need for rehabilitation measures such as soil and water conservation, soil cover improvement, or crop rejuvenation.

The results of zonal statistics as a table on the tea plant health map show the plant health class values for each plantation block. Based on this table, most plantation blocks are in the very healthy category (score 3), indicating relatively good and productive vegetation conditions. Several blocks with a score of 2 indicate the need for additional management measures, while blocks with a score of 1 are a priority for management intervention.

Overall, the spatial distribution of tea plant health levels provides a comprehensive picture of the variation in plant physiological conditions in the field. This information can be used as a basis for planning plantation management strategies. Areas with a very healthy class need to be maintained through intensive maintenance and control of disruptive factors. Areas that are quite healthy require optimization of management, such as adjusting fertilization and regulating water management. Meanwhile, unhealthy areas are the main focus for the implementation of corrective and conservative measures so that plantation productivity can be improved in a sustainable manner.

3.5 Validation of Tea Plants Health Map

Validation of tea plant health maps is carried out to ensure that the spatial analysis results obtained from satellite imagery represent actual conditions in the field. This validation process is carried out by comparing the plant health index values generated through the weighted overlay method (main parameters NDVI , LST , and slope) with the wet tea production data for each plantation block. Conceptually, a good level of plant health is expected to correlate positively with high wet production, while plants in poor health show lower production yields.

The initial validation step was carried out by calculating the average wet production value of tea plants for the 2020-2024 period for each plantation block. Next, to group plant health levels based on wet production, a statistical approach was used to determine the first quartile (Q1) and second quartile (Q2) values. The first quartile was used as the lower limit for the unhealthy plant category, while the second quartile was used as the upper limit for the moderately healthy plant category. The first quartile value was calculated using the following formula.

$$Kuartil\ 1\ (Q1) = \frac{(n+1)}{4} \quad (3)$$

Meanwhile, the calculation of the second quartile follows the following formula.

$$Kuartil\ 2\ (Q1) = \frac{(n+1)}{2} \quad (4)$$

Where n is the amount of production data. Based on the calculation results, Q1 is 60,889.98 kg and Q2 is 80,595.93 kg. Thus, the classification limits for production volume can be defined as follows: (1) unhealthy plants with a production value ≤ 60,889.98 kg, (2) moderately healthy plants with a production value between 60,889.98 and 80,595.93 kg, and (3) very healthy plants with a production value ≥ 80,595.93 kg.

After the production classification was determined, a comparison was made with the tea plant health index values from the satellite image analysis (value 1 = unhealthy, value 2 = moderately healthy, and value 3 = very healthy). The comparison was carried out for each plantation block. The validation process was carried out by calculating the conformity between the two data sources using the formula.

$$Validasi\ Data = \left(\frac{Sampel\ Benar}{Jumlah\ Sampel}\right) \times 100 \quad (5)$$

Based on the calculation results, out of a total of 103 samples, 48 samples showed consistency between the plant health class from satellite imagery and the category based on wet production data. Thus, the accuracy rate of the tea plant health map reached 46.60%. This accuracy value indicates that the satellite image analysis results do not fully reflect the actual conditions in the field. The 53.40% difference is likely due to various factors, both from the limitations of satellite data and variations in agronomic conditions in the field. Regarding remote sensing data, the limited spatial resolution of the images causes a single pixel to capture mixed surface features, including vegetation, open land, and other non-vegetative elements, thereby affecting the accuracy of the NDVI index. Atmospheric factors such as clouds, fog, or variations in air humidity can also reduce the quality of image reflectance and cause index value deviations. Meanwhile, in terms of field conditions, variations in maintenance practices between blocks, differences in plant age, canopy density, and environmental factors such as rainfall and light intensity also affect tea production. In addition to physiological factors, non-physiological aspects such as pest and disease attacks, the effects of extreme weather, and inaccuracies in production data recording also have the potential to cause differences between observation and spatial analysis results.

In general, these validation results show that tea plant health maps generated from satellite image analysis have moderate capabilities in describing the actual conditions of tea plantations. Although its accuracy is still relatively low, this approach still provides useful spatial information for the initial identification of plant conditions and can be further developed through improvements in image spatial resolution, more detailed field data calibration, and the integration of additional parameters such as rainfall, soil moisture, and plant phenology data.

4 Conclusion

This study demonstrates the potential use of Landsat 8 Level-2 imagery to assess the health level of tea plants in Afdeling Cinyiruan and Kertamanah, Kertamanah Unit, Malabar Tea Plantation, Bandung. The resulting thematic maps can depict spatial variations between tea plant blocks so that areas requiring agronomic attention can be precisely identified. One of the key findings shows that the very healthy class predominantly occupies the central to southern areas of the plantation, while the unhealthy class is generally distributed in peripheral areas and on steeper slopes. This spatial pattern indicates that topography (slope) and geographic location play a dominant role in influencing tea plant health conditions. The main parameters used in the weighted overlay analysis include the Normalized Difference Vegetation Index (NDVI), Land Surface Temperature (LST), and Digital Elevation Model (DEM). NDVI and LST were calculated using the Google Earth Engine platform, while DEM processing, weighting processes, and final map preparation were performed using ArcMap 10.8.2 software. The analysis yielded three classes of plant health: very healthy, moderately healthy, and unhealthy, which were spatially visualized to support plantation management decisions. Validation of the mapping results based on wet production data showed an overall accuracy (total accuracy) of 46.60%, calculated from 48 consistent samples out of a total of 103 samples, where the mapped classes corresponded with the wet production classes. The remaining 53.40% difference is primarily influenced by the mixed pixel effect caused by the 30-meter spatial resolution of Landsat imagery, which mixes vegetation and non-vegetation objects within a single pixel, in addition to atmospheric interference on reflectance values, and variations in field conditions such as differences in maintenance techniques, plant age, rainfall, the presence of pests and diseases, and inaccurate production records. Overall, the study shows that Landsat 8 Level-2 imagery can be used to monitor tea plant health at the plantation scale, particularly for identifying broad spatial patterns related to topography and geographic location, although its accuracy is still limited. Better mapping accuracy can be achieved through the use of higher-resolution satellite imagery, the integration of more detailed field data, and the development of weighting methods that are more adaptive to vegetation characteristics and local environmental conditions.

List of abbreviations:

- NDVI : Normalized Difference Vegetation Index
- LST : Land Surface Temperature
- DEM : Digital Elevation Model
- TM : Planting Area (Tanaman Menghasilkan)

Acknowledgement

The authors gratefully acknowledge financial support from the Institut Teknologi Sepuluh Nopember for this work, under project scheme of the Publication Writing and IPR Incentive Program (PPHKI) 2025.

Availability of data and materials:

All the data used in this research are available at the website Google Earth Engine (GEE) for NDVI and LST processing, website inageoportal to download DEM (https://tanahair.indonesia.go.id/portal-web/unduh), and wet tea production data can be obtained by direct interview with PT Perkebunan Nusantara 1 Regional 2 (this data is confidential).

Author's contributions:

- CRFPP formal analysis, methodology, validation, review & editing
- VC formal analysis, methodology, validation, review & editing
- MNC supervisor

References

[1] P. Wang et al., "Determining the critical threshold of meteorological heat damage to tea plants based on MODIS LST products for tea planting areas in China," Ecol. Inform., vol. 77, Nov. 2023, doi: 10.1016/j.ecoinf.2023.102235.

[2] S. N. Salimah, A. Junaedi, and Sudradjat, "Pengelolaan Pemetikan Tanaman Teh (Camellia sinensis (L.). O. Kuntze) di Wonosobo, Jawa Tengah," Buletin Agrohorti, vol. 11, no. 2, pp. 249–259, 2023, doi: 10.29244/agrob.v11i2.47161.

[3] L. P. Caro, Wages and working conditions in the tea sector: the case of India, Indonesia and Viet Nam Background note. 2020. [Online]. Available: www.ilo.org/publns.

[4] C. Wang, J. Han, Y. Pu, and X. Wang, "Tea (Camellia sinensis): A Review of Nutritional Composition, Potential Applications, and Omics Research," Jun. 01, 2022, MDPI. doi: 10.3390/app12125874.

[5] N. Sahu et al., "Analysis of Tea Plantation Suitability Using Geostatistical and Machine Learning Techniques: A Case of Darjeeling Himalaya, India," Sustainability (Switzerland), vol. 15, no. 13, Jul. 2023, doi: 10.3390/su151310101.

[6] D. D. P. Afner, A. Aprisal, and Y. Yulnafatmawita, "Indeks Stabilitas Agregat Tanah Pada Perkebunan Teh Berbasis Slope Dan Umur Tanaman Di Kecamatan Gunung Talang Kabupaten Solok," Jurnal Tanah dan Sumberdaya Lahan, vol. 8, no. 1, pp. 75–81, 2020, doi: 10.21776/ub.jtsl.2021.008.1.10.

[7] M. N. Cahyadi et al., "Analysis of the effect of the 2021 Semeru eruption on water vapor content and atmospheric particles using GNSS and remote sensing," Geod. Geodyn., vol. 15, no. 1, pp. 33–41, Jan. 2024, doi: 10.1016/j.geog.2023.04.005.

[8] M. N. Cahyadi, T. Asfihani, R. Mardiyanto, and R. Erfianti, "Loosely Coupled GNSS and IMU Integration for Accurate i-Boat Horizontal Navigation," International Journal of Geoinformatics, vol. 18, no. 3, pp. 111–122, Jun. 2022, doi: 10.52939/ijg.v18i3.2233.

[9] M. N. Cahyadi and I. Rwabudandi, "Integration of GNSS-IMU for increasing the observation accuracy in Condensed Areas (Infrastructure and Forest Canopies)," in E3S Web of Conferences, EDP Sciences, May 2019. doi: 10.1051/e3sconf/20199403015.

[10] M. N. Cahyadi et al., "Comparative Analysis of Low-Cost GNSS OEM Board K706 and BX316 (Case Study: Bulusidokare Village Sidoarjo Regency)," in IOP Conference Series: Earth and Environmental Science, IOP Publishing Ltd, Apr. 2021. doi: 10.1088/1755-1315/731/1/012024.

[11] M. A. Lazaridou and A. C. Karagianni, "Landsat 8 multispectral and pansharpened imagery processing on the study of civil engineering issues," in International Archives of the Photogrammetry, Remote Sensing and Spatial Information Sciences - ISPRS Archives, International Society for Photogrammetry and Remote Sensing, 2016, pp. 941–945. doi: 10.5194/isprsarchives-XLI-B8-941-2016.

[12] B. Mamaghani, M. G. Saunders, and C. Salvaggio, "Inherent reflectance variability of vegetation," Agriculture (Switzerland), vol. 9, no. 11, Nov. 2019, doi: 10.3390/agriculture9110246.

[13] M. B. Moisa, B. T. Gabissa, L. B. Hinkosa, I. N. Dejene, and D. O. Gemeda, "Analysis of land surface temperature using Geospatial technologies in Gida Kiremu, Limu, and Amuru District, Western Ethiopia," Artificial Intelligence in Agriculture, vol. 6, pp. 90–99, Jan. 2022, doi: 10.1016/j.aiia.2022.06.002.

[14] A. Sekertekin and S. Bonafoni, "Land surface temperature retrieval from Landsat 5, 7, and 8 over rural areas: Assessment of different retrieval algorithms and emissivity models and toolbox implementation," Remote Sens. (Basel)., vol. 12, no. 2, Jan. 2020, doi: 10.3390/rs12020294.

[15] R. Madugundu et al., "Impact of Field Topography and Soil Characteristics on the Productivity of Alfalfa and Rhodes Grass: RTK-GPS Survey and GIS Approach," Agronomy, vol. 12, no. 12, Dec. 2022, doi: 10.3390/agronomy12122918.

Management Food Waste in Indonesia: A Review

Nadya Ul Haq Mahmud[1], *IDAA* Warmadewanthi[1*], *Arseto* Yekti Bagastyo[1], and *Dang* Vu Bich Hanh[2]

[1]Department of Environmental Engineering, Faculty of Civil Engineering and Planning, Institut Teknologi Sepuluh Nopember, Surabaya, Indonesia
[2]Department of Environmental Engineering, Faculty of Environment and Resources, Ho Chi Minh City University of Technology, Ho Chi Minh, Vietnam

Abstract. Food waste is a global problem of considerable magnitude, and in Indonesia, the households account for 80% of the 23–48 million tons of food waste produced per year, which is equal to 155–184 kg/person. One of the main aims of this study is to investigate the problem of food waste generation in Indonesia in depth, both with regard to its patterns and causes. It is also an objective of the study to assess food waste management practices critically, as well as to determine the potential influence of the "Makan Bergizi Gratis" program on future waste accumulation. The research applies a rigorous scientific methodology and brings together the official records of the National Waste Management Information System of the Indonesian Ministry of Environment and Forestry and national databases. The study discusses both direct and indirect factors that contribute to food loss and waste, including poor harvesting practices, ineffective supply chains, consumer behavior, and policy limitations. The paper aims to support the transition towards a sustainable food system through essential policy areas and real initiatives by offering a comprehensive overview of the food waste dynamics in Indonesia.

1. Introduction

Food waste is a significant global challenge, with around 1.3 billion tons of edible food being lost or discarded each year, which accounts for roughly one-third of all food produced for human consumption (Saliem et al., 2021). This phenomenon results in approximately 3.3 billion tons of carbon dioxide equivalent emissions, accounting for 8-10% of human-caused greenhouse gas emissions, and leads to significant environmental challenges, such as climate change, eutrophication, water scarcity, and excessive land use (Zulkifli et al., 2025). From an economic perspective, it leads to annual losses surpassing one trillion US dollars. Socially, it worsens food insecurity for approximately 828 million individuals globally, directly opposing Sustainable Development Goal 12.3, which seeks to reduce per capita global food waste at the retail and consumer stages by half by the year 2030 (Martianto et al., 2024; Saliem et al., 2021).

In Indonesia, the issue of food waste is particularly severe, with households responsible for approximately 80% of the 23–48 million tons produced each year from 2000 to 2019. This amounts to 115–184 kilograms per person annually, placing the country second worldwide with 300 kilograms per capita, as reported by the Economist Intelligence Unit (Saliem et al., 2021; Cahyana et al., 2024). Recent data from the Sistem Informasi Pengelolaan Sampah Nasional (SIPSN; National Waste Management Information System) indicate that food waste comprised 41.7% of total solid waste in 2023, while organic waste, largely

* Corresponding author: warma@its.ac.id

food-related, constitutes 40–62.95% of municipal solid waste. In Depok City in 2021, food waste formed 62.95% of 310,443 tons generated from a population of 2.085 million (Waluyo & Kharisma, 2023; Zulkifli et al., 2025). These figures underscore households as the primary source, followed by food services and retail, with edible portions reaching 44% (Saliem et al., 2021).

Studying the patterns and causes of food waste in Indonesia is essential for devising effective interventions, as inefficiencies span the supply chain from poor harvesting and inadequate storage to consumer behaviors like over-portioning and policy shortcomings (Cahyana et al., 2024). Amid rapid urbanization and population growth exceeding 260 million, such analyses bridge data gaps in SIPSN reporting, align with national targets like Presidential Regulation Number 97 of 2017 for 30% waste reduction by 2025, and support circular economy transitions to mitigate environmental, economic, and social impacts (Zulkifli et al., 2025).

This review paper aims to comprehensively investigate food waste generation patterns, causes, current management practices, and the potential impacts of the "Makan Bergizi Gratis" program on waste accumulation in Indonesia, drawing from SIPSN records, national databases, and peer-reviewed international and national journals up to 2025. The scope focuses on quantitative data across supply chain stages, challenges, and sustainable strategies to inform evidence-based policies for a resilient food system.

2. Food Waste in Indonesia

Food waste in Indonesia predominantly arises at the downstream stages of the food supply chain, encompassing retail, food services, and particularly households, where it is characterized by the discard of safe and edible food portions resulting from behavioral decisions such as overbuying, excess cooking, inadequate storage, aesthetic rejections, and plate leftovers (Martianto et al., 2024). According to Food and Agriculture Organization (FAO) guidelines, this contrasts sharply with upstream food losses by emphasizing intentional human actions influenced by urbanization, rising disposable incomes, and cultural practices that promote prestige-driven overconsumption, with households globally responsible for 61% of total food waste and similarly dominating at 80% in Indonesia (Martianto et al., 2024). Indonesia's per capita food waste ranks it second globally at 115-300 kg/year, equivalent to 5-19 million tons annually from households alone between 2000-2019, rising from 39% to 55% of overall waste with a trend underscoring the urgency of SDG 12.3 to halve retail and consumer-level waste by 2030 through targeted education, planning, and nudges (Saliem et al., 2021; Martianto et al., 2024).

2.1 Key Food Waste Statistics in Indonesia

Indonesia faces a food waste problem of considerable magnitude, contributing significantly to global environmental concerns. Annually, 23–48 million tons of food waste are produced in the country, with households contributing a substantial 80% of this total, equating to an average of 155–184 kg per person per year (Martianto et al., 2024; Saliem et al., 2021). This massive volume represents not only a loss of valuable resources but also a significant environmental burden, as food waste constitutes a large portion of Indonesia's overall waste stream. For instance, in 2021, Indonesia generated 30 million tons of waste, with 40% being organic food waste. In a broader context, food waste can constitute as much as 41% of the nation's overall waste, amounting to roughly 12.89 million tons each year as of 2022 (Audrey & Rukmini, 2024). Jakarta, the capital city, is expected to produce around 4,050 tons of food waste each day, accounting for roughly 54% of its overall daily waste (Bhaskara, 2023).

Table 1. Summary of food waste data in indonesia

Description	Value	Source
Annual Food Waste Production	20.9–48 million tons	Cahyana et al., 2024; Martianto et al., 2024
Household Contribution to Food Waste	80%	Martianto et al., 2024
Food Waste Per Capita Per Year	77–184 kg	Saliem et al., 2021; Cahyana et al., 2024
Food Waste as Percentage of Total Waste	41% (12.89 million tons in 2022)	Audrey & Rukmini, 2024
Daily Food Waste in Jakarta	4,050 tons (54% of total waste)	Bhaskara, 2023

Through direct empirical measurements, a comprehensive quantification was achieved, as demonstrated by a cross-sectional study involving 215 households in Bogor Regency. This study utilized waste compositional analysis and beverage diaries, uncovering that the average household generates 77 kg of food waste per person annually. Of this, 37.7% is edible, amounting to 29 kg, while 62.3% is inedible, totaling 48 kg (Martianto et al., 2024). Urban households produced significantly more at 79.4 kg/capita/year compared to 45.8 kg in rural areas, reflecting disparities driven by accessibility and lifestyles, while city-specific data highlight hotspots: Surabaya generates 0.37 million tons annually, and Depok's 2021 municipal solid waste of 310,443 tons included 62.95% food waste (~195,000 tons) (Martianto et al., 2024; Zulkifli et al., 2025). These figures align with national estimates placing households at 80% of consumption-stage waste (74 kg/capita globally benchmarked), positioning Indonesia among top producers and necessitating data-driven prevention amid stalled 30% waste reduction targets by 2025 (Martianto et al., 2024).

Table 2. Household Food Waste Quantities

Household Food Waste Quantities (kg/capita/year)	Urban	Rural	Total	Edible %
Edible Portion	-	-	29	37.7%
Inedible Portion	-	-	48	62.3%
Total Waste	79.4	45.8	77	-

3. Patterns and Causes of Food Waste in Indonesia

Indonesia is one of the leading nations contributing to the significant global problem of food waste. Data from the SIPSN indicates food waste comprises about 40.5% of Indonesia's total waste, exceeding 29 million tons yearly (Kusumawardani et al., 2023; Waluyo & Kharisma, 2023). This massive volume represents a substantial loss of resources and a major driver of environmental degradation. The decomposition of this organic material, especially in landfills, releases potent greenhouse gases like methane (25 times more powerful than CO_2), significantly contributing to climate change (Kusumawardani et al., 2023).

Food waste patterns in Indonesia are primarily driven by households and inefficiencies throughout the food supply chain. Rice/cereals and vegetables are major edible waste contributors, while fruits mainly contribute inedible waste from peels and spoilage (Martianto et al., 2024). In cities such as Jakarta and Depok, food waste makes up a significant portion of their municipal solid waste, highlighting the substantial environmental challenge it poses for urban waste management.(Bhaskara, 2023; Ulfah et al., 2023).

Table 3. Summary of Key Food Waste Quantities and Proportions (Martianto et al., 2024)

Description	Quantity/Proportion
National Total (annual estimate)	23–48 million tons
Per Capita/Year	155–184 kg
Household Contribution	80% of total

Factors amplifying environmental damage include consumer behavior and policy limitations. Household waste is significantly influenced by consumer behaviors such as purchasing more than needed, making impulsive buys, serving overly large portions, and storing items incorrectly (Kusumawardani et al., 2023; Cahyana et al., 2024). Low public awareness about food waste's environmental consequences further compounds the issue (Audrey & Rukmini, 2024). Existing policies, like Presidential Regulation No. 97/2017, often lack specific targets, clear measurement indicators, and robust monitoring to effectively reduce food waste's GHG contribution (Waluyo & Kharisma, 2023). This regulatory gap leads to suboptimal waste management, causing environmental problems like water and soil pollution from open dumping (Cahyana et al., 2024).

Current food waste management, predominantly landfilling, has the highest environmental impact. However, there is high potential for valorizing food waste into bioenergy via Black Soldier Fly larvae processing, composting, and biodigesters, significantly reducing its environmental footprint (Ulfah et al., 2023). Adopting circular economy strategies in the food industry can cut greenhouse gas emissions by as much as 30%, curb carbon output, and prevent land degradation. Critical assessment of practices, including programs like "Makan Bergizi Gratis," is essential to ensure food security initiatives also align with environmental sustainability.

4. The Role of the "Makan Bergizi Gratis (MBG)" Program in Food Waste Management

The "Makan Bergizi Gratis" program represents a significant Indonesian government initiative, having been officially launched in early 2025 with far-reaching goals for national welfare (Agustini & Mulyani, 2025). The program's primary objectives are multifaceted, specifically aiming to address prevalent nutritional deficiencies, enhance the overall quality of human resources across the nation, and ensure greater educational equity by consistently providing nutritious meals in schools. This comprehensive initiative strategically targets vulnerable demographics, including schoolchildren, toddlers, pregnant, and lactating mothers, underscoring its broad social and public health impact. Through these crucial provisions, the program endeavors to significantly improve the health outcomes of its beneficiaries, boost learning concentration among students, elevate academic achievement, and effectively reduce existing social disparities that hinder national progress (Qomarrullah et al., 2025).

A critical analysis of the "Makan Bergizi Gratis" program reveals a notable discrepancy between its commendable stated objectives and the tangible outcomes observed during its real-world implementation, particularly concerning its pervasive impact on food waste accumulation. The initial phases of MBG implementation have unfortunately introduced a new and pressing problem: a significant increase in food waste. This rise in waste is primarily attributed to the perceived low quality and, consequently, the poor acceptance of the provided meal menus by recipients, leading to a substantial portion of the food not being fully consumed. Such a scenario underscores a critical disconnect between the program's noble goals of enhancing food security and its practical execution, as evidenced by persistent issues in food access, utilization, and stability within the target groups (Komara et al., 2025). Moreover, external factors such as the menu's suitability, the overall duration of the program, and the precise sizing of meal portions have been identified as crucial variables that significantly influence the ultimate amount of food waste.

The "Makan Bergizi Gratis" program, despite its positive intentions to improve nutrition, has regrettably not led to a reduction in food waste; instead, initial reports indicate an increase in discarded food. This counterintuitive outcome highlights a significant challenge for a program designed to enhance both nutrition and food security, suggesting that resources are not being utilized as effectively as intended (Komara et al., 2025). The phenomenon of increased waste generation underscores a critical area for program re-evaluation and adjustment, moving beyond mere provision to address the complexities of food consumption.

Several interconnected factors explain why the MBG program is currently contributing to food waste:

- **Low Quality and Poor Acceptance of Meals**: A primary reason for the increased food waste within the MBG program stems from the quality and palatability of the meals provided to students. If the food offered is not appealing or does not meet the taste preferences of the children, they are significantly less likely to consume it, leading directly to higher plate waste (Komara et al., 2025). Studies on school meal services emphasize that the sensory attributes of food, particularly taste and texture, play a crucial role in meal acceptance and directly influence the amount of food wasted. Enhancing food service management practices, such as careful menu planning, sourcing, preparation, and cooking, is seen as an effective approach to increase acceptance and minimize waste (Setyaningtyas et al., 2022).
- **Implementation and Operational Challenges**: Operational shortcomings and structural issues within the program's implementation significantly contribute to the accumulation of food waste. These challenges can encompass inconsistencies in how nutritional standards are met, uneven distribution of meals, and inadequate coordination among various stakeholders involved in the program's delivery (Agustini & Mulyani, 2025). Such logistical deficiencies can result in meals being unsuitable or unavailable when needed, further increasing the likelihood of food being wasted. Historically, school feeding programs in Indonesia have faced similar difficulties with resource allocation, diverse geographical needs, and management issues, which can lead to limited program coverage and overall ineffectiveness (Rimbawan et al., 2023).

5. Challenges and Obstacles in Food Waste Management in Indonesia

Indonesia faces significant challenges in managing food waste, stemming from technical, social, and policy constraints. A major technical hurdle is the inadequate waste management infrastructure, with limited recycling efforts often hampered by a lack of public knowledge and underutilized advanced technologies like anaerobic digestion for energy generation (Handayati & Widyanata, 2024). Socially, there is slow adoption of waste management practices by the public, with participation often discontinuing without consistent support, alongside a general lack of public understanding regarding effective food waste handling (Bhaskara, 2023; Handayati & Widyanata, 2024). From a policy perspective, existing regulations, such as Presidential Regulation 97/2017, are often insufficient to meet ambitious reduction targets, lacking specific metrics for measurement, monitoring, and evaluation, and offering no clear roadmap for reducing household food waste (Ulfah et al., 2023).

Furthermore, limitations in data collection and information systems pose significant obstacles to effective management. Waste generation data is often incomplete, primarily focusing on food leftovers and excluding other organic forms, which skews the overall picture. The National Waste Management Information System struggles with inconsistent reporting from various districts and cities, making comprehensive analysis difficult (Kusumawardani et al., 2023). Critically, the specific amounts of urban food waste from different sources (food services, markets, households and retail) remain unknown because data are often combined, hindering precise targeting of interventions (Cahyana et al., 2024). These data gaps prevent a thorough understanding of food waste dynamics and impede the development of evidence-based policies.

Implementation of food waste management programs also encounters various barriers. The government alone cannot effectively manage all waste responsibilities due to infrastructure deficits and logistical complexities (Handayati & Widyanata, 2024). There are also challenges in scaling and commercializing food waste valorization initiatives, such as converting waste into bioenergy, due to issues with technological feasibility and integrating supply chains. Without robust infrastructure and clear logistical pathways, even well-intentioned programs struggle to achieve their full potential, leading to inefficiencies and continued waste accumulation.

6. Recommendations and Strategies for Sustainable Food Waste Management

To foster sustainable food waste management in Indonesia, a multi-pronged approach involving robust policies, practical initiatives, and active stakeholder engagement is essential. Policy recommendations include formulating a national strategy with clear targets for food loss and waste reduction, complemented by a national data system to track progress accurately (Saliem et al., 2021). Adopting circular economy principles, as successfully implemented in other nations, can provide a comprehensive framework for addressing waste and promoting sustainability (Waluyo & Kharisma, 2023). Specific policies should also target household food waste reduction through awareness campaigns, behavioral change initiatives, and guidance on both preventive measures and waste management techniques (Martianto et al., 2024).

The active participation of various stakeholders is paramount for successful food waste management. The government should lead by implementing comprehensive education programs, improving solid waste services, and developing clear guidelines for all actors involved (Cahyana et al., 2024). They must also be responsive, adaptive, and committed to transparent and participatory teamwork. The community is urged to become more involved in waste management by actively participating in reduction initiatives with ongoing support and engaging in educational campaigns about food waste (Saliem et al., 2021; Handayati & Widyanata, 2024). The private sector can assist by supplying material resources or by serving as buyers for waste products that have been processed by communities. An integrated approach where all stakeholders collaborate effectively is vital for achieving national waste management goals.

7. Conclusion

The "Makan Bergizi Gratis" program, initiated by the Indonesian government to enhance nutrition and human resource quality, has unfortunately led to an increase in food waste during its initial implementation. This unexpected outcome stems primarily from issues such as the low quality and poor acceptance of the meals provided, coupled with various implementation and operational challenges including inconsistent nutritional standards and uneven distribution. Children's taste preferences and program design elements like portion sizes also contribute significantly to this waste.

This situation occurs within a broader context of significant food waste in Indonesia, where households are major contributors. To address this, effective food waste management requires overcoming technical, social, and policy hurdles, including improving data collection and strengthening program implementation. Future strategies must involve robust policies with clear targets, practical initiatives focusing on high-waste categories and behavioral change, and active collaboration between communities, government, and the private sector to ensure a sustainable and efficient food system.

Funding

This study was funded by the Directorate General of Science and Technology Resources and Higher Education, Kemendiktisaintek through the scholarship "Master's to Doctorate Education Program for Excellent Undergraduates (PMDSU) Batch VII" (038/E5/PG.02.00.PL/2024).

References

A. S. Cahyana, I. Vanany, N. I. Arvitrida, I. Gunawan, Modeling the relationships among the driving factors of food waste in indonesian city. Journal of Industrial Engineering and Management, **17**, 1 (2024).https://doi.org/https://doi.org/10.3926/jiem.6012

A. Ulfah, I. D. A. A. Warmadewanthi, Y. Fransiscus, A. U. Farahdiba, E. Rosyidah, J. Hermana, A. Yuniarto, The present and proposed sustainable food waste treatment technology in indonesia : a review. Environmental Technology & Innovation, **32**, 103256 (2023). https://doi.org/10.1016/j.eti.2023.103256

C. P. Komara, S. Pitriani, D. S. Sari, D. Yulianti, Free nutritious meal program amidst the food waste crisis. E-Sospol. Electronical Journal of Social and Political Sciences, **12**, 2 (2025). https://doi.org/https://doi.org/10.19184/e-sospol.v12i2.53789

D. Kusumawardani, N. A. Hidayati, A. Martina, K. S. Agusti, Y. Rahmawati, Y. Y. Amalia, N. F. Ramdaniyah, Household food waste in indonesia: macro analysis. Polish Journal of Environmental Studies, **32**, 6 (2023). https://doi.org/10.15244/pjoes/163157

D. Martianto, R. Diana, Y. F. Baliwati, D. Sukandar, A. Hendriadi, The quantity and composition of household food waste: Implications for policy. PLoS ONE, **19**, 6 (2024). https://doi.org/10.1371/journal.pone.0305087

G. I. Bhaskara, Religious values of chefs in bali and food waste awareness : a preliminary study food waste in developing countries. E-Journal of Tourism, **10**, 1 (2023). https://doi.org/https://doi.org/10.24922/eot.v10i1.100744

H. P. Saliem, S. Mardianto, Sumedi, E. Suryani, S. M. Widayanti, Policies and strategies for reducing food loss and waste in Indonesia Policies and strategies for reducing food loss and waste in Indonesia. Earth and Environmental Science, **892** (2021). https://doi.org/10.1088/1755-1315/892/1/012091

M. Audrey, E. Rukmini, Food waste warriors : Indonesian communities engage through workshops and social media education. International Journal of Public Health Science (IJPHS), **13**, 2 (2024). https://doi.org/10.11591/ijphs.v13i2.23660

R. Qomarrullah, Suratni, W. S. Lestari, M. Sawir, The long-term impact of the free nutritious meal program on health and educational sustainability. Indonesian Journal of Intellectual Publication, **5**, 2 (2025).

R. Rimbawan, R. Nurdiani, P. H. Rachman, Y. Kawamata, Y. Nozawa, School lunch programs and nutritional education improve knowledge, attitudes, and practices and reduce the prevalence of anemia: a pre-post intervention study in an indonesian islamic boarding school. Nutrients, **15** (2023). https://doi.org/https://doi.org/10.3390/nu15041055

S. W. Setyaningtyas, A. N. Arimbi, R. Diana, Exploration of school meal food service management : a case study of a full-day school. International Journal of Public Health Science, **11**, 3 (2022). https://doi.org/10.11591/ijphs.v11i3.21331

S. Zulkifli, A. Jayanegara, B. P. Noorachmat, M. R. Fahmi, T. Tandio, Sustainability analysis of environmental waste alleviation through bioconversion using black soldier fly larvae: a case study in depok city, indonesia. Journal of Natural Resources and Environmental Management, **15**, 1 (2025). https://doi.org/http://dx.doi.org/10.29244/jpsl.15.1.19

U. Agustini, S. Mulyani, Effectiveness and challenges of free nutritious meal program policy as an educational intervention in indonesia. Jurnal Kiprah Pendidikan, **4**, 3 (2025).

Waluyo, D. B. Kharisma, Circular economy and food waste problems in Indonesia: Lessons from the policies of leading countries. Cogent Social Sciences, **9**, 1 (2023). https://doi.org/10.1080/23311886.2023.2202938

Y. Handayati, C. Widyanata, Effective food waste management model for the sustainable agricultural food supply chain. Scientific Reports, **14**, 10290 (2024). https://doi.org/10.1038/s41598-024-59482-w

Automatic Extraction 3D Building Models from UAV LiDAR Point Cloud Data: A Case Study of Tunjungan Street, Surabaya

Joseph Frederick Marbun[1], *I* Gusti Ngurah Antaryama[2], *Hepi* Hapsari Handayani[1*]

[1] Department Geomatics Engineering, Faculty of Civil, Planning, and Geo Engineering, Sepuluh Nopember Institute of Technology, Surabaya, Indonesia

[2] Architecture Department, Faculty of Civil, Planning, and Geo Engineering, Sepuluh Nopember Institute Of Technology, Surabaya/ East Java, Indonesia

Abstract. High resolution 3D City Models are essential for the Smart City paradigm, yet acquiring accurate spatial data in dense tropical urban environments remains challenging due to the limitations of passive optical sensors. This study addresses these issues by validating a semi-automated workflow for generating Level of Detail 2 (LOD2) building models using UAV LiDAR data, specifically focusing on mitigating systematic strip misalignment errors. Using Surabaya's Tunjungan Street as a case study, the research implements a rigorous strip adjustment method followed by nDSM-based extrusion guided by 2D footprints. Results demonstrate that strip adjustment is indispensable, improving Z-axis consistency by nearly 50% (from 1.23 mm to 0.63 mm). Crucially, this improvement minimizes vertical discrepancies and outliers, ensuring the high data cohesion required for precise architectural reconstruction and preventing geometric misinterpretations. The final products achieved a Horizontal Accuracy (CE90) of 0.079 m and Vertical Accuracy (LE90) of 0.385 m, surpassing Indonesian Class 1 standards for 1:5,000 scale mapping. Furthermore, the LOD2 models exhibited a vertical RMSE of 0.268 m, confirming the workflow's reliability for precision critical urban planning.

[*] Corresponding author: hh.handayani@its.ac.id

1 Introduction

The "Smart City" paradigm has emerged as a fundamental tenet of modern urban management, particularly in the 21st century, where rapid urbanization necessitates the adoption of advanced technologies to foster efficiency and sustainability [1]. A pivotal element within this framework is the 3D City Model, which serves as a digital twin of the physical environment. These models enable complex simulations and analyses unattainable through traditional 2D mapping. Traditional 2D maps fail to represent the volumetric density and vertical complexity of urban canyons, which are critical for simulating wind flow, solar potential, and thermal comfort [2], [3], [4]. In dense corridors like Tunjungan Street, 3D models provide a digital twin that allows for vertical-space management and precise heritage conservation analysis. High-resolution spatial data is indispensable, particularly in complex urban environments like Surabaya, where areas such as the Tunjungan Street corridor dominated by high-rise buildings and dense vertical infrastructure require highly accurate data for effective management.

Conventional methods for high-accuracy spatial data acquisition, such as Total Stations and Terrestrial Laser Scanning (TLS), have long served as the standard approach [5]. While these methods yield high geometric accuracy, they are frequently constrained by logistical inefficiencies, time-consuming data collection processes, and difficulties in accessing busy commercial districts, rendering them impractical for large-scale applications. Conversely, aerial photogrammetry utilizing passive optical sensors struggles within narrow "urban canyons," where severe shadowing and texture homogeneity hinder the accurate reconstruction of vertical structures.

UAV LiDAR technology offers a robust solution to these challenges [6], [7]. As an active sensor, UAV LiDAR is capable of acquiring high-density 3D point clouds in complex urban environments, independent of lighting conditions, and can penetrate vegetation canopies to detect underlying structures. This makes UAV LiDAR particularly well-suited for mapping dense urban corridors, especially in tropical environments like Indonesia. However, a primary challenge in deriving accurate 3D building models from UAV LiDAR is the geometric inconsistency between adjacent flight lines, commonly referred to as strip misalignment.

These systematic errors arise from the integration of the onboard GNSS and IMU systems, causing vertical and horizontal shifts in overlapping point clouds [8], [9]. If not adequately addressed, these misalignments can severely degrade the quality of the resulting 3D building models. Furthermore, extracting building geometries to meet Level of Detail (LOD) 2 standards specifically roof structures poses a significant computational challenge. Unlike LOD1, which merely represents buildings as prismatic blocks with flat tops based on average height, LOD2 requires the interpretation of complex roof slopes and tiered structures from noisy point clouds. A hybrid workflow capable of reducing noise while preserving architectural detail is required to ensure model accuracy and viability. Consequently, the implementation of strip adjustment methods becomes imperative within the data processing workflow. This method functions by minimizing discrepancies between overlapping flight lines through mathematical transformations of the vehicle's trajectory [10], [11].

This study aims to: (1) evaluate the effectiveness of strip adjustment algorithms in minimizing systematic sensor errors within UAV LiDAR data, and (2) develop and validate a semi-automated workflow for generating LOD2 3D building models to ensure high-fidelity urban representation. To achieve these objectives, the research first implements geometric corrections to improve trajectory consistency, which is fundamental for assessing the accuracy of the resulting base map. Subsequently, LOD2 building models are constructed through an extrusion method based on a Normalized Digital Surface Model (nDSM), utilizing the 'Extract Roof Form Information' task in ArcGIS Pro and guided by 2D building footprints. Finally, the vertical accuracy of the generated models is validated against independent terrestrial measurements from a Total Station to confirm their reliability for precision-critical urban management.

2 Methods

2.1 Study Area

This research was conducted along the Tunjungan Corridor, located in Surabaya, East Java, Indonesia as shown in Fig 1. The study area is situated within the geographic coordinates of 7°15'30" S-7°16'00" S and 112°44'00" E-112°44'30" E. The corridor stretches approximately 0.7 km with an average road width of 20 meters. This location was selected due to its significance as a major commercial hub and the complex urban morphology it presents. The area is characterized by high-rise buildings, shopping centers, office buildings, and hotels, creating an "urban canyon" effect. This environment presents specific challenges for geospatial data acquisition, including the potential for GNSS signal multipath errors caused by reflections from glass buildings, as well as occlusion zones that block the sensor's line of sight to the ground surface. These conditions make the Tunjungan Corridor an ideal site for evaluating the limitations and accuracy of UAV LiDAR systems.

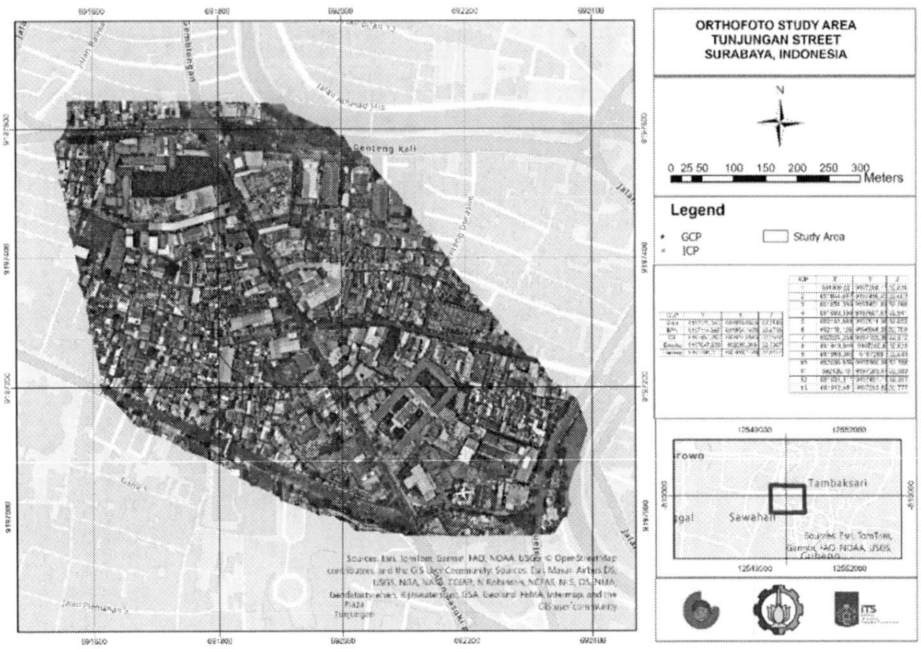

Fig. 1. Study Area

2.2 Dataset

The data utilized in this study comprises primary and secondary sources essential for obtaining precise geospatial information. The primary dataset consists of a UAV LiDAR point cloud collected using a Matrice 350 RTK platform equipped with a DJI L2 sensor, providing high-density 3D coordinates, reflectance intensity, and time-stamped trajectory data. For the building geometry, 2D building footprints were manually digitized from high-resolution orthophotos at a scale of 1:5,000. This vector data serves as the foundational geometry for the extrusion process. To ensure reliability, the positional consistency of these footprints was validated by cross-referencing them with the LiDAR-derived orthophoto, ensuring precise alignment with the physical roof extents. Furthermore, 5 Ground Control Points (GCPs) and 13 Independent Check Points (ICPs) were employed for georeferencing and accuracy assessment. The GCPs were strategically placed using a corridor-based distribution along Tunjungan Street to minimize potential model distortion in this narrow urban environment. Additional vertical validation was conducted through field measurements using an Electronic Total Station (ETS) to serve as an independent ground truth for the generated 3D models.

2.3 Method

The workflow of this research follows a systematic approach that includes several stages: data acquisition, preprocessing (which involves trajectory processing and strip adjustment), point cloud classification, elevation model creation, and 3D building modeling.

The first stage of the process involves the acquisition and preprocessing of UAV flight data. Raw GNSS/IMU data is processed utilizing the Post-Processed Kinematic (PPK) method to generate a Smoothed Best Estimate of Trajectory (SBET). This method was selected for its robustness in maintaining high trajectory accuracy in dense urban environments. This SBET is then integrated with raw laser range data to perform georeferencing of the point cloud. The next step is strip adjustment, which addresses the inevitable errors in the IMU boresight. Flight paths in adjacent areas often exhibit mismatches due to these errors. To resolve this, the study uses an automatic algorithm, such as TerraMatch, to identify "cut lines" or "section lines" on both vertical surfaces, like building walls, and horizontal surfaces in overlapping regions[10]. By examining the geometric differences between these features, particularly in terms of vertical misalignment (Z-mismatch), correction parameters for the sensor's orientation (heading, roll, and pitch) are computed. These adjustments are then applied across the entire dataset. The use of thousands of section lines as internal observations ensures the robustness and accuracy of the adjustment process.

In the point cloud classification and elevation model creation stage, the ground and non-ground points were interpolated into a DTM and DSM using the Inverse Distance Weighting (IDW) algorithm with a grid resolution of 0.68 meters. These parameters were chosen to preserve fine architectural details in the resulting nDSM. The main goal of this step is to generate accurate surface models. To achieve this, the ground points are used to create a Digital Terrain Model (DTM), and the non-ground points are used to generate a Digital Surface Model (DSM). Once both the DSM and DTM are created, the Normalized Digital Surface Model (nDSM) is derived by subtracting the DTM from the DSM[12], [13] (eq.1).

$$nDSM = DSM - DTM \quad (1)$$

The resulting nDSM isolates the height of objects above the ground surface by removing the terrain's topography. The pixel values in the nDSM represent the absolute height of objects, effectively distinguishing between the terrain surface and the structures above it, such as buildings. This model is critical for accurately capturing the heights of objects in a 3D space, which is essential for subsequent 3D building modeling.

For 3D building modeling, a hybrid approach based on classified data, known as Extrusion-Based Modeling, is employed. This technique combines the geometric accuracy of 2D cadastral building footprints with the vertical precision provided by LiDAR data. The main process starts with the Input Geometry step, where the 2D footprint provides the exact horizontal boundaries (X, Y) for the building. The next step involves Height Extraction, the LOD2 models were generated using the 'Extract Roof Form Information' method within the 'Create Buildings' task in ArcGIS Pro. This process utilizes procedural rules to analyze the nDSM and automatically differentiate between various roof types, such as flat, gabled, or hipped structures. Rather than relying on a single average height, the method performs zonal statistical analysis within each footprint to extract critical parameters: the maximum height is utilized to determine the ridge height (roof peak), while the median or mode height is used to estimate the eave height (wall top). This dual-parameter approach effectively captures the true volumetric shape of slanted or multi-tiered roofs while mitigating the influence of sensor noise and small rooftop outliers, such as chimneys or antennas.

Finally, the accuracy of the generated maps and models is rigorously assessed. According to Badan Informasi Geospasial (BIG) Regulation No. 15 of 2014, the base map accuracy is evaluated using 13 Independent Check Points (ICPs), as shown in Fig. 1. The Root Mean Square Error (RMSE) is calculated for both horizontal (X, Y) and vertical (Z) components, and these values are then converted into CE90 and LE90 to determine the map's accuracy scale[14]. For the 3D building model, the vertical accuracy is validated by comparing the extracted building heights from the model with independent field measurements taken using a Total Station. To confirm the reliability of the generated 3D models, a total of 13 buildings were used for validation. These buildings were selected to represent a wide range of vertical diversity, spanning from low-rise commercial shophouses (approx. 5-10 meters) to high-rise structures such as the Double Tree and Platinum Hotels (exceeding 50 meters). RMSE is calculated by comparing the heights from the model with the reference measurements, which provides an accurate assessment of the vertical performance of the 3D building model.

This multi-step methodology ensures high precision and accuracy in creating 3D building models, which are essential for urban planning, land management, and other applications that require detailed geospatial information.

3 Result and Discussion

3.1 Strip Adjustment

Initial point cloud inspections revealed visible vertical discrepancies between adjacent flight paths due to IMU boresight calibration errors. To resolve this, a strip adjustment process was implemented, which significantly enhanced geometric integrity by reducing the average Z-mismatch and RMSE by approximately 50%. Table 1 presents the comparative statistics, grounding the claim of improved data cohesion in quantitative results.

Table 1. Strip Adjustment Error Statistics

Parameter	Value Before Adjustment	Value After Adjustment (m)	Percentage Improvement
Average 3D Mismatch	0.00123	0.00063	48.78%
Average Z Mismatch (Vertical)	0.00063	0.00063	48.78%
RMSE (Sum Z)	0.002	0.001	50.00%
Maximum Mismatch	0.035	0.013	62.68%

The data in Table 1 indicates a substantial improvement in data cohesion after the strip adjustment process. The average vertical mismatch (Average Z Mismatch) decreased by nearly 50%, reaching a sub-millimeter level of 0.00063 meters, significantly enhancing the statistical quality of the dataset. More importantly, the maximum Z mismatch, which represents outliers, dropped dramatically from 3.5 cm to 1.3 cm. This reduction of outliers is crucial for applications requiring high precision, such as micro-drainage analysis, where such errors could lead to misinterpretations, such as incorrect identification of roof texture in the 3D model.

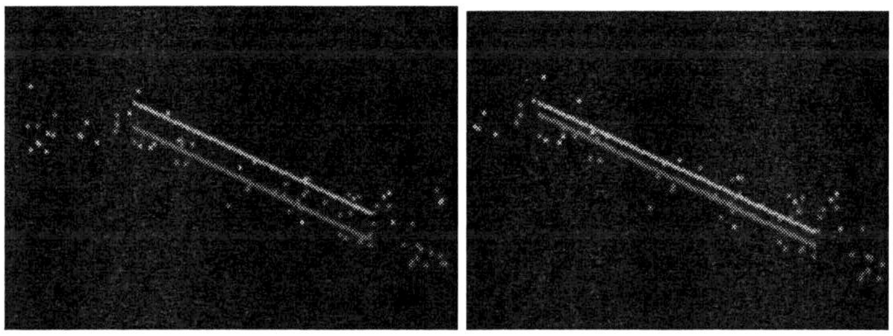

Fig. 2. Comparison of Point Cloud Alignment with Tie Line - Left: Before Strip Adjustment, Right: After Strip Adjustment

The strip adjustment process involved adjusting 6,034 section lines, providing a substantial internal observation dataset. The large number of observations ensures a robust statistical basis, confirming that the global representation of sensor errors, especially those due to IMU calibration, was effectively corrected. As a result, the strip adjustment minimized the RMSE compared to the laser system's inherent noise floor, improving the overall geometric quality of the model. Visually, as shown in Fig. 2, this adjustment resulted in a smoother surface and a significant reduction in "step edges" the vertical offsets previously visible between overlapping flight strips. The clearer alignment of profile lines on roof surfaces and walls demonstrates the substantial improvement in vertical accuracy, enabling the generation of a more refined 3D surface model.

3.2 Geometric Accuracy

The accuracy of the geospatial product from the LiDAR campaign was evaluated based on several key metrics, as shown in the Table 2. The spatial resolution of the product was measured at 0.681 meters, indicating the level of detail captured by the LiDAR system and reflecting the precision with which features in the landscape can be distinguished. The horizontal accuracy, represented by the CE90 value, was 0.079 meters, which significantly surpasses the tolerance of 2.5 meters for a scale of 1:5,000 as set by the Geospatial Information Agency (BIG) standards. This value also meets the Class 1 requirement for horizontal accuracy at a scale of 1:1,000, where the acceptable CE90 is 0.2 meters. In terms of vertical accuracy, the LE90 value was 0.385 meters. This result meets the Class 1 requirement (LE90 $\leq$ 1 meter) for a 1:5,000 scale but falls within the Class 3 standard (LE90 $\leq$ 5 meters) for a 1:1,000 scale. This level of vertical precision is highly relevant for LOD2 building modeling, as it ensures that the extracted roof heights and structural geometries remain within the acceptable error margins for high-fidelity urban simulations and infrastructure management. Overall, these results demonstrate that the LiDAR system provides highly accurate spatial data, particularly in horizontal positioning, making it suitable for 3D building model generation. However, further refinement of vertical accuracy is necessary for producing highly precise 3D models that require the strictest level of detail, especially in urban planning and detailed building design.

Table 2. Geometric Accuracy Assessment Results

Aspect	Measurement Result (m)
Spatial Resolution	0.681
Horizontal Accuracy (CE90)	0.079
Vertical Accuracy (LE90)	0.385

3.3 3D Model Buildings

The core output of this study is the LOD2 3D building model, which is generated through an extraction process that utilizes the Normalized Digital Surface Model (nDSM) to determine the vertical dimensions of structures, defined by the 2D footprint. The use of the nDSM effectively filters out the influence of terrain topography variations, allowing the algorithm to focus solely on the height of objects relative to the ground. This technical approach is especially critical along the 0.7 km Tunjungan Street corridor, where the ground elevation exhibits notable variations. By quantifying and removing these terrain slopes through the nDSM, the workflow prevents topographical distortions from affecting the building height calculations, ensuring that the generated models align accurately with the actual vertical dimensions of the structures. As demonstrated in the Fig.3, the application of the nDSM ensures accurate building height extraction, highlighting the vertical dimensions of the structures and ensuring they are represented accurately, free from the influence of varying terrain.

The generated model successfully captures the roof shapes, distinguishing them from the simpler LOD1 block. By analyzing the point cloud distribution within each footprint, the algorithm reconstructs roof geometries, such as differentiating between the flat roofs of modern buildings and the gabled roofs of heritage structures. This approach significantly improves the dataset's utility, allowing applications like solar radiation simulations that depend on roof slope and orientation. The 3D model clearly illustrates these roof forms and accurately represents building features, making the model more suitable for advanced planning and analysis.

The extrusion method used in this study results in a clean and geometrically consistent model. The model aligns consistently with the cadastral footprint, resolving issues such as jagged building edges. However, it is important to acknowledge the limitations of extrusion-based modeling, particularly when dealing with highly complex or non-traditional roof geometries that may require more advanced reconstruction techniques. The resulting 3D model demonstrates a smoother, more accurate urban representation, enhancing the model's reliability for urban planning and other applications. This 3D building model, derived from the nDSM-based extraction process, showcases the effectiveness of this methodology in generating high-quality, detailed 3D models, making them ideal for further urban analysis and planning applications.

Fig. 3. Visualization of LOD2 Building Models generated from nDSM-based extrusion along Tunjungan Street

3.4 Validation Vertical Height Of Model 3D

To validate the absolute vertical accuracy and roof height consistency of the generated 3D models, a rigorous ground-truthing campaign was undertaken utilizing an Electronic Total Station (ETS). A sample set of thirteen buildings, exhibiting a wide range of elevations, was selected to ensure the validation process adequately represented the vertical diversity within the study area.

The selection of these samples was designed to capture a broad spectrum of the urban topography. This included high-rise structures, such as the Double Tree and Platinum Hotels, as well as low-rise commercial clusters like shophouses. Integrating this variety was crucial to verify the model's consistency across different vertical tiers.

The analysis yielded a Root Mean Square Error (RMSE) of 0.268 meters (approximately 27 cm). Notably, this performance metric surpasses the LE90 value recorded for the general ground surface (0.385 m), demonstrating that the model is well within the acceptable decimeter level thresholds for LOD2 building modeling and urban planning applications. This discrepancy is likely attributable to the physical properties of the building roofs; as hard, reflective surfaces, they provide cleaner and less ambiguous laser returns. In contrast, ground-level measurements are often complicated by vegetation, uneven terrain, or shadow occlusion, which tend to reduce accuracy in standard LE90 calculations.

4 Conclusion

This study establishes a validated, automated workflow for extracting LOD2 3D building models from UAV LiDAR data, utilizing the Tunjungan Street corridor in Surabaya as a representative urban case study. A pivotal component of this success was the integration of a rigorous strip adjustment method based on section lines, which proved indispensable for high-fidelity modeling—specifically in improving roof height accuracy and reducing systematic strip misalignment. This correction process effectively mitigated vertical mismatches by nearly 50%, tightening Z-axis consistency from 1.23 mm to 0.63 mm, thereby ensuring a robust foundation for subsequent modeling steps.

In terms of geospatial compliance and model reliability, the generated outputs demonstrated exceptional precision. With a Horizontal Accuracy (CE90) of 0.079 m and a Vertical Accuracy (LE90) of 0.385 m, the data comfortably exceeds the Class 1 standards for 1:5,000 scale mapping mandated by the Badan Informasi Geospasial (BIG), confirming its suitability for formal planning purposes. Furthermore, the extrusion-based technique successfully produced LOD2 geometries that accurately capture roof structures surpassing the limitations of standard LOD1 block models with a vertical RMSE of 0.268 m when corroborated against independent Total Station surveys. This level of precision explicitly confirms the model's suitability for LOD2 applications and complex urban planning tasks, meeting standard vertical tolerances for 3D building city development.

Acknowledgment

The authors wish to express their sincere gratitude to the field survey team for their dedicated assistance during the data acquisition and ground-truthing campaign. We also extend our appreciation to the Genteng District Administration for their cooperation and for granting the necessary permissions to conduct the field study within their jurisdiction.

References

[1] M. Hämäläinen, "Smart city dvelopment with digital twin technology," dalam *33rd Bled eConference – Enabling Technology for a Sustainable Society: June 28 – 29, 2020, Online Conference Proceedings*, University of Maribor Press, 2020, hlm. 291–303. doi: 10.18690/978-961-286-362-3.20.

[2] B. Bryant, A. Agha Karimi, B. Atazadeh, dan A. Rajabifard, "Modelling 3-Dimensional cadastral boundaries in a BIM environment: a case study using volumetric format plans in Queensland, Australia," *Int. J. Digit. Earth*, vol. 18, no. 1, hlm. 2518578, Des 2025, doi: 10.1080/17538947.2025.2518578.

[3] G. Cremen, C. Galasso, dan J. McCloskey, "A Simulation-Based Framework for Earthquake Risk-Informed and People-Centered Decision Making on Future Urban

Planning," *Earths Future*, vol. 10, no. 1, hlm. e2021EF002388, 2022, doi: 10.1029/2021EF002388.

[4] M. Korkmaz, Y. E. Akyildiz, S. Demirkesen, S. Toprak, P. Nowak, dan B. Ciftci, "A Digital Twin Approach to Sustainable Disaster Management: Case of Cayirova," *Sustainability*, vol. 17, no. 21, hlm. 9626, Jan 2025, doi: 10.3390/su17219626.

[5] S. Hendriatiningsih, D. Suwardhi, dan J. Januragadi, "3D Model Based on Terrestrial Laser Scanning (TLS) Case study: The Cangkuang Temple, Garut District, West Java, Indonesia," *J. Eng. Technol. Sci.*, vol. 47, no. 1, hlm. 1–19, Feb 2015, doi: 10.5614/j.eng.technol.sci.2015.47.1.1.

[6] Y.-C. Lin, R. Manish, D. Bullock, dan A. Habib, "Comparative Analysis of Different Mobile LiDAR Mapping Systems for Ditch Line Characterization," *Remote Sens.*, vol. 13, no. 13, hlm. 2485, Jan 2021, doi: 10.3390/rs13132485.

[7] D. Hu dan J. Minner, "UAVs and 3D City Modeling to Aid Urban Planning and Historic Preservation: A Systematic Review," *Remote Sens.*, vol. 15, no. 23, hlm. 5507, Jan 2023, doi: 10.3390/rs15235507.

[8] F. Pöppl, G. Mandlburger, dan N. Pfeifer, "EVALUATION OF A GNSS/IMU/LIDAR-INTEGRATION FOR AIRBORNE LASER SCANNING USING RTKLIB PPK AND PPP GNSS SOLUTIONS," *Int. Arch. Photogramm. Remote Sens. Spat. Inf. Sci.*, vol. XLVIII-1-W3-2023, hlm. 161–166, Okt 2023, doi: 10.5194/isprs-archives-XLVIII-1-W3-2023-161-2023.

[9] F. Pöppl, A. Spitzer, A. Ullrich, dan N. Pfeifer, "Airborne Laser Scanning for Corridor Mapping: Georeferencing with Tightly-coupled Multi-view LiDAR," *Int. Arch. Photogramm. Remote Sens. Spat. Inf. Sci.*, vol. XLVIII-1-W4-2025, hlm. 109–114, Jun 2025, doi: 10.5194/isprs-archives-XLVIII-1-W4-2025-109-2025.

[10] K. Liu, H. Ma, L. Zhang, Z. Cai, dan H. Ma, "Strip Adjustment of Airborne LiDAR Data in Urban Scenes Using Planar Features by the Minimum Hausdorff Distance," *Sensors*, vol. 19, no. 23, hlm. 5131, Nov 2019, doi: 10.3390/s19235131.

[11] Z. Chen, J. Li, dan B. Yang, "A Strip Adjustment Method of UAV-Borne LiDAR Point Cloud Based on DEM Features for Mountainous Area," *Sensors*, vol. 21, no. 8, hlm. 2782, Jan 2021, doi: 10.3390/s21082782.

[12] F. Soleimani Vostikolaei dan S. Jabari, "Large-Scale LoD2 Building Modeling using Deep Multimodal Feature Fusion," *Can. J. Remote Sens.*, vol. 49, no. 1, hlm. 2236243, Jul 2023, doi: 10.1080/07038992.2023.2236243.

[13] R. Župan, A. Vinković, R. Nikçi, dan B. Pinjatela, "Automatic 3D Building Model Generation from Airborne LiDAR Data and OpenStreetMap Using Procedural Modeling," *Information*, vol. 14, no. 7, hlm. 394, Jul 2023, doi: 10.3390/info14070394.

[14] P. BIG, *Perka BIG No. 15 Tahun 2014*, Peraturan Perundang-undangan 15, Oktober 2014. doi: http://peraturan.bpk.go.id/Details/269446/perka-big-no-15-tahun-2014.

Analysis of Residential Land Changes Using Object Based Image Analysis (OBIA) Method with SPOT Imagery in 2014 and 2024 (Case Study: Banyumanik, Semarang City)

Adya Amalia Putri [1*], *Filsa Bioresita* [1]

[1]Geomatic Engineering, Faculty of Civil Engineering, Planning, and Geoscience, Sepuluh Nopember Institute of Technology, Surabaya, Indonesia

Abstract. Residential land expansion in urban areas, particularly in Banyumanik District, Semarang City, has increased significantly over the past decade. This study analyzes residential land-use changes between 2014 and 2024 using Object-Based Image Analysis (OBIA) applied to SPOT 7 satellite imagery and examines the influence of public and social facilities on these changes. An integrated spatial and statistical approach was employed. Residential land-use maps for 2014 and 2024 were generated and overlaid with public and social facility maps to assess spatial relationships, followed by statistical analysis to evaluate their influence on residential expansion.The results indicate a substantial increase in residential land area, predominantly in locations with high accessibility to public and social facilities. Statistical findings confirm that facility availability significantly influences residential land expansion, highlighting infrastructure provision as a key driver of urban growth. These findings imply that urban development can be more effectively managed by aligning the spatial distribution of public and social facilities with land-use planning and zoning regulations. Such integration is essential to promote sustainable urban growth and to mitigate uncontrolled residential expansion in Semarang City.

1 Introduction

According to the 2019 World Bank report, approximately 151 million Indonesians, or 56% of the total population, resided in urban areas. Urban regions in Indonesia have experienced significant growth, driven by the role of cities as centers of economic and social activities. This trend has encouraged population movement from rural to urban areas, a phenomenon commonly referred to as urbanization.

Urbanization in Indonesia occurs due to increasing settlement density, the expansion of infrastructure, and the development of various public facilities, which collectively contribute to the reclassification of rural areas into urban zones. Rapid urbanization often leads to unplanned land use changes. As cities expand, residential land use commonly shifts in

* Corresponding author: adya.amalia1502@gmail.com

response to economic, social, political, technological, environmental dynamics, and natural population growth. These factors collectively contribute to the expansion of urban areas [1].

Changes in residential land use, particularly in urban environments, pose major challenges for spatial planning and land management. This phenomenon is influenced by numerous drivers such as population growth, economic development, infrastructure expansion, and shifts in social behavior [2, 3, 4]. In Indonesia, Semarang City is among the areas experiencing significant residential land use changes.

Statistical data from BPS indicates that within a decade, residential land in Semarang City nearly doubled. In 2014, the residential land area was recorded at 17,768 hectares, increasing to 32,311 hectares in 2024. The rapid growth of residential areas has led to limited land availability in the city center, pushing development toward peripheral regions, including Banyumanik District [5].

Banyumanik District is one of the 16 administrative districts of Semarang City, consisting of 11 sub-districts: Pudakpayung, Gedawang, Jabungan, Padangsari, Banyumanik, Srondol Wetan, Pedalangan, Sumurboto, Srondol Kulon, Tinjomoyo, and Ngesrep. This area has undergone substantial land use changes, particularly the conversion of land into residential areas. These changes are influenced by urbanization processes and socio-economic dynamics. In addition, the development of infrastructure—such as the expansion of educational facilities and the growth of commercial centers—serves as a primary driver of residential land use change in Banyumanik District.

To analyze such land use dynamics, remote sensing technology plays a crucial role. Remote sensing enables rapid, efficient, and continuous acquisition of spatial data over large areas. Satellite imagery provides temporal perspectives that make it possible to monitor land use changes, making remote sensing particularly relevant for urbanization studies.

One widely applied analytical method in remote sensing is Object-Based Image Analysis (OBIA). Unlike traditional pixel-based approaches, OBIA groups pixels into meaningful objects based on spectral similarity, texture, shape, and spatial context. This method produces more accurate classification results, especially in urban areas characterized by complex spatial patterns.

In this study, SPOT satellite imagery is utilized to map and analyze residential land use changes in Banyumanik District. The high spatial resolution of SPOT imagery enables detailed and accurate detection of land use changes and allows examination of spatial relationships between residential areas and the development of supporting facilities such as educational and commercial centers. Therefore, this study aims to analyze residential land use changes in Banyumanik District between 2014 and 2024 and to identify the influence of public and social facilities on the rate of residential land conversion.

The objective of this research is to analyze residential land use changes in Banyumanik District between 2014 and 2024 using OBIA-based classification of SPOT imagery and to assess the influence of public and social facilities on the rate and spatial pattern of residential land expansion.

The research questions addressed in this study are:

1. How has residential land use in Banyumanik District changed between 2014 and 2024?
2. What is the spatial relationship between the distribution of public and social facilities and residential land expansion?
3. To what extent do public and social facilities influence the rate of residential land conversion in Banyumanik District?

2 Materials and method

2.1 Materials

In this study, the materials used include SPOT 7 imagery of Semarang City from 2014 and 2024 to identify land cover in Banyumanik District. In addition, Google Earth data and data from the Spatial Planning Agency regarding the distribution of public facilities and social facilities in Banyumanik District were also utilized.

2.2 Land covers

Land cover refers to the observable biophysical cover on the Earth's surface, which results from the management, activities, and treatment carried out by humans on specific types of land cover for production, modification, or maintenance purposes. Remote sensing data greatly supports the provision of spatial information, particularly related to land cover and land use [6]. Land cover reflects the relationship between natural processes and social processes. It provides essential information for modeling and for understanding natural phenomena that occur on the Earth's surface [7].

Land cover may consist of vegetation or artificial constructions that occupy the Earth's surface. It is related to the physical appearance of features such as buildings, lakes, and vegetation [8]. In contrast, land use refers to all types of human utilization of land, ranging from agriculture to sports fields, residential houses to restaurants, hospitals to cemeteries.

2.3 Settlements

Based on Law Number 1 of 2011, the definition of housing refers to a collection of houses intended solely as a place of residence. Meanwhile, settlements are residential areas consisting of multiple housing units and functioning not only as living spaces but also as areas that support various activities. Settlements can be understood as human-made or natural environments equipped with facilities and infrastructure that enable communities to carry out their daily activities [9].

The regulatory framework concerning housing, settlements, and the supporting environmental infrastructure refers to Law Number 26 of 2007. The objective of this law is to improve the quality of life of the community through well-organized housing and settlement development [10].

The development of settlements is closely related to urban land use planning, particularly regarding the functional division of the city into working areas, residential areas, and recreational areas. Settlement patterns, including natural physical conditions, can serve as indicators for assessing settlement development [11]. In this study, the term "settlement" refers specifically to the settlements located in Banyumanik District, Semarang City.

2.4 Public facilities and social facilities

Facilities refer to all elements considered as means to achieve certain objectives in fulfilling specific needs. In another sense, when related to settlements, facilities are activities or physical elements that serve the needs of individuals or groups within a living environment. Systematically, these activities or elements can be categorized into two main groups: physical facilities and social facilities. Although different, both types of facilities are interrelated.

Social facilities can be defined as services that fulfill the social, mental, and spiritual needs of the community. These include educational facilities, health facilities, recreational facilities, worship facilities, sports facilities, commercial facilities, and government facilities [12]. Social facilities play a crucial role and are essential in ensuring the quality of every residential environment.

Meanwhile, physical facilities refer to services that meet the physical needs of the community and include public utilities. Social facilities hold a highly significant role in regional development. The role of infrastructure and facilities in regional development is to support the functioning of social and economic systems in the daily lives of communities [13].

Both social facilities and public facilities are accessible for public use. In the context of environmental infrastructure, public utilities and social facilities are often abbreviated as fasos and fasum for convenience.

Physical facilities are defined as basic physical components of the environment that enable housing and settlements to function properly. Facilities are supporting elements that serve the implementation and development of economic, social, and cultural activities.

2.5 Remote sensing

The basic concept of remote sensing consists of several elements or components, including the energy source, atmosphere, interactions between energy and objects on the Earth's surface, sensors, data processing systems, and various data applications. Energy in remote sensing serves as the medium that carries information about objects to the sensor, and it may take the form of sound, magnetic force, gravity, or electromagnetic energy.

A remote sensing image is a representation of an object, area, or phenomenon resulting from the recording of reflected or emitted energy by a remote sensing sensor and may appear as a photograph or digital data [14].

Remote sensing data collection is carried out using detectors or data acquisition devices known as sensors. These sensors, which gather data from a distance, are generally mounted on platforms such as aircraft, balloons, satellites, or other carriers [15]. The remote sensing data produced from sensor recordings are then processed for various applications.

2.6 SPOT-7 image

SPOT is a series of Earth-observation satellites that capture high-resolution optical imagery. The satellite system is designed to provide wide-area coverage suitable for cartography and monitoring. SPOT-7 was launched by Airbus Defence & Space on June 30, 2014.

Table 1. SPOT-7 Satellite Image Spesification

Specification	Description
Spatial Resolution	1.5 m (panchromatic, nadir)
	6 m (multispectral VNIR, nadir)
Spectral Bands	1 Panchromatic (450-745 nm) 4 Multispectral: -Blue (455-525 nm) -Green (530-590 nm) -Red (625-695 nm) -Near Infrared (760-890 nm)
Altitude	694 km
Orbit Type	Sun-synchronous
Dynamic Range	12-bit per pixel
Acquisition Capacity	Up to 6 million km^2 per day

Based on these specifications, SPOT-7 imagery includes four multispectral bands—blue, green, red, and near-infrared—as well as one panchromatic band. The multispectral bands have a spatial resolution of 6 meters, while the panchromatic band has a resolution of 1.5 meters.

2.7 Image fusion

The purpose of image fusion is to integrate information from multiple images while preserving all essential visual details contained in the original imagery. Image fusion is the process of combining information from a set of images into a single, more informative image that is suitable for both human interpretation and machine analysis [14].

One commonly used image fusion method is pan-sharpening. Pan-sharpening is a technique that merges multispectral (MS) imagery with panchromatic (PAN) imagery. Multispectral images contain rich spectral information but have relatively low spatial resolution, whereas panchromatic images offer high spatial resolution but limited spectral detail. Through pan-sharpening, low-resolution multispectral data are fused with high-resolution panchromatic data from the same area to produce an image with enhanced spatial detail and preserved spectral characteristics.

The resulting fused image provides sharper geometric features and more complete spectral information, making it more suitable for identifying and analyzing targets with greater accuracy [14].

2.8 OBIA Image segmentation

Image segmentation is an essential component of image processing. It is primarily used as a preprocessing step in object recognition systems. Image segmentation involves dividing an image into homogeneous regions based on specific similarity criteria, particularly the similarities between the gray-level values of a pixel and those of its neighbouring pixels. The results of this segmentation process are then used for higher-level tasks, such as image classification and object identification [15].

There are various image segmentation techniques, which can generally be classified into two main types based on their working principles [15]. One commonly used approach is segmentation based on color intensity, which assumes that objects to be separated tend to have different color intensities, while each object exhibits relatively uniform color characteristics. One segmentation technique based on color intensity is mean clustering. In mean clustering, image segmentation is performed by dividing the image histogram. The procedure consists of determining the minimum and maximum intensity values in the image, dividing the intensity range into N intervals representing the expected number of objects, grouping pixels into the nearest clusters, and replacing pixel values in each cluster with the mean value of that cluster.

The segmentation process also involves several parameters, namely scale, shape, and compactness [15]. The scale parameter in multiresolution segmentation does not correspond to the scale definition in remote sensing related to spatial resolution or pixel size. Instead, it represents an abstract value that controls the level of heterogeneity allowed within an object. Higher scale values permit greater heterogeneity, resulting in coarser segmentation and larger object sizes.

The shape parameter is influenced by the weighting of the color parameter. The selection of shape weight should be adjusted according to the characteristics of the study area and the phenomenon being analyzed. A higher shape weight emphasizes spatial homogeneity over spectral homogeneity, which may overly stress texture and not always produce optimal segmentation results.

The compactness parameter is used to distinguish compact objects from non-compact objects with relatively similar spectral values. Higher compactness values result in objects with more compact shapes, representing a deviation from an ideal compact form.

The primary objective of remote sensing data processing is to extract objects on the Earth's surface for interpretation and analysis. Rapid object extraction methods have developed significantly in recent years. In general, remote sensing classification can be divided into two approaches: pixel-based classification and object-based classification, commonly referred to as Geographic Object-Based Image Analysis (GEOBIA). Pixel-based classification assigns each pixel to a class based on its digital number, whereas object-based classification extracts meaningful objects using spectral, spatial, and contextual characteristics. Once these objects are identified, expert knowledge can be applied in subsequent classification stages [15].

2.9 Classification Accuracy Assessment

The accuracy of the land cover classification results was evaluated using a confusion matrix. This method provides quantitative information on the agreement between the classified image and reference data. Two accuracy indicators were applied for each land cover class, namely Producer's Accuracy and User's Accuracy. Producer's Accuracy represents the probability that a reference pixel is correctly classified and is calculated as the ratio between correctly classified pixels and the total reference pixels of a given class, indicating omission errors.

User's Accuracy represents the probability that a pixel classified into a certain class actually belongs to that class on the ground. It is calculated as the ratio between correctly classified pixels and the total number of pixels assigned to that class in the classified map, indicating commission errors [8]. Image classification results are considered acceptable when the overall accuracy value reaches or exceeds 80%.

The accuracy assessment was calculated using the following equations:

User's Accuracy : $\frac{x11}{x+1} x100\%$ ....(1)

Producer's Accuracy : $\frac{x11}{x1+} x100\%$ ....(2)

Overall Accuracy : $\left(\frac{\sum_{i=1}^{r} xii}{N}\right) x\ 100\%$...............................(3)

Description:

N : The total number of pixels

X1+ : The total number of pixels in row-i

X+1 : the total number of pixels in column i

Xii : the diagonal element of the confusion matrix representing the correctly classified pixels for class i

2.10 Regression analysis

Regression analysis is a statistical technique used to examine the relationship between two or more variables. Based on its level of complexity, regression analysis is divided into two types: when the model uses only one independent variable, it is referred to as simple linear regression, whereas when it involves more than one independent variable, it is referred to as multiple linear regression. Regression analysis consists of the following components:

a. Dependent Variable (Response): the variable being predicted or explained, denoted as Y.
b. Independent Variable (Predictor): the variable(s) used to predict or explain changes in the dependent variable, denoted as X.

The multiple linear regression model can be expressed mathematically as follows [4]:

$$Y = \beta_0 + \beta_1 X_1 + \beta_2 X_2 + \beta_n X_n + \varepsilon \quad (4)$$

Description:

- Y: Dependent variable
- β_0: Intercept (constant term)
- β: Regression coefficient
- X: Independent variable
- ε: Error term (difference between the actual value and the estimated model)

3 Result and analysis

3.1 Land cover change

According to the land cover classification analysis for Banyumanik District in the period of 2014–2024 with the OBIA procedure, from the 2014 till 2024 land cover maps, the extent of each land cover class along with its percentage change can be stated as follows to illustrate that a great number of land cover types have been increased. Table 2 shows the percentages of land cover areas:

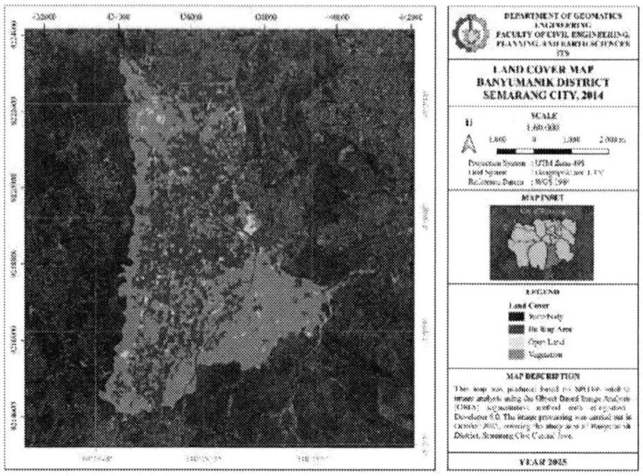

Fig. 1. Land Cover Map of Banyumanik District in 2014

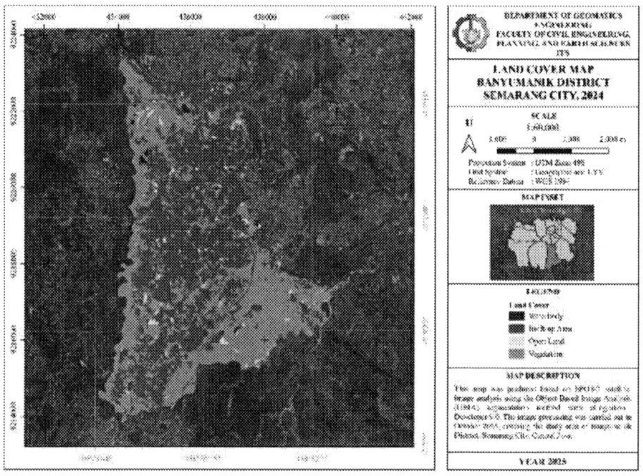

Fig. 2. Land Cover Map of Banyumanik District in 2024

Based on the land cover maps for 2014 and 2024, the area of each land cover class along with its percentage change shows that several land cover types have experienced an increase. The table of land cover area percentages is presented as follows

Table 2. Land Cover Area o Banyumanik District

Land Cover	2014		2024	
	Area (Ha)	%	Area (Ha)	%
Built-up Area	1209.414	37.93	1527.208	47.89
Open Land	109.109	3.42	79.288	2.49
Waterbody	9.923	0.31	18.113	0.57
Vegetation	1860.468	58.34	1564.305	49.05
Total	3188.914	100	3188.914	100

Based on the table above, it is shown that there is an increase in built-up land area and a decrease in open land and vegetation. Therefore, these land cover changes are interpreted through the following map and table:

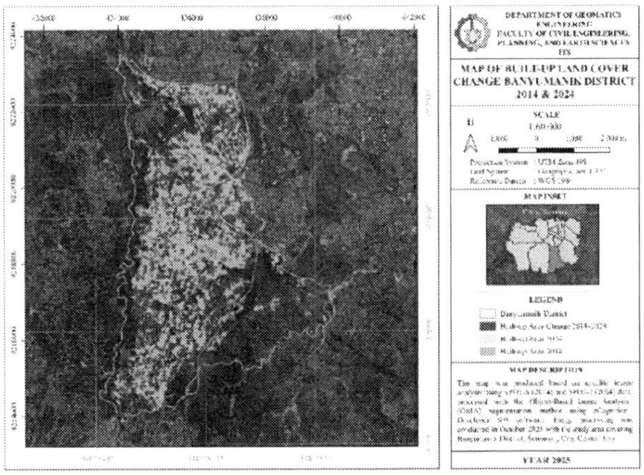

Fig. 3. Map of Land Cover Change in Banyumanik District

Table 3. Area of Land Cover Change in Banyumanik District

Land Cover	2014-2024	
	Area (Ha)	%
Built-up Area	317.794	9.97
Open Land	-29.821	-0.94
Waterbody	8.19	0.26
Vegetation	-296.163	-9.29
Total Land Cover Change	651.968	20.44

Based on the Land Cover Change Map of Banyumanik District from 2014 to 2024, the observed increase in built-up areas reflects an urban expansion process primarily driven by accessibility and infrastructure development. The concentration of land use change in the central and western parts of the district, as well as along major road corridors such as Jalan Setiabudi, indicates that residential growth has been strongly influenced by the availability of transportation networks. These corridors reduce travel time to the city center and the Tembalang education area, making adjacent land more attractive for residential and commercial development.

The dominant conversion of vegetation and open land into built-up areas is largely a consequence of limited land availability in Semarang's city center and rising land values in peripheral areas. As a result, development pressure has shifted toward Banyumanik District, where relatively large parcels of undeveloped land were still available. This process explains the substantial decrease in vegetation area (296.163 hectares) alongside the significant increase in built-up land (317.794 hectares) during the study period.

Furthermore, the clustering of land use change around public and social facilities suggests that facility provision acts as a catalyst for residential expansion. The presence of schools, healthcare facilities, commercial centers, and places of worship increases the functional attractiveness of surrounding areas, encouraging developers and households to concentrate near these services. Improved access to facilities reduces daily travel costs and enhances living convenience, which accelerates land conversion in nearby locations.

Overall, these findings indicate that residential land expansion in Banyumanik District is not random but follows a facility- and accessibility-driven development pattern. This pattern highlights the interaction between transportation infrastructure, public facility distribution, and land availability in shaping urban growth. Understanding these driving mechanisms is essential for spatial planning, particularly in controlling urban sprawl, protecting green open spaces, and guiding future development toward more sustainable urban forms.

3.2 Availability of public and social facilities

Based on the results of the analysis, field observations, and reviews of the availability of social facilities in previous years, there is a noticeable trend of increasing social facilities within the study area. As the number of social facilities increases, it affects the extent of their service coverage. The following are the maps showing the distribution of public and social facilities in Banyumanik District in 2014 and 2024.

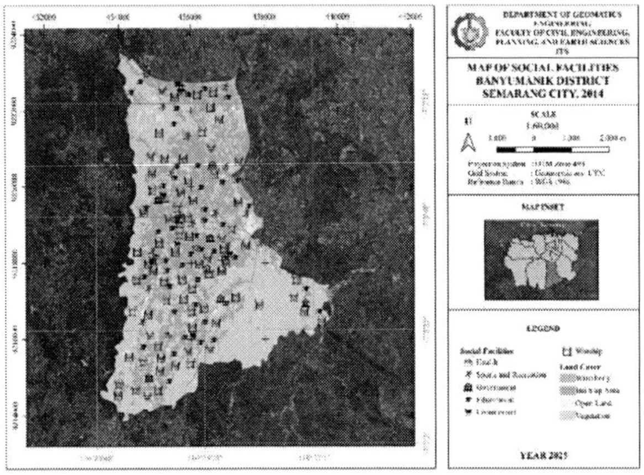

Fig. 4. Map of Distribution of Social Facilities in Banyumanik District in 2014

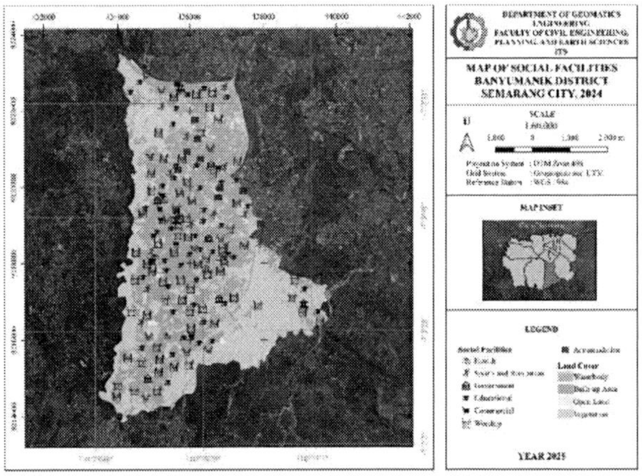

Fig. 5. Map of Distribution of Social Facilities in Banyumanik District in 2024

Table 4. Social Facilities in Banyumanik District

Public and Social Facilities	Categories	Year	
		2014	2024
Health Facilities	Hospital	2	2
	Community Health Center (Puskesmas)	4	4
Sports and Recreation Facilities	Sports Facilities	14	14
Government Facilities	District/Subdistrict Offices	10	10

Public and Social Facilities	Categories	Year	
		2014	2024
Educational Facilities	Elementary School (SD)	30	30
	Junior High School (SMP)	10	10
	Senior High School (SMA)	4	4
	University	2	2
Commercial Facilities	Markets and Malls	6	10
Worship Facilities	Mosque	67	67
	Church	7	7
	Vihara	1	1
Accommodation Facilities	Apartment	-	2
	Hotel/Guesthouse	1	10
Total		158	173

Based on tabular analysis and visual interpretation of the 2014 and 2024 social facility maps, there is an increase in the number and expansion of social facilities in Banyumanik District over a ten-year period. The number of social facilities increased from 158 units in 2014 to 173 units in 2024, indicating a growing trend in public services in line with population growth and residential development in the area.

Spatially, the 2024 map shows that the increase in facilities primarily occurred in areas experiencing residential development (built-up land). This is evident in the increasingly even distribution of facilities in the central and southern areas of the district, compared to the central and northern areas in 2014.

Several facility categories experienced significant growth, such as shopping facilities, which increased from 6 to 10 units, and accommodation, which expanded significantly from just one hotel/lodging in 2014 to 10 hotel/lodging units and two apartments in 2024. This increase aligns with the growth of residential areas and the need for commercial services in the area.

Meanwhile, educational, healthcare, religious, and government facilities remained unchanged in number, but their service coverage increased as the built-up area became denser. The distribution of facilities such as mosques, schools, and community health centers (Puskesmas) on the 2024 map appears to be more within new residential zones, which were previously vegetated in 2014.

Overall, the changes in the distribution of social facilities in the 2014–2024 period indicate that the development of built-up land is directly proportional to the need for and provision of social facilities. The increase in the availability of these facilities has the potential to improve the quality of public services, but must also be balanced with equitable distribution and sustainable planning to prevent disparities in service delivery between regions in Banyumanik District.

3.3 The influence of public facilities and social facilities on residential land-use change

Table 5. Correlation Between Social Facilities and Changes in Bulit-up Land Cover

Correlations

	Total Number of Public Facilities and Social Facilities	Residential Land Use Change

Total Number of Public Facilities and Social Facilities	Pearson Correlation	1	1.000**
	Sig. (2-tailed)		.
	N	2	2
Residential Land Use Change	Pearson Correlation	1.000**	1
	Sig. (2-tailed)	.	
	N	2	2

**. Correlation is significant at the 0.01 level (2-tailed).

Based on the results of the Pearson correlation test, a correlation coefficient of 1.000 was obtained with a significance level of 0.01 (2-tailed). A correlation value of 1.000 indicates a very strong and perfectly positive relationship between the number of public and social facilities and changes in residential land use. This means that any increase in the number of social and public facilities is directly proportional to an increase in changes in residential land use in the study area.

In other words, the increase in facilities such as shopping centers, hotels, places of worship, healthcare facilities, and educational facilities contribute to the expansion of residential areas. This can occur because the presence of social and public facilities increases the attractiveness of the area, strengthens service functions, and encourages the development of new activity centers. Areas with more complete access to facilities tend to experience accelerated housing development to meet the increasing housing needs of the community.

However, it should be noted that the sample size (N) consists of only two time points (2014 and 2024). Therefore, although the resulting correlation is perfect, it cannot yet comprehensively describe the causal relationship. This correlation further indicates that over the analyzed time period, the increase in social and public facilities coincided with the expansion of residential land, but cannot be concluded as a direct cause without further analysis.

Overall, these results indicate that the provision of public and social facilities is related to the dynamics of land use change, particularly in areas experiencing residential growth. This finding supports the importance of integrated facility planning to ensure that residential growth remains directed, equitable, and sustainable.

References

1. M. Roberts, F. Sander, S. Tiwari, Time to ACT: Realizing Indonesia's Urban Potential (World Bank Publications, Washington, 2019)
2. E.F. Lambin, H.J. Geist, E. Lepers, Dynamics of land-use and land-cover change in tropical regions, Annu. Rev. Environ. Resour. **28**, 205–241 (2003)
3. K.C. Seto, B. Güneralp, L.R. Hutyra, Global forecasts of urban expansion to 2030 and direct impacts on biodiversity and carbon pools, Proc. Natl. Acad. Sci. USA **109**, 16083–16088 (2011)
4. M. Batty, The New Science of Cities (MIT Press, 2013)
5. F. Wulandari, Analisis Perubahan Penggunaan Lahan di Kecamatan Ngemplak Kabupaten Boyolali Tahun 2004 dan 2015, Jurnal Universitas Muhammadiyah Surakarta (2017)
6. K. Darmoyuwono, Pedoman Penafsiran Liputan Lahan (Land Cover) dari Citra Landsat Skala 1:1.000.000–1:250.000 (BAKOSURTANAL, Cibinong, 1979)
7. R.M. Sampurno, A. Thoriq, Klasifikasi Tutupan Lahan Menggunakan Citra Landsat 8 Operational Land Imager (OLI) di Kabupaten Sumedang, Jurnal Teknotan **10**, 61–70 (2016)

8. T.M. Lillesand, W.J. Carper, R.W. Kiefer, The Use of Intensity–Hue–Saturation Transformations for Merging SPOT Panchromatic and Multispectral Image Data, Photogrammetric Engineering and Remote Sensing **56**, 461 (1990)
9. H.S. Yunus, Geografi Permukiman dan Beberapa Permasalahan Permukiman di Indonesia (Faculty of Geography, Universitas Gadjah Mada, Yogyakarta, 1987)
10. Republic of Indonesia, Law No. 26 of 2007 on Spatial Planning (2007)
11. Bintarto, Arsitektur Pembangunan dan Konservasi (Djambatan, Jakarta, 1979)
12. S. Burdah, cited in: S. Henlita, K.D.M.E. Handayeni, Tingkat Pelayanan Fasilitas Pendidikan Sekolah Menengah Tingkat Atas di Kabupaten Sidoarjo, Jurnal Institut Teknologi Sepuluh November 2, 2 (2013)
13. R.J. Kodoatie, Pengantar Manajemen Infrastruktur (Pustaka Pelajar, Yogyakarta, 2005)
14. F.S.H. Purwadhi, T.B. Sanjoto, Pengantar Interpretasi Citra Penginderaan Jauh (Universitas Negeri Semarang dan Lembaga Antariksa Nasional, Semarang, 2008)
15. L. Somantri, Teknologi Penginderaan Jauh (Remote Sensing) (Universitas Pendidikan Indonesia, Bandung, 2009)

Optimizing Waste Reduction Rate at The Integrated Waste Processing Facility in Ngipik Landfill, Gresik

Deqi Rizkivia Radita[1*], *Hanif* Ghifari[1], and *Susi* Agustina Wilujeng[1]

[1]Environmental Engineering Department, Institut Teknologi Sepuluh Nopember, Surabaya

Abstract. The average weight of municipal solid waste disposed of at the Ngipik Landfill was 212.2 tons per day, with 94.8% directly deposited in the landfill. Based on field observation, there was existing reduction carried out by Integrated Waste Processing Facility (*Tempat Pengolahan Sampah Terpadu,* TPST) before the waste was deposited in the landfill. However, the exact amount of waste that could be reduced at the TPST is unknown. Further research was needed on the amount of waste that can be reduced at TPST. This data could serve as additional information in the waste material flow of Ngipik Landfill. The objectives of this study were to assess the implementation of waste reduction by TPST operational activities, then to do the calculation of waste potential weight and composition that could be reduced by TPST. The weight and composition were measured using direct measurement methods. The potential reduction of the TPST was 7.74 tons per day. The results of the TPST optimization were an increase in the amount of waste entering to 18.8 tons per day with working hours from 7:00 a.m. until 4:00 p.m., the addition of 31 people in the sorting operation, and the expansion of land to 2,521.8 m^2.

1 Introduction

Municipal Solid waste issues is a problem faced by almost all cities in Indonesia. One of these issues is the municipal solid waste problem faced by Gresik Regency. The increase in population and number of industries in Gresik Regency has resulted in a significant increase in municipal solid waste generation. Based on the data obtained from the National Waste Management Information System (*Sistem Informasi Pengelolaan Sampah Nasional,* SIPSN), Gresik Regency produced 402.23 tons of municipal solid waste per day in 2024 [1]. This amount has increased by 22.72% or 74.46 tons per day from the amount of waste generated in 2020 [1].

The increasing amount of municipal solid waste produced per day is not followed by efforts to reduce waste from the source. Currently, most cities in Indonesia still adhere to old paradigm of collect, transport, and dispose. With this paradigm, the level of reduction at the source is very low, resulting in a huge burden for landfills. If this condition continues, it will cause the landfill to overload rapidly and must be closed.

A total of 212.2 tons of municipal solid waste is sent to Ngipik Landfill every day, or around 52.7% of the total daily municipal solid waste generated by Gresik Regency [1]. This data is also supported by research conducted by Rosariawari et al., which states that the average amount of municipal solid waste entering the Ngipik landfill is 210 tons per day [2]. The disposal method implemented at the Ngipik landfill is planned to be a sanitary landfill, with solid waste covered using soil by heavy equipment every day. However, this has not been implemented due to several constraints, one of which is the operational factors of vehicles and heavy equipment at the landfill [3]. Based on field observations on December 5, 2023, the municipal solid waste disposal system at the Ngipik landfill uses a controlled landfill. A total of 199 tons or 94.8% of the solid waste entering the Ngipik landfill is directly disposed of in the landfill. Currently, the amount of solid waste at the Ngipik landfill has exceeded its capacity [4]. If solid waste is not processed before being buried in the landfill, it is estimated that the Ngipik landfill will no longer be able to accommodate municipal solid waste from Gresik Regency in the next few years.

* Corresponding author: deqi@its.ac.id

Considering that the reduction rate at the source is very low, there are several actions that can be taken to reduce the amount of solid waste entering landfills. One action that can be taken is to conduct preliminary processing at the landfill before the solid waste is buried in the landfill zone. It is estimated that this preliminary processing will reduce the amount of waste buried in the landfill zone, thus extending the life of the landfill. One facility that can be applied for preliminary treatment by using Integrated Municipal Solid Waste Processing Facility or we called it TPST (*Tempat Pengolahan Sampah Terpadu*). From the study that conducted by Setyaningsih and Caroline in 2016 it obtained the results that TPST has potential to reduce almost 70% MSW entering the landfill [5].

Based on that reference, the Ngipik landfill management has make several efforts to reduce the amount of solid waste entering the landfill. Currently, there are processing efforts carried out by the informal sector, Integrated Waste Management Facilities (TPST), and composting houses. However, there is no data available on the amount of waste reduced by these sectors. From the waste processing data at the TPST, it will be possible to estimate the amount of solid waste reduction before it enters the landfill and the lifespan of the landfill. This will enable recommendations to be made on measures that can be taken to optimize solid waste reduction activities before it enters the landfill and efforts to extend the lifespan of the landfill. The research was conducted by examining solid waste reduction implementation at the Ngipik landfill from TPST operational activities, calculating the weight and composition of solid waste reduced by the TPST, and analyzing the material flow of solid waste at the Ngipik landfill. The expected outcome of this research is that the data produced can be used as a reference for landfill managers to calculate the lifespan of the landfill more accurately and evaluate existing activities at the Ngipik landfill.

2 Research method

2.1 Primary Data Collection

Primary data was obtained through direct observation, surveys, field calculations, and interviews with respondents at the research site. The data obtained in this study is as follows:

1. Data on the volume, density, and composition of daily solid waste entering the landfill.
   Volume data was obtained to determine the amount of solid waste entering the landfill per day. The data collection method used was weight volume analysis for 8 consecutive days by recording the weight of each truck entering and leaving the landfill at the weighbridge. The weight of the solid waste in the truck was obtained from the weight of the truck entering the landfill minus the weight of the truck leaving the landfill. In addition to the weight of the truck, the type of vehicle, vehicle license plate number, time of entry and exit of the truck, and the destination of the truck, whether directly to the landfill or to the TPST, were also recorded. The density was obtained by dividing the weight of the solid waste in the truck by the volume of the truck container. The composition of the solid waste is measured by taking 100 kg of waste that has just been unloaded from the truck, which is then sorted and weighed according to its composition.
2. Data on the weight and composition of waste reduced at the TPST.
   Data on TPST reduction was collected using direct measurement methods over a period of 4 days. The measurement of solid waste generation and composition was carried out in accordance with Indonesian National Standard (*Standar Nasional Indonesia,* SNI) number 19-3964-1994 concerning Methods for Collecting and Measuring Samples of Urban Waste Generation and Composition [6].
3. Data on the capacity of processing machines and the area of the TPST.
   The method of collecting data on machine capacity and TPST area was through interviews with TPST officers and direct calculations. Data from interviews with landfill and TPST management personnel included operating hours, the solid waste processing and management flow at the landfill and TPST.

2.2 Secondary Data Collection

In conducting this research, secondary data is needed to support the data analysis as follows:

1. Recapitulation of solid waste data entering the landfill.
2. Organizational structure and operational system at the TPST.

2.3 Data Analysis

Data processing and discussion were carried out on the primary and secondary data that had been collected. Research data analysis to determine reduction results, assessment of TPST operational activities, and material flow of solid waste at TPST was carried out in several ways, as follows:

1. Analysis of waste reduction from TPST operational activities at the Ngipik landfill.
   Analysis was carried out based on interviews with TPST officers and landfill management officers. Next, a descriptive analysis was conducted related to the operational system and organizational structure of the TPST, coordination between TPA and TPST activities, and the operating hours of the TPA and TPST. The existing conditions at the TPST were compared with the recovery factors of other TPSTs in order to optimize the recovery factor to match the ideal reduction. Then, the land area requirements per facility were calculated.
2. The analysis of TPST reduction results was carried out using direct measurement methods.
   The calculation of the average daily reduction of TPST from the processing results included the sorting of recyclable waste, plastic chips, compost, and briquettes.

3 Result and Discussion

3.1 Weight and Composition of Solid Waste Entering the Ngipik Landfill

Based on sampling conducted over 8 days, the average amount of waste entering the Ngipik Landfill was 234,262 kg per day or 234.2 tons per day. The largest amount of waste entered on Friday, amounting to 307 tons, and the smallest amount entered on Sunday, amounting to 160 tons. This occurred because the number of non-environmental agency trucks that disposed of waste on Monday-Friday was higher than on Saturday-Sunday, resulting in a significant decrease. Meanwhile, trucks from the Gresik environmental agency routinely disposed of waste every day. A summary of the weight of waste entering the landfill can be seen in Figure 1.

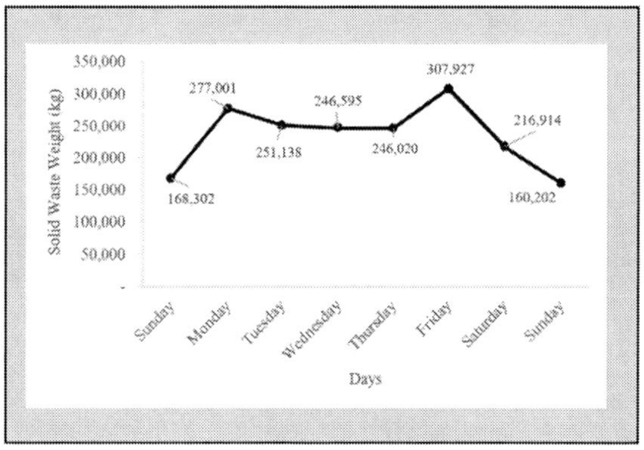

Fig. 1. Graph of Total Solid Waste Entering the Ngipik Landfill per Day

The composition of solid waste at the Ngipik Landfill is sorted based on SNI 19-3964-1994. The largest composition of municipal solid waste is food waste at 71.6%, while the smallest composition is other waste at 0.02% that mostly consist of sacks waste. The percentage composition of waste can be seen in Figure 2. Food waste and yard waste have high moisture content, which affects the weight of the waste and causes the percentage composition of these two types of waste to be the largest. Both types of waste have the potential for biological treatment because they can degrade naturally. The type of plastic waste with the highest percentage (47%) is Low Density Polyethylene (LDPE), which consists of disposable plastic bags, frozen food bags, bubble wrap, and food packaging plastic. This shows that LDPE plastic is still widely used by the community, especially for packaging. Other plastic compositions consist of multilayer plastics (19%), Polivinyl Chloride/PVC (12%), Polyethylene Terephthalate/PET (11%), Polypropylene/PP (5%), Polystyrene/PS (3.3%), and High Density Polyethylene/HDPE (2.3%). Residual waste consisting of diapers, menstrual pads, tissues, and other items are the third largest composition of waste entering the Ngipik landfill, with a percentage of 6.2%.

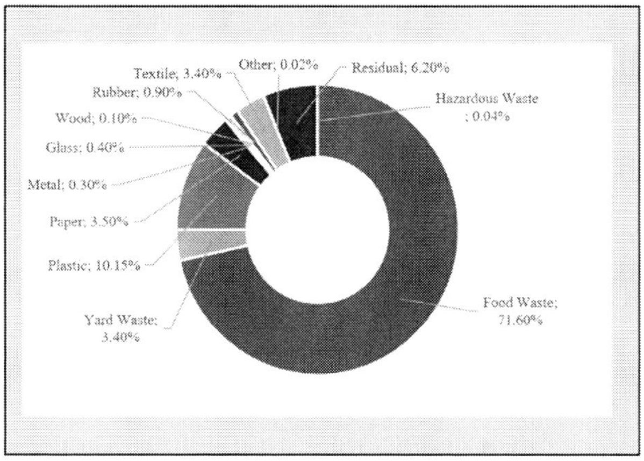

Fig. 2. Solid Waste Composition Entering the Ngipik Landfill

3.2 Existing Conditions of TPST

The TPST at the Ngipik landfill has a land area of 1,203 m^2 with a waste input capacity of 2.5 tons per hour. According to Minister of Public Works Regulation No. 3 of 2013, the requirement for establishing a TPST is to have a minimum building area of 20,000 m^2 [7]. In terms of land area, the TPST at the Ngipik landfill does not yet comply with the minimum building area requirement. The management office and weighbridge facilities are integrated with the Ngipik landfill. The layout of the existing TPST can be seen in Figure 3.

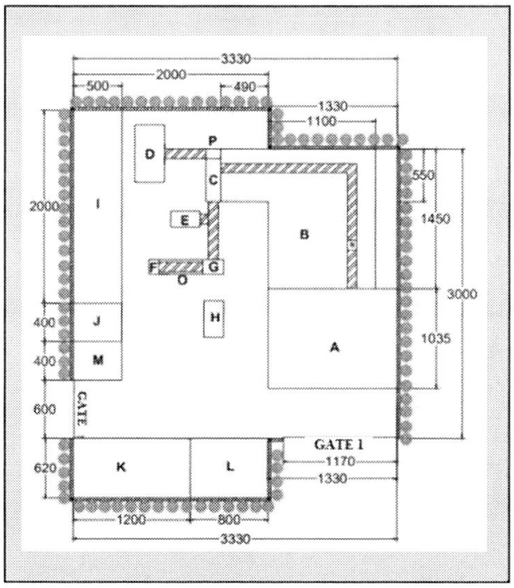

Description:
A Reception area
B Sorting area
C Sorting machine
D Compost shredding area
E Residue area
F Plastic shredding area
G Shredder
H Briquette production area
I Compost material storage area
J Mixed sorting output storage area
K Shredding output storage area
L Briquette output storage area
M Paper sorting output storage area
N Bag opener machine
O Plastic shredding area

Fig. 3. Existing TPST Layout

From the layout in Figure 3, there are areas for receiving, sorting, shredding, composting, and storing sorted waste, briquettes, and shredded waste. The processing machines at the TPST are a bag opener machine that tears open plastic wrappers from household waste, a plastic shredder machine that shreds plastic into small pieces, and a sorting machine that sorts compostable waste, residues, LDPE plastic, and others. The existing waste processing flow at the TPST can be seen in Figure 4.

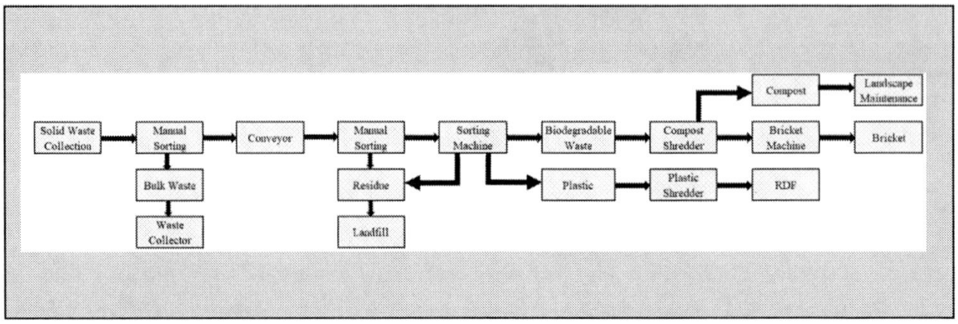

Fig. 4. Existing Waste Processing Flow at TPST

Based on an interview with the manager of the Ngipik landfill, the TPST at the Ngipik landfill began operating in July 2023. The TPST operates every Monday to Friday from 8:00 a.m. to 3:00 p.m. On average, 7-8 trucks enter the TPST every day, which are determined by the Environmental Agency. The final output of the TPST is plastic chips that will be used as Refuse Derived Fuel (RDF), as well as charcoal briquettes and compost materials that will be used for composting. At the TPST, waste that can be recycled, such as paper, cardboard, and certain types of plastic like HDPE, PET, PVC, and others, is sorted and later sold to waste collectors.

Recapitulation of TPST input weight and reduction can be seen in Figure 5. Based on the Figure 5, the largest reduction in TPST was on Wednesday, amounting to 8,169.3 kg or 8.16 tons, while the smallest reduction was on Thursday, amounting to 6,659.4 kg or 6.65 tons. The results of TPST processing, namely plastic chips, briquettes, compost, and sorted waste, can be recycled. From the field data, the average waste reduction rate at the TPST was 20%.

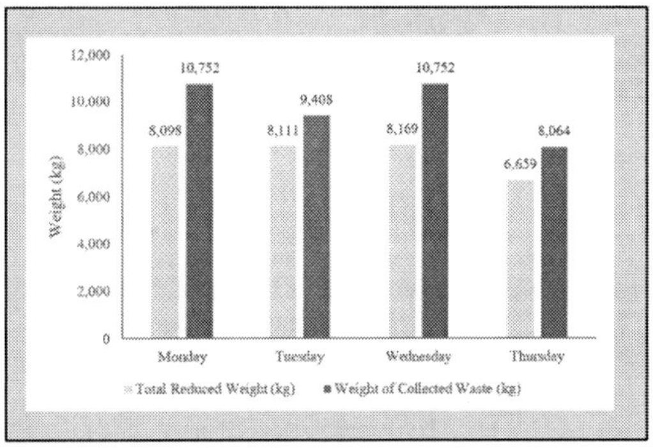

Fig. 5. Recapitulation of TPST Input Weight and Reduction

3.3 TPST Optimization

TPST optimization was carried out by increasing the weight of incoming waste from 7 trucks to 14 trucks per day with residential waste weighing a total of 18,819.5 kg. Additionally, working hours were extended from 6 hours to 8 hours in accordance with Law Number 11 of 2020 on Job Creation (Job Creation Law) [8]. The results of the calculation of the land area required for optimization can be seen in Table 1.

https://doi.org/10.1051/bioconf/202621606009

Table 1. Summary of Land Requirements for TPST Optimization

No.	Function	Existing Area (m²)	Optimized Area (m²)
1	Office	0	24
2	Receiving area	137.7	137.7
3	Sorting area	193.2	163.5
4	Plastic shredding area	11.4	27
5	Briquette production area	7.6	13
6	Residual waste handling area	4.98	10.6
7	Plastic shredding storage area	100	21
8	Mixed sorting results storage area	40	16
9	Paper sorting results storage area	40	9
10	Briquette storage area	49.6	13.8
11	Compostable waste shredding area	18	31.6
12	Composting area	100	567.8
13	Maturation area	0	197.5
14	Compost screening area	0	22.2
15	Compost storage	0	15.4
	TOTAL LAND AREA	662.4	1,270.1

From the Tabel 1, the total land area required for TPST optimization is 1,270.1 m². The results of the optimized building design for building 1 can be seen in Figure 6.

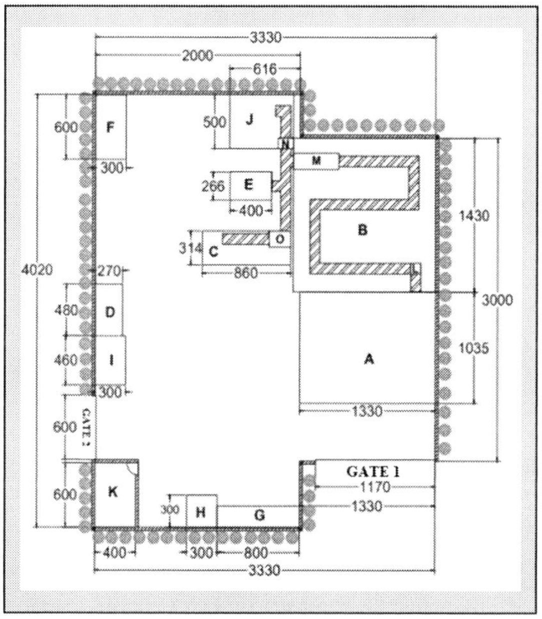

Description:
A Reception area
B Sorting area
C Plastic shredding area
D Briquette production area
E Residual waste handling area
F Plastic shredding storage area
G Mixed sorting storage area
H Paper sorting storage area
I Briquette storage area
J Compostable waste shredding area
K Administration office
L Bag opener machine
M Sorting machine
N Compost shredding machine
O Plastic shredding machine

Fig. 6. Layout Optimization of Building 1 TPST

The location for the composting facility is planned on unused landfill land 110 m from the waste transfer station. The location of the planned site can be seen in Figure 7.

Fig. 7. Composting Facility Planning Location

From Figure 7, the planned location of the composting facility is in the area marked with a yellow box, covering an area of 902.22 m^2, at distance of 110 m from the existing TPST building. The reason for choosing this location is that composting cannot be carried out on the existing TPST land, so a location around the landfill must be found for the planning of the composting building. The results of the optimized TPST building 2 design planning can be seen in Figure 8.

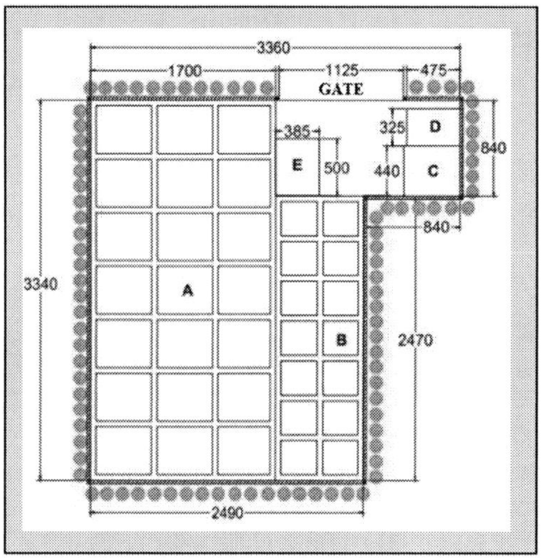

Description:
- A Composting area
- B Maturation area
- C Sifting area
- D Compost storage area
- E Compost material reception area.

Fig. 8. Optimization Layout of Building 2 TPST

4 Conclusion

The average weight of waste entering the Ngipik landfill is 234.26 tons per day. Of the 228.72 tons that go directly to the landfill, 9.74 tons go to the TPST. The TPST is estimated to reduce waste by 84.68% of the total waste entering the TPST, so that only 15.32% enters the landfill. To achieve this result, a scenario can be implemented by increasing the amount of waste entering the TPST from 7 trucks to 14 trucks per day. In addition, working hours will be increased from 6 hours to 8 hours. The processing results at the TPST include sorted recyclable waste, plastic chips, compost, and briquettes.

Funded by: Department of Environmental Engineering, Sepuluh Nopember Institute of Technology, Ministry of Education, Culture, Research, and Technology in accordance with the Implementation Agreement Department Research Batch 1 Year 2024 Researcher Contract Number: 1048/PKS/ITS/2024, dated February 29, 2024

References

1. Kementerian Lingkungan Hidup dan Kehutanan Republik Indonesia. (2023). Data Timbulan Sampah. Sistem Informasi Pengelolaan Sampah Nasional (SIPSN). https://sipsn.menlhk.go.id/sipsn/
2. Rosariawari, F., Rosydi, H. A., Achmad, A. S., Pratama, B. C., Pratiwi, S. W., & Rafi, M. I. (2023). Inovasi ramah lingkungan: Sosialisasi sampah plastik menjadi solusi inovatif untuk pengelolaan sampah eco-friendly innovation: Plastic waste socialization becomes an innovative solution for waste management. Kegiatan Positif, 1(4), 174–180. https://doi.org/10.61132/Kegiatan
3. Wardhani, M. K., & Harto, A. D. (2018). Studi komparasi pengurangan timbulan sampah berbasis masyarakat menggunakan prinsip bank sampah di Surabaya, Gresik, dan Sidoarjo. Jurnal Pamator, 11(1), 52–63.
4. Hertati, D., Nurhadi, & Tukiman. (2023). Effectiveness of zero-based waste management policy in realizing a clean Gresik Regency. JKMP (Jurnal Kebijakan Dan Manajemen Publik), 11(1), 1–7.
5. Setyaningsih, F. R., & Caroline, J. (2016). Tingkat efisiensi reduksi sampah di TPST Super Depo Sutorejo Kota Surabaya. In Prosiding Seminar Nasional Sains dan Teknologi Terapan (pp. 25–32).
6. Badan Standardisasi Nasional. (1994). Standar Nasional Indonesia 19-3964-1994 tentang Metode Pengambilan dan Pengukuran Contoh Timbulan dan Komposisi Sampah Perkotaan.
7. Kementerian Pekerjaan Umum dan Perumahan Rakyat. (2013). Peraturan Menteri Pekerjaan Umum dan Perumahan Rakyat Nomor 03 Tahun 2013 tentang Penyelenggara Prasarana dan Sarana Persampahan dalam Penanganan Sampah Rumah Tangga dan Sampah Sejenis Sampah Rumah Tangga.
8. Republik Indonesia. (2020). Undang-Undang No. 11 Tahun 2020 tentang Cipta Kerja.

Resistance Analysis of Remotely Operated Vehicle (ROV)

Rr. Niken Danartika Hadi Putri [1*], *I Ketut* Suastika [1].

[1]Department of Naval Architecture, Institut Teknologi Sepuluh Nopember, Surabaya, Indonesia

Abstract. Indonesia's vast underwater resources and infrastructure necessitate advanced technologies for exploration and maintenance. Remotely Operated Vehicles (ROVs) have emerged as essential tools, offering safe and efficient underwater operations. However, their performance is affected by hydrodynamic drag, influencing maneuverability and energy efficiency. This study may differ from previous research due to the ROV model used. The TKD Laboratory UUV model, named Segara Nauta with the type code SN-01, is currently under development in the TKD Laboratory of the Department of Naval Architecture, Faculty of Marine Technology (FTK) – ITS is used for this study. This study is focused on investigates the drag characteristics of the SN-01 ROV using Computational Fluid Dynamics (CFD) simulations. Grid independence tests identified an optimal mesh resolution of 1,163,064 elements, balancing computational efficiency and accuracy. Resistance tests at varying speeds (0.3–1.0 m/s) and movement directions revealed that resistance increases with speed. Forward-backward resistance ranged from 0.0019 N to 0.0190 N, while sideways resistance was significantly higher, from 0.0049 N to 0.0481 N, due to greater surface exposure. The resistance on the ROV increases as the speed of the ROV increases, therefore it is necessary to conduct a study on the service speed of the ROV SN-01 because it can affect the selection of the engine used so that the ROV can work effectively and efficiently. These findings underscore the influence of motion direction and speed on hydrodynamic resistance, emphasizing the need for optimized ROV designs. The results provide valuable insights for improving ROV efficiency, supporting Indonesia's sustainable underwater exploration and maritime infrastructure maintenance.

1. Introduction

Indonesia boasts an abundant wealth of underwater natural resources, including oil, natural gas, and minerals such as gold, copper, nickel, and iron ore. As one of the largest oil and gas producers in Southeast Asia, the country exploits these resources in offshore waters and mineral-rich regions such as Papua, Maluku, Sulawesi, and Kalimantan. Additionally, other potential underwater resources, such as marine minerals, gas hydrates, and marine biomass, are currently under exploration and research.

As an archipelagic nation, Indonesia has extensive underwater installations, including oil and gas pipelines (subsea pipelines), subsea risers, PLEMs (Pipe Line End Manifolds), as well as subsea communication (fibre optic) and power cables. These infrastructures require regular monitoring and inspection to identify potential damage or disruptions, ensure safety, and maintain their technical functionality. The growing activities in seabed mining and the construction of underwater pipelines and cables have driven an increase in underwater observation activities, both for surveys and inspections. By using unmanned technology and systems on underwater marine vehicles, namely Autonomous Underwater Vehicles (AUV), the underwater observation process will be assisted and safer (Sahir et al., 2017).

Remotely Operated Vehicle (ROV) technology has emerged as a primary solution to address the risks and limitations of human diving, particularly at depths of 100–300 meters. The use of ROVs offers safer, more cost-effective operations and allows for broader reach compared to manual diving. ROVs are equipped with tools such as cameras, sonar, and various sensors to support data collection. Additionally, Autonomous Underwater Vehicles (AUVs) are being developed as an alternative unmanned technology for underwater monitoring, although the mastery of such technology remains limited in Indonesia.

However, the performance of ROVs is influenced by hydrodynamic drag, which affects their maneuver ability, maximum speed, and energy efficiency. Therefore, studying the hydrodynamic drag on ROVs is essential for optimizing their design to enhance performance in supporting underwater surveys and inspections, particularly for pipeline infrastructure. Optimizing ROV designs with a focus on drag reduction will provide more efficient and safer technological solutions, supporting Indonesia's underwater exploration activities in a self-reliant and sustainable manner.

Research on Resistance Analysis of Remotely Operated Vehicle (ROV) may differ from previous research due to the ROV model used. The TKD Laboratory UUV model, named Segara Nauta with the type code SN-01, is currently under development in the TKD Laboratory of the Department of Naval Architecture, Faculty of Marine Technology (FTK) – ITS.

*Corresponding author: rrnikendanartika@gmail.com

Development of this model begins in 2023, focusing on design and electronics aspects as part of a final project (Ramadhani, 2023). This underwater vehicle features a fish-line body design, emphasizes hydrodynamic properties, and incorporates a modular structure.

2. Methodology

2.1. Remotely Operated Vehicle (ROV)

A Remotely Operated Underwater Vehicle (ROUV) is an underwater vehicle physically connected to an operator via a cable and controlled by an operator located on a vessel above the water surface (Gómez et al., 2017). This robot-like machine allows the operator to maneuver it from a relatively safe, dry, and comfortable location, typically on a ship or platform situated on the water's surface above the ROUV. The ROUV is classified as a tethered unmanned submersible robot. It is uninhabited, highly maneuverable, and operated by a human. The ROUV plays a critical role as an underwater robot, enabling operators to remain in a safe and comfortable environment while the ROUV performs tasks in hazardous underwater conditions (Azis et al., 2012).

2.2. ROV Model

The TKD Laboratory UUV model, named Segara Nauta with the type code SN-01, is currently under development in the TKD Laboratory of the Department of Naval Architecture, Faculty of Marine Technology (FTK) – ITS. Development of this model began in 2023, focusing on design and electronics aspects as part of a final project (Ramadhani, 2023). This underwater vehicle features a fish-line body design, emphasizing hydrodynamic properties, and incorporates a modular structure as shown in Figure 2.

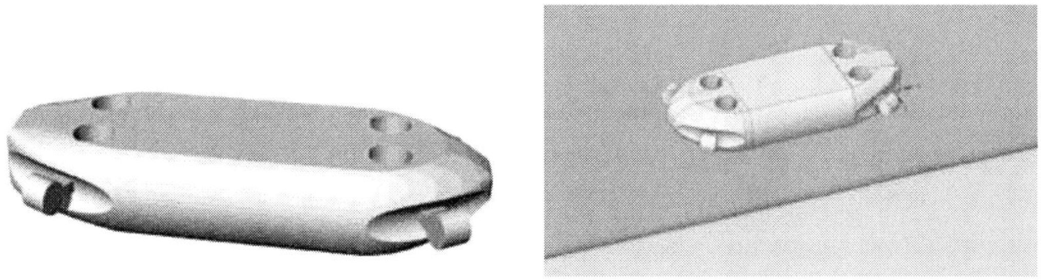

Figure 1 ROV Model

The modeling of the UUV SN-01 consists of components divided into the front, rear, and middle body sections, along with a thruster. Modifications were made to the components to simplify certain parts of the model for the meshing process. This simplification aims to reduce the geometric complexity of the model and minimize the size of the elements used to construct the geometry. The ROV model geometry specifications as a model are described in Table 1.

Table 1 Principal Dimension

Principal Dimension		
Length Overall	4,356	m
Bow length	1,016	m
PMB length	2,229	m
Aft length	1,111	m
Diameter	0,508	m
Wetted Surface Area	5,989	m^2
CG from bow	2,009	m

2.3. Computational Fluid Dynamics (CFD)

Computational Fluid Dynamics (CFD) is a field in science that, with the assistance of computers, provides quantitative predictions of fluid flow based on conservation laws (conservation of mass, momentum, and energy). CFD predictions refer to the defined flow geometry, physical properties, as well as boundary and initial conditions. The CFD computation principle begins by dividing the fluid domain into several integrated elements. Each element is governed by an equation using numerical calculations, which then produce results such as forces acting on the model or other information that describes the model's condition under specific boundary conditions. These boundary conditions are required as input. This principle is commonly employed in computational calculations using computer assistance. Another example of the application of this principle is Finite Element Analysis (FEA), which is used to calculate stress occurring in solid objects. (Kundu & Cohen, 2012).

2.4. Variation Determination

The test parameters in the CFD analysis of this study include control variables, constant variables, and free variables. The control variables consist of the operational speed and depth of the ROV, the constant variables include the speed and angles of the fluid inlet and outlet, while the free variables include the characteristics of the fluid flow passing through the ROV. The model was tested at 6 Speed Variations on Forward-Backward movement and 6 Speed Variations on Left-Right movement. As shown in Table 2.

Table 2 Variation Determination

No.	Forward - Backward	Right - Left
	Speed (m/s)	Speed (m/s)
1	0,3	0,3
2	0,4	0,4
3	0,55	0,55
4	0,6	0,6
5	0,7	0,7
6	1	1

2.5. Computational Domain, Boundary Conditions and Meshing

The simulation domain is established to represent the flow in the area around the model and is solved using numerical methods. Choosing the simulation domain is vital in demonstrating how the model's shape impacts the flow passing through it. The domain setup follows the guidelines recommended by the ITTC regarding the placement of the model in relation to the domain boundaries. This is important to ensure that the boundaries of the domain do not affect the flow around the model. The minimum distance between the domain boundaries and the model should be at least 1-2 times the length between perpendiculars (Lpp) (ITTC, 2011).

As seen in Figure 3, the computational domain is shaped like a box, with the boundaries located as follows. As seen in Figure 3, the computational domain is shaped like a box, with the boundaries located as follows. The outflow is situated 4000 downstream from the vessel, and the entry is situated 2000 upstream. The side walls are about 2000 feet away from the ship. The top wall is 2000 feet above the vessel, while the bottom wall is 2000 feet below it. In order to adequately capture the contour of the wave pattern behind the vessel, the domain length was made rather long (4000).

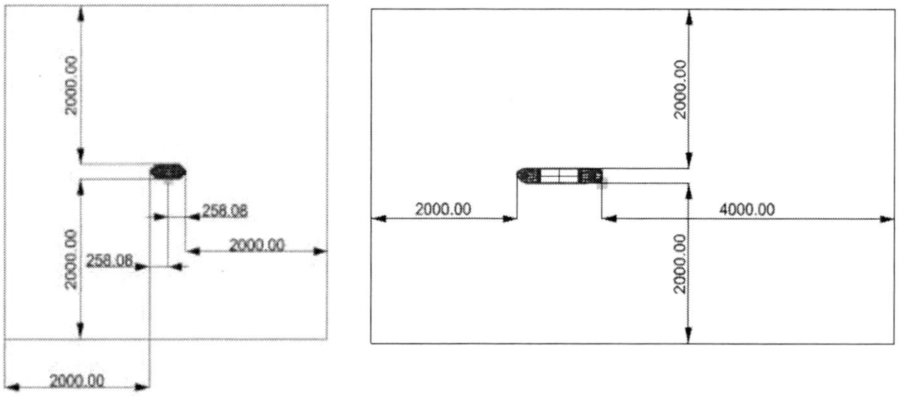

Figure 2 Ship Domain

Accurately depicting the fluid flow along the ship hull that is, the boundary layer flow that influences the vessel's drag is crucial. As a result, the mesh close to the hull was improved. In order to capture the free surface effects that is, the waves produced by the movement of the vessel on the water's surface the mesh close to the free surface was further improved. A side view of the computational mesh, using a multi-block structured grid, is displayed in Figure 4, with mesh improvements in the areas close to the hull and the free surface.

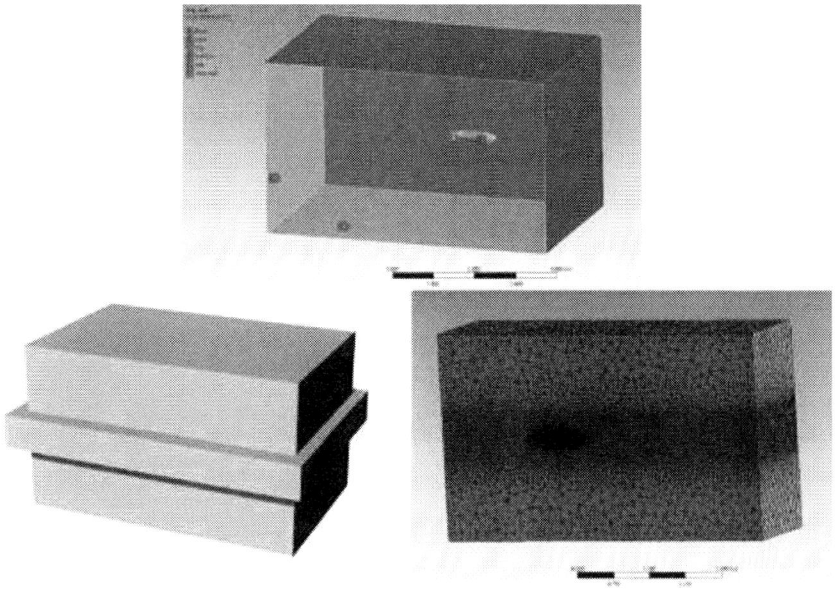

Figure 3 3D Side view of the computational mesh

2.6. Grid Independence Tests

To ascertain the ideal number of cells to be employed in the final setup of the CFD simulations and to confirm the convergence of the numerical computations, grid independence tests were conducted. The percentage rise or decrease in total resistance derived from these two consecutive simulations was calculated. In these experiments, the number of cells in a succeeding simulation was raised roughly 1.5 times until twice from the previous one. Table 3 tabulates the results, and plotted in Figure 5.

Table 3 Table Ship Resistance

Elements Qty.	Max Pressure (Pa)	Deviation	Ship Resistance (N)	Deviation
222855	0,05397		0,002013891	
568157	0,05401	0,07%	0,001969126	-2,22%
1163064	0,05294	-1,98%	0,001889085	-4,06%
2363086	0,05379	1,61%	0,001858333	-1,63%

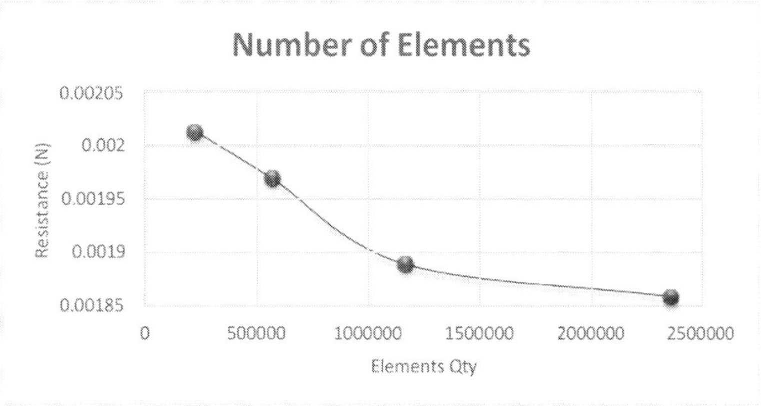

Figure 4 Result of Grid Independence Tests

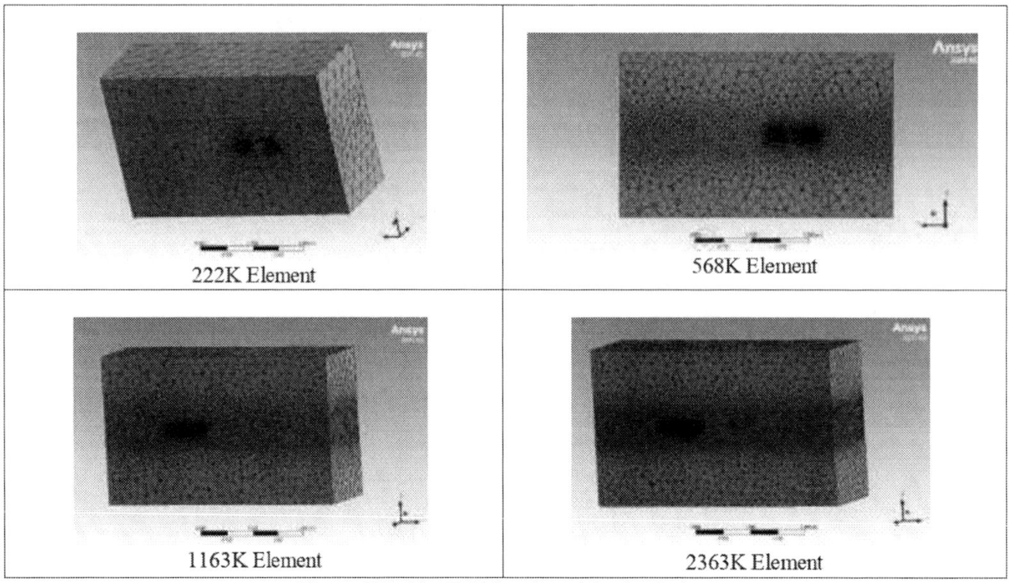

Figure 5 Meshing Trial Results at 0.3 m/s velocity.

From Table 3, the Deviation result with the most stable value is obtained, which is in the third row of the table or written in Table 4.

3. Results and Discussion

When the Deviation has been found, the next step is to look for Resistance in each Variation that has been determined. The table below is the result of Resistance in the Forward-Backward Variation. The results of the variation experiments carried out with the CFD method are Resistance which continues to grow along with the increasing speed on the ship both on the ship with the forward state and the ship in reverse state, as seen in Figure 6 and Table 4. When the ship is traveling at a speed of 0.3 m/s, the resistance is 0.0019 N until the ship increases the speed to 1.0 m/s, the resistance is 0.0190 N.

Table 4 Result of Resistance in the Forward-Backward

No	Speed (m/s)	Resistance (N)	
		Front - Back	**Back - Front**
1	0,30	0,001858333	0,001872720
2	0,40	0,003233097	0,003251852
3	0,55	0,005967690	0,006010472
4	0,60	0,007057926	0,007106079
5	0,70	0,009468552	0,009663272
6	1,00	0,019024893	0,019099682

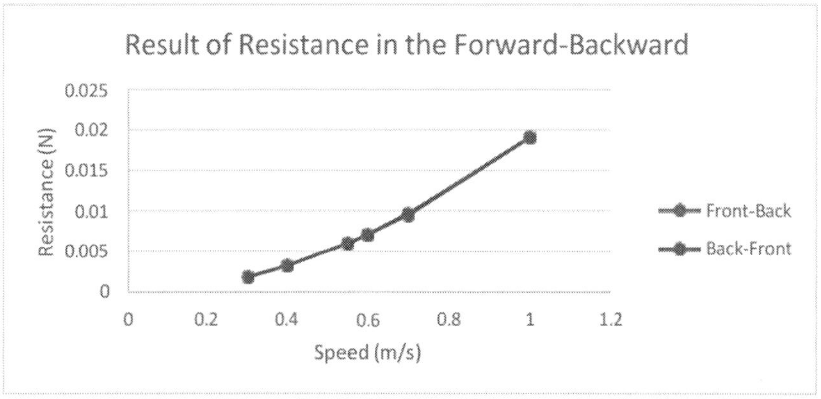

Figure 6 Diagram Result of Resistance in the Forward-Backward

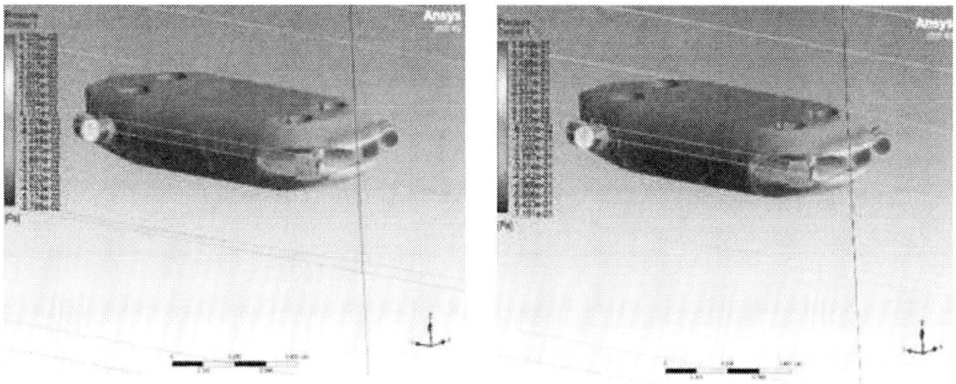

Figure 7 (left: Ship condition at 0.3 m/s), (right: Ship condition at 1.0 m/s)

After conducting resistance tests when the ship goes back and forth, the next is testing the ship's resistance to variations in ship speed when the ship moves to the right and left. The result is that the ship moving to the side makes the resistance greater than the movement of the ship forward and backward, due to several factors, namely the wider surface volume of the ship is one of the causes of greater ship resistance. The resistance of the ship moving sideways from right to left at a speed of 0.3 m/s is 0.004898206 N, and the resistance continues to increase when the speed is added up to 1 m/s with a resistance of 0.048133036 N. These results are not much different when the ship moves in the opposite direction, namely from left to right as shown in Table 5.

Table 5 Result of Resistance of the Ship Moving Sideways from Right to Left

No	Speed (m/s)	Resistance (N)	
		Right – Left	Left - Right
1	0,30	0,004898206	0,004883742
2	0,40	0,008398824	0,008385181
3	0,55	0,015313802	0,015321655
4	0,60	0,018043738	0,018073854
5	0,70	0,024251713	0,024234513
6	1,00	0,048133036	0,047969385

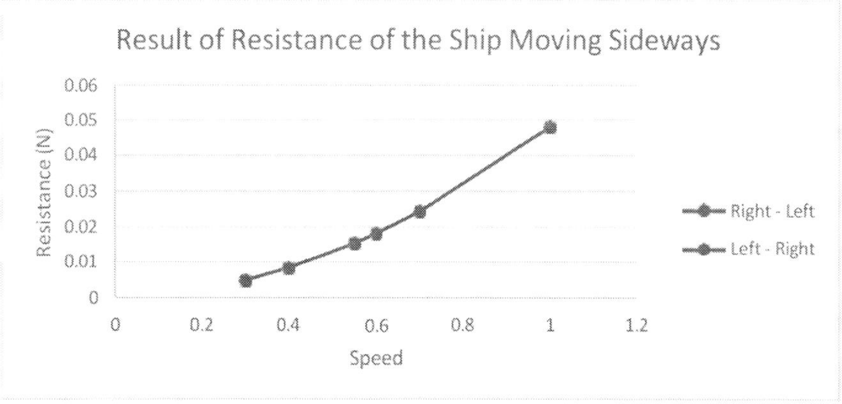

Figure 8 Diagram Result of Resistance of the Ship Moving Sideways from Right to Left.

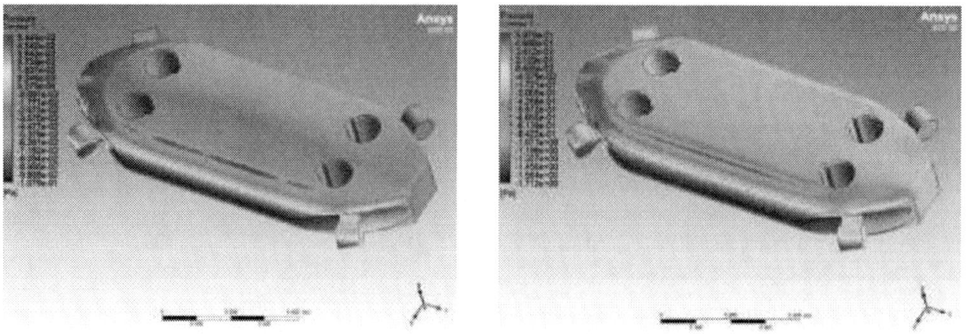

Figure 9 (left: Ship condition at 0.3 m/s), (right: Ship condition at 1.0 m/s)

4. Conclusion

The grid independence tests demonstrated that numerical convergence was achieved when the mesh density was sufficiently refined. Based on the results, the optimal mesh resolution was identified at 1,163,064 elements, where the deviation in ship resistance was minimal, ensuring a balance between computational efficiency and accuracy. In the resistance tests, it was observed that ship resistance increased with speed, consistent with theoretical expectations. For forward and backward movement, resistance ranged from 0.0019 N at 0.3 m/s to 0.0190 N at 1.0 m/s. Additionally, when the ship moved sideways, the resistance was significantly higher due to the increased surface area exposed to the fluid, starting at 0.0049 N at 0.3 m/s and reaching 0.0481 N at 1.0 m/s. The resistance on the ROV increases as the speed of the ROV increases, therefore it is necessary to conduct a study on the service speed of the ROV SN-01 because it can affect the selection of the engine used so that the ROV can work effectively and efficiently. These findings highlight the impact of motion direction and speed on hydrodynamic resistance, providing valuable insights for ship design and operational efficiency. These results can inform future analyses and improvements in maritime engineering.

References

[1] Aryawan, W.D., Utama, I.K.A.P., Hermawan, Y.A., & Sutiyo 2023. CFD Analysis into the Resistance Characteristics of Remotely Operated Vehicles when Submerges Under Water and Sails on the Surface. *CFD Letters*. 15(8). 166–178.

[2] ITTC. (2011). Practical Guidelines for Ship CFD Applications. *ITTC – Recommended Procedures and Guidelines ITTC*, 1–8.

[3] ITTC. (2021a). ITTC-Recommended Procedures and Guidelines Uncertainty Analysis in CFD Verification and Validation Methodology and Procedures ITTC Quality System Manual Recommended Procedures and Guidelines Procedure Uncertainty Analysis in CFD Verification and Validatio. *ITTC - Recommended Procedures and Guidelines*.

[4] ITTC. (2021b). ITTC - Recommended Procedures and Guidelines: Captive Model Test. *International Towing Tank Conference*, 1–20.

[5] Ramadhani, K. R. (2023). *Studi dan Rancang Bangun Unmanned Underwater Vehicle (UUV)Untuk Perairan Indonesia*. Institut Teknologi Sepuluh Nopember.

[6] Sahir, A. A., Munazid, A., & Suwasono, B. (2017). *Perancangan Kapal Selam Tanpa Awak (Unmanned Underwater Vehicle / Uuv) Sebagai Sarana Observasi Bawah Laut*. 72–83.

[7] Suastika, K 2023 *Dinamika Fluida Komputasi : Metode Volume* Hingga (Surabaya: ITS Press)

[8] Versteeg, H., & Malalasekera, W 1955 *An Introduction to Computational Fluid Dynamics* (England: Longman Scientific & Technical)

Smoothed Particle Hydrodynamics Simulations of Wave Run-Up over Hexaloc Armour on Rubble-Mound Breakwaters

Muhammad Brilliant Danu Ghazali Hafiz[1*], *Haryo Dwito Armono*[1], and *Muhammad Zikra*[1]

[1]Ocean Engineering, Faculty of Marine Technology, Sepuluh Nopember Institute of Technology, Surabaya, Indonesia

*Corresponding author: brilliantdanu06@gmail.com

Abstract. Indonesian beaches are increasingly vulnerable to abrasion and erosion, requiring reliable protection such as breakwaters. Since crest freeboard determines the level of safety, wave run-up predictions must be accurate. This study evaluates run-up on a 1:2 rubble-mound breakwater slope armored with Hexaloc using SPH (DualSPHysics) to improve design accuracy and efficiency. The geometry was constructed in AutoCAD and SketchUp. The simulation covered wave heights of 0.05–0.13 m, periods of 1.1–1.5 s, and single-layer and double-layer armor configurations. The numerical results were validated against analytical–empirical formulations, particularly the Ahrens relation. Ru/H increases with rising Iribarren number and then tends to plateau at high values. SPH produces Ru/H 0.538–1.076 (single-layer) and 0.588–1.216 (double-layer), while theoretical predictions are higher: 1.616–2.299 and 1.59–2.30. In the test range, single- and double-layer Hexaloc produced comparable run-ups; two layers were equivalent or slightly higher. The SPH–Ahrens fit was highly linear (R^2 = 0.9999 single-layer; 0.9948 double-layer), but SPH still predicted lower Ru/H than Ahrens.

1 Introduction

Environmental conditions in coastal areas are influenced by two main factors, namely human activities in coastal areas and natural processes, such as coastal erosion caused by ocean waves. Coastal changes are controlled by sediment transport, longshore currents, wave dynamics, and land-use practices in the surrounding coastal area. When sedimentation occurs, the shoreline shifts seaward, whereas abrasion drives the shoreline landward . These two processes are inherently connected, since erosion at one site may result in accretion at another. Coastal abrasion poses a major problem for Indonesia's shore regions, largely driven by wave forces that gradually wear away the shoreline and push it landward. This situation necessitates the installation of coastal protection structures to ensure shoreline stability and minimize the effects of wave action. To address erosion issues, coastal protection structures such as breakwaters are required. Breakwaters are primarily constructed to attenuate wave energy and provide calmer waters, particularly for harbor protection. For ports, breakwaters are designed to reduce wave agitation within the harbour basin, while also potentially influencing local sedimentation patterns near the entrance channel. In addition, detached/segmented breakwaters have been implemented as coastal defense measures that can modify the nearshore wave climate and consequently influence sediment transport patterns, potentially mitigating shoreline erosion and affecting local accretion/erosion. These structures are typically constructed from stacked concrete units with specific geometric forms, placed on the outermost layer to withstand the highest wave loads. Consequently, their shape and dimensions must be carefully designed to ensure structural stability. Breakwaters may be constructed as extensions of the shoreline to enhance port safety, or positioned offshore as detached structures to help mitigate coastal erosion. Their stability and performance largely depend on how they respond to wave interactions, particularly under extreme wave conditions.

In breakwater design, estimating the wave run-up height is a key parameter because it directly influences structural stability and the safety of the area sheltered behind the structure. Reliable run-up prediction supports a more efficient and robust design, reducing potential risks to both the breakwater and the adjacent coastline. Wave run-up refers to the vertical extension of water level along the surface of a coastal defense structure when impacted by waves [1]. The safety of a breakwater is governed by its crest elevation, which must exceed the predicted run-up height as well as the allowable level of wave overtopping. Wave run-up is defined as the vertical distance between the Still Water Level (SWL) and the highest point on the slope reached by the uprushing water. Wave run-up is governed by the incident wave direction, the geometry of the structure, and the roughness of the slope. This run-up occurs due to momentum transfer when waves impact the breakwater structure.

Numerous physical model studies on wave run-up have been carried out by various researchers at laboratory scale. Irribaren developed a formula to estimate run-up height on armor layers composed of different types of materials. Van der Meer and Stam investigated wave run-up on armor units made of fine and coarse stones, concluding that run-up is influenced by the significant wave height (Hs), structural slope angle (α), structural roughness, and wave period (Tp) [2]. Melito and Melby conducted similar experiments using core-loc armor units, demonstrating that the combined effects of high porosity and friction can reduce run-up height [3]. Schüttrumpf and Van Gent are conducted laboratory test. Laboratory experiments on a 1:6 slope showed that the run-up velocity exceeded the 2% exceedance level $\left(u_{2\%}\right)$ with values of 1.39 for the natural wave spectrum and 1.55 for the Texel Marsen Arsloe (TMA) spectrum. Furthermore, the water layer thickness flowing over the crest of the structure at the 2% exceedance condition $\left(c_{h,2\%}\right)$ was approximately 0.33, highlighting the strong influence of turbulence and aeration on the run-up and overtopping behavior of the breakwater [4].

Pang et al. subsequently carried out a comprehensive review of climate-change-driven coastal erosion using more than 200 global case studies. Their analysis suggests that a projected sea-level rise of 0.3–1.0 m by the end of the century could increase land loss by 20–40% in low-lying deltaic regions [5]. They also evaluated 95 case studies focusing on hydrodynamics and shoreline erosion worldwide, revealing that short-period waves (5–8 s) are capable of generating substantial sediment transport, leading to coastal retreat exceeding 5 m/year in highly energetic environments. The study highlights a methodological comparison: field measurements yield realistic data but are expensive and spatially limited, whereas numerical modeling offers greater flexibility but depends heavily on robust validation. Overall, the findings emphasize the need to integrate numerical simulations with field observations in coastal morphodynamic research.

Regarding coastal protection, geometric design criteria for berm breakwaters have been established based on a series of laboratory experiments. These tests yielded empirical formulas for assessing the stability of armour layers and indicated that berm breakwaters with milder slopes offer greater resistance to high wave run-up. Nonetheless, the findings are constrained by the laboratory scale, so site-specific calibration is still required before field implementation. Subsequently, numerical investigations were carried out using the Smoothed Particle Hydrodynamics (SPH) approach. DualSPHysics was employed to simulate wave overtopping on rubble-mound breakwaters, and the results showed that the SPH model can reproduce overtopping discharges with discrepancies of less than 15% relative to experimental data, especially under high-wave conditions and for permeable structures [6]. A key drawback, however, is the substantial computational cost when applied to large spatial domains. Wave run-up on breakwaters has also been analysed using a CFD-based RANS model. The principal outcome of this study was a validation curve comparing numerical predictions with laboratory measurements, revealing an average error of less than 10% for run-up heights on structures with a 1:2 slope. The results confirm the strength of RANS in representing turbulent processes, while also highlighting its limitations in accurately resolving extreme overtopping events with large runoff volumes.

The main aim of this study is to investigate wave run-up on sloped breakwaters using numerical modelling. The breakwater geometry is generated in SketchUp and simulated in DualSPHysics to examine hydrodynamic behaviour under varying wave and structural configurations. Key parameters analysed include wave height, wave period, water level conditions, and the slope angle of the breakwater. The outcomes of this work are expected to enhance understanding of how these wave and structural factors influence breakwater performance. Practically, the insights gained will offer coastal engineers and planners more reliable guidance for designing breakwaters that maintain stability and safety while keeping construction costs efficient. Moreover, by comparing the results with findings from earlier studies, this research underscores both the strengths and the limitations of SPH-based modelling for analysing coastal structures.

2 Methods

This research examines wave run-up on rubble-mound coastal structures equipped with Hexaloc armour units under irregular wave conditions. The study utilizes data gathered from literature reviews and digital geometric models developed in AutoCAD and SketchUp, which were subsequently imported into the DualSPHysics framework for numerical simulation.

2.1 Hexaloc Armour Unit

In coastal defense systems, one of the key elements is the armour rock layer. This outer layer serves to dissipate wave energy before it reaches the shore. These protective rocks, commonly referred to as armor units, are available in various forms such as dolos, tetrapods, A-jacks, and others. Typically cast from concrete, armor units are designed with specific geometries that promote stronger interlocking compared to conventional rock, thereby improving the overall stability of the structure.

Several countries have developed artificial armor units as substitutes for natural stones in coastal protection systems, featuring diverse shapes and stability characteristics. Alongside Europe and the United States, Indonesia has introduced its own artificial armor unit known as Hexaloc. This unit resembles the A-Jack and hexapod forms. The hexapod was first introduced in the United States in 1959, while the A-Jacks were developed later in 1998. Both units feature legs with tapered circular profiles. The A-Jacks are created by combining two identical components into one structure, resulting in six legs extending from a central filleted hub, each with a square cross-section. In contrast, the Hexaloc has six legs with a hexagonal cross-section that meet at a single joint without a filleted center [7]. Illustrations of these units are provided in Figure 1 and Table 1.

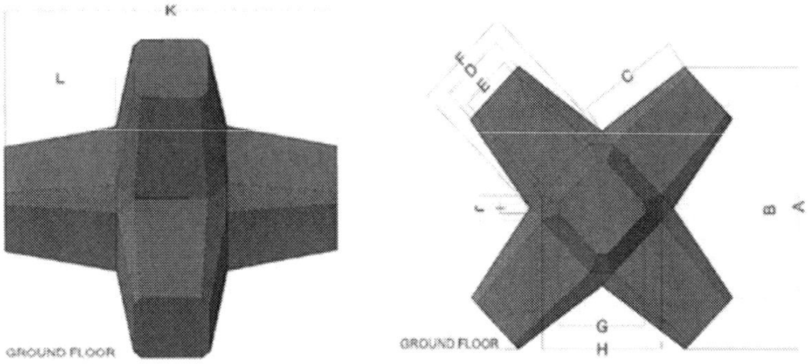

Fig. 1. Hexaloc

Table 1. Hexaloc Dimension

Symbol	A	B	C	D	E	F	G	H	I	J	K	L
Length (mm)	2106	1340	823	542	429	812	687	961	79	191	2437	808

2.2 Run-up and Irribaren Number

When waves strike a structure, their momentum drives the water to travel up along the surface of the structure. The maximum vertical elevation reached by this uprush is termed the wave run-up. This parameter is critical in the design of coastal protection works because it governs the required crest elevation of the structure. Wave run-up is controlled by several factors, including the geometry and slope of the structure, surface roughness, local water depth at the toe, and the incident wave characteristics. Run-up is commonly expressed in two forms: the mean run-up, defined as the average uprush height of all waves, and the 2% run-up, defined as the average of the highest 2% of run-up values from a sample of 100 waves [8]. The type of breaking that occurs can be classified using the Iribarren number (ξ), also known as the surf similarity parameter. Below are the equations used to calculate wave run-up.

$$\xi = \frac{tantan\, \alpha}{\sqrt{\frac{2\pi H_s}{gT^2}}} \quad (1)$$

where, ξ is the Irribaren number, α is the structure slope angle, H_s is the significant wave height, g is Earth's gravity, which has a value of 9.8 m/s^2, and T is the wave period.

Several studies have introduced empirical formulas for predicting wave run-up. Based on laboratory tests involving irregular waves on smooth, impermeable slopes, Mase developed an empirical relationship for estimating maximum run-up, valid for slope gradients between 1/30 and 1/5 and wave steepness values ranging from 0.007 to 0.07 [9]. The following formula is derived from the study.

$$\frac{R_{max}}{H_0} = 2.2\xi_0^{0.77} \tag{2}$$

Douglass examined wave run-up by eliminating the influence of beach slope from the governing equation. He reasoned that determining the slope is challenging and has only a minor effect on run-up, allowing the tan θ term in the Iribarren number to be substituted with a constant value [10]. The equation for calculating wave run-up is as follows:

$$\frac{R_{max}}{H_0} = \frac{0.12}{\sqrt{\frac{H_0}{L_0}}} \tag{3}$$

$$L_0 = \frac{gT^2}{2\pi} \tag{4}$$

where, R_{max} is the run-up maximum, L_0 is wave length, and H_0 is wave height.

Ahrens and Heimbaugh subsequently formulated an upper-limit equation to estimate the maximum run-up of irregular waves on riprap revetments, which represent rough and permeable slopes. Their formulation was derived from laboratory experiments conducted at a 1:16 scale [11]. In their study, run-up was defined as the elevation reached by green water only (excluding spray), and the resulting prediction is expressed as:

$$\frac{R_{max}}{H_{mo}} = \frac{1.022\xi}{1+0.247\xi} \tag{5}$$

2.3 Smoothed Particle Hydrodynamic

Gomez-Gesteira et al. characterize Smoothed Particle Hydrodynamics (SPH) as a meshless Lagrangian technique that has been widely adopted in Computational Fluid Dynamics (CFD) [12]. In SPH, the fluid is represented by discrete particles that can interact with solid boundaries and naturally follow large deformations in domains with moving boundaries. Over the years, continuous improvements to the SPH formulation have led to satisfactory levels of accuracy, stability, and robustness for engineering applications. The fluid field is discretized into a cloud of numerical particles, making SPH well suited for modelling a broad spectrum of problems, such as astrophysical flows, free-surface dynamics, and complex mixing processes. Unlike conventional grid-based CFD approaches, SPH does not require a fixed mesh, which offers certain computational advantages. In this framework, continuous media are represented by ensembles of particles, and for fluid flow, the Navier–Stokes equations are evaluated at each particle position based on interactions with neighboring particles. These neighbors are identified through a kernel function with a specified smoothing length, defined over a circular support in two dimensions or a spherical support in three dimensions. At every timestep, the physical variables of all particles are updated, and the particles are advected according to their newly computed properties.

Fig. 2. The Difference Between Mesh and Particle Discretization of Continous Free-Surface Flow [13]

2.4 DualSPHysics

DualSPHysics is an open-source computational tool designed specifically for modeling free-surface flow processes. It utilizes the Smoothed Particle Hydrodynamics (SPH) approach and incorporates hardware acceleration to efficiently compute particle motion within the simulation domain. Developed for real engineering applications, DualSPHysics extends the capabilities of the SPHysics model and is written in C++ with CUDA optimization. The SPH methodology in DualSPHysics is organized into three main stages: Neighbour List (NL), Particle Interaction (PI), and System Update (SU) [14]. To generate numerical outputs, users must proceed through a structured sequence of steps provided within the software. Workflow diagram illustrated in Figure 3.

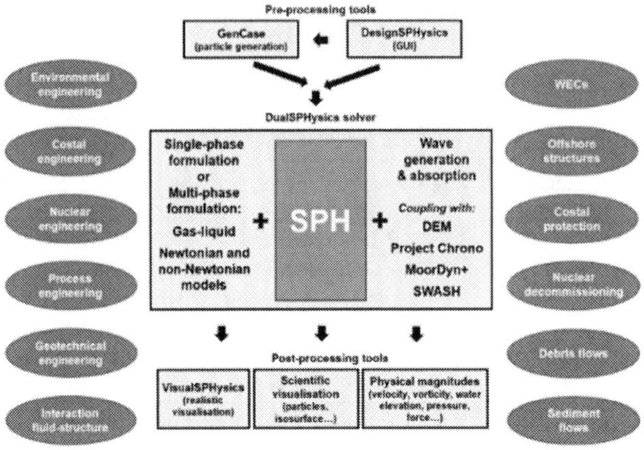

Fig. 3. Workflow of The DualSPHysics Program [14]

2.5 Data Analysis

In this study, the authors employed a breakwater with a length of 1.75 m, a height of 0.70 m, and a width of 2 m, placed at a water depth of 0.65 m. In this study, the breakwater slope was fixed at 1:2. The slope was not treated as a decision variable; instead, it was adopted as the baseline geometry of the investigated breakwater model. Accordingly, the analysis focuses on the effects of wave conditions and armour-unit arrangement on wave run-up for this fixed slope. Three variables were varied in the experiments, namely wave height, wave period, and the number of structural arrangements. Two wave gauges (WG1 and WG2) were used to monitor the free surface elevation. WG1 was located in the constant depth section downstream of the wave generator to record the generated wave conditions prior to the interaction with the structure. WG2 was positioned at the breakwater face to capture the near structure water surface evolution used for run-up estimation. Wave propagation was from left to right, from the wave generator toward the breakwater. For further details, see Figure 4 and Table 2.

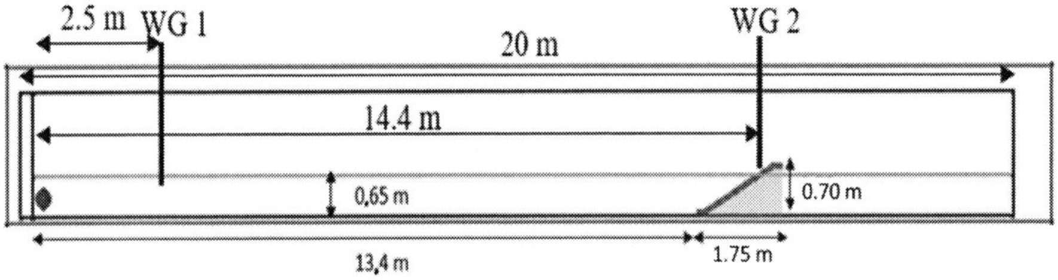

Fig. 4. Test Set-up in DualSPHysics

 https://doi.org/10.1051/bioconf/202621607001

Table 2. Test Scenario

Variations	Number Of Layers	Wave Height (m)	Wave Period (s)
1	1	0.05	1.5
2		0.07	1.2
3		0.09	1.2
4		0.11	1.1
5		0.13	1.1
6	2	0.05	1.5
7		0.07	1.2
8		0.09	1.2
9		0.11	1.1
10		0.13	1.1

In this study, calibration and validation were carried out by comparing numerical results for regular waves with analytical solutions derived from second-order wave theory. DualSPHysics applies the SPH method using discrete particles, where the distance particle (dp) can be tuned to control numerical resolution: a smaller dp increases accuracy but also raises computational demand. Considering hardware limitations, a value of dp = 0.03 m was adopted. Each simulation was run for 100 s, with wave gauges placed at x = 2.5 m and x = 14.4 m from the piston. The calibration and validation results, presented in Figure 5, show good correspondence with theory and support the use of this numerical wave tank for further simulations. The deviation between numerical and analytical results, quantified using the root-mean-square error (RMSE), is on the order of 0.07703.

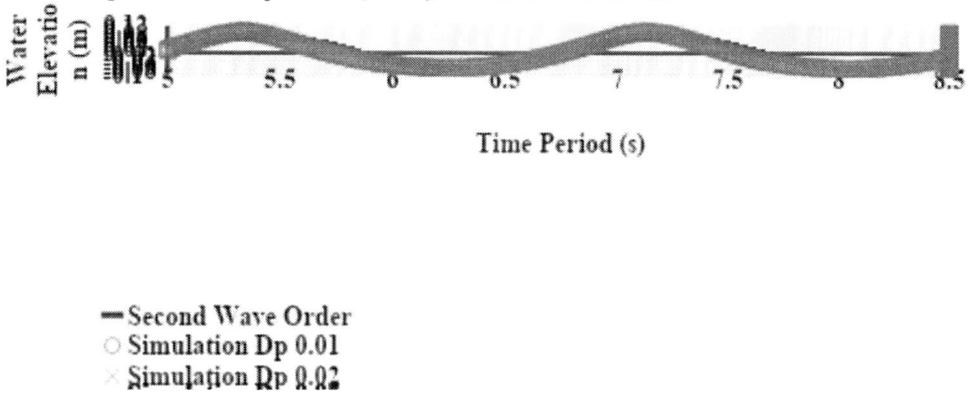

Fig. 5. DualSPHysics Model Validation Based on Second Wave Order Theory

3 Result and Discussion

3.1 Wave Run-up Based on Mase's Theory, Douglass's Theory, and Ahren's Theory

The evaluation of wave run-up on the non-overtopping breakwater in this study utilizes the formulations proposed by Mase, Douglass, and Ahrens, as presented in Eq. (2), Eq.(3), and Eq. (5). These models are selected because they align most closely with the conditions examined in this research. The resulting calculations serve as the foundation for validating wave run-up predictions generated using the Smoothed Particle Hydrodynamics (SPH) method. An example computation is included, beginning with the estimation of the wavelength (L) using Eq. (4) and the Iribarren number using Eq. (1). Table 3 presents the computed wave run-up values for each variation tested on the breakwater structure.

Table 3. Wave Run-up Height (m) Based on Theory

Variation	Wave Height, H (m)	Wave Height Significant, Hs (m)	Wave Period, T (s)	Wave Length, L (m)	Iribarren Number, ξ	*Ru*/H		
						Mase (1989)	Douglass (1992)	Ahrens (1988)
1	0.05	0.061	1.5	3.514	5.061	7.668	0.916	2.299
2	0.07	0.070	1.2	2.249	3.781	6.126	0.680	1.998
3	0.09	0.088	1.2	2.249	3.378	5.616	0.607	1.882
4	0.11	0.106	1.1	1.890	2.811	4.876	0.505	1.696
5	0.13	0.125	1.1	1.890	2.593	4.582	0.466	1.616
6	0.05	0.061	1.5	3.514	5.067	7.675	0.911	2.300
7	0.07	0.074	1.2	2.249	3.683	6.003	0.662	1.971
8	0.09	0.090	1.2	2.249	3.330	5.555	0.599	1.868
9	0.11	0.112	1.1	1.890	2.745	4.788	0.493	1.672
10	0.13	0.132	1.1	1.890	2.528	4.494	0.454	1.591

3.2 Wave Run-up Based on SPH Method

This research utilizes a numerical model based on the Smoothed Particle Hydrodynamics (SPH) method as implemented in DualSPHysics. As an example, a simulation output of wave run-up height is presented. The run-up is quantified as the maximum vertical position attained by free-surface particles when incident waves interact with the breakwater. The particle coordinates in the x, y, and z directions are illustrated in Figure 6. Consistent with the standard definition of wave run-up, the reported values are referenced to the still-water level, obtained by subtracting the initial water depth from the maximum free-surface elevation. The corresponding run-up values for each tested configuration of the breakwater are summarized in Table 4.

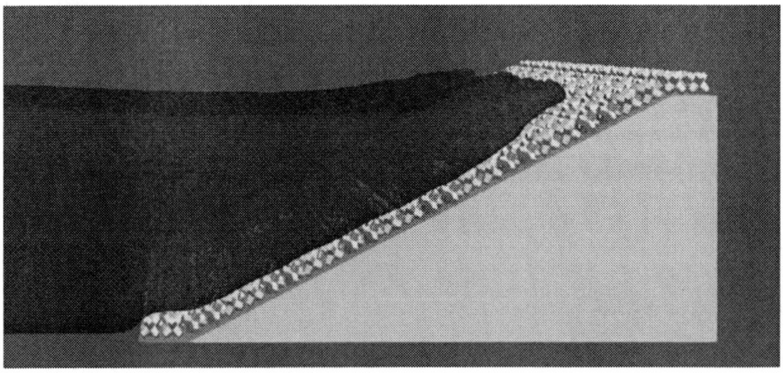

Fig. 6. Wave Run-up Height on Breakwater

Table 4. Wave Run-up Height (m) Based on SPH Method

Variation	Wave Height, H (m)	Wave Height Significant, Hs (m)	Wave Period, T (s)	Wave Length, L (m)	Iribarren Number, ξ	Wave Run-up, Ru (m)	*Ru*/H
1	0.05	0.061	1.5	3.514	5.061	0.066	1.076
2	0.07	0.070	1.2	2.249	3.781	0.062	0.891
3	0.09	0.088	1.2	2.249	3.378	0.069	0.783
4	0.11	0.106	1.1	1.890	2.811	0.066	0.618
5	0.13	0.125	1.1	1.890	2.593	0.067	0.538
6	0.05	0.061	1.5	3.514	5.067	0.074	1.216
7	0.07	0.074	1.2	2.249	3.683	0.068	0.925
8	0.09	0.090	1.2	2.249	3.330	0.079	0.872
9	0.11	0.112	1.1	1.890	2.745	0.078	0.696
10	0.13	0.132	1.1	1.890	2.528	0.077	0.588

3.3 Validation Wave Run-up Based on Previous Theory

In this section, the numerical results produced by the Smoothed Particle Hydrodynamics (SPH) model in DualSPHysics are compared with findings from earlier studies for validation purposes. Figure 7 presents the computed wave run-up (Ru/H) plotted against the Iribarren number, alongside data reported in previous research. In this plot, the green curve representing Ahrens' riprap experiments is used as the primary benchmark to assess the accuracy and consistency of the SPH-based run-up predictions in the present study.

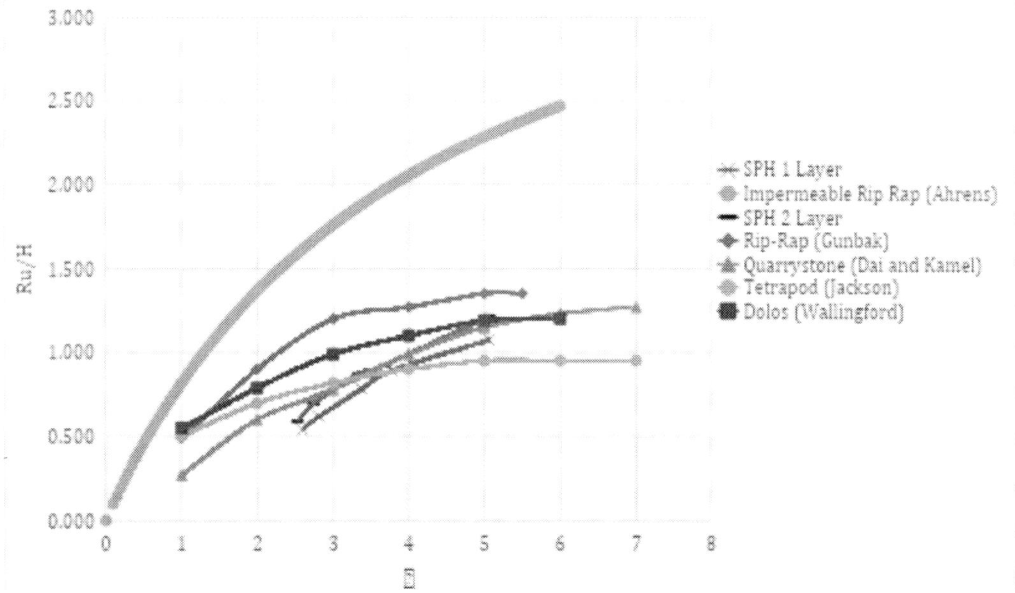

Fig. 7. The Influence of the Iribarren Number on the Wave Run-up

The graph illustrates the relationship between the non-dimensional wave run-up (Ru/H) and the Iribarren number (ξ) for various types of armour layers, together with numerical predictions from the SPH simulations for one-layer and two-layer Hexaloc configurations. Overall, all curves exhibit an increasing trend with ξ, consistent with classical run-up theory for rough and porous slopes. The Ahrens curve for impermeable riprap lies significantly above all other datasets, indicating that rough yet non-porous slopes tend to produce higher maximum run-up due to reduced energy dissipation. In contrast, concrete armour units such as tetrapods (Jackson) and dolos (Wallingford) display lower and more stable Ru/H values for $\xi > 4$, reflecting the enhanced dissipation caused by interlocking and higher porosity [15].

The SPH results both the single and double layer configurations generally fall below the riprap datasets of Gunbak and the quarry stone data of Dai & Kamel, suggesting that the Hexaloc armour in SPH simulations dissipates wave energy more effectively than natural stone armours. The differences between the one layer and two layer SPH results are minor, implying that adding a second layer does not significantly increase run-up, likely because internal permeability remains high enough to dissipate wave energy before reaching the upper slope. Overall, the graph demonstrates that SPH predictions fall within a physically consistent range compared with existing experimental studies. However, the systematic underestimation relative to the Ahrens upper bound formulation at high ξ values is notable and reflects known SPH behaviour, particularly the tendency for numerical over damping and sensitivity to particle resolution.

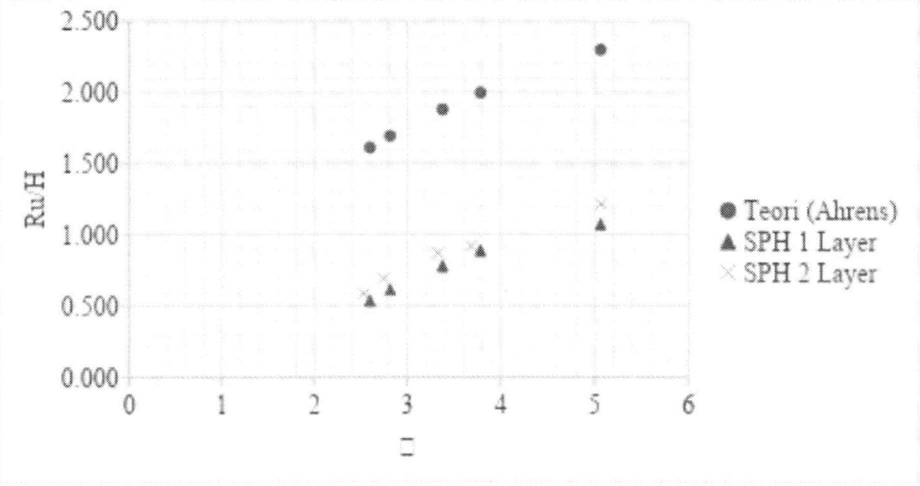

Fig. 8. The Relationship Between Wave Run-up Characteristic and The Iribarren Number

This graph compares the non-dimensional wave run-up (Ru/H) obtained from the SPH simulations with Ahrens' theoretical upper-bound curve for riprap revetments. The Ahrens data (blue markers) consistently exhibit higher run-up values across all Iribarren numbers (ξ), representing the behaviour of rough, low-permeability stone slopes. In contrast, the SPH results for both the one-layer and two-layer Hexaloc configurations are noticeably lower, although they follow a similar upward trend as ξ increases. This indicates that the numerical model successfully reproduces the fundamental dependence of run-up on the surf similarity parameter.

The separation between the SPH curves and the Ahrens reference highlights the stronger energy dissipation associated with Hexaloc armour, which has greater porosity and interlocking compared to conventional riprap. Moreover, the small difference between the one-layer and two-layer simulations suggests that adding an additional armour layer does not significantly alter the run-up response, likely because the overall permeability and dissipation characteristics of the structure remain similar. Overall, the figure demonstrates that the SPH model captures the qualitative behaviour of wave run-up while reflecting the unique hydrodynamic performance of Hexaloc armour relative to classical riprap systems.

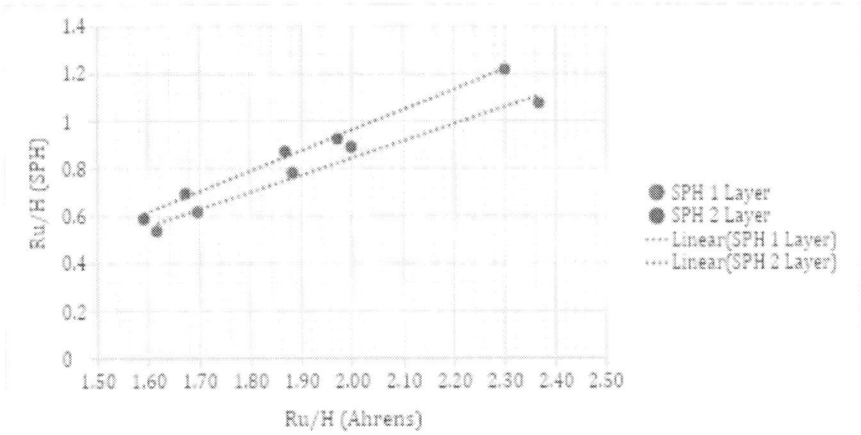

Fig. 9. Run-up Relationship Between SPH and Ahrens Theory

Figure 9 compares SPH-predicted run-up (Ru/H) with the theoretical values from Ahrens. Both the single-layer and double-layer Hexaloc configurations show a strong linear correlation with the Ahrens curve ($R^2 > 0.97$), indicating that the SPH model reproduces the same trend. However, the regression lines lie below the 1:1 line, confirming that SPH systematically predicts lower run-up than the Ahrens upper bound, with the two-layer configuration yielding slightly higher values than the single layer.

4 Conclusion

The comparison between SPH simulations and Ahrens' upper-bound formulation shows that the DualSPHysics model reproduces the expected increase of non-dimensional run-up (Ru/H) with the Iribarren number. Regression analysis between Ru/H(SPH) and Ru/H(Ahrens) yields very high coefficients of determination ($R^2 > 0.97$) for both single- and double-layer Hexaloc configurations, indicating that the SPH results are strongly correlated with the classical riprap run-up theory and capture its overall trend.

Despite this strong correlation, the SPH-predicted run-up values are consistently lower than Ahrens' theoretical curve, reflecting the higher porosity and energy dissipation capacity of Hexaloc armour compared with the impermeable riprap used in the reference formulation. The two-layer Hexaloc configuration produces slightly higher run-up than the single-layer case, but the difference is modest, suggesting that the additional layer does not fundamentally change the hydraulic response because the overall permeability of the armour system remains high.

These findings imply that SPH-based modelling with DualSPHysics is suitable for analysing relative trends in wave run-up and for comparing alternative armour configurations. However, for use in design practice, the systematic offset from Ahrens' upper bound indicates that case-specific calibration against laboratory or field data is still required to obtain conservative crest elevations for rubble-mound structures armoured with Hexaloc units.

References

[1] B. Triatmodjo, Teknik Pantai. Yogyakarta, Indonesia: Beta Offset, 1999.

[2] J. W. van der Meer and C. M. Stam, "Wave run-up on smooth and rock slopes of coastal structures," Ocean Engineering, 1992.

[3] I. Melito and J. A. Melby, "Wave runup, transmission, and reflection for structures armored with CORE LOC®," Coastal Engineering, vol. 45, no. 1, pp. 33–52, 2002, doi: 10.1016/S0378-3839(01)00044-8.

[4] H. Schüttrumpf and M. R. A. Van Gent, "Wave overtopping at seadikes," Coastal Structures 2003 - Proceedings of the Conference, vol. 40733, no. June, pp. 431–443, 2003, doi: 10.1061/40733(147)36.

[5] T. Pang, X. Wang, R. A. Nawaz, G. Keefe, and T. Adekanmbi, "Coastal erosion and climate change: A review on coastal-change process and modeling," Ambio, vol. 52, no. 12, pp. 2034–2052, 2023, doi: 10.1007/s13280 023-01901-9.

[6] A. J. C. Crespo et al., "DualSPHysics: Open-source parallel CFD solver based on Smoothed Particle Hydrodynamics (SPH)," Comput Phys Commun, vol. 187, pp. 204–216, Feb. 2015, doi: 10.1016/j.cpc.2014.10.004.

[7] C. Irawan and H. D. Armono, "Numerical Study of the of Proposed Hexaloc Concrete Armor Units under Static Loading," IOP Conf Ser Earth Environ Sci, vol. 1250, no. 1, 2023, doi: 10.1088/1755 1315/1250/1/012005.

[8] J. P. Ahrens, "Irregular wave runup on smooth slopes.," U.S. Army Coastal Engineering Research Center, Coastal Engineering Technical Aid, vol. 81–17, 1981.

[9] H. Mase, "Random Wave Runup Height on Gentle Slope," J Waterw Port Coast Ocean Eng, vol. 115, no. 5, pp. 649–661, 1989, doi: 10.1061/(asce)0733-950x(1989)115:5(649).

[10] S. L. Douglass, "Estimating extreme values of run-up on beaches," Journal of Waterway, Port, Coastal, and Ocean Engineering, vol. 118, no. 2, pp. 220–224, 1992.

[11] J. P. Ahrens, "Approximate upper limit of irregular wave runup on riprap," Journal of Waterway, Port, Coastal, and Ocean Engineering, vol. 114, no. 4, pp. 614–635, 1988.

[12] G. Argente, M. E. Gómez-Martín, and J. R. Medina, "Hydraulic stability of the armor layer of overtopped breakwaters," J Mar Sci Eng, vol. 6, no. 4, pp. 1–13, 2018, doi: 10.3390/jmse6040143.

[13] Ferryanto, O. Setyandito, Nizam, N. Yuwono, and W. Trihadi, "Wave run-up characteristic on armour layer breakwater using the smoothed particle hydrodynamics (SPH) method," in IOP Conference Series: Earth and Environmental Science, Institute of Physics, 2024. doi: 10.1088/1755-1315/1324/1/012031.

[14] J. M. Domínguez et al., "State-of-the-art SPH solver DualSPHysics: from fluid dynamics to multiphysics problems," pp. 1–43, 2021, doi: 10.1007/s40571-021-00404-2.

[15] M. A. Losada and L. A. Giménez-Curto, Mound Breakwaters Under Wave Attack. Santander, Spain: Dept. of Oceanographical and Ports Engineering, University of Santander, 1979.

Concept design of thin-steel purse seine fishing vessel for enhancing the safety of fishing operations

Andi Muhammad Almusadieq[1*], *Nazwa* Syafitri[1], *Reynard* Lisan[1], *Raden Sjarief* Widjaja[1]

[1] Department of Naval Architecture, Faculty of Marine Technology, Sepuluh Nopember Institute of Technology, Surabaya, Indonesia

* Corresponding author: 5018221117@student.its.ac.id

Abstract. Despite the continued use of traditional wooden fishing vessels by small-scale fishers in Indonesia, such vessels exhibit inherent limitations in durability, maintenance requirements, and seaworthiness under adverse marine conditions, which directly affect safety and operational efficiency. This study proposes a conceptual design of a 5–8 GT steel-hulled purse seiner fishing vessel constructed using thin steel plates to enhance safety, structural resilience, and sustainability. The research methodology involves the development of a conceptual vessel design with improved resistance to extreme weather conditions, while complying with applicable national and international standards, including classification rules and maritime safety regulations. The design process emphasizes overall vessel configuration, arrangement of main systems, and material selection to achieve efficient and robust construction. The use of thin steel plates is intended to support lightweight construction while maintaining structural strength and durability. The proposed design aims to reduce accident risks and improve operational safety at sea. The results of this study are expected to provide a practical reference for modern fishing vessel design and to support adoption by small- and medium-scale shipyards, contributing to improved safety and sustainability within Indonesian coastal fishing communities.

1 Introduction

Small-scale fishing vessels constitute the backbone of the global capture fisheries sector, particularly in archipelagic nations like Indonesia. However, despite their critical role in food security, this sector has historically suffered from a lack of formal engineering attention. Fyson [4] noted that the majority of fishing vessels under 30 meters have been neglected by naval architects, leaving their construction reliant on empirical methods rather than scientific design. This observation remains pertinent in the Indonesian context, where Susanto et al. [8] highlighted that domestic fishing vessels often evolve through artisanal tradition without rigorous stability evaluation. Consequently, this technical gap perpetuates the operation of fleets ill-equipped to withstand adverse marine conditions.

The continued reliance on timber for hull construction presents critical challenges regarding environmental sustainability and operational resilience. While wooden vessels have served traditional fleets for generations, they exhibit significant limitations in durability compared to modern materials. Bornmalm and Lagerqvist [2] emphasized that the historical shift from rustic boats to steel trawlers was pivotal in enhancing seaworthiness. From a safety perspective, recent quantitative assessments of marine accidents by Antão et al. [1] and Fu et al. [3] identify structural failure as a dominant Risk Influencing Factor (RIF) in severe incidents. Thus, the use of non-standardized wooden hulls poses a substantial risk to fisher safety.

Despite the structural advantages of steel, its adoption in the small-scale sector (<30 GT) is impeded by economic and manufacturing barriers. The transition from wood to steel is a complex shift in production management. Ross [6] highlights that accurate cost estimating is critical; for traditional communities, the high capital expenditure of conventional steel vessels is often prohibitive. Furthermore, conventional steel hulls require double-curvature plate bending, demanding heavy machinery scarce in small shipyards. Wibowo and Talahatu [9] argue that to make steel vessels accessible, the design must utilize simplified hull geometries, such as flat-plate designs, to reduce fabrication complexity.

Addressing these challenges, this research proposes a concept design for a 5-8 GT Purse Seine fishing vessel utilizing thin-steel plates. By integrating developable surface principles with hull form optimization, the design aims to minimize production complexity while maximizing structural integrity. The objective is to deliver a solution that adheres to Biro Klasifikasi Indonesia (BKI) and SOLAS standards, ensuring superior stability compared to wooden craft. This study seeks to provide a feasible model for small shipyards, fostering a sustainable transition towards resilient fishing fleets.

2 Method

2.1 Research Framework

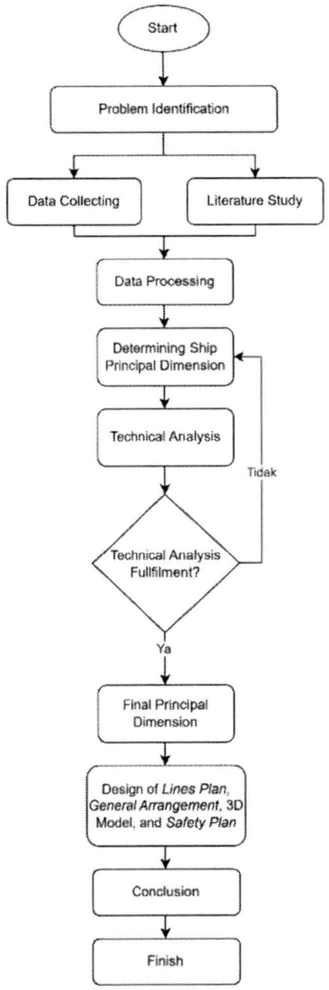

Fig. 1. Research Methodology

The methodology adopted in this study follows the iterative "Ship Design Spiral" approach as elaborated by Fyson [4]. Given the complex interdependence of naval architecture parameters, the design process is not linear but cyclical, where each decision regarding the hull form, weight distribution, and general arrangement requires continuous refinement to satisfy safety and operational constraints.

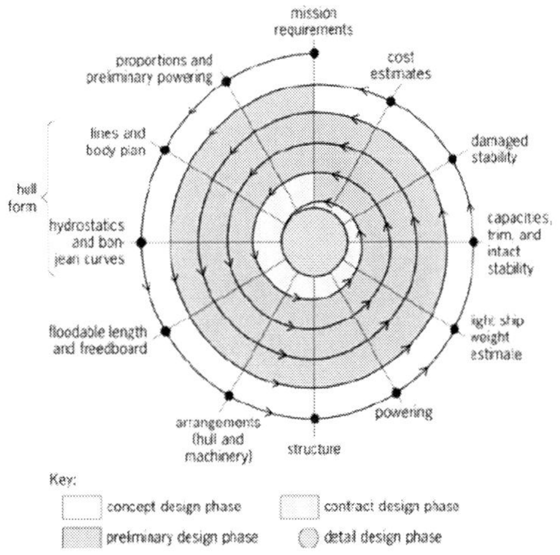

Fig. 2. Spiral Design Process

However, to address the specific challenges of adopting steel in small-scale shipyards, this research integrates the principle of Design for Production. Montwill et al. [5] emphasize that for efficient vessel lifecycle management, production processes must be considered during the early design phase. Consequently, the framework of this research is structured to ensure that the proposed "Thin-Steel" concept is not only hydrodynamically efficient but also technically feasible for construction in limited-resource environments. The research stages are systematically defined as follows:

1. Mission Profile Identification: Defining operational requirements based on the characteristics of Indonesian Purse Seiners.
2. Parametric Determination: Utilizing statistical regression of existing vessels to determine optimal principal dimensions (L, B, H, T).
3. Hull Form Development: Generating the lines plan using the developable surface method to accommodate thin-steel plates without complex bending.
4. Technical Assessment: Validating the design through scantling calculation (BKI Rules) and stability analysis (SOLAS criteria).

2.2 Determination of Principal Dimension

Instead of relying on a single parent ship design which may not reflect the diverse operational conditions of Indonesian waters, this study employs a statistical approach based on a comprehensive national database of fishing vessels. As highlighted by Susanto et al. [8], domestic fishing vessels often possess unique geometric characteristics evolved through traditional knowledge that must be evaluated and optimized rather than discarded. To derive the principal dimensions for the proposed vessel, a dataset of 3,732 existing fishing vessels in the 3–10 GT range was isolated from the national registry. Linear regression analysis utilizing the least squares method was performed to establish the correlation between Gross Tonnage (GT) and the main hull parameters: Length (Lpp), Breadth (B), Depth (H), and Draft (T). The regression aimed to identify the empirical geometric trends that have been proven reliable by the fishing community.

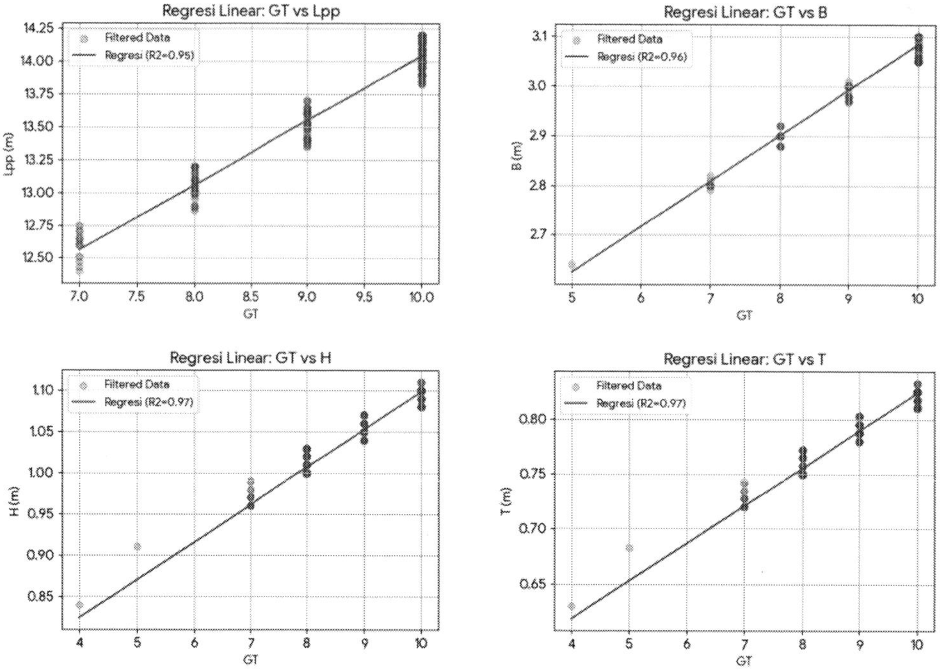

Fig. 3. Result of Linear Regression

The analysis employed an iterative outlier removal process to ensure high statistical significance ($R^2 > 0.95$). The resulting regression equations, which serve as the initial design baseline, are presented in Equations (1) to (4):

$$L = 0.4932(GT) + 9.1092 \quad (R^2 = 0.95) \tag{1}$$

$$B = 0.0919(GT) + 2.1648 \quad (R^2 = 0.96) \tag{2}$$

$$H = 0.0455(GT) + 0.6426 \quad (R^2 = 0.97) \tag{3}$$

$$T = 0.0341(GT) + 0.4820 \quad (R^2 = 0.97) \tag{4}$$

These equations provide the statistical foundation for the vessel's sizing. However, specific design interventions particularly regarding the vessel's breadth (B) are subsequently applied to enhance stability beyond these traditional averages, as detailed in the results section. Place the figure as close as possible after the point where it is first referenced in the text. If there is a large number of figures and tables, it might be necessary to place some before their text citation.

2.3 Hull Form Generation

The adoption of thin-steel plates (thickness < 4 mm) for the hull structure introduces significant constructability challenges, particularly regarding welding distortion and plate forming. Conventional fishing vessels typically employ compound curvature hull forms, which require extensive line heating or mechanical bending. As noted by Montwiłł et al. [5], the complexity of these manufacturing phases directly impacts the production cost and lifecycle efficiency. Furthermore, applying thermal forming to thin plates carries a high risk of material buckling and structural deformity.

To mitigate these risks, this research utilizes the Developable Surface Method as the core geometric strategy. Referencing the work of Wibowo and Talahatu [9], the hull is generated using ruled surfaces where the Gaussian curvature is zero. This geometric property allows the steel plates to be unrolled into 2D flat patterns without stretching or shrinking, thereby effectively eliminating the need for thermal forming processes.

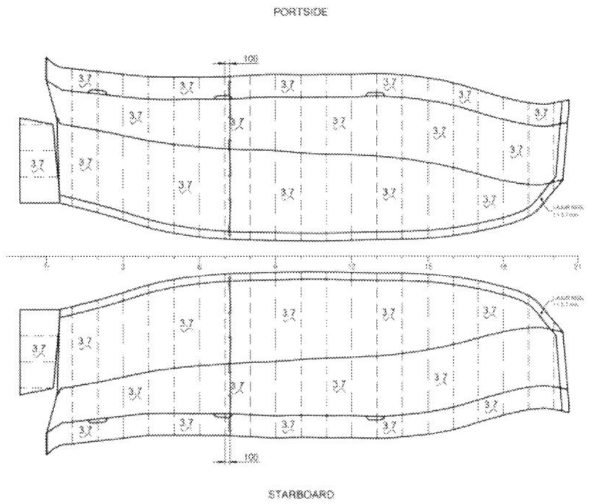

Fig. 4. Shell Expansion

The hull modeling was executed using computer-aided design (CAD) tools to optimize the developability of each panel. The design prioritizes a simplified chine configuration, enabling the integration of standard Flat Bars (FB) for stiffeners instead of the angle bars or bulb flats typically required for curved hulls. This integration ensures that the structural integrity of the thin-steel plating is maintained while significantly reducing fabrication man-hours and material scrap.

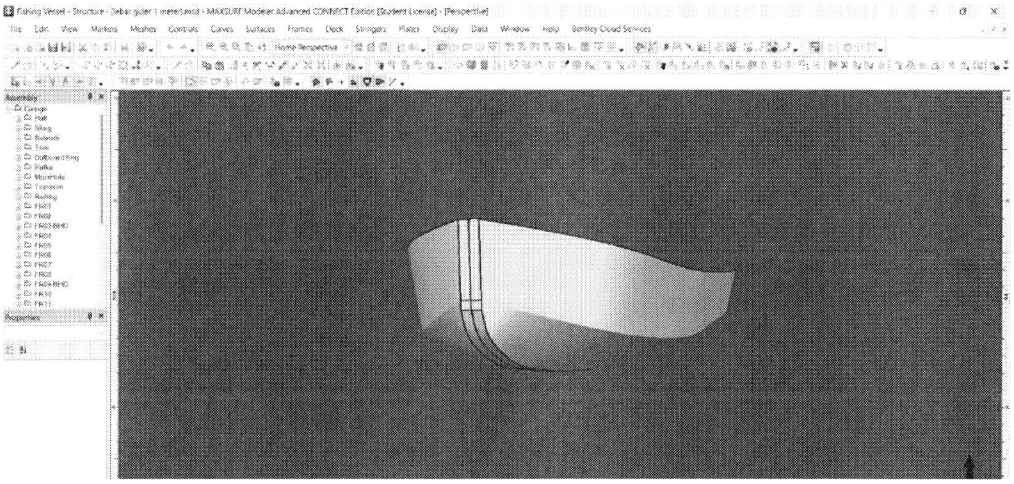

Fig. 5. 3D Hull Model

2.4 Evaluation Criteria

The proposed concept design is subjected to a rigorous technical assessment to ensure it meets the dual objectives of safety and manufacturing feasibility. Unlike traditional wooden vessel construction which often relies on empirical estimation, this study strictly adheres to deterministic regulations to mitigate the Risk Influencing Factors identified by Antão et al. [1]. To validate the feasibility of "Thin-Steel" application, the structural members—specifically the 3.7 mm hull plating and Flat Bar (FB) stiffeners—are calculated against the Rules for Small Vessels issued by Biro

Klasifikasi Indonesia (BKI). This structural assessment focuses on ensuring that the section modulus and plate thickness meet the minimum requirements for local strength, confirming that the significant reduction in material weight does not compromise hull girder integrity as emphasized in the historical transition analysis by Bornmalm and Lagerqvist [2].

Concurrently, given the operational profile of purse seiners which involves heavy lifting and dynamic loads, the vessel's intact stability is evaluated against the International Maritime Organization (IMO) Intact Stability Code (2008). The design must satisfy key parameters derived from the Righting Lever (GZ) curve, specifically the area under the curve at 0°–40°, the maximum GZ value occurring at an angle of no less than 25°, and a minimum Initial Metacentric Height (GM_0) of 0.35 m. Finally, aligning with the cost estimation principles outlined by Ross [6], the design is qualitatively evaluated for constructability, focusing on the minimization of double-curvature plates and the utilization of standard profiles to ensure the vessel remains economically accessible for small-scale fishers.

3 Result and Discussion

3.1 Evaluation Criteria

Based on the statistical regression methodology referenced from Susanto et al. [8], the final principal dimensions were established and subsequently adjusted to meet specific stability criteria. The comparison between the statistical regression baseline and the proposed design is presented in Table 1.

Table 1. Comparasion between Regression Result and Taken Design

Parameter	Regression Result (8 GT)	Proposed Design (7.89 GT)	Deviation & Analysis
Length (Lpp)	± 13.06 m	9.76 m	More compact to enhance maneuverability in crowded ports.
Breadth (B)	± 2.90 m	3.64 m	+25% Wider. Deliberately designed as a "beamy" hull to maximize the righting moment during hauling operations.
Depth (H)	± 1.01 m	1.79 m	+77% Higher. Provides substantial freeboard to ensure a dry deck condition and improve resilience against waves.
Draft (T)	± 0.76 m	1.00 m	Optimum draft for stability and cargo capacity balance.

Following the dimensional determination, the hull form was generated using the Developable Surface Method proposed by Wibowo and Talahatu [9]. This geometric strategy ensures that the 3.7 mm steel plates can be formed into the required hull shape without thermal bending, thereby preventing material buckling and significantly reducing production complexity, a critical factor emphasized by Montwiłł et al. [5]. The visualization of this hull form is presented in the Lines Plan (Fig. 6) and the 3D isometric view (Fig. 7).

https://doi.org/10.1051/bioconf/202621608001

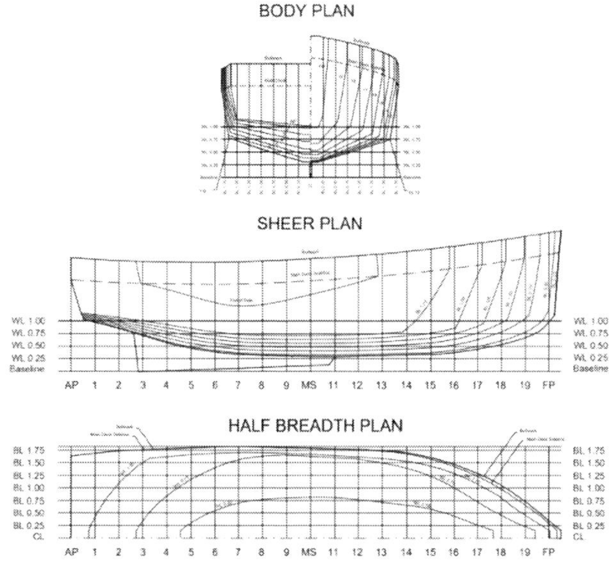

Fig. 6. Lines Plan

Fig. 7. 3D Isometric view

The analysis of the Lines Plan reveals a hard-chine (V-bottom) configuration. While this form theoretically exhibits slightly higher frictional resistance compared to a round bilge hull, it offers significantly better roll damping. This characteristic is crucial for crew safety and comfort, particularly when operating in the dynamic marine conditions typical of Indonesian fishing grounds.

3.2 General Arrangement and Capacity Configuration

The General Arrangement (GA) is strictly configured to maximize the working deck area required for purse seine operations. As illustrated in Fig. 8, the accommodation superstructure is positioned at the aft to provide the skipper with an unobstructed view of the working deck and net handling activities.

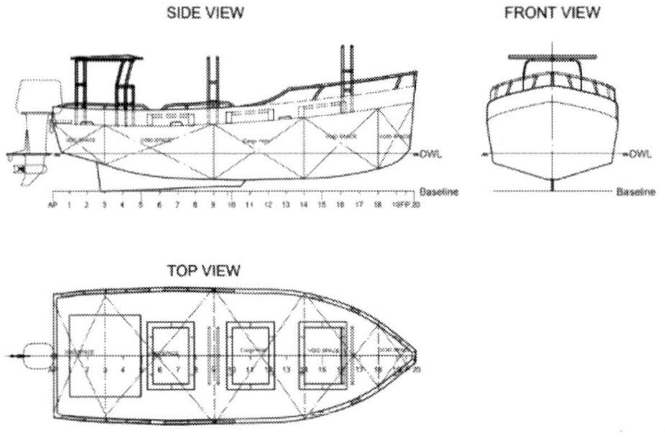

Fig. 8. General Arrangement

Regarding cargo capacity, the Capacity Plan shown in Fig. 9 details a fish hold with a total volume of 2.5 m^3, which translates to an effective capacity of approximately 2 tons of fish and ice. Furthermore, the portable fuel tanks and fresh water tanks are strategically positioned along the longitudinal axis to maintain an even keel trim condition under full load, ensuring optimal hydrostatics during transit.

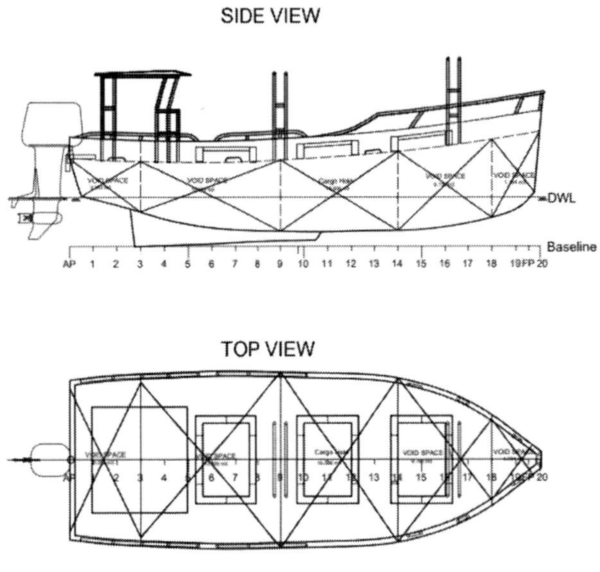

No	Name	Frame (aft)	Frame (fore)	Capacity (m^3)	Long Arm (m)	Trans Arm (m)	Vert Arm (m)
1	Storage	0	3	3.529	0.84	0	1.383
2	Void Space 1	3	9	11.503	3.067	0	1.171
3	Cargo Hold	9	14	11.059	5.754	0	1.189
4	Void Space 2	14	18	7.143	7.899	0	1.35
5	Void Space 3	18	20	1.257	9.357	0	1.594

Fig. 9. Capacity Plan

3.3 Propulsion Arrangement Strategy

Diverging from conventional steel vessels that typically employ inboard diesel engines, this design adopts a 20 HP Outboard Engine propulsion system, as illustrated in Fig. 10. The selection of this propulsion method is predicated on three strategic advantages. Firstly, regarding space efficiency, it eliminates the need for an internal Engine Room, thereby maximizing the hull volume available for the fish hold and directly contributing to revenue generation. Secondly, it simplifies maintenance routines; the engine can be easily detached for onshore repair, effectively addressing the scarcity of skilled diesel mechanics in remote coastal areas. Finally, aligning with the cost estimation principles outlined by Ross [6], this configuration eliminates the expensive shafting system, stern tube, and rudder assembly, significantly reducing the initial Capital Expenditure (CapEx).

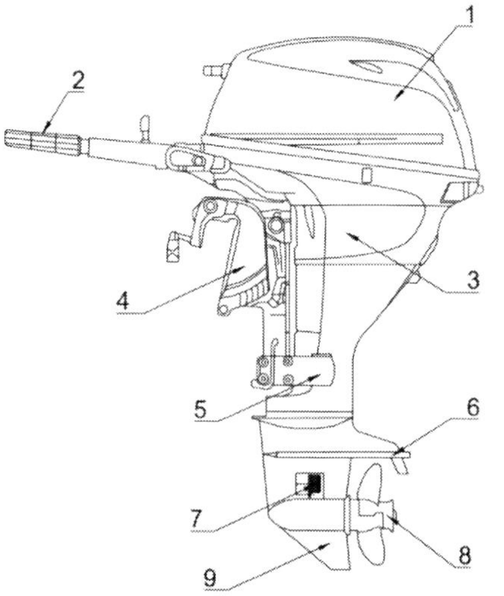

STEERING SYSTEM	
NO	DESCRIPTION
1	Engine
2	Tiller Handle
3	Upper Case
4	Bracket Unit
5	Tilt Adjustment
6	Anti Ventilation Plate
7	Water Intake Hole
8	Propeller
9	Skeg

Fig. 10. Steering System

3.4 Cargo Hold Insulation System: Structural and Thermal Analysis

Given the high thermal conductivity of the steel hull $K \approx 50\ W/mK$, the fish hold is designed with a composite sandwich system whose dimensions are determined through two deterministic calculation stages, integrating structural stiffness equivalence and thermal load analysis. The analysis begins by determining the thickness of the Fiberglass Reinforced Plastic (FRP) skins t_{frp} to ensure the inner hold walls possess impact resistance equivalent to the outer steel hull when subjected to heavy fish baskets and ice blocks. The Equivalent Stiffness principle is applied, expressed as

$$E_{st} \cdot I_{st} = E_{frp} \cdot I_{frp}$$

with the moment of inertia III proportional to t3t^3t3, leading to the required FRP thickness

$$t_{frp} = t_{st} \times \sqrt{3}\frac{E_{st}}{E_{frp}}$$

By substituting $t_{st} = 3.7\ mm,\ E_{st} = 206\ GPa,\ and\ E_{frp} = 10\ GPa$ the calculation becomes

$$t_{frp} = 3.7\times\sqrt{3}\frac{206}{10}t_{frp} = 3.7\times2.74 = 10.14\ mm$$

so the FRP skin thickness is rounded to 10 mm10~\text{mm}10 mm for both layers. This parameter is then used to determine the required Polyurethane (PU) foam core thickness t_{core} to meet thermal standards, using a design target heat transfer coefficient of $U_{design} = 0.46\, W/m^2k$ derived from the recommended limit of $0.60\, W/m^2k$ by applying a safety factor of 1.3. The corresponding required thermal resistance is

$$R_{req} = 2.17 m^2k/W$$

The two FRP skins contribute a thermal resistance of

$$R_{skins} = 2\times\left(\frac{t_{frp}}{k_{frp}}\right)R_{skins} = 2\times\left(\frac{0.01}{0.23}\right) = 0.087\, m^2k/W$$

so the required resistance from the PU core becomes

$$R_{core} = R_{req} - R_{skins} = 2.17 - 0.087 = 2.083\, m^2k/W$$

With PU foam conductivity $k_{pu} = 0.024\, W/mK$, the corresponding thickness is

$$t_{core} = 2.083\times 0.024 = 0.0499\, m$$

which is rounded to 50 mm for manufacturing efficiency. Thus, the final insulation configuration becomes 10 mm FRP − 50 mm PU Foam −10 mm FRP, providing structural protection equivalent to the steel hull while achieving thermal performance approximately 30% superior to the minimum FAO tropical standard.

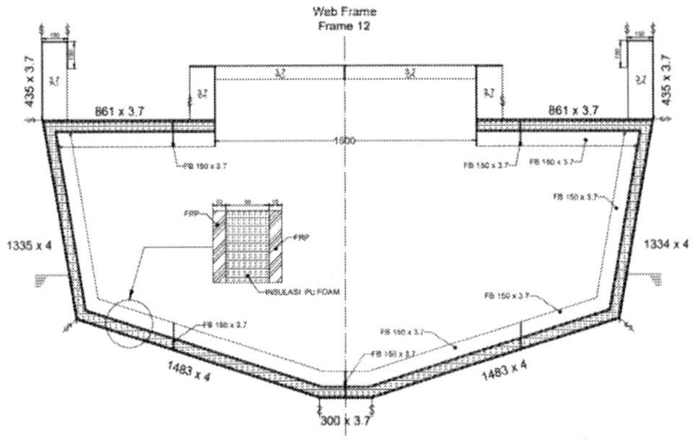

Fig. 11. Cargo Hold Constrction (Frame 12)

3.5 Safety Plan and SOLAS Compliance

The Safety Plan is designed in strict adherence to SOLAS regulations for vessels under 24 meters to mitigate the Risk Influencing Factors highlighted by Antão et al. [1]. As depicted in Fig. 12, the arrangement ensures rapid deployment of safety gear in emergencies. Standard equipment includes Life Saving Appliances such as a 4-person Life Float positioned on the wheelhouse top and Life Jackets stored in accommodation areas. Regarding fire safety, ABC-type Fire Extinguishers are strategically placed near the galley and portable fuel tank area, supplemented by a Fire Blanket. Furthermore, the vessel is equipped with standard COLREG navigation lights (Masthead, Sidelights, Sternlight) powered by a dedicated accumulator. This comprehensive safety arrangement addresses the structural and operational vulnerabilities of traditional vessels discussed by Bornmalm and Lagerqvist [2], ensuring a higher survival probability in adverse marine conditions.

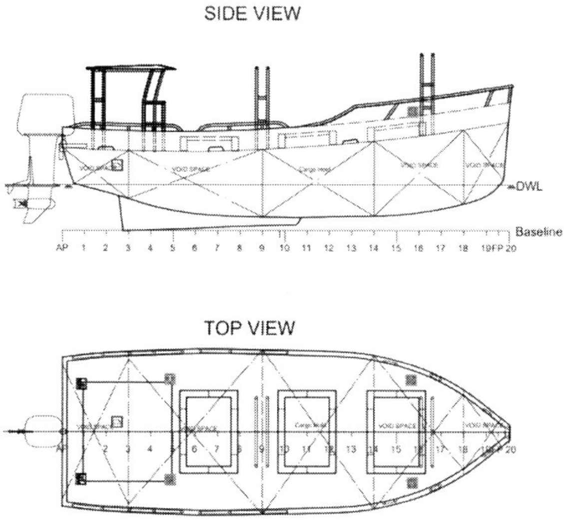

LIFE - SAVING			
SYMBOL	DESCRIPTION	QTY	LOCATION
	First Aid Box	1	Main Deck
	Life Jacket	3	Main Deck
	Lifebuoy	2	Main Deck

FIRE - FIGHTING EQUIPMENT			
SYMBOL	DESCRIPTION	QTY	LOCATION
	5 KG Powder Fire Extinguiser	2	Main Deck
	Fire Blanket	2	Storage

Fig. 12. Safety Plan

4 Conclusion

This research successfully proposes a resilient concept design for a 7.89 GT "Thin-Steel" Purse Seiner to modernize Indonesia's small-scale fishing fleet. Validated by a statistical regression of 3,732 existing vessels, the design features a distinctively stable hull geometry (L/B = 2.68) with a 25% increase in breadth to strictly satisfy IMO Intact Stability criteria for heavy lifting operations. Crucially, the application of the Developable Surface Method confirms the manufacturing feasibility of the 3.7 mm steel hull; by utilizing ruled surfaces and standard Flat Bar stiffeners, the design eliminates the need for complex thermal forming, thereby offering a cost-effective and constructible solution for local shipyards with limited infrastructure.

Operationally, the vessel integrates a strategic 20 HP outboard propulsion system to eliminate the inboard engine room, maximizing the cargo hold capacity to 2.5 m³. The deterministic thermal analysis establishes a specific sandwich insulation specification—comprising 10 mm FRP skins and a 50 mm Polyurethane foam core which achieves a Global Heat Transfer Coefficient (U-value) of 0.46 W/m²K, surpassing FAO tropical standards by 30%. Consequently, this study provides a feasible technical model for sustainable fleet modernization, balancing the rigorous safety standards of BKI and SOLAS with the practical constraints of production and cost.

References

[1] P. Antão, S. Sun, A.P. Teixeira, C.G. Soares, Quantitative assessment of ship collision risk influencing factors from worldwide accident and fleet data. Reliab. Eng. Sys. Saf. 234, 109166 (2023). https://doi.org/10.1016/j.ress.2023.109166

[2] L. Bornmalm, B. Lagerqvist, From Rustic Fishing Boats to Steel Trawlers: The development of fishing vessels on the west coast of Sweden, 1850-1980. Mar. Mirror 108, 66-89 (2022). https://doi.org/10.1080/00253359.2021.1978260

[3] S. Fu, M. Wu, Y. Zhang, B. Han, Z. Wu, Coupling and causation analysis of risk influencing factors for navigational accidents in ice-covered waters. Ocean Eng. 320, 120280 (2025). https://doi.org/10.1016/j.oceaneng.2024.120280

[4] J. Fyson, Design of Small Fishing Vessels (Fishing News Books Ltd / FAO, Farnham, 1985)

[5] A. Montwiłł, J. Kasińska, K. Pietrzaka, Importance of key phases of the ship manufacturing system for efficient vessel life cycle management. Procedia Manufacturing 19, 34-41 (2018). https://doi.org/10.1016/j.promfg.2018.01.006

[6] J.M. Ross, A Practical Approach for Ship Construction Cost Estimating, in Proceedings of the SNAME Annual Meeting (Boston, 2002)

[7] M. Shawyer, A.F. Medina Pizzali, The Use of Ice on Small Fishing Vessels (FAO Fisheries Technical Paper 436, Rome, 2003)

[8] A. Susanto, B.H. Iskandar, M. Imron, Evaluasi Desain Dan Stabilitas Kapal Penangkap Ikan Di Palabuhanratu. Mar. Fish. 2, 213-221 (2011)

[9] H.T. Wibowo, M.A. Talahatu, Pengembangan Desain Kapal Lambung Pelat Datar, in Proceedings of Seminar Nasional Tahunan Teknik Mesin (SNTTM) ke-9 (Palembang, 2010)

Big Data-Based Literature Study on Automatic Identifition System Data and Synthetic Aparture Radar Image Integration for Illegal Fishing in Maritime Awareness

Fifi Fitriah[1], *Adi* Novitarini Putri[1], and *Totok* Yulianto[1*]

[1]Department of Naval Architecture, Institut Teknologi Sepuluh Nopember, Surabaya, East Java, Indonesia

Abstract. Illegal, unreported, and unregulated (IUU) fishing remains a persistent threat in Indonesian waters, causing substantial economic losses and long-term ecological damage. This review synthesizes methods for fusing Automatic Identification System (AIS) data with Synthetic Aperture Radar (SAR) imagery to enhance maritime surveillance. AIS conveys vessel identity and reported position, whereas SAR detects vessels operating without AIS ("dark" vessels). The review covers approaches to spatiotemporal synchronization, data association, and machine-learning models that jointly exploit both modalities. In addition, this study provides a systematic mapping of recent AIS–SAR fusion methods and proposes a conceptual big data framework tailored to Indonesia's maritime surveillance context. According to the surveyed literature, AIS–SAR fusion has been reported to improve the identification of non-cooperative vessels, reduce false alarm and missed detection rates, and shorten response times. Effective implementation requires reliable spatiotemporal alignment, adequate computing resources for large-scale processing, and interagency data-sharing mechanisms. Collectively, the evidence indicates that large-scale AIS–SAR fusion can enhance maritime awareness and support Indonesia's efforts to counter IUU fishing.

1 Introduction

Illegal, unreported, and unregulated (IUU) fishing remains a major concern for coastal states because it erodes state revenue, distorts fishing efforts, and accelerates the degradation of marine ecosystems [2]. For Indonesia, an archipelagic country where more than two-thirds of the territory is ocean and national welfare is closely linked to the marine and fisheries sector, these pressures are particularly acute [12]. Monitoring such vast maritime areas using conventional patrols alone is costly and uneven, which motivates the use of technology-supported surveillance to complement existing enforcement tools [1].

In recent years, maritime monitoring has increasingly drawn on a combination of automatic vessel reporting systems and satellite remote sensing. The Automatic Identification System (AIS) provides vessel identity, position, and basic kinematic information broadcast at regular intervals, which can be used to reconstruct traffic patterns

* Corresponding author: tyulianto1970@its.ac.id

and fishing activity over time [6]. Synthetic Aperture Radar (SAR) imagery offers a complementary view: it records the sea surface day and night and under a wide range of weather conditions, enabling the detection of vessels that do not transmit AIS signals or purposely switch them off to avoid scrutiny [1], [9]. Together, AIS and SAR form a paired information source that can, in principle, reveal both declared and hidden activities in fishing grounds [1].

A growing body of work has explored different ways to use these data. Some studies focus mainly on AIS-based behavioral analysis, for example, by deriving indicators of fishing effort, loitering, or encounters at sea [6], [7]. Others concentrate on SAR-based ship detection and small-vessel monitoring, often highlighting the importance of sensor characteristics such as band, resolution, and polarization [4], [9]. Recent studies have combined AIS and SAR to spot non-cooperative vessels, characterize trawl activity, and follow offshore fleets in selected regions, and these works indicate that using both data sources together provides a clearer picture than treating them separately [1], [2], [3].

Despite this progress, several gaps remain in the literature. Many studies have been organized around particular case areas or sensor configurations, making it difficult to see how the methods relate to one another in the broader literature [1]. There is also limited synthesis of how choices about SAR band, polarization, revisit time, AIS reception (terrestrial versus satellite), and spatial–temporal matching windows interact in practice, especially for IUU fishing detection rather than general traffic analysis [3], [9]. In parallel, developments in large-scale SAR constellations, public benchmark datasets for dark vessels, and national initiatives on data sharing and interoperability suggest that technical and institutional landscapes are changing rapidly [5], [12].

This paper responds to these gaps by presenting a literature-based study of AIS and SAR integration for IUU fishing detection, with a focus on publications from 2021 to 2025. This review compiles and organizes recent research that combines AIS and SAR, examines how data sources, sensor settings, and association strategies are used in practice, and summarizes emerging patterns relevant to fisheries monitoring [1], [2], [3]. Building on this synthesis, this paper also outlines a conceptual big data framework for AIS–SAR fusion that is tailored to Indonesia's maritime surveillance context, including its emphasis on interagency coordination and accountable data governance [11], [12].

2 Theoretical and Literature Framework

2.1 Literature Search

This review focuses on publications from 2021 to 2025 that examine how AIS and SAR can be used together to detect illegal, unreported, and unregulated (IUU) fishing [1]. The time window was chosen to capture the recent growth in satellite constellations, public datasets, and operational interest in AIS–SAR fusion, while still keeping the number of studies manageable for detailed analysis. A small number of earlier, clearly foundational papers were retained when they helped clarify concepts or workflows that were still used in more recent work.

To identify relevant studies, we searched Google Scholar, Scopus, IEEE Xplore, MDPI (Remote Sensing), Frontiers, and arXiv.We combined keyword sets for AIS and SAR with terms for illegal or dark fishing, vessel or boat detection, and fishing behaviour when constructing the search queries [5].Boolean operators AND and OR were used to link these terms so that the search remained broad enough to capture different approaches, but narrow enough to avoid large volumes of obviously irrelevant material [5]. The main databases, time window, and keyword groups are summarized in Table 1.

The screening process was conducted in several stages. First, duplicate records across the databases were removed. Next, titles and abstracts were read to exclude studies that were clearly outside the scope, such as work that did not involve AIS or SAR, focused only on wake or speed without vessel detection, or relied on non-satellite sensors. For the remaining items, the full texts were examined to check whether AIS or SAR data were used, whether some form of fusion or spatiotemporal association was present, and whether the study contained a methodological contribution rather than being purely descriptive. Articles that appeared as both preprints and final journal versions were consolidated so that only the most complete version was retained.

Finally, additional eligibility criteria were applied to meet the objectives of this review. We included papers that discussed AIS–SAR integration, detection of IUU or dark vessels, or AIS-based behavior analysis with relevance to maritime surveillance in Indonesia or comparable settings [1]. Studies were excluded if the full text was not available in Indonesian or English, if they were review papers without new methodological content, or if they introduced a dataset without any analysis. The resulting set of studies obtained through this identification and screening process forms the basis for the synthesis presented in the following sections.

Table 1. Literature identification, screening, and eligibility criteria for AIS–SAR for IUU (2021-2025)

Section	Item	Content
Literature identification	Databases	Google Scholar; Scopus; IEEE Xplore; MDPI (Remote Sensing); Frontiers; arXiv
	Publication time period	2021–2025 (with exceptions for a small number of relevant foundational studies before 2021)
	Keywords A	AIS AND SAR AND "data fusion" AND ("illegal fishing" OR IUU OR "dark ships")
	Keywords B	("ship detection" OR "boat detection") AND SAR AND (VV OR VH OR polarimetric)
	Keywords C	AIS AND (behaviour OR behavior) AND (loitering OR transshipment)
	Keywords D	("xView3" OR "dark vessel") AND SAR AND ("machine learning" OR "deep learning")
Literature screening	Inclusion criteria (abstract reading)	Topics on AIS and SAR integration in remote sensing; detection of IUU

		or dark ships; or AIS behaviour analysis relevant to maritime surveillance in Indonesia or comparable regions.
Literature eligibility (full-text reading)	Exclusion criteria (general)	Full text not available; not written in Indonesian or English; duplicate preprint versus final version (only the most complete version retained).
Papers included in the systematic review	Exclusion criteria (detailed)	Review article without methodological contribution; no AIS or SAR data; dataset-only paper with no detection method; focus only on wake or speed without vessel detection; no fusion or spatiotemporal association; not based on satellite remote sensing.

2.2 Inclusion Criteria

The search results from all databases were first combined, and duplicate records were removed. Titles and abstracts were then screened to retain articles that dealt with the integration of AIS and SAR, or more broadly, satellite remote sensing, for the detection of illegal, unreported, and unregulated fishing [1]. At this stage, we set aside clearly unrelated material, such as work on non-maritime topics or studies that did not involve vessels at all.

The full text of the remaining studies was retrieved. Only papers available in Indonesian or English were retained so that the methods and results could be assessed in a consistent way. At the eligibility stage, we focused on studies that used satellite sensors, active or passive, such as SAR or optical imagery, together with AIS data, reflecting the idea that AIS and satellite observations provide complementary information on vessel identity, position, and visibility at sea [1], [9].

We further required that a study describe some form of fusion or association between AIS and satellite detections in space and time, report explicit vessel detection rather than only tracks or wake patterns, and provide at least one quantitative evaluation metric (for example, accuracy, precision, recall, F1, or mean average precision) [9]. Papers were excluded if they were purely review articles, if they released datasets without any accompanying detection method, or if they focused only on speed or wake characteristics without identifying vessels, as such work does not offer a technical basis for comparing AIS–SAR fusion approaches [5].

2.3 Data Analysis

The selected studies were then analyzed to identify broad patterns in methods, sensor use, and reported outcomes for AIS and SAR integration in the context of IUU fishing. For each study, we noted the satellite sensor families employed (for example, C-band versus X-band systems and the use of single or dual polarization), the type of AIS reception (terrestrial or satellite), the general strategy used to associate AIS tracks with satellite detections, and the evaluation measures reported for vessel detection or classification [9].

To make the comparison easier, the studies were grouped into a small number of categories. One set of groups reflected sensor characteristics (such as band and polarization), while another reflected application focus, for example, dark-vessel detection in open water, large-area association along shipping routes, or monitoring of dense coastal traffic. These groupings made it possible to relate design choices to practical drivers, such as image resolution, revisit rate, and traffic density [4], [9].

Where it was reasonable to do so, simple counts and summary tables were produced to show how often particular sensor types, fusion strategies, and evaluation practices appeared across the corpus. The narrative synthesis then drew attention to differences that matter for fisheries monitoring, including which configurations seem most promising for detecting small vessels, and which studies discuss issues that can be transferred to the Indonesian context, such as data latency, coverage gaps, and institutional arrangements for sharing information [3], [12].

3 Overview of The Research Items Retrieved

This section provides an overview of the studies retained in the review and shows how data sources, sensor choices, and fusion strategies are used in practice. Table 2 summaries the main characteristics of each article, including the year of publication, type of data, focus, method highlight, study region, and short notes on the contribution.

Across these studies, AIS is primarily used for vessel identification and tracking over time, allowing analysts to infer behavioral patterns such as loitering, hidden trawling, or encounters at sea [6]. SAR imagery, in turn, contributes direct evidence of vessel presence even when transponders are switched off, so that non-cooperative vessels remain observable in busy waters [1]. Several papers also make use of optical or multispectral sensors, but in most cases these play a supporting role compared with SAR.

On the sensor side, Sentinel-1 C-band is by far the most frequently used platform, reflecting its wide geographic coverage, regular revisit, and open data policy [9]. Small-vessel detection is a recurring challenge, and many authors report that using dual polarization (VV and VH) helps to increase the contrast between vessels and the surrounding sea surface in Sentinel-1 imagery, making it easier to separate small targets from sea clutter [4]. X-band sensors appear less often, but they tend to be selected in dense traffic or coastal areas where higher spatial resolution is required [9].

Table 2. Overview of retrieved research items (AIS–SAR for IUU)

Ref	Year	Data and sensors	Focus	Method highlight	Region or sea	Notes
[1] Galdelli et al.	2021	AIS and SAR	Suspicious activity and fisheries	AIS to SAR association, point to point	Adriatic Sea	Integration of AIS and SAR for non-cooperativ e vessels
[2] Marsaglia et al.	2025	AIS and optical or SAR	Hidden trawl fishing	Remote sensing fusion with AIS	Mediterranean	Distribution and intensity of fishing

[3] Li et al.	2024	Satellite AIS and remote sensing	Offshore fisheries monitoring	Satellite AIS analysis with imagery	Northern Indian Ocean	Offshore fleet monitoring
[4] Shin et al.	2024	SAR Sentinel-1 with VV and VH	Small vessels	Dual polarisation to raise target–sea contrast	General	Improved small-vessel detection
[5] Paolo et al. (xView3 SAR)	2022	SAR multi-mission dataset	Dark-vessel benchmark	Annotation scheme and evaluation protocol	Global	Public benchmark and dataset
[6] Zhou et al.	2025	AIS	IUU behaviour from AIS tracks	Behaviour features such as loitering and rendezvous	General	Indicators of IUU from AIS trajectories
[7] Han et al.	2025	AIS	Classification of behaviour and vessel type	Behaviour feature-based classifier	General	High accuracy for vessel type
[8] Li et al.	2021	AIS and single-channel SAR	Early integration	Single-channel SAR fusion with AIS	General	Foundational integration
[9] Lee et al.	2024	SAR X-band with AIS and V-Pass	Detection and tracking in dense traffic	Multi-sensor tracking	High-traffic areas	TerraSAR-X, KOMPSAT-5, Capella
[11] Nikitakos et al.	2021	Sentinel-1 dual polarisation with AIS	Large-scale monitoring	Dual polarisation with AIS	General	C-band with AIS
[12] BRIN, KKP, Bakamla	2025	Governance and policy	Open maritime data and interagency access	Framework for data exchange and audit	Indonesia	Supports interoperability for AIS and SAR integration

Figure 1 then groups the studies by sensor or data category and shows how often each category appears in the corpus. The distribution highlights the dominance of Sentinel-1 C-band and the complementary role of X-band SAR and other sensors, such as KOMPSAT-5 and Capella, in high-traffic environments.

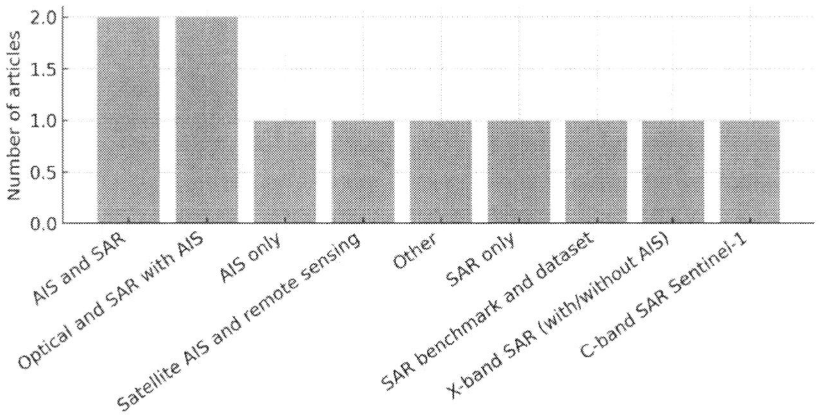

Fig. 1. Distribution of sensor categories in the collected articles

A synthesis of the retrieved literature points to several consistent patterns over the 2021–2025 period. First, cross-source fusion that combines AIS with SAR is now the dominant approach for monitoring fishing and other maritime activity, rather than relying on a single source alone [1], [3]. Second, two broad families of fusion strategy can be distinguished: point-to-point matching, which links individual SAR detections to AIS positions within a time–space window, and multi-feature association, which augments this with information such as target size or shape and is generally preferred in complex or crowded scenes [10].

Third, benchmark resources and shared protocols are beginning to emerge. The xView3 SAR mission, for example, is often used as a reference dataset for dark-vessel detection and as a way to compare annotation and evaluation practices across methods [5]. Performance is commonly reported using accuracy, precision, recall, F1 score, or mean average precision with Intersection-over-Union thresholds, which facilitates comparison among studies [9]. Taken together, these patterns provide a useful starting point for designing AIS–SAR fusion frameworks in Indonesia that emphasise robust time alignment, scalable processing for large data volumes, and coordinated data exchange across agencies [12].

4 Methods & Data Fusion

4.1 General Flow of AIS-SAR Integration

The overall workflow begins with two complementary data sources. AIS messages are processed to construct vessel tracks that record position, speed, and heading over time, whereas SAR images provide snapshots of vessel locations on the sea surface at specific acquisition times [1]. In the first stage, the AIS data are cleaned to remove duplicate messages and implausible position jumps, and the timestamps are converted to a common time standard so that they can later be aligned with satellite overpasses [6]. SAR scenes are calibrated and filtered to reduce speckle, producing stable backscatter values for ship detection [5]. The aim of this preparation step is to place both sources in a form that is suitable for matching, recognising that AIS represents continuous motion while SAR captures discrete scenes when the satellite passes over an area [1].

After preprocessing, analysts align AIS reports with SAR acquisitions within a practical time window so that vessel positions inferred from AIS remain plausible at the time the SAR image was recorded [1]. The spatial buffer around each AIS track segment is scaled to the image resolution and to local sea conditions rather than treated as a fixed distance because traffic density and message cadence differ between coastal and offshore regions and between terrestrial and satellite AIS reception [3]. Within these windows, candidate links between AIS tracks and SAR detections can be identified and passed to the association stage.

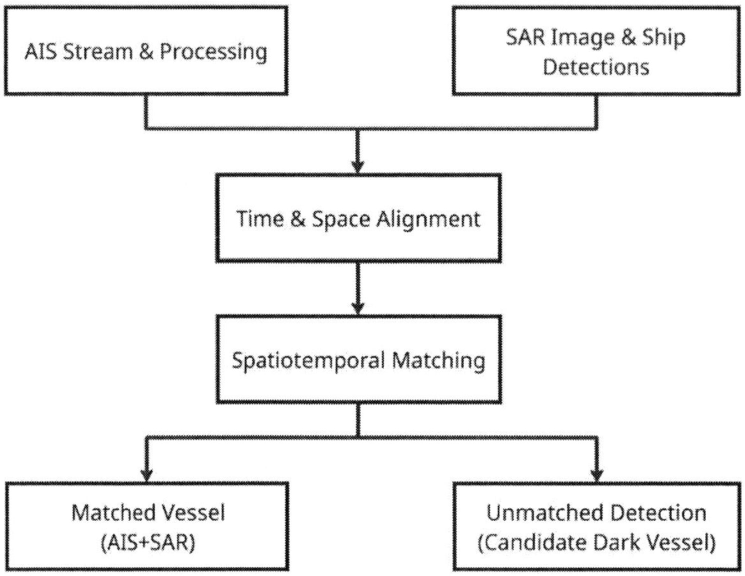

Fig. 2. Simplified AIS–SAR fusion workflow for identifying candidate dark vessels.

4.2 AIS Processing and SAR Detection

AIS processing consolidates the messages into coherent vessel trajectories. After duplicates and obvious outliers are removed, messages for each MMSI are ordered in time, and basic kinematic quantities, such as speed over ground, course, heading changes, and dwelling time, are computed [6]. In offshore areas, satellite AIS reception often results in larger gaps than terrestrial reception; therefore, short gaps may be bridged with simple interpolation or by using tighter time windows around SAR acquisitions [3]. Behavioral indicators such as prolonged loitering, rendezvous, and possible transshipment events can then be derived from these trajectories and later combined with SAR-based evidence [6].

On the SAR side, the imagery is radiometrically calibrated and filtered to reduce speckle so that the vessel targets stand out more clearly from the sea background [5]. Ship detection is then performed using a mixture of classical image-processing approaches and learning-based detectors to obtain bounding boxes and confidence scores for each object. The sensor configuration plays a central role in the quality of these detections. Dual polarization with VV and VH is frequently used because VV tends to capture strong reflections from vessel structures, while VH suppresses much of the sea clutter; together, they increase vessel-to-sea contrast and improve the detection of small boats [4]. X-band systems are often preferred in congested waters where finer spatial resolution is needed, with the Sentinel-1 C-band continuing to serve as a wide-area backbone sensor [4], [9]. The result of these preparations is two aligned products: a set of AIS trajectories enriched with behavioral features and a set of SAR-based vessel detections on a georeferenced scene. Both are ready for spatiotemporal association in the next step [1].

4.3 Data Association and Operational Evaluation

Spatiotemporal association links SAR detection with AIS position reports within the time window around each SAR acquisition, so that the inferred motion between successive AIS messages remains reasonable [1]. The spatial buffer used for matching is again scaled to image resolution and local conditions, for example, narrower buffers in calm, low-traffic waters, and slightly wider buffers in rough seas or busy shipping lanes [3].

In point-to-point association, candidate pairs are selected based on the nearest distance between an AIS track and SAR detection combined with the smallest time difference, after which simple physical checks on course and speed are applied to filter unlikely matches [1]. When traffic is dense or there are indications that AIS transponders may be switched off, studies increasingly rely on multifeature association. In this case, spatial and temporal proximity are combined with movement direction and attributes of SAR detection, such as bounding-box size and orientation, and, when available, supporting information from optical imagery [10]. These richer feature sets help reduce ambiguity when many vessels are present, or when dark targets are expected.

To support comparisons across studies, evaluation is usually reported using standard metrics such as accuracy, precision, recall, F1 score, and mean average precision with intersection-over-union thresholds [9]. Several papers have also reported the proportion of SAR detections that can be matched to AIS tracks within the chosen windows, which provides an operational indicator of how well the fusion works under different traffic and sensor conditions [3], [9]. Across the reviewed studies, initial parameter choices for time windows, spatial buffers, and validation checks are typically tuned to the sensing modality and traffic density, spanning low and medium traffic in C-band scenes, dense traffic in X-band scenes, and offshore situations dominated by satellite AIS [1], [3], [10].

4.4 Big Data for AIS-SAR Integraton

AIS and SAR integration benefits from big data architecture because the two data sources arrive in different forms. AIS messages form a continuous stream with a cadence of seconds to minutes, whereas SAR scenes arrive in batches that follow satellite revisit cycles [1]. In the ingestion layer, AIS streams are buffered and stored together with metadata such as the reception source, whereas SAR scenes are archived with orbit time, imaging mode, and polarization [4]. Both sources were written into a shared data lake using a common time standard and coordinate reference system so that they could later be merged, reprocessed, and audited [1].

The processing layer typically separates stream processing for AIS from batch processing for SAR. AIS streams are cleaned, trajectories are built, and behavioral features are extracted, while SAR batches undergo calibration, speckle reduction, and ship detection using the chosen sensor settings and polarization combinations [4]. Features from both sides are consolidated in a central repository of engineered variables, which allows experiments and models to be consistently reused across different tasks [9].

In the fusion layer, a spatiotemporal join is executed, and the association engine applies the point-to-point or multi-feature strategies described earlier [1], [10]. For training and validation, public benchmarks such as xView3 SAR are often used so that the methods can be tested under similar conditions before being adapted to local data [5]. Operationally, orchestration tools schedule various processing jobs, monitor latency and throughput, and log evaluation metrics, while the data lake maintains an audit trail so that each decision can be traced back to its inputs [9]. In the Indonesian context, this architecture must be complemented by governance arrangements and open maritime data policies that allow BRIN, KKP, and Bakamla to share AIS and SAR data transparently and accountably [12].

4.5 Method Summary

In summary, the methods reviewed in this paper combine behavioral information derived from AIS trajectories with visual evidence from SAR imagery to improve the detection of IUU-related activity. AIS processing yields clean tracks and indicators, such as loitering, rendezvous, and potential transshipment events, whereas SAR processing provides geo-located vessel detections under day-night and all-weather conditions [6]. Spatiotemporal association links these two views of maritime activity by aligning AIS and SAR in space and time and then applying point-to-point or multi-feature matching strategies adapted to local traffic and sensor settings [1], [10].

Embedding this workflow in a big data pipeline allows continuous AIS streams and batch SAR scenes to be handled at scale, with standard evaluation metrics and auditable processing steps. In the Indonesian setting, this type of architecture makes it easier to monitor activity almost in real time and gives the main maritime agencies a shared technical basis for working together in a transparent way [9], [12].

5 Findings

The reviewed studies converge on a consistent conclusion: fusing AIS tracks with SAR detections improves the ability to detect non-cooperative vessels (e.g., "dark" vessels) compared with relying on either data source alone. In low-traffic settings, straightforward point-to-point matching between AIS positions and SAR detections can be effective because the number of plausible candidates is small [1]. In higher-density scenes, however, association methods that incorporate additional cues—temporal offset, spatial distance, course/velocity consistency, and SAR-derived attributes—yield more robust matches and better separate true associations from false alarms [3], [10]. Benchmark datasets such as xView3 SAR further support method development by providing common data and evaluation protocols for dark-vessel detection before deployment on operational AIS streams [5].

From a sensor standpoint, Sentinel-1 C-band is the predominant backbone for AIS–SAR fusion due to its broad coverage, frequent revisit, and open-access policy [9]. Higher-resolution X-band sensors (e.g., TerraSAR-X, KOMPSAT-5, Capella) are used less often but are typically selected for congested coastal zones where resolving closely spaced targets is critical [9].

vessel–sea contrast, improving small-vessel detectability that is particularly relevant to fisheries monitoring [4]. These results indicate that sensor configuration materially affects the achievable performance of AIS–SAR fusion for IUU applications.

A complementary thread focuses on AIS-based behaviour analytics. Indicators such as sustained loitering, repeated returns to offshore hotspots, close-proximity encounters, and suspected transshipment are repeatedly highlighted as signals of potential IUU activity [6]. When coupled with SAR evidence, these behavioural flags help differentiate routine traffic from cases that merit follow-up. In several studies, behaviour-based classifiers are used to rank vessels by risk level, and the resulting candidates are cross-checked against SAR scenes to assess consistency between declared vessel type/activity and observed behaviour [7], [1].

Finally, the literature points to an operational blueprint: a big-data pipeline that ingests streaming AIS and batch SAR, performs cleaning and preprocessing, aligns both sources in space and time, and outputs fused vessel lists, risk scores, and maps of suspicious activity [1], [9]. For Indonesia, the reviewed work underscores that algorithmic advances must be paired with reliable data access and interagency governance; coordinated AIS and SAR sharing is as important as the detection method itself [12].

6 Conclusions

This review has examined recent work on the integration of AIS and SAR for monitoring illegal fishing, particularly illegal, unreported and unregulated (IUU) fishing, and related non-cooperative behavior at sea. Across the studies considered, AIS provides information on vessel identity and longer-term movement patterns, whereas SAR offers weather-independent snapshots of vessel presence at particular points in time [1]. When these two views are brought together through careful spatial and temporal alignment, dark vessels (vessels that disable or manipulate AIS) and other forms of unusual activity commonly associated with illegal fishing (e.g., AIS gaps, suspicious rendezvous/transshipment, loitering near boundaries, or irregular movement patterns) become easier to detect than with either data source alone [10]. The quality of the results depends on several factors: sensor choice and configuration, the way AIS and SAR are preprocessed, and the strategy used to associate detections with the tracks [9].

The literature suggests a gradual route towards implementation. In settings with moderate traffic, conservative point-to-point associations with well-chosen time windows and spatial buffers can provide useful starting points [1]. In more complex environments, multi-feature associations that also consider motion direction, bounding-box size, or supporting imagery appear to be more reliable [3], [10]. The Sentinel-1 C-band is likely to remain the backbone sensor for many countries, while X-band acquisitions and dual-polarization modes can be reserved for areas where small vessels and busy lanes create additional challenges [4], [9]. Reporting standard metrics such as accuracy, precision, recall, F1 score, and mean average precision with Intersection-over-Union thresholds helps to keep future work comparable and track progress over time [9].

For Indonesia, the main message from this review is that technical and institutional designs need to move together. A documented big data pipeline for AIS and SAR, paired with clear rules for data access and interagency cooperation, can support faster detection and enforcement of illegal fishing (IUU) by turning indicators such as dark vessels and unusual activity into actionable alerts for inspection and intervention.

References

1. Galdelli et al., A synergic integration of AIS data and SAR imagery to monitor fisheries and detect suspicious activities. Remote Sens. 13, 2455–2469 (2021).
2. M. Marsaglia et al., Shedding light on trawl fishing activity in the Mediterranean Sea with remote sensing data. ICES J. Mar. Sci. 82, 525–540 (2025).
3. J. Li, Q. Xing, X. Li, M. Arif, J. Li, Monitoring off-shore fishing in the Northern Indian Ocean based on satellite Automatic Identification System and remote sensing data. Sensors 24, 350–364 (2024).
4. H. Shin et al., Enhancement of small ship detection using polarimetric combination from Sentinel-1 imagery. Remote Sens. 16, 205–217 (2024).
5. F. Paolo et al., xView3-SAR: Detecting dark fishing activity using synthetic aperture radar imagery. arXiv (2022).
6. Y. Zhou et al., Identifying behaviors indicative of illegal fishing activities in Automatic Identification System data. J. Mar. Sci. Eng. 13, 34–49 (2025).
7. X. Han, Y. Zhou, J. Weng, L. Chen, K. Liu, Research on fishing vessel recognition based on vessel behavior characteristics from AIS data. Front. Mar. Sci. 12, 1–10 (2025).
8. J. Li et al., Integration of Automatic Identification System data and single-channel Synthetic Aperture Radar images. Ocean Coast. Res. 29, 189–202 (2021).
9. S. Lee et al., Assessment of maritime vessel detection and tracking using integrated SAR imagery and AIS/V-Pass data. J. Navig. 78, 305–321 (2024).
10. X. Li et al., Dark ship detection via optical and SAR collaboration: An improved multi-feature association method between remote sensing images and AIS data. Remote Sens. 17, 101–115 (2025).
11. N. Nikitakos et al., Large-scale automatic vessel monitoring based on dual-polarization Sentinel-1 and AIS data. Remote Sens. Lett. 14, 832–841 (2021).
12. BRIN, KKP, Bakamla, Open maritime data policy framework for Indonesia. Natl. Marit. Res. Bull. (2025).

MMSI Based Anomaly Classification in AIS: A Rules-First Baseline from International Numbering Standards

Dhya Aqilla Pramudhita[1], *Adi* Novitarini Putri[1], and *Totok* Yulianto[1*]

[1]Department of Naval Architecture, Institut Teknologi Sepuluh Nopember, Surabaya, Indonesia

Abstract. Research on Automatic Identification System (AIS) anomalies has largely focused on vessel trajectories and kinematics, while identifier validity is often assumed. This study fills that gap by using the nine-digit Maritime Mobile Service Identity (MMSI) to build a rules-first baseline for anomaly classification. Validation rules are derived from ITU-R M.585-9 and operational guidance (USCG NA VCEN, AMSA), covering format constraints, category and prefix patterns, and MID ranges. The same rules are applied to two public data sources (Global Fishing Watch and NOAA/Access AIS). The pipeline assigns per-record labels for validity, category, and diagnostic notes, and defines a taxonomy of identity anomalies: invalid format, misclassification or misuse, MID and policy inconsistencies, and spatiotemporal "cloned MMSI" detected via overlap tests when positions are available. Results indicate that identity screening reduces noise, highlights priority cases, and produces cleaner inputs for downstream behavioral models without relying on speed or trajectories. Contributions include a reproducible MMSI rule set, an anomaly taxonomy, and a per-source evaluation protocol to avoid misleading generalizations. The approach is transparent, computationally efficient, and easy to integrate as a first-stage filter in maritime analytics pipelines.

1 Introduction

The Automatic Identification System (AIS), which allows authorities to track vessel movements in almost real time, has emerged as a key component of modern maritime safety and monitoring. AIS data now serves a variety of purposes beyond its initial safety-of-life goal, such as environmental protection, fisheries management, and maritime security analytics. However, all of these applications implicitly rely on the dependability and consistency of the vessel identities encoded in Maritime Mobile Service Identities (MMSI) across platforms and jurisdictions. By analyzing the MMSI numbering standard and suggesting rules-first validation techniques that can be methodically applied to extensive AIS data, this paper challenges that assumption.

* Corresponding author: tyulianto1970@its.ac.id

1.1 Background

AIS is increasingly central to maritime safety, law enforcement at sea, and large-scale maritime analytics. Most mainstream AIS anomaly studies concentrate on trajectory and kinematics-based cues such as route deviations, speed anomalies, or loitering. In contrast, the validity of identifiers, especially the nine-digit MMSI, including its structure and consistency, is usually presumed rather than screened upfront. Without early checks, identity inconsistencies can cascade through later modeling stages; therefore, a transparent, lightweight, and reproducible identity-based screen should be the first step. [8]

The ITU-R M.585-9 standard formalizes the MMSI structure, including the definition of Maritime Identification Digits (MID) and numbering schemes for various categories, such as ship and shore stations, Aids to Navigation (AtoN), handheld DSC, and specialized devices such as AIS SART, MOB/MSLD, and EPIRB-AIS. This recommendation also regulates the assignment, use, and reuse of identities at the administrative level, including provisions on numbering consistency and management when devices or vessels change ownership or flag states. With this foundation, anomalies can be formulated deterministically, such as invalid digit patterns, inappropriate category prefixes, or MID inconsistencies, as strong and easily auditable validation rules [1].

At the operational level, the USCG NAVCEN underscores the imperative of timely MMSI registration to enable effective SAR operations and ensure regulatory compliance while providing a synoptic overview of the MMSI structure and applicable MID ranges. Correspondingly, the AMSA delineates the requirements and procedural pathways for MMSI allocation to AIS/DSC and AtoN units, emphasizing the centrality of MMSI to maritime safety and accountable identity management [2], [4].

On the data side, MMSI-based identities are documented and curated by two popular public sources, Global Fishing Watch (GFW) [5] and NOAA (U.S. National Oceanic and Atmospheric Administration) [6]. Variations in their curation practices are crucial for reliable cross-source evaluation and result interpretation. After filtering and harmonizing MMSI fields and vessel attributes, GFW releases a global static dataset of apparent fishing effort based on AIS [6]. On the other hand, downstream users bear the majority of the burden of MMSI validation since NOAA's Marine Cadastre AIS archive exposes raw message fields with little pre-screening [7]. To put GFW and NOAA records on an equal footing before analyses or cross-validation exercises, a clear rules-first approach is therefore required in this work.

1.2 Objectives

The objectives of this study are:

1. Formulated and implemented replicable MMSI validation rules derived from international standards (ITU-R M.585-9) and operational guidelines (USCG NAVCEN, AMSA) to establish a rule-based anomaly classification baseline for AIS data.
2. Identity-based anomaly indicators are defined as follows: (i) format anomalies, (ii) misclassification/misuse, (iii) "cloned MMSI" or cross-spatiotemporal identity collisions, and (iv) policy anomalies.
3. Evaluate the baseline on two data subsets (GFW and NOAA) to quantify detection rates, assess ease of replication, and gauge utility as a preprocessing step prior to trajectory- or kinematics-based modelling.

1.3 Problem Limitations

The problem limitations of this study are:

1. Rules and labels refer to ITU-R M.585-9 (format and categories) and NAVCEN/AMSA guidelines; variations in national policies outside these references are not explored in depth.
2. The case study data sources are subsets of GFW and NOAA, and the findings depend on the curation or definition policies of each provider (e.g., likely_gear, spoofing); therefore, generalization to other AIS providers needs to be tested.
3. Cloned or identity collision detection is treated as an identity-based indicator (requiring temporal overlap testing for the same MMSI in distant locations) and does not yet integrate cross-sensor fusion (e.g., VMS/VIIRS/SAR). Methodological reinforcement may follow identity verification practices in operational documentation.

2 Theoretical Basis

2.1 International Telecommunication Union (ITU)

The ITU is a specialized agency of the United Nations that oversees the radiocommunication sector (ITU-R) with a mandate to ensure the rational, equitable, efficient, and economical use of the radio frequency spectrum by all radiocommunication services. ITU-R technical recommendations are developed through studies in working groups and adopted by the Radiocommunication Assembly. In the context of maritime identity, ITU-R M.585-9 provides the scope, key terms, and appendix structure that describe the identity format and management guidelines.

Fig. 2.1 Cover of ITU-R document M.585-9: assignment and use of identities in the maritime mobile service identity (MMSI).

In essence, M.585-9 specifies the method of assignment, use, and conservation of identities in the maritime mobile service (MMS), covering formats for ship/coast/group, AIS

Aids to Navigation (AtoN), crafts associated with a parent ship, SAR aircraft, and special AIS SART/MOB/EPIRB-AIS devices. This document also emphasizes the availability of identity data for authorized entities such as Rescue Coordination Centers (RCC) and publication in the MARS database, while providing guidelines in Annex 3 for administration: assignment procedures, updates to the Radiocommunication Bureau, and reuse criteria (e.g., after two consecutive editions of List V or ≥2 years of absence) so that MID/MMSI resources are not depleted [1].

2.2 Australian Maritime Safety Authority (AMSA)

AMSA is Australia's national maritime safety authority that regulates accident prevention, SAR preparedness, and maritime radio equipment compliance. In the context of identity, AMSA manages the licensing or registration of devices that transmit identities on DSC/AIS, including the MMSI application process and checking the conformity of owner, vessel, and device category data. AMSA also explains the application of the MID block "503" (Australia) to various classes, providing readers with concrete examples of how ITU standards are translated into administrative practices in specific countries, from ship/coast/group numbering to specialized devices. The policy on the AMSA website helps researchers distinguish valid identities from misclassified cases, such as handheld devices being treated as vessels, and understand the relevant validation points for MMSI-based screening [4]. From a practical point of view, using AMSA's guidance in the rules-first baseline keeps the screening criteria close to real regulatory practice rather than arbitrary assumptions. Linking the ITU recommendations to how a national authority actually assigns and uses MMSI codes also makes the validation rules easier to interpret and more likely to be reused beyond the Australian context.

2.3 U.S. Coast Guard Navigation Center (NAVCEN)

NAVCEN is the official portal of the U.S. Coast Guard for maritime navigation information and radio communication identity services. In the context of MMSI, NAVCEN summarizes identity formats according to international recommendations, including class mapping such as ship/coast/group, SAR aircraft, handheld DSC, and AtoN, and links to national registration policies (e.g., through the FCC in the United States). NAVCEN also provides a range of valid MIDs (201–775) and explanatory materials to help vessel owners, radio operators, and authorities understand the relationship between MMSI and DSC/AIS and Search and Rescue (SAR) operations [2]. For data researchers/practitioners, the NAVCEN page is very useful for compiling summary tables of MMSI patterns and validation checklists prior to behavioral analysis. Examples of patterns that are often used as references include: coast 00MIDXXXX, group 0MIDXXXX, SAR aircraft 111MIDXXX, handheld 8MIDXXXXX, and AtoN pattern 99MIDXXXX. This material is practical to use as the basis for identity-based rules (format check, category check, MID consistency) in the pre-processing stage, so that administrative/format anomalies can be filtered out early before the kinematic model is applied [3].

2.4 MMSI Structure (9 digits) & Identity Class

The MMSI is a 9-digit numerical identity used by stations in maritime mobile services, such as ships, coastal stations, AtoN, SAR aircraft, and safety devices for identification purposes on AIS/DSC and integration with maritime safety numbering schemes. The first three digits are Maritime Identification Digits (MID) that represent the administration or authority that assigns the identity; the remaining six digits follow a pattern according to the station or device class. MMSIs are issued by national administrations or recognized authorities and are intended to be globally unique; they should be updated or retired when a vessel's flag, ownership, or equipment configuration changes. MID values indicate the assigning administration rather than nationality and are distinct from ISO country codes. Moreover, the leading digit constrains the station class (e.g., 0/00/99/8/9 prefixes for special use). Basic automated screening verifying nine-digit length, numeric content, permissible prefixes, and MID validity against allocation tables can eliminate many identity errors prior to behavioral analysis [1].

The official MMSI format details are specified by ITU-R M.585-9, including the following:

1. Ship station: MID + 6 digits (the first digit of the MMSI is in the range 2–7)
2. Group ship call: 0 + MID + XXXX
3. Coast/base station: 00 + MID + XXXX
4. Handheld VHF DSC: 8 + MID + XXXXX (handheld DSC/GNSS device)
5. Prefix "9" (special devices), including
6. AIS-SART 970…, MOB/MSLD 972…, EPIRB-AIS 974…
7. AIS AtoN 99 + MID + XXXX (with examples of grouping: 99MID1XXX physical, 99MID6XXX virtual, 99MID8XXX mobile)
8. Craft associated with a parent ship 918 + MID + XXXX

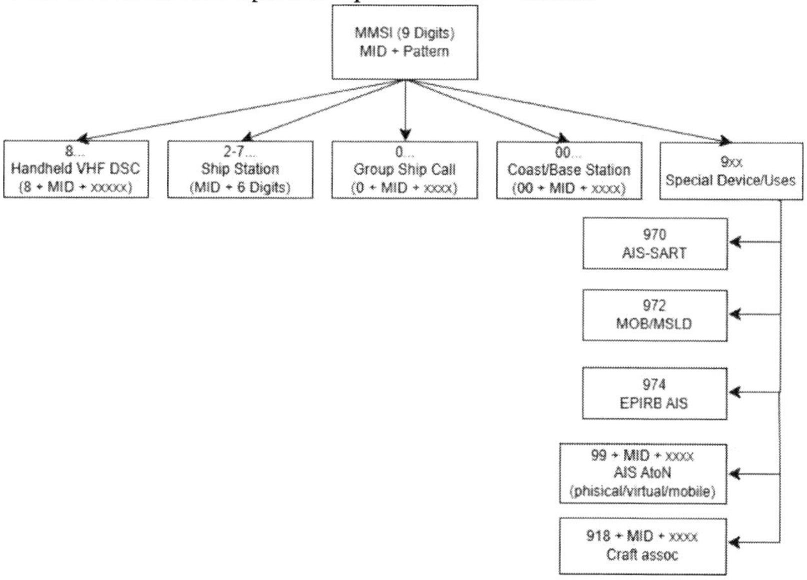

Fig. 2.2 MMSI Prefix Tree and Class Mapping (Derived from ITU R M.585-9)

Implications for validation or anomalies:

Because the numerical pattern of each class is deterministic (e.g., 0/00 = group/coast, 2–7 = ship, 8 = handheld, 9xx = special device), validation rules can capture (i) invalid format (non-digit or ≠9 digits), (ii) mis-categorization/misuse (e.g., prefix 8 but behaves like a commercial vessel), (iii) MID mismatch with flag/area of operation, and (iv) indication of identity collision/cloned MMSI (same identity appears simultaneously in distant locations; requires time overlap test). The management and reuse guidelines in Annex 3 (e.g., reuse after absence from two consecutive editions of List V or ≥2 years) can be used as a basis for policy anomalies[1].

3 Research Methodology

3.1 Research Design

This study adopted a computational survey method with a rules-first validation approach to MMSI identities in AIS data. The rules are derived from the MMSI numbering standard (9-digit format, class/prefix, and MID range 201–775) and then implemented as a validation script that generates validity labels, MMSI categories, and reasons or notations.

For the two data sources (NOAA and GFW), the design includes two validation paths that are consistent yet flexible with regard to column schemes.

1. GFW Path: utilizes validate_mmsi_indonesia_GFW.py, which contains the essential methods analyze_csv(), which exports to Excel with automatic flagging of invalid rows, and validate_mmsi(), which returns is_valid, category, and note.
2. NOAA Path: uses a generic validator package (validation module) via the validate_mmsi() function (boolean + reason), validate_rows(), and sort_mmsi_records() to generate separate CSV files (valid/invalid) and a. txt summary. These functions are exposed via _init ___. py and are also called by the CLI _main_. py.

The core cross-source process includes:

1. MMSI preprocessing (trim, numeric check, and 9-digit length);
2. Prefix pattern-based validation (ship/group/coast/handheld/9xx);
3. MID policy checks (201–775).
4. Category labeling + reason notation and export (Excel for GFW; CSV+summary for NOAA). The rule implementation, including the MID_MIN/MID_MAX limits, is uploaded in the module.

3.2 Research Time and Place

The computational experiment was conducted offline in a data processing environment (Python + pandas). Implementation period October 2025. The AIS datasets used in this study were obtained from the following sources:

1. GFW: previously downloaded validate_mmsi_indonesia_GFW.py. AIS subset; processed with validate_mmsi_indonesia_GFW.py.
2. NOAA: equivalent NOAA AIS archive (CSV format); processed with the validation module and mmsi_validator CLi.

3.3 Data Collection Methods

The data collection methods for this study are:

1. Primary data (computational results):
    - GFW: Excel output containing the columns mmsi_valid, mmsi_category, mmsi_note; invalid rows are automatically highlighted by analyze_csv() (openpyxl PatternFill).
    - NOAA: two sorted CSV files (*_valid.csv, *_invalid.csv) plus a summary*_mmsi_summary.txt from sort_mmsi_records().
2. (USCG) summary, and AMSA used as the basis for formulating validation rules (without changing the AIS data content).
3. AIS data source and execution:
    - GFWinput: AIS CSV (e.g., sar_vessel_detections_pipev3_20250922.csv), called with the --input argument and exported to Excel (--output).

- Input NOAA: CSV AIS NOAA (mis. AIS_YYYY_MM_DD.csv); via CLI mmsi_validator --input <file> [--output-dir <folder>]which prints the summary to stdout.

3.4 Required Data

The data required for this study is:

1. Key attributes: The minimum required input is the MMSI column, represented as a 9-digit string. During pre-processing, values are cleaned (trimming spaces and removing non-digit characters) and rejected if the length $\neq 9$ or if any non-digits remain. In this study, MMSI values were extracted from two AIS sources, Global Fishing Watch (GFW) AIS-based apparent fishing effort v3 and the NOAA Office for Coastal Management (MarineCadastre) AIS archive.
2. Validation parameters: Rules using parameters: MID_MIN = 201, MID_MAX = 775, dan VALID_FIRST_DIGITS = {2,3,4,5,6,7} for ship stations. The range and structure follow the international MMSI standard [1].
3. Category mapping (MMSI pattern):
   - Ship station: MID + 6 digit (first digit of MMSI ∈ {2–7}).
   - Group ship call: 0 + MID + XXXX.
   - Coast/Base station: 00 + MID + XXXX.
   - Handheld VHF DSC: 8 + MID + XXXXX.
   - Special devices/uses (prefix "9"): AIS-SART 970…, MOB/MSLD 972…, EPIRB-AIS 974…, AIS AtoN 99 + MID + XXXX (generally mapped physical/virtual/mobile), dan craft associated 918 + MID + XXXX.

   All of the above patterns are implemented in the validate_mmsi() function as a rule set so that each mmsi value can be labeled with a category and reason.
4. Input/output files:
   - GFW (Excel): …_mmsi_validation_GFW.xlsx berisi mmsi_valid, mmsi_category, mmsi_note + highlight invalid.
   - NOAA (CSV + txt): _valid.csv, _invalid.csv, dan _mmsi_summary.txt (total, valid, invalid, percentage valid).

3.5 Data Analysis

The data analysis for this study are:

1. Validation and labeling (the validate_mmsi() method).
   We process each mmsi in the following ways: (1) check the format (9 digits, all numbers); (2) check the category pattern based on the prefix (ship/group/coast/handheld/9xx); (3) check the MID policy within the valid range; and (4) arrange the output:
   - mmsi_valid ∈ {True, False};
   - mmsi_category ∈ {ship_station, group, coast, handheld, ais_sart, mob_msld, epirb_ais, aton, craft};
   - mmsi_note (brief reason, e.g.: "length not 9", "first digit not {2–7}", "prefix 9 does not match 970/972/974/98/99+MID", "MID 503", etc.).
2. Heuristic for detecting identity anomalies.
   From the three output columns above, we get four indicators:
   - Length $\neq 9$ or non-digit.
   - Misclassification/misuse, e.g., prefix 8 (handheld) used consistently like an operational vessel.

- Policy/MID outside 201–775 or inconsistent with administration/area of operation.
- Potential identity collision due to MMSI cloning, indicated by concurrent occurrences of a single MMSI in far apart areas (needs time-overlap testing of position data, prepared as a subsequent stage).

3. Summary in numbers.
   To describe data quality and the rule-based screening framework, the findings are summarized as follows:
   - Validity distribution: the percentage of records that are marked as valid compared to those that are not;
   - Category distribution: the breakdown of mmsi_category classes to show how items are used and stored.
   - The most common mmsi_note strings that point up systematic problems (such prefix violations that happen again and again or MIDs that are out of range).

   A concise summary table is provided to streamline the presentation of findings in the Results and Discussion section and to enable quicker, more transparent interpretation.
4. Export and look over: The full output is sent to Excel, where faulty rows are automatically highlighted to make manual auditing faster. In the Results & Discussion section, the summary table is used. All exports have consistent column definitions and provenance metadata (timestamp, rule version, and dataset source) to make sure they can be traced and reproduced.

3.6 Flowchart

The flowchart summarizes the rule-based MMSI validation and identity-anomaly screening workflow used in this study. The process starts with a focused literature review (ITU-R M.585-9, NAVCEN, AMSA) to ground the problem statement and the proposed solution. Three streams then proceed in parallel: (i) specifying anomaly-detection requirements, (ii) implementing the script (validate_mmsi() and related routines), and (iii) integrating domain rules (format, prefix/class, MID policy). These streams converge into a concise theoretical basis that formalizes the rule set.

A decision point is used to ensure that the rule set performs as intended. If the rules do not meet the expected criteria, the workflow iteratively applies targeted tests and quantitative evaluations to refine thresholds, pattern definitions, and diagnostic checks, followed by re-validation. This loop continues until the rules satisfy the required performance standards, at which point the validation is finalized. While the pipeline is executed on two AIS data streams (GFW and NOAA), it relies on a single reusable core of MMSI rules and generates consistent outputs (validity, category, and note). To ensure reproducibility, dataset provenance (source, access date, and input scope) is documented so that the entire screening process can be repeated and independently verified.

The flowchart aims to make the MMSI validation pipeline transparent and repeatable for other users, in addition to summarizing the steps. Each box makes clear what data is needed, what presumptions are being used, and how dataset-specific decisions fit into the workflow. This facilitates tracking the classification of specific identities, documenting any deviations, and transferring the same rule set to new AIS archives. Practically speaking, other maritime data projects can use the diagram as a checklist to implement and audit MMSI-based screening.

Lastly, the visual structure aids in distinguishing between dataset-specific and generic and reusable elements. The pipeline's fixed components (the MMSI rules) and adaptable components (data sources, thresholds, or local conventions) are easily visible to readers.

Because of this, the flowchart is a helpful tool for expanding the approach or contrasting it with other MMSI validation techniques.

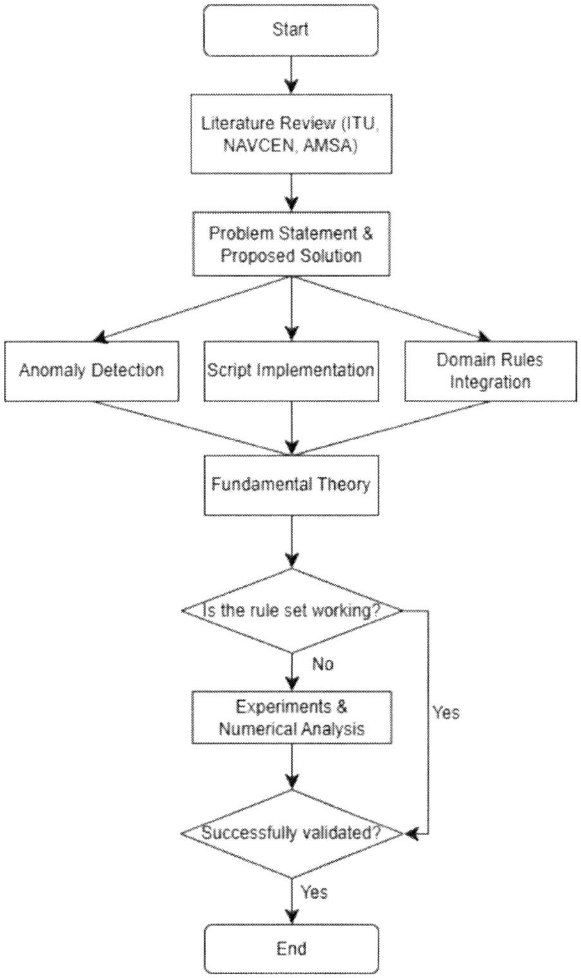

Fig. 3.1 Flowchart of the Rule-Based MMSI Validation and Identity-Anomaly Screening Workflow

4 Results and Discussion

4.1 Data Analysis Dataset Sources

The datasets used in result and discussion were obtained from two sources: Sentinel-1 SAR–based vessel detections from the Global Fishing Watch (GFW) Vessel Detections product and public AIS messages from the USCG Nationwide AIS (NAIS) redistributed via MarineCadastre NOAA AccessAIS. The GFW product supplies georeferenced detections (polygons/centroids) with confidence attributes derived from standardized processing of Sentinel-1 GRD scenes, while AccessAIS provides time-stamped AIS reports aggregated from terrestrial receivers across U.S. waters. For analysis, detection time, location, detection score (where available), and scene metadata (satellite, orbit, pass) were retained from GFW; the minimum required attribute from AccessAIS was MMSI, with additional fields (BaseDateTime, LAT/LON, SOG, COG, VesselType, CallSign, IMO) preserved for descriptive statistics. Both sources were accessed in October 2025, harmonized to UTC, and filtered to overlapping time windows; MMSI values were trimmed, non-digits removed, and 9-digit length enforced prior to identity screening [6], [7].

The MMSI validation pipeline was implemented in Python. The source code for the GFW AIS data is available in the MMSI Based GFW repository, and the code for the NOAA AIS data is available in the MMSI Based NOAA repository [9], [10].

4.2 Validity Summary

Rule-based filtering of the MMSI column produces two main labels: valid and invalid. The invalid category is mainly caused by (i) length $\neq$ 9 digits, (ii) non-digit characters, and (iii) prefix pattern mismatches (e.g., prefix 9 but not 970/972/974 or 98/99+MID). The complete distribution is shown in Table 4.1 for GFW and Table 4.2 for NOAA. The proportion of invalid entries confirms the high level of identity noise in field AIS and reinforces the importance of identity based pre filtering prior to trajectory analysis.

Table 4.1 Distribution of MMSI Validity in the GFW Dataset.

No	Status	Number (n)	Percentage (%)
1	True	156067	76.58
2	False	47741	23.42

Table 4.2 Distribution of MMSI Validity in the NOAA Dataset.

No	Status	Number (n)	Percentage (%)
1	True	8143389	99,84
2	False	13120	0,16

The MMSI validity screening results in Tables 4.1 and 4.2 reveal clear differences in identifier quality across the two AIS sources. For the GFW dataset (Table 4.1), 156,067 records (76.58%) conform to the rule base and are labeled as valid MMSI, while 47,741 records (23.42%) are flagged as invalid, indicating that a substantial portion of entries do not satisfy the enforced constraints, such as not being exactly 9 digits after cleaning, containing residual non digit characters, or failing expected prefix and MID patterns. This finding supports the need for a rules first quality gate, as nearly one in four records could otherwise propagate unreliable identity information into downstream analyses such as trajectory based anomaly detection or AIS SAR fusion. In contrast, the NOAA dataset (Table 4.2) shows near complete conformity to the same rules, with 8,143,389 records (99.84%) classified as valid and only 13,120 records (0.16%) invalid, suggesting a more consistent encoding of MMSI in NOAA while still reflecting a small number of corrupted or out of standard entries typical of real world archives. Taken together, the tables demonstrate that the proposed screening rules can deterministically separate conforming versus non conforming identifiers and that the extent of early stage identity cleaning is strongly source dependent, with GFW requiring substantially more filtering than NOAA before reliable identity driven analytics can be performed.

4.3 MMSI Category Distribution

Mapping of MMSI patterns shows the dominance of ship_station, consistent with the character of the AIS subset driven by ship activity. Note that 'X' is used as a wildcard for numeric digits (0–9) in the pattern notation (e.g., MID+XXXX), not as a literal character in MMSI values. The appearance of handheld (8...) and 9xx devices (e.g., AIS-SART/EPIRB-AIS) signals caution regarding mis-categorization/misuse (e.g., handhelds behaving like operational ships). Group/coast entries help identify non-vessel station records that sometimes enter the detection pipeline. Numerical summaries are shown in Table (see Table 4.3 for GFW and Table 4.4 for NOAA category derivatives from validation reasons).

Table 4.3 MMSI Category Distribution of GFW Dataset

No	Category	Number (n)	Percentage (%)
1	ship_station	156262	76.67
2	Unknown	45672	22.41
3	free_form	785	0.39
4	handheld_vhf_dsc	719	0.35
5	auxiliary_craft	358	0.18
6	ais_sart	8	0.0
7	epirb-ais	2	0.0
8	mob_msid	2	0.0

Table 4.4 Category Distribution of NOAA Dataset.

No	Category	Number (n)	Percentage (%)
1	unknown	10690	81.48
2	ship_station	2059	15.69
3	auxiliary_craft	371	2.83

Table 4.3 and Table 4.4 provide a breakdown of MMSI categories derived from the rule based pattern mapping, which helps distinguish vessel station identities from non vessel or special device identities that may appear in AIS streams. In the GFW dataset (Table 4.3), the distribution is dominated by ship_station with 156,262 records (76.67%), indicating that the subset largely represents conventional vessel transmissions, consistent with a ship activity oriented AIS sample. The second largest class is unknown with 45,672 records (22.41%), which suggests that a substantial portion of MMSI entries do not match the expected structural patterns used for category assignment, and therefore require caution in downstream analyses because identity type cannot be reliably inferred from MMSI alone. Smaller categories appear at low frequencies, including free_form (785; 0.39%), handheld_vhf_dsc (719; 0.35%), and auxiliary_craft (358; 0.18%), while special emergency and safety related devices such as AIS SART (8), EPIRB AIS (2), and MOB MSLD (2) occur only rarely (each

≈0.0% by proportion). Although these minor classes are numerically small, their presence is operationally important because handheld and 9xx device types can be misused or misconfigured and may produce tracks that resemble vessels if not filtered or handled separately.

In contrast, the NOAA dataset (Table 4.4) shows a markedly different profile, with unknown as the dominant category at 10,690 records (81.48%), followed by ship_station at 2,059 records (15.69%) and auxiliary_craft at 371 records (2.83%). This shift implies that, for the specific NOAA subset processed in this study, many MMSI values either fall outside the implemented category patterns or reflect data characteristics that make pattern based classification less straightforward, reinforcing the need to document dataset scope and pre processing choices when interpreting results. Taken together, the two tables show that MMSI category composition is strongly source dependent: GFW is primarily vessel station traffic, while NOAA contains a much higher proportion of identifiers that are not cleanly classifiable under the applied MMSI pattern rules. This insight is important for downstream pipelines, because category aware filtering can reduce false alarms by excluding non vessel stations and handling special devices separately before trajectory analysis or AIS SAR fusion.

4.4 Dominant Reasons for Invalid MMSI

The collection of reasons in the mmsi_note/mmsi_validation_reason column facilitates mass auditing and data cleansing prioritization. The most common reasons include: length not 9, non-digit, prefix 9 not compliant (not 970/972/974/98/99+MID), and MID outside 201–775. A summary of the top reasons is presented in Table (see Table 4.5 for GFW and Table 4.6 for NOAA).

Table 4.5 Dominant Reasons for Invalid MMSI in the GFW Dataset

No	Category	Number (n)	Percentage (%)
1	missing MMSI	41926	35.17
2	MID 412	23996	20.13
3	MID 413	9478	7.95
4	MID 525	8813	7.39
5	MID 636	8583	7.2
6	MID 538	6116	5.13
7	MID 352	4610	3.87
8	MID 574	3993	3.35
9	MID 431	3425	2.87
10	MID 257	3133	2.63
11	MID 563	2706	2.27
12	MID 477	2418	2.03

Table 4.6 Distribution of MMSI Validity in the NOAA Dataset.

No	Category	Number (n)	Percentage (%)
1	length not 9	8160	62.20
2	MID out of range	2240	17.07
3	invalid starting digit	2059	15.69
4	auxiliary MID out of range	371	2.83
5	invalid prefix 9 pattern	290	2.21

4.5 Dominant Reasons for Invalid MMSI

The foregoing results indicate that MMSI based screening functions effectively as a rules-first baseline, it attenuates noise, prioritizes invalid or suspicious records, and yields a cleaner corpus for downstream behavioral modeling. Detection of cloned identities can be incorporated by applying a spatiotemporal overlap test to records sharing the same MMSI, this indicator augments identity controls while imposing minimal computational overhead in early processing stages.

5 Results and Discussion

5.1 Summary of Findings

The analysis results demonstrate that the MMSI validation framework can be operationalized as a deterministic, auditable front end for AIS data quality control. Using rule references aligned with ITU R M.585 9 and NAVCEN and AMSA operational summaries, the implementation assigns each AIS record a structured set of outputs consisting of a validity label, an MMSI category, and an explicit rule based reason for any rejection or caution flag. Because the decision logic is expressed as transparent pattern and policy checks, the screening remains computationally lightweight and easy to audit, and it can be reproduced on different AIS sources using the same script. Importantly, the parameterizations that govern the screening behavior, such as MID bounds and prefix based patterns, can be versioned and documented, which supports faithful replication across experiments and facilitates controlled updates when numbering standards or operational interpretations evolve.

Beyond binary validity, the rule outputs also function as interpretable indicators of identity related risk. The observed error modes can be organized into four practical indicators: identity format violations, misclassification or misuse signals reflected by unexpected station patterns, MID policy violations, and potential identity collision or cloning cues. The first three indicators are derived directly from MMSI only checks and therefore can be applied even when positional fields are unavailable, while the cloning or collision cue is naturally treated as an extension that becomes testable once time and position attributes are included through a spatio temporal overlap analysis. This layered structure allows the pipeline to preserve the same core rules while expanding diagnostic depth as additional attributes become available, which is valuable for incremental deployment in real AIS processing environments.

When applied to the two datasets, GFW and NOAA, the screening behavior remains consistent in its logic while revealing source dependent quality characteristics and complementary reporting strengths. On the GFW stream, the rule base produces a concise profile of valid versus invalid identifiers, the composition of MMSI categories, and the most frequent rule based reasons, which directly supports audit triage by highlighting dominant error patterns and reducing identity noise prior to trajectory analytics or machine learning. On the NOAA and NAIS stream, the same rules remain stable at larger scale and generate separated valid and invalid outputs along with summary reports that are well suited for routine quality control and cleanup prioritization, particularly when executed reproducibly through a CLI workflow. Taken together, the category oriented view and the granular reason codes can be harmonized through simple standardization steps, such as consistent column naming and source tagging, enabling cross source comparisons of validity levels, mapping of error profiles, and preparation of cleaned datasets that are ready for subsequent trajectory based or multi sensor modules.

5.2 Recommendations

Based on the findings of this study, several recommendations are proposed to strengthen MMSI based identity screening and its downstream use in maritime analytics. First, the MID reference data should be enhanced by expanding and regularly synchronizing the MID dictionary against authoritative administrative lists and relevant national archives. Where appropriate, the screening can be reinforced with regionalization logic, such as fisheries zones, ALKI corridors, and EEZ boundaries, to improve policy conformance checks. Second, external registration cross checks should be automated whenever access is available, by linking screened MMSI records to official registration or assignment sources to validate MMSI ownership and class specific patterns, including identifiers associated with handheld devices, SAR aircraft, and AtoN units. Third, identity screening should be integrated upstream in kinematics and machine learning pipelines by deploying the rules first validation as an initial gating stage and benchmarking downstream model performance under ablation settings with and without pre filtering, using metrics such as precision and recall to quantify its operational benefit.

References

1. ITU-R, Recommendation ITU-R M.585-9, "Assignment and use of Maritime Mobile Service Identities (MMSI)," Geneva, Switzerland: International Telecommunication Union, 2023.
2. U.S. Coast Guard Navigation Center (NAVCEN), "Automatic Identification System (AIS): Overview," n.d. [Online]. Available: https://navcen.uscg.gov/automatic-identification-system-overview (accessed Oct. 2025).
3. U.S. Coast Guard Navigation Center (NAVCEN), "MMSI formats (including summary tables; MID 201–775)," n.d. [Online]. Available: https://www.navcen.uscg.gov/mmsi-formats (accessed Oct. 2025).

4. Australian Maritime Safety Authority (AMSA), "About maritime mobile service identity (MMSI) information," n.d. [Online]. Available: https://www.amsa.gov.au/mmsi (accessed Oct. 2025).
5. Australian Maritime Safety Authority (AMSA), "AIS aids to navigation," n.d. [Online]. Available: https://www.amsa.gov.au/safety-navigation/navigating-coastal-waters/ais-aids-navigation (accessed Oct. 2025).
6. Global Fishing Watch, "Global static dataset of AIS-based apparent fishing effort v3: Format changes (FAQ)," 2025. [Online]. Available: https://globalfishingwatch.org/faqs/2025-march-static-apparent-fishing-effort-v3-difference-with-map/ (accessed Oct. 2025).
7. NOAA Office for Coastal Management, "AIS data and tools — Frequently asked questions," 2023 (ver. Nov. 2023). [Online]. Available: https://coast.noaa.gov/data/marinecadastre/ais/faq.pdf (accessed Oct. 2025).
8. K. Wolsing, L. Roepert, J. Bauer and K. Wehrle, "Anomaly detection in maritime AIS tracks: A review of recent approaches," *J. Mar. Sci. Eng.* **10**, 112 (2022).
9. D. A. Pramudhita, "MMSI-Based-GFW: Rules-first MMSI validation for GFW AIS data," 2025. [Online]. Available: https://github.com/dhyaaqilla15-phy/MMSI-Based-GFW
10. D. A. Pramudhita, "MMSI-Based-NOAA: Rules-first MMSI validation for NOAA AIS data," 2025. [Online]. Available: https://github.com/dhyaaqilla15-phy/MMSI-Based-NOAA

Managing Biofouling from Ships in Indonesia: Regulation versus Implementation

I Ketut Aria Pria Utama[1,*], *Ahmad* Saikhu[2], *Mahendra Andiek* Maulana[3], *Ahmad* Nasirudin[1], *Januarti Jaya* Ekaputri[3]

[1]Department of Naval Architecture, Faculty of Marine Technology, Institut Teknologi Sepuluh Nopember, Surabaya, Indonesia.
[2]Department of Informatics, Faculty of Intelligent Electrical and Informatics Technology, Institut Teknologi Sepuluh Nopember, Surabaya, Indonesia.
[3]Department of Civil Engineering, Faculty of Civil Planning and Geo Engineering, Institut Teknologi Sepuluh Nopember, Surabaya, Indonesia.
*Corresponding author: kutama@its.ac.id

Abstract. The IMO Glofouling project, which was launched at the end of 2018, ended in early 2025. Indonesia, as one of the leads partnering countries (LPC), has responsibilities to implement the IMO guidelines in biofouling management. The present paper reviews the resolution MEPC.378 (80) together with the Indonesian Government Regulation as guidelines for the control and management of ships' biofouling to minimize the transfer of invasive aquatic species. Surveys are carried out in 4 major shipyards in Surabaya (2), Jakarta (1), and Balikpapan (1) as well as at the BKI Headquarters in Jakarta which responsible on ships checking maintenance and renew the classification of repaired ships. Investigation into traditional fisher communities is also included to comprehend their awareness about the very serious issue of biofouling attacks on their livelihoods. Based on the survey questionnaire, the discussions include the technical aspects of handling biofouling (such approaches and technologies used), social aspects (to cover the knowledge and understanding of the communities on such regulations), and economic aspects (e.g., spending cost on regular bottom cleaning). Finally, the appropriate method or solution is provided.

1 Introduction

Climate change has become one of the most critical challenges facing humanity, influencing almost every aspect of Biofouling is known as aquatic organism including microfouling such as bacteria, diatoms, and fungi, and macrofouling such as barnacles, mussels, and seaweeds [1]. Both microfouling and macrofouling can cause significant structural and economic damage, especially to ships and underwater equipment. In connection with ships and shipping, biofouling is well known as one of the main problems in the maritime sector due to its effect to increase the roughness of the ship hull [2]. Subsequently, this will increase frictional resistance and hence the fuel consumption and emission of toxic gas into the atmosphere.

In terms of environmental issues, biofouling can be invasive hence triggers foremost jeopardize to the sea and the conservation of biodiversity. Serious protection should be taken into consideration such as carried out by the Australian and New Zealand Governments; the two countries require every foreign vessel to have certificate of clean from biofouling before entering Australian and New Zealand waters [3]. Utama et al. [4] discussed the risk of Indonesia from biofouling attack because of being archipelagic country. The geographic position of Indonesia between two continents (Asia and Australia) and between two oceans (Pacific and Indian Oceans) has made Indonesia to be an open country for shipping worldwide. The potencies of danger may come from foreign flag fishing vessels, international recreational craft, offshore oil gas activities, domestic shipping, etc.

In terms of economy, biofouling strike has caused global economic loss up to 0.25% GDP in developed countries. According to OECD [5] Indonesia has to spend about 5.2 million USD annually to handle biofouling effect when there is no policy or regulation, whilst it becomes 2.6 million USD (or 50% reduction) when a proper policy is implemented. There is no clear policy before on handling biofouling waste from shipyards. The biofouling from a ship is washed away to the water without worrying about the risk of the life biofouling becomes invasive. By the regulation, the biofouling debris must be collected and treated further and cannot be dumped to water directly. So far, Indonesian Government has released regulation on antifouling control, ship hull, and tank cleaning under Ministry of Transportation Regulation Number 29/2014 [6]. Details of strategy to anticipate the attack of biofouling is not available, thus Indonesia follows the International Maritime Organization (IMO) guidelines with continuous efforts on strategy development under IMO Glofouling Project [7].

In order to understand in more details, the strategy and implementation of biofouling management in Indonesia, the current paper reviews the resolution MEPC.378 (80) [8] together with the Indonesian Government

Regulation as guidelines for the control and management of ships' biofouling to minimize its effects on Indonesian economy. Survey questionnaires are made and the participants include related parties in the maritime fields such as shipyards and classification societies.

2 Method

This study adopts a mixed-method approach that integrates literature review, data collection through regulatory analysis and field surveys, and targeted investigations across key maritime sectors in Indonesia. The overall objective is to evaluate the alignment between the regulatory framework for biofouling management and its actual implementation within shipyards, classification societies, and coastal communities.

First, a comprehensive literature review was undertaken to examine international and national regulatory instruments governing biofouling management. Core documents include IMO Resolution MEPC.378(80) [8] as the global guideline, complemented by relevant Indonesian government regulations that operationalize biofouling control and the prevention of invasive aquatic species. Academic publications, technical guidelines, and policy reports related to biofouling management practices, antifouling technologies, and invasive species pathways were also reviewed to establish a theoretical and regulatory foundation for the study.

Second, systematic data collection was conducted through two primary channels:

1. Survey-based data collection, which employed structured questionnaires distributed to shipyards, regulatory bodies, and classification societies. The survey aimed to capture stakeholders' understanding of biofouling management, their degree of compliance with IMO guidelines, and the technological or operational challenges encountered in practical implementation. Survey responses were analyzed to identify patterns of compliance, awareness levels, and gaps between regulatory expectations and industrial practices.
2. Regulatory data collection, involving the examination of national regulations, ministerial decrees, and technical standards that dictate the obligations of ship operators, shipyards, and inspection bodies in Indonesia.

Third, investigations were extended to traditional fishing communities to assess their awareness of biofouling and the extent to which fouling organisms affect the performance, safety, and economic sustainability of small fishing vessels. Surveys in some fisheries communities such as in Brondong (East Java) and Pengambengan (Bali) indicated that the fishermen do not really understand how to cope with biofouling on their fishing vessels. They just cleaned the surface and repainted it without checking if the biofouling has infiltrated inside the wooden boat hence can break the boat later. Consequently, this affected the safety of the boat and reduce its performance at sea. These investigations combined semi-structured interviews and observational assessments to document the experiences of fishers with hull fouling, their existing cleaning practices, and their knowledge of regulations associated with biofouling management.

The triangulation of these three methodological components allows the study to establish a holistic understanding of biofouling management in Indonesia. The literature review provides the regulatory and conceptual baseline; the survey data from shipyards and regulatory institutions quantify the level of implementation; and the community-level investigations offer insights into socio-economic impacts and local awareness. Combined, these methods, supported by thematic (qualitative) analysis of interviews, enable a rigorous assessment of the extent to which national implementation aligns with IMO's global framework and where gaps remain between regulation and practice.

3 Result and Discussions

3.1 The IMO Regulation (MEPC.378 (80)

IMO released MEPC.378 (80) adopted on 7 July 2023 as the guidelines for the control and management of ship's biofouling to minimize the transfer of invasive aquatic species (IAS). Continues studies by IMO members have identified the invasive aquatic species as one of the major threats for the well-being in the ocean. The guidelines have the intention to reduce the enlargement of biofouling growth on ships. Biofouling management practices may also improve a ship's hydrodynamic performance and can be effective at enhancing energy efficiency and reducing emission of toxic gases to the atmosphere from ships.

Fig. 1. Biofouling attack on ship hull: (a) barnacle dominated, and (b) barnacles, calcareous deposit, and soft fouling on ship's rudder

In the case of Indonesia, the Indonesian government determine outline for preventing the spreading of IAS and biofouling management in accordance with IMO regulations by national regulations, namely Minister of Transportation regulation on PM 29/2014. According to the regulation, an assessment and survey must be carried out to all ships of more than 400 GT or 24 m length operated in Indonesian waters; anti-fouling management and control are taken into consideration. An example of biofouling attack on a ship in Indonesia is shown in Figure 1. It is apparent that the biofouling (barnacles) attacks the surface of ship hull (Figure 1a) and barnacles combine with calcareous deposit and soft fouling attack ship's rudder.

The cleaning of biofouling can be carried out on land, e.g., using dry dock, and under water activities using professional divers together with appropriate tools as well as using robots of remotely controlled underwater vehicles. Furthermore, the debris must be treated properly and cannot be splashed out by water to dump the biofouling into the sea; collecting the debris with further treatment is recommended. Figure 2 shows the cleaning process of biofouling debris in a shipyard.

Fig. 2. Biofouling debris

3.2 Survey Investigation

In order to understand if the biofouling debris has been treated according to the regulation a series of surveys were carried out at shipyards in Balikpapan and Surabaya, together with special visit and survey to BKI (Indonesian Classification) which has the authority, see Figure 3. The survey findings are presented in Table 1.

PT. Galangan Balikpapan Utama

PT. Adiluhung Sarana Segara Indonesia

PT. Dumas Tanjung Perak Surabaya Shipyard PT. Biro Klasifikasi Indonesia

Fig. 3. Survey Activities Documentations

Table 1 presents a structured assessment of biofouling-related practices and regulatory conditions within Indonesian shipyards and classification societies, examined through technical, social, and economic dimensions. Collectively, these dimensions illustrate both the strengths and the systemic gaps that influence the overall effectiveness of biofouling management in the national maritime sector.

PT. Galangan Balikpapan Utama engages in meticulous hands-on inspections and supervision, focusing primarily on critical aspects such as hull integrity, the application of antifouling coatings, and the overall operations conducted within drydock facilities. This comprehensive approach ensures that every vessel undergoes thorough evaluations to maintain structural soundness and performance efficiency. The team employs advanced techniques and industry best practices to assess the condition of the hull, ensuring that any potential issues are identified and addressed promptly. Additionally, the careful application of antifouling materials is executed to enhance the longevity and operational efficiency of the vessels, preventing marine growth that can adversely affect performance. The drydock operations are meticulously managed to provide a safe and efficient environment for maintenance and repairs, reflecting the company's commitment to quality and excellence in maritime services.

PT. Adiluhung Sarana Segara Indonesia ensure adherence to various regulatory frameworks, such as the Minister of Transportation Regulation No. 29/2014, which pertains to antifouling measures and the cleaning of tanks. This compliance process could involve close collaboration with the Indonesian Classification Bureau (BKI) as well as the Ministry of Transportation to ensure that all operational practices meet the established safety and environmental standards. Engaging with these regulatory bodies is crucial for maintaining the integrity of maritime operations and ensuring that the company aligns with national policies aimed at promoting sustainable practices within the shipping industry.

PT. Dumas Tanjung Perak Surabaya Shipyard is intricately involved in the execution of antifouling systems, which are essential for maintaining the integrity and performance of marine vessels. The shipyard prioritizes hull maintenance to ensure that ships operate efficiently and safely, minimizing the risk of damage and enhancing their longevity. Additionally, the facility adheres to stringent pollution control procedures, reflecting its commitment to environmental stewardship and compliance with maritime regulations. By integrating these critical practices, PT. Dumas Tanjung Perak Surabaya Shipyard not only supports the operational needs of its clients but also contributes to the broader goal of sustainable shipping practices.

PT. Biro Klasifikasi Indonesia (BKI) serves as the national classification society tasked with ensuring adherence to various regulatory frameworks, including the Minister of Transportation Regulation No. 29 of 2014. This regulation specifically addresses standards related to the cleaning of hulls and tanks, which are critical for maintaining maritime safety and environmental protection. BKI's role involves rigorous verification processes to confirm that vessels comply with these established standards, thereby promoting best practices within the shipping industry and safeguarding marine ecosystems from potential pollutants. Through its oversight, BKI contributes significantly to the enhancement of maritime operations in Indonesia, ensuring that they align with both national and international safety and environmental regulations.

From a technical perspective, shipyards in Indonesia predominantly conduct hull-cleaning activities on repair vessels, while new-build facilities seldom undertake such procedures. This distinction suggests the existence of different operational priorities between maintenance-focused and production-oriented shipyards. The documented reliance on sandblasting, as well as the drying and mixing of biofouling residues, underscores the continued dependence on traditional surface-preparation techniques. The mention of licensed waste-management vendors and containment measures indicates an emerging awareness of environmental stewardship; however, the absence of detailed procedural standards suggests room for greater harmonization and regulatory enforcement. Ensuring consistent environmental controls is particularly important given the increasing international focus on hull fouling as a vector for invasive aquatic species.

The social aspects highlight a workforce that is generally trained for specialized cleaning operations, especially in sandblasting. The provision of standard personal protective equipment (PPE) reflects baseline compliance with occupational safety norms, though the availability of supplementary protective gear in some shipyards reveals variability in safety culture across facilities. Reports of minimal community impacts, with occasional odors noted in industrial zones, imply that current practices do not pose significant perceived social burdens. Nevertheless, this may reflect limited monitoring or underreporting rather than actual environmental neutrality. Comprehensive community-level assessments would help validate the extent of social acceptability and identify any latent concerns.

Economically, the cost structure of hull cleaning displays wide variability, ranging from relatively minor expenses during underwater or basic repair activities (0.5–5%) to substantially greater costs when painting and sandblasting are included (up to 45–50%). This spread indicates that hull-cleaning expenditures are highly sensitive to the scope and intensity of surface-renewal procedures. The behaviour of shipowners further contributes to this variability: some prefer frequent cleaning cycles without investing in antifouling paints, while passenger vessels typically adopt routine repainting. Such differences highlight the absence of a unified economic strategy for hull maintenance within the industry. Notably, the lack of community complaints regarding biofouling waste suggests either effective waste handling or limited public engagement with maritime environmental issues. Systematic socioeconomic impact assessments would be beneficial in validating these observations.

Finally, the section on classification societies underscores Indonesia's regulatory posture. Although the country has ratified international antifouling regulations through Perpres 66/2014, the corresponding national regulation on biofouling management remains under development. The involvement of BKI in a joint task force with the Ministry of Transportation reflects a proactive institutional approach toward aligning national frameworks with IMO MEPC.378(80). However, the ongoing drafting process highlights an existing regulatory void that may hinder uniform implementation across shipyards. Strengthening these frameworks is essential to ensure compliance, environmental protection, and alignment with global best practices.

Table. 1. Results of data collection

Category	Aspect	Description
Shipyard	Technical	- Hull cleaning is conducted on most repair vessels, while new-build yards rarely engage in such practices. - Waste management generally involves drying/mixing biofouling with sandblasting residues or disposal via licensed vendors. - Procedures ensure containment and prevent contamination of surrounding waters.
	Social	- Workers receive specialized training, primarily in blasting, and are equipped with standard PPE; some shipyards provide additional protective gear. - Community impacts are minimal, with occasional odor reports in industrial areas.
	Economic	- Hull cleaning costs range from 0.5–5% of repair/underwater work and may reach 45–50% when painting or sandblasting is included. - Shipowners either invest in antifouling coatings or prefer more frequent cleaning; passenger vessels typically undergo annual repainting. - No community complaints related to biofouling waste have been recorded.
Classification Society	-	Indonesia has ratified international antifouling regulations (Perpres 66/2014), but national rules on biofouling and IAS management aligned with IMO MEPC.378(80) are still being drafted. BKI is part of a joint task force with the Ministry of Transportation to develop these regulatory strategies.

Although Indonesia has ratified international antifouling regulations through Perpres 66/2014, the absence of a finalized national regulation on biofouling management presents a significant gap in the regulatory framework. This gap may lead to the inconsistent application of antifouling measures across different shipyards, potentially undermining efforts to control the spread of invasive aquatic species and protect marine ecosystems. Therefore, the development of comprehensive national biofouling management regulations is critical to ensure that international standards, such as those outlined in IMO MEPC.378(80), are effectively translated into enforceable domestic policies.

The involvement of the BKI in a joint task force alongside the Ministry of Transportation demonstrates a strategic and collaborative approach to addressing this regulatory shortfall. By aligning national policies with international mandates, the task force aims to establish uniform biofouling management practices that facilitate compliance. However, the ongoing drafting process underscores the complexity of integrating international guidelines into the national legal system, which requires coordination among multiple stakeholders and careful consideration of local maritime industry conditions. Until this process is complete, the regulatory void may pose challenges to consistent implementation and monitoring, highlighting the need for accelerated policy development and stakeholder involvement.

4 Conclusions

This study investigates the coherence between international and national regulations on biofouling management in Indonesia and their practical application in the maritime sector. Utilising a mixed-methods approach, this study integrates a comprehensive literature review, surveys conducted among shipyards and classification societies, and field investigations within traditional fishing communities. A critical analysis of the International Maritime Organization's MEPC.378(80) guidelines in conjunction with Indonesian regulations was undertaken. The survey findings indicate that shipyards predominantly engage in hull-cleaning activities for repair vessels, employing techniques such as sandblasting and various waste management practices, which exhibit inconsistent environmental controls. Although training programs for the workforce and safety protocols are in place, the economic implications of hull cleaning show significant variability across different operations. Further socialization and education should be given to fisheries communities due to their high numbers in population and the majority of them are very poor; the do not have adequate money to maintain their boats ad well as handling the biofouling and its effects.

Despite Indonesia's ratification of international antifouling regulations, the corresponding national legislation is still in the developmental phase, raising concerns about potential discrepancies in implementation. This study underscores the urgent need to establish comprehensive national regulations governing biofouling management and enhance regulatory frameworks to ensure alignment with global best practices. By addressing these gaps, Indonesia can better facilitate compliance and promote sustainable practices within its maritime industry, ultimately contributing to the protection of marine ecosystems and enhancing the sector's economic viability.

Acknowledgement

The research is funded under a research scheme called "ITS Centre Collaboration Research" Batch 1/2025 with contract number: 1952/PKS/ITS/2025.

References

[1] F. Weber and N. Esmaeili, "Marine biofouling and the role of biocidal coatings in balancing environmental impacts," *Biofouling*, vol. 39, no. 6, pp. 661–681, Jul. 2023, doi: 10.1080/08927014.2023.2246906.

[2] M. L. Hakim, I. K. Suastika, and I. K. A. P. Utama, "A practical empirical formula for the calculation of ship added friction-resistance due to (bio)fouling," *Ocean Engineering*, vol. 271, Mar. 2023, doi: 10.1016/j.oceaneng.2023.113744.

[3] IMO, "Biofouling." Accessed: Dec. 14, 2025. [Online]. Available: https://www.imo.org/en/ourwork/environment/pages/biofouling.aspx

[4] I. K. A. P. Utama, Y. A. Hermawan, R. C. Ariesta, S. Risdiyanto, M. Sitinjak, and W. Ardhiyanto, "Protecting the Country from Bio-invasion, a Case Study of Biofouling Management in Indonesia," *IOP Conf Ser Earth Environ Sci*, vol. 1250, no. 1, p. 012022, Oct. 2023, doi: 10.1088/1755-1315/1250/1/012022.

[5] OECD, *OECD Economic Outlook, Volume 2024 Issue 2*, vol. 2024. OECD Publishing, 2024. doi: 10.1787/d8814e8b-en.

[6] "Permenhub No. 29 Tahun 2014." Accessed: Dec. 14, 2025. [Online]. Available: https://peraturan.bpk.go.id/Details/103722/permenhub-no-29-tahun-2014

[7] "GloFouling Partnerships Project." Accessed: Dec. 14, 2025. [Online]. Available: https://www.imo.org/en/ourwork/partnershipsprojects/pages/glofouling-project.aspx

[8] IMO, "RESOLUTION MEPC.378(80) (adopted on 7 July 2023) 2023 GUIDELINES FOR THE CONTROL AND MANAGEMENT OF SHIPS' BIOFOULING TO MINIMIZE THE TRANSFER OF INVASIVE AQUATIC SPECIES," 2023.

BIO Web of Conferences **216**, 08005 (2026) https://doi.org/10.1051/bioconf/202621608005
SRCM 2025

Analysis of floating concrete breakwater mooring tension by DualSPHysics

Haryo Dwito Armono[1*], *Noris* Yolanda[1], and *Wahyudi* Tjitrosiswoyo[1]

[1]Ocean Engineering Department, Faculty of Marine Technology, Institut Teknologi Sepuluh Nopember, Surabaya, Indonesia

Abstract. An analysis of the effect of the distance between a Floating Concrete Breakwater (FCB) with dual configurations on the tension of the catenary mooring line system is presented. The Smoothed Particle Hydrodynamics method in DualSPHysics software was used to analyze the maximum tension on FCB mooring lines at the 1:10 model scale. The results indicate that the gap between FCBs influenced the mooring line's tension. In the narrow FCB (40cm width), the mooring line tension was reduced by 20% when the gap is reduced to 20 cm (50% width) from 30 cm (75% width) and increased by 1% when the gap was increased from 30 cm (75% width) to 40 cm (100% width). With a few exceptions, the wide FCB (102.5cm width) shows an 8% tension increase at a gap of 50cm (50%) to 75cm (75% width) and a 16% tension drop when the gap was reduced to 75cm (75% width) from 1 m (100% width). The wave's height and periods also influence the mooring line tension. At a wave height of 0.18 meters and 2.21 seconds wave period of 2.21 s, the maximum tensions for a narrow single FCB were 187.54 kN at the fairlead and 185.189 kN at the anchor. At a wave height of 0.28 m and a wave period of 2.84 s, the maximum tensions of the wide single FCB is 359.75 kN at the anchor and 362.42 kN at the fairlead. In the dual configuration, the maximum tension was recorded when the gap between the floating breakwaters was 50% of the width, either for the narrow or wide FCB.

1 Introduction

Ports are important infrastructures that support a country's economy. However, the use of remote ports in Indonesia is sometimes not optimal. High sea waves hinder port optimization because they can damage port infrastructures, disrupting the loading and unloading of goods and passengers. In general, these problems can be solved by developing coastal protection structures such as breakwaters[1]. However, not all ports in Indonesia can be built with rubble mound breakwater. Some ports in Indonesia have a sufficiently deep-water depth; therefore, if a rubble mound breakwater is built, the volume of material will incur expensive costs and take a long time. To overcome these problems, the construction of floating breakwaters is a viable solution

* Corresponding author: armono@its.ac.id

Floating concrete breakwaters (FCB) have brought significant innovations to protect ports from high waves in locations where conventional breakwaters have proven ineffective. Conventional breakwaters require complex planning and construction, and additional materials are required, resulting in high costs. Nevertheless, in deep waters, the construction of FCBs is less expensive than that of conventional breakwaters[2]. Other advantages of FCBs include their efficiency, which is not affected by tides and sea level changes, minimal environmental impact, low construction and demolition costs, short installation times, and the flexibility to adjust modules and layouts to accommodate future changes in use[3].

This study uses the SPH method with DualSPHysics software, which is proven accurate for simulating the movements of floating structures under the influence of regular waves[4,5]. FCB's performance is affected by immersion depth and wave conditions but is hardly affected by the density and gravity of the ballast water[6]. Dual FCBs are superior to single FCBs for wave reduction[7]. In addition, the distance between FCBs with dual configurations can affect the damping-wave performance and the tension on the mooring line[8,9]. This paper presents a model of a double-configuration FCB and catenary mooring system with distance variation. This study focused on a catenary mooring system so that the tension did not exceed the limit of breaking the load, as per the appropriate rules

2 Research Description

2.1 Research Data

The data needed in this study is the data of the numerical flume tank model used for the simulation. The existing FCB products were simulated in numerical flume tanks at 1: 10 scale for validation to ensure that the results of the numerical experiments reflect real-life situations as accurately as possible. Environmental data for the numerical flume tanks and existing FCB products are presented in Table 1.

Table 1. FCB dimension and environmental data.

Data Type	Variables	Scaled Value	Unit
FCB Structure Data	Length	0,745	m
	Width	0,5	m
	Depth	0,3	m
Environmental Data	Water Depth	0,514	m
	Wave Height	0,05	m
	Wave Period	1,2	s

Two types of FCB models were compared to the simulation results. The dimensions of each type were scaled 1:10 according to the Froude Scale Law shown in Table 2, and the shapes and dimension symbols of each model are presented in Figures 1 and 2.

Table 2. FCB's dimension

FCB Types	Variables		Actual Value	Scaled Value
Varian A (Narrow FCB)	Length	[L]	20 m	2 m
	Width	[B]	4 m	0,4 m
	Depth	[D]	1,8 m	0,18 m
	Weight	[W]	55 ton	55 kg
	Freeboard	[F]	0,5 m	0,05 m

FCB Types	Variables		Actual Value	Scaled Value
Varian B (Wide FCB)	Length	[L]	20 m	2 m
	Width	[B]	10,25 m	1,025 m
	Depth	[D]	3 m	0,3 m
	Weight	[W]	200 ton	200 kg
	Freeboard	[F]	0,8 m	0,08 m

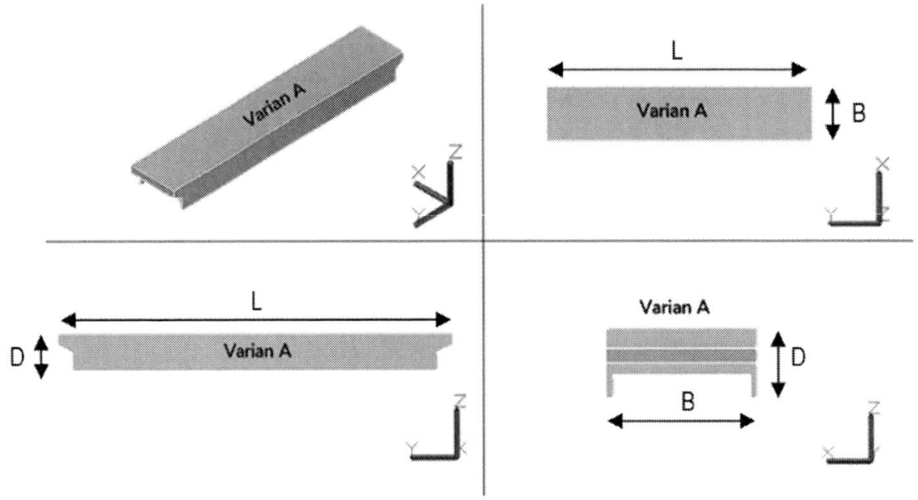

Fig.1. Isometric view of a narrow FCB (Type A).

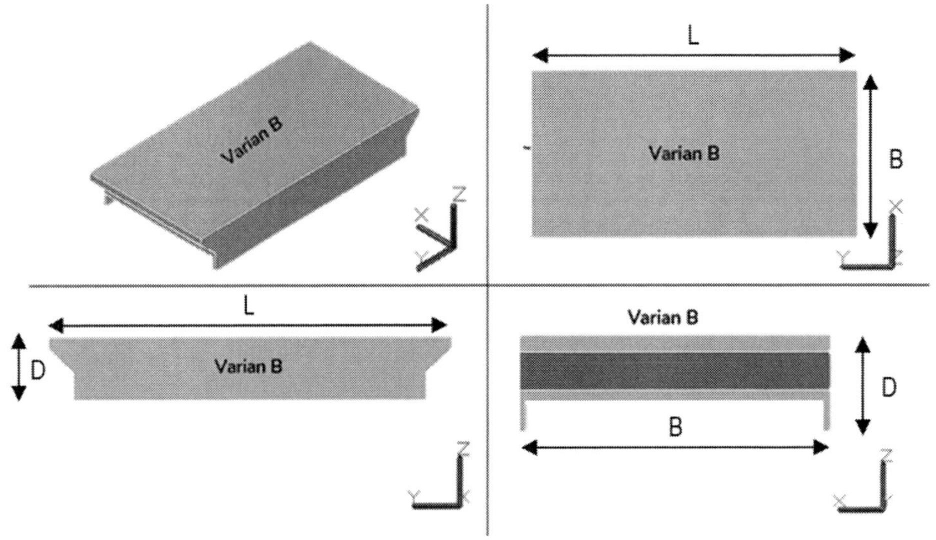

Fig.2. Isometric view of a wide FCB (Type B).

Environmental parameters were obtained and applied to potential FCB locations. The variation in the environmental data can be attributed to the type of FCB analyzed. The narrow-type, designed for calmer ocean conditions, will be tested using environmental data that tend to be smaller than those of the wide-type. The two main parameters to be applied are the wave height and the wave period as shown in Table 3.

Table 3. Environmental data.

FCB Types	Variabels	Prototype	Model
A (Narrow FCB)	Wave Height	1,4 m	0,14 m
		1,6 m	0,16 m
		1,8 m	0,18 m
	Wave Period	1,4 m	0,14 m
		1,6 m	0,16 m
		1,8 m	0,18 m
B (Wide FCB)	Wave Height	2 m	0,2 m
		2,4 m	0,24 m
		2,8 m	0,28 m
	Wave Period	4 s	1,26 s
		7 s	2,21 s
		9 s	2,84 s

The reference dimensions of the FCB units are given in Table 3, and the distance between the structures was also adjusted to 1:10 based on the Froude Scale. For each FCB unit, three spacing configurations were used according to the structure width as presented in Table 4. The distance between FCB unis and the water depth in the model is illustrated in Figure 3.

Table 4. FCB distance scenario.

FCB Types	Distance	S (m)
A (Narrow FCB)	0.5 B	0,2
	0.75 B	0,3
	B	0,4
B (Wide FCB)	0.5 B	0,5
	0.75 B	0,75
	B	1

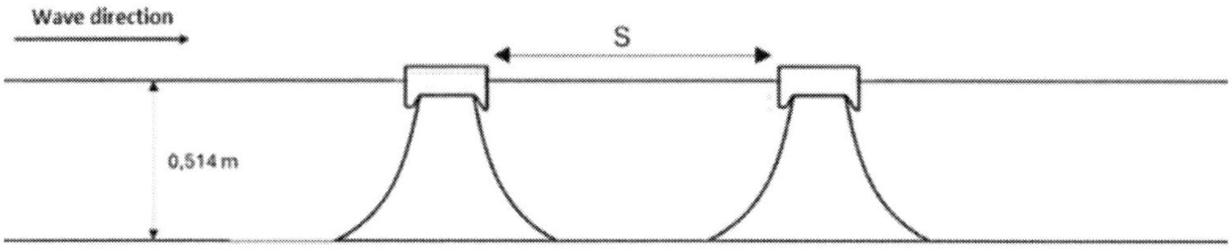

Fig. 3. The distance between FCB units

2.2 Numerical Flume Tank Model

A schematic of the numerical flume tank prepared by DualSPHysics is shown in Figure 4. After model preparation in DualSPHysics, a validation process was performed to determine whether the obtained model could be used for model exploration and further analysis. The FCB products to be simulated on the modeled tank flume are shown in Figure 5.

Fig. 4. Schematic of a numerical flume tank on DualSPHysics.

Fig. 5. FCB configuration model schematics on DualSPHysics.

The simulation results are presented in the form of the FCB structural movement and the response amplitude operator (RAO). The model's movement characteristics were compared to those of the RAO model reported by Chen et al. (2023). The modeling is based on 2 dimensional, so the resulting RAO only includes the surge, heave, and pitch. The tolerance value of the RAO comparison difference must be less than 10% [10]. The validation results in Tables 5–7 show that the RAO results differ by less than 10%. The RAO comparison is also shown in Figure 6. Therefore, numerical flume tank modeling can be used for further analysis.

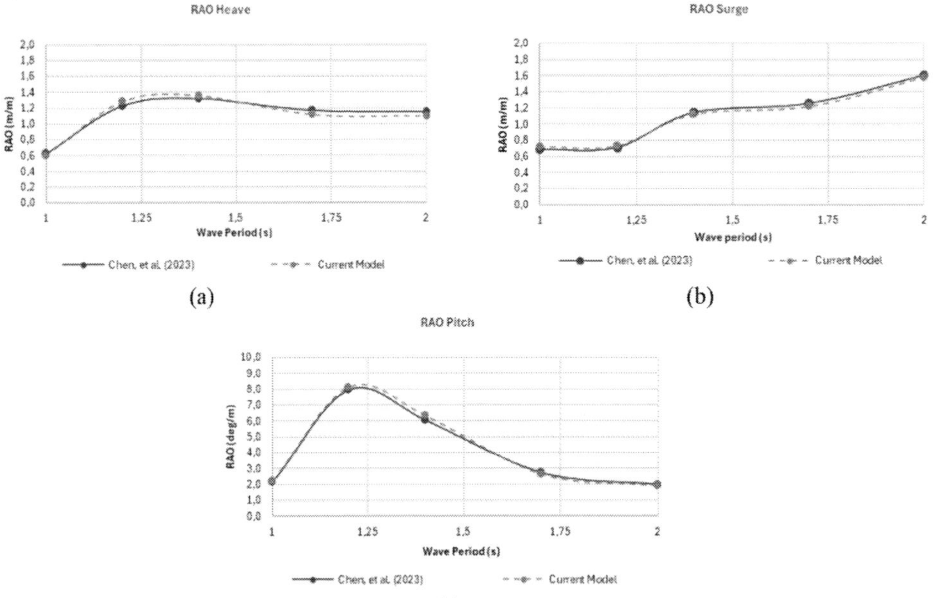

Fig. 6. RAO validation (a) heave (b) surge (c) pitch.

Table 5. Heave RAO validation results.

Wave Period (s)	RAO heave length (m/m)		Error	Remarks
	Chen et al. (2023)	Current Model		
1	0.628	**0.598**	4.81%	PASS
1.2	1.222	**1.281**	4.84%	PASS
1.4	1.314	**1.358**	3.34%	PASS
1.7	1.164	**1.115**	4.23%	PASS
2	1.147	**1.098**	4.27%	PASS

Table 6. Surge RAO validation results.

Wave Period (s)	RAO Surge (m/m)		Error	Remarks
	Chen et al. (2023)	Current Model		
1	0.691	**0.724**	4.71%	PASS
1.2	0.714	**0.736**	3.04%	PASS
1.4	1.143	**1.12**	2.00%	PASS
1.7	1.257	**1.218**	3.11%	PASS
2	1.6	**1.574**	1.63%	PASS

Table 7. Pitch RAO validation results.

Wave Period (s)	RAO Pitch (deg/m)		Error	Remarks
	Chen et al. (2023)	Current Model		
1	2.148	**2.236**	4.11%	PASS
1.2	7.977	**8.122**	1.81%	PASS
1.4	6.068	**6.347**	4.59%	PASS
1.7	2.727	**2.641**	3.16%	PASS
2	1.977	**1.921**	2.85%	PASS

2.3 Floating Concrete Breakwater (FCB) Modeling

The narrow and wide FCB geometric models were prepared using FreeCAD and further integrated with a previously validated numerical flume tank model. The dimensions, structure, and geometry of the scaled FCB are presented in Table 2. The models of the narrow and wide FCB structures are shown in Figures 7 and 8.

Fig. 7. Modeling scheme of narrow FCB on DualSPHysics.

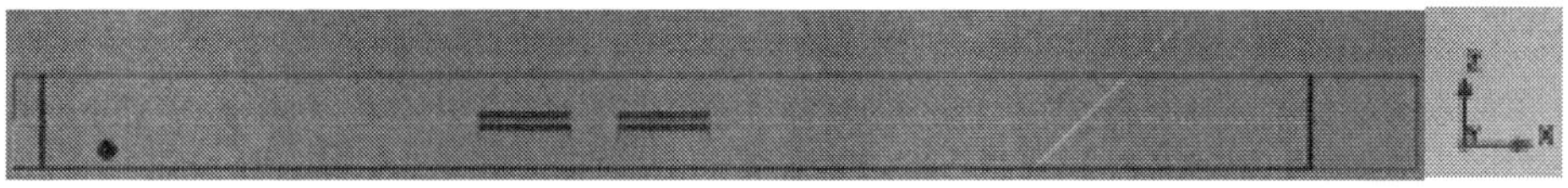

Fig. 8. Scheme for modeling a wide FCB on DualSPHysics.

2.4 Mooring System Modeling

The mooring lines on the FCB were modeled using DualSPHysics in a configuration consisting of two symmetrical mooring lines. The specifications of the mooring line used in this study are given in Table 8.

Table 8. Specification of mooring line.

Description	Scaled Value
Line Length	1,2 m
Diameter	3,6 mm
Mass per unit length	66 kg/100 m
Stiffness	2900 N

Description	Scaled Value
MBL	1020.2 N
Material	High Modulus Polyethylene 'Dynema ® SK75'

The configuration of the mooring line on the FCB is symmetric on each side. To recognize the tension lines during the analysis, each line was defined with a name. The naming details of the mooring line are shown in Figure 9.

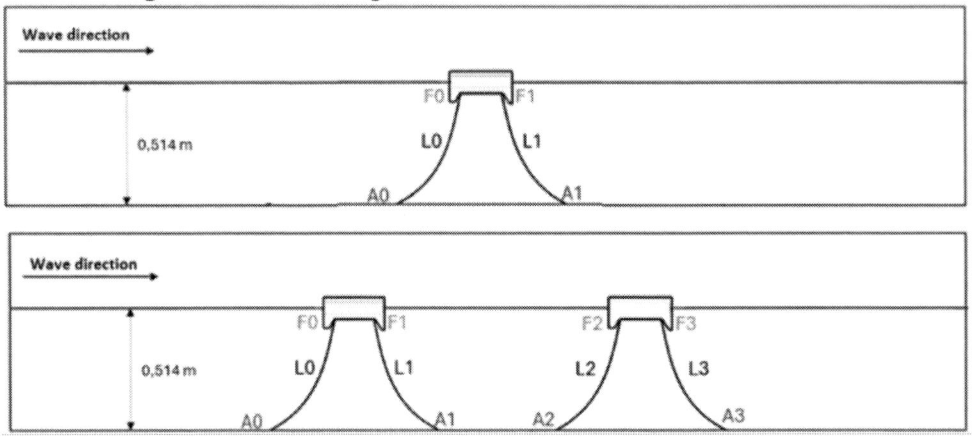

Figure 9. Single- and dual-configuration FCB mooring line labeling.

3 Results and Discussion

3.1 The influence of hydrodynamic parameters

The tension on the mooring line is affected by hydrodynamic variables, i.e., Wave period and height. The variations in the hydrodynamic variables resulted in different tension values. Regardless of the FCB type, hydrodynamic variables influence the maximum tension of the FCB mooring line. This can be observed through the change in the nondimensional magnitude of the wave steepness (H/gT2) with respect to the mooring line tension.

Figure 10 plots the relationship between the wave steepness and the mooring tension in the narrow FCB model. The results show that the mooring line tension tends to decrease as the wave steepness increases.However, when examined in more detail, the mooring line tension exhibits a similar pattern. The pattern of increasing the maximum tension value of the mooring line can be observed for each wave height variation and a given wave period. This suggests that although the mooring line tension generally decreases with increasing wave steepness, there are specific moments when the maximum tension exhibits a consistent increase, reflecting the complexity of the interaction between the wave and the mooring line.

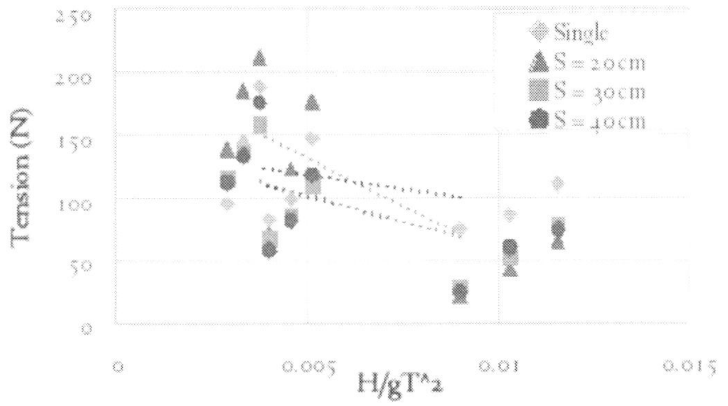

Fig. 10. Relationship between wave steepness and mooring tension in a narrow FCB (type A).

The wide FCB version has the same pattern as the narrow version, in which the trend of the mooring tension decreases with increasing wave steepness, as shown in Figure 11. However, there is still a pattern of increasing the maximum tension value of the mooring line at each wave height and for a certain period. This result shows that the steepness of the wave influences the tension of the mooring line. Thus, hydrodynamic variables (wave height and period) and parameters (wave steepness) affected the magnitude of the mooring line tension in both FCB types.

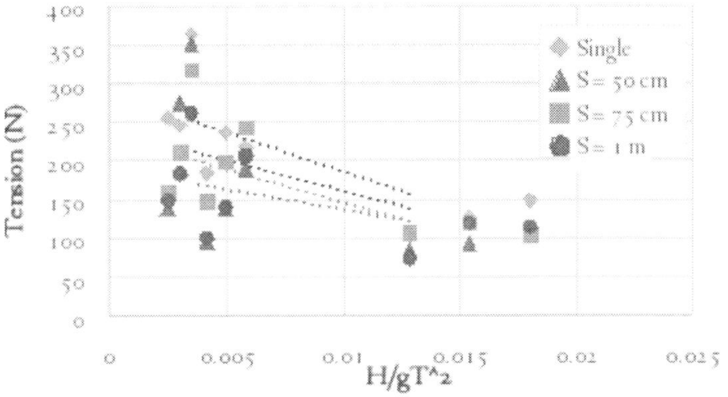

Fig. 11. Relationship between wave steepness and mooring tension for a wide FCB (type B).

3.2 Effect of FCB distance

In addition to the wave height and wave period parameters, this study also discusses the distance between the FCBs, specifically analyzing its effect on the mooring line tension in a dual configuration. The comparison of the distance between the FCB structure and the mooring line tension is shown in Figures 12 and 13. Both types result in a trend toward the mooring-line style, which tends to be inconsistent with the distance between structures.

The narrow version of the FCB exhibits the same pattern at periods of 1.89 s and 2.21 s for every Hs condition, which decreases the mooring line tension by 6%–38%. from a change distance from 20 cm to 30 cm. Then, from a gap increment of 30 cm to 40 cm, the tension of the mooring line also decreased by 2%–13%, except for Hs = 1.8 m, which increased by 8%–11%. Meanwhile, the trend for T = 1.26 s differs from that of the other T variations. At a distance of 20–30 cm, the tension of the mooring line tends to increase by 21%–30% and decreases at a distance of 30 cm to 40 cm by 6%–12%.

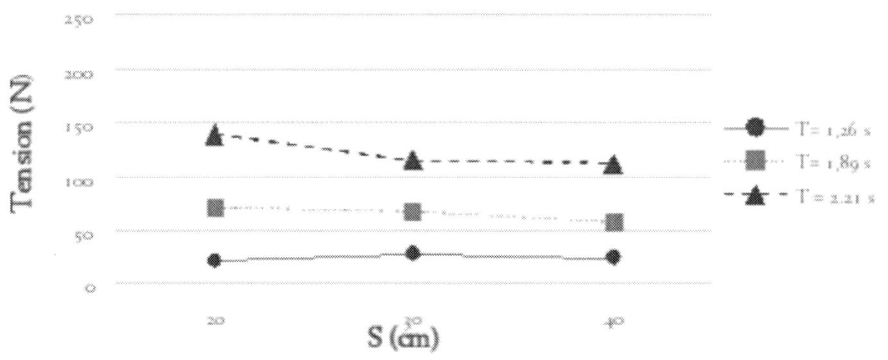

(a). Hs = 0.14 m

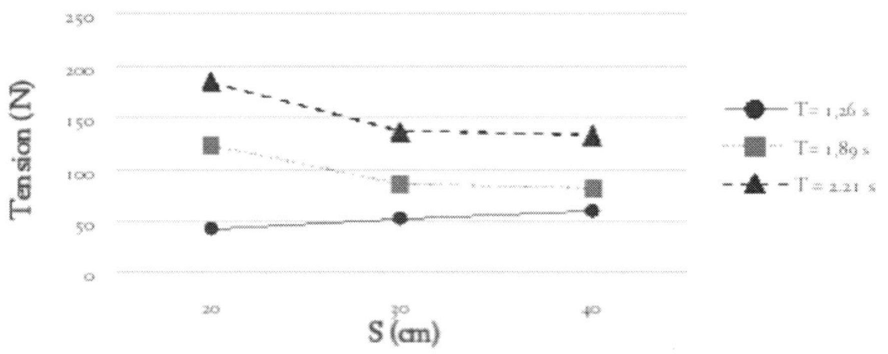

(b). Hs = 0.16 m

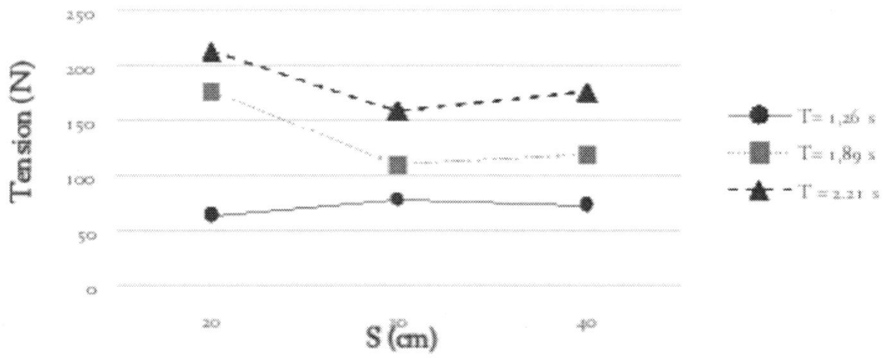

(c). Hs= 0.18m

Fig. 12. Relationship between FCB distance and mooring tension of narrow FCBs (type A) (a) Hs = 0.14 m, (b) 0.16 m, and (c) 0.18 m.

The wide FCB version exhibits a different trend from the narrow. At a wave period of 1.26 s, the mooring line tension increased by 25%–28% as the gap increased from 50 to 75 cm and decreased by 32% when the gap increased from 75 to 1 m. There was an anomaly that occurred in the wave period of 1.26s, when the wave height was 2.8 m, where the mooring line tension decreased by 10% from a distance of 50 cm to 75 cm and an increase of 11% from a distance of 75 cm to 1 m. At wave period 2.21 s, the mooring line tension showed a similar trend: a 28%–56% increase in tension at gap increment from 50 cm to 75 cm and a 15%–32% decrease in tension when the gap increased from 75 cm to 1 m.

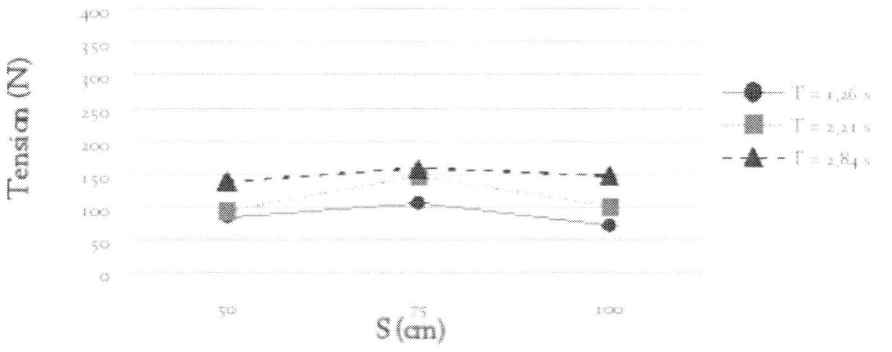

(a). Hs = 0.20 m

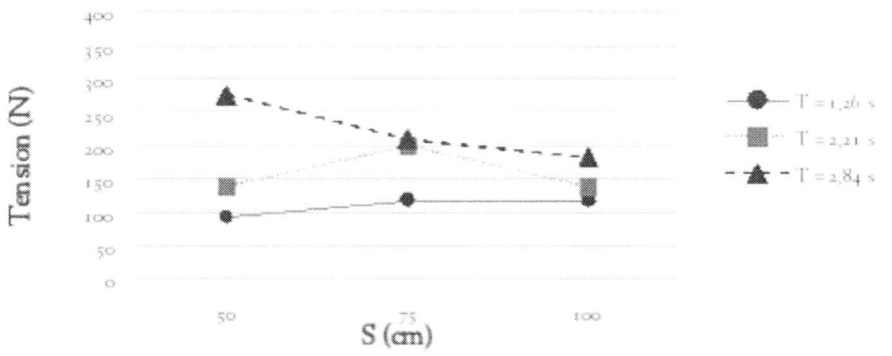

(b). Hs = 0.24 m

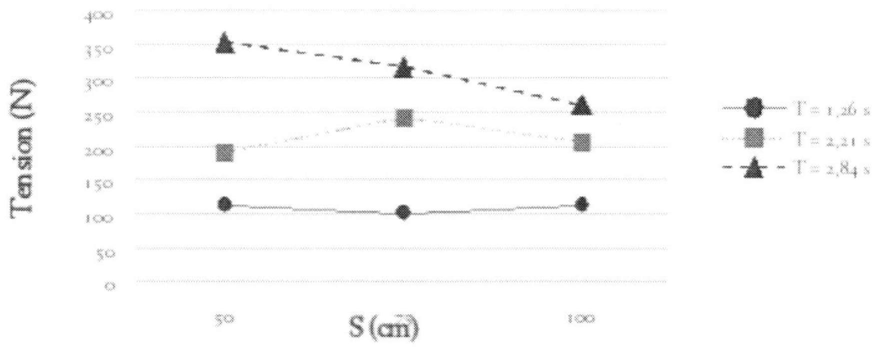

(c). Hs = 0.28 m

Fig. 13. Relationship between the FCB distance and the mooring tension of a wide FCB (type B): Hs = (a) 0.2 m (b) 0.24 m, and (c) 0.28 m.

Overall, the difference in distance significantly influenced the magnitude of the mooring line tension. In the narrow FCB, the average decrease in the mooring line tension was 20% from when the distance of 20 changed to 30 cm and an increase of 1% from the distance change of 30 to 40 cm. In the wide FCB version, the average increase in the mooring line tension was 8%, and the decrease was 16% from a distance of 75 cm to 1 m. However, the large number of anomalies suggests that more research is needed to better understand this phenomenon and reduce the uncertainty of the results. It is important to ensure that the analysis reflects real conditions and can be used as a basis for more accurate decision-making in the future.

3.3 Comparison between single and dual configurations

The mooring tensions of the single and dual configurations are given in Tables 9 and 10.

Table 9. Comparison between single and dual configurations of narrow FCBs (type A).

Configuration Type	Average Maximum Tension (N)	Comparison to Single Configuration
Single	**113.54**	-
Dual, S = 20 cm	**114.80**	**1.11%**
Dual, S = 30 cm	**91.95**	**-19.01%**
Dual, S = 40 cm	**92.80**	**-18.27%**

Table 10. Comparison between single and dual configurations of a wide FCB (type B).

Configuration Type	Average Maximum Tension (N)	Comparison to Single Configuration
Single	**209.05**	-
Dual, S = 50 cm	**164.28**	**-21.42%**
Dual, S = 75 cm	**177.79**	**-14.95%**
Dual, S = 100 cm	**148.99**	**-28.73%**

Table 9 shows that the mooring line tension was almost always lower in the dual configuration than in the single configuration. The distance gap that best reduced the tension of the mooring line for the narrow breakwater was 30 cm (75% of the width), resulting in a reduction of 19.01%. However, when the distance between structures was 20 cm (50% of the width), the mooring line tension increased by 1%. Although this number is still relatively small, this increase indicates the variability in the response of the mooring line to changes in the distance between structures. Overall, the average mooring line tension reduction by the dual configuration of the narrow FCB was 12.06% that of the single configuration.

The mooring line tension of wide FCB's dual configuration was always lower than that in the single configuration, as shown in Table 10. The best configuration for reducing the tension of the mooring line was 100 cm (full width), which reduced the tension by 28.73%. The average mooring line tension reduction by the dual configuration of wide FCBs was 21.7% compared to that of the single configuration.

3.4 Safety Factor

In the mooring analysis of the narrow and wide FCBs, the maximum tension results were used to calculate the safety factor under all the loading conditions. According to the API [11], the safety factor of the mooring line configuration under dynamic conditions is 1.67. The safety factors for all FCB-testing scenarios are presented in Tables 11–14. Based on the results of the maximum tension of the mooring line, its safety factor was found to be greater than 1.67, which means that the configuration of the mooring system and its configuration are suitable for operation.

Table 11. Safety factors for mooring line tension in a single configuration of a narrow FCB (type A).

Wave Height (m)	Wave Period (s)	Max Tension (N)	MBL (N)	Safety Factor	Status
0.14	1.26	73.990	1020.2	13.79	PASS
	1.89	82.075		12.43	PASS
	2.21	94.191		10.83	PASS
0.16	1.26	85.404	1020.2	11.95	PASS
	1.89	98.747		10.33	PASS
	2.21	143.531		7.11	PASS
0.18	1.26	109.832	1020.2	9.29	PASS
	1.89	146.574		6.96	PASS
	2.21	187.540		5.44	PASS

Table 12. A mooring line tension safety factor in a single configuration of a wide FCB (type B).

Wave Height (m)	Wave Period (s)	Max Tension (N)	MBL (N)	Safety Factor	Status
0.2	1.26	107.50	1020.2	9.49	PASS
	2.21	184.10		5.54	PASS
	2.84	254.41		4.01	PASS
0.24	1.26	126.94	1020.2	8.04	PASS
	2.21	236.25		4.32	PASS
	2.84	245.12		4.16	PASS
0.28	1.26	148.36	1020.2	6.88	PASS
	2.21	216.40		4.71	PASS
	2.84	362.42		2.81	PASS

Table 13. Safety factors for mooring line tension in a dual configuration of a narrow FCB (type A).

No.	Description	Wave Period (s)	Max. Tension (N)	MBL (N)	Safety Factor	Status
A.	**Hs = 0.14 m**					
1	S = 0.2 m	1.26	35.95	1020.2	28.38	PASS
		1.89	70.75		14.42	PASS
		2.21	138.82		7.35	PASS
2	S = 0.3 m	1.26	44.33	1020.2	23.01	PASS
		1.89	66.33		15.38	PASS
		2.21	114.94		8.88	PASS
3	S=0,4 m	1.26	40.41	1020.2	25.25	PASS
		1.89	57.64		17.70	PASS
		2.21	111.73		9.13	PASS
B.	**Hs = 0.16 m**					
1	S = 0.2 m	1.26	48.54	1020.2	21.02	PASS
		1.89	123.10		8.29	PASS
		2.21	184.36		5.53	PASS
2	S = 0.3 m	1.26	61.32	1020.2	16.64	PASS
		1.89	85.31		11.96	PASS
		2.21	135.83		7.51	PASS
3	S = 0.4 m	1.26	60.04	1020.2	16.99	PASS
		1.89	81.60		12.50	PASS
		2.21	132.67		7.69	PASS

No.	Description	Wave Period (s)	Max. Tension (N)	MBL (N)	Safety Factor	Status
C.	**Hs = 0.18 m**					
		1.26	61.27		16.65	PASS
1	S = 0.2 m	1.89	175.43	1020.2	5.82	PASS
		2.21	211.93		4.81	PASS
		1.26	78.08		13.07	PASS
2	S = 0.3 m	1.89	109.12	1020.2	9.35	PASS
		2.21	157.98		6.46	PASS
		1.26	73.68		13.85	PASS
3	S= 0.4 m	1.89	118.02	1020.2	8.64	PASS
		2.21	174.86		5.83	PASS

Table 14. A mooring line tension safety factor in a dual configuration of a narrow FCB (type B).

No.	Description	Wave Period (s)	Max. Tension (N)	MBL (N)	Safety Factor	Status
A.	**Hs = 0.2 m**					
		1.26	94.46		10.80	PASS
1	S = 0.5 m	2.21	85.32	1020.2	11.96	PASS
		2.84	131.44		7.76	PASS
		1.26	106.51		9.58	PASS
2	S= 0.75m	2.21	147.20	1020.2	6.93	PASS
		2.84	158.00		6.46	PASS
		1.26	99.46		10.26	PASS
3	S= 1 m	2.21	72.47	1020.2	14.08	PASS
		2.84	147.99		6.89	PASS
B.	**Hs = 0.24 m**					
		1.26	139.56		7.31	PASS
1	S = 0.5 m	2.21	92.69	1020.2	11.01	PASS
		2.84	275.01		3.71	PASS
		1.26	118.99		8.57	PASS
2	S= 0.75 m	2.21	198.80	1020.2	5.13	PASS
		2.84	210.08		4.86	PASS
		1.26	118.75		8.59	PASS
3	S= 1m	2.21	139.64	1020.2	7.31	PASS
		2.84	182.70		5.58	PASS
C.	**Hs = 0.28 m**					
		1.26	112.90		9.04	PASS
1	S = 0.5 m	2.21	188.90	1020.2	5.40	PASS
		2.84	323.27		3.16	PASS
		1.26	101.98		10.00	PASS
2	S= 0.75 m	2.21	242.12	1020.2	4.21	PASS
		2.84	315.63		3.23	PASS
		1.26	113.24		9.01	PASS
3	S= 1 m	2.21	206.10	1020.2	4.95	PASS
		2.84	256.24		3.98	PASS

4 Conclusion/Summary

Based on the analysis of the effect of distance between Floating Concrete Breakwaters (FCBs) with dual configurations on the tension of the catenary mooring line system, it is evident that the distance between FCB units significantly influences the magnitude of the mooring line tension. For a narrow FCB (40 cm width), the mooring line tension decreased by 20% when the gap between the FCBs was reduced from 20 cm to 30 cm and increased by 1% when the gap extended from 30 to 40 cm. In contrast, the wide version of the FCB (102.5 cm width) shows an 8% increase in tension at a gap of 50–75 cm, followed by a 16% decrease when the gap was increased from 75cm to 1 m. The wave height and period also have a notable effect on the mooring line behavior for both FCB configurations, with significant variations observed between the single and dual configurations despite similar overall trends. The maximum tension recorded for the single configuration of the narrow FCB was 185.189 kN at the anchor and 187.54 kN at the fairlead, with a wave height of 0.18 m and a period of 2.21 s. For the wide FCB, the single configuration reached 359.75 kN at the anchor and 362.42 kN at the fairlead at a wave height of 0.28 m and a wave period of 2.84 s. In the dual configuration, the narrow FCB exhibited a maximum tension of 221.93 kN on the L2 mooring line at the fairlead with a gap of 0.2 m, while the wide FCB reached 323.27 kN on mooring lines L0 and L2 with a gap of 0.5 m. The average reduction in the mooring line tension for the dual configurations was 12.06% for the narrow FCB and 21.7% for the wide FCB. Furthermore, the safety factors for all configurations remained above 1.67, indicating that the mooring system and its properties are adequate for operational purposes. These findings highlight the critical role of FCB spacing and environmental conditions in optimizing mooring line performance, suggesting areas for future research to address observed anomalies and enhance the understanding of complex interactions in floating breakwater systems.

References

1. I. A. Hakiki and L. E. Sembiring, "Uji Model Hidrolika Kinerja Pemecah Gelombang Apung Beton Dengan Konfigurasi Ganda," *Jurnal Teknik Hidraulik*, vol. 11, no. 2, pp. 83–92, 2020; doi: 10.32679/jth.v11i2.629.
2. M. W. Fousert, "Floating Breakwater," Delft University of Technology; 2006. [Online]. Available: http://resolver.tudelft.nl/uuid:87d7e889-8aaf-410b-9502-495412c59308
3. P. Ruol, L. M. Artinelli and P. Pezzutto, "Experimental and Numerical Investigation of the Effect of Mooring Stiffness on the Behavior of II-Type Floating Breakwaters," 2012. Accessed: Dec. 15, 2024. [Online]. Available: https://www.researchgate.net/publication/270566886_Experimental_and_numerical_investigation_of_the_effect_of_mooring_stiffness_on_the_behaviour_of_p-type_floating_breakwaters
4. J. M. Domínguez *et al.*, "SPH simulation of floating structures with moorings," *Coastal Engineering*, vol. 153, 2019, doi: 10.1016/j.coastaleng.2019.103560.
5. X. Han and S. Dong, "Interaction between regular waves and floating breakwater with protruding plates: Laboratory experiments and SPH simulations," *Ocean Engineering*, vol. 287, Nov. 2023, doi: 10.1016/j.oceaneng.2023.115906.
6. Z. Liu and Y. Wang, "Numerical studies of submerged moored box-type floating breakwaters with different shapes of cross-sections using SPH," *Coastal Engineering*, vol. 158, no. February, p. 103687, 2020; doi: 10.1016/j.coastaleng.2020.103687.
7. Y. Kun Chen, Y. Liu, D. D. Meringolo, and J. Ming Hu, "Study on the hydrodynamics of a twin floating breakwater by using SPH method," *Coastal Engineering*, vol. 179, 2023; doi: 10.1016/j.coastaleng.2022.104230.

8. C. Ji, X. Deng and Y. Cheng, "An experimental study of double-row floating breakwaters," *Journal of Marine Science and Technology (Japan)*, vol. 24, no. 2, 2019, doi: 10.1007/s00773-018-0554-2.
9. A. N. Williams, "Dual Floating Breakwaters," *Ocean Engineering*, vol. 20, no. 3, pp. 215–232, 1993, doi: 10.1016/0029-8018(93)90021-9.
10. D. Skandali, E. Lourens, and R. H. M. Ogink, "Calibration of response amplitude operators based on measurements of vessel motions and directional wave spectra," *Marine Structures*, vol. 72, p. 102774, Jul. 2020, doi: 10.1016/J.MARSTRUC.2020.102774.
11. API, "API RP 2SK: Design and Analysis of Stationkeeping Systems for Floating Offshore Structures," Dec. 2024.

Distribution and Environmental Parameters of Solen spp around the Pamekasan Coastal Water

Eva Ari Wahyuni[1*], Dwi Nurhayati Adhani[2], Lisa Lisdiana[3]

[1]Department of Natural Science Education, Faculty of Teaching and Education Science, Universitas Trunojoyo Madura, Indonesia; *email: evaariw@trunojoyo.ac.id

[2]Department of Early Childhood Teacher Education, Faculty of Teaching and Education Science, Universitas Trunojoyo Madura, Indonesia; email: dwi.adhani@trunojoyo.ac.id

[3]Department of Biology, Faculty of Mathematics and Natural Science, Universitas Negeri Surabaya, Indonesia; email: lisalisdiana@unesa.ac.id

Abstract. *Solen* sp. has grown to be a significant and economically valuable resource, particularly for the residents of Pamekasan, Madura Island; therefore, overfishing occurs if it is only used for natural purposes. The purpose of this research is to ascertain the density and environmental parameters of *Solen* sp around the Pamekasan southern coast. In September 2024, the study was required a descriptive methodology. *Solen* sp and environmental factors (temperature, salinity, pH, and substrate) served as the research materials. The gravimetric approach was used for the substrate analysis at the Oceanography Laboratory, Department of Marine and Fisheries, University Trunojoyo Madura. The results showed 1) *Solen* sp density of 8-12 individuals/m2 and 2) environmental parameters, including temperature 29-33^{0}C, salinity 31-33 ppt, pH 7.9-8.2, and substrate type of sandy mud. The density of *Solen* sp is thought to be related to environmental parameters that are still in accordance with sea water quality standards for marine biota, so that it becomes a habitat for growth and development of *Solen* sp.

Keywords: distribution, environmental parameters, Pamekasan, Solen sp, substrate

1. Introduction

One of the bivalve species with high economic value is Solen spp. [1, 2], including on the island of Madura [1, 3, 4]. As a result, it has become a significant commodity [5, 6, 7], with a tendency to be high and rising demand each year [4, 5]. An important reason to seek cultivation [8] is the high risk of overfishing [3, 4, 7, 9, 10, 11] due to people's habit of depending on the availability of Solen in nature [12], not least on Madura Island [4, 5]. There has been quite a lot of research on bivalve aquaculture [9, 13] although it has not been detailed because it focuses on one particular study, such as spawning characteristics [14]. Generally, the condition of Solen in some waters of the world, including in Pamekasan Regency [15], is very vulnerable to overfishing [9] as an implication of the high dependence of the community on the availability of Solen in nature [12] which is always utilized for various needs, one of which is community consumption [1, 2] although still on a traditional scale, thus affecting fluctuations in the price of solen commodities [3, 4, 5]. Solen sp was found on Madura Island with a variety of sizes and showed a fairly wide distribution of Solen sp [3]. Environmental characteristics suitable for Solen sp habitat include moderately warm temperatures [16], muddy substrates and/or muddy sand [17], and small changes in salinity [16]. Characteristics of the Madura Strait and Java Sea are quite similar to the general coastal waters around the world, two waters flanking Madura Island, include waves [18, 19] and weak currents [20, 21], temperatures tend to be warm throughout the year [16], and salinity is quite high (generally tropical areas) [16] and dominated by muddy substrates [3, 4, 22]. The water and environmental characteristics as described above are suitable habitats for Solen growth and development [23]. Therefore, the study of environmental parameters and distribution of Solen distribution is important as a first step to create a laboratory-scale experimental design to match the natural habitat of Solen as an effort to cultivate, so as to minimize the potential for overfishing Solen.

2. Methodology

Measurements of environmental parameters including DO, temperature, salinity, pH, and substrate, were conducted at the coastal village of Pragaan (Pamekasan) (Figure 1) in September 2024. The method of measuring environmental parameters is in accordance with [24]. Solen samples were morphologically analyzed to determine their density, surface water and sediment were sampled at the same time using physics parameters supported to explain the relationship between biotic and abiotic factors. DO, temperature, pH, salinity, and substrate have been measured. Water sampling uses seawater and a grab sampler. Sediment substrate are analyzed to estimate the grain size used the granulometric method (6, 7). Grain size separation was carried out. with sieves sized: >2; 1.4; 1; 0.5; 0.250; 0.150; 0.090; 0.063; and <0.063mm. Grain size classification is done by classification [25]. Type determination sediment based on classification Shepard's Triangle Diagram [26].

3. Result and Discussion

Environmental parameters measured during the study included DO, temperature, salinity, and pH as well as sediment substrate (Table 1). The DO value in 2023 was not measured [23], so it cannot be compared with the measurement value in 2024. The range of DO values shows 5.6 - 6.1 mg/L and is suitable for Solen sp habitat [24]. The range of sea surface temperature in 2023 was 28 - 30^{0}C, while in 2024 it was 29 - 31^{0}C, thus showing an increase in sea surface temperature by 1^{0}C and still reflecting the general characteristics of sea surface temperature patterns in the tropics [27] which tend to be warm throughout the year [28]. The salinity range in 2023 is 32 - 33 ppt, while in 2024 it is 31 - 33 ppt. In general, the value of the salinity range only differs at the lower interval, and is still suitable for Solen habitat [3] and reflects salinity conditions in the tropics [4]. The pH range shows the same range in 2023 and 2024, namely 8.0 - 8.2. Generally, in the tropics, the pH range is quite small and is suitable for Solen habitat [3, 23]. The substrate type shows no change (in 2023 and 2024), which is a muddy substrate.

The density of Solen during the measurement of environmental parameters was 8-12 species (week 1), 10-12 (week 2), 9-12 (week 3), and 11-12 (week 4), respectively. The relatively similar species density indicates that the differences in environmental parameter measurement values (Table 1) are not significant enough to affect the distribution of Solen. A decrease in salinity value will affect the survival of Solen [29], because Solen cannot adapt to low salinity [30, 31]. Sea surface temperature at the study site tends to be warm [32] and very small fluctuations, so it does not affect the survival of Solen [33]. Small fluctuations in all measured environmental parameters are not expected to affect the abundance or survival of Solen. However, in laboratory-scale experiments, exposure to temperature [33] and salinity [30, 31] significantly affected Solen survival.

4. Conclusion

For the environment of Solen sp., the range of environmental characteristics and circumstances is adequate and reasonable (DO 5.6 – 6.1 mg/L, temperature 29–31 °C, pH 8.0–8.2, salinity 32–33 ppt). According to the in-situ measurement, the condition is quite good. The Solen abundance is 8 – 12 species/m2. Temperature and salinity became the important environmental parameters (in laboratories scales) effecting on survival of *Solen* spp. We highly suggest to analysis the relationship between the environmental parameters and the abundance of *Solen* spp. based on statistical methods.

Acknowledgment

We thank to our research partners in Molecular Laboratory at Faculty of Mathematic and Natural Science, Surabaya State University.

References

[1] KURNIAWAN, E. R., AMBARWATI, R., & ISNANINGSIH, N. R. (2024). Diversity, abundance, and utilization of bivalves on the south coast of Pamekasan, Madura Island, Indonesia. *Biodiversitas Journal of Biological Diversity, 25*(6).

[2] Rusyadi, S. (2006). Karakteristik gizi dan potensi pengembangan kerang pisau (*Solen* spp) di perairan Kabupaten Pamekasan Madura. Skripsi. Program Studi Teknologi Hasil Perikanan. FPIK. IPB.
[3] Wahyuni, E. A., Insafitri, I., Ihsan, M. N., & Ciptadi, G. (2017). Biodiversity Solen sp. in Madura Island. *Applied Mechanics and Materials, 862*, 115-120.
[4] Wahyuni, E. A., Insafitri, I., Ciptadi, G., & Ihsan, M. N. (2016). Distribusi Solen sp di Perairan Kabupaten Bangkalan. *Jurnal Kelautan: Indonesian Journal of Marine Science and Technology*, *9*(1), 17-22.
[5] Darriba, D., 2006. Razor clams and aquaculture: studies in Galicia (NW of Spain). Abstracts Aquaculture Europe 2006. Firenze, Italy, 9–13 May.
[6] Barón, P.J., Real, L.E., Ciocco, N.F., Ré, M.E., 2004. Morphometry, growth and reproduction of an Atlantic population of the razor clam Ensis macha (Molina, 1782). Sci. Mar. 68, 211–217.
[7] Tuck, I.D., Bailey, N., Harding, M., Sangster, G., Howell, T., Graham, N., Breen, M., 2000. The impact of water jet dredging for razor clams, Ensis spp., in a shallow sandy subtidal environment. J. Sea Res. 43 (1), 65–81.
[8] Rahim, S. A. K. A. (2011). Razor clam (Solen spp.) fishery in Sarawak, Malaysia. *Kuroshio Science*, *5*(1), 87-94.
[9] Trisyani, N. (2020, February). Population growth and fishery status of the Lorjuk shellfish (Solen sp.) on Pamekasan beaches, Indonesia. In *IOP Conference Series: Earth and Environmental Science* (Vol. 441, No. 1, p. 012074). IOP Publishing.
[10] Fahy, E., Gaffney, J., 2001. Growth statistics of an exploited razor clam (Ensis siliqua) bed at Gormanstown, Co Meath, Ireland. Hydrobiologia 465, 139–151.
[11] Gaspar, M.B., Castro, M., Monteiro, C.C., 1998. The influence of tow duration and tooth length on the number of damaged razor clams, Ensis siliqua. Mar. Ecol., Prog. Ser.169, 303–305.
[12] Nadia, A., Leila, H., FMF, C. J., Zohra, H., & Da, C. F. (2016). Reproductive Cycle of the Razor Clam Solen marginatus (Pulteney, 1799) in the Southern Mediterranean Sea (Gulf of Gabes, South Tunisia).
[13] Da Costa, F., & Martínez-Patiño, D. (2009). Culture potential of the razor clam Solen marginatus (Pennánt, 1777). *Aquaculture*, *288*(1-2), 57-64.
[14] Trisyani, N., & Insafitri, I. (2012). METODE PEMIJAHAN PADA KERANG LORJUK (SOLEN sp).
[15] Ambarwati, R., & Irawan, B. (2020). The population of solen sp. (bivalvia: Solenidae) from Pamekasan, Indonesia. *Ecology, Environment and Conservation*, *26*, S199-S204.
[16] Medina-Lopez, E., & Ureña-Fuentes, L. (2019). High-resolution sea surface temperature and salinity in coastal areas worldwide from raw satellite data. *Remote Sensing*, *11*(19), 2191.
[17] Wang, C., Chen, M., Qi, H., Intasen, W., & Kanchanapant, A. (2020). Grain-size distribution of surface sediments in the chanthaburi coast, Thailand and implications for the sedimentary dynamic environment. *Journal of Marine Science and Engineering*, *8*(4), 242.
[18] Møller Jess, R., & Norstedt, F. (2023). Comparison of Particle Tracking and ADCP Including Construction of the Rating Curve at Islandsfallet.
[19] Bradtke, K., Brodziński, W., & Herman, A. (2025). Signatures of Breaking Waves in a Coastal Polynya Covered with Frazil Ice: A High-Resolution Satellite Image Case Study of Terra Nova Bay Polynya. *Remote Sensing*, *17*(18), 3198.
[20] Siswanto, A. D. (2015). Longshore current characteristics in Madura Strait. *Procedia Environmental Sciences*, *23*, 34-38.
[21] Utari, P. A., Setiabudidaya, D., Khakim, M., & Iskandar, I. (2019). Dynamics of the South Java Coastal Current revealed by RAMA observing network. *Terrestrial, Atmospheric & Oceanic Sciences*, *30*(2).
[22] Nurainie, I., & Wiyanto, D. B. (2021). Karakteristik sebaran sedimen dasar di Perairan Kalianget Kabupaten Sumenep. *Juvenil: Jurnal Ilmiah Kelautan Dan Perikanan*, *2*(3), 243-254.
[23] Wahyuni, E. A., Qomaria, N., & Lisdiana, L. (2023). Anatomical respiratory organ of Solen sp. In *BIO Web of Conferences* (Vol. 74, p. 03007). EDP Sciences.
[24] Asadi, M. A., Iranawati, F., & Andini, A. W. (2018). Ecology of bivalves in the intertidal area of Gili Ketapang Island, East Java, Indonesia. *Aquaculture, Aquarium, Conservation & Legislation*, *11*(1), 55-65.
[25] Doeglas, D. J. (1968). Grain-size indices, classification and environment. *Sedimentology*, *10*(2), 83-100.
[26] Shepard, F. P. (1954). Nomenclature based on sand-silt-clay ratios. *Journal of sedimentary Research*, *24*(3), 151-158.
[27] Zainab, S., Handajani, N., & Wibisana, H. (2020, May). Mapping of Sea Surface Temperature and its Correlation with Salinity Concentration in The Coastal Beach of Bangkalan Madura District. In *IOP Conference Series: Earth and Environmental Science* (Vol. 506, No. 1, p. 012050). IOP Publishing.
[28] Lin, J., Tang, M., & Wijaya, F.S. (2022). Satellite Oceanographic Data Processing and Analysis: Correlation Between Nino 3.4 Sea Surface Temperature & Sea Surface High and Wind. *Indonesian Journal of Earth Sciences*.

[29] Nel, H. A., Perissinotto, R., Taylor, R. H., & Carrasco, N. K. (2011). Salinity tolerance of the bivalve Solen cylindraceus (Hanley, 1843) (Mollusca: Euheterodonta: Solenidae) in the St Lucia estuary. *African Invertebrates*, *52*(2), 1-12.
[30] Tegelberg, H. C. (1964). Growth and ring formation of Washington razor clams. *Wash. Dep. Fish. Fish. Res. Pap*, *2*(3), 69-103.
[31] Souissi, N., Boughriba, S., Abdelhedi, O., Hamdi, M., Jridi, M., Li, S., & Nasri, M. (2019). Extraction, structural characterization, and thermal and biomedical properties of sulfated polysaccharides from razor clam Solen marginatus. *RSC advances*, *9*(20), 11538-11551.
[32] Niamaimandi, N. (2012). Biological parameters and abundance of the razor clam, (Solen brevis), from the Bushehr area of the Persian Gulf. *Agriculture, Forestry and Fisheries*, *1*(1), 1-6.

Compile all figure and tables

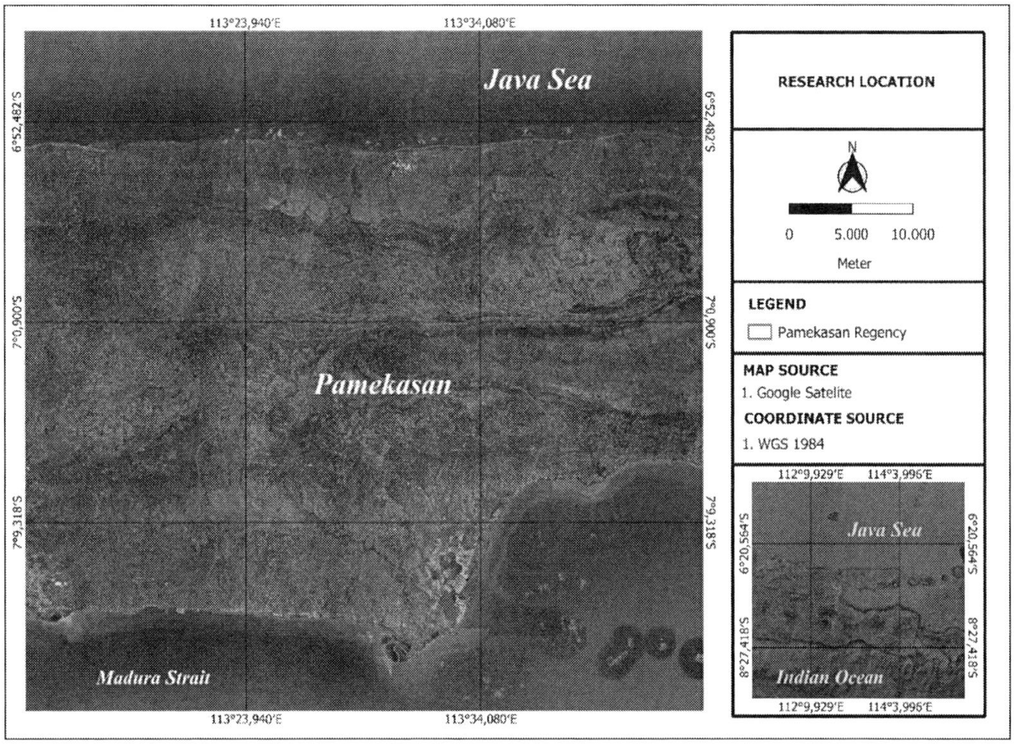

Figure 1. Research location

Table 1. Environmental Parameters of Solen in Pamekasan

Time	Weeks	DO (mg/L)	Temperature (°C)	Salinity (ppt)	pH	Substrate	References
July 2023	1	-	30	32	8.1	Muddy Sand	[23]
	2		30	33	8.1		
	3		31	32	8.0		
	4		31	32	8.0		
August 2023	1		31	33	8.0		
	2		30	32	8.2		
	3		29	33	8.1		
	4		30	33	8.0		
September 2024	1	5.6	31	33	8.0		Current research
	2	5.8	31	32	8.0		
	3	6.1	31	32	8.1		
	4	6.1	31	32	8.0		

Compile all figure and tables

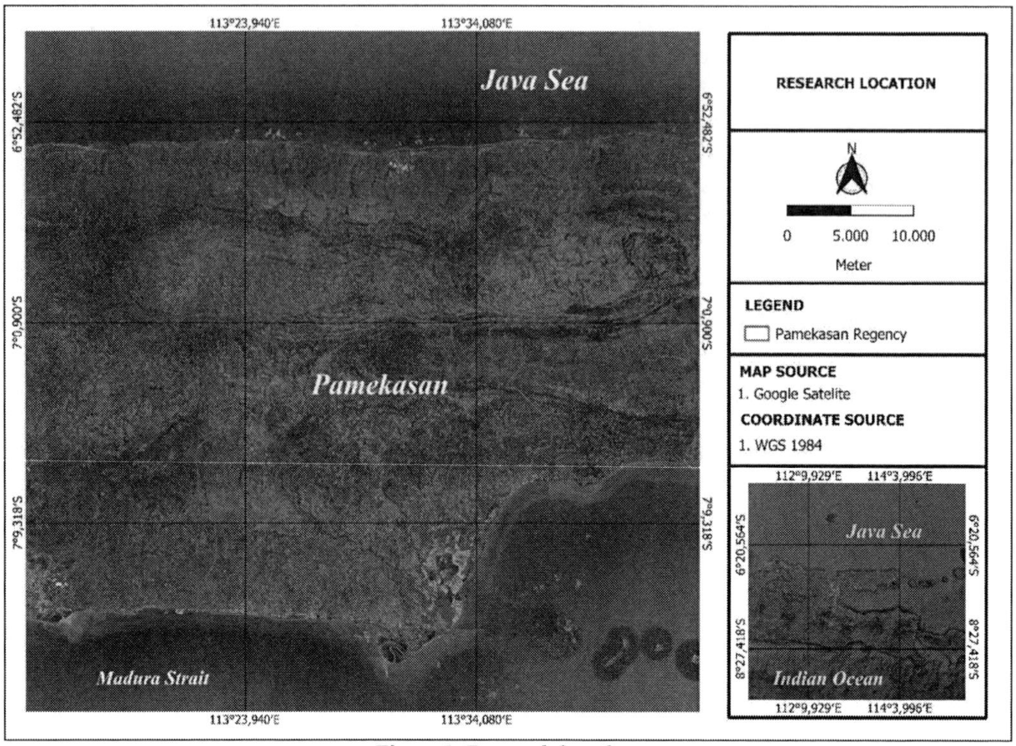

Figure 1. Research location

Table 1. Environmental Parameters of Solen in Pamekasan

Time	Weeks	DO (mg/L)	Temperature (°C)	Salinity (ppt)	pH	Substrate	References
July 2023	1	-	30	32	8.1	Muddy Sand	[23]
	2		30	33	8.1		
	3		31	32	8.0		
	4		31	32	8.0		
August 2023	1		31	33	8.0		
	2		30	32	8.2		
	3		29	33	8.1		
	4		30	33	8.0		
September 2024	1	5.6	31	33	8.0		Current research
	2	5.8	31	32	8.0		
	3	6.1	31	32	8.1		
	4	6.1	31	32	8.0		

Wind variability across in the Tomini Bay

Ilham Ainur Rahmat Hidayah[1], *Ashari* Wicaksono[1,3], *Putri* Meissarah[2], *Aries Dwi* Siswanto[1,2], *Ary Giri Dwi* Kartika[1,3]

[1]Study Program of Marine Science, Department of Marine and Fisheries, Universitas Trunojoyo Madura, Bangkalan, Indonesia.
[2]Parangtritis Geomaritime Science Park, Indonesia Geospatial Information Agency, Yogyakarta, Indonesia.
[3]Laboratory of Oceanography, Universitas Trunojoyo Madura, Bangkalan, Indonesia.

Abstract. This study aimed to investigate the wind variability and spectral analysis from wind data, specially focusing on Tomini Bay. We accessed data from Copernicus at https://marine.copernicus.eu/. We collected wind data in 2023 with longitude and latitude in the study area from 119° W to 130° E and 3° N to 2° S. The wind product includes hourly Level-4 sea surface wind and stress fields, with a horizontal spatial resolution of 0.125 degrees. In the Tomini Bay, the factors that are responsible for these one-of-a-kind spatial characteristics are the upwelling and downwelling conditions. Coastal areas exhibit consistent positive or negative anomalies, probably associated with upwelling and downwelling processes. Open-ocean areas have significant patterns of anomalies, indicating large-scale meteorological phenomena such as trade winds or monsoonal influences. The PSD graph focuses on the variations in wind pressure anomalies in Tomini Bay throughout 2023. Persistent oscillations around values close to zero suggest that the anomalies originate from either the long-term average or climate factors. The graph's peaks and through signify periods of volatility in wind pressure anomalies, either from seasonal variability, atmospheric dynamics, or particular climatic variables affecting regional wind patterns.

1. Introduction

In the investigation of atmosphere-ocean interactions [1-4], wind variability is particularly critical. Wind is a component of the weather that affects existence, including the utilization of renewable energy sources [7, 8], as well as the ecosystem as a whole [9]. Solar radiation, Earth's rotation [10], and uneven heating pressure gradients [11] interact to shape wind variability on a global scale. Surface roughness [12] and terrain [11] are examples of regional and local scale factors that can affect wind properties, leading to intricate and variable patterns of variability. This dynamic system has the potential to substantially influence the formation of monsoon and trade winds [13, 14], as well as the circulation of the atmosphere and ocean currents [4, 6]. Recent study has enhanced understanding of the elements affecting wind variability in both temporal and spatial dimensions, highlighting the influence of climate events [15] and anthropogenic [14, 16].

Current instruments and methods have developed rapidly, such as the use of supercomputers [4, 17] in computing so that they will be able to predict wind variability that has an impact on climate. With these instruments and technologies, researchers can find out the causes of wind variability in region that has high accuracy on a spatial and temporal scale [4-6].

The European Union initiated the Copernicus programme, which has become a critical initiative for Earth observation, providing a comprehensive collection of environmental data. Copernicus Atmosphere Monitoring Service (CAMS) and Copernicus Climate Change Service (C3S) are exceptional sources of wind and climate data, enabling researches to conduct highly precise studies of atmospheric and oceanic dynamics. These data are instrumental in the advancement of meteorology, oceanography, and energy applications, as well as the support of global initiatives to comprehend and mitigate the effects of climate change [18-21].

Tomini Bay is one of Indonesia's major bays; the closing line between the two mouths is more than 24 nautical miles long. The island of Sulawesi officially divides Tomini Bay into three provinces: Gorontalo, North Sulawesi, and Central Sulawesi. The Molucca Sea borders the exposed eastern edge of Tomini Bay [22, 23]. This study aimed to investigate the wind variability and spectral analysis from wind data, specially focusing on Tomini Bay.

2. Materials and Methods

We accessed data from Copernicus at https://marine.copernicus.eu/. We collected wind data in 2023 with longitude and latitude in the study area from 119° W to 130° E and 3° N to 2° S. The wind product includes hourly Level-4 sea surface wind and stress fields, with a horizontal spatial resolution of 0.125 degrees. We compute temporally-averaged difference fields [24] using scatterometer data from Metop-B and Metop-C ASCAT and corresponding variables from the European Centre for Medium-Range Weather Forecast (ECMWF) operational model. Copernicus's development and uses have made many important steps forward in coastal oceanography by using sentinel improvements and land effluents for shores and oceans. These changes have also made it easier for both old and new stakeholders to use CMEMS-based products [4, 21]. In this study we used variables latitude, longitude, air density, northward and eastward wind for u and v component, respectively. Analysis of wind using scatterometer are active sensors in that they transmit a microwave radar signal to the water surface and monitor the received energy [25].

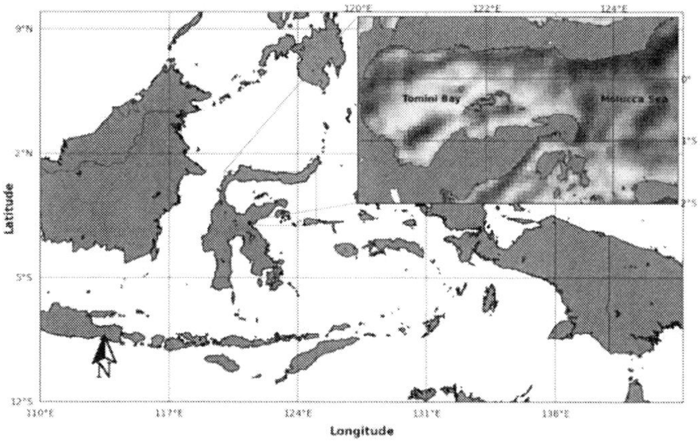

Fig 1. Research Location

3. Results and Discussion

3.1. Wind Variability

The extent of the wind variance that can be seen in Tomini Bay is hown in **Figure 2**, which can be found over here. The months of May through August were usually slow, and September to December were usually significant. However, the months of January through April were very different from these patterns. According to the findings of a spatial study, the primary portion of Tomini Bay was subjected to more intense wind conditions during the whole year. Both the topographic and oceanographic qualities have an impact on the development of spatial patterns that are either strengthened or weakened.

In the Tomini Bay, the factors that are responsible for these one-of-a-kind spatial characteristics are the upwelling and downwelling conditions. In Tomini Bay, the nutrient mixing and sea surface temperature (SST) variability may be determined by the effect of ocean circulation, such as the Indonesia Throughflow (ITF), which is both affected and produced by the wind.

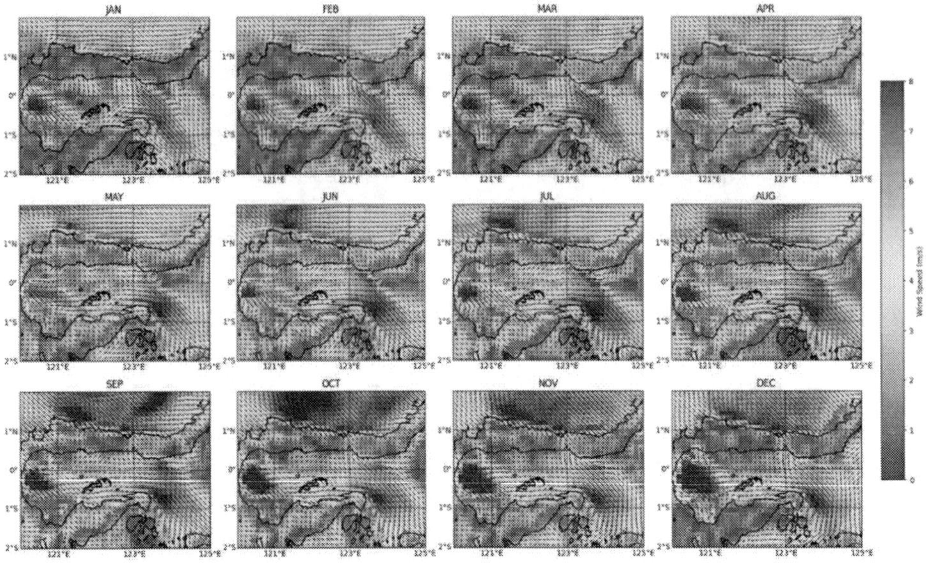

Fig 2. Wind variability in the Tomini Bay Periods 2023

3.2. Wind Stress Curl and Ekman Pumping Velocity

The monthly plots January to December (**Fig. 3**) exhibit distinct geographical and temporal fluctuations in wind stress curl anomalies throughout the research area. Positive anomalies (red hues) and negative anomalies (blue hues) alternate throughout the year, indicating strong seasonal fluctuation. Significant positive anomalies are noted in certain months, indicating heightened cyclonic activity or circumstances conducive to upwelling at these times. In contrast, pronounced negative anomalies, indicative of anticyclonic activity, prevail in other months. This indicates that the atmospheric circulation in this area is influenced by both seasonal dynamics and local wind patterns.

The geographical patterns indicate specific areas with significant abnormalities, perhaps associated with topography and oceanographic characteristics such as coasts, bathymetry, and ocean currents. Coastal areas exhibit consistent positive or negative anomalies, probably associated with upwelling and downwelling processes. Open-ocean areas have significant patterns of anomalies, indicating large-scale meteorological phenomena such as trade winds or monsoonal influences.

Seasonal variations are evident, particularly in March, April, and October, when anomaly patterns see significant alterations. Changes in large-scale atmospheric circulation, such as the beginning or decrease of monsoons or variations in the location of atmospheric pressure systems, may be a sign that the transition months are now occurring. It is possible that prevailing summer wind regimes, which may either exacerbate or attenuate wind stress curl anomalies, are responsible for the persistence of some patterns during the summer months (for example, from June to August).

The abnormalities in wind stress curl presumably significantly influence ocean-atmosphere interactions in this area. Positive curl anomalies may augment upward Ekman pumping, affecting thermocline dynamics and nutrient accessibility in the upper ocean. This affects biological productivity and regional climatic variability. Negative curl anomalies indicate downward Ekman pumping, which inhibits vertical nutrient transfer and influences surface ocean conditions. The identified anomalies correspond with established atmospheric and oceanic processes, such monsoon systems, ENSO effects, or regional circulation patterns. Further inquiry is required to elucidate the factors behind these anomalies, especially their association with large-scale climatic patterns.

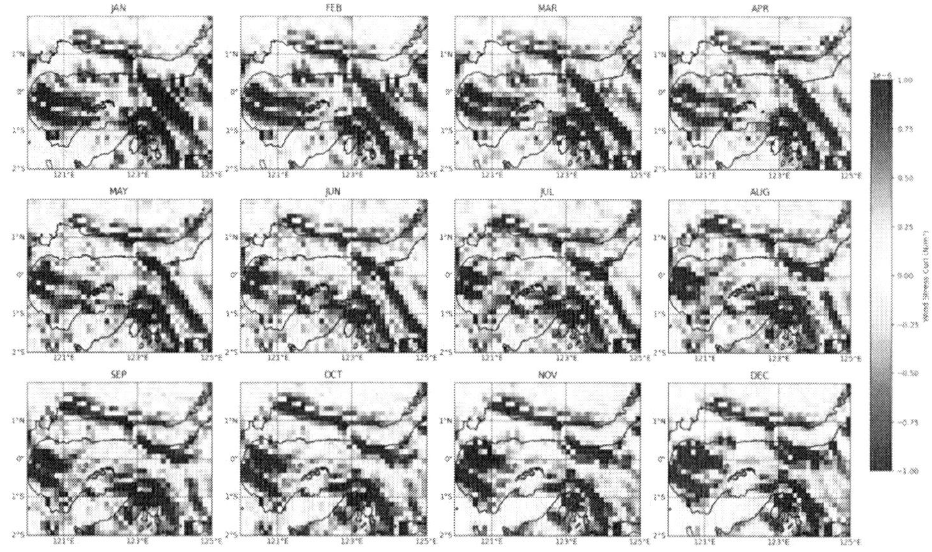

Fig 3. Wind Stress Curl

The distribution of EPV during January to May (**Fig. 4**), Tomini Bay has negative EPV values reaching -15 to -5 m/s which indicates the direction of vertical velocity is moving upwards. The results of EPV values show the same intensity and wind speed during the 5 months. The peak of negative EPV occurred in May (transitional season 1), where the EPV speed reached -15 m/s upwards. Negative EPV can cause water masses to move towards the surface through ekman pumping. According to Hori et al., [26] negative EPV can change sea surface conditions, because the results of the study show an indication of upwelling when EPV is negative. In June to August (east season), EPV begins to increase because in that season the wind speed also increases from the previous season. Wind speed has an influence on ocean dynamics, especially in the process of vertical movement of water masses [30]. EPV is part of the vertical circulation of seawater that occurs due to the interaction between the coriolis force and wind stress at the sea surface [30].

Entering the second transitional season (September - November), the intensity of EPV is getting stronger and spreads throughout Tomini Bay, because the variation of wind stress along Tomini Bay causes divergence or convergence of water masses, thus triggering upwelling or downwelling [27]. Ekman pumping velocity is influenced by higher wind speeds that tend to produce greater friction at the sea surface. The friction can affect the pressure gradient, which in turn determines the intensity of Ekman transport [28]. Strong and stable winds can cause Ekman pumping velocities to tend to be higher, compared to low wind speeds because they

BIO Web of Conferences **216**, 08007 (2026) https://doi.org/10.1051/bioconf/202621608007
SRCM 2025

only produce smaller pressure gradients, thus weakening Ekman pumping [29]. Research Berdandes et al., [29] shows that EPV with negative values in the Arafura Sea indicates an upwelling phenomenon, but Ekman pumping is not only influenced by wind stress, but is influenced by depth factors that also affect the value of EPV in an area.

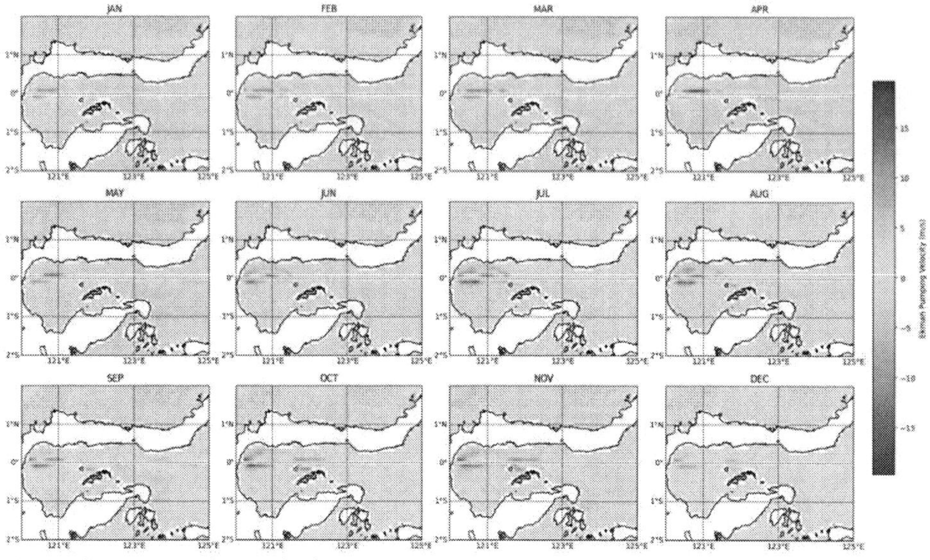

Fig 4. Ekman Pumping Velocity

3.3. Spectral Analysis

Figure 5 displays the results of the spectrum analysis. The initial graphic (above) illustrates the time series of the wind stress anomaly, recorded at coordinates (121° N, 0°E) throughout the year 2023. The second figure (below) depicts the power spectral density (PSD), which signifies the frequency domain attributes of a wind stress anomaly.

The PSD graph focuses on the variations in wind pressure anomalies in Tomini Bay throughout 2023. Persistent oscillations around values close to zero suggest that the anomalies originate from either the long-term average or climate factors. The graph's peaks and through signify periods of volatility in wind pressure anomalies, either from seasonal variability, atmospheric dynamics, or particular climatic variables affecting regional wind patterns.

The PSD approach enables the identification of atmospheric phenomena such as tropical cyclones or local winds through time series observations. The reason for this is the high-frequency fluctuations that result from ocean-atmosphere interactions, such as energy transfer in marine environments.

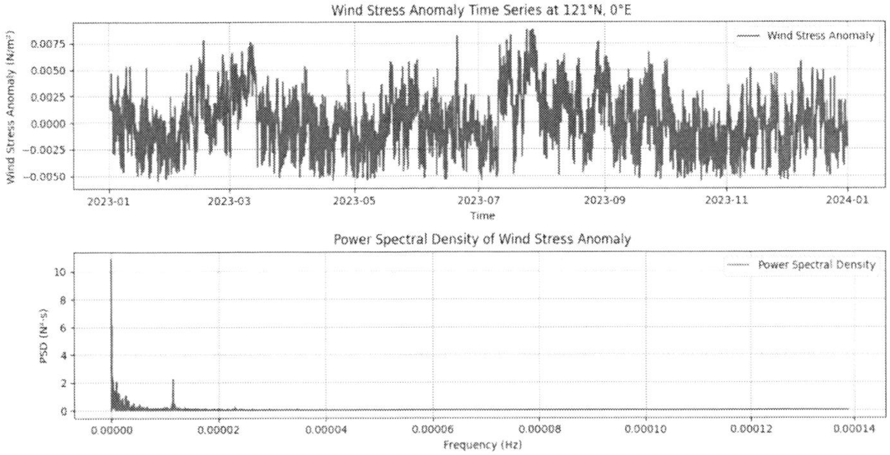

Fig 5. Spectral analysis and power spectral analysis

4. Conclussion

The monsoon that blow across in the area have an effect on the wind variability in the Tomini Bay. It is during the months of October and March that the transition and western season take place. The PSD analysis of CMEMS wind data, using coordinates 121° N, 0° E, indicated substantial variations in the data. Spectral analysis is a technique used to detect anomaly in the data.

Acknowledgement: We would like thanks to the blind reviewers and people for support for this article.

References

1. Capotondi, Antonietta, Shayne McGregor, Michael J. McPhaden, Sophie Cravatte, Neil J. Holbrook, Yukiko Imada, Sara C. Sanchez et al. "Mechanisms of tropical Pacific decadal variability." *Nature Reviews Earth & Environment* 4, no. 11 (2023): 754-769.
2. Trindade, Ana Filipa Mestre. "Development of high-resolution L4 ocean wind products." (2023).
3. Mayer, Michael, Magdalena Alonso Balmaseda, Stephanie Johnson, and Frederic Vitart. "Assessment of seasonal forecasting errors of the ECMWF system in the eastern Indian Ocean." *Climate Dynamics* 62, no. 2 (2024): 1391-1406.
4. Yari, Sadegh, Volker Mohrholz, and Mohammad Hadi Bordbar. "Wind variability across the North Humboldt Upwelling System." *Frontiers in Marine Science* 10 (2023): 1087980.
5. Mockert, Fabian, Christian M. Grams, Tom Brown, and Fabian Neumann. "Meteorological conditions during periods of low wind speed and insolation in Germany: The role of weather regimes." *Meteorological Applications* 30, no. 4 (2023): e2141.
6. Wixtrom, Tyler. "High-Resolution Near-Surface Wind Analysis with the Weather Research and Forecasting Model." (2023).

7. Antonini, Enrico GA, Edgar Virgüez, Sara Ashfaq, Lei Duan, Tyler H. Ruggles, and Ken Caldeira. "Identification of reliable locations for wind power generation through a global analysis of wind droughts." *Communications Earth & Environment* 5, no. 1 (2024): 103.
8. Richardson, D., A. J. Pitman, and N. N. Ridder. "Climate influence on compound solar and wind droughts in Australia." *npj Climate and Atmospheric Science* 6, no. 1 (2023): 184.
9. Brunet, Dominique, Reza Valipour, and Yerubandi R. Rao. "Wind variability over a large lake with complex topography: Lake of the Woods." *Journal of Great Lakes Research* 49, no. 1 (2023): 112-121.
10. Li, Peiyao, Jijian Lian, Chao Ma, and Jinliang Zhang. "Complementarity and development potential assessment of offshore wind and solar resources in China seas." *Energy Conversion and Management* 296 (2023): 117705.
11. Allouche, Mohammad, Elie Bou-Zeid, and Juho Iipponen. "The influence of synoptic wind on land–sea breezes." *Quarterly Journal of the Royal Meteorological Society* 149, no. 757 (2023): 3198-3219.
12. Luu, Linh N., Erik van Meijgaard, Sjoukje Y. Philip, Sarah F. Kew, Jouke HS de Baar, and Andrew Stepek. "Impact of surface roughness changes on surface wind speed over western Europe: A study with the regional climate model RACMO." *Journal of Geophysical Research: Atmospheres* 128, no. 12 (2023): e2022JD038426.
13. Zhang, Gang-Feng, Cesar Azorin-Molina, Deliang Chen, Tim R. McVicar, Jose A. Guijarro, Kai-Qiang Deng, Lorenzo Minola et al. "Variability and trends of near-surface wind speed over the Tibetan Plateau: The role played by the westerly and Asian monsoon." *Advances in Climate Change Research* (2024).
14. Jia, Zixuan, Massimo A. Bollasina, Wenjun Zhang, and Ying Xiang. "Response of the link between ENSO and the East Asian winter monsoon to Asian anthropogenic aerosols." *EGUsphere* 2024 (2024): 1-28.
15. Zhou, Yongchao, Xin Gao, and Jiaqiang Lei. "Characteristics of dust weather in the Tarim Basin from 1989 to 2021 and its impact on the atmospheric environment." *Remote Sensing* 15, no. 7 (2023): 1804.
16. Ullah, Irfan, Sourav Mukherjee, Sidra Syed, Ashok Kumar Mishra, Brian Odhiambo Ayugi, and Saran Aadhar. "Anthropogenic and atmospheric variability intensifies flash drought episodes in South Asia." *Communications Earth & Environment* 5, no. 1 (2024): 267.
17. Jiang, Jianhong, Yongjin Yu, Yang Zhou, Shimeng Qian, Hao Deng, Jianning Tao, and Wei Hua. "Influence of Model Resolution on Wind Energy Simulations over Tibetan Plateau Using CMIP6 HighResMIP." *Atmosphere* 15, no. 11 (2024): 1323.
18. Ricker, Marcel, Arno Behrens, and Joanna Staneva. "The operational CMEMS wind wave forecasting system of the Black Sea." *Journal of Operational Oceanography* (2024): 1-22.
19. Buontempo, Carlo, Samantha N. Burgess, Dick Dee, Bernard Pinty, Jean-Noël Thépaut, Michel Rixen, Samuel Almond et al. "The Copernicus climate change service: climate science in action." *Bulletin of the American Meteorological Society* 103, no. 12 (2022): E2669-E2687.
20. Garrigues, Sebastien, Julien Chimot, Melanie Ades, Antje Inness, Johannes Flemming, Zak Kipling, Istvan Laszlo et al. "Monitoring multiple satellite aerosol optical depth (AOD) products within the Copernicus Atmosphere Monitoring Service (CAMS) data assimilation system." *Atmospheric Chemistry and Physics* 22, no. 22 (2022): 14657-14692.

21. Sanchez-Arcilla, Agustin, Joanna Staneva, Luigi Cavaleri, Merete Badger, Jean Bidlot, Jacob T. Sorensen, Lars B. Hansen et al. "CMEMS-based coastal analyses: conditioning, coupling and limits for applications." *Frontiers in marine science* 8 (2021): 604741.
22. Rachma, Tia Rizka N., and Rizka Windiastuti. "Spatial Study of Indonesia's Historic Bay: A Case Study in Tomini Bay." In *IOP Conference Series: Earth and Environmental Science*, vol. 618, no. 1, p. 012038. IOP Publishing, 2020.
23. Tita, Aprilia Da Cruz, Anindya Wirasatriya, Denny N. Sugianto, Lilik Maslukah, Gentur Handoyo, Muhammad Helmi, and Praditya Avianto. "Persistence of high sea surface temperature (> 30° C) in Tomini Bay." In *IOP Conference Series: Earth and Environmental Science*, vol. 530, no. 1, p. 012038. IOP Publishing, 2020.
24. Access on https://marine.copernicus.eu/ with https://doi.org/10.48670/moi-00305.
25. Young, Ian R., Ebru Kirezci, and Agustinus Ribal. "The global wind resource observed by scatterometer." *Remote Sensing* 12, no. 18 (2020): 2920.
26. T. Horii, I. Ueki, and K. Ando, Coastal upwelling events along the southern coast of Java during the 2008 positive Indian Ocean Dipole. *J. Oceanogr.* **74**, 499–508 (2018).
27. F. Veron and W. K. Melville, Experiments on the stability and transition of wind-driven water surfaces, *J. Fluid Mech.*, 446, 25–65 (2001).
28. D. Li, P. Chang, S. Ramachandran, Z. Jing, and Q. Zhang, Contribution of the Two Types of Ekman Pumping Induced Eddy Heat Flux to the Total Vertical Eddy Heat Flux. *Geophys. Res. Lett.* **48**, (2021).
29. H. Bernades, A. A. D. Bernades, A. Wirasatriya, G. Handoyo, A. Rifai, L. Maslukah, and D. N. Sugianto, The effect of Ekman mass transport and Ekman pumping velocity on the variability of sea surface temperature in the Arafura Sea. *IOP Conf. Ser. Earth Environ. Sci.* **919**, 012026 (2021).
30. M. B. Rivas, A. Stoffelen, and A. Bentamy, Sea surface winds and Ekman pumping. *Journal Of Operational Oceanography*, **12**, (2019).

Study on the Reliability of Electric Motor Boat Propeller Structure Due to Axial Load

Admarch Gery Haryoseno[1] and Aries Sulisetyono[1*]

[1]Department of Naval Architecture, Faculty of Marine Technology, Sepuluh Nopember Institute of Technology
*Email: sulisea@na.its.ac.id

Abstract. This study investigates the structural reliability of B-Series propellers used in electric motor-driven ships under axial loads induced by fluid flow. The distinct operating characteristics of electric propulsion systems require dedicated reliability assessment to ensure safe and efficient performance. Computational Fluid Dynamics (CFD) simulations were conducted under uniform viscous flow conditions to evaluate hydrodynamic loading, yielding a maximum axial load of 32.74 kN at a rotational speed of 1500 RPM. The resulting loads were applied in Finite Element Method (FEM) simulations to assess structural stress and deformation, with maximum values of 273.27 MPa and 5.10 mm, respectively. Structural reliability analysis was performed using probability density functions of hydrodynamic loads and aluminum alloy material strength. The results indicate a structural reliability of 99.90% at a ship speed of 10 knots, which decreases significantly to 33.31% at 20 knots. The inclusion of safety factors of 10% and 40% effectively increased reliability by 9.99% and 100%, respectively. These findings emphasize the importance of appropriate safety factors and material selection to improve the structural reliability of electric motor ship propellers under high-speed operating conditions.

Keywords: B-Series Propeller, Computational Fluid Dynamics, Electric Motor Ship, Structural Reliability, Finite Element Method.

1. Introduction

The propeller is a critical component of electric motor boat propulsion systems, directly governing thrust generation, efficiency, and operational reliability. B-Series propellers are widely adopted due to their well-documented hydrodynamic characteristics, making them suitable benchmarks for numerical and experimental studies [2]. During operation, propeller blades are subjected to complex hydrodynamic loads, primarily axial and tangential forces, which significantly affect structural integrity and service life.

Axial loads arise from thrust generation along the shaft direction and may induce high stress levels, increasing the risk of deformation or structural failure if not properly considered during design [6]. Tangential forces generated by rotational motion contribute to torque and cyclic stresses, and their interaction with axial loads further intensifies structural demands on the blades [3]. Consequently, both load components play a decisive role in determining propeller reliability.

Structural performance is commonly evaluated through pressure distribution, deformation, and equivalent stress, while material selection is essential to withstand repeated loading cycles and seawater corrosion [4]. From a hydrodynamic perspective, achieving high thrust with minimal torque remains a primary objective, governed by parameters such as rotational speed, advance ratio, diameter, pitch ratio, and blade area ratio [11]. Therefore, propeller design must balance hydrodynamic efficiency and structural strength.

Computational Fluid Dynamics (CFD) is extensively applied to predict flow characteristics and hydrodynamic loads, whereas the Finite Element Method (FEM) is used to analyze stress distribution and deformation under axial and rotational loading. Their integration provides a comprehensive framework for fluid–structure interaction analysis, improving prediction accuracy and structural assessment [3], [12]. However, deterministic approaches cannot fully represent uncertainties in operating conditions, load variations, and material properties. Reliability-based methods have therefore been introduced to quantify failure probability and evaluate structural safety under uncertain scenarios [5].

Despite extensive research on hydrodynamic performance and structural response, limited attention has been given to the structural reliability of B-Series propellers for electric motor boats under axial loading. Accordingly, this study investigates the structural reliability of a B-Series propeller by integrating CFD and FEM to evaluate hydrodynamic loads, stress distribution, and deformation across varying operating speeds, followed by reliability analysis to estimate failure probability and assess safety margins. This work contributes to the development of reliable and efficient propeller designs for electric marine propulsion, supporting safe maritime operations [15].

2. Literature Review

Structural reliability of marine propellers is critical for electrically powered vessels due to high axial loads and complex hydrodynamic environments. Propeller performance and durability are strongly influenced by rotational speed, axial loading, and flow conditions [6], [7]. Hydrodynamic forces generated during operation directly affect propulsion efficiency and structural integrity.

To evaluate these effects, Computational Fluid Dynamics (CFD) is commonly used to predict flow behavior, while the Finite Element Method (FEM) analyzes stress and deformation on propeller blades [3]. This coupled approach enables identification of high-stress regions and potential failure locations. Reliability-based optimization further accounts for uncertainties in loading and material properties, leading to safer and optimized designs [5].

Hydrodynamic interaction between the propeller and surrounding water governs thrust generation and efficiency. Axial loads produced by pressure distribution along blade surfaces can cause deformation and accelerate material degradation if not properly considered in design. CFD is applied to determine pressure distribution and critical loading zones, whereas FEM evaluates structural responses under these loads [12], [13].

Excessive axial loading increases structural stress and wear, reducing operational lifespan. Numerical simulations are therefore widely adopted to model fluid–structure interaction, assess load variations, and optimize propeller geometry under different operating conditions. These approaches contribute to improved efficiency, reliability, and long-term durability of marine propellers [11], [14].

3. Methodology

The overall methodology of this study is illustrated in Figure 1, showing the integration of Computational Fluid Dynamics (CFD), Finite Element Method (FEM), and reliability analysis to evaluate the structural performance of a B-Series propeller. The process begins with collecting propeller geometry and material properties, followed by 3D modeling and meshing, hydrodynamic simulation, structural analysis, and probabilistic reliability evaluation.

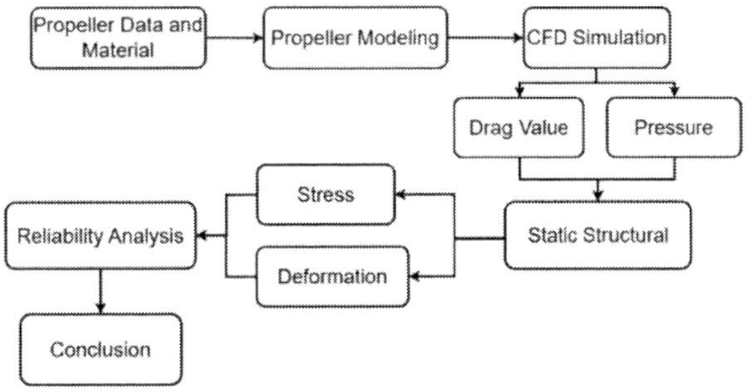

Figure 1 Flowchart

Propeller dimensions and material data are used to construct a three-dimensional model, which is discretized into finite elements. Mesh refinement is applied on blade surfaces to improve numerical accuracy, particularly in capturing pressure gradients and stress concentrations. CFD simulations are performed in ANSYS Fluent to obtain pressure distribution and axial loads under realistic operating conditions, including inflow velocity and rotational speed. These hydrodynamic loads are subsequently transferred to FEM for structural assessment, enabling evaluation of stress and deformation caused by axial loading [12].

Table 1 Principal Dimensions Propeller

PRINCIPAL DIMENSIONS		
Nama	**Dimension**	**Unit**
Diameter	0.7	M
Pitch	2.2	M
Blade Area	0.056	M^2
Number of Blades	4	-

To account for uncertainties in loading and material properties, a probabilistic framework is adopted. Axial loads are represented using probability density functions (PDFs), while material variability is obtained from tensile test data. Structural failure is defined when applied stress exceeds material strength, and failure probability is determined from the overlap of load and strength distributions. Table 1 summarizes the principal propeller dimensions, while Figure 2 presents the 3D geometry used for numerical analysis. Full-scale simulations are conducted to ensure realistic representation of operational conditions.

Figure 2 Propeller Design

3.1. CFD Simulation

The CFD simulation is performed to analyze the hydrodynamic behavior of the marine propeller at ten rotational speeds ranging from 1050 to 1500 rpm, representing typical operating conditions of electric motor boats. The analysis focuses on flow characteristics and pressure distribution on the propeller blades.

The simulations employ the Finite Volume Method (FVM) to discretize the Navier–Stokes equations, ensuring mass and momentum conservation [3]. Turbulence is modeled using the standard k–ε model, which is widely applied for submerged marine propellers due to its robustness in predicting turbulent flows with significant pressure gradients [2], and its suitability for steady operating conditions.

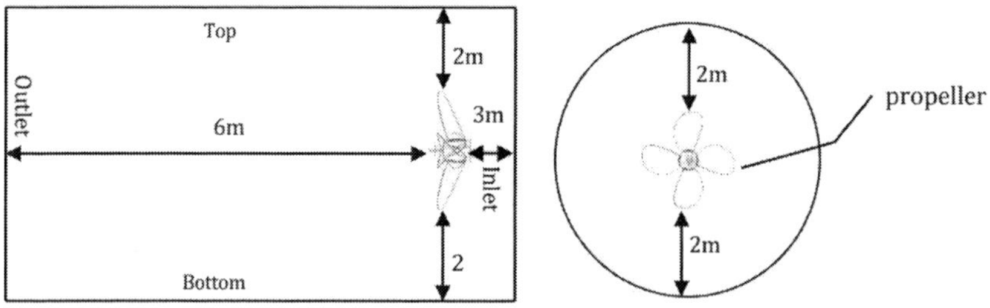

Figure 3 Boundary Condition

Pressure distribution on the blades is obtained by solving the momentum equations while considering fluid density and viscosity. Boundary conditions include velocity inlet, pressure outlet, and no-slip walls on the propeller and domain surfaces. The resulting pressure data are transferred to FEM for structural analysis, establishing fluid-structure coupling.

The computational domain and boundary conditions are shown in Figure 3, with the propeller placed 3 m from the inlet, 6 m from the outlet, and 2 m from surrounding boundaries to reduce numerical interference, following previous recommendations [6]. A mesh independence study yields an optimal grid of 27,347 elements (Table 2). Local mesh refinement is applied near blade surfaces, especially at leading and trailing edges, as illustrated in Figure 4, to capture high pressure gradients efficiently.

Table 2 Grid Independence Study

Grid Independence Study			
Element size	**Number of Mesh**	**Max Stress, MPa**	**Gap**
10 mm	27347	62.634	3%
20 mm	10862	64.332	6%
30 mm	8393	68.682	2%
40 mm	6874	67.242	-

The CFD results provide detailed pressure distributions required to evaluate hydrodynamic performance and identify high-load regions, including potential cavitation zones. These pressure fields serve as input for subsequent FEM-based stress and deformation analyses, following established numerical procedures [4].

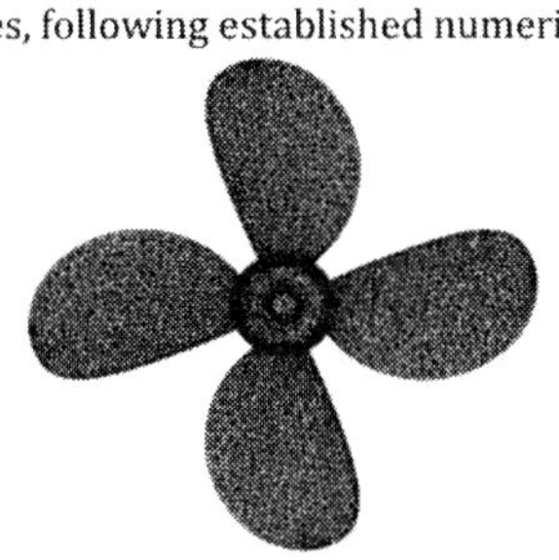

Figure 4 Meshing

3.2. Static Structural Analysis

In the Static Structural analysis in Figure 5, Static structural analysis is performed using FEM to evaluate stress distribution and deformation under hydrodynamic loading. Fixed support is applied at the front hub surface (point C), representing shaft attachment, while axial load is applied at the rear hub (point B). CFD-derived pressure is imposed on the blade surface (point A) to represent fluid-structure interaction [1].

The von Mises criterion is employed to assess equivalent stress, assuming linear elastic material behavior based on Hooke's law. This coupled CFD-FEM approach enables identification of critical regions susceptible to structural failure under axial loading [15].

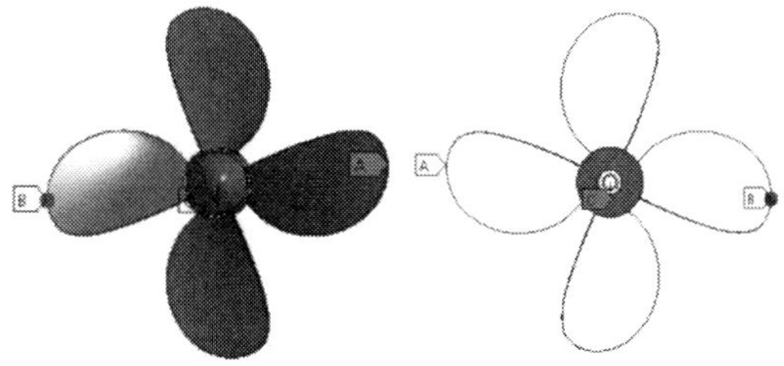

Figure 5 Static Structural

3.3. Strength of Material

Structural resistance is evaluated using yield strength as the primary failure criterion [9]. Aluminum is selected as the propeller material, with mechanical properties obtained from tensile test results (Table 3) [8]. The results show consistent yield and ultimate tensile strengths across different thicknesses, with maximum elongation of approximately 1.5%, indicating limited ductility. Material variability is incorporated into the probabilistic framework to account for uncertainty in structural strength.

Table 3 Tensile properties of aluminium samples from different thicknesses and heat treatments [Kim et al., 2021]

Wedge Thickness (mm)	YS (MPa)	UTS (MPa)	Elongation (%)	BHN
14	465	473	1.0	144
33	396	402	0.9	124
21	453	453	0.8	138
25	450	480	1.5	140
30	428	450	1.2	142
40	360	360	0.7	132
43	343	355	0.9	132
47	327	332	0.8	120
18	446	446	0.8	145
37	370	370	0.8	126

3.4. Reliability Analysis

This study applies a reliability-based approach to evaluate propeller structural integrity under axial loading. A probabilistic method is adopted using two stochastic variables: structural strength and applied load. Two Probability Density Functions (PDFs), representing strength $f_r(R_0)$ and load $f_s(S)$, are defined as shown in Figure 6, where R_0 and S are expressed in stress units. The probability of failure is determined from the overlap region of these PDFs, corresponding to the joint probability density function [1].

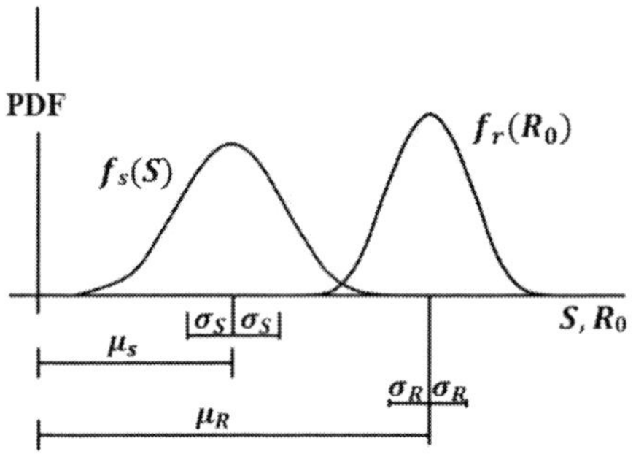

Figure 6 PDF Variabel R_0 dan S

A simple function, $g(S, R_0)$, is defined to represent the limit state function, which describes the safety margin SM, as the difference between the structural strength $f_r(R_0)$ and the applied load $f_s(S)$, expressed in Equation (1).

$$SM = g(S, R_0) = f_r(R_0) - f_s(S) \tag{1}$$

Based Based on the above equation, three criteria can be identified for each value of $g(S, R_0)$ as follows:

- $g(S, R_0) < 0$, Failure region, where the structural load, $f_s(S)$, exceeds the structural strength $f_r(R_0)$.
- $g(S, R_0) > 0$, the safe area, where the load, $f_s(S)$, is less than the strength of the structure $f_r(R_0)$.
- $g(S, R_0) = 0$, the boundary area between safety and failure, where the load, $f_s(S)$, is equal to $f_r(R_0)$.

The probability of failure can be calculated by equation (2).

$$P_f = P[M = g(S, R_0) \leq 0] = \iint_{g(S,R_0)\leq 0} f_s(S) \cdot f_r(R_0)\, ds\, dR_0 \tag{2}$$

Where $f_{s,r}(S, R_0)$ is the joint probability function of $f_r(R_0)$ and $f_s(S)$, and the limits of the integral area encompass all values of R_0 and S, where M is non-positive, indicating an unsafe condition. In the case where the statistical strength and structural load are independent, the probability of failure is expressed in Equation (3).

$$P_f = \Phi(-\beta) \tag{3}$$

Where $\Phi(\cdot)$ is the standard normal cumulative distribution function and β is the safety index, which can be calculated by equation (4), where μ is the average, and σ is the standard deviation.

$$\beta = \frac{\mu_{R_0} - \mu_S}{\sqrt{\sigma {R_0}^2 + \sigma S^2}} \tag{4}$$

The probability of structural failure (P_f) in some cases involving many variables can be calculated using the Monte Carlo simulation method [8]. Furthermore, the reliability of the structure (R) is determined by equation (5).

$$R = 1 - Pf \tag{5}$$

4. Results and Discussion

Structural reliability of the propeller under axial loads at ten different rotational speeds was evaluated using coupled CFD and static structural FEM analyses. The results include pressure distribution, stress, and deformation, which are discussed in the following subsections.

4.1. Pressure Distribution dan Axial Drag Analysis

Figure 7 shows the pressure distribution from CFD simulations at ten rotational speeds under constant inflow velocity. High-pressure regions appear mainly at the leading edge near the blade tip, while low pressure is observed toward the trailing edge and mid-span. As listed in Table 4, both minimum and maximum pressures increase with rotational speed, ranging from 0.026–0.041 MPa at 1050 RPM to 0.043–0.066 MPa at 1500 RPM. This pressure gradient governs the axial load acting on the blades and is critical for structural reliability assessment [11].

Table 5 presents axial drag forces at a ship speed of 5.14 m/s, increasing from 10.3 kN at 1050 RPM to 19.9 kN at 1500 RPM. These results confirm that higher rotational speeds generate greater axial loads, increasing the potential for stress concentration and structural degradation, consistent with previous studies [13].

Table 4 Pressure Data Table

RPM	Minimum (Mpa)	Maximum (Mpa)
1050	0.026031	0.04129
1100	0.0290164	0.0443551
1150	0.0311521	0.0492308
1200	0.0328566	0.0518323
1250	0.0344271	0.0542286
1300	0.0360817	0.059595
1350	0.037105	0.0581774
1400	0.0393519	0.0614345
1450	0.0411436	0.0639996
1500	0.0426101	0.0660518

Table 5 Axial Drag

Rpm	Axial Drag (N)	Vs(m/s)	ω
1050	19406.6	8.084	0.573
1100	23750.4	8.099	0.576
1150	23845.5	8.216	0.598
1200	24598.6	8.417	0.638
1250	25918.6	8.624	0.678
1300	27309.8	8.714	0.695
1350	27925.3	9.020	0.755
1400	30061.2	9.226	0.795
1450	31542.0	9.389	0.827
1500	32743.6	2.267	0.559

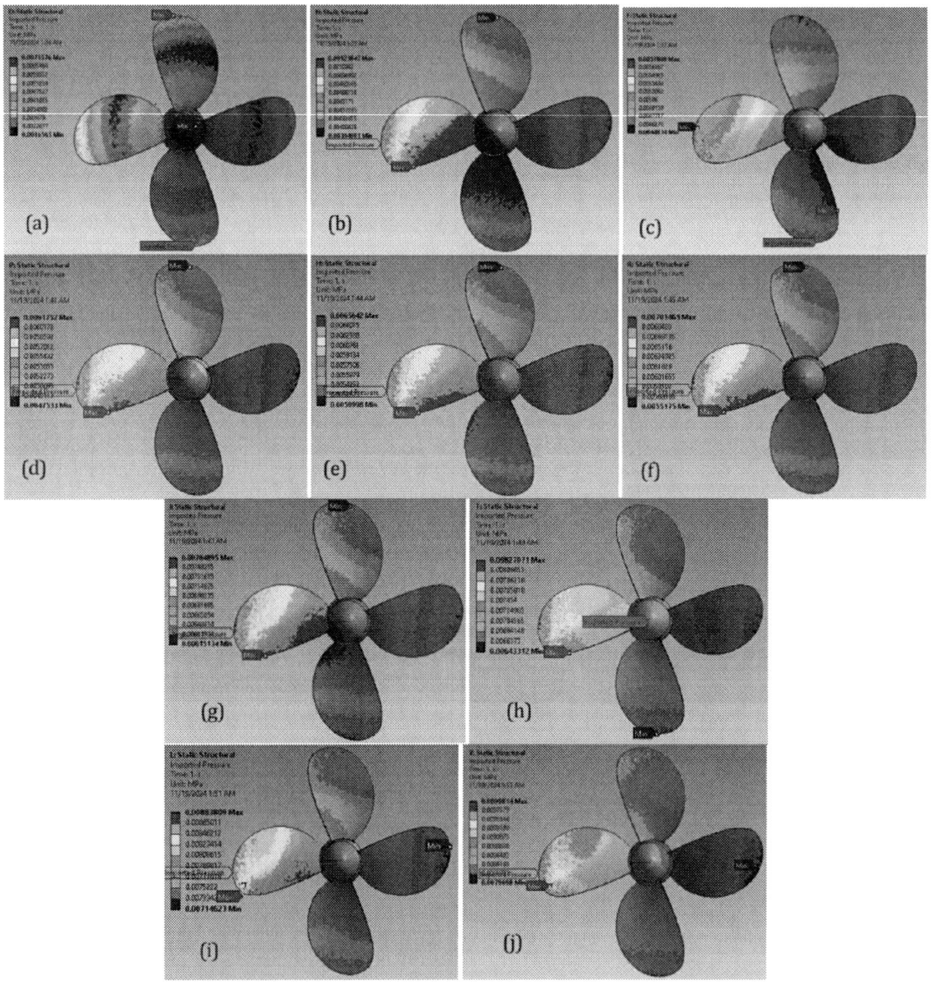

Figure 7 Imported Pressure by (a) 1050 Rpm, (b) 1100 Rpm, (c) 1150 Rpm, (d) 1200 Rpm, (e) 1250 Rpm, (f) 1300 Rpm, (g) 1350 Rpm, (h) 1400 Rpm, (i) 1450 Rpm, (j) 1500 Rpm

4.2. Stress Distribution Analysis

Static structural FEM analysis was performed using axial loads obtained from CFD to evaluate von Mises stress distribution on the propeller blades. Figure 8 shows stress contours at different rotational speeds, where maximum stress appears mainly near the blade root and along the load transfer path toward the hub.

Stress magnitude increases with rotational speed, reflecting the rise in axial force and bending moment. As listed in Table 6, the maximum von Mises stress reaches 125.36 MPa at 1500 RPM, representing the most critical operating condition. These results highlight the importance of stress evaluation under axial loading to ensure structural reliability during high-speed operation.

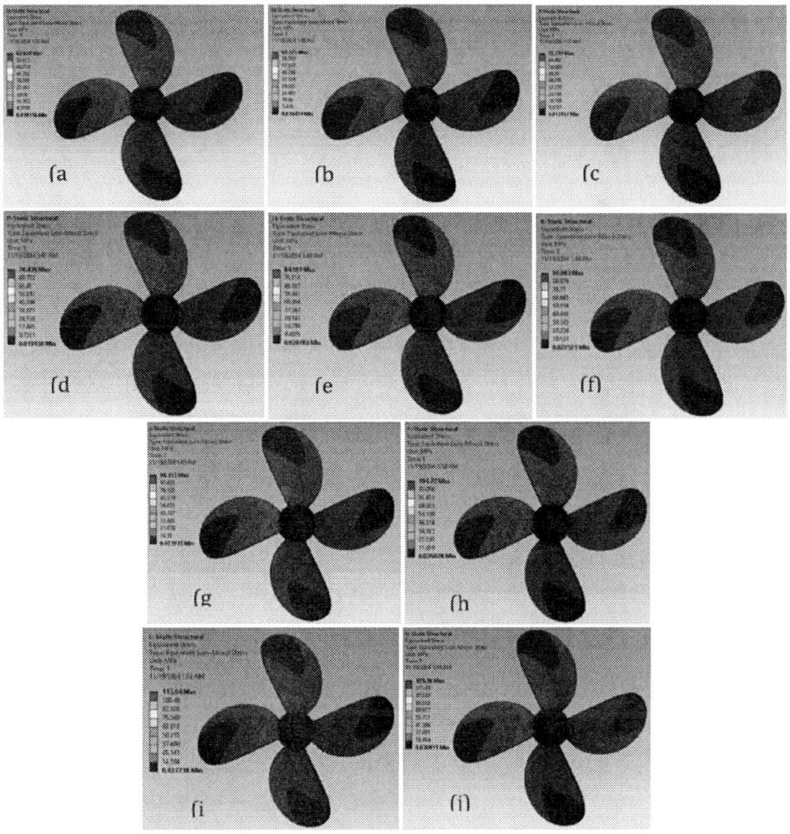

Figure 8 Equivalent Stress by (a) 1050 Rpm, (b) 1100 Rpm, (c) 1150 Rpm, (d) 1200 Rpm, (e) 1250 Rpm, (f) 1300 Rpm, (g) 1350 Rpm, (h) 1400 Rpm, (i) 1450 Rpm, (j) 1500 Rpm

Table 6 Stress & Deformation

RPM	Max Stress (Mpa)	Max Deformation (mm)
1050	163.84	3.0524
1100	193.16	3.5920
1150	199.40	3.7173
1200	207.26	3.8643
1250	218.00	4.0666
1300	229.20	4.2779
1350	234.94	4.3853
1400	251.54	4.6983
1450	263.56	4.9239

1500	273.27	5.1022

4.3. Deformation Structure

Figure 9 presents the deformation contours of the propeller at different rotational speeds, with maximum deformation concentrated near the blade tips. Deformation increases with rotational speed due to higher axial loads and bending effects, as summarized in Table 6. This evaluation ensures that blade displacement remains within acceptable limits to prevent geometric distortion and performance degradation. Similar trends have been reported in previous studies, highlighting the influence of rotational speed and hydrodynamic loading on propeller deformation [14].

Figure 9 Total Deformation by (a) 1050 Rpm, (b) 1100 Rpm, (c) 1150 Rpm, (d) 1200 Rpm, (e) 1250 Rpm, (f) 1300 Rpm, (g) 1350 Rpm, (h) 1400 Rpm, (i) 1450 Rpm, (j) 1500 Rpm

4.4. Structural Reliability Analysis with Safety Factor

Stress results from FEM and axial loads from CFD were used to evaluate structural reliability. Mean stress and standard deviation for each rotational speed are summarized in Table 7, and stress PDFs were constructed using a normal distribution. Figure 10 presents the joint PDF between structural stress and material strength, where yield strength from tensile tests (Table 3) defines the failure limit.

Safety factors of 20% and 40% were applied (Figures 10b and 10c), showing a significant reduction in failure probability. The safety index was calculated using Equation (4) (Table 8), followed by failure probability (Equation (3)) and structural reliability $R = 1 - P_f$ (Equation (5)). The results confirm high structural reliability, especially with safety factors applied, indicating that the propeller is suitable for axial loading conditions.

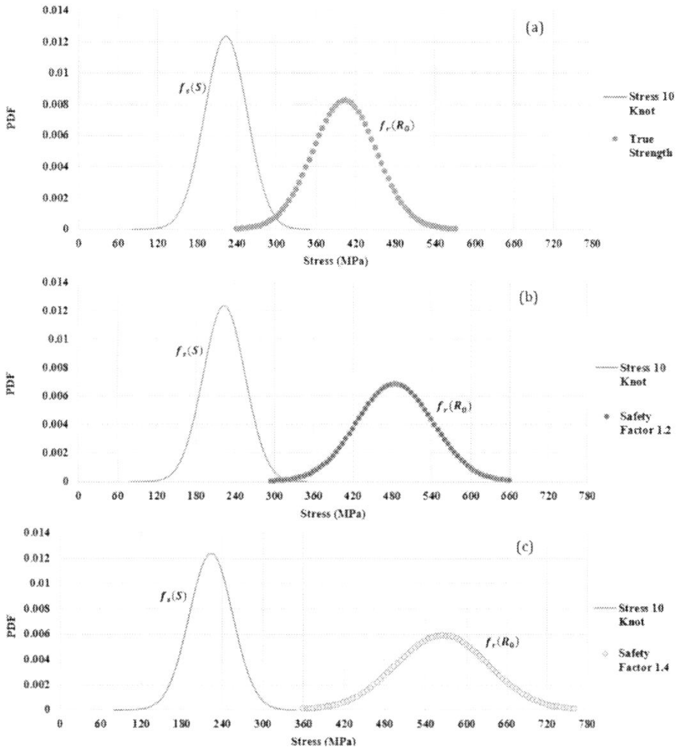

Figure 10 : Joint Probability Density Function (PDF) with Safety Factor

Table 7 : Stress of loads and compared to material strength at various safety factors.

Parameter	Stress of loads (μ_s)	Stress of Tensile Test		
		True Strength (μ_{R_0})	Safety Factor 1.2 (μ_{R_0})	Safety Factor 1.4 (μ_{R_0})
Mean (μ)	223.42	403.8	484.56	565.32
Sigma (σ)	32.23	48.36	58.03	67.70

Table 8 : Analysis of safety index, failure probability, and structural reliability with safety factors.

Parameter	True Strength	Safety Factory	Safety Factor 1.4
Safety Index (β)	3.104	3.934	4.561
Failure (P_f)	0.10%	0.004%	0.00%
Reliability ($1 - p_f$)	99.90%	99.996%	100%

4.5. Structural Reliability Analysis with Various Loads

Figure 11 shows the joint PDF between structural stress and material strength under axial loading at inlet velocities of 10, 15, and 20 knots. Increasing inlet velocity intensifies hydrodynamic loads, resulting in higher stress levels. The corresponding mean and standard deviation values are summarized in Table 9 and used to calculate the safety index Table 10 via Equation (4), followed by failure probability Equation (3) and structural reliability Equation (5).

The results indicate that structural reliability decreases with increasing ship speed due to higher hydrodynamic loading. Operation above 15 knots significantly raises failure risk; therefore, a maximum speed of 15 knots is recommended to maintain acceptable reliability. The use of materials with higher yield strength than aluminum is also suggested to improve propeller durability under elevated loading conditions.

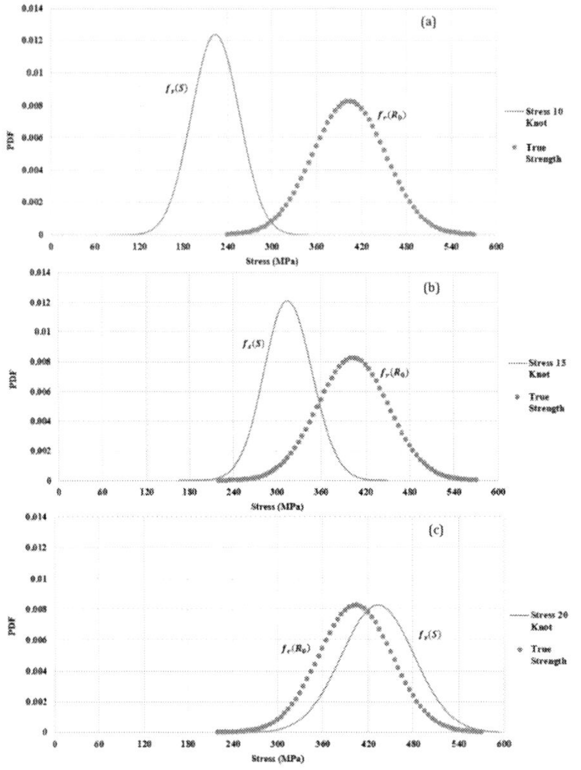

Figure 11 : Curve Probability Density Function (PDF) with Various Load

Table 9 : Stress of Loads with Various Flow Speed and Tensile Test

	Stress of Loads with Various Flow Speed			**Stress of Tensile Test**
Parameter	**10 Knot (μ_s)**	**15 Knot (μ_s)**	**20 Knot (μ_s)**	**True Strength (μ_{R_0})**
Mean (μ)	223.417	315.272	433.244	403.8
Sigma (σ)	32.227	33.002	48.154	48.357

Table 10 : Analysis of safety index, failure probability, and reliability with Various Flow Speed

Parameter	**Stress 10 Knot**	**Stress 15 Knot**	**Stress 20 Knot**
Safety Index (β)	3.104077	1.51213643	-0.4314569
Failure (p_f)	0.10%	6.525%	66.693%
Reliability ($1 - p_f$)	99.90%	93.475%	33.307%

5. Conclusion

This This study investigated the structural reliability of an electric motor boat propeller subjected to axial loads using an integrated approach combining Computational Fluid Dynamics (CFD), Finite Element Method (FEM), and reliability analysis. CFD simulations were employed to evaluate hydrodynamic loads at various propeller rotational speeds, while FEM was used to assess the resulting stress and deformation responses. Reliability analysis was subsequently conducted to quantify failure probability and the influence of safety factors under different operating conditions.

The CFD results indicate that axial loads increase with rising propeller rotational speed, reaching a maximum of 32.74 kN at 1500 RPM. Correspondingly, FEM analysis at this condition produced a maximum von Mises stress of 273.27 MPa and a deformation of 5.10 mm. Reliability evaluation across all rotational speed variations yielded a low failure probability of 0.1%, corresponding to a reliability level of 99.90%. The application of safety factors of 10% and 40% further improved the reliability, increasing it by 9.99% and up to full reliability, respectively.

Additional reliability analysis considering variations in ship speed revealed that structural reliability remained high at 10 knots (99.90%) but decreased significantly to 33.31% at 20 knots due to increased hydrodynamic loading. These findings demonstrate that higher operational speeds substantially reduce structural reliability. Therefore, it is recommended to limit ship speed to acceptable operational ranges and consider alternative propeller materials with higher yield strength than aluminum to enhance structural durability and reliability under elevated loading conditions.

References

[1] Truelock, D., Lavroff, J., Pearson, D., Czaban, Z., Luo, H., Wang, F., ... & Nicholls-Lee, R. (2022, September). Committee v. 5: Special vessels. In International Ship and Offshore Structures Congress (p. D011S001R006). SNAME.

[2] Purwana A, Ariana IM, Handani DW, Wardhana W (2018) Performance and noise prediction of marine propeller using numerical simulation. IPTEK J. Proceedings Series 1:20–25.

[3] Guan G, Zhang X, Wang P, Yang Q (2022) Multi-objective optimization design method of marine propeller based on fluid-structure interaction. Ocean Eng 252:111222.

[4] Kim SW, Lee SJ, Kim DU, Kim MS (2021) Experimental investigation on tensile properties and yield strength modeling of T5 heat-treated counter pressure cast A356 aluminum alloys. Metals 11:1192.

[5] Gholinezhad H, Torabi SH (2022) Reliability-based multidisciplinary design optimization of an underwater vehicle including cost analysis. J Mar Sci Technol 27:11–26.

[6] Ortolani F, Dubbioso G, Muscari R, Mauro S, Di Mascio A (2018) Experimental and numerical investigation of propeller loads in off-design conditions. J Mar Sci Eng 6:45.

[7] Guo HP, Zou ZJ, Wang F, Liu Y (2019) Numerical investigation on the hydrodynamic characteristics of a marine propeller operating in oblique inflow. Appl Ocean Res 93:101969.

[8] Vizentin G, Vukelic G, Murawski L, Recho N, Orovic J (2020) Marine propulsion system failures—A review. J Mar Sci Eng 8:662.

[9] Iqbal M, Gafur A, Antoko B (2023) Analisis Optimalisasi Nilai Thrust, Torque, dan Efficiency Propeller B Series dan Kaplan Series pada Studi Kasus Kapal Pencalang 15 GT Menggunakan Metode CFD. In: Proceedings of the Conference on Marine Engineering and Its Application 6(1):1–6.

[10] Özsayan S, Köksal ÇS, Usta O, Çelik C, Yılmaz N, Korkut E (2022) An experimental and numerical investigation of the cavitation effects on propeller performance and erosion on DTMB-5415 propeller. In: Proceedings of the 7th International Symposium of Marine Propulsors, Wuxi, China.

[11] Marques CH, Belchior CR, Caprace JD (2019) Marine propeller parametric optimization and matching to electric motor. J Braz Soc Mech Sci Eng 41:1–11.

[12] Rochmad AN, Sulisetyono A (2024) Fluid–Structure Interaction of a Flat-Rudder Floater in N219 Floatplane Maneuvers at the Water's Surface. SpringerBriefs in Applied Sciences and Technology:21–30. https://doi.org/10.1007/978-3-031-67788-5_3.

[13] Zou, D., Zhang, J., Ta, N., & Rao, Z. (2019). Study on the axial exciting force characteristics of marine propellers considered the effect of the shaft and blade elasticity. Applied Ocean Research, 89, 141-153.

[14] Magionesi, F., Dubbioso, G., & Muscari, R. (2024). Fluid–structure interaction of a marine rudder at incidence in the wake of a propeller. Physics of Fluids, 36(6).

[15] Stambaugh, K. A. (2020). On ship structure risk and total ownership Cost management assisted by prognostic hull structure monitoring. *Delft University of Technology*.

The Adsorptive Power of Nylon–Gelatin–Hematite Composite: Study on NaCl Solution Adsorption Through Equilibrium, Kinetic, and Thermodynamic Approach

Lilik Miftahul Khoiroh[1,2], *Nurul* Widiastuti[1*], *Ananda* Anggelita Aprina Puri[2], *Triyanda* Gunawan[1], and *Moch* Purwanto[3]

[1]Department of Chemistry, Faculty of Science and Data Analytics, Institut Teknologi Sepuluh Nopember, Jl. Arif Rahman Hakim, Kampus ITS Keputih-Sukolilo, Surabaya 60111, Indonesia

[2]Chemistry Study Program, Faculty of Science and Technology, Universitas Islam Negeri Maulana Malik Ibrahim Malang, 65144, Indonesia

[3]Department of Chemical Engineering, Institut Teknologi Kalimantan, Balikpapan 76127, Indonesia

[*]Corresponding author: nurul_widiastuti@its.ac.id

Abstract. The development of composite adsorbent materials is a promising strategy for saline water treatment. In this study, a nylon–gelatin–hematite composite with a porous structure was investigated for NaCl ion removal. The porous matrix and functional groups on the polymer and hematite surfaces facilitated Na^+ and Cl^- adsorption through electrostatic interactions and surface adsorption mechanisms. Adsorption performance was evaluated using a batch method with variations in adsorbent dosage, contact time, and temperature. Functional groups were confirmed by Fourier-transform infrared (FTIR) spectroscopy. Adsorption performance was evaluated through measurements of salinity, total dissolved solids, and electrical conductivity. The adsorption behavior was examined through equilibrium, kinetic, and thermodynamic analyses to elucidate the underlying mechanism and efficiency. The FTIR spectra confirmed characteristic vibrations of nylon, gelatin, and hematite, indicating the successful formation of the composite. Equilibrium data fitted well with the Freundlich isotherm models (R^2 > 0.99) at dosage variation, suggesting favorable multilayer adsorption on heterogeneous surfaces. The maximum adsorption capacity q_{max} reached 1.626 mg/g. Kinetic studies revealed that the adsorption process followed a pseudo-first-order model, indicating a dominant physisorption mechanism. Negative Gibbs free energy values (ΔG = -1.905 to -2.139 kJ/mol) confirmed the spontaneous adsorption process.

1 Introduction

Water remains a vital necessity for sustaining life. Global water demand is projected to reach 6.9 trillion cubic meters by 2030. However, water scarcity and increasing salinity are causing serious global problems due to high levels of industrialization and seawater intrusion [1, 2]. Among the various methods to address the problem of high salinity in water, Reverse Osmosis (RO) has been practiced on a large scale due to its high level of separation efficiency [3]. Although RO is very effective in providing clean water on a large scale, this technology still requires high pressures of 10–100 bar, high power consumption, and a high potential for membrane fouling. Therefore, this technology often uses chemical cleaning agents. Adsorption is a promising alternative approach due to its ease of operation, efficiency, and environmental friendliness in removing dissolved salts [4-6].

In recent years, gelatin-iron oxide nanocomposites, especially gelatin–hematite (α-Fe_2O_3), have gained attention as adsorbents due to the cumulative interaction of biopolymer functional groups ($-NH_2$,$-COOH$) and metal oxide surfaces [7, 8] and observed that gelatin-conjugated hematite nanoparticles had a high adsorption capacity of

169.49 mg g^{-1} Pb(II) with a removal efficiency of over 90%. The gelatin and α-Fe_2O_3 composite could adsorb Pb^{2+} with a high capacity of 800 mg/g and a removal efficiency of 82% based on pseudo-first-order kinetics [4]. Gelatin-Pluronic F127-hematite patterned flakes demonstrated adsorption activity towards ibuprofen, with an adsorption capacity of 25.5-55 mg g^{-1} and a removal efficiency of 22%. These studies validate that gelatin has a high adsorption capacity; however, mechanical instability and particle aggregation remain limitations of gelatin-based composites in aqueous conditions.

Nylon has been incorporated as a structural reinforcement polymer within the gelatin–hematite composite framework to overcome these challenges. The presence of nylon provides a robust and flexible matrix that enhances the adsorbent's mechanical integrity, stability, and reusability. Moreover, its fibrous morphology facilitates improved ion diffusion and transport throughout the adsorbent structure [6, 9]. Additionally, nylon minimizes hema-tite particle aggregation and stabilizes gelatin under aqueous conditions, thereby enhancing adsorption capacity and operational durability. According to Hamad [9], nylon-based materials exhibit significant adsorption performance, achieving up to an 86% removal of Congo red dye with a 300 mg/L capacity. Similarly, a nylon–Fe_3O_4 composite has demonstrated efficient Cu^{2+} ion removal, with a removal efficiency of 94.67% and an adsorption capacity of 11.57 mg/g. Nylon–gelatin–hematite composites represent a promising class of low-cost adsorbent materials for desalination applications. The interconnectedness of the components provides advantageous physicochemical characteristics, while the nylon matrix offers an inherently porous structure capable of accommodating small molecules. Moreover, the potential to utilize nylon derived from waste further enhances the economic feasibility and sustainability of this composite.

This study serves as a preliminary investigation of the membrane system aimed at gaining a fundamental understanding of the adsorption mechanisms involved in membrane-based processes. Specifically, the research focuses on the adsorption behavior of a nylon–gelatin–hematite composite for sodium chloride (NaCl) removal. The adsorption performance was systematically evaluated using equilibrium, kinetic, and thermodynamic analyses to clarify the controlling mechanisms, adsorption efficiency, and associated energy characteristics. The findings offer new insights into the development of composite adsorbent systems for saline water treatment, demonstrating enhanced ion removal capability, improved stability, and greater sustainability of the gelatin–hematite composite.

2 Method

2.1 Material and instrumentation

The materials used in this research include nylon yarn, hematite, gelatin, 37% hydrochloric acid (HCl) (Merck), glutaraldehyde, demineralized water, and sodium chloride (NaCl) (Central Drug House Ltd.). The instruments used in this research were Fourier-transform Infrared (FTIR) spectroscopy, utilizing a Varian spectrometer, to determine the functional groups of hematite and composites. Water tester meter (EZ-9909) to determine the total dissolved solids (TDS), electrical conductivity (EC), and salinity of samples.

2.2 Method

2.2.1 Preparation and characterization of nylon-gelatine-hematite composite

The nylon solution was prepared by dissolving 3 g of nylon yarn in 15% hydrochloric acid (HCl). Subsequently, 1 g of gelatin and 10–50 mg of hematite was added to the nylon solution. The mixture was heated at 50°C for 30 minutes under continuous stirring to ensure homogeneity. The homogeneous mixture was then cast onto a glass plate to form a thin film and left to stand for several minutes before being immersed in a 1% glutaraldehyde solution within a coagulant bath for cross-linking. After coagulation, the composite film was carefully removed, washed, and dried at room temperature. The composite film was characterized by FTIR using the wavenumber 400 – 4000 cm^{-1}.

2.2.2 *Adsorption of nylon-gelatine-hematite composite to saline water*

The prepared composite film measuring 7x7 cm with thicknes range 0.125-0.2 mm which was cut into small pieces with weight of adsorbent range 10-50 mg according the variation of the dosage of adsorbent used. The composite samples were tested as adsorbents for a 3500ppm saline (NaCl) solution. Adsorption experiments were conducted under varying contact times (1–24 h) and temperatures (60–100 °C) to investigate the effects of time and thermal conditions on adsorption performance. The adsorption data were fitted into the adsorption isotherm model, as shown in Tables 1 and 4, to determine the nature of the adsorption phenomenon and find the adsorption capacity of the adsorbent. The removal rate is calculated according to the formula [6] :

$$\%R = \frac{Co-Ct}{Co} \text{ x } 100 \quad (1)$$

C_o is the initial concentration of the solution, and C_t is the concentration after Adsorption of the solution.
The adsorption capacity (mg/g) is calculated according to the equation:

$$Qe = \frac{(Co-Ct)V}{m} \quad (2)$$

The Langmuir and the Freundlich isotherms evaluate the equilibrium adsorption as described by the following equations (3-4). The kinetics of the reaction were evaluated by applying pseudo-zero-order, pseudo-first-order, and pseudo-second-order models, as described by the following [5, 6, 10, 11]:

$$\frac{1}{qe} = \frac{1}{Kl\, x\, qmax} + \frac{1}{qmax} \text{ x } \frac{1}{Ce} \quad (3)$$

$$\ln qe = \ln Kf + \frac{1}{n} \ln Ce \quad (4)$$

$$\ln (qe - qt) = \ln qe - k_1 t \quad (5)$$

$$\frac{t}{q_t} = \frac{1}{k_2(q_e)^2} - \frac{1}{q_e} t \quad (6)$$

3 Result and Discussion

3.1 Characterization of nylon-gelatine-hematite composite

The structural characteristics of the nylon–gelatin–hematite composite were examined using Fourier Transform Infrared (FTIR) spectroscopy, and the corresponding spectrum is presented in Fig. 1. The FTIR bands observed at 3294, 2932, 1637, 1540, 1461, and 1263 cm^{-1} are attributed to N–H stretching, C–H stretching, and the amide I (C=O stretching), amide II (N–H bending), and amide III (C–N stretching) vibrations of the amide groups derived mainly from nylon [12]. Similar absorption bands at 1637 cm^{-1} amide I (C=O stretching), 1540 cm^{-1} (amide II), and 1263 cm^{-1} (C–O stretching) are also characteristic of gelatin [4, 13]. In addition, a distinct absorption band around 576-520 cm^{-1} corresponds to the Fe–O stretching vibration of hematite [14], confirming the incorporation of iron oxide within the composite matrix. No significant difference or noticeable shift in the absorption bands was observed among the three composite variations. This indicates that the small amount of hematite incorporated into the polymer matrix was insufficient to alter the characteristic absorption intensity of nylon or gelatin. Based on characterization, FTIR interpreted that the interaction between the polar groups of gelatin/nylon and the hematite surface is reflected in the stable position in the composite network, supporting the interpretation that hematite particles are not only physically embedded, but also associated through weak electrostatic or coordinative interactions, as shown in Fig. 2.

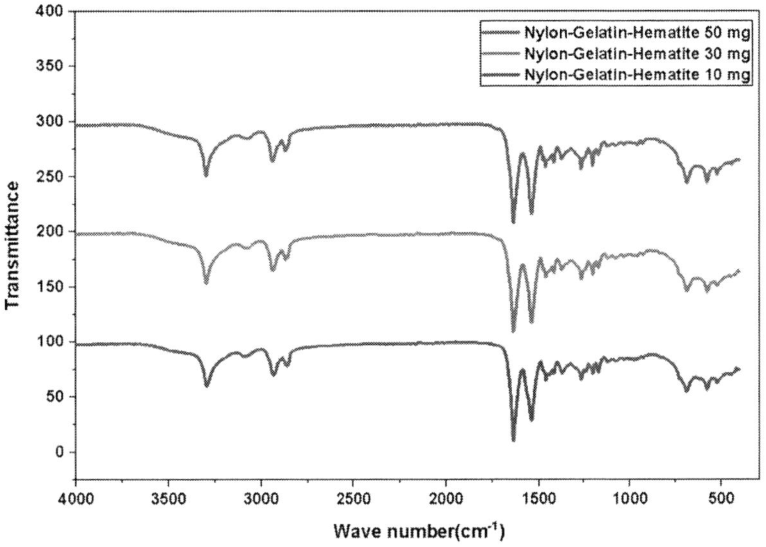

Fig 1. The Infrared Spectra FTIR of nylon-gelatin-hematite composite with variation of hematite composition.

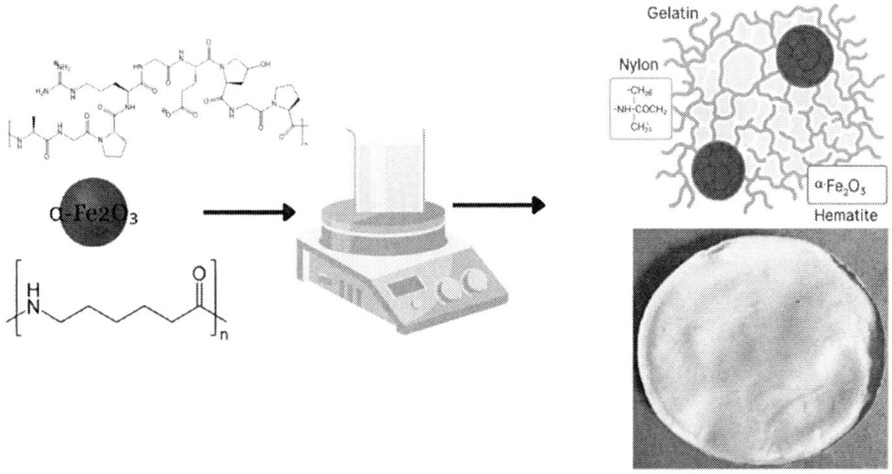

Fig 2. Schematic representation of the fabrication process nylon-gelatin-hematite composite

3.2 Effect of Adsorbent Dosage on Saline Solution

The effect of adsorbent dosage on NaCl removal was evaluated using 10, 30, and 50 mg of nylon–gelatin–hematite composite at a fixed concentration of 3500 ppm and a contact time of 1.5 hours. Figure 3 shows that the percentage of electrical conductivity removal increased, but total dissolved solids and salinity decreased with increasing adsorbent mass. The removal efficiency of electrical conductivity increased from 3.9% to 6.4%, but the percentage of total dissolved solids removal decreased from 3.1% to 2.1%, and salinity decreased from 3.5% to 2.1% with increasing dosage from 10 to 50 mg. The increase in adsorption performance on electroconductivity values is due to the increasing number of active surface sites, so that more Na^+ and Cl^- ions can be adsorbed on the adsorbent surface through electrostatic interactions or ion exchange. However, the decrease in the percentage of TDS and salinity removal is due to the remaining adsorbent mass without full interaction due to the limited availability of more ions and the increase in the higher surface area of the adsorbent [5].

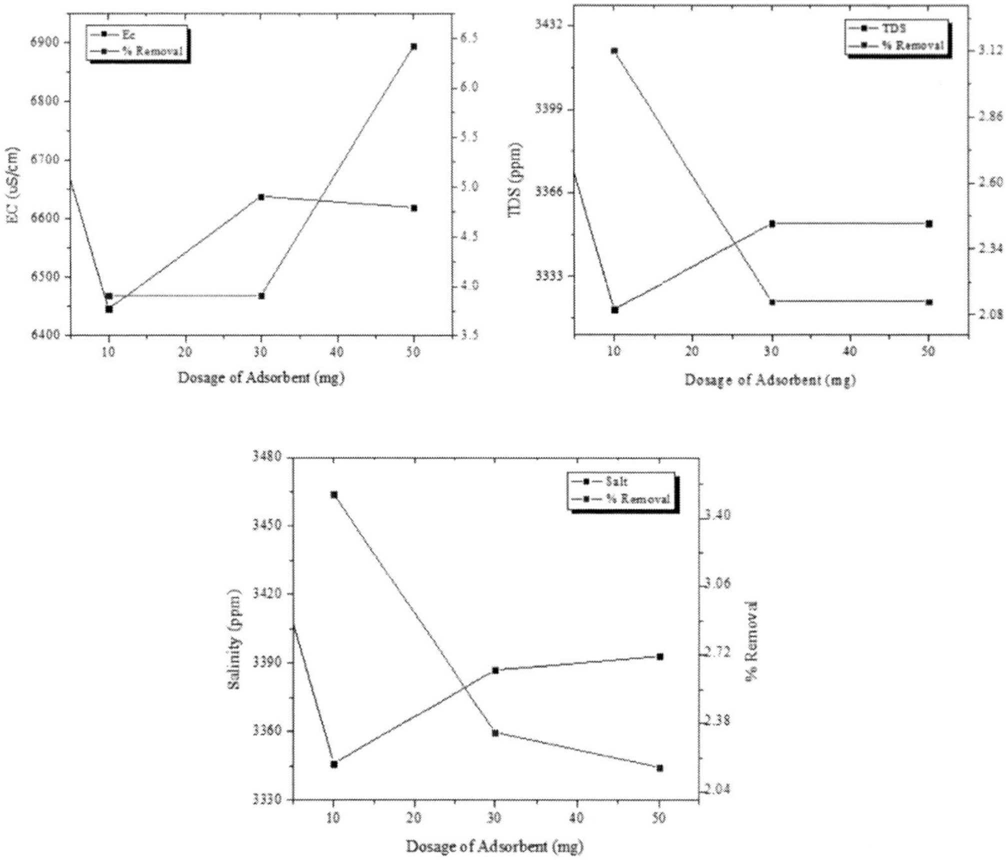

Fig 3. The Effect of the dosage of nylon-gelatin-hematite on NaCl Solution

The Salinity and TDS removal decreased when the dosage increases. At a dosage of 50 mg, the adsorption capacity tends to decrease (Fig. 3), indicating that particle agglomeration may occur at higher dosages, which reduces the effective surface area and leads to the overlap of adsorption sites. This phenomenon is consistent with observations reported for iron oxide–polymer composites by [6, 9], where excessive adsorbent loading diminished active site accessibility due to interparticle interactions.

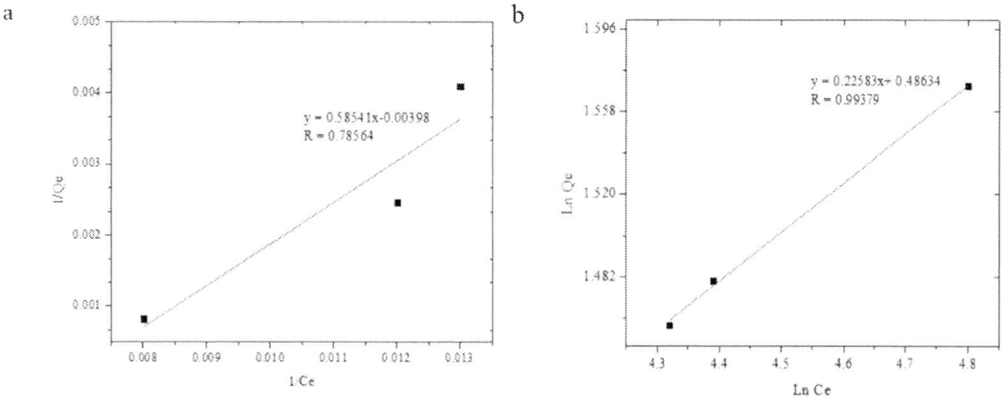

Fig 4. Linear isotherm of saline solution adsorption based on composite nylon-gelatin-hematite from data on salinity

Table 1. Isothermal adsorption-desorption model fitting parameters

Adsorption Model			
Langmuir		**Freundlich**	
q_{max} (mg/g)	1.708	K_F (mg/g)(L/mg)$^{1/n}$	1.626
R	0.78564	**R**	0.99379

Based on Figure 4, the equilibrium interaction between NaCl ions and the composite surface follows the Freundlich adsorption isotherm, with a correlation coefficient (R^2) of approximately 0.99, and based on the Langmuir adsorption isotherm, a maximum absorption capacity of 1.708 mg/g, as shown in Table 1. This result suggests that the adsorption process occurs through multilayer adsorption mechanisms on the heterogeneous surface of the composite. The Adsorption of NaCl onto the nylon–gelatin–hematite composite involves a combination of physical interactions. Electrostatic attraction occurs between Na^+ and Cl^- ions and the polar functional groups (–OH, –NH, and –COOH) in gelatin and nylon, initiating Adsorption. Subsequently, partial ion exchange occurs between Na^+ ions and the Fe–OH sites on the hematite surface. Another possible mechanism for reducing salinity is the interaction between amino functional groups and Na^+ ions through Lewis acid–base interactions, suggesting that nitrogen-containing groups play a crucial role in enhancing adsorption performance [7].

Table 2. A comparison of q_{max} of the prepared composite with other previously reported

Reported adsorbent	**Molecul adsorbate**	**Q_{max} (mg/g)**	**References**
Gelatin - α-Fe_2O_3 composite	Pb (II) solution	800	(4)
Nylon fiber waste (porous nylon sheet)	Congo red	188	(9)
Gelatin-conjugated hematite (α-Fe_2O_3) nanoparticles	Pb(II) solution	169.49	(7)
Modified gelatin / montmorillonite nanocomposite (gelatin-based)	Malachite Green (MG)	950.5	(10)
Gelatin-Pluronic F127-hexagonal hematite	Ibuprofen	55	(8)
Nylon@Fe_3O_4	Copper ion	11.57	(6)
Nylon-gelatin-hematite	NaCl Solution	1.626	This work

3.3 Effect of Time on Saline Solution Adsorption

The adsorption behavior of NaCl on nylon–gelatin–hematite composites was investigated at various contact times, ranging from 1 to 24 hours. As illustrated in Figure 5, the adsorption rate increased sharply during the initial stage (first two hours), with a removal efficiency of approximately 3%. Subsequently, the adsorption rate gradually increased at a low rate. It reached equilibrium after 24 hours, achieving removal efficiencies of approximately 10% for electrical conductivity (EC), 13% for total dissolved solids (TDS), and 13% for salinity. The initial rapid adsorption can be attributed to the many vacant active sites on the composite surface, which allow Na^+ and Cl^- ions to be readily adsorbed through electrostatic attraction and ion exchange processes. As adsorption progresses, these sites become occupied, leading to a slower diffusion-controlled phase where intraparticle diffusion governs ion transport to the interior of the adsorbent pores [9].

This kinetic profile, characterized by an initial adsorption followed by a slower intraparticle diffusion stage, is typical of heterogeneous polymer–oxide composites. The equilibrium time of 24 hours observed in this study aligns with reports on similar materials, such as gelatin–hematite hydrogels [7] and gelatin–αFe_2O_3 composites [10], confirming that the overall adsorption process is primarily governed by electrostatic interaction and ion-exchange between Na^+/Cl^- ions and the surface functional groups (–OH, –COOH, and –NH). These results support the mixed physisorption mechanism previously identified through isotherm.

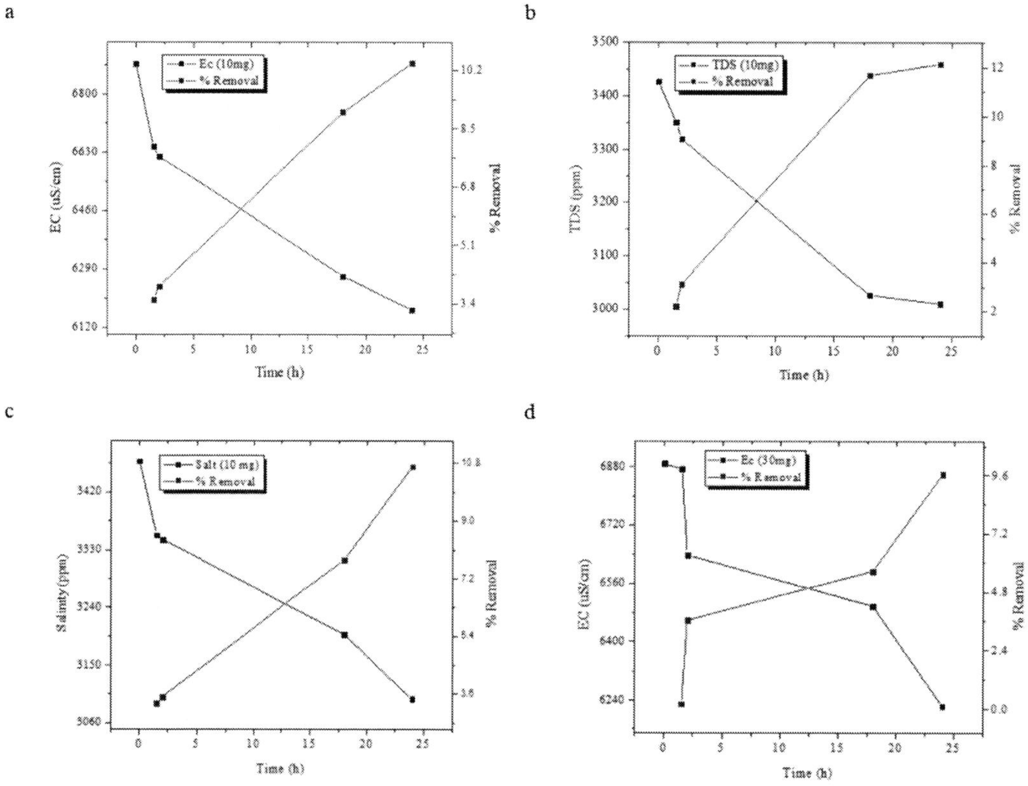

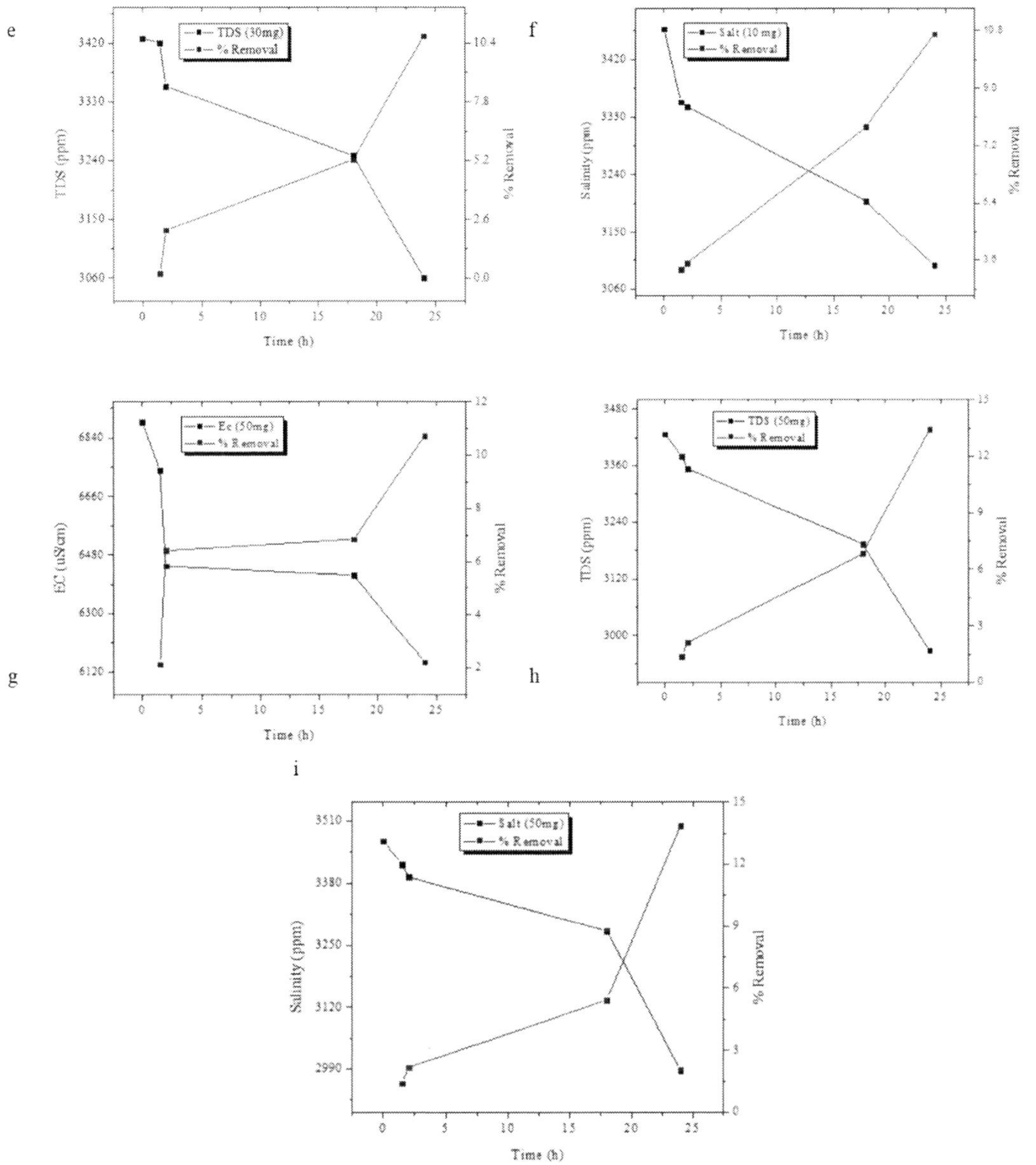

Fig 5. The Effect of dosage of nylon-gelatin-hematite on NaCl Solution

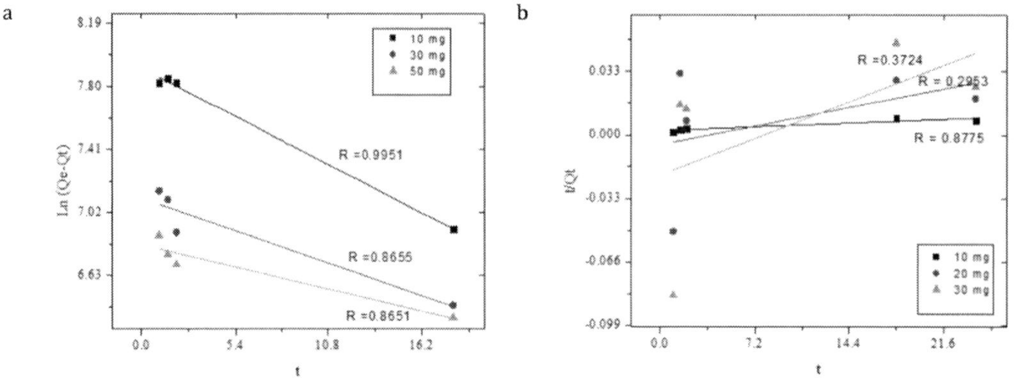

Fig 6. Linear isotherm of saline solution adsorption based on composite nylon-gelatin-hematite at 100 °C (from data salinity) a) Pseudo-first-order b) Pseudo-second-order

Table 3. Kinetic parameters for the pseudo-first and second order

Kinetic Model	Parameter	Dosage (10 mg)	Dosage (30 mg)	Dosage (50 mg)
Pseudo-first-order	R^2	0.9951	0.8656	0.8651
Pseudo-second-order	R^2	0.8775	0.2953	0.3724

The adsorption kinetics of NaCl onto the nylon–gelatin–hematite composite were evaluated using the pseudo-first-order (PFO) and pseudo-second-order (PSO) kinetic models, and the corresponding parameters are presented in Table 3. The pseudo-first-order model exhibits consistently higher correlation coefficients (R^2 = 0.86–0.99) compared to the pseudo-second-order model, indicating that the adsorption process is better described by the PFO kinetics. This finding suggests that the adsorption of Na^+ and Cl^- ions is predominantly governed by physical interactions, such as electrostatic attraction and surface diffusion, rather than by chemisorption. The dominance of the PFO model implies that the rate-limiting step is associated with mass transfer and ion diffusion to the available surface sites of the composite adsorbent.

At an adsorbent dosage of 10 mg, the pseudo-second-order model shows a moderate linear correlation (R^2 = 0.87) with a relatively low rate constant, indicating a slow adsorption rate. This behavior can be attributed to the limited number of accessible active sites and the low probability of effective ion–surface interactions during the initial adsorption stage. As the adsorbent dosage increases to 30 mg and 50 mg, a pronounced decrease in the correlation coefficient (R^2), reaching 0.2953, is observed along with fluctuations in the k_2 values. This trend can be explained by particle aggregation and surface overlapping at higher adsorbent concentrations, which reduce the effective surface area and hinder ion diffusion toward active sites. Consequently, the applicability of the pseudo-second-order model becomes less significant at higher dosages, further confirming that chemisorption is not the dominant mechanism governing NaCl adsorption under the studied conditions [5].

3.4 Effect of Temperature on Saline Solution Adsorption

Temperature variation experiments were conducted at 60°C, 80°C, and 100°C to evaluate the thermodynamic properties of NaCl adsorption. Figure 7 shows that the higher the adsorption temperature, the greater the decrease in electrical conductivity, TDS, and salinity, with a sharp decrease observed at 100°C. This is likely due to increased entropy at the molecular interface, as evidenced by the low value of Gibbs delta, especially at 100°C. The frequent decrease in conductivity, TDS, and salinity values with increasing temperature indicates that the adsorption process is endothermic.

The adsorption capacity increases with increasing temperature, indicating that the process is endothermic. Thermodynamic parameters were determined from equilibrium data and summarized in Table 5. The Gibbs free energy change (ΔG) values were -1.905, -2.019, and -2.139 kJ mol^{-1} for temperatures of 60, 80, and 100 °C, respectively. Negative ΔG values indicate spontaneous Adsorption, while their magnitude increases with increasing temperature, indicating increased feasibility at higher thermal energies.

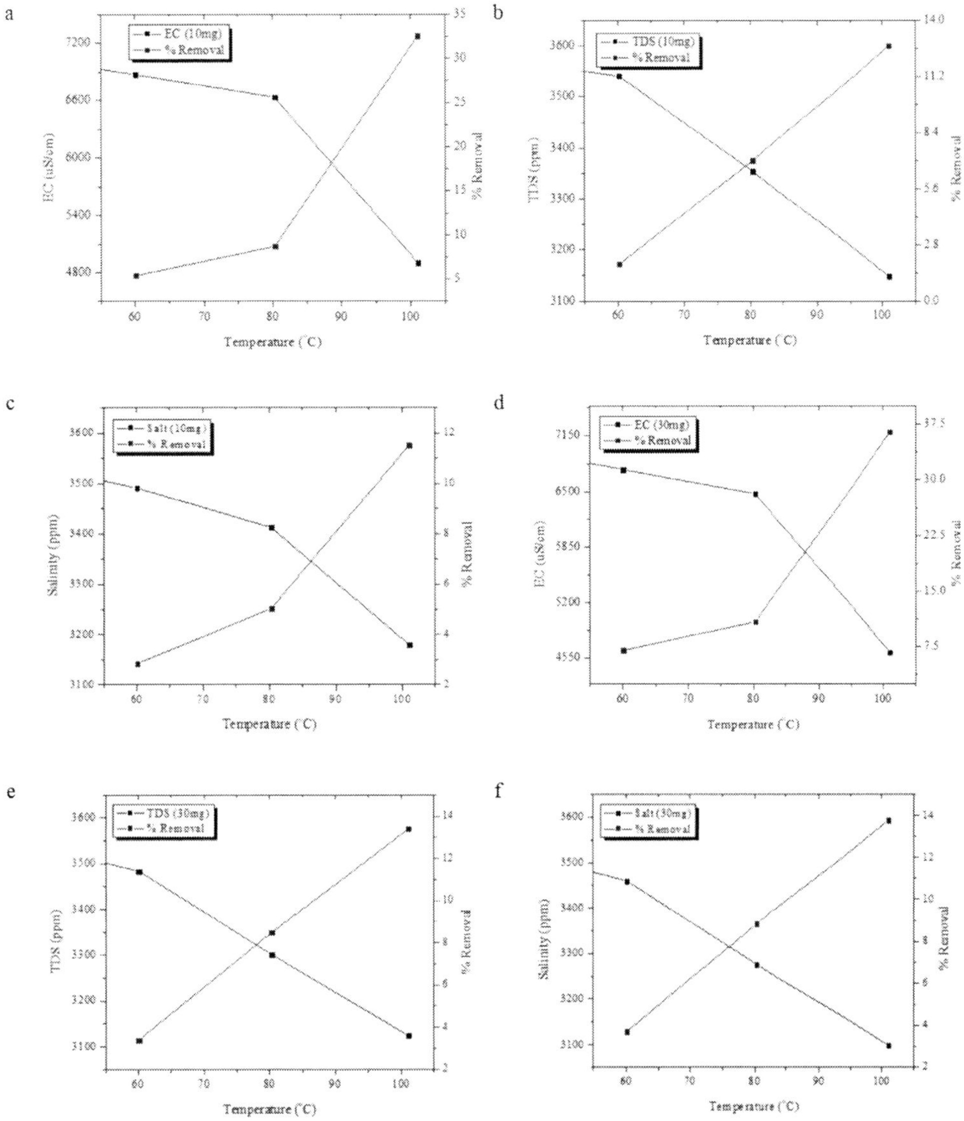

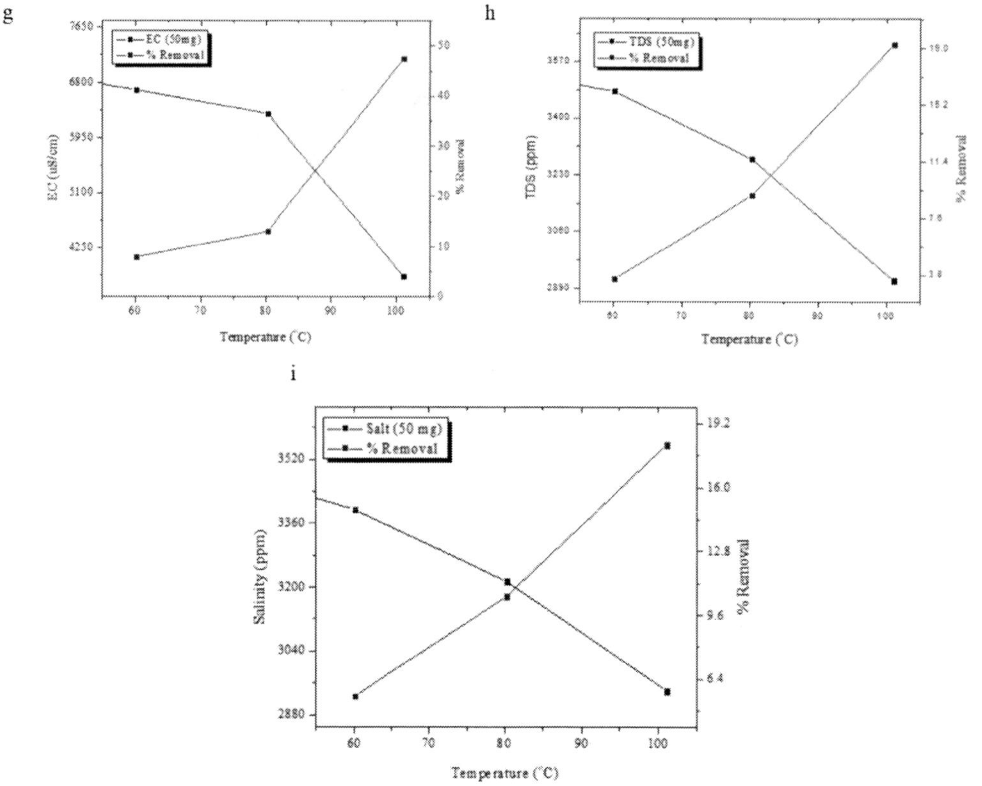

Fig 7. The Effect of temperature on NaCl solution by nylon-gelatin-hematite composite

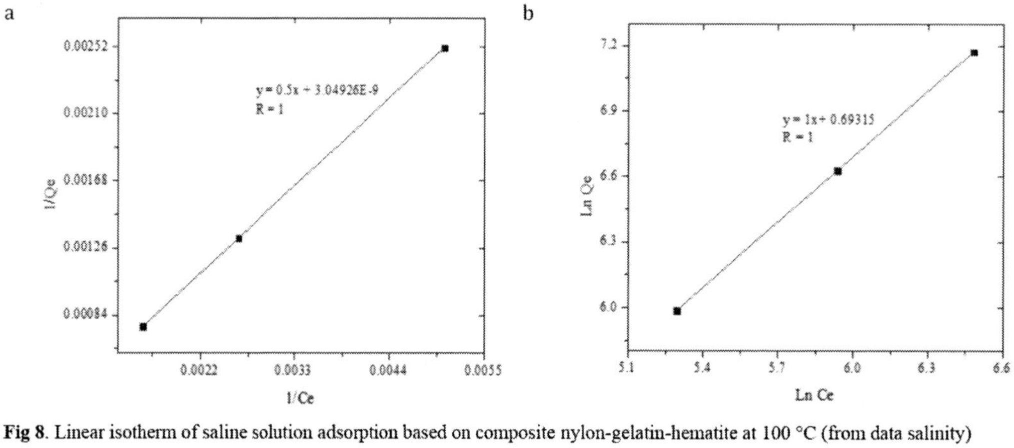

Fig 8. Linear isotherm of saline solution adsorption based on composite nylon-gelatin-hematite at 100 °C (from data salinity)

Table 4. Isothermal adsorption-desorption model fitting parameters

Adsorption Model			
Langmuir		**Freundlich**	
q_{max}	2	**1/n**	1
K_L(L/mg)	1.64E+08	**K_F(mg/g)(L/mg)$^{1/n}$**	1.99
R^2	1	**R^2**	1

Table 5. Thermodynamic parameters for NaCl solution removal on nylon-gelatin-hematite at various temperatures

ΔG (kJ/mol)		
T = 60 °C	T = 80 °C	T = 100 °C
-1.905	-2.019	-2.139

The negative ΔG values indicate spontaneous Adsorption, while the increasing magnitude with temperature suggests enhanced feasibility at higher thermal energy. Similar thermodynamic trends have been reported for Na^+ and Cl^- Adsorption on biopolymer–iron oxide composites. This behavior suggests that the adsorption mechanism involves electrostatic attraction and diffusion-assisted ion penetration into the composite matrix.

4 Conclusion

The nylon–gelatin–hematite composite demonstrated relatively good adsorption performance for removing NaCl from saline solutions. The adsorption data fit well with the Freundlich isotherm models, indicating multilayer and heterogeneous adsorption behavior. Kinetic analysis revealed that the process followed a pseudo-first-order model with an R^2 value of 0.99, confirming physisorption as the dominant mechanism. Thermodynamic results revealed negative ΔG values, ranging from -1.905 to -2.139 kJ/mol, indicating spontaneous Adsorption. Increasing temperature and adsorbent dosage enhanced the adsorption capacity. Overall, the synergistic combination of nylon, gelatin, and hematite significantly improved surface activity and ion-binding affinity, making this composite a promising material for efficient saline water treatment.

References

1. Venkatesh SS, Vellaichamy P, Thirumalachari S, Ramalingam V, Doraiswamy Raju M. Experimental investigation and comparison of PBI/MWCNT and PSF/MWCNT membranes for recovering water from RO reject of brackish water by FO. Heliyon. 2024 Apr;**10**(7):e28455. https://doi.org/10.1016/j.heliyon.2024.e28455
2. Elma M, Mahmud M, Mustalifah FR, Akhbar A, Suryani L, Pratiwi AE, et al. Evaluasi Kinerja Membran Silika Pektin Untuk Desalinasi Air Payau Terhadap Suhu Kalsinasi Membran. Jukung J Tek Lingkung [Internet]. 2021 Jun 12 [cited 2025 Mar 11];**7**(1). Available from: https://ppjp.ulm.ac.id/journal/index.php/jukung/article/view/10816 https://doi.org/10.20527/jukung.v7i1.10816
3. Curto D, Franzitta V, Guercio A. A Review of the Water Desalination Technologies. Appl Sci. 2021 Jan 12;**11**(2):670. https://doi.org/10.3390/app11020670
4. Khoiroh LM, Barroroh H, Taufik M, Royyanuddin M, Akbar FA, Amaliyah NR, et al. Adsorptive Performance Of Gelatin-Hematite Composite Synthesized From Iron Lathe Waste Using Guava Leaf Extract For Pb(II) Removal From Aqueous Solutions. **12**(1).
5. Tran HN, You SJ, Chao HP. Thermodynamic parameters of cadmium adsorption onto orange peel calculated from various methods: A comparison study. J Environ Chem Eng. 2016 Sep;**4**(3):2671–82. https://doi.org/10.1016/j.jece.2016.05.009

6. Hou C, Zhang Y, Liao Q, Chen Q. Ultra-cheap and ultra-stable adsorption materials based on nylon and removal performance of copper ion. Sep Purif Technol. 2024 May;336:126346. https://doi.org/10.1016/j.seppur.2024.126346
7. Kim HJ, Lee JM, Choi JH, Kim DH, Han GS, Jung HS. Synthesis and adsorption properties of gelatin-conjugated hematite (α-Fe2O3) nanoparticles for lead removal from wastewater. J Hazard Mater. 2021 Aug;416:125696. https://doi.org/10.1016/j.jhazmat.2021.125696
8. Ulfa M, Prasetyoko D, Bahruji H, Nugraha RE. Green Synthesis of Hexagonal Hematite (α-Fe2O3) Flakes Using Pluronic F127-Gelatin Template for Adsorption and Photodegradation of Ibuprofen. Materials. 2021 Nov 10;**14**(22):6779. https://doi.org/10.3390/ma14226779
9. Hamad KH, Yasser AM, Nabil R, Tarek R, Hesham E, El-telbany A, et al. Nylon fiber waste as a prominent adsorbent for Congo red dye removal. Sci Rep. 2024 Jan 11;**14**(1):1088. https://doi.org/10.1038/s41598-023-51105-0
10. Abu Elella MH, Aamer N, Abdallah HM, López-Maldonado EA, Mohamed YMA, El Nazer HA, et al. Novel high-efficient adsorbent based on modified gelatin/montmorillonite nanocomposite for removal of malachite green dye. Sci Rep. 2024 Jan 12;**14**(1):1228. https://doi.org/10.1038/s41598-024-51321-2
11. Dimbo D, Abewaa M, Adino E, Mengistu A, Takele T, Oro A, et al. Methylene blue adsorption from aqueous solution using activated carbon of spathodea campanulata. Results Eng. 2024 Mar;21:101910. https://doi.org/10.1016/j.rineng.2024.101910
12. Oshiro M, Takashima K, Furukawa Y. Infrared Stark spectra for a Nylon 6 film. Chem Phys Lett. 2019 Aug;728:32–6. https://doi.org/10.1016/j.cplett.2019.04.068
13. Cebi N, Dogan CE, Mese AE, Ozdemir D, Arıcı M, Sagdic O. A rapid ATR-FTIR spectroscopic method for classification of gelatin gummy candies in relation to the gelatin source. Food Chem. 2019 Mar;277:373–81. https://doi.org/10.1016/j.foodchem.2018.10.125
14. Bakhtiar K, Azhdar B. Influence of pH and annealing temperature on hematite (α-Fe2O3) nanoparticle synthesis with pine needle extract and investigation of structural, magnetic, and dielectric properties. J Sci Adv Mater Devices. 2025 Jun;**10**(2):100876. https://doi.org/10.1016/j.jsamd.2025.100876

Analysis of a Single Squared Wingsail Design as an Auxiliary Propulsion System on a Tanker Ship

Hilman Ikmalul Iman[1], *Ahmad* Nasirudin[1], *Hesty* Anita Kurniawati[1], *Wasis* Dwi Aryawan[1], *Hasanudin*[1], *Erzad* Iskandar Putra[1], and *Ardi* Nugroho Yulianto[1]

[1]Department of Naval Architecture, Faculty of Marine Technology, Institut Teknologi Sepuluh Nopember (ITS), Surabaya, Indonesia

Abstract. This study presents an analysis of a single squared wingsail used as an auxiliary propulsion system for an 8,700 DWT oil tanker. The objective is to evaluate the aerodynamic performance and stability feasibility of the wingsail and to assess its contribution to engine power reduction. Numerical simulations were conducted to obtain lift and drag coefficients for three aspect ratios (AR 2, 4, and 6) with a constant sail area of 289 m^2. Stability assessment was performed at multiple voyage waypoints along the Gresik–Bontang round trip using the weather criterion ($K \geq 1$) and metacentric height ($GM \geq 0.3$ m). The results indicate that all configurations satisfy the stability requirements, while AR 4 provides the most effective performance in terms of thrust generation and power reduction. The AR 4 configuration, with a span of 34.0 m and a chord of 8.5 m, achieves the highest average engine power reduction of 5.3%, compared to 4.9% for AR 2 and 4.1% for AR 6. These findings demonstrate that the selected wingsail configuration is aerodynamically effective and stable, offering a practical solution for auxiliary wind-assisted propulsion on tanker vessels.

1 Introduction

Climate change has become one of the most critical challenges facing humanity, influencing almost every aspect of the environment and human activity. Its effects are increasingly visible through extreme temperature variations, rising sea levels, more frequent floods and droughts, powerful storms, wildfires, landslides, and the accelerated melting of polar ice caps [1,2]. Recognizing the seriousness of this global issue, the United Nations (UN) established the Intergovernmental Panel on Climate Change (IPCC) in 1988 to provide comprehensive scientific assessments of climate-related phenomena [3]. This was followed by the establishment of the United Nations Framework Convention on Climate Change (UNFCCC) in 1992, which brought nations together to discuss and coordinate climate policies. Subsequent milestones, such as the Kyoto Protocol in 1997 and the Paris Agreement in 2015, have reinforced international commitments to maintaining global temperature increases below 1.5°C above pre-industrial levels [4].

Within this global framework, the maritime sector plays an essential role in world trade and transportation, serving as an integral component of international supply chains and economic connectivity [5]. Despite its efficiency in moving large volumes of goods, the industry continues to evolve in response to technological and environmental developments. The International Maritime Organization (IMO), through its series of Greenhouse Gas (GHG) Studies, has highlighted the significant contribution of the maritime sector to global emissions. IMO as the global authority on maritime safety and performance standards, has introduced several frameworks to promote the modernization of ship systems and design. The Energy Efficiency Design Index (EEDI) and the Energy Efficiency Existing Ship Index (EEXI) are two key indices established to ensure ships adopt innovative design concepts and operational improvements [6]. These frameworks have encouraged ship designers and researchers to explore the use of advanced technologies that can improve overall vessel performance while maintaining structural and operational safety [7].

One promising approach that has attracted renewed attention is wind-assisted ship propulsion (WASP). The concept of using wind power for marine propulsion has existed for millennia, particularly in the early stages of maritime trade when sails were the primary means of propulsion [8]. In modern times, this concept has evolved with advances in aerodynamic design, control systems, and material technology. The development of modern devices such as Flettner rotors, suction wings, and rigid wingsails has made it possible to utilize wind energy as an auxiliary source of thrust in large ocean-going vessels [9,10]. By generating aerodynamic lift and drag forces, these systems provide additional thrust that can supplement engine-driven propulsion, thereby improving the vessel's hydrodynamic behaviour and potentially reduce engine load.

Among the different types of WASP technologies, the rigid or squared wingsail configuration has shown promise for integration into large ships. Unlike flexible fabric sails, rigid wingsails use airfoil-shaped structures like airplane wings to generate lift more effectively [11]. This configuration allows for higher lift-to-drag ratios, better control of aerodynamic forces, and greater efficiency at a wide range of wind angles. Furthermore, the squared design simplifies the mechanical system, making it easier to construct, maintain, and retrofit on existing vessels without major modifications to the hull or deck structure . The wingsail can also be automated for deployment and adjustment, which enhances operational safety and consistency under varying weather conditions [12].

However, the installation of a wingsail system introduces several engineering challenges. The aerodynamic forces acting on the sail can induce heeling and overturning moments that alter the ship's hydrostatic balance. These effects become particularly significant in large displacement vessels, where the center of effort of the wingsail is located high above the waterline. Such forces can influence the metacentric height and overall stability of the vessel. Therefore, it is crucial to evaluate the aerodynamic loading in conjunction with the ship's stability criteria to ensure safe and reliable operation under different wind directions and speeds.

2 Method

2.1 Case Study

2.1.1 Main Dimension

The ship selected for this case study is a 8700-ton DWT tanker operating in Indonesia, with the specification as listed in Table 1.

Table 1. Main Dimension

Item	Value
Length Between Perpendiculars (Lpp)	106.4 m
Breadth (B)	18.2 m
Height (H)	10 m
Draught (T)	7.8 m
Volume Displacement (V)	11500 m^3
Service Speed (Vs)	13.5 knots
Engine Power	5310 HP

2.1.2 Ship's Route

In this study, wind data were gathered along the ship's operational route from July 2023 to July 2024, from reference [13,14]. The data include true wind speed and wind direction recorded during the vessel's voyages over a one-year period. These parameters were collected to represent the actual wind conditions experienced by the ship during its operations, serving as input for the aerodynamic and performance analyses of the wingsail.

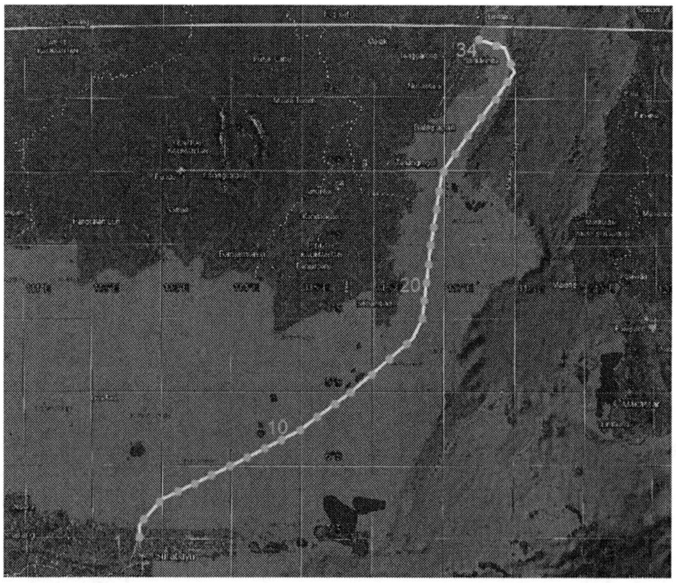

Fig. 1. Ship's operating route and waypoints

The ship operates along a domestic trade route connecting Gresik and Bontang, located in the Java Sea and Makassar Strait. This route represents a typical east-west shipping corridor frequently used by tanker vessels for fuel and chemical transport. The voyage between Gresik and Bontang is conducted regularly in a round-trip pattern, providing consistent environmental conditions suitable for evaluating wind characteristics and potential wingsail performance over the course of one year.

2.1.3 Ship Voyage Data

To analyze the potential performance and applicability of the wingsail system, this study utilized wind data collected along the ship's operational route for a continuous one-year period, from July 2023 to July 2024. The dataset comprises both wind speed and wind direction information recorded throughout the vessel's typical voyage path. This data serves as a fundamental input to evaluate the aerodynamic behavior of the wingsail and to estimate its possible contribution under realistic sea conditions.

Fig. 3.(a) illustrates the variation of true wind speed along the route during the observation period. The wind data were recorded at several coordinate points along the voyage. Fig. 2.(b) presents the wind direction. The directional variability suggests that the route experiences a combination of favorable and crosswind conditions, which are essential considerations in estimating the aerodynamic performance and thrust generation of the wingsail. Overall, the wind characteristics within this period represent realistic environmental conditions for analyzing the wingsail's effectiveness in assisting propulsion and reducing fuel consumption.

(a)

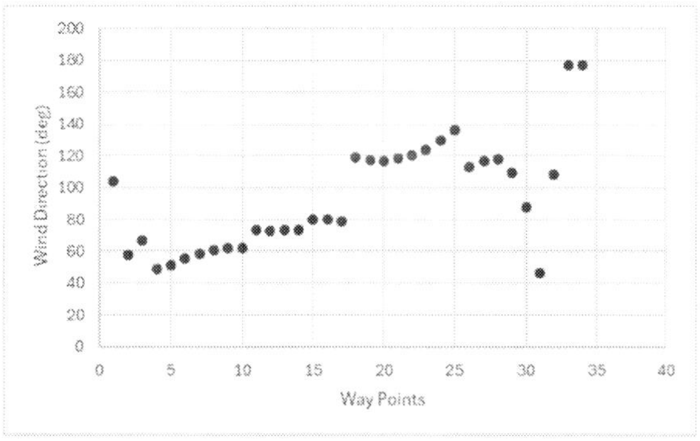

(b)

Fig. 2. Data way points for (a) true wind speed (b) wind direction

Fig. 3 presents the true wind speed and wind direction distributions for the entire observation period. The data show that the average wind speed along the ship's operational route generally ranges between 6 and 7 knots, with occasional peaks above 8 knots during certain months. These variations indicate that the vessel encounters a wide range of relative wind conditions depending on its heading and seasonal changes in the Indonesian maritime region.

2.2 Wingsail Geometries

In this study, the wingsail employed is a rigid squared sail using the NACA 0012 airfoil profile. The total sail area used in this work is 289 m², selected as the reference area for further performance evaluation. To investigate the influence of aspect ratio on aerodynamic behavior, the sail area is divided into three geometries with aspect ratios (AR) of 2, 4, and 6. For the AR 2 configuration, the sail dimensions consist of a span of 24 m and a chord of 12 m. The AR 4 configuration has a span of 34 m and a chord of 8.5 m, while the AR 6 configuration is defined by a span of 41.6 m and a chord of 6.9 m. These variations allow for comparative analysis of aerodynamic coefficients and performance trends across different geometrical arrangements of the wingsail.

2.2.1 Configuration of Wingsail

In this study, the wingsail employed is a single squared rigid wingsail designed as an auxiliary propulsion device for an 8,700 DWT oil tanker. The wingsail has a rectangular planform with an aspect ratio (AR) of 2, 4, and 6 and a total sail area (A_S) of 289 m². The airfoil section used for the design is based on the NACA 0012 symmetrical profile. The wingsail geometry is selected to achieve an optimal balance between aerodynamic efficiency and stability performance while maintaining structural simplicity for potential retrofit applications. The wingsail is installed at a longitudinal position of 79.5 m from the aft perpendicular (AP), corresponding to a location between the midship section and the forecastle deck. This placement is selected to provide sufficient exposure to incoming wind while maintaining an acceptable longitudinal weight distribution and minimizing adverse effects on the vessel's trim and stability characteristics.

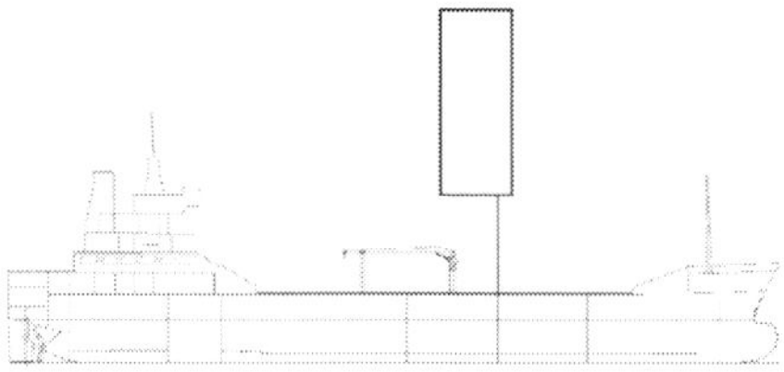

Fig. 3. Wingsail configuration

The rectangular configuration was chosen because of its straightforward manufacturing process and consistent aerodynamic performance across the span. With span length and chord length, the designed wingsail offers sufficient lift generation for thrust assistance without exceeding the vessel's stability limits. The aerodynamic performance parameters, including lift and drag, are evaluated through numerical simulations to determine the effectiveness of the proposed design.

2.3 Simulation Set-Up

The numerical simulations were conducted to evaluate the aerodynamic performance of a squared rigid wingsail for three different aspect ratios, namely AR 2, AR 4, and AR 6. These configurations were selected to represent variations in wingsail geometry while maintaining a constant total sail area of 289 m². The geometrical modifications were applied by adjusting the span and chord dimensions according to the specified aspect ratios, ensuring that only the influence of aspect ratio on aerodynamic behavior was investigated.

Each wingsail model was developed using the NACA 0012 airfoil profile and analyzed under identical operational and environmental conditions. The numerical approach was employed to capture the flow characteristics around the wingsail surface and to determine the corresponding aerodynamic forces generated by each configuration. The primary aerodynamic parameters obtained from the simulations were the lift coefficient (C_l) and drag coefficient (C_d), which were then used as inputs in the subsequent thrust and performance analysis.

This simulation framework enables a consistent comparison between the three aspect ratios, allowing the influence of geometric variation on aerodynamic efficiency to be clearly identified. The results of this analysis provide a reliable basis for selecting the most appropriate wingsail configuration in terms of aerodynamic performance for integration as an auxiliary propulsion system on the tanker vessel.

2.4 Wingsail Thrust Calculations

The forces acting on the sail are illustrated in Fig. 1. The thrust force (T) generated by the sail is the component in ship direction of the resultant force (R) of lift (L) and drag (D). Additionally, the R also produces a side force (S).

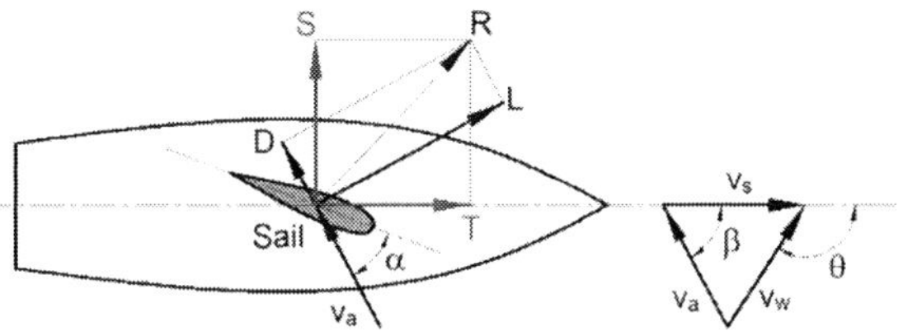

Fig. 4. Sailing vessel forces

L and D are generated by the sail at an angle α relative to the direction of the apparent wind with velocity V_A. The apparent wind velocity V_A is the resultant of the true wind velocity (V_w) and the ship's velocity (V_S). The angle between V_A and V_S is denoted as β, while the angle between V_w and V_S is represented as θ.

T and S can be calculated by using (1) and (2).

$$\mathbf{T} = \mathbf{L} \sin \beta - \mathbf{D} \cos \beta \tag{1}$$

$$\mathbf{S} = \mathbf{L} \cos \beta + \mathbf{D} \sin \beta \tag{2}$$

where,

$$\beta = \text{acos}\left[(V_w \cos \theta + V_s) / V_A\right] \tag{3}$$

$$V_a = (V_w^2 + V_s^2 + 2\, V_w\, V_s \cos \theta)^{0.5} \tag{4}$$

$$\mathbf{L} = 0.5\, \rho\, V_A^2\, A_s\, C_l \tag{5}$$

$$\mathbf{D} = 0.5\, \rho\, V_A^2\, A_s\, C_d \tag{6}$$

In this formulation, C_l and C_d are obtained from the numerical simulations for each wingsail configuration, while A□ represents the sail area and AR denotes the aspect ratio. These parameters are used as fundamental inputs for calculating the aerodynamic forces and thrust generated by the wingsail. The aerodynamic forces acting on the wingsail are governed by the apparent wind velocity (V_A) and the wind angle relative to the ship's course (β). These parameters determine the magnitude of lift (L) and drag (D) generated by the airfoil-shaped sail. The resulting forces are resolved into thrust (T), acting in the longitudinal direction of the ship, and side force (S), acting perpendicular to the ship's motion. While the thrust force contributes directly to reducing the propulsion demand of the main engine, the side force induces heeling moments that affect ship stability. Therefore, the aerodynamic performance of the wingsail must be evaluated in conjunction with stability criteria to ensure that the generated thrust does not compromise the vessel's safety.

2.5 Stability Criteria and Calculations

The stability assessment was conducted to evaluate the influence of the wingsail installation on the vessel's transverse stability. Two main criteria were applied: the weather criterion and the metacentric height (GM) requirement. The weather criterion defines the ratio between the overturning moment generated by the aerodynamic force on wingsail (M_q) and wind heeling moment (M_f). This relationship can be expressed as:

$$K = \mathbf{M_q}/\mathbf{M_f} \tag{7}$$

The vessel is considered stable when K ≥ 1, which indicates that the restoring moment is sufficient to counteract the heeling effect produced by the wingsail. The weather criterion, defined as the ratio between the overturning moment (M_q) induced by aerodynamic wind forces and the restoring moment (M_f) provided by the ship, indicates the vessel's ability to resist wind-induced heeling. When K is lower than unity ($K < 1$), the overturning moment exceeds the available restoring moment, implying that the ship may experience excessive heel or be unable to recover to an upright position under wind loading conditions. The overturning moment (M_q) is obtained from the product of the aerodynamic force acting on the sail. The wind heeling moment (M_f) consists of two main components, the moment acting on the sail structure (M_{fs}) and the moment acting on the hull body (M_{fb}). M_q and M_f can be calculated by using (9) and (10).

$$\mathbf{M_q} = l_q\, \Delta \tag{8}$$

$$\mathbf{M_f} = \mathbf{M_{fs}} + \mathbf{M_{fb}} \tag{9}$$

The second stability criterion involves the metacentric height (GM), which characterizes the ship's initial transverse stability. The vessel should maintain a minimum GM value of 0.3 m to ensure adequate righting capability under external heeling forces. A GM value below the minimum requirement (GM < 0.3 m) indicates insufficient righting leverage, resulting in slow roll response and increased vulnerability to capsizing when subjected to external disturbances. Therefore, both criteria must be satisfied to ensure that the wingsail-generated aerodynamic forces do not compromise the vessel's stability during operation. Both criteria were employed in this study to determine the acceptable wingsail dimensions that maintain compliance with IMO stability requirements.

3 Result and Discussions

3.1 Lift and Drag Coefficient

The lift coefficient (C_l) and drag coefficient (C_d) were obtained from the numerical simulation results for each wingsail configuration with aspect ratios of 2, 4, and 6. These coefficients represent the aerodynamic response of the sail under varying true wind angles along the ship's operational route. The extracted C_l and C_d values are subsequently used as primary inputs in the thrust calculation to evaluate the aerodynamic performance of the wingsail and its contribution to ship propulsion. The variation of these coefficients across different aspect ratios provides a basis for comparing the thrust and performance trend of each wingsail configuration.

A grid independence study was conducted to ensure that the numerical results obtained from the numerical simulations are not significantly influenced by mesh resolution. Several mesh densities were tested by progressively increasing the total number of elements, and the corresponding lift coefficient (C_L) and drag coefficient (C_D) were monitored. As shown in Fig. X, an increase in mesh density leads to noticeable variations in both C_L and C_D at lower element counts; however, beyond approximately 5×10^6 elements, the changes in aerodynamic coefficients become marginal. The lift coefficient shows a gradual convergence trend, while the drag coefficient stabilizes with only negligible differences for finer meshes. This indicates that further mesh refinement does not significantly improve solution accuracy. Therefore, the selected mesh resolution provides a reasonable balance between computational efficiency and numerical accuracy and is considered sufficient for subsequent aerodynamic analyses.

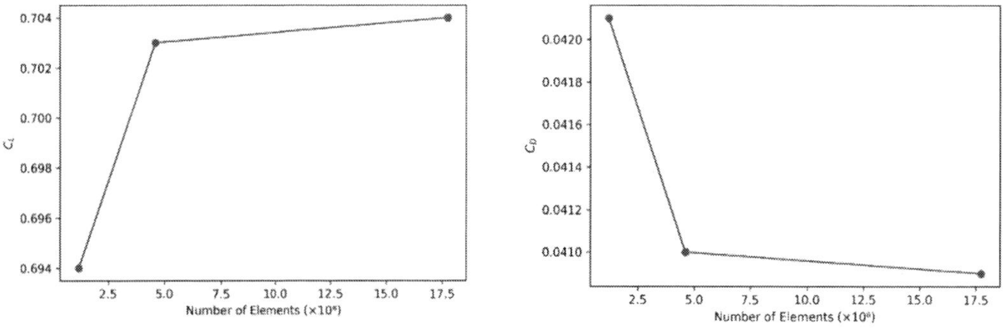

(a) (b)

Fig. 5. Grid independence study (a) C_l (b) C_d

The lift coefficient (C_l) and drag coefficient (C_d) were obtained from the numerical simulations for each aspect ratio (AR = 2, 4, and 6) under three Reynolds numbers, namely Re = 3.3M, 6.0M, and 8.9M. Figures X and Y present the variation of C_l and C_d to aspect ratio, illustrating the aerodynamic response of the squared wingsail geometry.

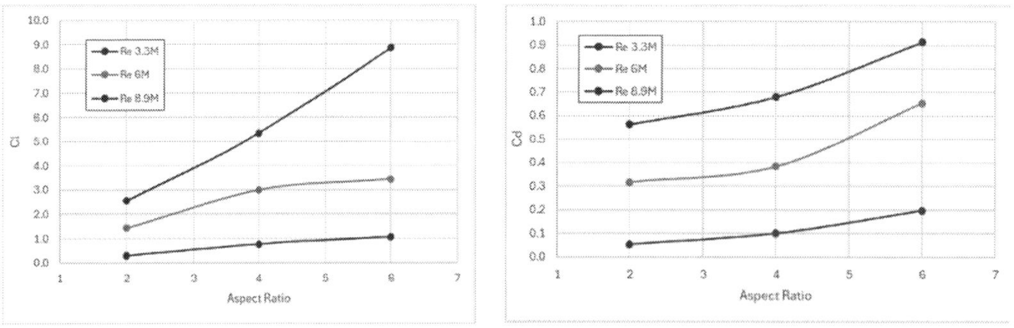

(a) (b)

Fig. 6. (a) C_l maximum and (b) C_d maximum for different aspect ratio and reynold number

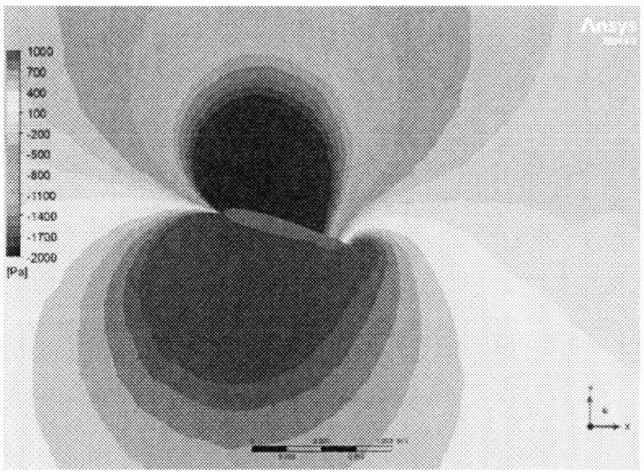

(a)

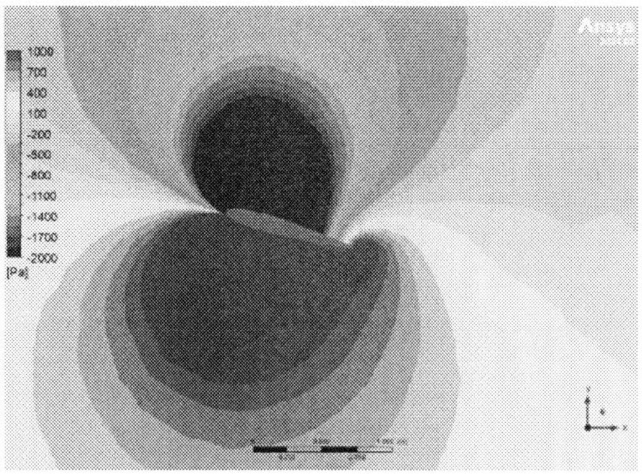

(b)

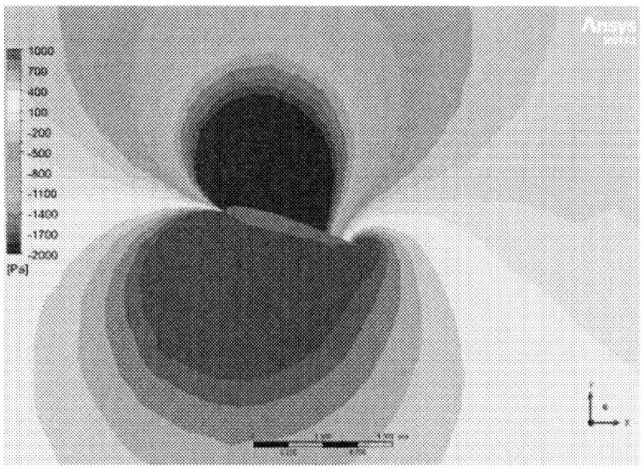

(c)

Fig. 7. Contour pressure (a) AR 2 (b) AR 4 (c) AR 6

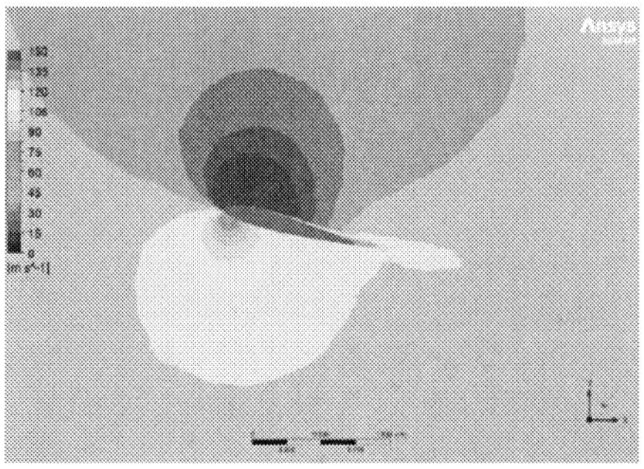

(a)

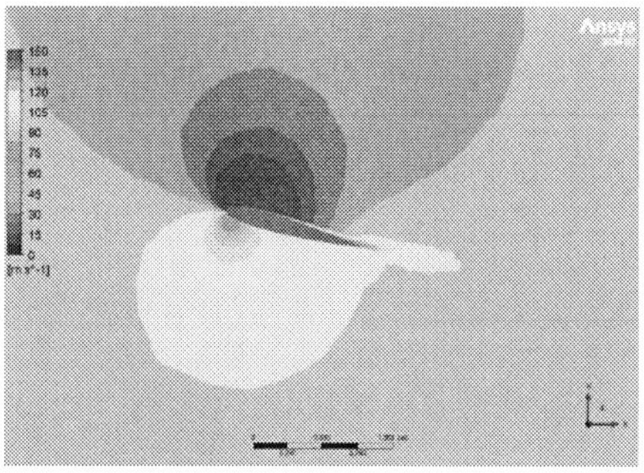

(b)

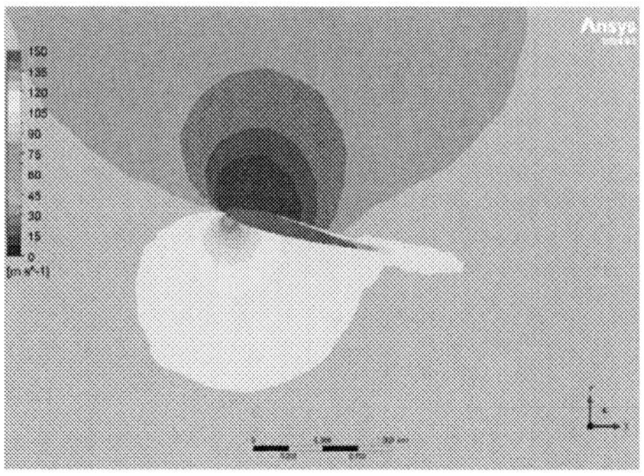

(c)

Fig. 8. Contour velocity (a) AR 2 (b) AR 4 (c) AR 6

In Fig. 5, demonstrate that increasing aspect ratio leads to higher lift and drag coefficients across all Reynolds numbers. This trend indicates that while larger aspect ratios enhance lift generation, they also introduce greater aerodynamic resistance.

Based on the results, the AR 4 configuration produced the highest lift coefficient, with a C_l value of 1.50, indicating superior lift generation capability compared to AR 2 and AR 6. The AR 2 configuration yielded a C_l of 1.40, while AR 6 showed the lowest lift coefficient at 1.16. This trend suggests that increasing AR does not always result in higher lift for the squared rigid wingsail, as geometric efficiency and flow behavior also influence aerodynamic performance. In terms of drag, AR 2 exhibited the highest drag coefficient with a C_d value of 0.310, followed by AR 6 with 0.217. The lowest drag coefficient was obtained by the AR 4 configuration at 0.162. This indicates that AR 4 offers a more favorable balance between lift and drag, resulting in a higher lift-to-drag ratio compared to the other configurations.

3.2 Stability Calculations

The stability calculations were conducted to evaluate the effect of the wingsail installation on the ship's overall stability performance. Based on the analysis, the wingsail generates an overturning moment (M_q) and a wind heeling moment (M_f) that were used to determine the stability ratio (K) and the metacentric height (GM).

The stability evaluation was conducted to assess the feasibility of the selected wingsail configuration under realistic operational conditions along the ship's entire voyage route. The route considered in this study consists of 34 waypoints for a single direction, resulting in a total of 68 evaluation points for a complete round trip between Gresik and Bontang. At each waypoint, the stability parameters were calculated based on the prevailing wind conditions and the corresponding aerodynamic forces generated by the wingsail.

Initially, three aspect ratios (AR = 2, 4, and 6) were examined to investigate their influence on both aerodynamic performance and ship stability. The results indicated that the AR 2 configuration successfully satisfied the stability criteria at all evaluated points, with the weather criterion ratio K remaining above 1 and the metacentric height GM exceeding 0.3 m throughout the entire route. Conversely, the AR 4 configuration failed to meet the stability requirements at several locations, particularly at 34 evaluated points where the K value dropped below the acceptable limit, indicating insufficient restoring capability under certain wind conditions. The AR 6 configuration exhibited an even higher tendency to violate the stability constraints.

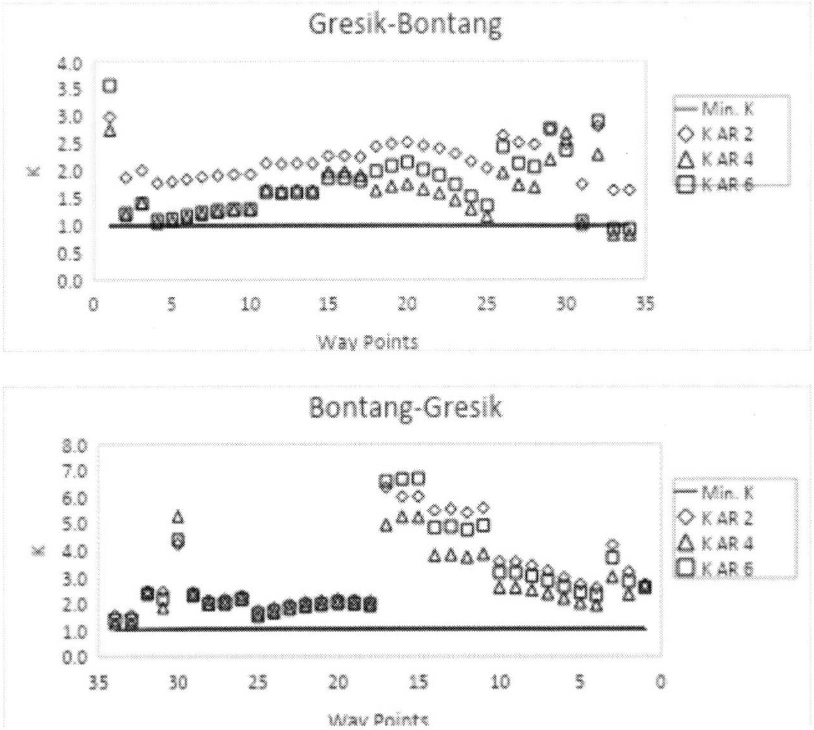

Fig. 9. Weather criterion (K) at each waypoints

https://doi.org/10.1051/bioconf/202621609002

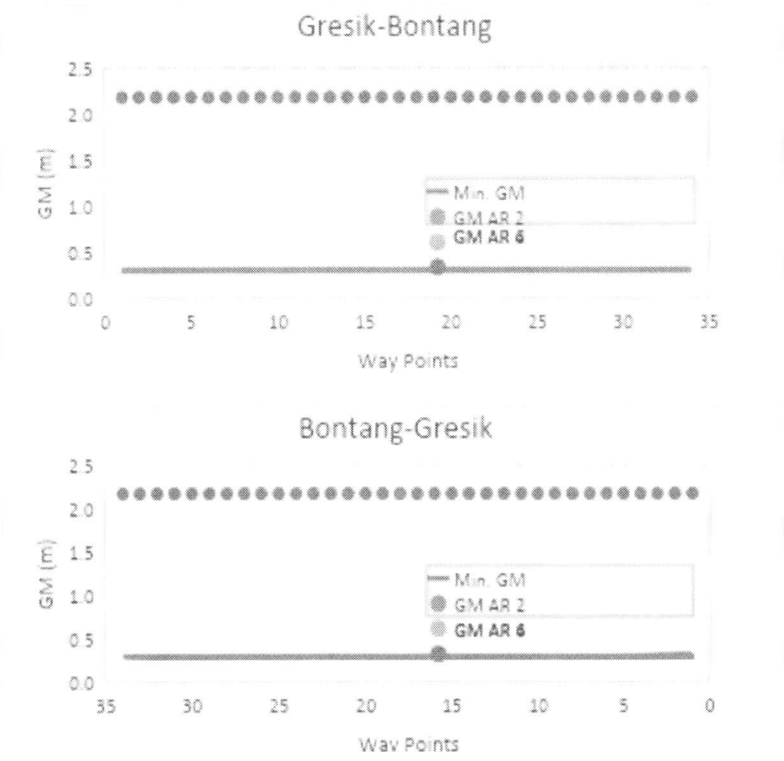

Fig. 10. GM criterion at each waypoints

To balance between aerodynamic performance and stability safety, a compromise aspect ratio of AR 4 was selected as the final configuration. The stability ratio (K) was further evaluated at each waypoint along the ship's operational route using Equation (7) for the selected wingsail configuration with AR = 4. The analysis shows that all calculated K values along the Gresik–Bontang segment satisfy the weather criterion requirement, with values exceeding 1 and reaching up to approximately 4. For the route from Bontang to Gresik, the stability ratio exhibits even higher margins, with peak values approaching 6.3 at specific waypoints. This consistent compliance across all evaluated points confirms that the proposed wingsail configuration provides sufficient restoring capability against wind-induced overturning moments throughout the entire round-trip operation.

The metacentric height (GM) was obtained from stability analysis using Maxsurf Stability to evaluate the effect of the wingsail installation on the vessel's transverse stability. For the Gresik–Bontang route under ballast condition, the GM values were 2.187 m for AR 2, 2.184 m for AR 4, and 2.180 m for AR 6. Meanwhile, for the Bontang–Gresik route under full load condition, the corresponding GM values were 2.194 m for AR 2, 2.187 m for AR 4, and 2.178 m for AR 6.

3.3 Evaluation of Aspect Ratio and Thrust Performance

The comparative assessment of aspect ratios AR 2, AR 4, and AR 6 were conducted based on both aerodynamic performance and stability compliance along the vessel's operational route. Although AR 2 demonstrated full compliance with the stability criteria at all evaluated waypoints, its aerodynamic performance, particularly in terms of thrust generation, was comparatively lower due to its reduced effective span. Conversely, AR 4 exhibited improved aerodynamic characteristics; however, stability analysis revealed that this configuration did not satisfy the weather criterion at several waypoints, indicating potential operational limitations under varying wind conditions. Considering these findings, an intermediate aspect ratio of AR 4 was adopted as a compromise configuration, providing a balanced combination of enhanced thrust potential and acceptable stability performance. This selected aspect ratio ensures that the wingsail operates within the required stability limits while achieving higher propulsive contribution compared to AR 2, thereby supporting a more effective integration of the wingsail system as an auxiliary propulsion device.

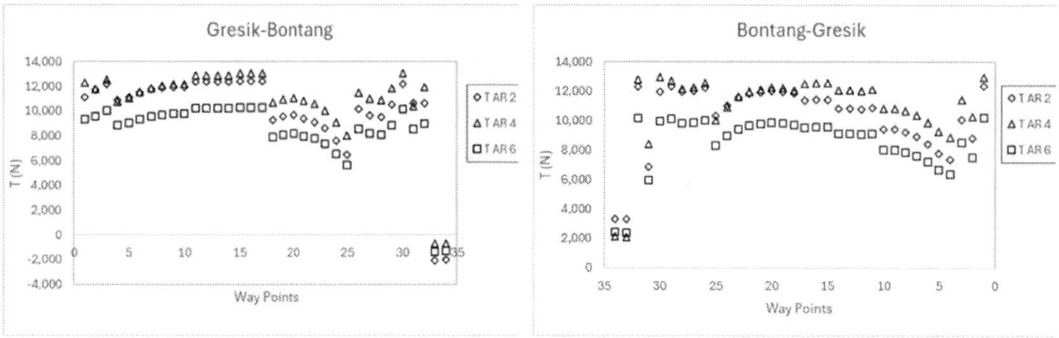

Fig. 11. Wingsail thrust

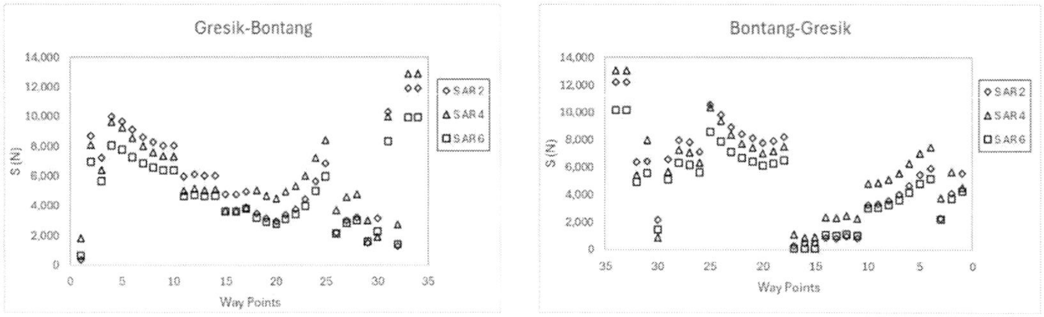

Fig. 12. Wingsail side force

3.4 Contribution of Wingsail

The contribution of the wingsail system to propulsion performance was evaluated by comparing the average aerodynamic thrust generated by the sail with the ship's resistance at the service speed of 13.5 knots. The analysis was performed separately for each voyage direction to account for differences in wind characteristics and hydrodynamic resistance.

For the Gresik–Bontang voyage, the total ship resistance at the service speed of 13.5 knots was evaluated to be 205.4 kN. Under the corresponding wind conditions along this route, the AR 4 wingsail configuration generated sufficient aerodynamic thrust to achieve an engine power reduction of 5.29%. This reduction indicates that the wingsail effectively assisted the main propulsion system by partially offsetting the hydrodynamic resistance, thereby decreasing the required brake power output while maintaining the vessel's operational speed.

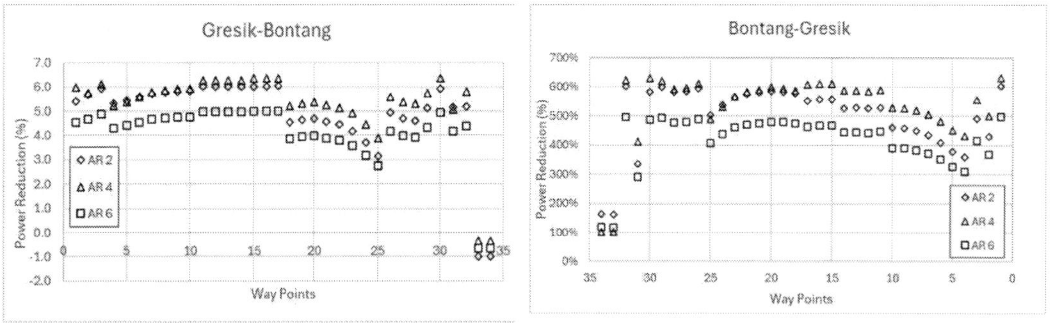

Fig. 13. Power reduction

Similarly, for the Bontang to Gresik, the total ship resistance increased to 229.6 kN due to variations in environmental and operational conditions. Despite the higher resistance, the AR 4 wingsail maintained robust aerodynamic performance, resulting in a slightly greater engine power reduction of 5.34%. The consistency of these power savings across both voyage directions demonstrates the reliability of the wingsail system in providing auxiliary propulsive contribution under varying wind and resistance conditions. Considering both voyage directions, the overall average engine power reduction provided by the AR 4 wingsail for a complete round trip was therefore determined to be 5.3%, confirming the consistent contribution of the wingsail system in reducing propulsion power demand under varying operational conditions.

4 Conclusions

The results of this study indicate that a wingsail area of 289 m^2 represents the maximum feasible configuration that fully satisfies the applied stability criteria for all evaluated aspect ratios (AR 2, AR 4, and AR 6). This area defines the upper operational limit at which the wingsail system can be implemented without compromising the vessel's transverse stability and safety margins. Among the investigated configurations, the wingsail with an aspect ratio of 4, corresponding to a span of 34 m and a chord of 8.5 m, demonstrates the most favorable overall performance. This configuration generates the highest average thrust contribution, resulting in an average engine power reduction of approximately 5.3% over the evaluated voyage. The reduction in required engine power is achieved through the longitudinal thrust component produced by aerodynamic lift and drag forces acting on the wingsail, which partially offsets the ship resistance and decreases the propulsion demand of the main engine. Consequently, the AR 4 configuration provides an effective balance between aerodynamic performance and stability compliance and is therefore identified as the optimal design for integration on the selected tanker vessel within the scope of this study. Future research should extend the present analysis by incorporating structural design and strength assessment of the wingsail, evaluating dynamic wind and wave effects, and investigating operational control strategies and alternative placement configurations to further improve propulsion efficiency and ensure reliable performance under realistic operating conditions.

References

[1] W. I. Okonkwo, "Impacts of Climate Change on Environment and the Remedies @ ECRTD-UK: https://www.eajournals.org/," vol. 8, no. 2, pp. 1–9, 2022.

[2] E. A. Singh and M. R. Shindikar, "A Comprehensive Review on Climate Change and Its Effects," vol. 13, no. 11, pp. 924–931, 2023, doi: 10.9734/IJECC/2023/v13i113240

[3] F. Committee, "Protection of global climate for present and future generations of mankind: Programme budget implications of draft resolution A/C. 2/58/L. 14/Rev. 1: Report of the 5th Committee: General Assembly, 58th session," no. January 2004, pp. 2004–2005, 2005.

[4] P. R. S. Masson-Delmotte, V., P. Zhai, H.-O. Pörtner, D. Roberts, J. Skea, M. I. G. A. Pirani, W. Moufouma-Okia, C. Péan, R. Pidcock, S. Connors, J.B.R. Matthews, Y. Chen, X. Zhou, and and T. W. E. Lonnoy, T. Maycock, M. Tignor, *IPCC, 2018: Summary for Policymakers. In: Global Warming of 1.5°C. An IPCC Special Report on the impacts of global warming of 1.5°C above pre-industrial levels and related global greenhouse gas emission pathways, in the context of strengthening the global.* Cambridge and New York: Cambridge University Press, 2018. doi: https://doi.org/10.1017/9781009157940.001.

[5] Melnyk Oleksiy Mykolayovych, "STRATEGIES FOR THE DEVELOPMENT OF MARITIME TRANSPORTATION AND METHODS FOR EFFICIENT OPERATION," vol. 8, no. 22, pp. 324–335, 2023.

[6] M. Tadros, M. Ventura, and C. G. Soares, "Review of the IMO Initiatives for Ship Energy Efficiency and Their Implications," pp. 662–680, 2023.

[7] A. T. A. Wijaya, I. M. Ariana, H. Prastowo, and T. Pitana, "Consideration of Energy Efficiency Operational Index evaluation Consideration of Energy Efficiency Operational Index evaluation".

[8] T. Chou, V. Kosmas, M. Acciaro, and K. Renken, "A Comeback of Wind Power in Shipping : An Economic and Operational Review on the Wind-Assisted Ship Propulsion Technology," 2021.

[9] G. Atkinson, H. Nguyen, and J. Binns, "Considerations regarding the use of rigid sails on modern powered ships," *Cogent Eng.*, vol. 5, no. 1, pp. 1–20, 2018, doi: 10.1080/23311916.2018.1543564

[10] P. F. Rynne, S. Member, and K. D. Von Ellenrieder, "Development and Preliminary Experimental Validation of a Wind- and Solar-Powered Autonomous Surface Vehicle," *IEEE J. Ocean. Eng.*, vol. 35, no. 4, pp. 971–983, 2010, doi: 10.1109/JOE.2010.2078311

[11] A. Hillenbrand, A. Oceanbird, and J. Kuttenkeuler, "Wind Tunnel Tests of a Two-Element Wingsail with Focus on Near-Stall Aerodynamics," vol. 9, no. 1, pp. 110–127, 2025.

[12] R. Zhang *et al.*, "Numerical investigation on the effects of heel on the aerodynamic performance of wing sails," *Ocean Eng.*, vol. 305, no. April, p. 117897, 2024, doi: 10.1016/j.oceaneng.2024.117897. Available: https://doi.org/10.1016/j.oceaneng.2024.117897

[13] "Global Fishing Watch." Available: https://globalfishingwatch.org/. [Accessed: Sep. 20, 2025]

[14] R. Schlitzer, "Ocean Data View," 2016. Available: http://odv.awi.de. [Accessed: Sep. 20, 2025]

Innovative Concrete Mold and Hexaloc Production System for Sustainable Coastal Protection in Indonesia

Candra Irawan[1*], *Haryo* Dwito Armono[2], *Made* Yujiro Semita[1] *Nurul* Istiqomah[2] and *Muhammad* Alvin A. P.[1]

[1]Department of Civil Engineering, Institut Teknologi Sepuluh Nopember, Surabaya, Indonesia

[2]Department of Ocean Engineering, Institut Teknologi Sepuluh Nopember, Surabaya, Indonesia

Abstract. Coastal erosion is a major challenge for Indonesia, where extensive shorelines are exposed to high-energy waves and anthropogenic pressures. This study introduces Hexaloc, a novel concrete armor unit with six interlocking arms designed to enhance stability and ease of installation in breakwater structures. The research focuses on the innovation of concrete molds and production systems, enabling scalable fabrication of Hexaloc units with consistent quality. Laboratory-scale trials were conducted to validate compressive strength, density, and durability using Ultrasonic Pulse Velocity (UPV), hammer test, and compressive strength testing. Results confirm that Hexaloc units achieve the targeted strength of 30 MPa and demonstrate effective interlocking performance under simulated wave conditions. The prototype development reached Technology Readiness Level (TRL) 5, with potential for industrial-scale application and community-based coastal protection programs. This innovation contributes to sustainable coastal infrastructure by offering a practical, durable, and environmentally adaptive solution for shoreline defense in tropical regions.

1 Introduction

Indonesia's coastal regions are increasingly vulnerable to environmental pressures such as shoreline erosion, storm surges, and large-scale reclamation activities [1], [2], [3]. These phenomena not only threaten ecological balance but also pose risks to infrastructure, livelihoods, and long-term coastal resilience. Conventional breakwater systems, typically constructed from natural rocks or standard concrete armor units, have been widely employed to mitigate these hazards. However, such traditional solutions often exhibit limitations in terms of hydraulic stability and structural efficiency, particularly when subjected to high-energy wave environments [4]. The lack of optimized interlocking capacity among conventional units can result in displacement, reduced durability, and higher maintenance costs over time.

Fig. 1. Interlocking property of the Hexaloc concrete armor units

* Corresponding author: chandra@ce.its.ac.id

In response to these challenges, the Hexaloc unit was conceived as an innovative concrete armor block specifically engineered to enhance coastal protection. Characterized by six symmetrical arms, the Hexaloc design (shown in Fig. 1) aims to maximize interlocking potential and improve overall stability under dynamic wave action [5]. Its geometric configuration facilitates energy dissipation while minimizing movement, thereby offering superior performance compared to conventional armor units. Recent innovations such as BRINpod further demonstrate the potential of novel interlocking designs to strengthen coastal defense systems [6]. The development of Hexaloc represents a significant advancement in coastal engineering, combining structural robustness with practical applicability. By integrating novel design principles into coastal defense systems, this unit contributes to more sustainable and resilient shoreline management strategies in Indonesia and beyond.

Building on prior investigations of artificial roots, armor units, and preliminary numerical analyses of Hexaloc [5], [7], [8], this study seeks to bridge the gap between conceptual design and practical implementation. The primary objective is to establish a reliable production methodology for Hexaloc units, including mold fabrication, concrete mix optimization, and laboratory testing of mechanical properties. By integrating design innovation with scalable manufacturing processes, the research aims to demonstrate the feasibility of Hexaloc as a coastal protection solution that is both structurally robust and environmentally resilient. Ultimately, this paper presents the development, testing, and evaluation of Hexaloc units as a novel contribution to sustainable coastal infrastructure.

2 Methodology

2.1 Mold Innovation

The research successfully produced designs of Hexaloc molds corresponding to unit weights of 0.14 ton (a) and 1.0 ton (b) as shown in Fig. 2. The Hexaloc mold consists of two main parts, the upper and lower sections, which are connected at the center using bolted joints. The lower section is supported by steel columns positioned at the four sides of the mold, ensuring stability and alignment during casting. Openings are provided on all four sides as well as on the top surface of the mold to facilitate the placement of concrete, allowing efficient filling and compaction. On each tip of the arms, openings are provided to aid in the demolding process. In addition, the corners of the mold are deliberately designed at angles greater than 90 degrees to ensure that the concrete surface remains smooth and free of damage when the mold is removed. This design innovation reduces the risk of stress concentration at the edges, prevents chipping during demolding, and eases the release process, thereby improving both the visual quality and durability of the finished Hexaloc units.

By minimizing surface defects and simplifying the demolding process, the >90° corner design directly contributes to lower maintenance costs, reduced rejection rates, and longer service life of the Hexaloc units in coastal applications. The improved surface finish not only enhances durability against abrasion and weathering but also ensures better interlocking performance when units are deployed in breakwater systems. Consequently, the mold design provides both technical reliability and economic efficiency, strengthening the case for industrial-scale production and practical implementation in coastal defense projects.

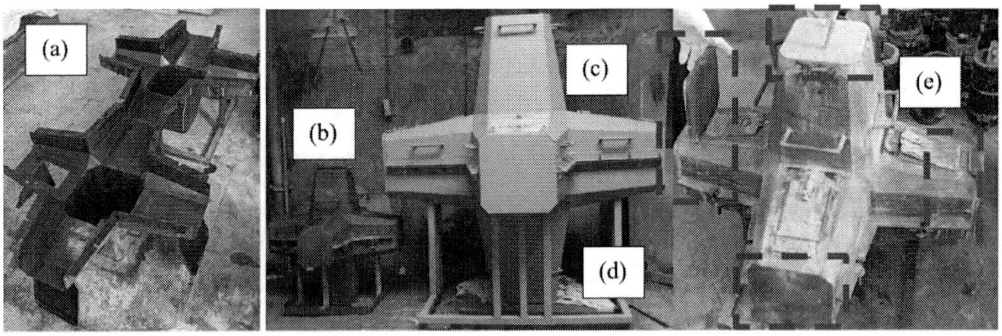

Fig. 2. (a) Two-part Hexaloc molds for unit weights of 0.14 ton (b) and 1.0 ton (c) equipped with (d) steel supports and (e) openings at all six tips of the arms

2.2 Production Process

The detailed mix proportions are summarized in Table 1, which outlines the material quantities used per cubic meter of concrete. The mix was proportioned to achieve both high strength and durability, with a cement content of 444.44 kg/m³ and using a 0.45 water/cement ratio. The relatively high cement dosage, combined with carefully balanced aggregates (716.22 kg/m³ of sand and 1074.33 kg/m³ of crushed stone) ensured adequate workability while maintaining dense particle packing. The addition of 200 liters of water provided sufficient hydration without compromising strength, as reflected in the low w/c ratio. This mix design was intentionally selected to produce concrete with superior mechanical properties, particularly compressive strength exceeding standard requirements, and to guarantee durability under aggressive coastal exposure conditions. The systematic proportioning not only facilitated consistent production of Hexaloc units but also ensured that the resulting concrete could withstand long-term service in marine environments, where resistance to erosion, abrasion, and chloride penetration is critical.

Table 1. Mix Composition of Hexaloc Concrete f'c 30 MPa

Material	Quantity per m³ of concrete
Cement	444.44 kg
Sand	716.22 kg
Crushed stone	1074.33 kg
Water	200 L
w/c ratio	0.45

2.3 Technology Readiness Level (TRL)

The project successfully achieved Technology Readiness Level (TRL) 5, demonstrating that the Hexaloc prototype could be validated under relevant laboratory conditions. At this stage, the design was tested in controlled environments that closely replicate real coastal scenarios, confirming both its structural performance and stability. Achieving TRL 5 indicates that the concept has moved beyond theoretical modeling and small-scale experimentation, reaching a level where the prototype can reliably function in simulated operational settings. Looking ahead, the next milestone involves scaling the innovation to TRL 6, which requires demonstration in a relevant industrial environment. This progression will be pursued through collaboration with precast concrete manufacturers, enabling the transition from laboratory molds to industrial production systems. Such partnerships are expected to ensure consistent quality, reproducibility, and cost-effectiveness in large-scale fabrication. By advancing to TRL 6, the project aims to bridge the gap between laboratory validation and real-world application, positioning Hexaloc as a viable solution for coastal infrastructure projects that demand durability, scalability, and resilience [9], [10].

3 Results and Discussion

As shown in Fig. 3, the mold was successfully tested for producing Hexaloc units with a weight of 0.14 ton, demonstrating the feasibility of the design at a smaller scale. Visual inspection of the finished product confirmed that the concrete surface exhibited good quality, with smooth texture, uniform finish, and minimal surface defects such as honeycombing or segregation. These observations suggest that the mold design ensures proper compaction and effective concrete flow during casting, which are critical factors for achieving durability and long-term performance in marine environments. Although the assessment was primarily visual, the results provide encouraging evidence that the production system can deliver consistent quality, thereby supporting the reliability of Hexaloc units for coastal protection applications.

https://doi.org/10.1051/bioconf/202621609003

Fig. 3. Laboratory process for producing Hexaloc units

Concrete specimen testing is shown in Fig. 4, where the Hexaloc armor unit's compressive strength was evaluated to determine their ability to withstand structural loads in coastal environments. Test results indicated values ranging between 47–52 MPa, which significantly exceeded the design target of 30 MPa. This performance demonstrates excellent bonding between cement paste and aggregates, as well as effective compaction during casting. The higher strength values confirm that Hexaloc units are capable of resisting both static and dynamic forces generated by wave action and interlocking stresses, ensuring long-term durability in service.

Fig. 4. UPV test, rebound hammer test, and compressive strength test of Hexaloc concrete samples.

The Ultrasonic Pulse Velocity (UPV) test results, as shown in Fig. 5, reveal notable variations in concrete quality across the testing sequence. Two distinct peaks are observed around testing numbers 8–10 and 28–30, where the UPV values approach 4700 m/s, indicating zones of high material density and internal cohesion. According to widely accepted standards such as those referenced in ASTM C597 (as shown in Table 2) and international guidelines, concrete quality can be classified based on UPV values as follows:

Table 2. Grading of Concrete Quality Based on Ultrasonic Pulse Velocity (UPV) [11]

UPV Range (m/s)	Concrete Quality
> 4500	Excellent
3500 – 4500	Good
3000 – 3500	Medium
< 3000	Poor

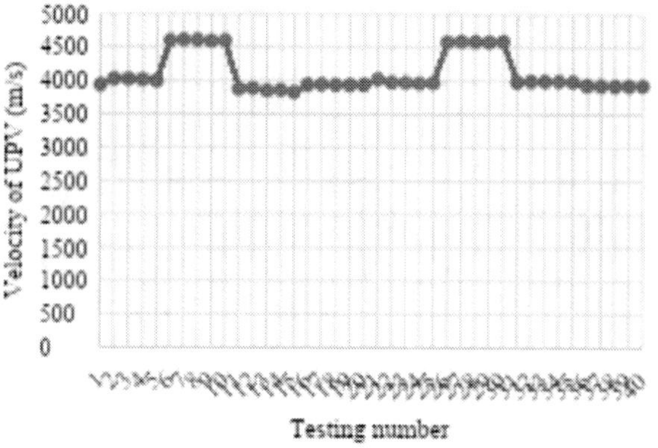

Fig. 5. Variation in UPV Velocity Across Testing Sequence for Hexaloc Concrete Units

Outside the peak intervals, the velocity fluctuates around 4000 m/s, placing most samples within the "Good" quality category, with select batches reaching "Excellent" classification. These fluctuations may reflect differences in curing conditions, compaction quality, or aggregate distribution during the casting process. The presence of high UPV values in specific batches confirms the potential for achieving consistent mechanical integrity in Hexaloc units, while the variability highlights the importance of process control during production. Overall, the UPV data supports the structural viability of the Hexaloc design and provides a non-destructive benchmark for quality assurance in future scaling efforts.

The application of Hexaloc units for coastal protection at Dalegan Beach, Gresik further illustrates the practical relevance of this innovation (see Fig. 6). The deployment of the interlocking concrete blocks demonstrates how the system can be effectively utilized to mitigate shoreline erosion and reduce the destructive impacts of wave action in real coastal settings. By stabilizing the beach profile and dissipating wave energy, Hexaloc units contribute to maintaining coastal integrity while offering a sustainable alternative to conventional protection methods such as tetrapods or rubble mounds. Field observations confirmed that the modular design facilitated ease of installation and adaptability to varying site conditions, while the high mechanical strength and surface quality of the units ensured durability under continuous marine exposure. This case study not only validates the engineering performance of Hexaloc but also highlights its potential for broader application in Indonesia's coastal management programs, where erosion and habitat loss remain pressing challenges.

Fig. 6. Application of Hexaloc for coastal protection at Dalegan Beach, Gresik.

4 Conclusion

This study successfully developed and validated innovative molds and production systems for Hexaloc concrete armor units, marking a significant advancement in sustainable coastal engineering. The mold design, characterized by bolted connections, steel column supports, and corner angles greater than 90°, ensured smooth surfaces, reduced stress concentration, and facilitated efficient demolding. Laboratory trials confirmed that the optimized concrete mix achieved compressive strengths of 47–52 MPa, exceeding the design target of 30 MPa, while Ultrasonic Pulse Velocity (UPV) results classified the units within "Good" to "Excellent" quality standards. These findings demonstrate the structural reliability and durability of Hexaloc under relevant laboratory conditions, thereby achieving Technology Readiness Level (TRL) 5.

The pilot application at Dalegan Beach, Gresik further illustrated the practical relevance of Hexaloc, showing its effectiveness in mitigating shoreline erosion and dissipating wave energy. The modular design facilitated ease of installation and adaptability to site conditions, confirming its potential for broader adoption in Indonesia's coastal management programs. Looking ahead, collaboration with precast concrete manufacturers will enable scaling to TRL 6, ensuring reproducibility, cost-effectiveness, and industrial readiness. Overall, Hexaloc presents a viable, durable, and scalable solution for coastal infrastructure projects, bridging laboratory innovation with real-world application and contributing to long-term shoreline resilience.

This work was financially supported by the Prototype Output Assistance Program, Fiscal Year 2024, under the Directorate General of Higher Education, Research, and Technology through the Directorate of Research, Technology, and Community Service. The support was provided based on the Cooperation Agreement (PKS) No. 2639/PKS/ITS/2024. We sincerely thank the management of Dalegan Beach, Gresik, for their support during this study.

References

[1] H. Warsilah, Resilience of coastal cities in facing climate change: Reclamation case of Benoa Bay Bali and North Jakarta Bay. In *Climate Change Research, Policy and Actions in Indonesia: Science, Adaptation and Mitigation*, R. Djalante, J. Jupesta, E. Aldrian (Eds.), Springer International Publishing, Cham, 121–147 (2021). https://doi.org/10.1007/978-3-030-55536-8_7

[2] G. Argente, M. E. Gómez-Martín, J. R. Medina, Hydraulic stability of the armor layer of overtopped breakwaters. *Journal of Marine Science and Engineering*, 6(4), 0143 (2018). https://doi.org/10.3390/jmse6040143

[3] M. E. Gómez-Martín, J. R. Medina, Heterogeneous packing and hydraulic stability of cube and cubipod armor units. *Journal of Waterway, Port, Coastal, and Ocean Engineering*, 140(1), 100–108 (2014). https://doi.org/10.1061/(ASCE)WW.1943-5460.0000223

[4] M. Sarfaraz, A. Pak, Numerical investigation of the stability of armour units in low-crested breakwaters using combined SPH–polyhedral DEM method. *Journal of Fluids and Structures*, 81, 14–35 (2018). https://doi.org/10.1016/j.jfluidstructs.2018.04.016

[5] C. Irawan, H. D. Armono, Numerical study of the proposed Hexaloc concrete armor units under static loading. *IOP Conference Series: Earth and Environmental Science*, 1250, 012005 (2023). https://doi.org/10.1088/1755-1315/1250/1/012005

[6] I. Wulandari, et al., Development and performance evaluation of BRINpod: A sustainable breakwater armor unit. *Marine Systems & Ocean Technology*, 20(4), 42 (2025). https://doi.org/10.1007/s40868-025-00187-8

[7] H. D. Armono, A. Manurung, Sujantoko, I. K. Suastika, Wave transmission model over hexagonal artificial reefs using DualSPHysics. *IOP Conference Series: Earth and Environmental Science*, 1095, 012025 (2022). https://doi.org/10.1088/1755-1315/1095/1/012025

[8] H. D. Armono, et al., Wave reflection of concrete armour unit: A 2D physical model study. *IOP Conference Series: Earth and Environmental Science*, 1250, 012002 (2023). https://doi.org/10.1088/1755-1315/1250/1/012002

[9] NASA, *Technology Readiness Assessment: Best Practices Guide*, Washington, DC (2011).

[10] European Commission, *Technology Readiness Levels (TRL)*, Brussels (2014).

[11] ASTM International, *ASTM C597-16: Standard test method for pulse velocity through concrete*, West Conshohocken, PA (2016).

Potential of Edu-Ecotourism Based on OTEC Utilizing Indonesian Post-Operation Offshore Oil/Gas Facilities

Fadrian Athaya Muzhaffar[1], **Mukhtasor[1*]**, *Dendy* Satrio[1], *Shade* Rahmawati[1], *Widi Agoes* Pratikto[1], *Hasan* Ikhwani [1], and *Arfie Ikhsan* Firmansyah[1,2]

[1] Department of Ocean Engineering, Institut Teknologi Sepuluh Nopember, Surabaya, 60111, Indonesia
Email: rianathaya@gmail.com; **mukhtasor@oe.its.ac.id***; dendy.satrio@its.ac.id; shade@oe.its.ac.id; pratikto@oe.its.ac.id; hikhwani@oe.its.ac.id
[2] National Research and Innovation Agency (BRIN), South Tangerang, 15314, Indonesia
Email: arfi004@brin.go.id

Abstract Indonesia is striving to meet its growing electricity demands in a sustainable way, by moving away from oil and gas sources. The government, particularly in East Kalimantan, is considering the repurposing of the offshore facilities for Ocean Thermal Energy Conversion (OTEC), which not only provides a clean and sustainable electricity, but also offers additional benefits such as agricultural and aquaculture products. This study will look at how OTEC may be integrated into an edu-ecotourism concept, which aims to promote renewable energy technology and create a sustainable economy that benefits local communities and stakeholders. To meet the study's aims, research of the possible use of OTEC as an edu-ecotourism attraction will be conducted, through papers that are reviewed from previous studies to further the understanding and the potential implementation of OTEC for edu-ecotourism. The OTEC itself is planned to have the capability of producing around 20 kW of net power. Therefore, analysis and review of potential uses and benefit of OTEC is crucial to integrate the multi-functionality of OTEC to cater not just to the goal of creating an edu-ecotourism attraction, but also to the various needs of the various stakeholders, that have a part or affected by this project.

Keywords: Edu-ecotourism; decommissioning; sustainable development; ocean renewable energy; OTEC.

1 Introduction

The Indonesian government is currently trying to meet its electricity demand, which is increasing by 4.9% per year, in an environmentally friendly way. The fulfilment of electricity demand in Indonesia, which was previously mostly done with oil and gas fuels, has led to a high number of offshore oil and gas facilities needed to exploit these fuels. The large number of offshore oil and gas facilities creates problems related to decommissioning in the post-operation phase which requires a large amount of sum. At the same time, Indonesia is starting to develop multiple researches for ocean renewable energy with one of them being ocean thermal energy or OTEC [1], [2]. In addition, the Indonesian government is trying to deal with problems related to the large number of abandoned offshore oil and gas facilities, because if left unattended these facilities can cause abandoned offshore facilities to cause environmental pollution, disruption of the fishing industry, and also shipping.

One of the efforts made by the government, in this case through the initiative of the Regional Government of East Kalimantan Province, is a study of the utilization of offshore oil and gas facilities in the form of oil and gas export pipelines that are converted to function as pipes carrying cooling water for marine thermal power plants known as Ocean Thermal Energy Conversion (OTEC). This research is a development of technical feasibility studies that have been carried out previously [1], [2]. The advantage of using OTEC systems for power generation is their potential to provide non-intermittent electricity from an environmentally friendly energy source [3]. In addition, OTEC can also provide various derivative products such as agricultural products, aquaculture, and other derivative products that can be attractive because of their benefits to various groups of people with various backgrounds [4].

2 Ocean Thermal Energy Conversion

2.1 OTEC Working Principles

Ocean Thermal Energy Conversion (OTEC) is one of the ocean potentials that can be utilized to provide electrical energy. In OTEC, electrical energy is generated by utilizing the temperature difference between near-surface waters that have higher temperatures and deep waters that have lower temperatures, this temperature difference must have at least a 20°C difference [3]. The higher-temperature surface water is pumped into the system to create a cycle of steam, either by heating another working fluid in an closed cycle OTEC (Figure 1a), or by the hot seawater itself in an open cycle OTEC, which drives the turbine while the lower-temperature deep-sea water is used to cool and re-condense the steam back into liquid, creating a cycle (Figure 1b) [3].

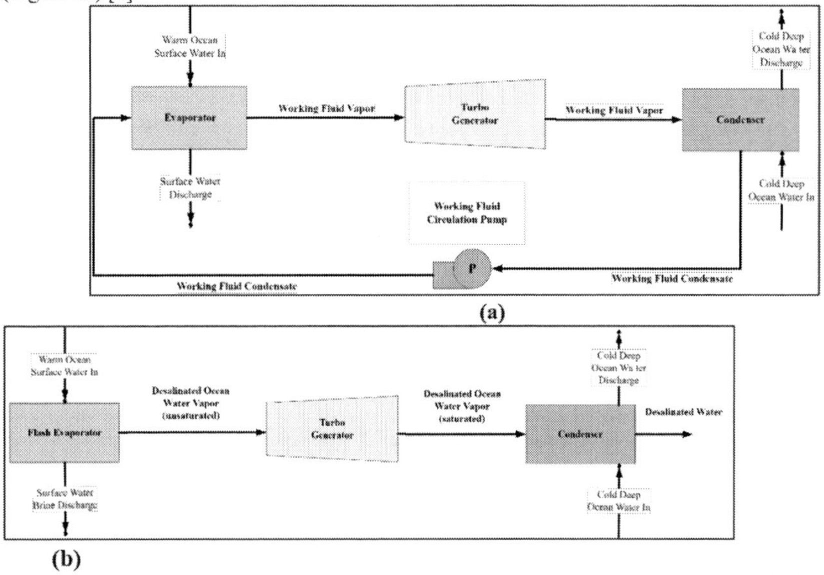

Figure 1. (a) Closed Cycle OTEC Principle (b) Open Cycle OTEC Principle [3]

One example of an OTEC currently in operation is a closed-cycle OTEC in Kumejima, Okinawa, Japan with a generating capacity of 100 kW The advantage of using OTEC systems as power plants is their potential to provide electricity on a continuous (non-intermittent) basis [3]. Another advantage the utiliziation of an OTEC system (Figure 2) is the potential of OTEC to provide cooling for the generating system independently as well as clean fresh water (directly for Open Cycle OTEC or through a desalination system for a Closed Cycle OTEC) in addition to providing electricity, thus making OTEC a viable alternative energy sources for coastal areas or remote islands that are off the grid or have a lack of clean water for human consumption, wether for the OTEC plant staff or the community. In addition to its fresh water production, cold water from the deep sea that are utilized in an OTEC system also brought additional benefits such as nutrients that are critical and can be used in aquaculture, not to mention its potential to provide a natural air conditioning system for buildings and community through a method termed as Sea Water Air Conditioning (SWAC).

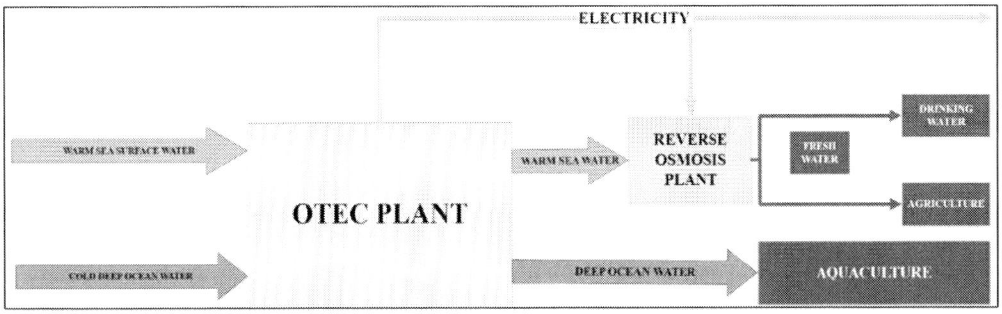

Figure 2. Multifunctionality of an OTEC system adapted from [3]

2.2 Multifunctionality of OTEC

OTEC has a distinct advantage over other marine renewable energy sources especially for utilization in the tropics. As mentioned in the researches that have been published, OTEC have an ability to produce electricity constantly while also having the capability of producing additional products that can be utilized by mankind [5]. With a stable temperature gradient between the surface and deep water in the tropics, OTEC plants could produce stable and continuous electricity throughout the year. In addition, OTEC could also produce additional by products and services (Figure 2). For example:

2.2.1 Aquaculture

In its operational state an OTEC plant utilized large amounts of Deep Ocean Water (DOW) for its electricity generation. These DOW contains large amounts of nutrients while also generally being relatively clean either from pathogens or other contaminants such as microplastics or industrial wastes [6] (Figure 3). These characteristics of a clean and nutrient rich water source are what the aquaculture industry such as for those who breeds oysters; clams; prawns; and other species are needs in their industry. By utilizing the DOW that are already being pumped to the surface by the OTEC plants, the Aquaculture industry could produce such products at a lower cost by lowering the need to purchase chemical enrichments and water treatment products.

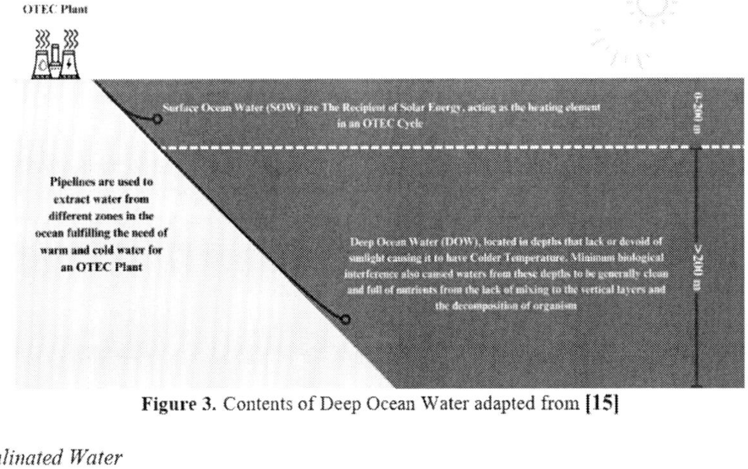

Figure 3. Contents of Deep Ocean Water adapted from [15]

2.2.2 Desalinated Water

According to research, an open-cycled or hybrid OTEC plant can directly produce desalinated water by converting seawater vapour that have seperated from its salt content that is produced during it's electricity generating cycle to drinkable fresh water. A 1 MW hybrid OTEC could be taken as an example where it has the ability to generate around 4500 m^3 of drinking water daily [5]. Meanwhile a closed cycle OTEC system would have to introduce a dedicated desalination facilities which needs additional powers to convert its seawater intake due to its seawater cycle lack of change in its material state as its stays as liquid thus retaining its salt water content througout the loop. Economically speaking, it's production cost and the quality of produced drinking water, would have similar characteristics to its conventional counterparts thus making it quite economically and technically feasible especially since its raw resource (i.e., pumped water from the ocean for the operation of an OTEC plant).

2.2.3 Agriculture

In the agricultural sectors, fresh waters produced by an OTEC system could be used to supply waters that could be used to water fresh produce. This could be beneficial in coastal area where clean supply of fresh waters are limited and agricultural products remains as one of the main source of income [5]. Other than the usage of fresh waters, DOW could also help cultivate plants or produce usually grown in a more temperate climate, in areas with warmer temperature such as the tropics and subtropics. In this utilization, the DOWs are used to decrease the temperatures of the soil where the plants are planted. This method is known as refrigerated soil agriculture, where normally other methods of refrigeration are used to decrease the temperature of the soil and that can be costly when viewed in a technological aspect or from its electricity consumption alone.

3 Edu-Ecotourism

Edu-ecotourism is a form of sustainable tourism that have an highlight in it's educational value. Thus, it consist of two main aspect which are Ecotourism and Edutourism.

3.1 Ecotourism

Ecotourism is a type of tourism that focuses on environmentally friendly and responsible travel to natural areas. The goal of an ecotourism is to protect the environment, improve the well-being of local residents, and provide an educational experience for travellers. Ecotourism involves activities related to nature, such as observing and appreciating natural landscapes, wildlife, and local cultures. It also emphasizes the importance of minimizing environmental impacts, supporting local communities, and promoting conservation efforts [7].

In a sustainable development strategy for increasing tourism in a specific area, ecotourism is a viable strategy due to its having this following aspects:

- Have minimal negative impacts on the natural resources of an area;
- Involvment of multiple stakeholders (individuals, communities, tourists, operators, and government agencies) in the planning, development, implementation, and monitoring phases;
- Respect local cultures and traditions;
- Creates sustainable income for local communities and stakeholders;
- Creates revenue for the conservation area;
- Educate all stakeholders on their role in conservation.

3.2 Edutourism

An educational tourism activity emphasizes the development of knowledge, added value to the economy, and brings long-term socio-economic benefits in the form of entertaining and interesting tourism activities. Taken from the book titled "Managing Educational tourism", age, gender, income, and occupation are one of the many key factors of a succsessfull educational tourism that influences a tourist attraction in certain areas or activities [8]. The book also explained some components of an edu-tourism, first of all there are a primary ccomponent which are:

- Attractions and activities that provide the means of providing learning experiences (e.g. parks, historical sites, natural environments);
- Resource Specialists responsible for delivering the learning component of tourism activities (Such as curators, lecturers, researchers, academics);
- Tour Planners who are responsible for planning and developing learning programs for tourists (such as conservation organizations, universities, interest groups);
- Tour Operators who have responsibility for packaging and marketing tours to tourists.

In addition, the secondary component consists of supporting services consisting of:

- Transportation such as cruise, bus, train or other modes of transportation that play a role in bringing tourists to and from tourist sites.
- Hospitality services such as the provision of meals, recreation, social activities, and accommodation options.
- Travel Services such as travel agents and insurance providers.
- Marketing Organizations that operate on a national, regional or local scale to promote educational tourism to tourists.

4 OTEC Based Edu-Ecotourism

4.1 Worldwide Example of OTEC Based Edu-Ecotourism

Around the world a couple of OTEC demonstration plants have been built as a pilot project not just to promote the usage of OTEC as a new source of renewable energy, but also as a model on how the multifunctionality of OTEC could be harnessed to benefit the local community. Such examples are as follows:

4.1.1 Okinawa Deep Ocean Water Research Center (ODRC)

The ODRC in Kumejima, Okinawa, Japan had created a pilot project in the form of a 100 kW OTEC plant that aims to not just produce electricity but also utilize it's multifunctionality under the name the "Kumejima Model" [9]. The "Kumejima Model" is an innovative approach by utilizing Deep Ocean Water (DOW) and Ocean Thermal Energy Conversion (OTEC) technologies for sustainable development on Kumejima Island. This model integrates multiple uses of DOW, allowing it to be utilized not just in power generation, but also in the local aquaculture, agriculture, and even cosmetics industry, thereby maximizing resource efficiency and minimizing waste. For instance, after being used for OTEC power generation, the cold DOW can be repurposed for cooled-soil agriculture and aquaculture, demonstrating its versatility and potential for reuse up to four times or more depending on the application.

Kumejima itself have the target of achieving 100% renewable energy by 2040, with OTEC expected to supply a large portion of the island's electricity needs. The projected installation of a 1MW-class OTEC facility is predicted to provide approximately 15% of Kumejima's electricity supply, which is crucial given the island's current reliance on diesel power generation for its day to day electricity generation. Furthermore, the local economic impact of DOW industries has been notable, with revenues increasing from roughly 20 million USD in 2011 to 24 million USD in 2016, highlighting the economic benefits of integrating these technologies that support additional products from an OTEC plant into the community.

The OTEC plant in Kumejima is utilized for various applications and not just power production. One significant use is in aquaculture, where the cold deep ocean water (DOW) is utilized to produce healthy baby prawns free from polutants at the Okinawa Kuruma Prawn Hatchery, ensuring a stable supply throughout the year. Additionally, DOW is used in the farming of sea grapes, that's already an established local produce, which now benefits from the nutrient-rich and temperature-controlled environment provided by the ocean water.

Moreover, the Kumejima Model integrates DOW for chilled-soil agriculture, where fresh water is cooled using DOW in heat exchangers, allowing for temperature regulation near plant roots, which is essential for growing foods and food coloring extracts from sensitive plants. The cosmetics industry also use DOW, as local companies use desalinated DOW as a base for their products, benefiting from its low-cost access to minerals and bacteria-free quality [9]. The companies manufactures a range of cosmetic products using desalinated DOW as a base, which allows them to incorporate local ingredients and minerals into their formulations, providing a consistent and bacteria-free resource for production. This approach not only enhances the quality of their products but also offers marketing advantages due to the unique properties of DOW. Additionally, there's also an ongoing development of bioreactor systems within typhoon-resistant greenhouses, which support the commercialization of chilled-soil agriculture techniques, further integrating DOW into the local industries These approaches in multi-produce utilization of OTEC not only enhances local industries but also promotes sustainable development on the island.

4.1.2 Cozumel Island OTEC Ecopark

 https://doi.org/10.1051/bioconf/202621609004

The Ocean Thermal Energy Conversion (OTEC) Ecopark proposed off Cozumel Island aims to address the local needs for water, energy, and food in coastal communities [10]. This system is designed to produce 60 MW of base load power, which is expected to be integrated into the local electrical grid through the Chankanaab electrical node, located around 44 km away. In addition to energy generation, the OTEC system will also produce over 77,000 m^3 of desalinated water daily, sufficient to meet the needs of a population of 730,000.Furthermore, the desalination submodule will generate 3 million m^3 of nutrient-rich deep ocean water (DOW) each day, which can be utilized for sustainable aquaculture, particularly in cultivating Ulva spp. seaweed.

The integration of OTEC technology not only provides clean energy but also supports the local economy by creating opportunities in the blue economy sector. Not to mention, the cultivation of "*Ulva* spp." is projected to yield approximately 69.75 tons per day, contributing to food security and carbon sequestration efforts.. Overall, the OTEC Ecopark represents a comprehensive approach to enhancing the resilience and sustainability of coastal communities in Mexico, while also addressing the challenges posed by climate change and resource scarcity.

4.2 OTEC Based Edu-Ecotourism Potential in Indonesia

A study had been conducted in Indonesia regarding the utilization of Post-Operation oil and gas facilities located in East Kalimantan [1]. The results of the study indicate that reusing post-operation offshore oil and gas pipelines (POGP) as cold-water pipes (CWP) for a 20 kW Closed-Cycle Ocean Thermal Energy Conversion (OTEC) systems could significantly reduce investment costs, as the POGP has a remaining service life of over 20 years, making it a viable option for such applications . The analysis revealed that the temperature of cold seawater in the CWP could increase by approximately 3 to 6°C due to exposure to ambient temperatures along the 60 km pipeline length . This temperature change is deemed as critical, as it affects the performance of the OTEC system, particularly in the condenser where the working fluid interacts with the cooling medium. The study emphasizes the importance of conducting a comprehensive assessment of the temperature changes in the cold seawater flowing through the CWP, as these changes can impact the efficiency of the OTEC system.

The findings suggest that careful design and feasibility studies are essential to optimize the efficiency of the OTEC system while utilizing the existing POGP infrastructure. Overall, the potential for cost reductions in both decommissioning and OTEC development presents a viable potential for the reutilization of POGP in Indonesia's energy transition efforts by utilizing as a part of the OTEC System. The location chosen are taking into account the 200 m, exclusion zone of PT.Pertamina Hulu Kalimantan Timur (PHKT) which forbids any construction inside or near the pipeline canal and also the planned Coastal Tourism Development Plan of Semangko Village. The location and also nearby Zone of Interest (ZoI) of a potential OTEC Edu-Ecoutourism could be seen in Figure 4, where each colored zone represent a different ZoI located in Semangko Village, Marangkayu District, Kutai Kartanegara Regency, East Kalimantan, Indonesia.

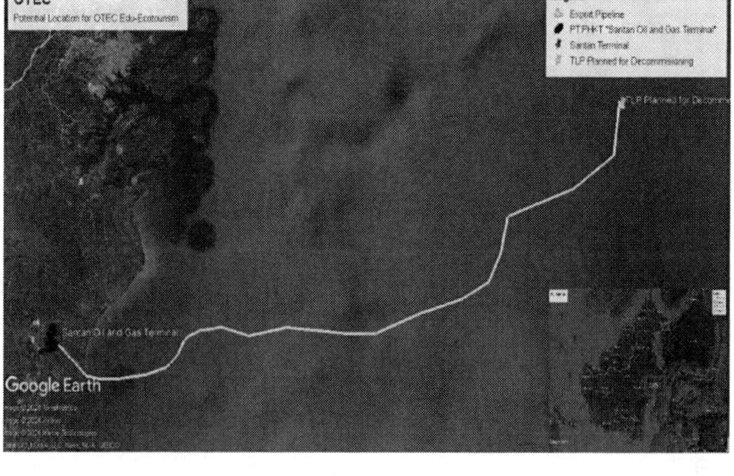

 https://doi.org/10.1051/bioconf/202621609004

(a)

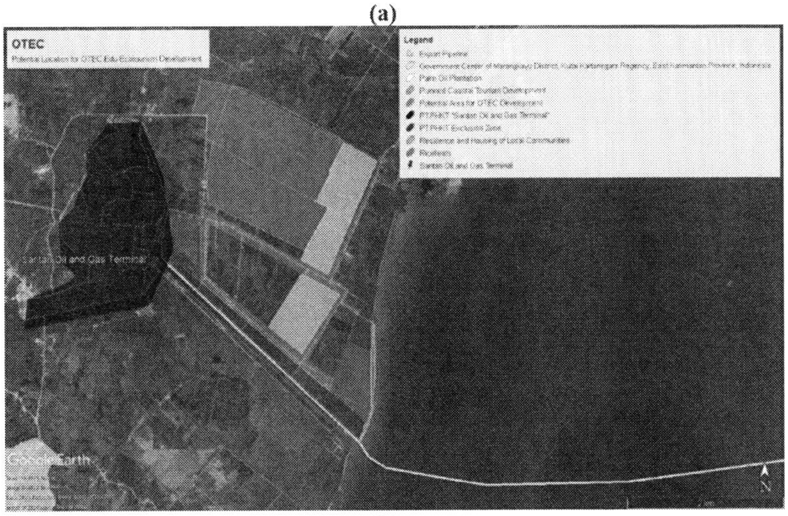

(b)

Figure 4. (a) General Area of the Planned OTEC in East Kalimantan, Indonesia. (b) Location of Planned OTEC site, Marangkayu, East Kalimantan. Image adapted from Google Earth.

If we based the potential of OTEC, in East Kalimantan, that have been researched for a development of an OTEC Plant, to the OTEC utilization concept developed around the world. The development of OTEC in said location will not only help the electricity demand of the area in a sustainable manner, but also can be utilized to develop the local community (Figure 5).

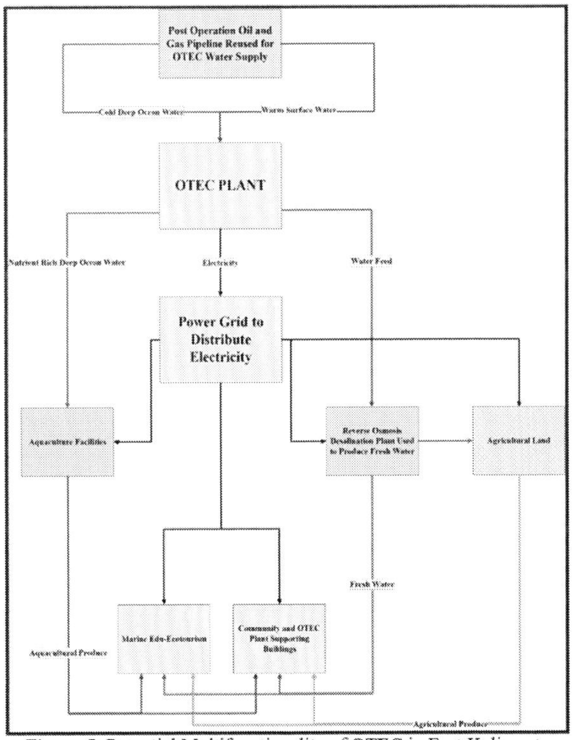

Figure 5. Potential Multifunctionality of OTEC in East Kalimantan

4.2.1 Aquacultural Products

Based on a report from the Marine and Fisheries Service of the Province of East Kalimantan, the report showed that farming and aquaculture of Tiger Prawns and Seaweeds remained as one of the main sources of income for the local communities as per seen in Table 1 [11]. Tiger Prawn as a species is quite sensitive to Water Quality [12]. As per the research done by Pantjara et al, the lack of a supperior seed in addition to inferior water quality could lead to the decrease in survival and growth rate of juvenile tiger prawns. This study is also supported by assessment done by the local community, where the heavily physical parameters of the water used in prawn farming either due to the high content of sediments or other pollution such as parasite has declined the existence of prawn farming in Marangkayu. As proven in Kumejima, Japan, DOW can be utilized to enrich the waters used for aquaculture in place of chemical enrichments, thus providing a sustainable and environmentally friendly method of increasing the production of aquaculture products such as prawn and seaweeds.

Table 1. Recapitulation of Fishery Activities in Marangkayu, East Kalimantan [11].

No	Fishery Activities	Investment Criteria			Profit Month
		NPV	IRR	Net BCR	
1	Dogol	87,549,442	99%	5.11	Rp2,593,271.00
2	Bubu	35,039,057	89%	4.64	Rp1,075,833.00
3	Fishing	41,881,943	87%	4.54	Rp1,549,000.00
4	Gill Net	48,591,959	91%	4.6	Rp1,770,583.00
5	Gondrong	58,964,216	94%	4.85	Rp2,116,083.00
6	Spear Fishing	37,892,102	84%	3.91	Rp1,437,063.00
7	Rakkang	21,365,208	50%	2.88	Rp1,100,417.00
8	Bagan	295,529,738	89%	4.62	Rp4,925,496.00
9	Bandeng Presto	60,817,821	108%	5.29	Rp2,078,333.00
10	Salted Fish Production	28,016,508	90%	4.41	Rp1,020,333.00
11	Prawn Aquaculture	421,450,773	80%	4.11	Rp11,779,106.00
12	Seaweed Aquaculture	275,924,762	88%	4.59	Rp10,148,958.00

4.2.2 Fresh Water Production

According to research, sustaining a clean water supply remains as one of the problem faced in East Kalimantan [13]. Based on research conducted in 2020, the local water service cannot fulfil the total needs in Balikpapan alone, thus opening the potential for a new freshwater production source in East Kalimantan. OTEC through either Open Cycle flash evaporation or the usage of reverse osmosis desalination, for a closed cycle OTEC, could prove to be one of the steady supplier of fresh waters, due to its 24/7 operational capability all year long. This will prove beneficial to the community, especially those in coastal areas where a steady supply of clean fresh water might be difficult to be provided.

4.2.3 Agricultural Product

In 2021, a research have been conducted on the agricultural needs and land development in the Marang Kayu, Sub-district [14]. Said research has shown that agricultural and plantations are the second most common land usage in the area, after fisheries and aquaculture (Table 2). This has meant that other than being a fishery-based community, Semangko Village, in Marangkayu District was also an agricultural one. The potential of OTEC in not only supporting the local energy demands, but also in providing a way to support its agriculture industry are one of the untapped potentials from developing an OTEC plant in the region. From Figure 8, we could see those agricultural lands of paddy fields (Dark Greenish-Yellow), located north-east of The Santan Oil and Gas Terminal encompass an area of around 240 Ha. Currently the paddy fields and it's farmers are facing a decline due to either flooding in rainy season or the lack of fresh water during drought based on observation and interview done towards the local community. As proven in Kumejima and also the research done for the planned OTEC plant in Cozumel Island, OTEC could benefit the agricultural sector by providing a steady supply of fresh water, especially during drought, and also in the diversification of local produce, that otherwise needed more temperate temperatures.

Table 2. Land Usage in East Kalimantan [14].

Space Pattern			**Slope**			**Grand Total**
Sub-District	**<2%**	**2-8%**	**16-25%**	**26-40%**	**40%**	
Bontang Selatan		**204.60**	**406.31**		**197.79**	**808.70**
Protection Forest		166.68	257.46		151.37	575.51
Corversion Production Forest		8.08	97.90			106.06
Plantation		27.51	50.87		46.42	124.80
Settlement		2.33				2.33
Marangkyu	**10,659.14**	**12,830.09**	**37,124.25**	**11,280.53**	**13,858.76**	**85,752.77**
Protection Forest	137.26	1,240.40	380.83		57.29	1,815.78
Conversion Production Forrest	565.74	844.46	269.65		229.41	2,909.26
Permanent Production Forest	154.54	6,301.41	22,868.55	7,768.01	4,630.64	41,723.35
Fisheries Area	4,206.37	204.96				4,411.33
Plantation	1,786.33	1,374.69	10,296.08	3,434.01	7,183.62	24,073.73
Settlement	879.69	1,657.29	2,896.62	78.51	756.59	6,268.60
Food crops and Horticulture	2,929.21	1,206.68	412.62		1.01	4,549.72
Muara Kaman			**602.79**	**366.94**	**15.47**	**985.20**
Permanent Production Forest			598.38	3,694.00	11.63	976.95
Nature Reserve Area			4.41		3.84	8.25
Sangatta Satan		**173.31**	**900.98**	**773.11**	**1,631.93**	**3,479.33**
Nature Reserve Area		173.31	900.98	773.11	1,631.93	3,479.33
Teluk Pandan	**396.14**	**5,832.48**	**16,400.17**	**10,206.24**	**30,378.89**	**63,218.92**
Protection Forest	315.95	227.54	245.76		7,121.37	7,910.62
Conversion Production Forrest	68.83	487.55				556.38
Permanent Production Forest		451,151.00	11,785.68	7,018.81	16,362.60	39,676.60
Nature Reserve Area		605.45	4,378.72	3,187.43	6,867.01	18,038.61
Plantation	11.36	0.43	0.01		22.91	34.71
Grand Total	**11,055.28**	**19,040.48**	**55,444.50**	**22,666.82**	**46,077.84**	**154,244.92**
Percentage	**7%**	**30%**	**36%**	**15%**	**12%**	

4.2.4 Edu-ecotourism Attraction

A pilot project or OTEC facilities, such as those in Kumejima, Japan could increase visitors from outside of the region, either for academical reasons or tourism. The potential for said OTEC plant in attracting visitors can be predicted in a couple of ways, for one the OTEC plant itself could prove as a pilot project of the viability and benefit of OTEC as a source of renewable energy in Southeast Asia and the Tropics, and also shows that a transition from fossil based fuels to renewable energy could be done in an economically beneficial method. Currently the local government are committed to find a way to utilize planned or already decommissioned oil and gas facilities, so that it could be utilized for the transition to renewable energy, while also serving as an Edu-Ecotourism attraction by creating a pilot plant and an energy transition museum.

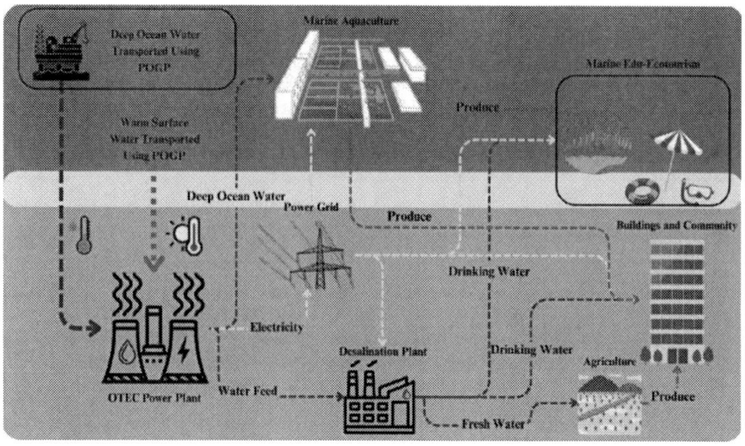

Figure 6. Illustration of an OTEC Edu-Ecotourism Concept in East Kalimantan

As an Edu-Ecotourism concept, OTEC could provide an integrated area where various products and electricity could be produced sustainably (Figure 6). While mainly the area is used for research purposes regarding an OTEC Pilot Projects that attracts researchers and visitors, as per seen in Kumejima the various products that are produced from the multifunctionality of OTEC could support the local community and tourism. Where electricity could be produced sustainability and continuous, desalination plants could produce fresh waters both for drinking and agriculture, and various marine produces from aquacultures can be used to feed tourist and be sold to generate additional income all integrated into one another. This new and unique method of an integrated multifunctional facility in South East Asia, could potentially boost visitors that aims to research the OTEC and it's multifunctionality while also attracts visitors for recreational reason to experience the uniqueness of the produce and it's production method, to enjoy the natural landscape of its coastal area, and even learn regarding on how OTEC could be utilized to sustainably produce electricity and it's other products as a method of transitioning from oil and gas so that the nature could be safeguarded. In addition, local knowledge and cultural centre could also be established as a way to increase awareness and spread knowledge regarding conservation efforts and local cultures. This concept of sustainable tourism while also providing a way of improving one's knowledge and awareness are exactly what made an OTEC as a potential Edu-Ecotourism attraction, which are by utilizing its uniqueness as a power plant that have various capabilities and also providing integration with local resources to promote sustainable tourism.

Other than that, the given improvement of road access and local infrastructures needed to support OTEC plant will also prove beneficial in the long term for the community, even after the plant have been built by improving road access to the region and developing the various local infrastructures and facilities. Thus this proposed method of sustainable tourism could co-exist and support the planned increase of Marine Ecotourism planned by the local government to improve the tourism rate in Semangko Village, Marangkayu District, Kutai Kartanegara Regency, East Kalimantan as per shown by the pink area *coastal area* in Figure 5.

4 Conclusion

In conclusion, the integration of Ocean Thermal Energy Conversion (OTEC) technology into Indonesia's energy landscape presents a unique opportunity for sustainable development. The potential to repurpose decommissioned offshore oil and gas facilities for OTEC systems not only addresses the growing electricity demands in a sustainable manner, but also offers significant economic benefits to local communities through the production of fresh water, aquaculture, and agricultural products. The successful examples of OTEC implementation in places like Kumejima, Japan, demonstrate the viability of such projects, showcasing their ability to generate clean energy while simultaneously supporting local industries and promoting eco-tourism.

Moreover, the establishment of OTEC facilities can serve as educational and ecotourism attractions, enhancing public awareness and appreciation for renewable energy technologies. This dual focus on energy production and community engagement aligns with global sustainability goals, making OTEC a promising endeavour for Indonesia's transition towards a greener economy. Overall, the the importance of strategic planning and stakeholder involvement in harnessing the full potential of OTEC for the benefit of both the environment and local communities and also the technical analysis of OTEC and it's multifunctionality if developed in East Kalimantan remains paramount for future research.

Acknowledgement

The authors would like to convey their appreciation to Institut Teknologi Sepuluh Nopember (ITS), which funded the current project under a scheme called Departmental Research Scheme under-contract number 2023/PKS/ITS/2024. An appreciation is also expressed to Institut Teknologi Kalimantan (ITK), who had helped accommodate and assist of the field research.

References

[1] M. Mukhtasor, A. I. Firmansyah, D. Satrio, S. Rahmawati, R. Shadiq, and O. Yaakob, "On Assessing a Potential Reuse of the Indonesian Post-Operation Offshore Oil/Gas Pipelines as A Cold-Water Pipe for an Ocean Thermal Energy Conversion (OTEC) System: A Thermal and Fluid Dynamic Perspective," *Journal of Advanced Research in Fluid Mechanics and Thermal Sciences*, vol. 123, no. 1, pp. 53–68, Oct. 2024, doi: 10.37934/arfmts.123.1.5368.

[2] A. I. Firmansyah, Mukhtasor, D. Satrio, S. Rahmawati, H. Ikhwani, and W. A. Pratikto, "A study of the temperature distribution in the OTEC cold water pipe using a heat and mass transfer approach," in *IOP Conf. Series: Earth and Environmental Science*, IOP Publishing Ltd, 2024, pp. 1–11. doi: 10.1088/1755-1315/1372/1/012018.

[3] R. Kempener and F. Neumann, "Ocean Thermal Energy Conversion: Technology Brief," 2014.

[4] A. Copping, H. Farr, C. Rumple, K. Park, and Z. Yang, "Feasibility of Multi-Use Ocean Thermal Energy Conversion (OTEC) Platforms," *J. Mar. Sci. Eng.*, vol. 14, no. 1, Jan. 2026, doi: 10.3390/jmse14010064.

[5] J. Herrera, S. Sierra, and A. Ibeas, "Ocean thermal energy conversion and other uses of deep sea water: A review," Apr. 01, 2021, *MDPI AG*. doi: 10.3390/jmse9040356.

[6] M. Mac Takahashi and P.-Y. Huang, "Novel Renewable Natural Resource of Deep Ocean Water (DOW) and Their Current and Future Practical Applications," *Koroshio Science*, vol. 6, no. 1, 2012.

[7] N. K. Arismiyanti, "Development Strategy of Sustainable Marine Ecotourism in Indonesia," *ASEAN Journal on Hospitality and Tourism*, vol. 15, no. 2, pp. 118–138, 2017.

[8] W. B. Ritchie, N. Carr, and C. P. Cooper, *Managing Educational Tourism*, 1st ed. Clevedon: Channel View Publication, 2003.

[9] B. Martin, S. Okamura, T. Yasunaga, Y. Ikegami, and N. Ota, "OTEC and Advanced Deep Ocean Water Use for Kumejima: An Introduction," in *Oceans Conference Record (IEEE)*, Institute of Electrical and Electronics Engineers Inc., 2022. doi: 10.1109/OCEANSChennai45887.2022.9775240.

[10] J. G. Tobal-Cupul *et al.*, "An Assessment of the Financial Feasibility of an OTEC Ecopark: A Case Study at Cozumel Island," *Sustainability (Switzerland)*, vol. 14, no. 8, Apr. 2022, doi: 10.3390/su14084654.

[11] Juliani *et al.*, "Identifikasi dan Pemetaan Zonasi Kawasan Konservasi Perairan dan Pulau-Pulau Kecil Kabupaten Kutai Kartanegara," Samarinda, 2022.

[12] B. Pantjara, H. S. Suwoyo, and I. Rusdi, "The production of tiger prawn (Penaeus monodon) juveniles using the hapas on brackishwater pond in Sidoarjo Regency," in *IOP Conference Series: Earth and Environmental Science*, IOP Publishing Ltd, Oct. 2021. doi: 10.1088/1755-1315/860/1/012032.

[13] U. Fitriati, H. Kurdi, A. Isramaulana, and M. Mayang, "Study of Clean Water Needs Balikpapan, East Kalimantan, Indonesia," European Alliance for Innovation n.o., Mar. 2020. doi: 10.4108/eai.22-10-2019.2291474.

[14] A. Sopian, S. Hardwinarto, M. Sumaryono, and M. I. Aipassa, "Analysis of Land Availability and Needs with Potential Agricultural Land-Development Within Santan Watershed and Marangkayu, East Kalimantan," 2021.

[15] Y. Ikegami and B. Martin, "OTEC Viability as a Catalyst for Transformative Island Development," Kumejima, Sep. 2022.

Developing a rainfall estimation model using XGBoost with Himawari-8/9 satellite and atmospheric data in East Java

Gede Gangga Wisnawa[1,2*], *Fajar Setiawan*[1,2]

[1]Department of Geomatics Engineering, Institut Teknologi Sepuluh Nopember, Surabaya, Indonesia

[2]Marine Meteorological Station of Tanjung Perak, Badan Meteorologi Klimatologi dan Geofisika, Surabaya, Indonesia

Abstract. Accurate rainfall estimation in tropical regions is often hindered by non-linear atmospheric interactions and extreme data imbalance. This study develops a multi-stage precipitation estimation framework—comprising binary classification, multi-class classification, and regression—using an optimized Extreme Gradient Boosting (XGBoost) architecture. Applied to East Java, Indonesia, the model integrates Himawari-8/9 satellite brightness temperatures, global atmospheric indices, and high-resolution topography. To mitigate the dominance of non-rain events (91.6% of the dataset), Stratified Random Under-sampling (RUS) was employed. Hyperparameters were tuned using Bayesian Optimization and evaluated via 10-fold site-based cross-validation to prevent spatial data leakage. Results show that the optimized model significantly outperforms the baseline. In the regression stage, MAE and RMSE decreased by 21.5% and 23.0%, respectively, while the Pearson correlation coefficient improved by 43.1%. In classification, the Critical Success Index (CSI) rose by 16.2% for binary and 34.5% for multi-class stages, indicating an enhanced capability to detect rare rainfall events. Performance gains were most pronounced in mountainous regions, suggesting improved representation of orographic effects. The proposed hierarchical framework demonstrates potential as an effective approach for satellite-based rainfall estimation in topographically diverse tropical regions.

1 Introduction

East Java Province, Indonesia, is a region characterized by high vulnerability to disasters triggered by extreme weather conditions. Data from the National Disaster Management Agency of Indonesia (BNPB) reveals that floods accounted for approximately 48.8% of all natural disaster events in this region, culminating in material losses of IDR 275.5 billion in 2024 [1]. This situation highlights the urgent need for an accurate precipitation estimation system with high spatial and temporal resolution to support more effective mitigation and early warning efforts in East Java.

Numerical Weather Prediction (NWP) models are commonly used to generate weather forecasts. Although physically robust, NWP demands intensive computational resources and

*Corresponding author: gede.wisnawa@bmkg.go.id

is operationally expensive. Furthermore, NWP models face challenges in the parameterization of sub-grid scale processes, like cloud microphysics and turbulence, which are fundamental to understanding convective precipitation in tropical regions. These high costs and uncertainties at the physical process level create a distinct opportunity gap. This gap can be filled by data-driven empirical models leveraging machine learning (ML), which offer a potentially more efficient and accurate alternative [2].

The availability of high-resolution geostationary satellite data from Himawari-8/9, provides a strong foundation for developing precipitation estimation models. The Advanced Himawari Imager (AHI) sensor provides 16 observation channels with 10-minute temporal and 2-km spatial resolution, enabling the continuous capture of atmospheric dynamics. The utilization of predictor variables from this satellite is not merely based on arbitrary numerical values, they represent physically relevant atmospheric processes. For example, low Brightness Temperature (BT) values in Band 13 (10.4 μm) indicate high and cold cloud tops, which are associated with strong convective activity. Similarly, Brightness Temperature Difference (BTD) between channels offers additional information regarding cloud phase, particle size, and convective growth rates [3]. This combination of information establishes satellite infrared data as a valid predictor for precipitation, even though the relationship between satellite signals and precipitation is indirect and highly non-linear [4].

Algorithm selection is crucial given the complexity of remote sensing data and the variability of precipitation in tropical regions. Conventional ML models, such as Decision Trees and Support Vector Machines, generally struggle to accommodate high-dimensional, non-linear data characteristics influenced by complex feature interactions [5] In this context, ensemble methods offer more stable and accurate performance [6]. XGBoost was selected due to its numerous advantages, including high scalability, the ability to handle complex non-linearities, and built-in regularization (L1/L2) that effectively reduces overfitting [7]. Moreover, various meteorological studies have demonstrated that XGBoost consistently outperforms classical ML models in satellite-based precipitation estimation tasks [4,8]. Nevertheless, studies indicate that XGBoost's performance is highly sensitive to its hyperparameter configuration [4]. Therefore, the model's success depends not only on the algorithm itself but also on an appropriate hyperparameter optimization strategy. Accordingly, this study aims to develop a 3-hourly precipitation estimation model for the East Java region using XGBoost by integrating Himawari-8/9 data with atmospheric and topographic variables. The study will also evaluate the result of hyperparameter optimization on the model's accuracy, error patterns, and its capability to detect operationally significant heavy rainfall events.

2. Materials and methods

This section details the data sources, pre-processing stages, and analytical methods employed in this study. The process commences with the determination of the study area and the collection of primary data, followed by missing data handling, temporal and spatial resolution alignment, and data resampling to balance the occurrences of rain and no-rain events. Subsequently, model design, hyperparameter tuning, and evaluation of the results were conducted to obtain an accurate and reliable analysis.

2.1 Study area

The study area is East Java Province, Indonesia, geographically bounded by 110°–115° E longitude and 5°–9° S latitude. This region features complex topography, extending from coastal lowlands to highland areas, and is highly vulnerable to hydrometeorological hazards. The ground-truth data consist of 3-hourly precipitation accumulation, aggregated from 10-

minute observations from a network of 79 Automatic Weather Stations and Automatic Rain Gauges (AWS/ARG). This network is operated by the Meteorology, Climatology, and Geophysics Agency of Indonesia (BMKG), with provides data covering the period from 2020 to 2024

2.2 Data

This study utilizes a multi-scale and multi-source dataset compiled to capture patterns for generating accurate precipitation predictions. The predictor variables consist of Himawari-8/9 satellite observations (Brightness Temperature/BT and Brightness Temperature Difference /BTD), global and regional atmospheric indices including large-scale indices such as Sea Surface Temperature Anomaly (SSTA) Nino 3.4 and Dipole Mode Index (DMI) from The Bureau of Meteorology (BoM), and regional indices such as Western North Pacific Monsoon Index (WNPMI) and Australian Summer Monsoon Index (AUSMI) from European Centre for Medium-Range Weather Forecasts (ECMWF), and a Digital Elevation Model (DEM) from United States Geological Survey (USGS).

Table 1. Summary of Target and Predictor Variables.

Variable Name	Source	Spatial Resolution	Temporal Resolution	Purpose
Precipitation	BMKG (AWS/ARG)	Point (79 station)	10 min	(Target)
BT (Band 7-16)	JMA (via BMKG)	2 km x 2 km	10 min	Predictor
BTD	Derived features from BT	2 km x 2 km	10 min	Predictor
SSTA Nino 3.4	BoM	Global	7 days	Predictor
Dipole Mode Index (DMI)	BoM	Global	7 Hari	Predictor
WNPMI	ECMWF ERA5	Regional	3 Jam	Predictor
AUSMI	ECMWF ERA5	Regional	3 Jam	Predictor
Digital Elevation Model (DEM)	GMTED2010	500 m	Statis	Predictor
Julian Date	calculated	-	day	Predictor

Brightness Temperature Difference (BTD) data, which represent the difference in BT values between two infrared (IR) channels. These BTD features, as shown in Table 2, can provide additional information regarding cloud microphysical characteristics, which helps to differentiate cloud types and identify clouds with a higher precipitation potential [3, 8]. Specifically, this study using several key BTD combinations to capture distinct atmospheric and cloud properties. For example, the difference value between Band-8 and Band-14 is used to assess upper tropospheric water vapor content. Other combinations, such as Band-7 – Band-14 and Band-7 – Band-10, provide information on cloud phase and cloud thickness, respectively. Critically for precipitation analysis, Band-11 and Band-14 difference helps detect low cloud growth, while Band-14 and Band-15 difference is used to identify the development of convective clouds.

Temporal and spatial resolution alignment was performed to ensure all input variables possessed a uniform data structure prior to model training. As the estimation target is on a 3-hourly scale, data with a finer temporal resolution, such as the 10-minute BMKG observed rainfall and Himawari-8/9 BT/BTD, were harmonized. Specifically, rainfall was accumulated into 3-hour totals, while the BT/BTD values were selected at the timestamp nearest to the target interval (e.g., 00:00, 03:00, 06:00 UTC, up to 21:00 UTC). Conversely, data with a coarser duration, such as the weekly Sea Surface Temperature Anomaly (SSTA) of Nino 3.4 and Dipole Mode Index (DMI), were applied constantly across all 3-hour slots within that specific week. Time variables, like the Julian Date, were calculated directly from each sample's timestamp. Spatially, gridded data, including Himawari-8/9 and the DEM, were extracted at the station coordinate points using the nearest neighbor method to ensure an accurate local representation.

2.3 Missing value handling

Data quality control was performed by applying a data completeness threshold. Observation stations with missing data rates exceeding 25% were excluded from the model development process to prevent potential bias caused by excessive data gaps. Incomplete data points for the remaining stations were omitted to ensure the training set consisted solely of reliable and valid observations. This rigorous filtering process maintains the overall integrity of the analysis while limiting performance degradation associated with input data inconsistencies.

2.4 Data resampling

The distribution of rainfall data as the target variable in this study exhibits a significant imbalance, where the frequency of no-rain events vastly outnumbers that of rain events. This condition poses a risk of biasing the model toward the majority class, thereby compromising its accuracy in detecting lower-frequency events. To handle this disparity, the study utilizes Random Under-sampling (RUS) with stratification specifically targeting the majority class. Unlike oversampling techniques such as SMOTE, which generate synthetic data and risk introducing noise or non-physical weather patterns, RUS operates by randomly reducing the number of samples in the majority class (CR) until its proportion is balanced with the minority class. This approach aims to proportionally reduce the dominance of the majority data without discarding critical patterns associated with global climate variability and monsoon systems. By utilizing stratification, the majority data is down-sampled evenly across each climate phase, ensuring that the representative characteristics of each phase are preserved [9]. Consequently, the RUS process yields a more proportional inter-class distribution prior to the application of augmentation techniques on the minority class.

2.5 Model design

The methodology incorporates Extreme Gradient Boosting (XGBoost), a sequential ensemble algorithm that improves predictive accuracy by iteratively generating decision trees to address the deficiencies of prior iterations. Model performance is optimized by minimizing an objective function that integrates a loss function with a regularization term, ensuring high accuracy while effectively mitigating overfitting [10]. Regularization is achieved by penalizing the number of leaves and leaf weights within each tree; this enhances XGBoost's generalization capabilities, particularly for large-scale and sparse datasets through its sparsity-aware learning mechanism. In the context of spatiotemporal rainfall modeling, computational efficiency is further augmented by histogram-based algorithms and parallel processing support, which accelerate the identification of optimal splits for each feature.

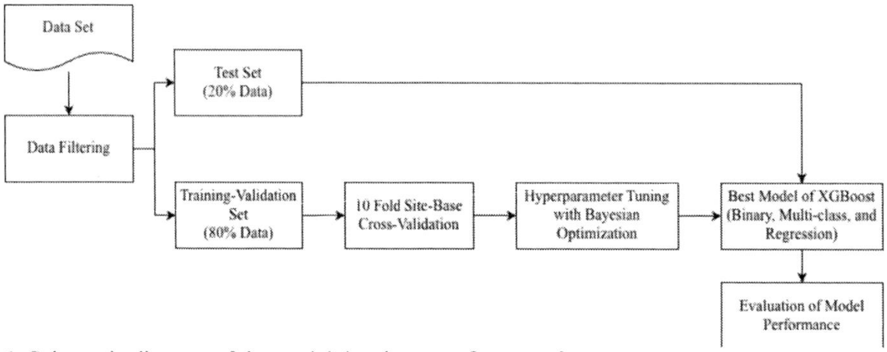

Fig. 1. Schematic diagram of the model development framework.

The proposed framework is designed as a multi-stage XGBoost architecture comprising three distinct phases: binary classification, multi-classification, and regression. This structure aims to generate more accurate and informative precipitation estimates. In the first phase, a binary classification model is utilized to discriminate between rain and no-rain events based on a specific precipitation threshold. The second phase subsequently focuses exclusively on samples identified as 'rain,' employing multi-classification to categorize rainfall intensity (e.g., light, moderate, and heavy). Finally, the third phase applies a regression model solely to the rain samples to estimate the precipitation quantity. By isolating the positive rain instances, this hierarchical strategy effectively mitigates the bias typically introduced by the zero-inflated nature of precipitation data, allowing the model to focus its learning capacity exclusively on the complex variability of rainfall intensities [11].

2.6 Hyperparameter optimization

Model performance is critically dependent on the selection of appropriate hyperparameters [4]. This study examines several pivotal hyperparameters, including: *learning_rate* (0.01–1) to control the weight update step size; *max_depth* (0–20) to define the maximum tree depth; *n_estimators* (100–1000) representing the total number of trees in the ensemble; *subsample* (0.1–1) to determine the fraction of data used per tree; *min_child_weight* (0.1–2) to set the minimum weight required for a node split; *gamma* was utilized (set between 0 and 1) to specify the minimum loss decrease needed to permit further partitioning , while *colsample_bytree* (spanning 0.1 to 1) was adjusted to manage the ratio of available features included in each sequential tree. These parameter ranges were utilized in the search for the optimal configuration across the XGBoost models (binary classification, multi-classification, and regression).

Hyperparameter optimization was conducted using Bayesian Optimization, which efficiently explores the parameter space by probabilistically modeling combinations likely to enhance model performance [7]. Performance evaluation was carried out using 10-Fold

Site-Based Cross-Validation (CVSI), a validation strategy that partitions data according to station locations. This approach prevents data leakage, which can occur when data from a single station appears simultaneously in both training and validation sets across different timestamps. Consequently, CVSI ensures that the model is rigorously evaluated based on its spatial generalization capability [8]. The research dataset was divided into 80% for training and validation, and 20% as an unseen test set. All reported evaluation results are derived from models using the optimal hyperparameters obtained during this validation process.

2.7 Evaluation of model result

Several statistical indicators were used to analyze the effectiveness of the regression stage. Specifically, Mean Absolute Error (MAE) and Root Mean Square Error (RMSE) served as the primary tools for quantifying the extent of prediction inaccuracies. Furthermore, the degree of linear agreement and the directional relationship between predicted outputs and observed data were evaluated through the Pearson Correlation Coefficient (R).

$$RMSE = \sqrt{\frac{1}{n}\sum_{n}^{n}(\hat{y}_i - y_i)^2} \quad (1)$$

$$MAE = \frac{1}{n}\sum_{n}^{n}|\hat{y}_i - y_i| \quad (2)$$

$$R = \frac{\sum_{i=1}^{n}(y_i - \bar{y})(\hat{y}_i - \bar{\hat{y}})}{\sqrt{\sum_{i=1}^{n}(y_i - \bar{y})^2} \cdot \sqrt{\sum_{i=1}^{n}(\hat{y}_i - \bar{\hat{y}})^2}} \quad (3)$$

The developed XGBoost model also generates multi-classification predictions categorized into four rainfall intensity classes: No Rain (< 0.1 mm/3h), Light Rain (0.1–10 mm/3h), Moderate Rain (10–20 mm/3h), and Heavy Rain (> 20 mm/3h). Classification performance was evaluated via a confusion matrix, enabling the calculation of overall accuracy (ACC). However, as accuracy can be inflated by the dominance of True Negatives, this study also utilized the Critical Success Index (CSI). CSI provides a more objective assessment of rare events and is therefore considered the most representative metric for evaluating the model's capability to accurately detect rainfall occurrences.

$$Accuracy = \frac{Hits + True\ Negative}{Hits + False\ Alarm + Miss + True\ Negative} \quad (4)$$

$$CSI = \frac{Hits}{Hits + False\ Alarm + Miss} \quad (5)$$

3. Result and discussion

A comprehensive analysis of the data's inherent properties is a mandatory prerequisite to ensure that the subsequent model construction phase is built upon a valid understanding of the dataset's complexities. This analysis serves to validate methodological assumptions and identify specific modeling challenges, particularly those arising from the data's inherent class imbalance and the non-linear relationships among variables.

3.1 Data distribution and correlation

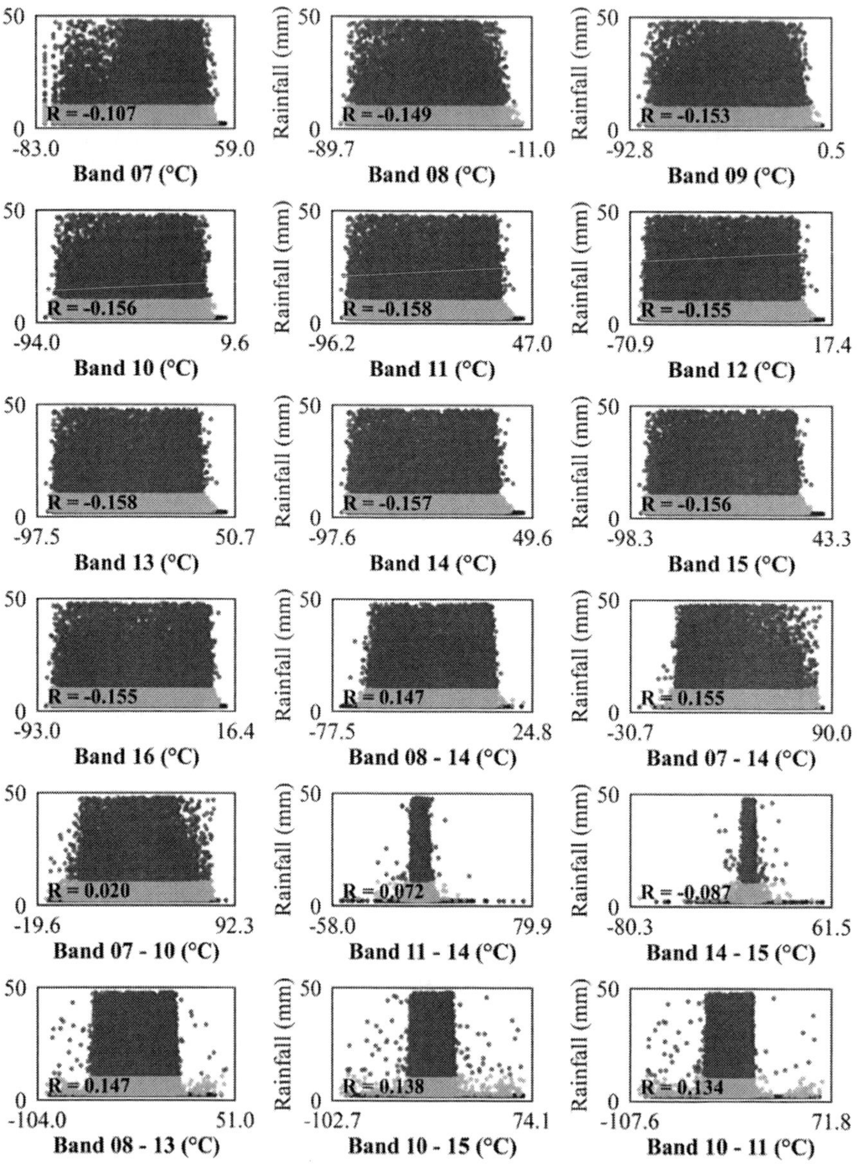

Fig. 2. Scatter plots illustrating the relationship between Himawari-8/9 predictors (BT and BTD) against observed 3-hourly rainfall intensity, overlayed with linear regression lines and Pearson correlation coefficients (*R*).

Scatterplots illustrating the relationship between various BT of Himawari channels and BTD against rainfall exhibit a highly dispersed pattern (Fig. 2). No discernible strong linear correlation (*R*) was identified. The Pearson correlation coefficients (*R*) for all BT channels and BTD against surface rainfall range from -0.158 to 0.155. The majority of data points are clustered near zero rainfall, with the variance widening as predictor values increase. Furthermore, Fig. 3 reveals that similar patterns are observed in the auxiliary data, including climate indices and topographic parameters. Relationships between these predictors and the target precipitation also demonstrate substantial scatter and lack a distinct linear trend. These findings have important implications for model selection. The lack of linear correlation indicates that linear regression cannot effectively handle the complex dynamics of precipitation. Therefore, the use of XGBoost is validated. Unlike traditional models, XGBoost is designed to handle complex, non-linear, and multivariate relationships.

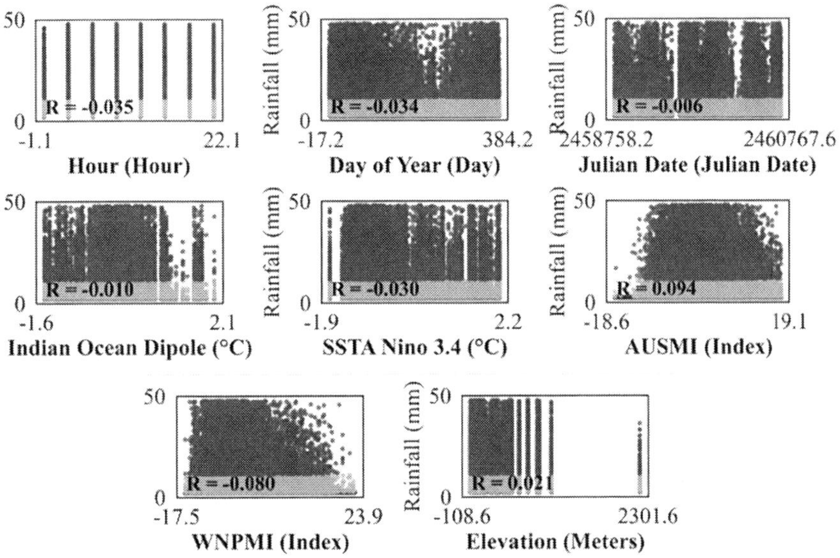

Fig. 3. Scatter plot matrix of auxiliary predictors (Time, Climate Indices, and Elevation) against observed 3-hourly rainfall intensity, overlayed with linear regression lines and correlation coefficients (*R*).

Beyond the challenge of non-linearity, the dataset also exhibits extreme class imbalance. Fig.4 illustrates the initial probability mass function (PMF) of the data from 71 observation stations following the filtering process. As evidenced in the figure, the Clear (CR) class overwhelmingly dominates the dataset, comprising 693,572 samples (91.6%). Conversely, rainfall events constitute the minority classes, with Light Rain (HR) accounting for 50,731 samples (6.7%), followed by Moderate Rain (HS) with 6,815 samples (0.9%), and Heavy Rain (HL) with a mere 6,057 samples (0.8%). Visually, the distribution exhibits a sharp leptokurtic pattern near 0 mm, confirming the prevalence of non-rainy conditions. In contrast, the probability curve for Heavy Rain (HL) appears nearly flat (platykurtic), reflecting the scarcity of extreme events. This severe imbalance, compounded by the distribution overlap between classes, poses a significant risk of model bias, where the algorithm may favor the majority class (CR) and fail to detect extreme precipitation events.

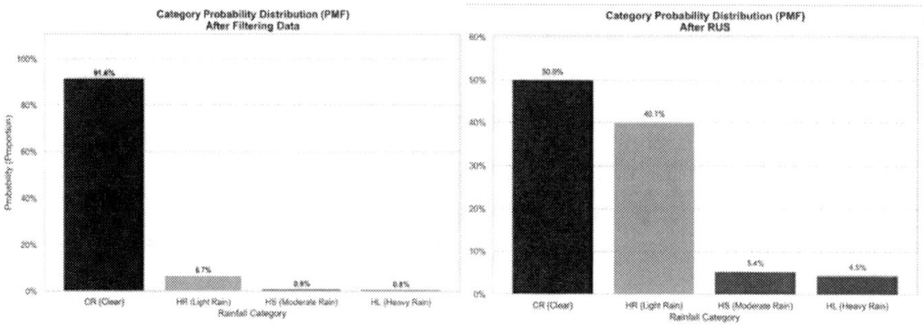

Fig. 4. Comparison of rainfall category probability distributions before (left) and after (right) applying the Stratified Random Under-sampling (RUS) technique.

To mitigate this issue, Stratified Random Under-sampling (RUS) was implemented. This method downsamples the majority class (CR) while preserving the climatological structure associated with IOD, ENSO phases, and seasonal variability. The under-sampling process targeted a reduction of the CR class to approximately 63,603 samples, aligning it with the total accumulated count of rainfall samples (HR + HS + HL). The outcome of the RUS application is presented in Figure 4.4. As shown, the proportion of the CR class was successfully adjusted to 50.0%, achieving a balanced 1:1 ratio between rain and no-rain events. However, it is crucial to note that an internal imbalance within the rainfall categories persists. Light Rain (HR) continues to dominate the rain category at 40.1%, far exceeding HS (5.4%) and HL (4.5%). This residual imbalance underscores that model evaluation cannot rely solely on Accuracy. Consequently, metrics sensitive to minority classes, such as the Critical Success Index (CSI), are essential for a robust performance assessment [11].

3.2 Hyperparameter optimization result

The hyperparameter optimization results presented in Table 2 reveal that each modeling stage (binary, multi-classification, and regression) requires a distinct configuration tailored to its specific function within the multi-stage framework. For the binary classification stage, a notably high *max_depth* (17) and relatively high *min_child_weight* indicate the necessity for a complex tree structure to effectively separate the initial two classes (rain and no-rain). This complexity is primarily driven by the overwhelming dominance of the Clear (CR) class at 0 mm rainfall [8]. Outputs identified as no-rain are immediately assigned to the CR class with a value of 0 mm and undergo no further processing. Conversely, samples detected as "rain" are forwarded to the multi-classification stage.

In multi-calssification model, the optimal configuration is more moderate, characterized by a lower *max_depth* (7), a higher *colsample_bytree* (0.76), and a substantial *min_child_weight*. These settings suggest that the model must constrain its complexity to prevent overfitting, particularly given the significant distributional overlap between HR, HS, and HL rainfall classes. Subsequently, the regression stage designed to estimate the specific rainfall magnitude for samples labeled HR, HS, or HL yielded a lighter hyperparameter combination, such as significantly fewer *n_estimators* (117) and a higher *learning_rate* (0.33). This implies that the regression task operates more efficiently on continuous value patterns and does not require the extensive tree complexity demanded by the classification stages. Overall, these configurational variations are consistent with the hierarchical nature of the proposed modeling framework, where each stage addresses a distinct statistical challenge.

Table 2. Optimal hyperparameter configurations for binary classification, multiclass classification, and regression tasks.

Hyperparameter	Best Hyperparameter of XGBoost		
	Binary	Multi-classification	Regression
colsample_bytree	0.23	0.76	0.57
gamma	0.63	0.36	0.08
learning_rate	0.01	0.01	0.33
max_depth	17.00	7.00	6.00
min_child_weight	3.82	8.47	1.17
n_estimator	426.00	354.00	117.00
subsample	0.66	0.77	0.98

3.3 Regression model performance evaluation

Quantitative evidence from Table 3 illustrates how the optimized XGBoost architecture consistently outperforms the default configuration across all evaluation criteria. The Mean Absolute Error (MAE) decreased by 21.5%, dropping from 3.93 to 3.08 mm/3hours, indicating a reduction in the model's average absolute error. A more substantial improvement was observed in the Root Mean Square Error (RMSE), which fell by 23.0% from 8.66 to 6.67 mm/3hours. This performance boost signals that optimization effectively mitigates large prediction errors. The sharper decline in RMSE compared to MAE suggests that the optimization successfully suppressed errors in predicting rainfall events, for example cases where the model predicts 0 mm during heavy observed rainfall or the opposite condition. Furthermore, the Pearson correlation coefficient (R) increased by 43.1%, rising from 0.33 to 0.48. This represents an improvement in the linear relationship between predictions and observations, shifting from a weak category to a moderate one.

Table 3. Performance comparison between the optimized and non-optimized (default) XGBoost models across regression, binary, and multi-classification stages.

Scheme	Regression			Binary		Multi-classification	
	RMSE	MAE	R	ACC	CSI	ACC	CSI
Optimized	6.68	3.08	0.48	0.76	0.59	0.90	0.49
Non-Optimized	8.66	3.92	0.33	0.72	0.51	0.84	0.37
Improvement (%)	-23.0	-21.5	43.1	5.5	16.2	6.8	34.5

These improvements align with the spatial findings illustrated in Fig. 5, which reveal significant changes in the distribution of regression performance across observation stations in East Java. In the default model, the MAE map displays extensive areas of high error, particularly in the central and southern regions dominated by mountainous topography. Following optimization, the red areas on the map are visibly reduced and replaced by lower MAE values. A similar enhancement is evident in the correlation map. While the default model generally exhibited low correlation ($R \approx 0.3$), the optimized model demonstrates a clear shift toward higher correlation ($R > 0.4$), especially in the coastal areas and the eastern part of East Java.

This spatial analysis provides direct insight into the influence of elevation on model performance, where RMSE and MAE values tend to increase, and correlation coefficients tend to decrease, as regional elevation rises [12]. The default model proved less capable of representing rainfall dynamics in topographically complex regions, as reflected by the high MAE and low correlation in highlands and mountainous areas. Conversely, the optimized model showed the most substantial performance gains specifically in these rugged terrains. These findings indicate that while the optimized model outperforms the default version, the challenge of orographic rainfall remains a significant barrier to accurate detection.

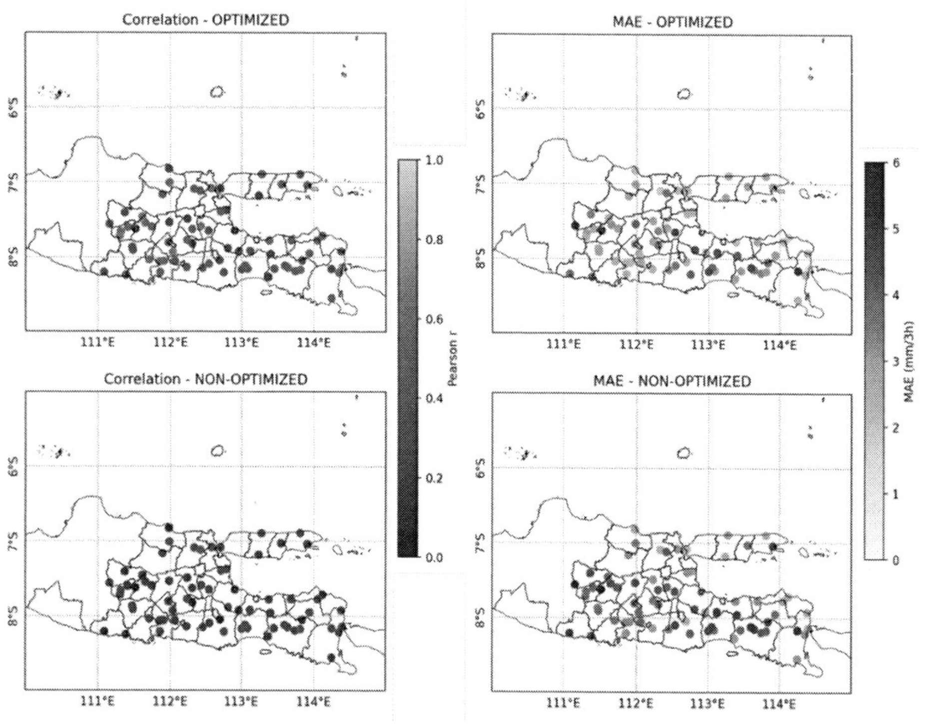

Fig. 5. Spatial distribution of regression performance metrics comparing the Optimized (top row) and Non-Optimized (bottom row) XGBoost models across observation stations in East Java. The left panels display Pearson Correlation coefficients (R), while the right panels display Mean Absolute Error (MAE in mm/3h).

The quantitative evaluation of model performance across varying topographic gradients reveals a distinct degradation in predictive performance at higher elevations, as evidenced by the systematic shift in error metrics and correlation coefficients. This trend is substantiated by the positive linear regression slopes observed for both Root Mean Square Error (RMSE) and Mean Absolute Error (MAE) relative to station elevation, alongside a corresponding decline in the Pearson correlation coefficient (R) (Fig. 6). To facilitate this assessment, observation stations were partitioned into three distinct elevation groups: lowlands (stations below 60 m above sea level [asl]), mid-elevation (61–200 m asl), and high-altitude regions (stations exceeding 200 m asl). Specifically, categorical analysis shows that the mean RMSE increases progressively from 6.23 in the lowland group to 7.28 in the high-altitude areas, while the mean MAE rises from 2.67 to 3.65, indicating that the model's predictive uncertainty intensifies in complex mountainous terrains. Simultaneously, the mean correlation drops from 0.50 to 0.45 at higher elevations, suggesting a weakened linear agreement between the satellite-derived predictors and surface observations. Such performance discrepancies likely stem from the intricate orographic effects and localized convective processes inherent to high-altitude topography, which present significant challenges for infrared-based brightness temperature (BT) and difference (BTD) channels to resolve accurately compared to more uniform lowland conditions. Consequently, these findings highlight the necessity for enhanced topographic integration or localized recalibration to mitigate elevation-dependent biases in satellite-based rainfall estimation models.

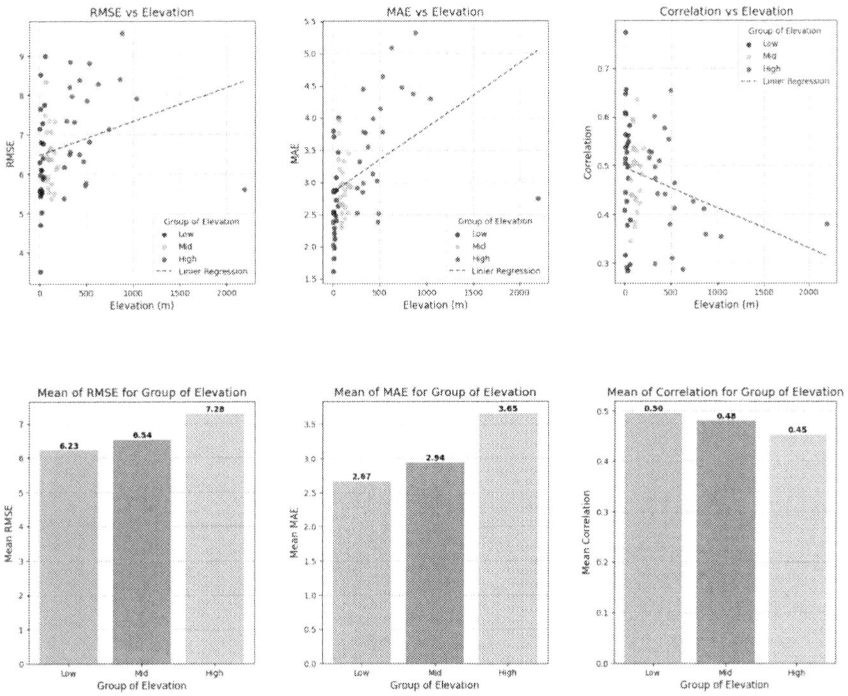

Fig. 6. Analysis of model performance metrics (RMSE, MAE, and Correlation) relative to station elevation. The upper panels show scatter plots with linear regression trends, while the lower panels display the mean error values for low, mid, and high-elevation groups.

3.4 Classification model performance evaluation

The evaluation results presented in Table 3 also demonstrate that hyperparameter optimization yields consistent performance improvements across both classification modeling stages. In binary classification, ACC increased by 5.5%; however, the most significant improvement was observed in CSI, which rose by 16.2%, signaling a tangible enhancement in the model's ability to detect rainfall events. A similar pattern is evident in multi-classification, where ACC increased by 6.8% to 0.90, while CSI experienced a much larger surge of 34.5%, rising from 0.38 to 0.49. The dominance of CSI improvement over ACC indicates that optimization was particularly effective in reducing misses and false alarms, thereby improving the model's precision in identifying rainfall classes that are difficult to distinguish due to class imbalance. Thus, hyperparameter optimization is proven to not only enhance general accuracy but, more importantly, to improve the model's sensitivity toward rare yet critical rainfall events [11].

The spatial evaluation illustrated in Fig. 7 and Fig. 8 provides essential context regarding the multi-classification model's performance. The Accuracy map in Fig. 7 displays very high values for all rainfall classes across nearly all stations in East Java; however, this pattern reflects the Accuracy Paradox inherent in imbalanced datasets. Accuracy appears high not because the model is genuinely superior, but because the 'Clear' class is the dominant event, meaning a prediction of 'no rain' is almost always correct.

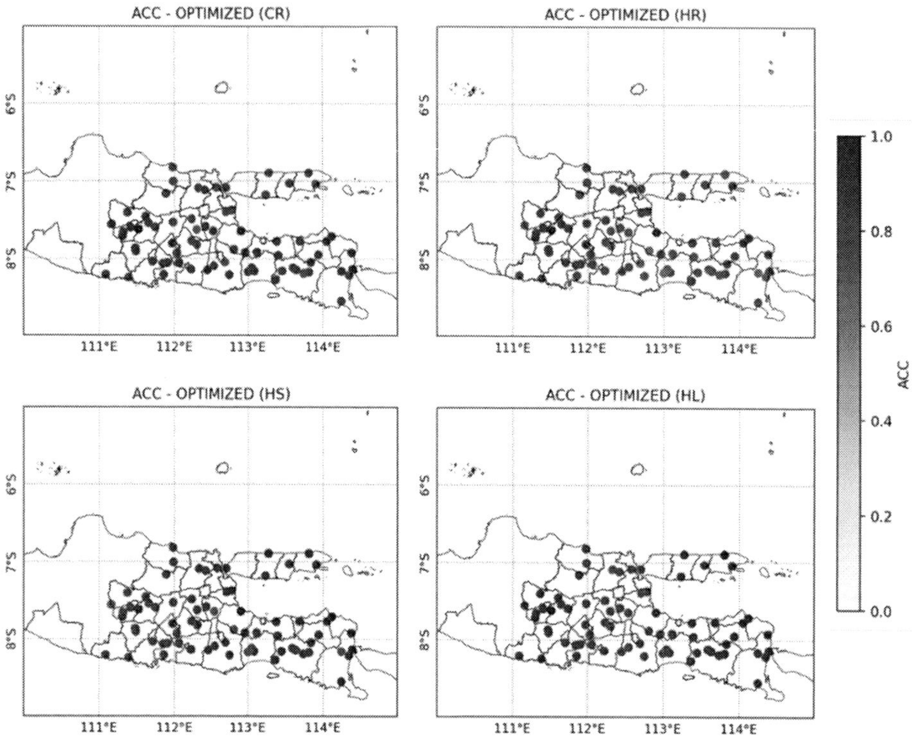

Fig. 7. Spatial distribution of Accuracy (ACC) scores for the optimized XGBoost model across four rainfall categories: Clear (CR), Light Rain (HR), Moderate Rain (HS), and Heavy Rain (HL).

The CSI maps provide a clear spatial overview of model performance, revealing distinct behavioral patterns across each rainfall class (Fig. 8). For the Clear class, CSI values remain high and uniform across the entire region, indicating that the binary classification stage excels at identifying no-rain conditions. In contrast, the CSI distribution for Light and Moderate Rain classes exhibits significantly greater variability. Coastal areas and the northern lowlands demonstrate superior CSI values, whereas the central mountainous regions consistently display a decline in performance. This pattern reinforces the understanding that topographic complexity, particularly orographic rainfall mechanisms and local convective processes, serves as a primary factor diminishing prediction accuracy in highland areas. Regarding the Heavy Rain class, CSI values are low across nearly all stations, confirming that these rare high-intensity events remain the model's most significant challenge. This finding aligns with the underestimation observed in the PDF analysis and illustrates that the scarcity of extreme events lies at the root of the data imbalance problem in precipitation modeling. However, the difficulty in detecting heavy rainfall is not attributable solely to data imbalance, but also stems from the inherent physical limitations of infrared sensors. In tropical regions, heavy precipitation is frequently driven by 'warm rain' processes where cloud tops are not sufficiently cold to be flagged by IR sensors [13], while cold cirrus clouds often trigger false alarms [14]. Overall, these spatial patterns confirm that CSI provides a more representative measure of the model's detection capability than accuracy.

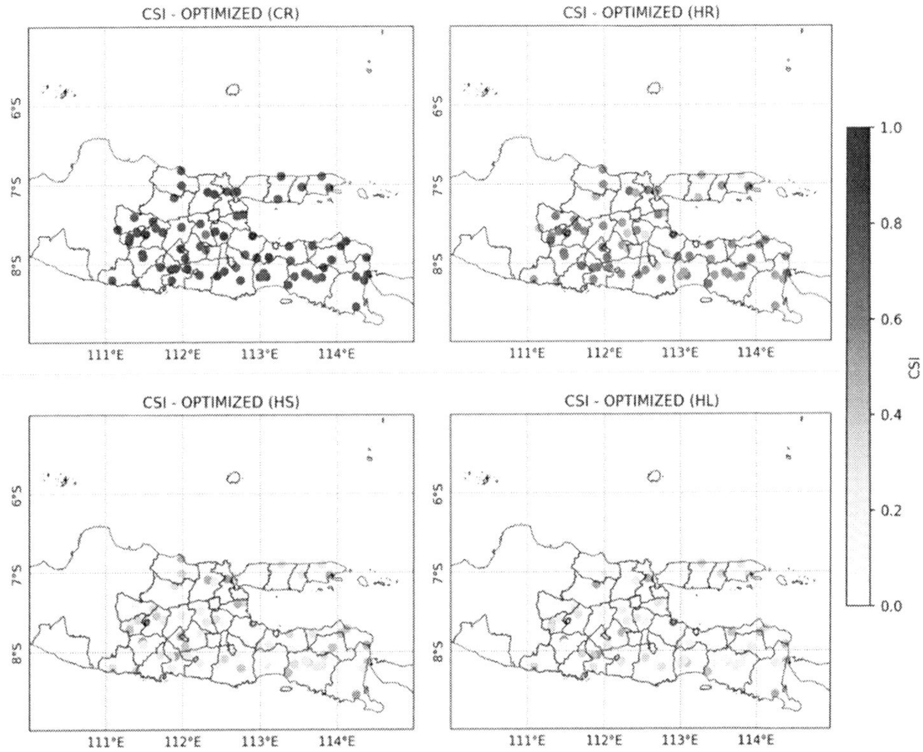

Fig. 8. Spatial distribution of Critical Success Index (CSI) scores for the optimized XGBoost model across four rainfall categories: Clear (CR), Light Rain (HR), Moderate Rain (HS), and Heavy Rain (HL).

4. Conclusion

This study demonstrates that integrating multi-source data (comprising Himawari-8/9 Brightness Temperature (BT) and Brightness Temperature Difference (BTD), atmospheric indices, and topography) with the XGBoost algorithm yields an accurate and operationally relevant 3-hourly rainfall estimation system for East Java. Initial analysis confirmed that the relationship between satellite predictors and precipitation is highly non-linear, further complicated by extreme class imbalance and distributional overlap among rainfall categories.

The optimization of XGBoost hyperparameters proved effective, resulting in significant performance enhancements. In the regression stage, Mean Absolute Error (MAE) and Root Mean Square Error (RMSE) decreased by 21.5% and 23.0%, respectively, while correlation increased by 43.1%. This indicates that optimization effectively suppresses large errors and improves prediction precision. Spatial evaluation indicates that the most substantial improvements occurred in mountainous regions, signaling the optimized model's capability to capture complex orographic rainfall dynamics. Regarding classification, the Critical Success Index (CSI) improved by 16.2% for the binary stage and 34.5% for the multi-classification stage, confirming the model's enhanced sensitivity in detecting rare rainfall events. Nevertheless, performance for the heavy rainfall class remains limited, reflecting challenges stemming from the scarcity of observational data.

Overall, this study confirms that the optimized XGBoost based model delivers significant performance gains and successfully mitigates the spatial heterogeneity of predictions. The model demonstrates strong potential as an operational component within a satellite-based early warning system for the East Java region. However, the detection of rare heavy rainfall remains a critical limitation. This opens avenues for future research utilizing image-based deep learning approaches, radar data integration, or hybrid physical-statistical models to further enhance sensitivity toward extreme events.

The Acknowledgements: The authors would like to express their sincere appreciation to the Meteorology, Climatology, and Geophysics Agency (BMKG) for providing the data and necessary facilities that supported this research.,
Fundings: This research was financially supported by the Indonesia Endowment Fund for Education (LPDP),
Data availability: This study has been conducted using BMKG Data.

References

1. BNPB, Jumlah kejadian bencana menurut jenis bencana. (2025). Diakses dari https://data.bnpb.go.id/dataset/data-bencana-indonesia/resource/9b41007e-c998-456b-8cbc-385b17986e46
2. S. Berkhahn, L. Fuchs, I. Neuweiler, An ensemble neural network model for real-time prediction of urban floods, *J. Hydrol.* **575**, 743 (2019).https://doi.org/10.1016/J.JHYDROL.2019.05.066
3. M. Min, C. Bai, J. Guo, F. Sun, C. Liu, F. Wang, H. Xu, S. Tang, B. Li, D. Di, L. Dong, J. Li, Estimating Summertime Precipitation from Himawari-8 and Global Forecast System Based on Machine Learning, *IEEE Trans. Geosci. Remote Sens.* **57**, 2557 (2019).https://doi.org/10.1109/TGRS.2018.2874950
4. M. Putra, M.S. Rosid, D. Handoko, High-Resolution Rainfall Estimation Using Ensemble Learning Techniques and Multisensor Data Integration, *Sensors* **24**, 5030 (2024). https://doi.org/10.3390/s24155030
5. S. Kundu, S.K. Biswas, D. Tripathi, R. Karmakar, S. Majumdar, S. Mandal, A review on rainfall forecasting using ensemble learning techniques, *e-Prime* **6**, 100296 (2023). https://doi.org/10.1016/j.prime.2023.100296
6. H. Hang, J. Mallick, S. Alqadhi, A.A. Bindajam, H.G. Abdo, Exploring forest fire susceptibility and management strategies in Western Himalaya: Integrating ensemble machine learning and explainable AI for accurate prediction and comprehensive analysis, *Environ. Technol. Innov.* **35**, 103655 (2024). https://doi.org/10.1016/j.eti.2024.103655
7. B. Wu, P. Chen, M. Wei, Bayesian optimization-based XGBoost for performance Prediction of Carbon Nanotube Membranes. (2024).https://doi.org/10.21203/RS.3.RS-4562640/V1
8. S. Zhou, Y. Wang, Q. Yuan, L. Yue, L. Zhang, Spatiotemporal estimation of 6-hour high-resolution precipitation across China based on Himawari-8 using a stacking ensemble machine learning model, *J. Hydrol.* **609**, 127718 (2022). https://doi.org/10.1016/j.jhydrol.2022.127718

9. G.E.A.P.A. Batista, R.C. Prati, M.C. Monard, A study of the behavior of several methods for balancing machine learning training data, *ACM SIGKDD Explor. Newsl.* **6**, 20 (2004). https://doi.org/10.1145/1007730.1007735
10. T. Chen, C. Guestrin, XGBoost: A scalable tree boosting system, in *Proceedings of the 22nd ACM SIGKDD International Conference on Knowledge Discovery and Data Mining* (2016), pp. 785–794. https://doi.org/10.1145/2939672.2939785
11. A.U.G. Senocak, M.T. Yilmaz, S. Kalkan, I. Yucel, M. Amjad, An explainable two-stage machine learning approach for precipitation forecast, *Journal of Hydrology.* **627**, 130375 (2023). https://doi.org/10.1016/j.jhydrol.2023.130375
12. J. Ko, K. Lee, H. Hwang, S. G. Oh, S. W. Son, K. Shin, Effective training strategies for deep-learning-based precipitation nowcasting and estimation, *Computers & Geosciences* **161**, 105072 (2022). https://doi.org/10.1016/J.CAGEO.2022.105072
13. B. Sohn, G. Ryu, H. Song, Observational Characteristics of Warm-Type Heavy Rainfall, *Advances in Global Change Research* **69**, pp. 745-759 (2020). https://doi.org/10.1007/978-3-030-35798-6_15
14. C. Kidd, V. Levizzani, Status of satellite precipitation retrievals, *Hydrology and Earth System Sciences* **15**, pp. 1109-1116 (2011). https://doi.org/10.5194/hess-15-1109-2011

Iot-Based Monitoring and Alert System for Carrageenan Production in Coastal Village

Yuda Apri Hermawan[1,*], *Totok* Yulianto[1], *Dedi Budi* Purwanto[1], *Irfan Syarif* Arief[2], *Muhammad Ikhsan* Suryadarma[1] *and Muhammad* Farhan[3]

[1]Department of Naval Architecture, Faculty of Marine Technology, Institut Teknologi Sepuluh Nopember, Surabaya, Indonesia
[2]Department of Marine Engineering, Faculty of Marine Technology, Institut Teknologi Sepuluh Nopember, Surabaya, Indonesia
[3]Department of Electrical Engineering, Faculty of Intelligent Electrical and Informatics Technology, Institut Teknologi Sepuluh Nopember, Surabaya, Indonesia

Abstract. Carrageenan is a polysaccharide compound extracted from seaweed that is widely cultivated in Sumenep waters, has a higher economic value. The process of processing seaweed into carrageenan is carried out by the extraction process at a temperature of about 80-90° C which takes several hours until the carrageenan is completely dissolved and the drying process in the oven until it becomes dry material. Both processes greatly affect the quality of the processed carrageenan. Therefore, real-time and continuous monitoring is required so that the quality can be maintained properly. With the processing site located in a coastal area far from the city centre, an Internet of Things (IoT)-based monitoring and alert system is proposed that can send real-time information. The data obtained can be used as a reference for alerts to operators in the control room through a web that can be accessed easily. The system employs ESP32 board, Thermocouple Type K sensor, SCT-013-010 sensor, which transmits data to Firebase real-time database where the data is stored. A web-based User Interface is created to display data and alerts with easier access. The system applied to carrageenan processing in coastal areas shows the usefulness of monitoring and warning against excessive temperatures.

1 Introduction

Indonesia, as an archipelagic nation, possesses immense marine and fisheries potential, with seaweed being one of its leading commodities, contributing 38 % of the global seaweed market [1]. The waters of Sumenep, Madura, are a significant center for the cultivation of *Eucheuma cottonii*, a type of red seaweed that is the primary raw material for producing carrageenan [2]. Carrageenan is a high-value hydrocolloid polysaccharide widely used in the food, pharmaceutical, and cosmetic industries as a gelling agent, thickener, and stabilizer [3]. The economic value of seaweed increases substantially when it is processed into carrageenan, making this downstream industry vital for local economies [4].

The problem with Indonesian seaweed is the low perception of product quality. The production of high-quality carrageenan is heavily dependent on precise control over the manufacturing process, particularly during the extraction and drying stages [5]. The extraction phase requires heating a seaweed slurry to a specific temperature range, typically between 80°C and 90°C, for several hours to dissolve the polysaccharide completely. Subsequently, the resulting gel is dried until it reaches a specific moisture content. Both temperature and duration are critical parameters in these stages. lower heat can lead to no gel produced because the chemical process to produce carrageenan did not take place very well, thus leaving seaweed uncooked [6]. Conversely, insufficient heat or time can lead to low yields.

Traditionally, monitoring these processes relies on manual checks by operators, which are often intermittent and prone to human error. Manual checks cannot provide the continuous data stream necessary to identify "spikes" or "drops" in temperature that occur between logging intervals [7]. Uncontrolled heat treatment for seaweed process directly affects the color and sulfate content of the final product [8]. The application of technology in the extraction

* Corresponding author: yuda.apri@its.ac.id

and drying process, particularly in the heating stage, has been proven to improve the quality of the resulting carrageenan [9]. The challenge is compounded by the fact that many processing facilities are located in remote coastal areas, far from urban centers, making consistent supervision and access to advanced control technology difficult [10]. This lack of real-time monitoring not only results in suboptimal processing conditions but also eliminates the potential for data-driven evaluation, leading to variations in product quality and potential economic losses for manufacturers.

To address these challenges, modern technological solutions are required. The emergence of the Internet of Things (IoT) offers a powerful tool for real-time, continuous, and remote monitoring of industrial processes. By integrating sensors with microcontrollers and cloud-based platforms, an IoT system can provide operators with live data and automated alerts, enabling them to maintain optimal conditions and respond immediately to any deviations. Therefore, this study proposes the development of an IoT-based monitoring and alert system specifically designed for carrageenan processing. This system aims to improve quality control and operational efficiency by providing a reliable, accessible, and real-time data feed to operators, ensuring the production of high-quality carrageenan.

2 Requirement Analysis

2.1 Carrageenan Processing Workflow

Carrageenan is a natural polysaccharide extracted from the cell walls of red seaweeds. Fig. 1 show production process. The processing begins with washing the harvested seaweed using clean freshwater. This step is repeated several times to remove sand, residual sea salt, small organisms, and other contaminants. After washing, the seaweed undergoes soaking for several hours to soften its texture, facilitating easier cutting. Next, the seaweed is cut into pieces measuring 2–5 cm to increase the contact surface area during extraction. This preparation accelerates carrageenan release and optimizes heat transfer efficiency.

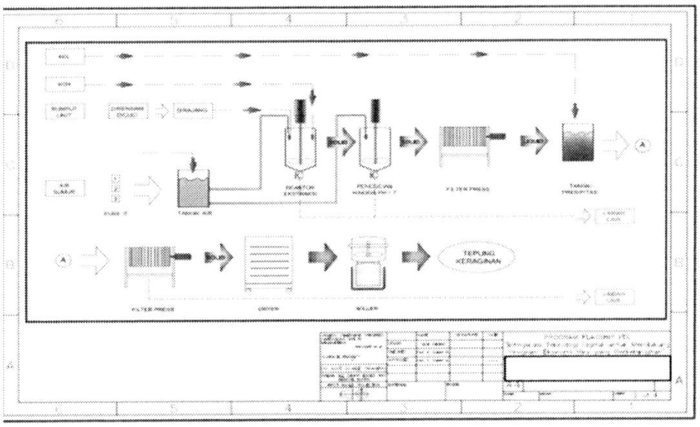

Fig. 1. Carrageenan flour processing scheme.

Extraction is carried out in stainless steel reactors to prevent corrosion and metal contamination. Water is added at a ratio of 1:10–1:20 relative to the dry weight of the raw material. Heating is maintained for approximately one hour at 80–85 °C, the optimal range for dissolving carrageenan without damaging its molecular structure. The extraction yields a mixture of carrageenan solution and solid residue. This mixture is rinsed with clean water until a neutral pH (7) is achieved. A pressing stage is then performed to separate the carrageenan-rich solution from the pulp. The solution subsequently undergoes precipitation using a designated precipitating medium to isolate the solid phase containing concentrated carrageenan. Precipitation is followed by a second pressing to significantly reduce moisture content before drying [11].

Drying is carried out in an oven at 60 °C for 48–72 hours, depending on the initial water content and ambient humidity. Strict temperature control is essential to prevent thermal degradation of the polysaccharide. Once completely dry, the carrageenan solids are ground in a high-speed milling machine into fine powder for packaging.

2.2 Monitoring System Design

Based on the production process described in the previous chapter, monitoring requirements for carrageenan production can be established. Two main units are prioritized for monitoring:

- Extraction unit – monitored parameters include extraction temperature and stirrer motor rotation speed (RPM). Temperature directly affects carrageenan solubility, while RPM influences heat distribution and prevents material clumping.
- Drying oven – monitored for internal temperature stability. Temperature fluctuations can cause uneven drying, affecting the quality, color, and gel strength of carrageenan.

Fig. 2 shows the architectural scheme of the monitoring system based on the requirements that have been established. The system architecture is divided into three main systems, namely the oven monitoring system, the extraction monitoring system, and the visual monitoring system. The oven monitoring system consists of devices for monitoring temperature and for visualizing temperature. Meanwhile, the extraction monitoring system is designed to measure the speed of the mixer machine and the temperature of the extraction chamber. Visual monitoring is also added to monitor conditions inside the seaweed processing chamber.

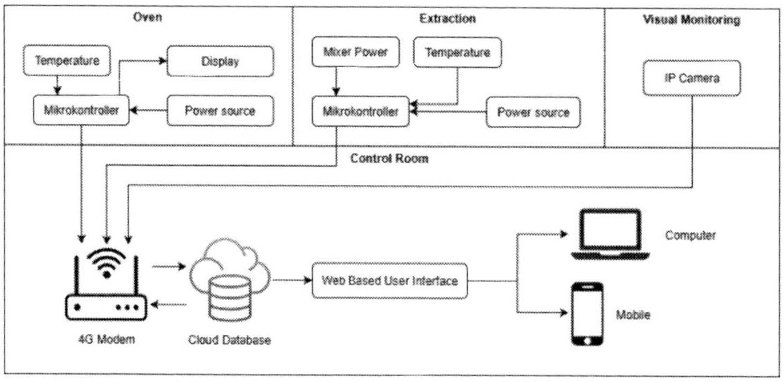

Fig. 2. Architecture of monitoring system for carrageenan process.

3 Hardware and Software Implementation

3.1 Monitoring System Design

The microcontroller used is the ESP32 DevkitC developed by Espressif Systems. One of the main advantages of the ESP32 DevkitC is its ability to connect to WiFi networks and Bluetooth devices. This allows IoT devices to communicate wirelessly with other devices, cloud servers, or smartphone applications.

Additionally, the ESP32 features a dual-core processor with high processing power for a microcontroller. However, the ESP32 is designed with a low-power mode, enabling IoT devices to operate in battery-powered or energy-constrained scenarios, making it suitable for simple applications. Another advantage is its support for various development environments, including the Arduino IDE and the Espressif IoT Development Framework (ESP-IDF). This makes it easy to use for developers with varying levels of experience [12].

3.2 Sensors

The selection of sensors is adjusted to the function based on the placement of sensors in the production process. Temperature sensor for monitoring system, a type K thermocouple sensor is used due to its wide temperature range, durability, and cost-effectiveness. This sensor was chosen because it performs well at temperatures between 25 and 100 degrees Celsius, which is suitable for drying and extraction processes [13]. The signal from the temperature sensor is first received by the max6675 module for calibration and stability of the analog signal reading from the sensor compared to the direct reading by analog-to-digital conversion on the microcontroller [14]. Meanwhile, to measure

https://doi.org/10.1051/bioconf/202621611001

the mixer's power, an SCT-013-000 sensor is used to read the amount of current passing through a conductor to a load. The use of sensors can measure current without the need to interrupt the current path or make modifications to the main circuit. Finally, after the sensor is determined, an electronic system circuit is formed, which can be seen in Fig. 3.

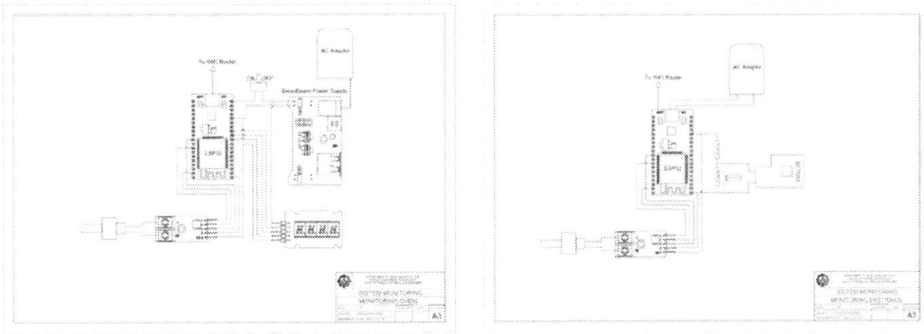

(a) (b)

Fig. 3. Wiring diagram of the electronic system used; (a) oven monitoring, (b) extraction monitoring.

3.3 Database and Web Monitoring Design

The system's design is centered on an Internet of Things (IoT) framework to integrate and monitor data from various sensors in real-time. This architecture (shown in Fig. 4) involves three key stages: data transmission from the equipment, storage and processing in the cloud, and data visualization. The entire process leverages Google's Firebase platform, chosen for its powerful Realtime Database, robust authentication for security, and integrated hosting services. On the hardware side, each monitoring device is equipped with a microcontroller and a Wi-Fi module that connects to the internet via a 4G router. The microcontroller is programmed to acquire sensor data, authenticate with the Firebase system, and then transmit the information to a designated directory in the Realtime Database. Finally, this live data is presented on a simple and easy-to-understand web-based User Interface (UI), built with HTML and deployed using Firebase Hosting. This dashboard, which displays critical parameters like temperature and RPM, is designed to automatically refresh whenever new data is received, allowing all stakeholders to conveniently monitor the seaweed processing conditions.

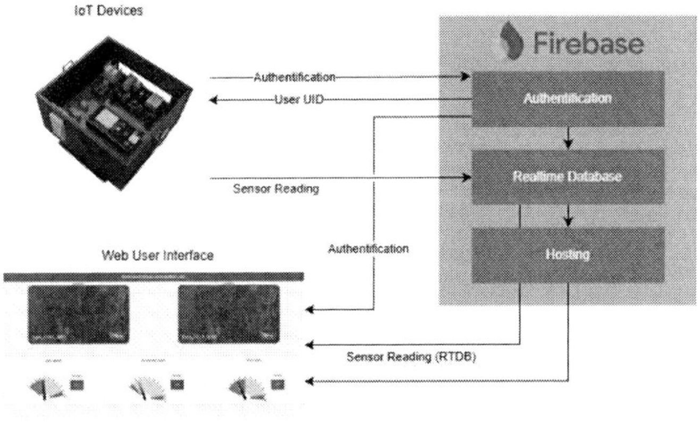

Fig. 4. Architecture of monitoring system for carrageenan process.

4 Results and Discussion

https://doi.org/10.1051/bioconf/202621611001

Preliminary testing was conducted to ensure that all electronic circuits, software components, and system connectivity operated according to the design specifications. The results indicated that the microcontroller, temperature sensor, and current sensor consistently performed accurate data acquisition. Collected data was successfully transmitted to the Firebase Realtime Database without significant delays or packet loss. The serial monitor displayed a "PASSED" status for each data transmission cycle, confirming that data integrity was maintained (Fig. 5 (a)).

The oven temperature sensor provided stable readings, with the maximum recorded value reaching 87 °C (Fig. 5 (b)). The 4-digit 7-segment display showed identical values to those transmitted to the database, demonstrating the accuracy of the microcontroller's analog-to-digital conversion process. The SCT-013-000 current sensor successfully detected variations in the mixer's motor load, which correlated with changes in RPM. While minor fluctuations occurred due to the instability of the drill motor used for simulation, the system was still able to record and track the overall trend with high accuracysurface.

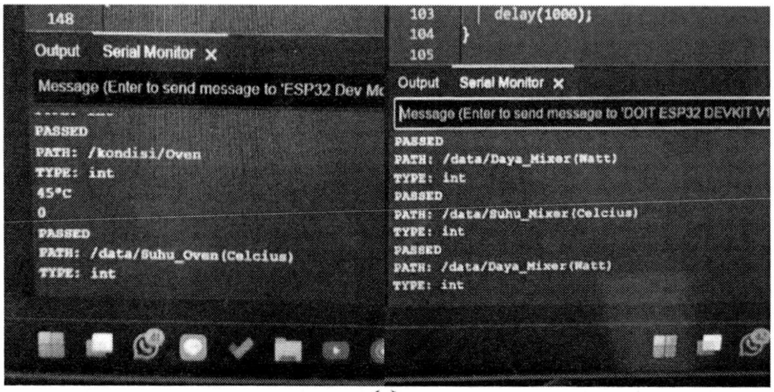

(a)

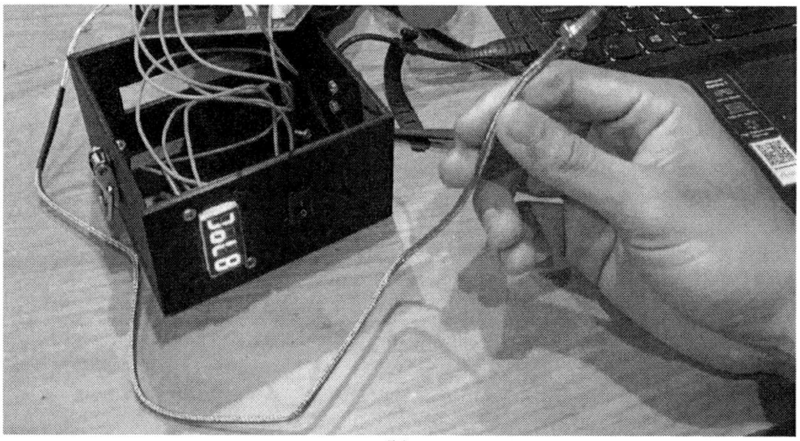

(b)

Fig. 5. IoT Device Testing; (a) passed sending data test, (b) temperature sensor test.

Data transmission was carried out via a Wi-Fi connection paired with a 4G router. A load test was performed by continuously sending data for one hour, during which no anomalies or connection failures were observed. The average update latency in the database was less than one second, meeting the requirements for real-time monitoring.

https://doi.org/10.1051/bioconf/202621611001

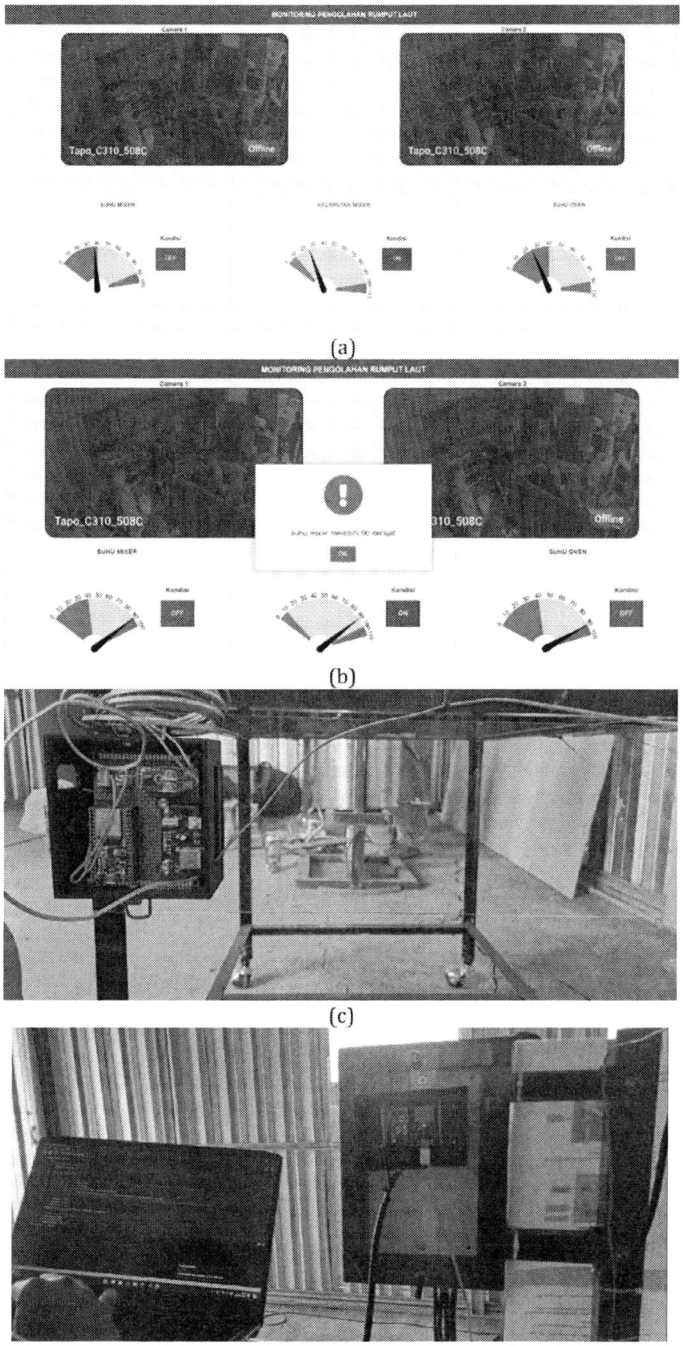

Fig. 6. Monitoring system result; (a) web user interface, (b) web alert system, (c) device installation at drying oven, (d) device installation at extraction unit.

The web-based dashboard, hosted on Firebase, successfully displayed real-time updates of oven temperature and mixer speed readings in synchronization with sensor measurements (Fig. 6 (a)). Value changes were visible almost instantly, without the need for manual page refresh. The interface's minimalistic design allowed operators to interpret results easily, even without extensive technical background. Each sensor in every process has its own safety limit determined based on the processing stage. The alert system is designed as a popup window that will appear with a notification sound if any sensor exceeds the predetermined limits (Fig. 6 (b)). This system is expected to help operators become aware of errors, thus avoiding accidents.

During implementation at the carrageenan processing facility, all sensors and electronic modules were installed without requiring major modifications to the existing machinery, seen in Fig. 6 (c, d). The oven temperature sensor was mounted through a drilled opening in the oven wall, with the electronic module positioned externally to prevent component degradation from heat exposure. The current sensor was installed non-invasively on the extraction machine's electrical panel, ensuring that the original functionality was unaffected.

5 Conclusion

The IoT monitoring system developed in this study performed successfully during both laboratory and on-site testing at the carrageenan facility. The system proved capable of accurately collecting data from temperature and current sensors and transmitting it to a cloud database in real-time with minimal delay. The web-based interface provided a clear and automatically updating display of key operational data, such as temperature and RPM. Easy to install and stable during 24-hour continuous operation, the system significantly improved the production process by enabling remote monitoring, allowing for early detection of anomalies, and enhancing overall quality control and operational efficiency.

References

[1] M. Basyuni, M. Puspita, R. Rahmania, H. Albasri, I. Pratama, D. Purbani, A. Aznawi, A. Mubaraq, S. S. Al Mustaniroh, F. Menne, Y. I. Rahmila, S. G. Salmo III, A. Susilowati, S. H. Larekeng, E. Ardli and T. Kajita, "Current biodiversity status, distribution, and prospects of seaweed in Indonesia: A systematic review," Heliyon, vol. 10, no. 10, p. e31073, 2024.

[2] S. Widjaja, E. N. Prasetyo, S. Nurkhamidah, I. S. Arif and S. Megawati, "An Integrated Business Model Based on Linear Programming for Carrageenan Production Chain in Coastal Village," in SRCM 2023, Surabaya, 2024.

[3] F. Jabeen, Zil-e-Aimen, R. Ahmad, S. Mir, N. S. Awwad and H. A. Ibrahium, "Carrageenan: structure, properties and applications with special emphasis on food science," RSC Adv., vol. 15, no. 27, pp. 22035-22062, 2025.

[4] M. Mulyati, J. Geldermann and T. Kusumastanto, "Carrageenan supply chains in Indonesia," IOP Conf. Series: Earth and Environmental Science, vol. 414, p. 012013, 2020.

[5] S. Salengke, A. Hasizah, Supratomo, A. Waris, M. Mahendradatta, A. Laga and Metusalach, "Optimization of Carrageenan Extraction from Eucheuma Spinosum Using Pilot Scale Ohmic Technology," in 15th ASEAN Food Conference, Ho Chi Minh, 2017.

[6] S. Mustapha, H. Chandar, R. Saghravani and M. Y. Harun, "Production of semi- refined carrageenan from Eucheuma cotonii," Journal of Scientific & Industrial Research, vol. 70, pp. 865-870, 2011.

[7] D. Witczak and S. Szymoniak, "Review of Monitoring and Control Systems Based on Internet of Things," Appl. Sci, vol. 14, no. 19, p. 8943, 2024.

[8] M. Cascais, P. Monteiro, D. Pacheco, J. Cotas, L. Pereira, J. Marques and A. Gonçalves, "Effects of Heat Treatment Processes: Health Benefits and Risks to the Consumer," Appl. Sci., vol. 11, no. 18, p. 8740, 2021.

[9] A. Hasizah, M. Mahendradatta, A. Laga, M. Metusalach, Supratomo, A. Waris and S. Salengke, "A novel ohmic-based technology for seaweed processing," International Food Research Journal, vol. 25, no. 4, pp. 1341-1348, 2018.

[10] W. A. Teniwut and R. M. K. Teniwut, "Minimizing the instability of seaweed cultivation productivity on rural coastal area: a case study from Indonesia," AACL Bioflux, vol. 11, no. 1, pp. 259-271, 2018.

[11] S. Widjaja, E. N. Prasetyo, S. Nurkhamidah, I. S. Arif and S. Megawati, "An Integrated Business Model Based on Linear Programming for Carrageenan Production Chain in Coastal Village," in The 4th Sustainability and Resilience of Coastal Management (SRCM 2023), Surabaya, 2024.

[12] Espressif Systems, "ESP32WROOM32 Datasheet," 2023. [Online]. Available: https://www.espressif.com/sites/default/files/documentation/esp32-wroom-32_datasheet_en.pdf.

[13] O. Igwilo, Mathurine.G, I. Onyegbadue and Azike.R.U, "Experimental Design, Characterization, coupling and calibration of type k thermocouple," Journal of Engineering and Applied Sciences, vol. 2, no. 3, pp. 388-399, 2023.

[14] R. Septiana, I. Roihan, J. Karnadi and R. A. Koestoer, "Calibration of K-Type Thermocouple and MAX6675 Module With Reference DS18B20 Thermistor Based on Arduino DAQ," Prosiding SNTTM, vol. XVIII, pp. 1-6, 2019.

Determinants of Enhanced Trained Human Resources in Flood Disaster Risk Reduction (A Case Study in Sidomulyo Village, Lamongan)

Muhammad Rafli Al Ghiffari Ferani[1], and *Adjie* Pamungkas[1,*]

[1]Department of Urban and Regional Planning, Institut Teknologi Sepuluh Nopember, Surabaya, Indonesia

Abstract. In many developing regions, top-down infrastructural solutions fail to mitigate recurrent disasters, such as the annual Bengawan Jero flood in East Java, Indonesia, which causes estimated losses of IDR 29 billion (approximately USD 1.8 million). This failure underscores the critical need for community-level capacity, yet Sidomulyo Village remains a "cold spot" with a severe deficit in trained human resources. This research aims to identify the key determinants for enhancing these resources. Employing Partial Least Squares-Structural Equation Modelling (PLS-SEM), the roles of local institutions, social capital and human capital were assessed. The analysis shows that human capital has a strong, direct positive effect on trained human resources ($\beta = 0.644$, p value 0.000), while local institutions and social capital have no significant direct effect. However, the study uncovers a critical indirect path wherein social capital significantly impacts trained human resources by mediating the development of human capital ($\beta = 0.343$, p value 0.013). The clear conclusion is that effective strategies for enhancing trained human resources must focus on building human capital while using social capital as the fundamental supporting mechanism.

1 Introduction

The integration of disaster risk management into spatial development is a critical necessity, particularly in regions where hazard exposure and systemic vulnerabilities intersect [1]. In East Java, the persistent inundation of the Bengawan Jero area exemplifies a high-risk profile characterized by complex hydrological imbalances and expanding spatial impacts [2]. Despite ongoing governmental interventions, the region's overall resilience remains constrained, with "low human resources" identified as a pivotal vulnerability [2]. The flooding impacts are substantial and protracted. A clear example occurred in early 2023, when a six-month flood event inundated 6,672 homes and 5,970 hectares of farmland, leading to significant economic losses [2]. Evidence suggests an escalating crisis, with the spatial extent of the flooding expanding from 42 to 59 villages in a single year [3]. The Bengawan Jero floods are attributed to complex hydrological deficits, including river

* Corresponding author: adjie@its.ac.id

geometry imbalances, low flow velocity [2], and overflow from the primary Bengawan Solo watershed [3].

Current assessments categorize the overall resilience of the Bengawan Jero area as "medium" (67.7%), yet explicitly identify "low human resources" as a significant vulnerability [2]. Specifically, Desa Sidomulyo has emerged as a "cold spot," suffering from a concentrated deficit in trained personnel, a demographic essential for ensuring adaptive and effective community-level response [3]. Addressing this gap requires a comprehensive understanding of community resilience through a capital-based approach, where the interplay between social and human capital serves as a foundation for preparedness [4]. While social capital facilitates community mobilization and mutual aid, human capital acts as a turnkey resource for the technical competence required in high-risk environments [5]. Therefore, this study aims to identify the determinants of enhanced trained human resources for flood disaster risk reduction in Sidomulyo Village, establishing the necessary empirical foundation using Partial Least Squares-Structural Equation Modelling (PLS-SEM) prior to the formulation of a concrete DRR framework.

2 Literature review

2.1 Trained human resources

Trained human resources, defined by specialized competencies in disaster mitigation and emergency response, serve as a fundamental determinant of community preparedness [3]. While disaster management necessitates multi-stakeholder engagement, volunteers both from national bodies and local organizations within the Bengawan Jero area remain central to operationalizing response and sustainable prevention efforts. Despite the active presence of these groups, systemic capacity in the region remains deficient, with Desa Sidomulyo critically identified as a "cold spot" [3]. Rectifying this deficit requires a capital-based approach to resilience, emphasizing the interplay between foundational social structures and individual capabilities [4,6].

Local institutions play a vital role in this framework, acting as pillars for governance, preparedness education, and the development of resilient livelihoods [7]. Furthermore, contemporary literature posits a formative theoretical link wherein social capital facilitated through norms, trust, and networks, drives the development of human capital [5]. Consequently, this study evaluates three interconnected indicators hypothesized to enhance trained human resources: local institutions; social capital, for its capacity to mobilize mutual aid and collective action; and human capital, which encompasses the specialized education, skills, and health baseline inherent to effective disaster personnel.

2.2 Floods

Flooding remains a dominant global hazard, defined by the United Nations Office for Disaster Risk Reduction (UNDRR) as the submergence of land by water that exceeds the drainage capacity of either natural or artificial systems [8]. These events result in significant multi-dimensional losses, encompassing casualties, property damage, and long-term environmental degradation [8]. A primary exemplification of this risk profile is found in East Java, where recurrent inundations are formally classified as high-priority threats within regional disaster management frameworks [2]. In the specific context of the Bengawan Jero area in Lamongan Regency, flooding is characterized as an annual, episodic, and predictable phenomenon [3]. These inundations are driven by the overflow of the Bengawan Solo River and further exacerbated by the region's distinctive low-lying topography [2,3]. The impact is notably protracted, often lasting between four to six months with substantial water depths, consistently affecting a wide spatial extent of villages across the regency [2,3]. Such protracted events underscore the necessity of moving beyond structural solutions toward adaptive, community-based frameworks that address localized vulnerabilities and capacity deficits.

2.3 Disaster Risk Management

Disaster Risk Management (DRM) is conceptualized as a systematic process aimed at minimizing hazard probabilities and mitigating unavoidable damages through a continuum of identification and governance [8,9]. This management paradigm is operationalized via risk assessments that analyze the critical interplay between Hazards, Vulnerability, and Capacity. According to the United Nations Office for Disaster Risk Reduction (UNDRR), these components represent the potential for harm, the conditions heightening susceptibility, and the internal strengths available to enhance resilience, respectively [8]. Strategically, interventions are categorized as prospective, corrective, or compensatory, the latter focusing on strengthening resilience against residual risks [8]. In the Bengawan Jero area, where the flood risk index is driven by land use and soil characteristics [2], implementing an effective DRM framework is imperative. While the region maintains a medium overall capacity, the specific deficit in "trained human resources" has emerged as a critical vulnerability, particularly in "cold spot" areas like Desa Sidomulyo [3]. This critical gap in trained human resources constitutes a residual risk, necessitating the identification of its most influential determinants. Identifying these determinants will provide a foundation for compensatory DRM, which can then be integrated into future strategic and development plans.

2.4 Local institution

Local institutions (LIs), comprising community associations, village-level entities, and civil society organizations, are recognized as fundamental drivers for enhancing disaster risk management and localized resilience. In the context of the Bengawan Jero area, where LIs such as TAGANA and RBJ are active, their responsibilities span the full disaster cycle, from pre-disaster training and awareness to during-disaster rescue and post-disaster recovery [2,3]. Despite the implementation of national resilience programs, critical institutional gaps persist in specific locations such as Sidomulyo Village. These deficiencies require targeted capacity enhancement through both formal technical training and informal self-study to bridge the "cold spot" resource deficit [3]. The capability of these institutions is contingent upon a multifaceted interplay of governance, influence, and tangible existence [10]. Drawing from adaptive governance frameworks, this study focuses on several critical institutional variables essential for operational capability: the existence of the institution as a facilitator for resilience; adequacy of personnel to prevent capacity shortages; development initiatives that foster community-level education; and the establishment of regulations, specifically Standard Operating Procedures (SOP) and external rules, which are vital for effective collaboration during crisis scenarios.

2.5 Social Capital

Social capital (SC) is conceptually defined as the access to and participation in formal and informal networks that incorporate trust, norms, and collective institutions [4,7]. Theoretically, it is bifurcated into structural dimensions, such as participation and rules, and cognitive dimensions, including shared values and reciprocity [11]. This research operationalizes social capital through four key variables to determine its influence on trained human resources [7,11]. These variables are: Norms, the behavioral standards for community coordination [7,11]; Participation is examined through three sub-variables number of participants, frequency of participation, and level of participation as engagement in development programs directly enhances community capabilities [4,11]; Networks are assessed across reliable person, their bonding social (intra-community), bridging social (inter-group), and linking social (community-to-group) [4,7]; Finally, trust, comprising interpersonal trust and shared collective values, is evaluated for its role in fostering beneficial relationships and ensuring adherence to risk management protocols, thereby contributing to the enhancement of trained human resources [7,11].

2.6 Human Capital

Human capital (HC), defined as the acquired expertise, knowledge, and physical health of individuals, serves as a "turnkey" resource in enhancing community resilience to hydrometeorological hazards [12]. Within the context of disaster management, HC is operationalized through several critical variables that collectively determine the effectiveness of local resources. First, formal education is assessed via two sub-variables: education level [12] and disaster education material [4,7]. Second, skills are evaluated across the pre-disaster skills, during-disaster skills, and post-disaster skills [7,12]. Third, health is analyzed via two sub-variables: disability population and prevalence of severe disease, which identify underlying population vulnerabilities [12]. Finally, disaster knowledge, often acquired informally [7,12], is examined to establish the existing baseline capability of local human resources.

3 Study area and data sources

3.1 Study area

This research selects Desa Sidomulyo, a village within the Bengawan Jero Region, as the study area. The village, comprising four dusun (hamlets), covers 222 Ha and supports a population of 2,442 (density: 1,100/km^2). The rationale for its selection is its significant and recurrent vulnerability to flooding, primarily caused by overflow from the Bengawan Solo watershed and supplemental runoff from Lamongan Kota. These flood events generate profound socio-economic consequences, as evidenced by a 2023 disaster that submerged hundreds of homes and resulted in substantial agricultural losses valued at hundreds of billions of rupiah. Inundation depths are significant, reaching 20-50 cm in residential zones and 2 m in agricultural areas during the 2023 event, while 2024-2025 levels remained high (1-2 m) in agricultural zones despite being lower (10-20 cm) in settlements.

The persistent flood risk in the region is attributed to two key factors: land use and soil texture [13]. The village's land use is dominated by food crops (186 Ha, 78%) and rural settlements (12%), with residential zones situated near water bodies, thereby increasing exposure. This vulnerability is compounded by the area's soil composition, which predominantly features fine and medium textures [13]. This fine texture inhibits soil permeability, consequently elevating the probability of surface inundation [13]. As illustrated in **Figure 1**, illustrates the flood disaster evacuation network in Sidomulyo Village, detailing the strategic distribution of assembly points and designated evacuation routes across its four hamlets to facilitate organized community displacement.

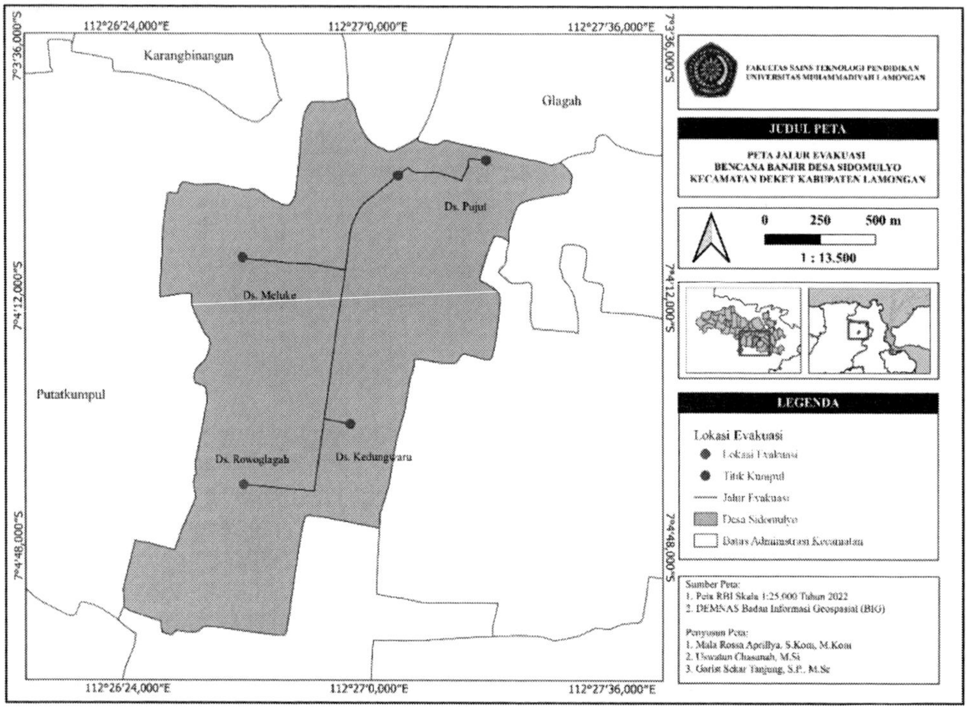

Fig. 1. Map of flood disaster evacuation routes in Sidomulyo Village

3.2 Data sources

This research is based on a systematic data collection methodology, utilizing both primary and secondary sources. The primary data acquisition strategy is centered on assessing the condition of trained resources within Desa Sidomulyo. To achieve this objective, empirical data were gathered directly from the public of the study area through the administration of structured questionnaires and in-depth interviews. To complement and contextualize the primary findings, secondary data collection was concurrently conducted. This process involved a multi-pronged approach; wherein relevant data was systematically sourced through institutional surveys and media surveys.

4 Methodology

This study employs a multi-stage quantitative framework, initiating with a Literature Review to identify hypothesized indicators, variables, and subvariables for the enhancement of trained human resources. Subsequently, Questionnaire and In-depth Interview data are utilized to establish respondent-derived weights and profiles for these selected factors. The analytical phase bifurcates: Descriptive and Inferential Statistics (Correlation) are applied to analyze the existing condition of trained human resources in

Desa Sidomulyo, while Partial Least Squares-Structural Equation Modeling (PLS-SEM) is implemented to identify the statistically significant determinants influencing their enhancement. The entire research methodology is illustrated in **Figure 2**.

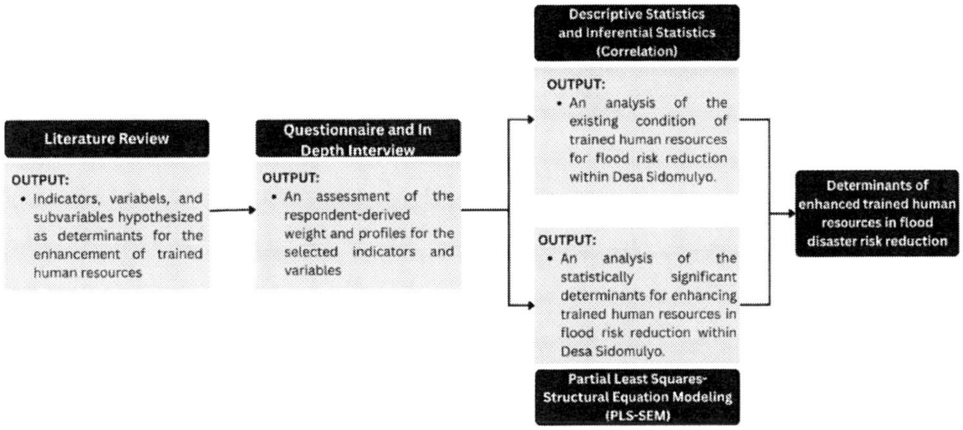

Fig. 2. Research methodology

4.1 Questionnaire and in-depth interview

Primary data for this research were acquired through the administration of a structured questionnaire and supplementary in-depth interviews. The sample comprised 42 community respondents from Desa Sidomulyo, selected via a proportional stratified sampling technique to ensure representation across the four hamlets (dusun). This sample size (N=42) satisfies the minimum requirements for Partial Least Squares-Structural Equation Modeling (PLS-SEM) analysis. The data collection instruments were designed to capture respondent-derived assessments and weighting for the selected indicators and variables. A 1-to-5 Likert scale was utilized to provide quantitative inputs for the subsequent statistical modeling.

4.2 Descriptive statistics and inferential statistics (correlation)

This study employs a dual analytical approach for its initial objective. Descriptive statistics are utilized to systematically gather, summarize, and outline the data, providing a clear overview of its primary characteristics and establishing a baseline profile of the existing condition of trained human resources for flood risk reduction in Desa Sidomulyo [14]. Following this, inferential statistics are applied to draw broader conclusions about the population from the sample data [14]. Specifically, correlation analysis (Pearson) is implemented to measure the strength and direction of the relationship between the variables under investigation.

4.3 Partial least squares-structural equation modeling (pls-sem)

To identify the statistically significant determinants of trained human resources, this research employs Partial Least Squares-Structural Equation Modeling (PLS-SEM). This analytical method is utilized to test and estimate the causal relationships between latent constructs [15]. It is particularly well-suited for this study as it effectively handles complex models without convergence issues, non-normally distributed data, and smaller sample sizes. The analysis follows a two-stage process [15]: first, the specification of the outer model (measurement model) and the inner model (structural model); and second, the systematic evaluation of these models. The outer model is assessed for construct reliability and validity using criteria such as outer loadings, Average Variance Extracted (AVE), and the Heterotrait-Monotrait Ratio (HTMT), while the inner model is assessed by examining path coefficients, T-statistics (via bootstrapping), and *p*-values (<0.05) to determine the significance of the hypothesized relationships [15].

5 Result and discussion

5.1 Characteristics of each indicator

5.1.1 Characteristics of local institution

The Local Institution (X02) indicator was assessed to determine the perceived effectiveness of formal local institutions (e.g., Village Government, PKK, Karang Taruna) in developing trained human resources. The analysis reveals a significant disconnect between the community's high regard for institutional governance and its dissatisfaction with practical program implementation. A summary of the dominant respondent perceptions for each variable is presented in **Table 1.**

Table 1. Summary of dominant respondent perception for local institution

Variable/ Sub-variable	**Code**	**Dominant Respondent Perception**	**% (n)**
Local Institution	X02	**Score 2**: Efforts are minimal; limited to occasional verbal information/warnings (e.g., at mosque, via WhatsApp).	42.9% (18)
Existence	X03	**Score 3**: Gained basic practical skills (e.g., turning off power, basic first aid) from institutional programs.	42.9% (18)
Adequacy of Personnel	X04	**Split Perception:** **Score 3:** Sufficient (most involved residents were served well).	35.7% (15)
		Score 2: Lacking (service was limited and uneven).	35.7% (15)
Development Initiatives	X05	**Score 2:** Training was provided but offered only "fleeting insight" and did not build lasting, practical skills.	50.0% (21)
Regulations	X06	**Score 4:** Regulations are "very clear and accessible" and actively encourage compliance.	42.9% (18)
Standard Operating Procedures	X07	**Score 4:** Information is "very clear and consistent" and disseminated through accessible channels.	40.5% (17)
External Rules	X08	**Score 4:** Respondents "obey and actively encourage others" to comply with regulations.	61.9% (26)

The local institution (X02) is defined by a critical contradiction. On one hand, the Regulations (X06) variable was perceived positively, with 42.9% (n=18) finding rules "very clear and accessible". This was strongly supported by its sub-variables; 61.9% (n=26) reported high compliance with External Rules (X08), and 40.5% (n=17) found the Standard Operating Procedures (X07) "very clear and consistent". Conversely, this high regard for governance did not translate to practical execution. Development Initiatives (X05) were viewed as superficial, with 50.0% (n=21) stating that training offered only "fleeting insight" rather than building lasting skills. This implementation gap was further reflected in Adequacy of Personnel (X04), which was perfectly split between sufficient (35.7%, n=15) and lacking (35.7%, n=15), indicating uneven service delivery. Despite these criticisms, the Existence (X03) variable indicated some tangible benefits, as 42.9% (n=18) of respondents acknowledged gaining basic practical skills from past programs.

5.1.2 Characteristics of social capital

The Social Capital (X09) indicator, measuring the capacity derived from community interaction, was perceived positively. A significant majority (85.7%, n=36) rated social interaction as either having a "positive influence" (Score 3) or being a "strong driver" (Score 4) in motivating residents to participate in training. The construct's performance was analyzed across four key variables—norms, participation, networks, and trust—with a summary of dominant respondent perceptions presented in **Table 2**.

Table 2. Summary of dominant respondent perception for social capital

Variable/ Sub-variable	**Code**	**Dominant Respondent Perception**	**% (n)**
Social Capital	X09	**Score 3:** Social interaction has a positive influence, motivating residents to learn.	47.6% (20)
Norms	X10	**Score 3:** A "sufficiently strong" habit exists to participate in training; residents respond well.	40.5% (17)
Participation	X11	**Score 2:** Participation is limited to giving input or practicing basic skills (e.g., First Aid).	31.0% (13)
Number of Participants	X12	**Score 2:** Involvement is mostly passive (e.g., listening to socialization).	47.6% (20)
Frequency of Participation	X13	**Score 2:** Attended once; gained knowledge but not practical skills.	45.2% (19)
Level of Participation	X14	**Score 2:** Involvement was limited to providing opinions in a forum.	35.7% (15)
Network	X15	**Score 1:** Cooperation is limited and functions only when there is an official order.	33.3% (14)
Reliable Person	X16	**Score 2:** Only a "handful" of the same few individuals can be relied upon.	45.2% (19)
Bonding Social	X17	**Score 4:** Many residents take the initiative to help each other in training programs.	35.7% (15)
Bridging Social	X18	**Score 4:** Groups (e.g., IKSPI Kera Sakti) actively and voluntarily seek collaboration.	35.7% (15)
Linking Social	X19	**Score 3:** Communication is "good enough" and provides space for discussion.	40.5% (17)
Trust	X20	**Score 3:** Trust is "good enough" for residents to discuss and conduct activities together.	64.3% (27)
Interpersonal Trust	X21	**Score 3:** Residents trust each other sufficiently for active discussion and small agreements.	52.4% (22)

Shared Collective Values	X22	**Score 3:** Residents voluntarily take real action (e.g., clearing evacuation routes) based on shared values.	52.4% (22)

The analysis of Norms (X10) indicated a positive foundation, with 40.5% (n=17) perceiving a "sufficiently strong" habit among residents to participate in training when invited. However, the Participation (X11) variable revealed this involvement is largely passive, with the dominant perception (31.0%, n=13) being limited to giving input rather than taking an active role. This is supported by the sub-variables, where Number of Participants (X12) was characterized by listening to socialization (47.6%, n=20), and Frequency of Participation (X13) was limited to a single attendance, which provided knowledge but not practical skill (45.2%, n=19). Reinforcing this, the Level of Participation (X14) was also predominantly consultative, with 35.7% (n=15) reporting their highest involvement as providing opinions in a forum, rather than taking an active role in implementation.

The Network (X15) variable highlighted a significant structural weakness, as 33.3% (n=14) reported that cooperation is limited and functions only when there is an official order. This is largely due to Reliable People (X16) being limited to a "handful" of the same individuals (45.2%, n=19). Despite this, Bonding Social (X17) and Bridging Social (X18) were strong, with 35.7% of respondents for both variables reporting high levels of initiative and voluntary collaboration. Linking Social (X19) was deemed "good enough" for discussion (40.5%, n=17). Finally, the Trust (X20) indicator was strong, with 64.3% (n=27) describing it as "good enough" for joint discussion and action. Interpersonal Trust (X21) was sufficient for active discussion and small-scale agreements (52.4%, n=22). This trust manifested as strong Shared Collective Values (X22), where 52.4% (n=22) confirmed that residents voluntarily take real action, such as clearing evacuation routes, based on shared values.

5.1.3 Characteristics of human capital

The Human Capital (X23) indicator, which assesses the knowledge, skills, and health capacity of the community, was perceived as being in a developmental stage. The dominant perception (50.0%, n=21) was that residents capacity was limited to passive knowledge gained from socialization (Score 2) , while 40.5% (n=17) reported active skill development through formal training or simulations (Scores 3 and 4). A detailed breakdown of the dominant perceptions for each variable within this construct is presented in **Table 3.**

Table 3. Summary of dominant respondent perception for human capital

Variable/ Sub-variable	**Code**	**Dominant Respondent Perception (Score & Description)**	**% (n)**
Human Capital	X23	**Score 2:** Attended socialization, gaining knowledge but not skills.	50.0% (21)
Formal Education	X24	**Score 3:** "Sufficiently equipped" to follow planned steps.	42.9% (18)
Education Level	X25	**Score 4:** Will assess the situation and prioritize actions.	38.1% (16)

Variable/ Sub-variable	**Code**	**Dominant Respondent Perception (Score & Description)**	**% (n)**
Disaster Education Material	X26	**Score 3:** Taught "basic steps" (e.g., go to assembly point).	38.1% (16)
Skills	X27	**Score 4:** Actively seek info, help neighbors, join community work.	45.2% (19)
Pre-disaster Skills	X28	**Score 2:** Clean drains and move items to higher ground.	35.7% (15)
During-disaster Skills	X29	**Score 4:** Check on neighbors and share critical information.	59.5% (25)
Post-Disaster Skills	X30	**Score 4:** Join community work (*kerja bakti*) to help neighbors.	54.8% (23)
Health	X31	**Score 2:** Participated in data collection of vulnerable residents.	42.9% (18)
Disability Population	X32	**Score 2:** Participated in data collection of residents w/ disabilities.	42.9% (18)
Prevalence of Severe Disease	X33	**Score 2:** Participated in data collection of residents w/ severe illness.	42.9% (18)
Disaster Knowledge	X34	**Score 4:** Check warnings, coordinate w/ neighbors, join post-flood work.	50.0% (21)

The analysis of Formal Education (X24) indicated it "sufficiently equipped" residents to follow planned steps (42.9%, n=18). This is supported by the high Education Level (X25) (47.2% High School, 36.1% College), which correlates with respondents confidence in assessing situations rather than acting spontaneously (38.1%, n=16). However, Disaster Education Material (X26) in schools was perceived as basic or purely theoretical (35.7%, n=15), focusing on procedures like moving to assembly points (38.1%, n=16). The Skills (X27) variable was rated highly, with 45.2% (n=19) actively helping neighbors and joining community recovery efforts. While Pre-Disaster Skills (X28) were largely limited to household preparations like clearing drains (35.7%, n=15), During-Disaster Skills (X29) and Post-Disaster Skills (X30) were exceptionally strong and community-oriented, with majorities (59.5% and 54.8%, respectively) focused on checking on neighbors and collective recovery. A significant weakness was found in the Health (X31) variable, which assesses planning for vulnerable populations. Actions for both Disability (X32) and Severe Illness (X33) populations were predominantly passive, limited to "data collection" (42.9% for all three variables). Conversely, Disaster Knowledge (X34) was very high, with 50.0% (n=21) demonstrating a comprehensive understanding of the full disaster cycle, from checking warnings to post-flood coordination.

5.2 Validity and reliability test

To ensure the quality of the research instrument, the validity of the data was rigorously assessed. Validity, which confirms the instrument measures its intended constructs, was evaluated using the Pearson correlation test at a 5% significance level. For the sample size of 42 respondents, the critical r value was 0.304. The analysis demonstrated that all 3 indicators, 12 variables, and 18 sub-variables achieved a calculated r value greater than this critical r threshold. Consequently, all items were deemed valid, confirming that the instrument effectively and accurately measures the concepts under investigation. The detailed results of this validity test are presented in **Table 4.**

Table 4. Result of validity test

Indicators	**Variable/**Sub-variable	**Item**	**Validity Test**						
			Calculated r value		**Critical r values**	**Sig.**		***α***	**Description**
Trained Human Resources		Y01	0,334	>	0,304	0,031	<	0,05	**Valid**
Local Institution		X02	0,627	>	0,304	0,000	<	0,05	**Valid**
	Existence	X03	0,611	>	0,304	0,000	<	0,05	**Valid**
	Adequacy of Personnel	X04	0,441	>	0,304	0,003	<	0,05	**Valid**
	Development Initiatives	X05	0,546	>	0,304	0,000	<	0,05	**Valid**
	Regulations	X06	0,539	>	0,304	0,000	<	0,05	**Valid**
	Standard Operating Procedures	X07	0,626	>	0,304	0,000	<	0,05	**Valid**
	External Rules	X08	0,397	>	0,304	0,009	<	0,05	**Valid**
Social Capital		X09	0,539	>	0,304	0,000	<	0,05	**Valid**
	Norms	X10	0,710	>	0,304	0,000	<	0,05	**Valid**
	Participation	X11	0,684	>	0,304	0,000	<	0,05	**Valid**
	Number of Participants	X12	0,367	>	0,304	0,017	<	0,05	**Valid**
	Frequency of Participation	X13	0,563	>	0,304	0,000	<	0,05	**Valid**
	Level of Participation	X14	0,538	>	0,304	0,000	<	0,05	**Valid**
	Network	X15	0,575	>	0,304	0,000	<	0,05	**Valid**
	Reliable Person	X16	0,661	>	0,304	0,000	<	0,05	**Valid**
	Bonding Social	X17	0,597	>	0,304	0,000	<	0,05	**Valid**
	Bridging Social	X18	0,594	>	0,304	0,000	<	0,05	**Valid**
	Linking Social	X19	0,720	>	0,304	0,000	<	0,05	**Valid**
	Trust	X20	0,640	>	0,304	0,000	<	0,05	**Valid**
	Interpersonal Trust	X21	0,642	>	0,304	0,000	<	0,05	**Valid**
	Shared Collective Values	X22	0,608	>	0,304	0,000	<	0,05	**Valid**
Human Capital		X23	0,504	>	0,304	0,001	<	0,05	**Valid**

Indicators	Variable/Sub-variable	Item	Validity Test						
			Calculated r value		Critical r values	Sig.		α	Description
	Formal Education	X24	0,532	>	0,304	0,000	<	0,05	**Valid**
	Education Level	X25	0,561	>	0,304	0,000	<	0,05	**Valid**
	Disaster Education Material	X26	0,567	>	0,304	0,000	<	0,05	**Valid**
	Skills	X27	0,393	>	0,304	0,010	<	0,05	**Valid**
	Pre-disaster Skills	X28	0,445	>	0,304	0,003	<	0,05	**Valid**
	During-disaster Skills	X29	0,430	>	0,304	0,004	<	0,05	**Valid**
	Post-disaster Skills	X30	0,506	>	0,304	0,001	<	0,05	**Valid**
	Health	X31	0,818	>	0,304	0,000	<	0,05	**Valid**
	Disability Population	X32	0,818	>	0,304	0,000	<	0,05	**Valid**
	Prevalence of Severe Disease	X33	0,818	>	0,304	0,000	<	0,05	**Valid**
	Disaster Knowledge	X34	0,475	>	0,304	0,001	<	0,05	**Valid**

Subsequently, the data was subjected to a reliability analysis, which yielded a Cronbach's Alpha of 0.936. As this value exceeds the accepted threshold of 0.60, it indicates high reliability. This signifies that the consistency of the answers provided by the 42 respondents is classified as high. The complete results of this reliability analysis are available in **Table 5.**

Table 5. Result of reliability test

Case Processing Summary			
		N	**%**
Cases	Valid	42	100.0
	Excluded	0	.0
	Total	42	100.0
Cronbach's Alpha		**N of Items**	
.936		34	

5.3 Correlation test

A Pearson correlation analysis was conducted as an initial verification of the relationships between the independent variables (constructs) and their respective indicators. This step was essential for preliminary data screening and to identify the strength and direction of the relationships, utilizing a significance threshold of $p < 0.05$. A summary of the significant correlations and their respective strengths is presented in **Table 6.**

Table 6. Summary of pearson correlation strength

Indicator	Variable / Sub-Variable	Code	Correlation Strength
Local Institution	**Existence**	X03	High
	Adequacy of Personnel	X04	Low
	Development Initiatives	X05	High
	Regulations	X06	Low
	Standard Operating Procedures	X07	Low
	External Rules	X08	Low
Social Capital	**Norms**	X10	Medium
	Participation	X11	Low
	Number of Participants	X12	Low
	Frequency of Participation	X13	Low
	Level of Participation	X14	Low
	Network	X15	Low
	Reliable Person	X16	Low
	Bonding Social	X17	Medium
	Bridging Social	X18	Medium
	Linking Social	X19	Medium
	Trust	X20	Medium
	Interpersonal Trust	X21	Medium
	Shared Collective Values	X22	Medium
Human Capital	**Formal Education**	X24	Low
	Education Level	X25	Low
	Disaster Education Material	X26	Low
	Skills	X27	Low
	Pre-disaster Skills	X28	Low
	Health	X31	Medium
	Disability Population	X32	Medium
	Prevalence of Severe Disease	X33	Medium
	Disaster Knowledge	X34	Low

For the Local Institution (X02) construct, Existence (X03) and Development Initiatives (X05) showed a high correlation, indicating that tangible program impact is strongly linked to development initiatives. Conversely, Regulations (X06) and External Rules (X08) were found to be non-significant ($p > 0.05$), suggesting that formal rules alone do not currently drive the perception of institutional capacity. Analysis of Social Capital (X09) revealed that Norms (X10), Network (X15), and Trust (X20) including all sub-variables for Bonding Social (X17), Bridging Social (X18), and Linking Social (X19) demonstrated significant, medium-strength correlations. Critically, the entire Participation

(X11) construct (including sub-variables X12, X13, X14) and the Reliable Person (X16) variable were non-significant, implying that general participation levels are not strongly associated with the overall indicator. In the Human Capital (X23) indicator, Health (X31) (and its sub-variables X32, X33) showed a medium correlation, while Pre-disaster Skills (X28) and Disaster Knowledge (X34) were also significant. Notably, Formal Education (X24) and the general Skills (X27) construct (specifically 'during' and 'post-disaster' skills) were non-significant, suggesting preparedness is influenced more by specific health-related awareness and pre-disaster knowledge than by formal education levels.

5.4 Determinants of enhanced trained human resources

To answer the primary research objective, the hypothesized framework was tested using Partial Least Squares-Structural Equation Modeling (PLS-SEM) with SmartPLS 4 software. This method was selected for its suitability in handling smaller sample sizes (N=42) and complex models with non-normally distributed data.

5.4.1 Hypothesis Development

Based on the theoretical framework synthesized from the literature review, five primary hypotheses were developed. The framework posits that Local Institutions (X02), Social Capital (X09), and Human Capital (X23) are direct determinants of trained human resources (Y01). Furthermore, based on established theory that social structures influence human skill development, the model also hypothesizes that Local Institutions (H4) and Social Capital (H5) have a significant positive influence on Human Capital, positioning Human Capital as a potential mediator. The complete set of hypotheses is detailed in **Table 7** and the specified structural model, illustrating the hypothesized paths between the constructs, is presented in **Figure 3**.

Table 7. Hypothesized relationships in the model

Hypothesis	Path	Description
H1	X02 -> Y01	Local Institution (X02) has a significant positive effect on Trained Human Resources (Y01).
H2	X09 -> Y01	Social Capital (X09) has a significant positive effect on Trained Human Resources (Y01).
H3	X23 -> Y01	Human Capital (X23) has a significant positive effect on Trained Human Resources (Y01).
H4	X02 -> X23	Local Institution (X02) has a significant positive effect on Human Capital (X23).
H5	X09 -> X23	Social Capital (X09) has a significant positive effect on Human Capital (X23).

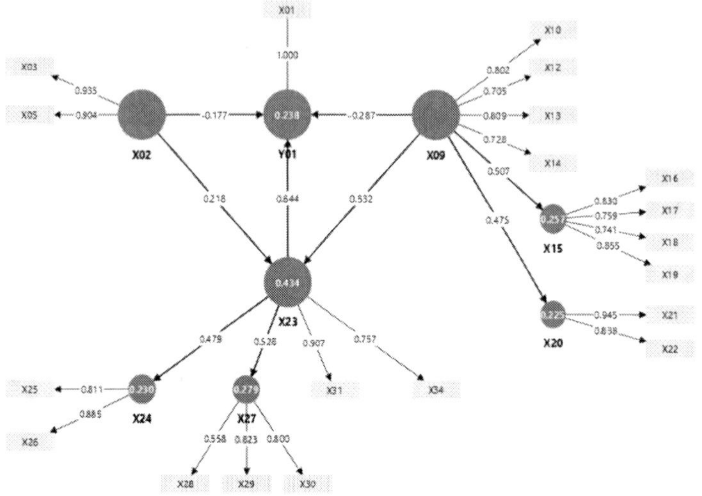

Fig. 3. Model Specification

5.4.2 Measurement Model Assessment

The analysis began with a rigorous assessment of the measurement model's reliability and validity. The initial model was refined based on these statistical criteria; Adequacy of Personnel (X04) and Regulations (X06) were removed from the Local Institution construct for failing to meet outer loading and AVE thresholds. Similarly, the Participation (X11) construct was reformulated, elevating its indicators (X12, X13, X14) to first-order variables to achieve model validity. The refined model demonstrated strong psychometric properties.

Reliability was assessed using Composite Reliability (ρA), which is considered a more robust measure than Cronbach's Alpha in PLS-SEM. All constructs met the required thresholds. Local Institution (X02), Social Capital (X09), Network (X15), and Trust (X20) all demonstrated high reliability (ρA > 0.80). Human Capital (X23), Formal Education (X24), and Skills (X27) were within the 0.60–0.70 range, which is acceptable for exploratory research. Validity was established by confirming both convergent and discriminant validity.

1. Convergent Validity was confirmed using two metrics. First, Outer Loadings for most indicators exceeded the 0.70 threshold. An exception was made for Pre-Disaster Skills (X28) (Loading = 0.558), which was retained due to its theoretical importance in the local context, where recurrent flooding has normalized the hazard. Second, the Average Variance Extracted (AVE) for all constructs surpassed the 0.50 minimum, confirming each construct explained over 50% of its indicators variance.
2. Discriminant Validity was established using two modern standards. First, the Heterotrait-Monotrait Ratio (HTMT) values for all constructs were below the 0.90 threshold, confirming empirical distinctiveness. Second, the Fornell-Larcker criterion was met, as the square root of the AVE for each construct was greater than its correlation with all other constructs.
3. Finally, a collinearity analysis found that all indicator Variance Inflation Factor (VIF) values were ≤ 2.140, well below the threshold of 5, confirming no

multicollinearity issues. Based on these tests, the measurement model was deemed robust, reliable, and valid.

5.4.3 Structural Model Assessment

Following validation of the measurement model, the structural model was assessed by running a bootstrapping procedure (5,000 samples) to test the hypothesized paths ($p < 0.05$). The results, summarized in **Table 8**, revealed that Human Capital (X23) was the only construct with a significant, strong positive direct effect on Trained Human Resources (Y01) ($\beta = 0.644$, $p < 0.001$). Contrary to H1 and H2, both Local Institution (X02) and Social Capital (X09) had no significant direct influence on Trained Human Resources. Further analysis showed that Social Capital (X09) had a significant, strong positive influence on Human Capital (X23) ($\beta = 0.532$, $p < 0.001$).

Table 8. Summary of structural model hypothesis testing

Hypothesis	Path	Original Sample (β)	T-Statistics	P-Values	Result
H1	X02 -> Y01	-0.177	1.203	0.229	Not Supported
H2	X09 -> Y01	-0.287	1.241	0.215	Not Supported
H3	**X23 -> Y01**	**0.644**	**3.830**	**0.000**	**Supported**
H4	X02 -> X23	0.218	1.437	0.151	Not Supported
H5	**X09 -> X23**	**0.532**	**4.439**	**0.000**	**Supported**

The non-significance of Local Institutions (X02) and Social Capital (X09) in directly enhancing Trained Human Resources (Y01) reveals a critical "implementation gap" within the specific context of Sidomulyo Village. While institutional governance is highly regarded with 42.9% of respondents finding Regulations (X06) "very clear" and 40.5% perceiving Standard Operating Procedures (X07) as "consistent", this does not translate into effective resource development. Empirical evidence shows that 50.0% of residents view Development Initiatives (X05) as superficial, providing only "fleeting insights" rather than sustainable technical skills. Furthermore, the adequacy of personnel (X04) is perfectly split, with 35.7% citing a lack of service that leads to uneven delivery. Similarly, the direct influence of Social Capital remains non-significant because Participation (X11) is predominantly passive and consultative (31.0%), with 45.2% of participants attending only once and gaining no practical expertise. Although the village maintains strong Interpersonal Trust (X21) and Shared Collective Values (X22) both at 52.4% these social bonds do not translate into immediate technical action for flood mitigation without a targeted mechanism for skill acquisition.

Conversely, the indirect effect of Social Capital (X09) on Trained Human Resources (Y01) through Human Capital (X23) was statistically significant ($\beta = 0.343$, $p = 0.013$). This central finding of the model illustrates that social dimensions such as norms, participation, networks, and trust function as foundational supporting mechanisms that first foster the development of Human Capital, which subsequently leads to the emergence of trained human resources. This mediated framework proves that enhancing Trained Human Resources in high-risk "cold spot" areas is dependent on a specific conversion process, where social connectivity is leveraged to build the tangible human capabilities necessary for effective and resilient disaster risk reduction. These significant determinants are summarized in **Table 9** for further strategic consideration.

Table 9. Recapitulation of variables influencing trained human resources

Indicator	Variable	Sub-variable
Human Capital (X23)	Formal Education (X24)	• Education Level (X25) • Disaster Education Material (X26)
	Skills (X27)	• Pre-disaster Skills (X28) • During-disaster Skills (X29) • Post-disaster Skills (X30)
	Health (X31)	
	Disaster Knowledge (X34)	
Social Capital (X09)	Norms (X10)	
	Number of Participants (X12)	
	Frequency of Participation (X13)	
	Level of Participation (X14)	
	Network (X15)	• Reliable Person (X16) • Bonding Social (X17) • Bridging Social (X18) • Linking Social (X19)
	Trust (X20)	• Interpersonal Trust (X21) • Shared Collective Values (X22)

6 Conclusion

This study concludes that enhancing trained human resources in high-risk, low-capacity areas like Desa Sidomulyo is a nuanced process, not of direct institutional action, but of multi-level capital conversion. The Partial Least Squares-Structural Equation Modeling (PLS-SEM) analysis confirmed that Human Capital (X23) is the only construct with a direct, strong, and statistically significant positive influence on the development of Trained Human Resources (Y01). Contrary to initial hypotheses, Local Institutions (X02) and Social Capital (X09) were found to have no significant direct effect on Trained Human Resources. However, the mediation analysis uncovered a critical pathway: Social Capital (X09) acts as a powerful predictor of Human Capital (X23), fully mediating the relationship between community bonds and trained resources.

Consequently, a fundamental paradigm shift is required: we must transform the current model of top-down instruction into one of bottom-up, sustained skill development. Strategic frameworks must stop treating social capital as the solution itself; instead, community bonds should be leveraged strictly as a driving tool to encourage residents to participate in training. This necessitates a hard pivot from abstract general socialization to concrete, applied training such as drills and mapping which is operationalized through four actionable strategies. First, regarding Variable of Formal Education, we must bridge the gap between theory and reality by integrating Bengawan Jero case studies into local curriculum to contextualize flood science. Second, Variable of Disaster Knowledge must evolve into operational competence by teaching residents to translate generic warning levels into specific checklists. Third, Variable of Practical Skills must embrace the shift towards simulation, incorporating route mapping workshops and boat operation drills. Finally, addressing Variable of Health Vulnerability requires a data-driven "Health Guardian" system that assigns neighbors to assist specific individuals with disabilities during evacuations, effectively using social ties to protect the most vulnerable.

Acknowledgement

The authors express their sincere gratitude to the residents and the village government of Sidomulyo for their participation and to the Department of Urban and Regional Planning at Institut Teknologi Sepuluh Nopember for their academic support. Strict measures were implemented to ensure that participant anonymity and confidentiality were maintained throughout the data collection process. However, this research is subject to certain limitations, including a relatively small sample size (n=42) and a specific focus on a single geographic "cold spot". Furthermore, because the analysis relies on respondent-derived perceptions, the findings may not be fully generalizable to regions with distinct territorial risk profiles or different socio-economic characteristics.

References

[1] BPBD Jawa Timur, "Dokumen Kajian Risiko Bencana Jawa Timur 2022-2026" 2021, Surabaya. Accessed: Feb. 24, 2025 [Online]. Available: https://files.bpbd.jatimprov.go.id/DOKUMEN/DOKUMEN%20KRB%20NASIONAL%20JAWA%20TIMUR%202022-2026.pdf
[2] A. Pamungkas, P. Atmodiwirjo, A.R. Harani, "Pengembangan Model Mitigasi Bencana Banjir Sungai Bengawan Jero Berbasis Praktik Adaptasi dan Partisipasi Komunitas", 2023
[3] A.N. Rahma, Strategi peningkatan kapasitas masyarakat dalam pengurangan risiko bencana banjir di Bengawan Jero, Kabupaten Lamongan (Institut Teknologi Sepuluh Nopember, Surabaya, 2024)
[4] H. Tariq, C. Pathirage, T. Fernando, Measuring community disaster resilience at local levels: an adaptable resilience framework. Int. J. Disaster Risk Reduct., 1-16 (2021). https://doi.org/10.1016/j.ijdrr.2021.102358
[5] R.D. Weaver, N. Habibov, Social capital, human capital, and economic well-being in knowledge economy: result from Canada's general social survey. J. Sociol. Soc. Welf. 39, 31-53 (2012). https://doi.org/10.15453/0191-5096.3665
[6] J.S. Mayunga, Understanding and Applying the Concept of Community Disaster Resilience: A Capital-Based Approach (USA, 2007)
[7] M. Choudhury, C. Haque, B. Doberstein, Adaptive governance and community resilience to cyclones in coastal Bangladesh: addressing the problem of fit, social learning, and institutional collaboration. Environ. Sci. Policy 126, 174-183 (2021). https://doi.org/10.1016/j.envsci.2021.08.007
[8] UNDRR, "Terminology" 2017, Geneva. Accessed: Feb. 24, 2025 [Online]. Available: https://www.undrr.org/terminology/
[9] Badan Nasional Penanggulangan Bencana, "Peraturan Kepala BNPB Nomor 2 Tahun 2012 tentang Pedoman Umum Pengkajian Risiko Bencana," 2012.
[10] A. Munawir, S. Rewa, M. Taufik, Capability of local government organizations in handling natural disasters in Maros Regency. KnE Soc. Sci. 8, 95-107 (2023). https://doi.org/10.18502/kss.v8i17.14107

[11] S. Sanyal, J.K. Routray, Social capital for disaster risk reduction and management with empirical evidences from Sundarbans of India. Int. J. Disaster Risk Reduct. 19, 101-111 (2016). https://doi.org/10.1016/j.ijdrr.2016.08.010

[12] A. Al-Maruf, J.C. Jenkins, A. Bernzen, B. Braun, Human capital as a turnkey resource in resilience to cyclones and storm surges: empirical evidence from coastal Bangladesh. Mar. Pollut. Bull. 188, 1-15 (2023). https://doi.org/10.1016/j.marpolbul.2023.115721

[13] M.F. Hasan, S. Prasetya, Analisis tingkat kerawanan banjir di Bengawan Jero Kabupaten Lamongan. Swara Bhumi, 239-247 (2015)

[14] Z. Iba, A. Wardhana, Metode Penelitian (CV. Eureka Media Aksara, Purbalingga, 2023)

[15] J.F. Hair, A. Alamer, Partial least squares structural equation modeling (PLS-SEM) in second language and education research: guidelines using an applied example. Res. Methods Appl. Linguist. 1, 1-16 (2022). https://doi.org/10.1016/j.rmal.2022.100027

Implementation of Preparedness Efforts for Rip Current Hazard at Pangandaran Beach

Yessi Nirwana Kurniadi[1*], Mia Wimala[2], Resa Ardian Rivaldi[1], Muhamad Ridhan[1]

[1] Civil Engineering Department, Institut Teknologi Nasional Bandung. Jl. PHH Mustopa No 23 Bandung, Indonesia. Email: **yessi@itenas.ac.id*** , muhamadridhan0@gmail.com

[2] Civil Engineering Department, Universitas Katolik Parahyangan. Jl. Ciumbuleuit No.94, Bandung, Indonesia. Email: miasoejoso@unpar.ac.id

Abstract. Rip currents are powerful current that flow from the shore through the surf zone and out to deeper waters. These currents play a critical role in coastal dynamics but they are also of significant concern for beach safety, as they pose a threat to swimmers and beachgoers. The number of fatalities due to rip current every year at Pangandaran Beach raises the question of whether the beach management is ready to face this hazard. Therefore, this study aims to assess the preparedness efforts against rip current hazard in Pangandaran Beach. The benchmarks of preparedness that will be studied include trainings, campaigns, supporting instruments, and beach closure. This qualitative research will use data from interviews with BPBD, Balawista, and Basarnas of Pangandaran Beach, supported by data from field observation. The findings indicate four criteria and eighteen metrics for evaluating rip current preparedness, applicable to beach stakeholders. The stakeholders engaged in rip current hazard management at Pangandaran Beach have undertaken preparedness initiatives that can be categorized as adequately effective. The execution of both routine and non-routine training, the distribution of campaigns via several media, the acquisition of supporting equipment, and the closure of the beach have been completed to date. Nevertheless, feedback concerning the quantity and condition of warning signs, together with the enhancement of materials and platforms for distributing information specifically about rip currents, requires further refinement.

Keywords: Rip current, preparedness, hazard, beach, Pangandaran.

1 Introduction

Rip current is one of the most dangerous and unpredictable coastal hazards, leading to numerous drowning and injury annually. According to National Weather Service (NWS), rip currents are the most serious hazard to beach safety in the United States, as well as causing more deaths on average than lightning, tornadoes or hurricanes [1]. A rip current is a narrow, swift, seaward moving current that flows from the shore to the ocean. Triatmodjo (1999) states that rip currents arise from discrepancies along the shoreline and the elevation of breaking waves. A rip current comprises multiple components: the feeder current, the neck of the current, and the head of the current. The velocity of a rip current can attain 1 m/s and may potentially reach 2 m/s. This current can extend up to 61-762 meters length but typically has a width of less than 9 meters. The combination of waves, currents, and coastal morphology can generate rip currents. Rip currents frequently arise in shallow waters, where individuals are more like to swim. Nevertheless, the lack of awareness among several domestic and international visitors of beach safety and their lack of knowledge about rip currents leads to their ability to understand the hazard system or warning signs accurately. This increases the number of rip currents victim, particularly during the vacation season when tourist numbers peak. Rip currents have carried thousands of individuals out to sea, and in Australia, they are regarded as the primary hazard for beachgoers, accounting for 90% of maritime rescue incidents and instances of individuals being swept away. In Florida, rip currents account for 80% of sea rescues and are deemed a significant natural hazard [2], resulting in over 100 fatalities annually in the United States [3]. In England, the Royal National Lifeboat Institution (RNLI), which offers rescue services for 69% of the country's beaches, documented that 68% of maritime incidents from 2005 to 2007 were attributable to rip currents. During the period from 2005 to 2007, it was documented that 80% of beach incidents along the Atlantic coast of England were attributable to rip currents. Rip currents pose a significant threat due to their unpredictable patterns, which render them challenging to anticipate. Multiple elements contribute to the formation of rip currents, such as wave height, wave period, wind data, and tidal conditions.

Pangandaran Beach is a famous tourism spot located on a peninsula on the south coast of West Java, Indonesia. This beautiful beach is an attractive place for visitors due to natural landscape and water sport area such as swimming, surfing and other water activities. However, the currents and waves in Pangandaran coastal area at certain periods can be a potential danger for visitors. According to the Regional Board for Disaster Management (Badan Penanggulangan Bencana Daerah, BPBD) in Pangandaran Regency, there were 102 drowning victims during the holiday season in 2007 [4], and 10 beachgoers drown during the New Year holiday in 2010 [5] . During the 2016 holiday season, there were 50 accidents; in 2017, there were 20 accidents, resulting in one fatality. This underscores the significant threat posed by rip currents. Rip current accidents continue to occur in designated "safe areas" of the beach at Pangandaran, as these currents are unpredictable, and visitors either remain uninformed about this phenomenon or ignore the numerous warning signs posted along the beach [6]. The comprehension of rip currents by beachgoers and the readiness of beach managers regarding rip currents significantly contribute to reducing casualties at Pangandaran Beach.

Mucerino et al [7] stated that large and rapidly growing tourism activities in coastal environments impose an urgent need for both beach safety and risk zone at the sea. Local managers should provide information to beachgoers about rip current hazards to reduce the level of danger for visitors to Alassio beach, Italy. Locknick et al. [8] advocate for the regular annual documentation of rescues, aid, and drowning incidents, including precise times and places, to enhance data regarding beachgoers behaviour. The data can further be assessed to forecast the degree of rip current hazard [9]. The Korea Meteorological Administration has enhanced beach user safety by anticipating rip currents and supplying data from numerical models like HAECUM. Nevertheless, these initiatives are still inadequate to guarantee the protection of beachgoers. Effective hazard management strategies should focus on areas susceptible to rip currents, and comprehensive life-saving actions must be executed.

However no attempts have been made to assess the preparedness of stakeholders for rip current risk management in Pangandaran beach. Therefore, this research is focused to evaluate the current initiatives undertaken by stakeholders about the significance of management in hazard mitigation, as evidenced by multiple prior studies, as well as to develop various improvement strategies for future advantages.

2 Methodology

The qualitative research was mainly conducted through in-depth interviews with many stakeholders, including BPBD, Water Tourism Rescue Agency-Lifeguard (Badan Penyelamat Wisata Tirta, BALAWISTA) as the beach manager, and National Search and Rescue Agency (Badan Nasional Pencarian dan Pertolongan, BASARNAS). Six respondents comprise two individuals from each of the stakeholders. The interviews addressed the following subjects: job descriptions, current initiatives or programs, as well as the barriers and challenges encountered. Furthermore, a literature review of previous studies was employed to support this research, in conjunction with field observations conducted between July 2022 and December 2024 in prominent areas to achieve a more thorough understanding of the daily situations. The observations was performed at those two intervals due to the rise in visitors during the school holiday season, thereby enhancing the understanding of the preparedness of the stakeholders and tourists involved.

3 Case Study

Pangandaran Beach is a favored location for both local and international tourists that can be accessed by vehicle, train, or airplane. The journey from Jakarta, the capital of Indonesia, requires up to 7 hours by car or 1 hour by airline. The tourism infrastructure of Pangandaran Beach has been enhanced with numerous hotel alternatives, hostels, and pleasant homestays to ensure a comfortable stay for visitors. The coastline measures around 91 kilometers and the beach is oriented towards the Indian Ocean. The morphological structure along the southwestern coast of Java Island is highly intricate, comprising bays, headlands, shallow waters, and steep shorelines. The tourist areas are situated in many beach locations, namely Pantai Barat, Pantai Timur, Batu Karas, and Pasir Putih. Each beaches exhibits distinct characteristics, resulting in different functions. Pantai Barat is regarded as a secure location for swimming, Pantai Timur is allocated for fishing, Batu Karas is an optimal site for surfing, while Pasir Putih is characterized by its white sandy beach. Lifeguards are stationed at two locations which are in Pantai Barat Beach and Batu Karas Beach. This research will concentrate on Pantai Barat Beach, as it attracts more visitors than other beach areas. There are five lifeguard observation stations as show in **Figure 1**, positioned 400 meters apart from one another, except for the distance between posts 4 and 5, which is around 500-600 meters.

Figure 1 Lifeguard Stations at Pangandaran Beach (Google Earth, 2023)

4 Result and Discussion

4.1 Stakeholders of Hazard Management at Pangandaran Beach

In contrast to Search and Rescue (SAR) Barakuda, which oversees the fishermen of Pangandaran Beach, or the Pangandaran Marine and Air Police Unit that concentrates on criminal activities, the BPBD of Pangandaran Regency, BASARNAS, and BALAWISTA of Pangandaran Beach are the stakeholders that directly engage with tourists. Consequently, this research will concentrate on the last three parties. These three stakeholders play an important role related to hazzard preparedness, especially rip current. The policy makers have their respective roles whose ultimate goal is to provide safety to the public or tourists.

BPBD is a governmental body responsible for disaster management in certain regency or city in Indonesia. The vision is basically to create resilient communities to perpetually prepared to confront a disaster. Specifically, the objectives of BPBD of Pangandaran Regency are: 1) to enhance the disaster risk reduction system, 2) to improve disaster emergency services and outreach, and 3) to expedite post-disaster recovery services. In other hand, BASARNAS envisions the execution of efficient and dependable search and rescue operations. This concept seeks to instil a sense of security throughout the community regarding their actions. To achieve this aim, the BASARNAS has established numerous missions, including: 1) to maintain continuous vigilance in search and rescue operations, as well as in the evacuation of the public during accidents, disasters, and situations that pose a threat to human safety, with reliability, efficacy, speed, efficiency, and security, 2) to collaborate with national and international agencies/organizations to execute SAR operations and promote

SAR awareness to enhance SAR capabilities, 3) to enhance the organization's technical and managerial competencies and to perpetually advance, evolve, and refine all facets, 4) to implement capacity building and enhance human resource preparedness, together with ongoing coordination, to ensure optimal execution of search and rescue operations at all times, and 5) to supply operational facilities and infrastructure, communication apparatus, and SAR information systems as required to achieve the vision and goal. Meanwhile, BALAWISTA of Pangandaran Beach is an organization affliated with the Pangandaran Regency Tourism and Culture Office, and is directly responsible to the Pangandaran Regent. The vision of Balawista itself is *We Are There For You - We Are Nothing Without You*. The mission of Balawista itself is to save lives on the beach and other marine tourism through education, prevention and rescue services. Members of Balawista are lifeguards tasked with supervising and rescuing, ensuring a sense of security and comfort for tourists at Pangandaran Beach. BASARNAS and BALAWISTA has distinct focuses and responsibilities, although both are vital to disaster response and emergency relief, complementing one another. BASARNAS oversees extensive search and rescue operations encompassing natural disasters, transportation incidents, and various emergencies across Indonesia, whereas BALAWISTA concentrates on the rescue and monitoring of tourists at beaches and recreational areas, primarily aimed at mitigating accident risks in these locations. BALAWISTA also engages in socialization and safety education for tourists to mitigate the risk of accidents in tourist locales, particularly those associated with water activities or dangerous outdoor activities.

4.2 Identification of Parameter for Hazard Preparedness

Table 1 presents a few of hazard preparedness parameters that have been effectively determined from diverse sources or experts, including from Indonesian National Agency for for Disaster Management (Badan Nasional Penanggulangan Bencana, BNPB) (2008), Mucerino et al. (2021) [7], Litzinger and Leatherman (2021) , Llopis et al. (2018) [10], Locknick and Houser (2021) [8], and Lee et al. (2016) [9].

Table 1 Parameter *for Hazard Preparedness*

No.	Criteria	Parameter	Reference
1	Trainings	a) Activation of disaster preparedness units along with all its supporting components	BNPB (2008)
		b) Preparedness training, simulations, drills, and technical exercises for each sector, including disaster response (SAR, social services, health, infrastructure, and public works)	
		c) Inventory of emergency support supplies	
2	Campaigns	a) Preparation of assistance and mobilization of resources and logistics	
		b) Development of a swift and cohesive information and communication system to facilitate disaster operations	
		c) Establishment and deployment of early warning system instruments	
		d) Formulation of a contingency plan	
		e) Mobilization of resources (people and infrastructure/equipment).	
		f) Provision of information on the dangers of rip currents to beachgoers	Mucerino et al (2021)
3	Supporting Instruments	a) Procurement of safety signs	Mucerino et al (2021)
		b) Improvement of lifeguard skills	

Table 1 Parameter *for Hazard Preparedness*

No.	Criteria	Parameter	Reference
		c) Provision of ring buoys and guidelines for their deployment	Litzinger & Leatherman, (2021)
		d) Warning signs redesign to include instructions for escape	Llopis Et al (2018)
4	Beach Closure	a) Consistent annual reporting on rescues, assistance, and sinkings with information on specific times and locations	Locknick & Houser (2021)
		b) Allocation of warning signs and flags	
		c) Placement of beach access points for swimming and surfing	
		d) Alert regarding the rip current hazards in beach promotion	
		e) Beach closure (due to potential hurricane and elevated tidal conditions)	Lee et al., (2016)

4.2.1 Training

The American Academy of Paediatrics (AAP) Committee on Injury and Poison Prevention Drowning, an autonomous organization dedicated to water safety and security, asserts that drowning ranks as the fourth leading cause of unintentional death. Annually, about 4,000 individuals succumb to drowning, with one-third being children under 14 years of age. Proficiency in swimming is crucial for mitigating the risk of drowning. Likewise, for individuals engaged in rip current preparedness, swimming proficiency is essential for rescuing people at sea. Consequently, a primary training activity undertaken by the three parties is swimming training. Given BPBD's primary function of supporting BASARNAS and BALAWISTA, swimming training and additional programs will be organized upon request. In contrast to BPBD, the training has frequently been executed by BASARNAS and BALAWISTA in a systematic manner and is planned on an annual basis. Trainings are largely conducted to enhance skill and ability in managing various emergency situations, especially those related to rescue, assistance, and disasters. Standard training sessions encompass SAR training, natural disaster management, maritime and terrestrial rescue operations, SAR operation management, health and first aid, mapping, and the application of technology including drones and GIS, as well as simulations and collaborative exercises between the two teams. Training is compulsory for all new members, whereas standard training, such as swimming and first aid, is advisable for volunteers.

4.2.2 Rescue and Evacuation

Knowledge regarding the rescue and evacuation of victims is a key sign of preparedness against rip currents. The subsequent document delineates the Standard Operating Procedures (SOP) for the rescue phase executed by the three interconnected entities:

1. Rapid response
   Lifeguards will promptly intervene to rescue drowning victims using supportive equipment, including a rescue tube and spine board. Should the victim lose consciousness, the lifeguard personnel will collaborate with one another.
2. A jet ski is employed when the sufferer is situated at a significant distance.
3. In the event of a reported drowning victim, the lifeguard will inquire about the reporter's name and the sequence of events pertaining to the occurrence. This reporting also pertains to BALAWISTA (lifeguards) who execute rescue operations.
4. If the victim sustains minor injuries, the BALAWISTA team will provide urgent assistance; if the victim incurs major injuries, they will be swiftly transported to the nearest hospital.

5. Should the victim remain unfound after 7 days, the BALAWISTA will collaborate with BASARNAS to initiate a search for the victim. BASARNAS will work alongside with BPBD to secure evacuation routes, provide emergency medical assistance, and implement broader disaster management if required.
6. In the event of a drowning occurrence occurring outside of working hours, the BALAWISTA will still perform a rescue if notified by the public.
7. Throughout the vacation season, the BALAWISTA will partner with BASARNAS and volunteers for beach patrols. Volunteers must have undergone training and possess swimming proficiency.

BALAWISTA must possess the capacity for self-recognition while rendering support. In the event of an accident involving a victim being carried away by a rip current, lifeguards, namely BALAWISTA members, must assess their capabilities to see if they can directly assist the victim. If they are unable to execute the rescue or feel apprehensive, a lifeguard must initiate an emergency call or solicit immediate aid from fellow lifeguards. Furthermore, lifeguards must be capable of recognizing indicators that a victim requires aid. Five lifeguard stations positioned around the Pangandaran Beach are essential to provide guidance to guests swimming in the designated area. In the event of an accident or emergency, BALAWISTA members coordinate among watch posts, activate the danger siren, and notify people via loudspeakers to vacate the beach, thereby aiding the rescue operation. Direct statements from the BALAWISTA indicate that the readiness and cooperation for drowning victims are commendable. The lifeguards will collaborate to rescue the victim. The collaboration among the lookout guards dictates whether they can independently execute the rescue. When it cannot be managed independently, BALAWISTA will facilitate the involvement of BASARNAS to go to the site. Subsequently, first aid and evacuation of the casualties must be executed in accordance with the existing conditions. If the victim is beyond rescue, which is often the case, BASARNAS will place them in a corpse bag provided at the guard post and transport them to the hospital for an autopsy. The assertions from both parties suggest that they are deemed to have comprehended the management of drowning victims through the training provided.

4.2.3 Supporting Equipment

Interviews with the three parties reveal a summary of their supporting equipment, as detailed in **Table 2**.

Table 2 Supporting Equipment

No	Supporting Equipment	BPBD	BASARNAS	BALAWISTA
1	Buoy	✓	✓	✓
2	Lifeguard Car	✓	✓	✓
3	Speed Boat	✓	✓	
4	First Aid Kit	✓		
5	Drone	✓		
6	Safety Cap		✓	
7	Aqua Eye		✓	
8	Body Bag		✓	
9	SAR Boat		✓	
10	Snorkle Set		✓	✓
11	Whistle			✓
12	Surfboard			✓
13	Inflatable Boat	✓	✓	

4.2.4 Rip current Hazard Sign

The BPBD, BASARNAS, and BALAWISTA collaborated to develop a unified program aimed at conducting a campaign and disseminating information to the public. They recognize the significance of this campaign to inform the public and visitors about the attributes and hazards of this stream. Online campaigns can be executed using social media, print media, and electronic media. Simultaneously, direct campaigns may be executed through various initiatives aimed at educating beachgoers or the community. The interviews with BPBD, BASARNAS, and BALAWISTA indicate that the web campaign has been insufficiently effective due to its failure to reach all media outlets. Furthermore, the educational materials supplied are inadequate for effectively reaching the audience. Campaigns performed directly are distinct. They will be able to immediately access several beachgoers present at the place during that time. The campaign's readiness requirements involve redesigning warning signs to include escape instructions and highlighting the dangers of rip currents.

Indonesia National Disaster Management Agency (BNPB) in has a regulation on signs and disaster information boards, disaster signs consist of guidance signs, warning signs, prohibition signs and information signs. The signs explain the types of disaster signs used for types of disasters such as earthquakes, tsunamis, volcanic eruptions, land movements, floods, and forest fires. However, for rip current hazard, there are no regulated signs because it has not been categorized as a disaster.

At Pantai Barat Pangandaran there are several signs for tsunami evacuation, assembly point sign if tsunami occur, red flag to indicate rip current and "no-swimming warning" board. The red flag is a non permanent sign consist of long stick and red flag with "Danger current" sign. Lifeguard (Balawista) will place this red flag if there is a rip current at their observation post. **Figure 2** shows red flag and warning sign condition

Figure 2 Red flag and warning sign condition

From the interviews with lifeguards that prohibitory signs and warning signs are needed more than information signs. Therefore, the design of prohibition signs will be recommended as well as warning signs. These warning signs were developed from rip current information image in National Weather Service, National Oceanic and Atmospheric Administration (NOAA) . The images present multiple indicators for recognizing rip currents such as 1) A narrow channel of darker, calmer water surrounded by breaking waves and foam; 2) A rotating water current distinct from the surrounding water; 3) A difference in water color, such as an area of muddy-appearing water, and 4) A consistent area of foam or seaweed being carried through the surf. This design also adapt from traffic sign to ensure people about the importance of this information.

a. "No Swimming" Warning Sign
This recommendation warning sign based on BNPB regulation No 7 Year 2015 about sign and disaster information sign and Ministry of transportation regulation No 13 Year 2014 on traffic sign. These information consist of color, dimension, font type and also material with warning sign in bahasa Indonesia "Dilarang Berenang, Arus Kuat" which means "No swimming, strong current" as shown in **Table 3** and **Figure 3**.

b. Rip Current Hazard Sign
This sign based on NOAA rip current sign which indicates rip current can moves people away from shore. The yellow color contains warnings of possible dangers or vulnerable places. To make it easier to read, yellow is the basic color of this sign. For the text and images use black color so that it is easier to see.

c. Rip Current Information Sign
This recommended sign consist of fact about rip current, how rip current form in the nearshore, rip current speeds and the danger of this current in Bahasa Indonesia.

https://doi.org/10.1051/bioconf/202621612002

Table 3 "No Swimming" Warning Sign Dimension

"No Swimming" Warning Sign	BNPB Regulation No. 07 Year 2015	Ministry of Transportation Regulation No. 13 Year 2014	Recommendation design
Color	Red	Red	Red
Board Sign Length	120 – 160 cm	120 – 160 cm	120 cm
Board sign width	160 – 400 cm	160 – 400 cm	200 cm
The sign pole height	-	175 - 265 cm	175 cm
Font type	Clearview Highway	Clearview Highway	Clearview Highway
Sign pole material	-	-	Galvanized Iron
Sign material	-	Almunium sheet	Almunium sheet

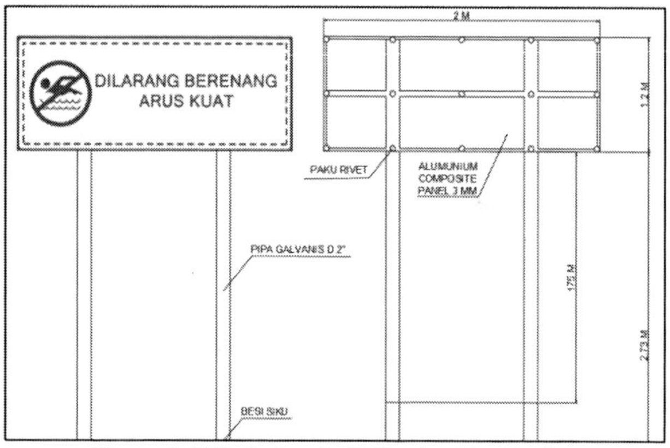

Figure 3 ""No Swimming" Warning Sign Design

Table 4 Rip Current Hazard Sign Dimension

Rip Current Hazard Sign	BNPB Regulation No. 07 Year 2015	Ministry of Transportation Regulation No. 13 Year 2014	Recommendation design
Color	Yellow	Yellow	Yellow
Board Sign Dimension	45 – 90 cm	45 – 90 cm	60 cm
The sign pole height	-	175 – 265 cm	175 cm
Font type	Clearview Highway	Clearview Highway	Clearview Highway
Sign pole material	-	-	Galvanized Iron
Sign material	-	Almunium sheet	Almunium sheet

Table 5 Rip Current Information Sign Dimension

Rip Current Information Sign	BNPB Regulation No. 07 Year 2015	Ministry of Transportation Regulation No. 13 Year 2014	Recommendation design
Color	Yellow	Yellow	Yellow
Board Sign Length	140 cm	140 cm	140 cm
Board Sign Width	105 cm	105 cm	105 cm
The sign pole height	-	175 - 265 cm	175 cm
Font type	Clearview Highway	Clearview Highway	Clearview Highway
Sign pole material	-	-	Galvanized Iron
Sign material	-	Aluminium sheet	Aluminium sheet

(a) (b)

Figure 4 (a) Rip Current Hazard Sign and (b) Rip Current Information Sign

4.2.5 Beach Closure

Beach closure is an action to closing the beach that must be taken if there are hazard in coastal areas. This is done to maintain public safety. The proposed beach closing is to consider the possibility of high waves that result in many rip currents along the coast even for the beach that is considered safe will become dangerous. This beach closing is necessary to urge tourists not to swim in the beach area due to high wave conditions. Wave height is a key factor in the probable formation of a rip current. Wave height can be measured directly or assessed using existing wind data. The quantity of wave reflections manifested as bays and depressions serves as the primary indicator of rip current formation. Concerning the beach closure, the three involved parties often coordinate with multiple media outlets, including print, online, and electronic, to notify the public about the closure within a specified timeframe, supported by compelling justifications. It is anticipated that the public would exercise greater caution when visiting the beach in the future, as natural occurrences can arise unexpectedly.

In general, the implementation of rip current preparedness efforts at Pangandaran Beach can be classified as adequate. The evaluation results are summarized in **Table 6**.

Table 6 Evaluation Results

No.	Criteria	Parameter	Description
1	Trainings	a) Activation of hazard preparedness units along with all its supporting components	Available
		b) Preparedness training, simulations, drills, and technical exercises for each sector, including disaster response (SAR, social services, health, infrastructure, and public works)	Available
		c) Inventory of emergency support supplies	Available
2	Campaigns	a) Preparation of assistance and mobilization of resources and logistics	Available
		b) Development of a swift and cohesive information and communication system to facilitate disaster operations	Re-design
		c) Establishment and deployment of early warning system instruments	Re-design
		d) Formulation of a contingency plan	Available
		e) Mobilization of resources (people and infrastructure/equipment).	Available
		f) Provision of information on the dangers of rip currents to beachgoers	Re-design
3	Supporting Instruments	a) Procurement of safety signs	Re-design
		b) Improvement of lifeguard skills	Available
		c) Provision of ring buoys and guidelines for their deployment	Available
		d) Warning signs redesign to include instructions for escape	Re-design
4	Beach Closure	a) Consistent annual reporting on rescues, assistance, and sinkings with information on specific times and locations	Available
		b) Allocation of warning signs and flags	Re-design
		c) Placement of beach access points for swimming and surfing	Available
		d) Alert regarding the rip current hazards in beach promotion	Available
		e) Beach closure (due to potential hurricane and elevated tidal conditions)	Available

5 Conclusion

This research successfully formulated four criteria and 18 assessment parameters for rip current preparedness by relevant stakeholders, namely BPBD, BASARNAS, and BALAWISTA. Furthermore, an overview of the execution of rip current hazard management can be acquired through interviews with the three relevant parties. All parties are cognizant of and comprehend their individual roles and the interrelations among them. They have undertaken several initiatives to date, including fundamental training, engagement in self-improvement programs, acquisition of supportive materials, and installation of warning signs along the beach, despite their state being suboptimal or partially compromised. General campaigns about beach safety efforts have also been publicized through various print and electronic media. Unfortunately, the program has failed to address the hazards of rip currents in specific way. Consequently, several enhancement initiatives may focus on improving the campaign materials and their distribution channels, in addition to increasing the number of warning signs along the beach. The assessment of preparedness initiatives against rip currents conducted by all stakeholders can be categorized as adequate.

Acknowledgement

This research is fully funded by Lembaga Penelitian dan Pengabdian kepada Masyarakat (LP2M) ITENAS and in collaboration with Parahyangan University (UNPAR) Bandung. Many thanks to Pangandaran Coast Guard (BALAWISTA), Department of Tourism and Culture West Java (Disparbud), and Regional Disaster Management Agency Pangandaran (Badan Penanggulangan Bencana Daerah) for research permit in Pangandaran area.

References

[1] S. B. L. Stephen P. Leatherman, "Rip Currents: A Major Coastal Hazard and Public Safety Challenge," in *Solution to Coastal Disaster*, 2011.

[2] E. B. T. A. J. R. Jamie H. MacMahan, "Rip current review," *Coastal Engineering*, vol. 53, no. 2-3, pp. 191-208, 2006.

[3] A. R. J. B. Jamie H. Macmahan, "An Introduction to Rip Currents Based on Field Observations," *Journal of Coastal Research*, p. 27, 2011.

[4] "102 Pengunjung Pangandaran Terseret Ombak," Republika, 2007.

[5] Krisdiantoro, " 10 Wisatawan Pangandaran Terseret Ombak".

[6] W. B. S. a. E. Kusmanto, "Arus Rip di Perairan Pesisir Pangandaran, Jawa Barat," *Ilmu Kelautan*, 2013.

[7] L. C. L. S. C. F. P. E. S. E. &. F. M. Mucerino, "Rip current hazard assessment on a sandy beach in Liguria, NW Mediterranean," *Natural Hazards*, pp. 137-156, 2021.

[8] S. &. H. C. Locknick, "Correspondence of Beach User Perception, Lifesaving Strategies and Rip Currents: Implications for Beach Management," *Coastal Management*, vol. 49, 2021.

[9] G. H. S. L. C. K. J. &. L. J. Lee, "Rip current zoning map to manage safety at Haeundae Beach, South Korea," *Journal of Coastal Research*, 2016.

[10] I. A. E. A. G. T. S. B. C. &. H. C. Llopis, "Determining beach user knowledge of rip currents in Costa Rica," *Journal of Coastal Research*, vol. 34, 2018.

Evaluating the Effect of Constructed Wetland and Microbial Fuel Cell (CW-MFC) in Wastewater Treatment with Floating and Emergent Plant-Based Systems

Bara Awanda Marhendra[1], *Isni* Arliyani[1*], *Afrinda* Dwi Wahyuni[1], *Fahmi* Ikhlasul Amalludin[1], *Eddy* Setiadi Soedjono[1], *Adhi* Yuniarto[1], *Bowo* Djoko Marsono[1]

[1]Department of Environmental Engineering, Faculty of Civil, Planning, and Geo-Engineering, Institute of Technology Sepuluh Nopember Surabaya, Surabaya 60111/ East Java, Indonesia

Abstract. Constructed Wetlands and Microbial Fuel Cells represent two promising low-energy technologies for domestic wastewater treatment. Their integration has the potential to enhance both pollutant removal and bioelectricity generation through combined microbial, plant-based, and electrochemical processes. This study evaluated the performance of an integrated CWMFC system operated with two plant configurations, namely the emergent species Equisetum hyemale and the floating species Pistia stratiotes. Five reactor configurations were examined: CW MFC with aeration and bacterial augmentation (CWMFC AB), CW MFC with bacterial augmentation (CWMFC B), CW MFC without augmentation (CWMFC), MFC only system, and CW only system. Each reactor with a volume of 9 liters was operated in batch mode for 18 d using domestic greywater, with monitoring conducted every three days. The CWMFC AB configuration consistently produced the greatest reductions in BOD, COD, and TAN. Emergent plant systems demonstrated stronger TAN removal than floating plant systems. Bioelectricity generation reached a maximum of 794.7 milliwatts per square meter in the emergent CWMFC AB reactor, whereas floating systems produced up to 196.5 milliwatts per square meter. These results indicate that the combined CW MFC approach supported by aeration and Lactobacillus plantarum augmentation, offers an effective strategy for decentralized wastewater treatment and energy recovery.
Keyword: Constructed wetland; Microbial fuel cell; Emergent plants; Floating plants; Domestic wastewater; Bioelectricity generation

1 Introduction

Domestic wastewater in coastal cities, such as Surabaya in Indonesia, contains a complex mixture of pollutants originating from household activities. Typical contaminant groups include organic matter, nutrients, detergents, suspended solids, and microbial loads, which collectively elevate key parameters such as Chemical Oxygen Demand, Biochemical Oxygen Demand, Total Ammonia Nitrogen, and fluctuations in pH [1]. When discharged without adequate treatment, these pollutants accelerate eutrophication, degrade aquatic ecosystems, and pose risks to coastal environments that are already vulnerable to tidal influences and limited freshwater exchanges. The increasing population density in suburban and peri-urban districts further exacerbates these challenges, creating a pressing need for decentralized, low-energy, and adaptive treatment systems suitable for domestic wastewater management in coastal urban settings [2].

Constructed Wetlands have emerged as an attractive alternative because they operate using natural treatment mechanisms involving plants, microbial biofilms, and supporting media. CWs can reduce organic matter, stabilize pH, and transform ammonia through plant-assisted nitrification and microbial pathways. Their advantages include low operational costs, minimal energy demand, and high resilience to variable influent quality. However, CWs can be

* Corresponding author: isniarliyani@its.ac.id

limited by slow degradation rates, low oxygen transfer, and inconsistent performance when treating wastewater with elevated organic or nutrient loads, which are common in domestic greywater.

Microbial Fuel Cells represent another innovative low-energy treatment option, in which electroactive bacteria oxidize organic matter and generate an electric current as electrons move through an external circuit. MFCs have demonstrated the ability to reduce COD and BOD while simultaneously producing renewable electricity. Despite this potential, standalone MFC systems often suffer from limited power output, sensitivity to environmental conditions, and unstable, long-term operation. Consequently, their application in the treatment of real domestic wastewater remains constrained.

The integration of Constructed Wetlands (CWs) with Microbial Fuel Cells (MFCs) has garnered significant attention as a strategy to leverage the complementary strengths of both technologies. As suggested by several studies, the combined CW-MFC system can enhance pollutant degradation by coupling plant-mediated oxygenation with electrochemical microbial processes, potentially improving Chemical Oxygen Demand (COD), Biochemical Oxygen Demand (BOD), Total Ammonia Nitrogen (TAN), and pH stabilization simultaneously [3]. Consequently, CW-MFC modules with vegetation demonstrated higher treatment efficiency than unplanted wetland modules, with average COD, NH4 +, and TP removal efficiencies ranging from around 80-90% [4] [5]. However, it remains unclear whether these systems consistently enhance each other or whether operational interactions may restrict performance. The presence of plant roots may alter electrode aeration, biofilm development, and electron transfer patterns, whereas the electrochemical environment in the MFC may influence rhizosphere microbial communities. These uncertainties underline the need for a systematic evaluation of integrated system dynamics under real wastewater conditions.

Furthermore, while emergent and floating plants are frequently utilized in constructed wetland (CW) applications, their specific impact on the performance of integrated CW microbial fuel cells (MFCs) has not been thoroughly examined. Research suggests that various types of vegetation demonstrate differing effectiveness in the removal of contaminants from wastewater, with the emergent plant Typha angustifolia exhibiting superior removal capabilities compared to three other emergent plant species [2]. Emergent plants with solid root structures may provide better oxygen diffusion, whereas floating plants offer larger surface area coverage but limited rhizosphere depth. In addition, the role of bacterial augmentation in enhancing CW processes, MFC activity, or both components collectively remains inadequately explored. Determining whether augmentation enhances treatment efficiency or primarily benefits only one subsystem is crucial for optimizing the design of the combined CW MFC reactor.

This study evaluated these knowledge gaps by assessing organic matter removal, ammonia transformation, pH regulation, and electricity production in an integrated CW MFC system operated with emergent and floating plant-based configurations. The experimental approach compared individual and combined systems to determine whether integration yields a synergistic effect or introduces operational constraints under real domestic wastewater conditions.

2 Materials and Methods

This study was conducted at the Environmental Engineering Laboratory of Institut Teknologi Sepuluh Nopember (ITS), Surabaya, in October 2024. The methodological framework was designed to evaluate the treatment efficiency and electricity generation of combined constructed wetland and Microbial Fuel Cell systems under controlled laboratory conditions. The procedures encompassed domestic wastewater characterization, plant propagation and acclimatization, bacterial culture preparation, and the construction and operation of CW MFC reactors. Each component of the methodology was structured to ensure consistent operating conditions and enable systematic comparisons across reactor configurations, including both individual and integrated systems.

2.1 Domestic Wastewater Characterisation

Domestic greywater was collected from a residential complex in Sedati District, Sidoarjo, East Java. Sampling followed the Indonesian National Standard SNI 8990-2021 regarding methods for sampling wastewater for physical and chemical analysis. Initial characterization showed that the wastewater exceeded the quality standards set by the Regulation of the Minister of Environment and Forestry (Permen LHK) No. 68 of 2016. The key parameters measured were pH, Biochemical Oxygen Demand (BOD), Chemical Oxygen Demand (COD), Total Suspended Solids (TSS), and Total Ammonia Nitrogen (TAN). The wastewater exhibited high organic loads and alkaline conditions, with COD and BOD values far above the permissible limits, indicating the presence of both biodegradable and recalcitrant pollutants. Ammonia concentrations also exceeded regulatory thresholds, suggesting a high risk of eutrophication if it is discharged without treatment. These characteristics underscore the need for effective biological and electrochemical treatment approaches.

Table 1. Domestic Wastewater Concentration.

Parameters	Unit	Domestic Wastewater Concentration	Maximum Concentration	Description
BOD_5	mg/L	912	30	Non Compliant
COD	mg/L	4.096	100	Non Compliant
TSS	mg/L	50,55	30	Non Compliant
Ammonia	mg/L	36	10	Non Compliant
pH	-	9,62	6-7	Non Compliant

2.2 Plant Propagation and Acclimatization

Prior to their use in the reactors, the selected plant species were propagated and acclimatized to controlled laboratory conditions. *Pistia stratiotes* (floating plant) and *Equisetum hyemale* (emergent plant) were chosen based on their robustness, pollutant uptake capacity, and aesthetic potential for future landscape-integrated CW-MFC applications. Propagation involved encouraging vegetative growth until sufficient healthy shoots were obtained for the experiment. Acclimatization was conducted in clean water for one week, during which the plant condition was monitored based on leaf turgidity, stem structure, color, and absence of necrotic symptoms. A range-finding test using wastewater concentrations of 10%, 25%, 50%, 75%, and 100% was conducted to evaluate plant tolerance. Both species exhibited optimal performance at a wastewater concentration of 25 %, which was selected as the influent concentration for reactor operation.

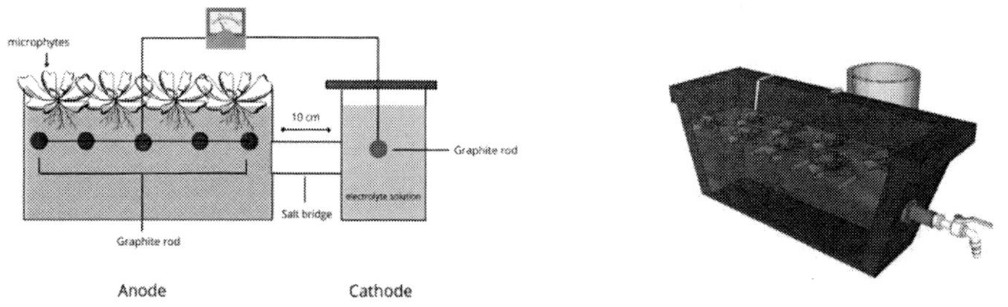

Fig. 1. Reactor Design; (left) Schematic Configuration of CW-MFC Reactor; 3D configurations of CW-MFC Reactors

2.3 Bacterial Culture

Lactobacillus plantarum was used as the bacterial augmentation agent. Cultivation began by inoculating sterile MRS broth with bacterial colonies and incubating under anaerobic conditions at 37°C for 18–24 h. The culture was then scaled up by transferring the starter into a larger volume of MRS broth and incubating further until the concentration reached approximately 1x108 CFU/mL, as verified through optical density measurements and a standard calibration curve. This inoculum was added to the designated reactors to evaluate its influence on pollutant removal and power generation.

2.4 Reactor Configuration

The CW MFC system consisted of laboratory-scale batch reactors, as shown in fig. 1. to evaluate the combined and individual performance of the CW and MFC components. Each reactor measured 50 cm × 24 cm × 20 cm with a working volume of 9 L for floating systems and 7.5 L for emergent systems, the latter supplemented with a 2.5 L gravel substrate layer. The substrate comprised two gravel layers: a 4 cm layer with a particle size of 1–2 cm and a 3 cm layer with a particle size of 3–5 cm. Gravel was used to support biofilm development and root–microbe interaction.

Graphite electrodes (15 cm length) and a 10 cm salt bridge containing 1 M KCl in agar were used as the electrochemical components. The electrodes were pretreated with 1 M HCl, followed by 1 M NaOH, to remove contaminants. The anode and cathode chambers were separated by a PVC pipe housing the salt bridge, and NaCl served as an electrolyte. The voltage and current outputs were monitored using a digital multimeter.

Five reactor configurations were used for each plant type (floating and emergent), resulting in ten reactors.

1. CWMFC AB: Integrated CW MFC with aeration and bacterial augmentation
2. CWMFC B: Integrated CW MFC with bacterial augmentation but no aeration
3. CWMFC: Integrated CW MFC without aeration and without bacterial augmentation
4. MFC: MFC-only system (control 1)
5. CW: CW-only system (control 2)

The reactors were operated in batch mode for 18 d, with sampling conducted every three days (Days 0, 3, 6, 9, 12, 15, and 18). Environmental conditions were maintained in an indoor greenhouse with a stable temperature, natural light exposure, and no rainfall.

All wastewater quality analyses in this study were conducted using nationally recognized Indonesian standards, referring to the Standar Nasional Indonesia (SNI) for physical and chemical testing. These methods were applied consistently across all sampling intervals to ensure the accuracy and comparability of the data. The measured parameters and their corresponding analytical standards are as follows:

The pH of the wastewater was measured according to SNI 06-6989.11-2004 using a calibrated digital pH meter equipped with a glass electrode. The Biochemical Oxygen Demand (BOD_5) test followed SNI 6989.72:2009, with samples incubated at 20°C for five days before measuring the dissolved oxygen (DO). Chemical Oxygen Demand (COD) analysis was performed following SNI 6989.73:2009, using the closed reflux dichromate method with spectrophotometric detection. The concentration of Total Ammonia Nitrogen (TAN) was determined based on SNI 6989.30:2005. Electrical performance, including voltage and current generation in the MFC components, were recorded twice daily using a digital multimeter, and power density (mW/m^2) was calculated using the formula,

$$P=V^2/(R\times A) \quad (1)$$

where V is the measured voltage, R is the external resistance, and A is the projected anode surface area.

3 Result and Discussion

3.1 pH Dynamics Across Reactor Configurations

The initial pH of the domestic wastewater was highly alkaline at 9.62, which exceeded the optimal range for most biological treatment processes. Across all reactors, the pH declined progressively during the eighteen day operation period, reflecting biological acidification associated with organic degradation and nitrogen transformation processes [4]. However, the rate and magnitude of this decline differed substantially between the reactor configurations and plant types.

In the emergent plant systems, the pH decreased rapidly during the first nine days, reaching values between 6.77 and 7.92 by day 12. The lowest pH was observed in CWMFC B, which dropped from 9.62 to 6.41 on day 18. This substantial reduction reflects the strong oxygen transport capacity of *Equisetum hyemale* as an emergent plant that promotes nitrifying bacterial activity through root zone aeration. Enhanced nitrification leads to proton generation and subsequent pH reduction, explaining the steeper decline compared to that in floating systems. The CWMFC AB reactor also showed a stabilized pH close to neutrality (6.78) by day 18, indicating that aeration combined with microbial augmentation supported balanced microbial metabolism and mitigated excessive alkalinity.

The MFC reactor exhibited the highest pH fluctuation, consistent with earlier studies showing that MFC performance is strongly dependent on electron acceptor availability at the cathode, which can vary significantly in plant-free systems [6]. In the absence of plant-mediated oxygenation, cathodic reactions may proceed less consistently, and electroactive biofilms develop more slowly, contributing to less stable internal biochemical conditions. In contrast, the integrated CW-MFC systems benefited from plant-supported oxygen transfer and enhanced microbial activity, which promoted more stable pH trajectories.

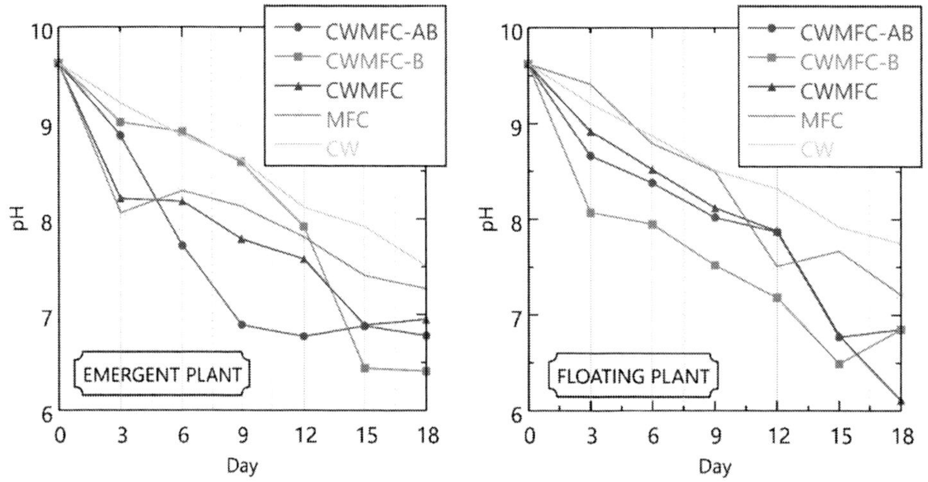

Fig. 2. pH dynamics over time for (left) emergent plants and (right) floating plants.

3.2 Biochemical Oxygen Demand Removal

BOD concentrations declined steadily in all reactor configurations during the 18 days of operation, although the extent of reduction varied across the treatment systems and plant types. In the emergent plant units, the integrated CW-MFC configurations achieved the highest levels of organic pollutant removal. The CWMFC AB system reduced BOD from 228 mg/L to 20 mg/L by day 18, representing approximately 91% removal. This superior performance is associated with the combined influence of aeration, bacterial augmentation, and the development of electroactive biofilms, which together strengthen both aerobic and anaerobic degradation pathways, as previously reported in related studies [7]. The CWMFC B reactor followed with a final concentration of 24 mg/L, while the non-augmented CWMFC system exhibited slower degradation, ending at 116 mg/L. The individual MFC and CW reactors showed comparatively limited reductions, reaching 112 mg/L and 100 mg/L, respectively, further indicating that neither technology alone was able to achieve substantial BOD removal under the tested conditions.

In the floating plant systems, all reactors also showed declining BOD trends, although the overall performance was lower than that observed in the emergent systems. The CWMFC AB configuration again produced the most pronounced decrease, reaching a final BOD concentration of 28 mg/L. This result was followed by CWMFC B at 40 mg/L and the non-augmented CWMFC system at 112 mg/L, respectively. The floating MFC and CW controls remained above 100 mg/L at the end of the experiment, confirming their limited efficiency in the absence of system integration. The slower rates observed in floating systems may be linked to the lack of gravel media, which restricts biofilm formation, and the relatively low radial oxygen release from *Pistia stratiotes* roots. These limitations reduce the availability of aerobic microhabitats and weaken the early establishment of effective microbial communities.

Emergent plant systems consistently outperformed floating systems in all reactor variants. This improved performance is likely supported by the extensive rhizosphere of *Equisetum hyemale*, which promotes greater oxygen transfer, provides larger microbial attachment surfaces, and facilitates diverse pollutant degradation pathways. Floating plants produce a shallower and less oxygenated root zone, which reduces the intensity of plant–microbe interactions and slows organic matter degradation. Despite these constraints, the floating CWMFC AB configuration achieved substantial improvement compared to the floating control systems, demonstrating that aeration and bacterial augmentation can compensate for the physiological limitations associated with floating vegetation.

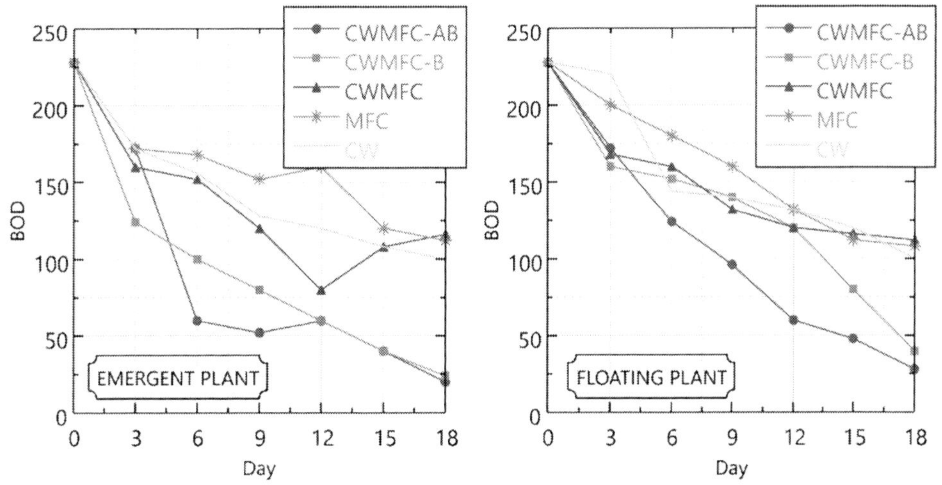

Fig. 3. BOD removal over time for (left) Emergent Plant and (right) Floating Plant.

The treatment hierarchy was consistent for both plant categories, with CWMFC AB providing the strongest removal efficiency, followed by CWMFC B, CWMFC, and finally, the single technology reactors. This performance pattern reinforces the positive interaction between constructed wetland processes and the mechanisms of microbial fuel cells. Constructed wetlands enhance oxygen diffusion and root exudate supply, whereas microbial fuel cells support electron transfer and sustained bioelectrochemical oxidation. The improved outcomes observed in the augmented and aerated reactors align with the existing literature, showing that enriched microbial communities and improved oxygen availability accelerate organic pollutant degradation and strengthen electroactive consortia [8]. The findings confirm that the integrated CW-MFC concept delivers significantly higher BOD removal than either system operating independently, and that appropriate operational enhancements can further maximize treatment performance.

3.3 Chemical Oxygen Demand Removal

COD concentrations decreased steadily in all reactor configurations across the 18-day operational period. The initial COD value of 1024 mg/L was identical for every reactor, which allowed a direct comparison of the treatment performance between the individual and integrated systems. The observed trends highlight the contributions of plant-mediated processes, microbial activity, and bioelectrochemical reactions to organic pollutant degradation in both emergent and floating plant configurations.

In the emergent plant systems, the most substantial COD reduction occurred in the CWMFC AB. The COD declined from 1024 mg/L to 89.6 mg/L on day 18, indicating an overall removal efficiency of approximately 91 percent. This outcome demonstrates the combined effect of aeration, bacterial augmentation, and MFC-mediated electron transfer in accelerating the breakdown of organic matter, which has also been reported in previous coupled CW and MFC studies. The CWMFC B reactor had a final COD concentration of 128 mg/L, confirming that bacterial enrichment notably contributed to the system performance. The non-augmented CWMFC reactor achieved a slower decline, reaching 396.8 mg/L, whereas the MFC only and CW only systems ended at 371.2 mg/L and 358.4 mg/L, respectively. These results highlight that neither wetlands nor MFCs alone are sufficient to achieve high COD removal within the tested time frame.

A similar trend was observed in the floating plant reactors. The CWMFC AB configuration again produced the highest COD reduction, with a final concentration of 89.6 mg/L. The CWMFC B reactor followed with 153.6 mg/L on day 18, reflecting the influence of bacterial augmentation on strengthening organic matter oxidation. The non-augmented CWMFC reactor remained at 409.6 mg/L, whereas the MFC only and CW only systems performed comparably, finishing at 396.8 mg/L and 384 mg/L, respectively. These findings indicate that floating systems are capable of substantial pollutant removal, although the overall rate of COD reduction is slightly lower than that achieved by emergent systems. The reduced performance may be due to the absence of a gravel layer in the floating reactors, which limits biofilm establishment and slows microbial pathway diversification, as suggested in earlier wetland studies [9].

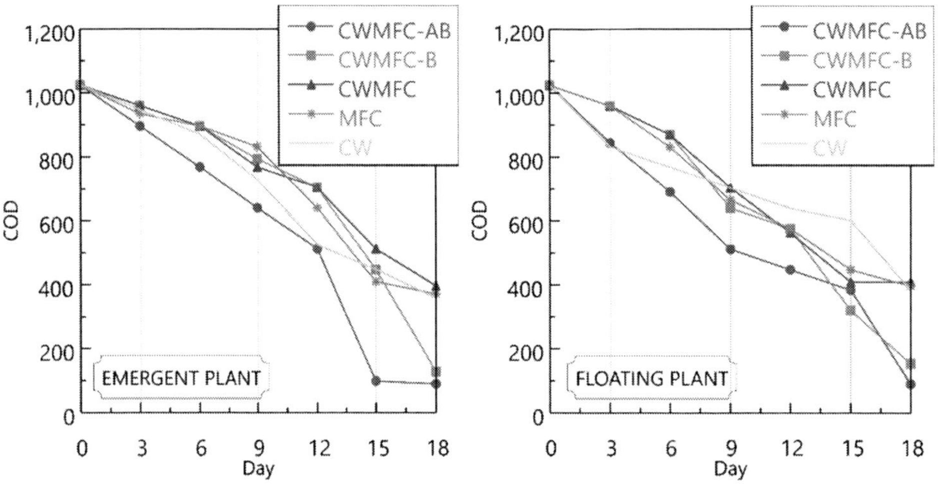

Fig. 4. COD removal over time for (left) emergent plants and (right) floating plants.

The performance gap between the emergent and floating plants remained relatively small for COD removal. In both configurations, the reactors that included both CW and MFC components consistently demonstrated superior removal efficiencies. The statistical analysis performed through ANOVA revealed no significant difference between the floating and emergent configurations on day 18, indicated by an F statistic of 0.029 and a p value of 0.867. This confirms that both plant types support comparable levels of COD reduction and that system performance is influenced more strongly by bacterial augmentation and MFC integration than by plant morphology alone.

The overall pattern across all treatments confirmed that the CWMFC AB configuration provided the most effective treatment performance. The synergy between plant-mediated oxygen transfer, enriched microbial populations, and electrochemical oxidation results in rapid COD degradation. The removal hierarchy observed in this study is consistent with previous investigations of integrated CW and MFC systems, where coupling these processes produces more efficient treatment than either technology operating independently [9]. The results of the present study confirm that the combination of biological, microbial, and electrochemical mechanisms provides a strong foundation for improving decentralized domestic wastewater treatment.

3.4 Ammonia Nitrogen Removal

Ammonia concentrations decreased across all reactors during the eighteen-day experiment, although the magnitude and rate of removal varied among the treatment configurations. The results demonstrated that systems incorporating plants, bacterial augmentation, and microbial fuel cell components consistently achieved the greatest decrease in total ammonia nitrogen, whereas reactors that lacked one or more of these elements showed slower conversion.

In the emergent plant systems, the CWMFC AB reactor achieved the lowest total ammonia nitrogen concentration by the end of the experiment, reaching 7 mg/L total ammonia nitrogen. The CWMFC B reactor produced a final concentration of 9 mg/L, reflecting the contribution of bacterial supplementation to nitrogen conversion processes. The non-augmented CWMFC reactor demonstrated a moderate decrease to 11 mg/L. The MFC only and CW only reactors showed the weakest performance, with final concentrations of 17 mg/L and 14 mg/L, respectively. These results indicate that the combined influence of plant-rooted zones, bacterial activity, and electrochemical reactions provides conditions that strongly promote ammonia transformation in emergent plant systems [10].

A similar pattern was observed in the floating plant reactors. The CWMFC AB system exhibited the fastest removal, with a final concentration of 9 mg/L. The CWMFC B reactor followed closely, reaching 10 mg/L. The non-augmented CWMFC system decreased to 17 mg/L by day 18. The MFC-only and CW-only reactors showed the slowest reductions, ending at 18 mg/L and 20 mg/L, respectively. Therefore, floating systems require the combined action of plants, bacteria, and microbial fuel cells to achieve the strongest removal, while systems lacking any single component performed noticeably less effectively than emergent plants with substrates.

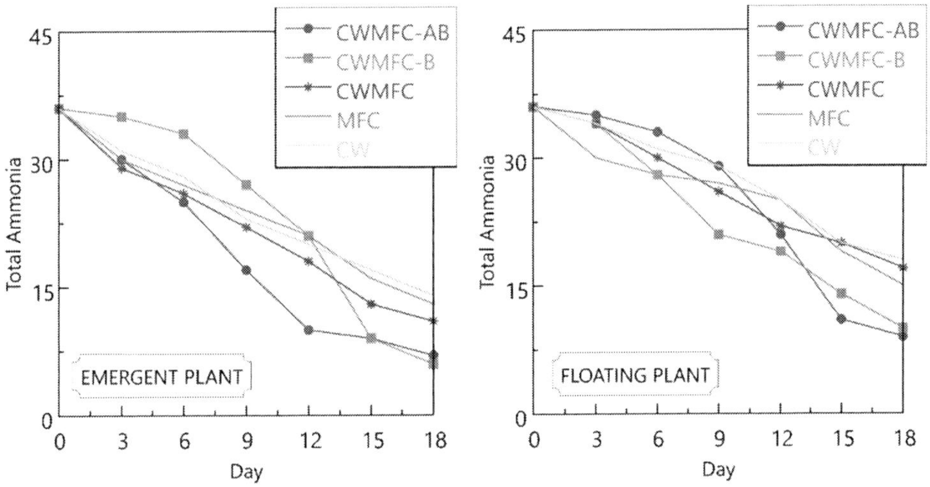

Fig. 5. TAN removal over time for (left) emergent plants and (right) floating plants.

A comparison between the emergent and floating subsystems revealed that the emergent plants achieved slightly greater ammonia removal. The presence of gravel substrate in the emergent reactors may have supported more stable microbial colonization and provided favorable conditions for nitrification and denitrification. Although floating plant reactors are responsive and generally effective, they depend more heavily on bacterial augmentation and microbial fuel cell integration to achieve comparable performance levels. Despite these differences, both plant-based systems generated substantially higher removal rates than the non-planted control reactors, confirming that plant physiological processes contribute meaningfully to nitrogen reduction.

Nitrogen removal observed in the integrated reactors may involve multiple biological and electrochemical pathways, including microbial nitrification and denitrification, plant-mediated uptake, and redox enhancement facilitated by microbial fuel cells. Previous studies have noted that hybrid constructed wetland and microbial fuel cell systems can accelerate nitrogen transformation by enabling interactions between aerobic and anoxic microenvironments and supporting electroactive microbial communities [10]. Overall, the reactors that incorporated plants, bacteria, and microbial fuel cell components produced the most substantial total ammonia nitrogen reductions in both the emergent and floating configurations. The control reactors displayed slower declines, confirming that each component plays an important role in the treatment process. Therefore, the results demonstrate that the integrated constructed wetland and microbial fuel cell configuration provides the most favorable conditions for effective nitrogen conversion in domestic wastewater.

3.5 Electricity Production and Electrochemical Performance

Electricity generation increased progressively in all CW MFC reactors during the eighteen-day operational period, although the magnitude of the increase varied considerably across the treatment configurations. The observed trends reflect the combined influence of plant physiology, microbial activity, and electrochemical processes occurring within the microbial fuel cell. Control reactors without one or more essential components produced negligible electricity, confirming that each removed element plays a critical role in creating electrochemical gradients and enabling current generation.

In the floating plant system, the CWMFC reactors produced very low current densities throughout the operational period. Power generation in CWMFC increased gradually from 7.03×10^{-4} mW/m^2 to 1.40×10^{2} mW/m^2, while CWMFC B increased from 1.68×10^{0} mW/m^2 to 1.96×10^{2} mW/m^2. These values indicate limited electroactive biofilm development, which is likely related to the reduced oxygen transfer and weaker rhizosphere support provided by the floating roots. The floating CWMFC AB reactor showed a substantially higher output, reaching 1.96×10^{2} mW/m^2. Electricity generation was significantly higher in the emergent plant system. The CWMFC reactor increased from 1.15×10^{-1} mW/m^2 to 4.93×10^{1} mW/m^2, and the CWMFC B reactor followed a similar trend with moderate outputs, indicating that aeration or bacterial augmentation alone did not substantially enhance performance. The highest power densities were observed in the emergent CWMFC AB reactor, which reached a maximum of 7.95×10^{2} mW/m^2. The elevated performance of emergent systems can be attributed to stronger oxygen release into the rhizosphere, improved structural support for biofilm growth, and better integration of electrodes with plant-mediated microenvironments that favor electron transfer [11].

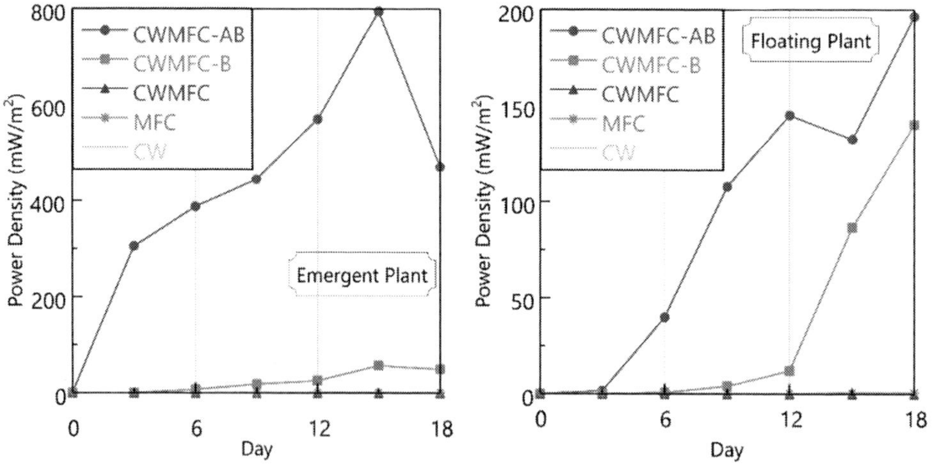

Fig. 6. Power density generation over time for (left) Emergent Plant and (right) Floating Plant.

Electricity generation patterns in both plant systems indicated that the combined application of plants, bacterial augmentation, and aeration in the CWMFC AB configuration produced the most favorable conditions for electrogenic activity. Root exudates stimulate microbial metabolism, and oxygen released by emergent plants supports cathodic reactions that facilitate electron transfer. This interaction lowers the internal resistance and improves the bioelectrochemical conversion efficiency. Previous studies reported that integrated CW MFC systems provide an extensive surface for microbial colonization and promote the growth of electrochemically active bacteria, resulting in increased bioelectricity generation in integrated constructed wetland microbial fuel cells [12]. The electricity generated in this study was also consistent with previous CW-MFC studies that reported power densities as low as 9.4 mW/m^2 [13] and as high as 4.5-7.19x10^2 mW/m^2 [14]. Overall, the emergent plant reactors produced markedly higher power densities than the floating reactors, and the CWMFC AB configuration consistently outperformed all other configurations. However, the electricity generation results clearly show that the integration of plants, bacterial augmentation, and the microbial fuel cell system produced strong improvements in the electrochemical performance of both floating and emergent plants. Reactors that combined all three components exhibited the highest increase in power density, whereas the control reactors lacking any component produced minimal or unstable outputs.

3.6 Parameter Relation

The combined trends of pH, BOD, COD, TAN, and power density show that pollutant removal and electricity generation in CW MFC systems are interdependent processes governed by microbial activity, plant-mediated functions, and electrochemical reactions. Reactors that incorporated plants, bacterial augmentation, and MFC components consistently produced the most stable operating conditions and highest performance across all parameters. pH stability was a key factor supporting both organic degradation and electricity generation. The reactors containing plants showed more consistent pH profiles, which helped to maintain optimal microbial activity. This stability contributed to faster reductions in BOD and COD because electrogenic and heterotrophic bacteria perform more efficiently under stable pH conditions. In contrast, MFC-only reactors exhibited larger pH fluctuations, which likely reduced the establishment of electroactive biofilms and resulted in both slower organic removal and lower power densities.

The patterns of BOD and COD reduction were closely linked to the electricity generation. Reactors with rapid decreases in organic load also produced higher power densities, indicating that the oxidation of organic matter directly supplied electrons for current production [12]. This relationship was most apparent in the CWMFC AB reactors, which demonstrated the highest organic matter removal and the strongest power output. Systems lacking either plant or bacterial augmentation showed weaker removal rates and correspondingly lower power generation, confirming that efficient biodegradation supports more effective electron transfer within the MFC. TAN removal also aligns with broader system behavior. Reactors with complete configurations exhibited faster ammonia reduction, which is consistent with enhanced microbial nitrification and plant uptake. These reactors produced greater COD and BOD removal, as well as higher power densities, suggesting that the microbial communities responsible for nitrogen transformation are part of a broader set of biological processes that improve electrochemical performance [15]. In contrast, limited TAN removal in the control reactors paralleled weaker organic degradation and minimal electricity generation.

Overall, the results show that pH stability, organic matter removal, ammonia reduction, and electricity generation are interrelated. Systems that achieved strong performance for one parameter tended to perform well for all other parameters. This pattern indicates that CW MFC processes reinforce each other when plants, bacteria, and the MFC unit operate concurrently. Therefore, the integrated CWMFC AB configuration provides the most balanced biochemical and electrochemical environments, resulting in superior pollutant removal and power production compared to partial or single-component systems.

4 Conclusion

This study demonstrates that the integration of constructed wetlands with microbial fuel cells (CWMFC) significantly enhances the treatment of domestic wastewater and electricity generation compared to standalone constructed wetland (CW) or microbial fuel cell (MFC) systems. The CWMFC AB configuration, which combines plants, bacterial augmentation, and electrochemical processes, achieved the highest and most consistent removal efficiencies for biochemical oxygen demand (BOD), chemical oxygen demand (COD), and total ammonia nitrogen (TAN) while maintaining a stable pH conducive to microbial and electrogenic activity. Both floating and emergent plants contributed to pollutant removal, with emergent plants yielding superior power densities, likely due to increased root surface area and improved rhizosphere conditions that foster electrogenic bacteria. The observed positive correlation between pollutant degradation rates and electrical output confirms that efficient substrate utilization by electroactive microbes simultaneously drives treatment performance and bioelectricity generation. Compared to benchmark studies, the removal efficiencies and power densities achieved here (BOD removal: 91%; COD removal: 91.25%; TAN removal:81%; power density: 7.95×10^2 mW/m^2) are comparable or superior to reported values in similar CWMFC configurations, highlighting the effectiveness of combined biological and electrochemical enhancements. These results align with broader findings in the field, reinforcing the potential of CWMFC systems as sustainable, multifunctional solutions for decentralized wastewater management and renewable energy recovery.

References

[1] Widyarani, D. R. Wulan, U. Hamidah, A. Komarulzaman, R. T. Rosmalina, and N. Sintawardani, Domestic wastewater in Indonesia: generation, characteristics and treatment, Environ Sci Pollut Res **29**, 32397 (2022).

[2] C. Sha, S. Shen, J. Zhang, C. Zhou, X. Lu, and H. Zhang, A Review of Strategies and Technologies for Sustainable Decentralized Wastewater Treatment, Water **16**, 3003 (2024).

[3] D. Sheoran and S. Singh, Examining the efficacy of constructed wetland coupled microbial fuel cell to treat textile wastewater using local wetlands plant species, Environ Sci Pollut Res **31**, 66136 (2024).

[4] G. S. Colares et al., Hybrid constructed wetlands integrated with microbial fuel cells and reactive bed filter for wastewater treatment and bioelectricity generation, Environ Sci Pollut Res **29**, 22223 (2022).

[5] X. Xie, J. Hu, X. Cao, S. Zhang, T. Sakamaki, and X. Li, In Situ Utilization of Electron-Enhanced Degradation of Azo Dyes in a Constructed Wetland–Microbial Fuel Cell Coupling System, Sustainability **16**, 3181 (2024).

[6] D. Ucar, Y. Zhang, and I. Angelidaki, An Overview of Electron Acceptors in Microbial Fuel Cells, Front. Microbiol. **8**, (2017).

[7] X. Wang, M. Xue, Z. Wang, W. Xia, and C. Zhang, Integrated Constructed Wetland–Microbial Fuel Cell Systems Using Activated Carbon: Structure-Activity Relationship of Activated Carbon, Removal Performance of Organics and Nitrogen, Water **16**, 278 (2024).

[8] S. Kesarwani, D. Panwar, J. Mal, N. Pradhan, and R. Rani, Constructed Wetland Coupled Microbial Fuel Cell: A Clean Technology for Sustainable Treatment of Wastewater and Bioelectricity Generation, Fermentation **9**, 6 (2022).

[9] N. Paucar and C. Sato, An Overview of Microbial Fuel Cells within Constructed Wetland for Simultaneous Nutrient Removal and Power Generation, Energies **15**, 6841 (2022).

[10] J. Fang, R. Zhao, Q. Cao, Q. Quan, R. Sun, and J. Liu, Effects of emergent aquatic plants on nitrogen transformation processes and related microorganisms in a constructed wetland in northern China, Plant Soil **443**, 473 (2019).

[11] I. Araneda, N. F. Tapia, K. Lizama Allende, and I. T. Vargas, Constructed Wetland-Microbial Fuel Cells for Sustainable Greywater Treatment, Water **10**, 940 (2018).

[12] F. Liu, L. Sun, J. Wan, A. Tang, M. Deng, and R. Wu, Organic matter and ammonia removal by a novel integrated process of constructed wetland and microbial fuel cells, RSC Adv. **9**, 5384 (2019).

[13] S. Bhaduri and M. Behera, Advancement in constructed wetland microbial fuel cell process for wastewater treatment and electricity generation: a review, Environ Sci Pollut Res **31**, 50056 (2024).

[14] D. G. Jacobs, L. O. Kachienga, M. C. Rikhotso, A. L. K. Abia, A. N. Traoré, and N. Potgieter, Assessing the current situation of constructed wetland-microbial fuel cells as an alternative power generation and wastewater treatment in developing countries, Front. Energy Res. **12**, 1448730 (2024).

[15] K. Angassa, T. Getu, and M. Abewaa, Removal of nutrients and bioelectricity generation from institutional wastewater using constructed wetland-microbial fuel cell, Results in Engineering **24**, 103254 (2024).

Distribution of Cromium (Cr) in the Bottom Substrate around Kebintik and Batu Belubang Coastal Waters, Bangka Island

Umroh[1], Aditya Pamungkas[1], Irma Akhrianti[1], Aries Dwi Siswanto[2], Nuryanto[3*]

[1]Department of Marine Science, Faculty of Agriculture, Fisheries and Marine, Universitas Bangka Belitung, Bangka, 33172, Indonesia, [2]Department of Marine Science, Faculty of Agriculture, Universitas Trunojoyo, Madura Island, East Java, 69169, Indonesia, [3]Semarang Polytechnic of Health, Ministry of Health, Republic of Indonesia, Central Java Indonesia

*Email: nuryanto@poltekkes-smg.ac.id

Abstract. Within the next five years, the post-tin mining coastal waters of Kebintik and Batu Belubang in Central Bangka are expected to run out. Habitats for commercially significant and community-consumed benthos (bivalve class, such as *Anadara granosa* and *Meretrix meretrix*) can be found in both coastal seas. Consequently, there are worries regarding the effects of heavy metals (such Cr) from tin mining on marine resources since they build up in the biota, which is eventually ingested by humans. This work uses quantitative descriptive methods and Cr measurement with ICP-EOS to evaluate the concentration and distribution of Cr in sediments surrounding Kebintik and Batu Belubang coastal waters. According to the findings, the sediments in both coastal waterways had Cr concentrations ranging from 0.0210 to 0.044 ppm (the category is still within the quality threshold). The area of post-tin mining tailings pilings at station 6 had the largest dispersion of Cr. The distribution and concentration of heavy metals in the sediments of Kebintik and Batu Belubang's coastal waters are still below the quality standards, meaning that the waters are still suitable for benthic life—which is significant to the local economy and does not pose a health risk to humans—and are therefore generally in good condition.

Keywords : Batu Belubang, Kebintik, Cr, sediment

1 Introduction

On Bangka Island, heavy metal pollution in tin mining and post-mining areas has grown to be a significant problem. According to previous research, heavy metals found in tin mining waste [1, 2] have raised public concerns, particularly about the possible disruption of marine resources. Because it is poisonous to both humans and animals, chromium (Cr) is one of the heavy metals that has grown to be a significant problem in mining [3]. Cr buildup will disrupt the function of enzymes and result in death if exposed for an extended period of time [4]. The loss of marine larval organisms in the water, which lowers the marine biological resources in Kebintik and Batu Belubang, is the effect that the community is most worried about. According to [4], Cr metal will lower the quality of embryo hatching and larval growth, which will lower the organisms' survival rate. In the waters of Kebintik and Batu Belubang, non-fish resources like shellfish are still plentiful during the season and could be a possible catch for the local population.

Numerous studies have examined the heavy metal content of Bangka Island's waters, including those on Pb, Cd, Zn, and Cu pollution in Matras Waters, Bangka Induk [5], Pb metal in Kelabat Bay [6], Pb, Cd, Zn, Cu, Sn, and Al metals in the central Bangka waters [1, 7], and Cd, Pb, Co, and Ni metals in the central Bangka waters [2]. It is imperative to address public concerns because the extensive distribution of Cr metal has not been investigated in any of the studies carried out in the Bangka Island maritime area.

Therefore, this study aims to assess the concentration and distribution of Cr metal in sediments in Kebintik and Batu Belubang waters, which are post-tin mining areas. The area is also a habitat for benthic, bivalve classes such as blood clams (*Anadara granosa*) and crab clams (*Meretrix meretrix*) which are abundant in season. With this concentration and distribution data, it is very useful in habitat security for fish and non-fish biological resources on the Kebintik and Batu Belubang coasts.

2 Methodology

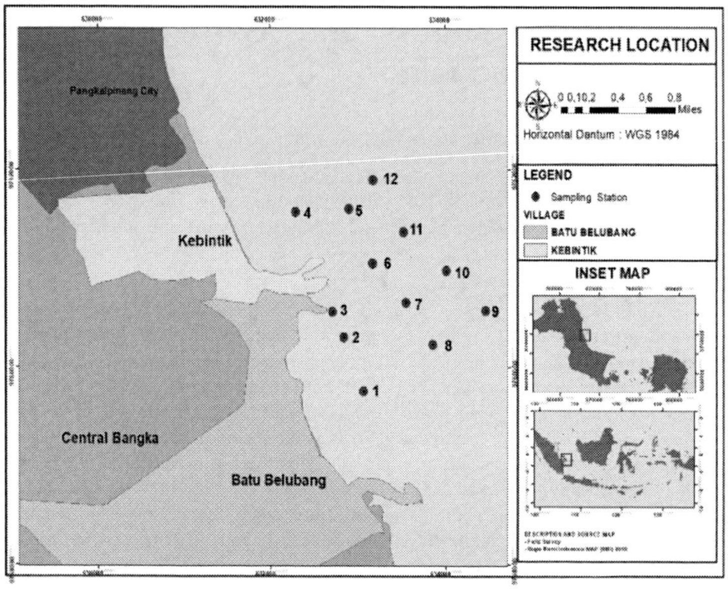

Fig. 1. Sampling stations for Cr metal in water and sediment at Kebintik and Batu Belubang Coast

Sampling locations for sediment, water, and aquatic physico-chemical parameters were conducted at 12 points (**Figure 1**). Sampling was conducted in the waters of Kebintik and Batu Belubang in March 2023. The research method is descriptive quantitative, consisting of field surveys (determining the location and time of research), sediment sampling and physical parameters - water chemistry, sample analysis in the laboratory and data analysis.

Preparation of Cr metal in sediments based on [1, 2, 7], and measurement of heavy metal concentrations using Inductively Coupled Plasma - Optical Emission Spectroscopy (ICP-EOS).

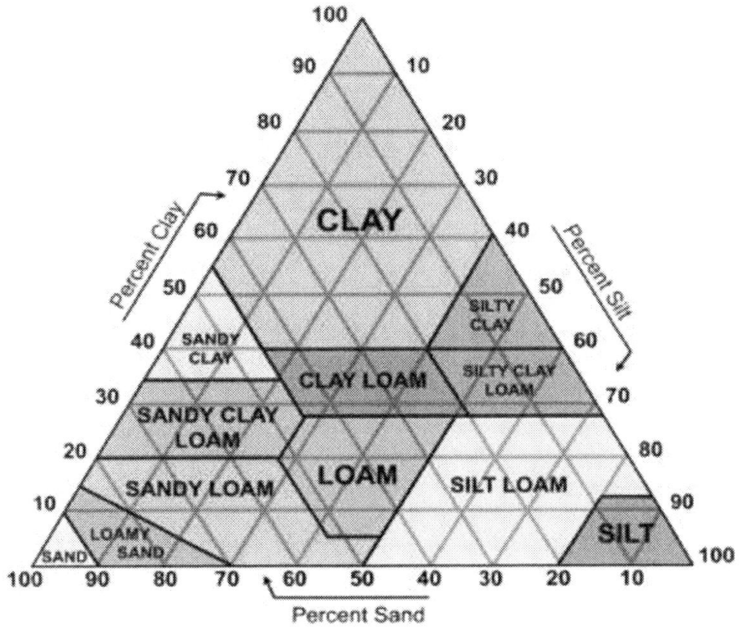

Fig. 2. The Triangle of Shepard [10]

Sediment Texture Analysis

100 grams of sediment samples were collected, and they were dried at 1000C for 24 hours. 25 grams of the dried sediment samples, with diameters of 4 mm, 1.00 mm, 0.150 mm, and 0.038 mm, were put in a sieve shaker and sieved for 20 minutes after being ground up with a mortar. Following each sieve level, the products were weighed. Additionally, Shepard's triangle was used to examine sediment samples [8]. Following that, each fraction's percentage value was determined, displayed, and given a name based on the Wentworth categorization [9].

Analysis of Cr Concentration Distribution Map in Sediment

Maps of Cr heavy metal distribution in sediments from Kebintik and Batu Belubang waters were processed using Arcgis 10.4.1 and interpreted using different color scales in the form of displays with the Inverse Distance Weighted (IDW) method, then analyzed descriptively [7].

3 Result and Discussion

Table 1. Classification of Sediment Grain Size using Wenworth Scales [11]

Large Grain Size (mm)	Name
>256	Boulder
64-256	Couble
4-64	Pebble
2-4	Granule
1-2	Very course sand
½-1	Coarse sand
¼-½	Medium sand
1/8 -1/4	Fine sand
Large Grain Size	
1/16-1/8	Very find sand
1/256 – 1/16	Silt
<1/256	Clay

Table 2. Concentration of Cr in the sediment and water (ppm) at Pantai Kebintik and Batu Belubang, Bangka Tengah

Stations	S	E	$Cr_{sediment}$	Cr_{air}
1	02°10'23,1	106°11'47,8	0,0260	<0.001
2	02°10'05,4	106°11'40,5	0,0340	<0.001
3	02°09'56,9	106°11'36,1	0,0210	<0.001
4	02°09'24,0	106°11'22,4	0,0330	<0.001
5	02°9'22,77	106°11'41,8	0,0360	<0.001
6	02°09'40,9	106°11'51,0	0,0440	<0.001
7	02°09'53,8	106°12'03,2	0,0320	<0.001
8	02°10'07,7	106°12'13,1	0,0340	<0.001
9	02°09'56,5	106°12'32,8	0,0220	<0.001
10	02°09'43,2	106°12'18,3	0,0220	<0.00
11	02°09'30,5	106°12'02,3	0,0340	<0.001
12	02°09'13,4	106°11'51,2	0,0230	<0.001
[11] Guidelines		Low	80	
		High	370	

The range of 0.0210 - 0.044 ppm (**Table 2**) for the concentration of Cr metal in the sediments of the Kebintik and Balu Belubang waters indicates that the category is still below the [12] quality level. After more than five years of tin mining, both streams are post-tin. The concentration of Cr metal in water was likewise very low, at less than 0.001. This is because the metal is soluble, changing from carbonate to hydrocarbonate and hydroside, before joining other particles to form a bond and sinking to the bottom of the water's perticles.

Table 3. Physical – Chemical Parameters at the Pantai Kebintik dan Batu Belubang

Stations	Water						Sediment Texture (%)		
	Temperature (°C)	pH	Salinity (°/₀₀)	DO (mg/L)	Depth (m)	TSS (mg/L)	Sand	Silt	Clay
1	28,3	5,33	28	7,3	2,40	0,13	73,15	20,95	5,91

Stations	Water						Sediment Texture (%)		
	Temperature (°C)	pH	Salinity (°/₀₀)	DO (mg/L)	Depth (m)	TSS (mg/L)	Sand	Silt	Clay
2	28,8	5.32	28	7,1	2,20	0,18	91,39	7,73	0,95
3	29,6	5,33	28	6,8	1,63	0,14	89,40	8,49	2,19
4	28,8	5,34	31	6,8	2,00	0,10	78,99	20,02	0.98
5	29,6	5,32	29	7,0	2,40	0,13	99,41	0,47	0,15
6	29,8	5,28	29	7,0	2,67	0,24	88,76	11,05	0,19
7	29,9	5,31	31	6,9	2,92	0,12	97,39	2,10	0,53
8	30,0	5,32	29	6,9	2,00	0,25	98,54	1,35	0,16
9	30,5	5,26	29	6,8	4,63	0,11	87,19	12,17	0,74
10	30,4	5,25	31	6,9	4,18	0,14	98,43	1,51	0,08
11	30,2	5,32	30	7,0	2,62	0,16	98,89	1,07	0,07
12	30,9	5,28	30	6,7	2,82	0,19	94,83	4,93	0,26
Quality standard	28-30*	-	33-34*	>5*	-	<20*			

*Quality standard based on [18]

Compared to Benoa Bay sediments [13]; Trimulyo Waters, Semarang [14]; and Muara Sungai Mati [15], the Cr metal content in the sediment and water column in Kebintik and Batu Belubang waters is lower. Sand dominates the sediment texture characteristics from stations 1 through 12, with 73.15% being the lowest and 99.41% being the highest (**Table 3**). This may result in a low concentration of Cr metal in the sediment. According to [2], sand has a restricted surface area and a decreased ability to bond with metals. In the sediments of stations 1 through 12, there was a link between textural type and Cr metal ranging from -0.610 to 0.600, according to a Pearson statistical analysis. Low Cr metal and sand texture had a positive correlation (r = 0.600), but silt and clay had negative correlations (r = -0.512 and -0.292, respectively). According to studies [13], there is a positive association (r=0.434) between sand texture and Cr metal in Benoa sediments in Bali. These findings are consistent with that finding. In addition to being caused by the sediment texture factor, the concentration of Cr metal in the sediment is relatively low. It is also known that the primary source of activity that flows waste containing Cr metal is still uncommon. Many sources of waste containing Cr metal are generated from waste from the textile and screen-printing industries [16 17]. The results of research by [17] in the sediments of the Muara Sungai Mati show that Cr metal has exceeded the quality standard limit, due to input from the waste stream of the screen-printing industry.

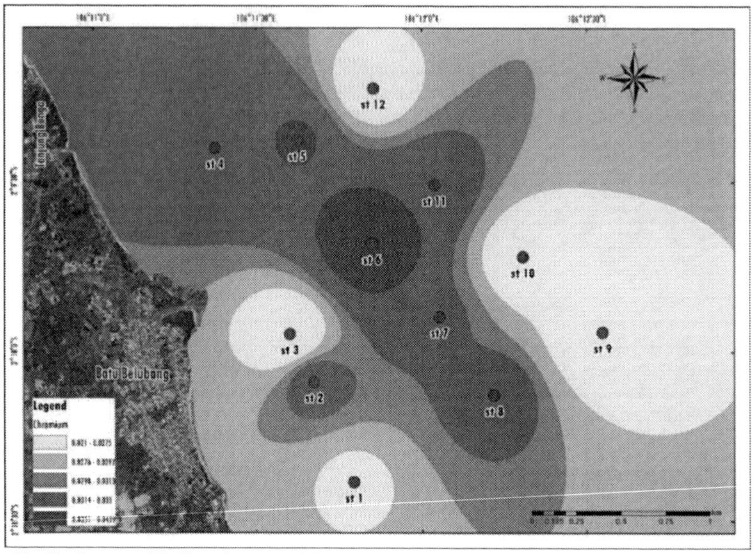

Fig. 3. Map of Cr distribution at Kebintik and Batu Belubang waters, Central Bangka

The area of post-tin mining tailings accumulation, station 6, exhibits a tendency towards the largest distribution of Cr metal (**Figure 3**). Furthermore, because there has already been sedimentation, the current conditions force particles that have bonded with metals to stop and gather in the vicinity of station 6. Since sediment sampling was done in the transitional season with moderate currents (0.098–0.183 m/s), there was also little variation in the distribution of Cr metal. The findings of the investigation of the distribution and concentration of heavy metals in the sediments of Kebintik and Batu Belubang demonstrate that the environmental conditions are still favorable and conducive to the existence of benthos, which have significant economic value and are consumed by the local population.

4 Conclusion

The concentration of Cr metal in the sediments of Kebintik and Batu Belubang Waters ranged from 0.0210 - 0.044 ppm, and was classified as still below the [12]. The distribution of Cr metal in the Kebintik and Batu Belubang water areas shows no significant difference due to moderate current conditions.

Acknowledgment

Thanks to the University of Bangka Belitung, the Institute for Research and Community Service (LPPM), lecturers and students of Marine Science, University of Bangka Belitung.

References

[1] Umroh, Bengen, D. G., Prartono, T., & Riani, E. (2022). Heavy Metals Source Identification by Enrichment Factor in Bangka Island Sediments, Indonesia. *EnvironmentAsia*, *15*(2), 120–131. https://doi.org/10.14456/ea.2022.39

[2] Umroh, Wahidin, L. O., Hudatwi, M., & Pryambada, A. (2024). Exploring Ecological Dynamics: A Case Study of Post-Unconventional Tin Mining in Batu Belubang Waters, Bangka Island. *EnvironmentAsia*, *17*(3), 151–161. https://doi.org/10.14456/ea.2024.43

[3] Casalegno, C., Schifanella, O., Zennaro, E., Marroncelli, S., & Briant, R. (2017). Collate literature data on toxicity of Chromium (Cr) and Nickel (Ni) in experimental animals and humans. *EFSA Supporting Publications*, *12*(2). https://doi.org/10.2903/sp.efsa.2015.en-478

[4] Aslam, S., & Yousafzai, A. M. (2017). Chromium toxicity in fish: A review article. *~ 1483 ~ Journal of Entomology and Zoology Studies*, *5*(3), 1483–1488.

[5] Nugraha, M. A., Pamungkas, A., Syari, I. A., Sari, S. P., Umroh, U., Hudatwi, M., Utami, E., Akhrianti, I., & Priyambada, A. (2022). Penilaian Pencemaran Logam Berat Cd, Pb, Cu, dan Zn pada Sedimen Permukaan Perairan Matras, Sungailiat, Bangka. *Jurnal Kelautan Tropis*, *25*(1), 70–78. https://doi.org/10.14710/jkt.v25i1.12317

[6] Komalasari, A., Afriyansyah, B., Ihsan, M., & Nugraha, M. A. (2019). Bioakumulasi Logam Berat Pb dan Cu terhadap Penaeus merguiensis di Perairan Teluk Kelabat Bagian Dalam. *Jurnal Kelautan Tropis*, *22*(1), 1. https://doi.org/10.14710/jkt.v22i1.3727

[7] Umroh, Sari, S. P., Fabiani, V. A., Ariyanto, D., Siswanto, A. D., & Nuryanto. (2023). Distribution and Contamination Level of Cuprum (Cu) and Plumbum (Pb) in Bulk Sediments of the Bangka Island. *Ilmu Kelautan: Indonesian Journal of Marine Sciences*, *28*(3), 278–288. https://doi.org/10.14710/ik.ijms.28.3.278-288

[8] Ardiansyah, A., Paryono, P., & Waspodo, S. (2023). Diversity of Gastropods and Bivalvia in Jukung Bay Waters, Pemongkong Village, Jerowaru District, East Lombok Regency. *Jurnal Biologi Tropis*, *23*(4), 27–36. https://doi.org/10.29303/jbt.v23i4.5568

[9] Setiawan, I., Miswar, E., Zulfahmi, I., & Yuni, S. M. (2019). Distribution of sediment grain size in Kajhu coastal area, Aceh Besar District, Indonesia. *IOP Conference Series: Earth and Environmental Science*, *348*(1). https://doi.org/10.1088/1755-1315/348/1/012107

[10] Folk, R. L. (1954). The distinction between grain size and mineral composition in sedimentary-rock nomenclature. *The Journal of Geology*, *62*(4), 344-359.

[11] Wentworth, C. K. (1922). A scale of grade and class terms for clastic sediments. *The journal of geology*, *30*(5), 377-392.

[12] ANZECC, & ARMCANZ. (2000). Australian and New Zealand Guidelines for Fresh and Marine Water Quality - Aquatic Ecosystems - Rationale and Background Information- Volume 2. *Australian and New Zealand Environment and Conservation Council, Agriculture and Resource Management Council of Australia and New Zealand*, *2*(4), 678

[13] Suteja, Y., Dirgayusa, I. G. N. P., & Purwiyanto, A. I. S. (2020). Chromium in Benoa Bay, Bali - Indonesia. *Marine Pollution Bulletin*, *153*(February), 111017. https://doi.org/10.1016/j.marpolbul.2020.111017

[14] Nuraini, R. A. T., Endrawati, H., & Maulana, I. R. (2017). Analisis Kandungan Logam Berat Kromium (Cr) Pada Air, Sedimen Dan Kerang Hijau (Perna viridis) Di Perairan Trimulyo Semarang. *Jurnal Kelautan Tropis*, *20*(1), 48. https://doi.org/10.14710/jkt.v20i1.1104

[15] Suteja, Y., & Dirgayusa, I. G. N. P. (2018). Bioaccumulation and translocation of chromium on crabs and mangroves in Mati River estuary, Bali, Indonesia. *AACL Bioflux*, *11*(2), 469–475.

[16] Motzer, W.E., 2016. Chemistry, geochemistry, and geology of chromium and chromium compounds. In: Chromium (VI) Handbook, pp. 23–91. https://doi.org/10.1201/ 9780203487969.ch2.
[17] Dirgayusa, I. G. N. P., Suteja, Y., & Adnyana, I. B. P. (2017). Mangroves Sediment Ability as a Traps of Heavy Metal Chrome in Tukad Mati Estuary, Bali – Republic of Indonesia. *Journal of Sustainable Development*, *10*(4), 1. https://doi.org/10.5539/jsd.v10n4p1
[18] PPRI (Peraturan Pemerintah Republik Indonesia). (2021) Penyelenggaraan Perlindungan dan Pengelolaan Lingkungan Hidup. Sekretariat Negara Republik Indonesia. http://www.jdih.setjen. kemendagri.go.id/

AUTHOR INDEX